AF769410

DESIGNING DENDRIMERS

DESIGNING DENDRIMERS

Edited by

SEBASTIANO CAMPAGNA
PAOLA CERONI
FAUSTO PUNTORIERO

WILEY

A JOHN WILEY & SONS, INC., PUBLICATION

Copyright © 2012 by John Wiley & Sons, Inc. All rights reserved

Published by John Wiley & Sons, Inc., Hoboken, New Jersey
Published simultaneously in Canada

No part of this publication may be reproduced, stored in a retrieval system, or transmitted in any form or by any means, electronic, mechanical, photocopying, recording, scanning, or otherwise, except as permitted under Section 107 or 108 of the 1976 United States Copyright Act, without either the prior written permission of the Publisher, or authorization through payment of the appropriate per-copy fee to the Copyright Clearance Center, Inc., 222 Rosewood Drive, Danvers, MA 01923, (978) 750-8400, fax (978) 750-4470, or on the web at www.copyright.com. Requests to the Publisher for permission should be addressed to the Permissions Department, John Wiley & Sons, Inc., 111 River Street, Hoboken, NJ 07030, (201) 748-6011, fax (201) 748-6008, or online at http://www.wiley.com/go/permission.

Limit of Liability/Disclaimer of Warranty: While the publisher and author have used their best efforts in preparing this book, they make no representations or warranties with respect to the accuracy or completeness of the contents of this book and specifically disclaim any implied warranties of merchantability or fitness for a particular purpose. No warranty may be created or extended by sales representatives or written sales materials. The advice and strategies contained herein may not be suitable for your situation. You should consult with a professional where appropriate. Neither the publisher nor author shall be liable for any loss of profit or any other commercial damages, including but not limited to special, incidental, consequential, or other damages.

For general information on our other products and services or for technical support, please contact our Customer Care Department within the United States at (800) 762-2974, outside the United States at (317) 572-3993 or fax (317) 572-4002.

Wiley also publishes its books in a variety of electronic formats. Some content that appears in print may not be available in electronic formats. For more information about Wiley products, visit our web site at www.wiley.com.

Library of Congress Cataloging-in-Publication Data:

Designing dendrimers / edited by Sebastiano Campagna, Paola Ceroni, Fausto Puntoriero.
p. cm.
Includes index.
ISBN 978-0-470-43355-3 (hardback)
1. Dendrimers. I. Campagna, Sebastiano. II. Ceroni, Paola. III. Puntoriero, Fausto, 1974
TP1180.D45D48 2012
620.1′92–dc23

2011021419

Printed in Singapore

oBook ISBN: 9781118111086
ePDF ISBN: 9781118111055
ePub ISBN: 9781118111062
eMobi ISBN: 9781118111079

10 9 8 7 6 5 4 3 2 1

CONTENTS

PREFACE

The term dendrimer derives from the Greek words *dendron* (tree) and *meros* (part) and refers to its regularly and highly branched structure. Many dendritic architectures are present in Nature: from dendritic cells to ice crystals. In chemistry, the first examples of iterative synthetic procedure (now recognized as a trademark of dendrimers) was described by Vögtle in 1978, while the first thoroughly investigated dendrimers were reported by Tomalia and Newkome in the mid-1980s and by Fréchet in the early 1990s. In the beginning, the term *arborol* (from Latin word means tree) was sometimes used, but finally the word dendrimer conquered the scene. Still in the early 1990s other distinguished researchers, such as Astruc, Balzani, Meijer, Moore, Müllen, and the Majoral/Caminade group, joined the field and contributed to the early successes of the dendrimer chemistry. At that time, synthesis of a dendrimer was a real challenge both for the synthetic procedure and for the analytical characterization. The interest in dendrimer chemistry has enormously increased since then, also because it was immediately clear that new properties emerged from the dendritic structures, properties, which could not be featured by the individual subunits or from alternative assembly organizations. The great potential of dendrimers is now well recognized by the huge growth in the number of papers (from few dozens in 1993 to more than 2000 in 2007, see Figure P.1) and patents, and by the advent of companies selling dendrimer-based products. This research stimulated developments in other areas of knowledge, including new characterization techniques.

Like trees, dendrimers usually exhibit aesthetically pleasant structures, but the interest in a specific dendrimer does not depend, of course, only on its beauty; rather, it depends on the "fruit" (i.e., the specific function) that the dendrimer is able to produce. Indeed, they are currently attracting the interest of a great number of scientists because of their unusual chemical and physical properties and the wide range of potential

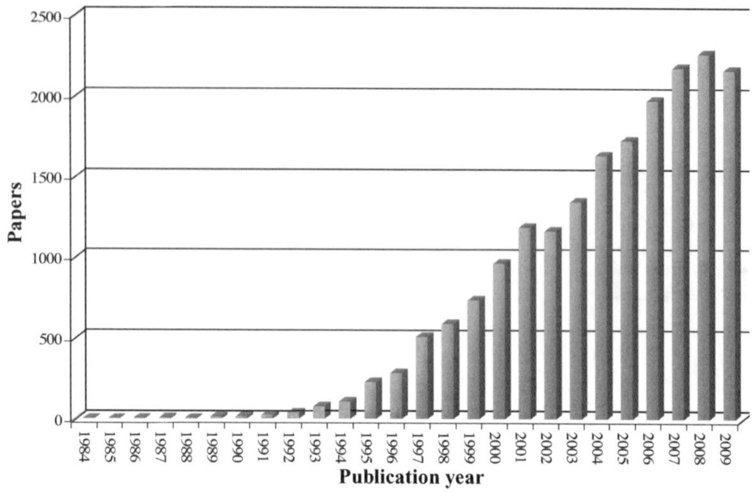

FIGURE 1 The increasing number of published papers devoted to dendrimer chemistry along the years. The 2009 data are limited to November 10, 2009.

applications in such diverse fields as medicine, biology, chemistry, physics, and engineering.

Other books on dendrimers have been published during the past 15 years, but very recent progress in new areas stimulated us to edit the present volume to provide an overview of the field's progress and future perspectives.

In this book, leading scientists in this field report their up-to-date contributions covering a broad range of applications and suggesting future possibilities. The book is not intended to be a comprehensive text on dendrimer science, but rather it is intended to guide the newcomers in the complex and multi-faceted fields of dendrimer synthesis and applications and to provide the experts with the state-of-the-art studies. We hope that the present book will be useful not only for scientists engaged in the dendrimer research but also in nanotechnology, materials science, and molecular biology since much space is devoted to the recent dendrimer applications, ranging from optoelectronic devices to drug delivery and artificial proteins. The volume is also aimed at graduate students and postdoctoral fellows interested in exploring the frontiers of dendrimer chemistry as an inspiration for new research topics. It may also serve as a basic text or a complementary reading source for graduate and postgraduate students attending nanoscience and nanotechnology courses. Indeed, although the book does not include introductory or tutorial sections, most chapters begin with a discussion of the basic concepts that are relevant for the presented topics.

The book consists of 16 chapters, which are not grouped into sections, but nevertheless are logically ordered. The first chapters provide an introductory view—even including perspectives for alternative approaches to the dendrimer realm—mainly centered on synthetic methods/approaches. The following ones are more application-oriented, exploring electrochemically based applications, photochemical energy conversion, single-molecule detection, biomedical applications, and future perspectives.

Our sincere acknowledgments are devoted to the distinguished colleagues and friends who have contributed the chapters. We would also like to express our gratitude to the Wiley staff members for their help in the various phases of the editorial work.
Color figures are available at ftp://ftp.wiley.com/public/sci_tech_med/designing_dendrimers.

Sebastiano Campagna
Paola Ceroni
Fausto Puntoriero

Messina and Bologna, September 2011

LIST OF CONTRIBUTORS

Beatriz Alonso, Departamento de Química Inorgánica, Universidad Autónoma de Madrid, Madrid, Spain

Vincenzo Balzani, Dipartimento di Chimica "G. Ciamician", Università di Bologna, via Selmi 2, I-40126 Bologna, Italy

Martin Baumgarten, Max-Planck-Institut für Polymerforschung, Ackermannweg 10, D-55128 Mainz, Germany

Giacomo Bergamini, Dipartimento di Chimica "G. Ciamician", Università di Bologna, via Selmi 2, I-40126 Bologna, Italy

Anne-Marie Caminade, Laboratoire de Chimie de Coordination (LCC) CNRS, 205 route de Narbonne, 31077 Toulouse Cedex 4, France; Université de Toulouse, UPS, INPT, LCC, F-31077 Toulouse, France

Carmen M. Casado, Departamento de Química Inorgánica, Universidad Autónoma de Madrid, Madrid, Spain

Paola Ceroni, Dipartimento di Chimica "G. Ciamician", Università di Bologna, via Selmi 2, I-40126 Bologna, Italy

Jørn B. Christensen, Department of Chemistry, University of Copenhagen, DK-2100 Copenhagen, Denmark

Tamis Darbre, Department of Chemistry and Biochemistry, University of Berne, Freiestrasse 3, CH-3012, Berne, Switzerland

Fan Gao, Beijing National Laboratory for Molecular Sciences (BNLMS), Key Laboratories of Polymer Chemistry and Physics and of Bioorganic Chemistry and Molecular Engineering of Ministry of Education, College of Chemistry and Molecular Engineering, Peking University, Beijing 100871, China

María Pilar García-Armada, Departamento de Ingeniería Química Industrial, Universidad Politécnica de Madrid, Madrid, Spain

Marc Gingras, CNRS, Aix-Marseille II University, UPR 3118 Interdisciplinary Center on Nanosciences of Marseille, 163 Av. de Luminy, Case 913, 13288 Marseille Cedex 09, France

Takane Imaoka, Chemical Resources Laboratory, Tokyo Institute of Technology, Yokohama 226-8503, Japan

Angel E. Kaifer, Center for Supramolecular Science and Department of Chemistry, University of Miami, Coral Gables, FL 33124-0431, USA

José Losada, Departamento de Ingeniería Química Industrial, Universidad Politécnica de Madrid, Madrid, Spain

Yuguo Ma, Beijing National Laboratory for Molecular Sciences (BNLMS), Key Laboratories of Polymer Chemistry and Physics and of Bioorganic Chemistry and Molecular Engineering of Ministry of Education, College of Chemistry and Molecular Engineering, Peking University, Beijing 100871, China

Jean-Pierre Majoral, Laboratoire de Chimie de Coordination (LCC) CNRS, 205 route de Narbonne, 31077 Toulouse Cedex 4, France; Université de Toulouse, UPS, INPT, LCC, F-31077 Toulouse, France

Enrico Marchi, Dipartimento di Chimica "G. Ciamician", Università di Bologna, via Selmi 2, I-40126 Bologna, Italy

Klaus Müllen, Max-Planck-Institut für Polymerforschung, Ackermannweg 10, D-55128 Mainz, Germany

George R. Newkome, Departments of Polymer Science and Chemistry, University of Akron, Akron, Ohio 44325-4717, USA

Jean-François Nierengarten, Laboratoire de Chimie des Matériaux Moléculaires, Université de Strasbourg et CNRS, Ecole Européenne de Chimie, Polymères et Matériaux (ECPM), 25 rue Becquerel, 67087 Strasbourg Cedex 2, France

Jian Pei, Beijing National Laboratory for Molecular Sciences (BNLMS), Key Laboratories of Polymer Chemistry and Physics and of Bioorganic Chemistry and Molecular Engineering of Ministry of Education, College of Chemistry and Molecular Engineering, Peking University, Beijing 100871, China

Fausto Puntoriero, Università di Messina, Dipartimento di Chimica Inorganica, 98166 Vill. S. Agata – Messina, Italy

Tianshi Qin, Max-Planck-Institut für Polymerforschung, Ackermannweg 10, D-55128 Mainz, Germany

Jean-Louis Reymond, Department of Chemistry and Biochemistry, University of Berne, Freiestrasse 3, CH-3012, Berne, Switzerland

A. Dieter Schlüter, Laboratory of Polymer Chemistry, Department of Materials, ETH Zurich, Wolfgang-Pauli Strasse 10, HCI J 541, 8093 Zurich, Switzerland

Carol Shreiner, Department of Chemistry, Hiram College, Hiram, Ohio 44234, USA

Donald A. Tomalia, The National Dendrimer & Nanotechnology Center, Department of Chemistry, Central Michigan University, Mt. Pleasant, MI 48859, USA

Mitsuru Ueda, Department of Organic & Polymeric Materials, Graduate School of Science & Engineering, Tokyo Institute of Technology, 2-12-1, O-okayama, Meguro, Tokyo 152-8552, Japan

Sergei A. Vinogradov, Department of Biochemistry and Biophysics, University of Pennsylvania, Philadelphia, PA 19104, USA

Tom Vosch, Nano-Science Center, Department of Chemistry, University of Copenhagen, Universitetsparken 5, 2100 Copenhagen, Denmark

Isao Washio, Material Design Unit, Material Science Laboratory, Mitsui Chemicals, Inc., 580-32 Nagaura, Sodegaura, Chiba 299-0265, Japan

David F. Wilson, Department of Biochemistry and Biophysics, University of Pennsylvania, Philadelphia, PA 19104, USA

Kimihisa Yamamoto, Chemical Resources Laboratory, Tokyo Institute of Technology, Yokohama 226-8503, Japan

Jing Yan, Beijing National Laboratory for Molecular Sciences (BNLMS), Key Laboratories of Polymer Chemistry and Physics and of Bioorganic Chemistry and Molecular Engineering of Ministry of Education, College of Chemistry and Molecular Engineering, Peking University, Beijing 100871, China

Afang Zhang, Laboratory of Polymer Chemistry, Department of Materials, ETH Zurich, Wolfgang-Pauli Strasse 10, HCI J 541, 8093 Zurich, Switzerland

1

DENDRIMERS AS QUANTIZED NANO-MODULES IN THE NANOTECHNOLOGY FIELD

JØRN B. CHRISTENSEN[1] AND DONALD A. TOMALIA[2,3]

[1]*Department of Chemistry, University of Copenhagen, DK-2100 Copenhagen, Denmark*

[2]*The National Dendrimer & Nanotechnology Center, Department of Chemistry, Central Michigan University, Mt. Pleasant, MI, 48859 USA*

[3]*NanoSynthons LLC, 1200 N. Fancher Ave., Mt. Pleasant, MI, 48858 USA*

1.1 INTRODUCTION

According to an insightful opinion expressed by Nobel Laureate, R. Hoffmann [1], chemistry and physics meet in the solid state where these two disciplines share a significant mutual interest concerning the structure of matter. That withstanding, chemists are usually most concerned with the use of atom building blocks and the manipulation of their electrons to produce more complex molecular structure. Historically, much of this earlier activity focused on the evolution of first principles and a "central paradigm" as briefly outlined in Figure 1.1.

These first principles led to the emergence of "elemental periodicity" that ultimately produced Mendeleev's Periodic Table (1869), and a systematic framework for unifying and defining the very rich contemporary science of traditional small molecule and macromolecular chemistry. One of the greatest challenges facing today's interdisciplinary field of nanoscience is the absence of a "central paradigm," and a nanomaterials classification roadmap for organizing and defining the growing number of nanostructures and assemblies that are being reported in the literature. These issues were examined as a critical theme for a recent National Science

Designing Dendrimers, Edited by Sebastiano Campagna, Paola Ceroni, Fausto Puntoriero.
© 2012 John Wiley & Sons, Inc. Published 2012 by John Wiley & Sons, Inc.

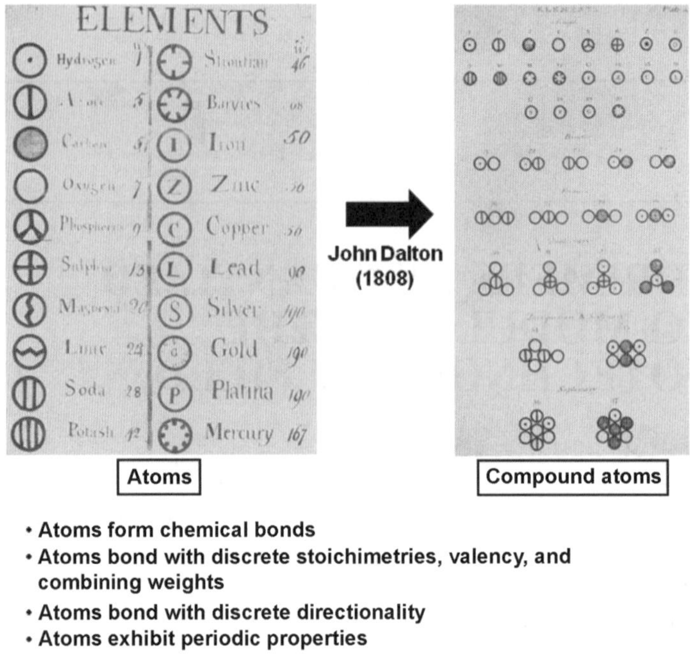

FIGURE 1.1 Dalton's first table of elemental atoms and their conversion to compound atoms according to his atom/molecular hypotheses. Key components of traditional chemistry central dogma based on Dalton's hypothesis [2].

Foundation (NSF) workshop entitled: "Periodic Patterns, Relationships and Categories of Well-Defined Nanoscale Building Blocks" [3]. An embryonic consensus of principles, classifications, and fundamentals emerged from this workshop that has now evolved into a systematic framework for unifying and defining nanoscience. This conceptual framework was based on historical principles, which were first used for traditional chemistry and included: (a) a nanomaterials classification roadmap, (b) a table of well-defined nano-module (element) categories, (c) combinatorial libraries of nano-compounds/-assemblies, and (d) the observation of nano-periodic property patterns. This concept was recently published [4]. It focused exclusively on well-defined monodisperse (0-D/1-D) nanoscale materials and the division of these well-defined materials into hard and soft nanoparticles (i.e., modules) following elemental compositional and architectural criteria invoked for traditional inorganic and organic materials (Figure 1.2). These well-defined nano-modules were selected based on criteria associated with traditional features/behavior of atoms (i.e., atom mimicry) and were referred to as hard and soft particle nano-element categories. As such, these nano-element categories mimic atoms by reacting or self-assembling to form combinatorial libraries of hard–hard, hard–soft, or soft–soft nano-compounds or nano-assemblies. Finally, due to the conserved nature of the precursor building blocks (i.e., atoms and/or small molecules) leading to these well-defined nano-element categories and their nano-compounds/-assemblies, many nano-periodic property

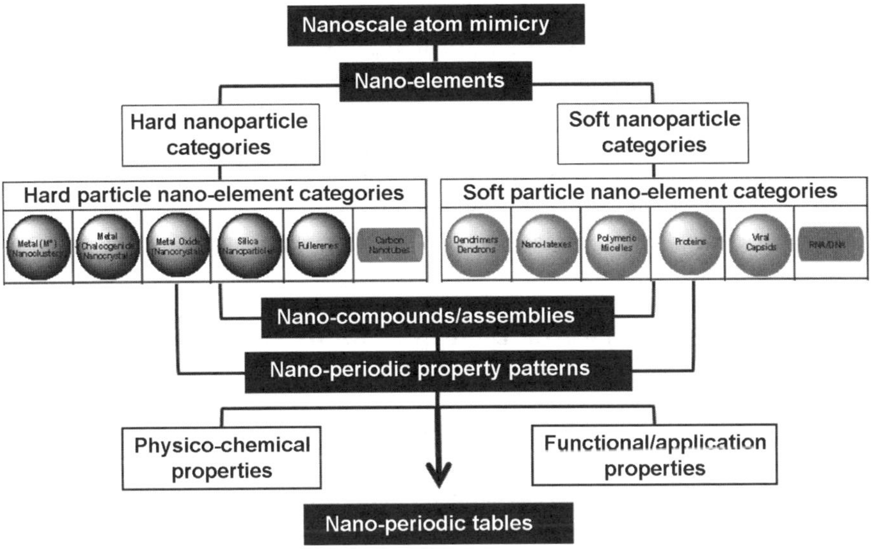

FIGURE 1.2 Concept overview. Using first principles and step logic that led to the "central dogma" for traditional chemistry, the criteria of nanoscale atom mimicry was applied to Category I-type, well-defined nanoparticles. This produced 12 proposed nano-element categories, which were classified into six hard particle and six soft particle nano-element categories. Chemically bonding or assembling these hard and soft nano-elements leads to hard:hard, soft: hard, or soft:soft type nano-compound categories, many of which have been reported in the literature. Based on the discrete, quantized features associated with the proposed nano-elements and their compounds, an abundance of nano-periodic property patterns related to their intrinsic physicochemical and functional/application properties have been observed and reported in the literature.

patterns have been observed and reported in the literature. These nano-property trends/periodic property patterns may be classified as either intrinsic physicochemical or functional/application type properties. In all cases these unique nano-property trends/patterns appear to be inextricably related to well controlled *c*ritical *n*anoscale *d*esign *p*arameters (CNDPs), such as size, shape, surface chemistry, flexibility, and architecture as overviewed in Figure 1.2.

This account will focus only on dendrons/dendrimers that have been previously reported to exhibit many features associated with traditional elemental atoms (i.e., *atom mimicry*) [5,6]. Inspired by these early examples of dendrimer-based heuristic atom mimicry, a broader concept was evolved which embraced both hard and soft particle nano-element categories. Presently, six hard nano-element and six soft nano-element categories have been proposed. Dendrons/dendrimers constitute one of the six proposed *soft particle, nano-element categories, namely the [S-1]; core–shell type category in this* nano-periodic system. As such, dendrimers behave as *soft particle nano-elements* to produce well-characterized core–shell (tecto) dendrimers (i.e., *dendrimer–dendrimer type nano-compounds*). Similarly, reactions with fullerenes have been reported to produce stoichiometric *dendrimer–fullerene type nano-compounds* [7], whereas, dendrimers bearing either surface or interior ligation sites reacted with metal salts to produce stoichiometric *dendrimer–metal ligation type nano-compounds* [8,9]. Some of these constructs could be further reduced to yield *core–shell; dendrimer–metal nano-cluster type nano-compounds.* These results optimistically portend the extension of this concept to a wide variety of other well-defined soft particle nano-element categories (i.e., proteins, DNA/RNA, viral capsids, etc.), and hard particle nano-element categories (i.e., metal oxide nanocrystals, metal chalcogenide nanocrystals, fullerenes, etc.).

1.2 CRITICAL NANOSCALE DESIGN PARAMETERS (CNDPs): STRUCTURAL CONTROL IN DENDRIMERS

Dendrimers are widely recognized as an important subclass within the fourth major polymer architecture category referred to as (IV) *dendritic polymers* [10] (see Figure 1.3). Hierarchically speaking, dendrimers are collections of 10^3–10^4 atoms with relative molar masses of 10^4–10^5 Da and hydrodynamic dimensions ranging from 1 to 30 nm. Among the most important features of dendrimers is that they can be synthesized with precise structural control over their CNDP's, which include nanoscale: (a) *sizes*, (b) *shapes*, (c) *surface chemistries*, (d) *flexibilities/ rigidities*, and (e) *architecture* [11]. This is in sharp contrast to statistical poly-dispersed features which are normally associated with traditional polymer architectures, namely: (I) linear, (II) cross-linked polymers, and (III) branched types as shown in Figure 1.3.

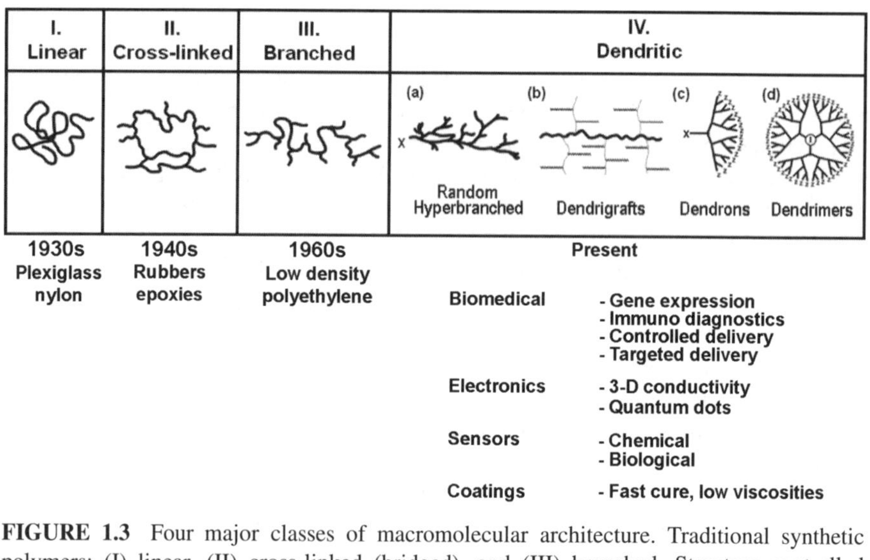

FIGURE 1.3 Four major classes of macromolecular architecture. Traditional synthetic polymers: (I) linear, (II) cross-linked (bridged), and (III) branched. Structure controlled polymers (IV) dendritic.

1.2.1 Dendrimer Synthesis: Divergent and Convergent Methods

In contrast to traditional polymers, dendrimers are monodisperse, macromolecular core–shell structures, consisting of three basic architectural components, namely: (1) a core, (2) an interior of monomer shells (generations) consisting of repeating branch-cell units, and (3) terminal functional groups (i.e., the outer shell or periphery). Dendrimer synthesis strategies may be divided into two major approaches and are referred to as either a *divergent* or *convergent* assembly strategies. The required building blocks and synthetic pathways are as outlined in Figure 1.4.

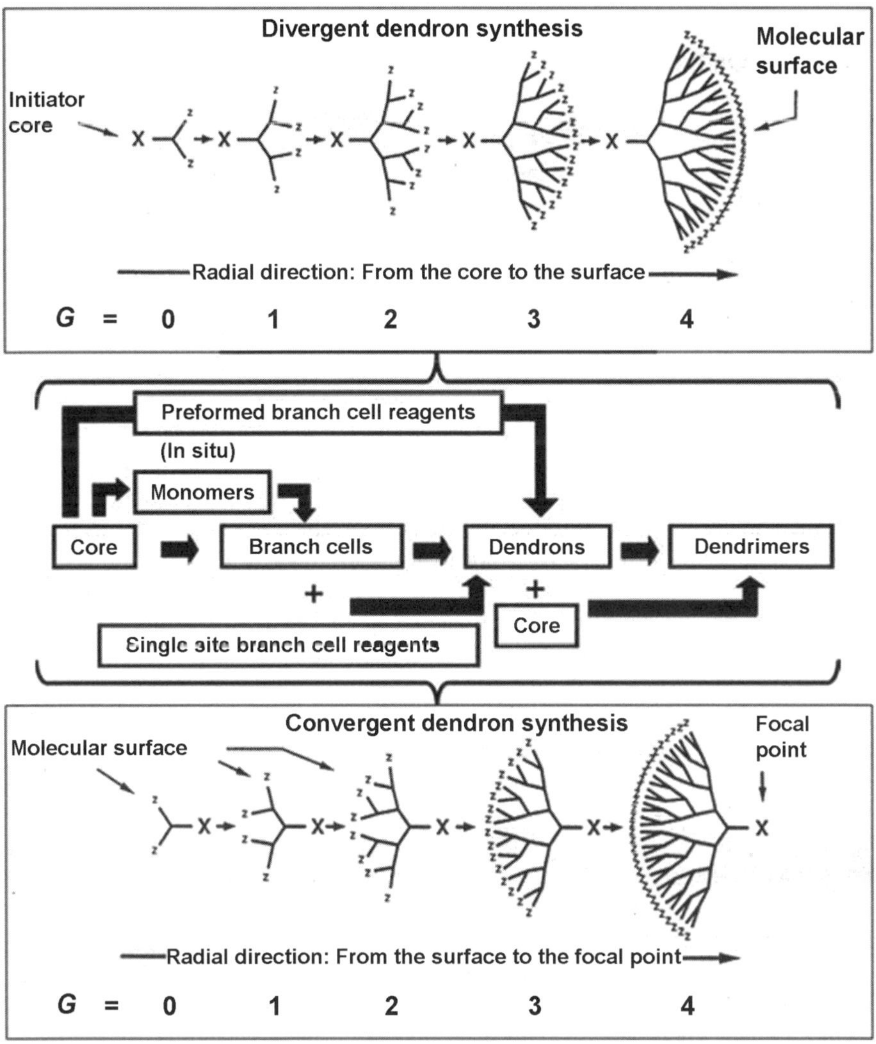

FIGURE 1.4 Hierarchical assembly scheme illustrating the options for constructing dendrimers by either divergent (Tomalia-type) or convergent (Frechet-type) synthetic strategies.

In general, dendrimers produced by the convergent strategy are nearly free of defects. This is in contrast to dendrimers obtained by the divergent approach; wherein, low defect levels are observed in the early generations with substantially higher defect levels occurring at higher generations (Figure 1.4). A comparison is made between *convergent*—Yamamoto-type dendrimers and *divergent*—Meijer-type dendrimers as shown in Figures 1.6 and 1.7, respectively.

Within each of these major approaches, there may be variations in methodology for branch-cell or dendron construction, (i.e., supramolecular assembly [12], click chemistry [13], etc.). Many of these issues, together with experimental laboratory procedures, have been extensively reviewed elsewhere [14–19].

1.2.2 Quantized Component Relationships to Produce Core–Shell, Soft Nano-Matter Dendrimer Structures

The steps required for the divergent assembly of dendrimers may be thought of as beginning with quantized, core information. This core information that defines initiator size, shape, multiplicity, and directionality then orchestrates subsequent iterative monomer shell polymerization events. The resulting monomer shells (i.e., generations) consist of mathematically defined numbers of monomers and branch-cell domains, which are required to produce these discrete nanoscale core–shell structures. As such, dendrimers may be viewed as nanoscale processing devices; wherein, their precise sizes and shapes are directed and controlled as illustrated in Figure 1.5.

Each architectural component manifests a specific function, and at the same time defines important periodic properties for these nanostructures as they are grown generation by generation. For example, as illustrated in Figure 1.5 the *core* may be thought of as the molecular information center from which *size*, *shape*, *directionality, and multiplicity* are expressed *via* covalent connectivity to the outer shells [11]. Within the *interior*, one finds the *branch-cell amplification region*, which defines the type and amount of interior void space that may be enclosed by the terminal groups as the dendrimer is grown to a surface congested level (Figure 1.9).

Branch-cell multiplicity (N_b) determines the density and degree of amplification as an exponential function of generation (G) (Figure 1.8). The interior composition and amount of solvent-filled void space determines the extent and nature of guest–host (endo-receptor) properties (i.e., for encapsulation of metals, pharma, or small guest molecules) that is possible with a particular dendrimer family and generation (Figure 1.24). Finally, the surface may consist of either reactive or passive terminal groups that may perform a variety of functions. With appropriate function, they serve as a *template polymerization region*; wherein, each generation is amplified and covalently attached to the precursor generation.

Secondly, the surface groups may function as passive or reactive gates controlling entry or departure of guest molecules from the dendrimer interior. These three architectural components essentially determine a rich range of nano-physicochemical properties, and the overall sizes, shapes, reactivity, and flexibility of dendrimers. It is important to note that dendrimer diameters increase linearly as a function of shells or

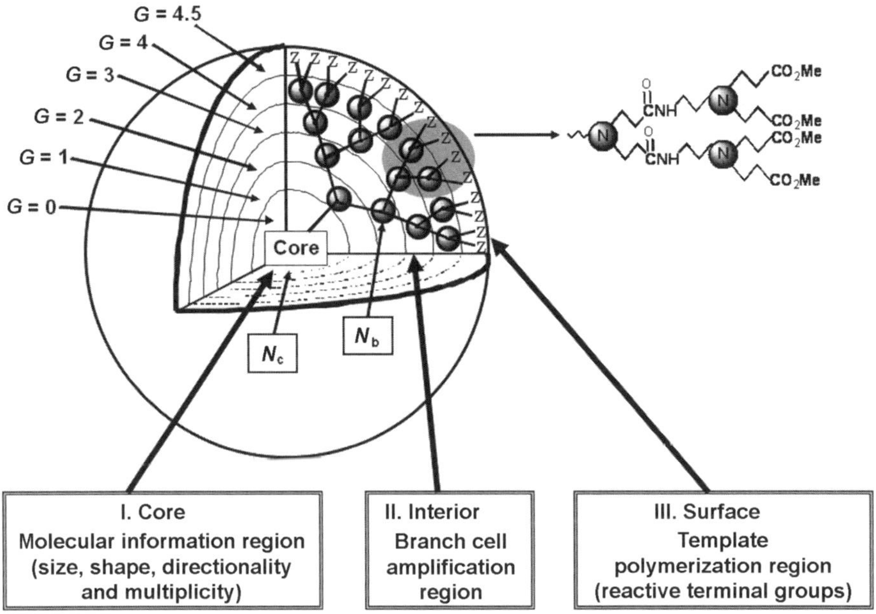

FIGURE 1.5 Tomalia type ($G = 4.5$); {*dendri*-poly(amidoamine)-$(CO_2Me)_Z$}; (PAMAM) dendrimers. A 3-D projection of such a dendrimer core–shell architecture for $G = 4.5$ poly (amidoamine) (PAMAM) dendrimer with principal architectural components (I) core, (II) interior, and (III) surface [20].

generations added; whereas, the terminal functional groups increase exponentially (Figure 1.8) as a function of generation. This dilemma enhances "tethered congestion" of the anchored dendrons, as a function of generation, due to the steric crowding of the end groups. As a consequence, lower generations are generally open, floppy structures; whereas, higher generations become robust, less deformable spheroids, ellipsoids, or cylinders [23] depending on the shape and multiplicity of the core.

1.3 EXPERIMENTALLY PROVEN STRUCTURAL CONTROL OF CRITICAL NANOSCALE DESIGN PARAMETERS (CNDPs)

In this section, we focus on experimentally proven examples that illustrate structural control of CNDP's when synthesizing dendrons/dendrimers. This brief overview illustrates examples of atom mimicry based on the synthetic ability to control; (a) *sizes*, (b) *shapes*, (c) *surface chemistry*, (d) *flexibility*, and (e) *architecture at the nanoscale level*. It is apparent, that both the core multiplicity (N_c) and branch-cell multiplicity (N_b) determine the precise number of terminal groups (Z) and mass amplification as a function of generation (G). One may view those generation sequences as quantized polymerization events that adhere to the dendritic aufbau

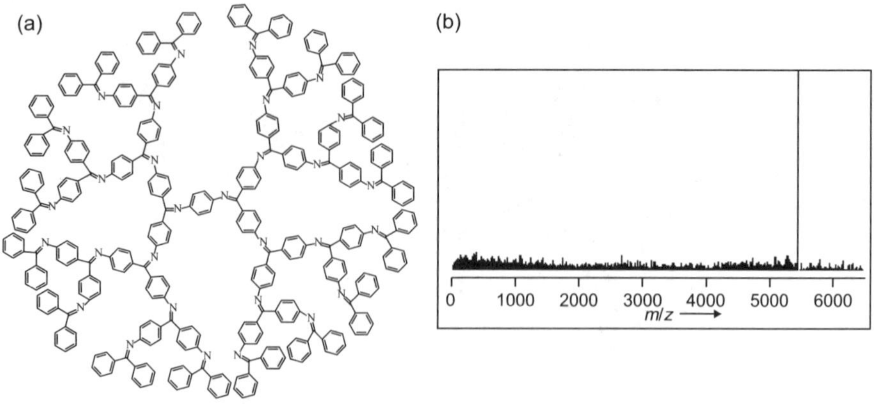

FIGURE 1.6 (a) Yamamoto-type [core:*p*-phenylene]; ($G=4$); {*dendri*-poly(phenylazomethine)} (DPA) dendrimers, (b) MALDI-TOF of the G4; DPA dendrimer [21]. Copyright 2001 American Chemical Society.

mathematics described in Figure 1.9. The assembly of reactive monomers [11,24], branch cells [11,25,26], or dendrons [12,27] around atomic or molecular cores to produce dendrimers according to divergent/convergent dendritic branching principles has been well demonstrated (Figure 1.4). This systematic filling of nano-space around cores with branch cells in generational growth stages (branch-cell shells) produces discrete, quantized bundles of mass that are mathematically predictable [5]. Predicted molecular weights are routinely confirmed by mass spectroscopy [22,28,29] (Figures 1.6 and 1.7) and other analytical methods [27,30–32]. Predicted numbers of branch cells, terminal groups (*Z*), and molecular weights, as a function of generation, are as described in Figure 1.8. It should be noted, that the molecular weights approximately double as one progresses from one generation to the next. The surface groups (*Z*) and branch cells (BC) amplify mathematically according to a power function; thus, producing discrete, monodisperse structures with precise molecular weights. These predicted values are routinely verified for convergent synthesized dendrimers by mass spectroscopy (Figure 1.6). However, for divergent synthesized dendrimers, minor mass defects are often observed for higher generations as congestion induced *de Gennes dense packing* and nano-steric effects begin to occur [11,33] (Figure 1.7).

1.3.1 Quantized Size-Control of Monodisperity

The bottom-up synthesis of dendrons/dendrimers provides one of the most precise and tunable strategies known for constructing a systematic continuum of reproducible soft matter nanoscale structures. Size and structure control observed for dendron/dendrimer synthesis rival that expected for proteins and DNA/RNA. In fact, dendrimers are often referred to as *artificial proteins* [34–36]. Based on the close mimicry of globular protein size scaling and their monodispersity, they are often used as protein replacements/substitutes in many nanomedicinal applications [34,35,37,38].

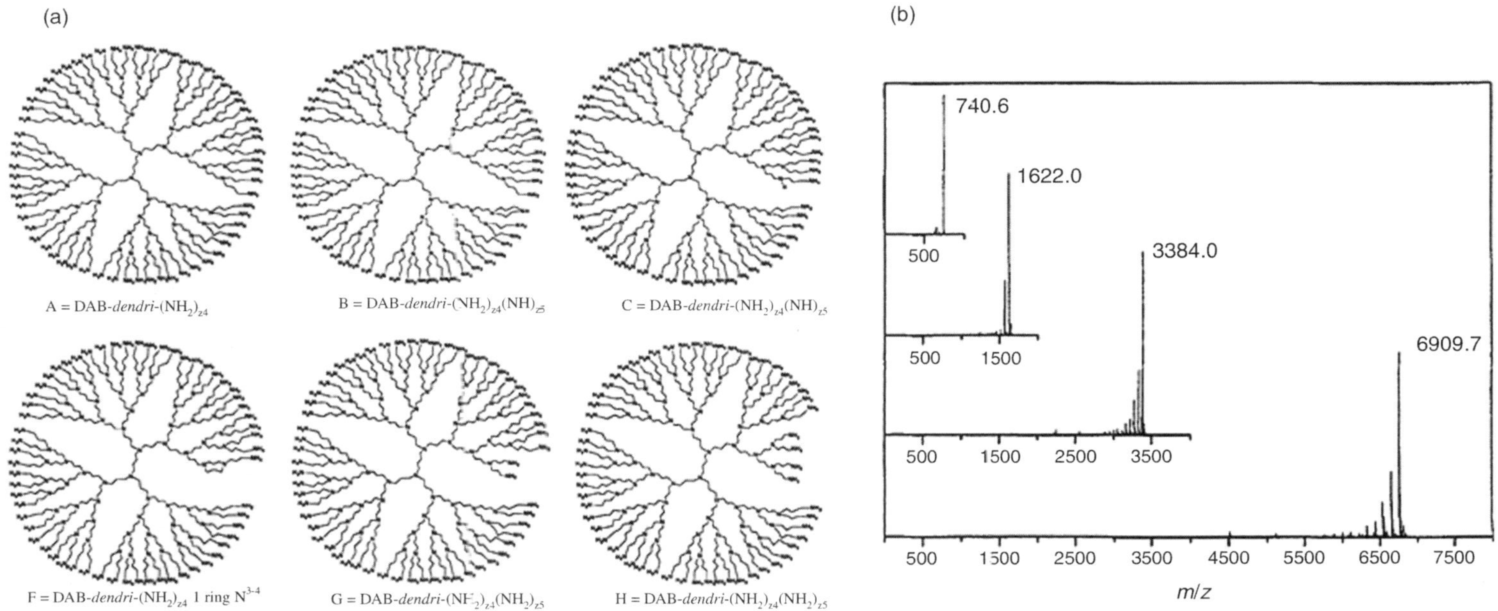

FIGURE 1.7 (a) Meijer-type [core:diaminobutane]; ($G = 4$); {*dendri*-poly(propyleneimine)-$(NH_2)_Z$} (PPI) dendrimers illustrating perfect structure and defective structures observed by mass spectrometry, (b) deconvoluted ESI-MS data showing masses for perfect structures and defective structures for PPI dendrimers; $G = 1$–4 [22]. Copyright Wiley-VCH Verlag Gmbh & Co. KGaA.

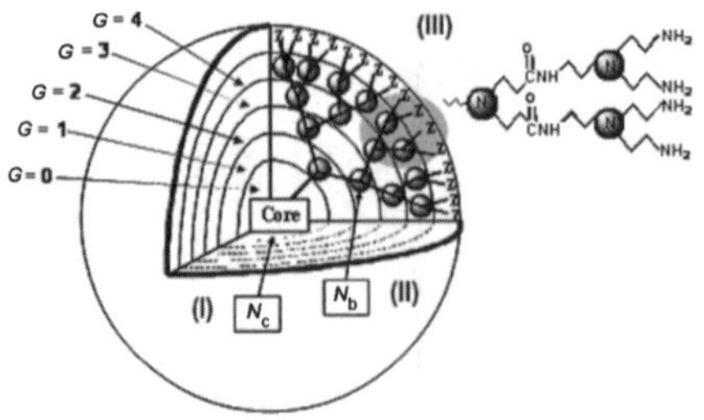

Number of surface groups	Z	$= N_c N_b^{G}$	Surface group amplification per generation
Number of branch cells	BC	$= N_c N_b^{G-1}$	
Molecular weights	MW	$= M_c + N_c[M_{Ru} N_b^{G-1} + M_t N_b^{G}]$	

G	No. of NH_2 Surface Groups	Molecular Formula	MW	Hydrodynamic Diameter (nm)
0	4	$C_{24}H_{52}N_{10}O_4S_2$	609	1.5
1	8	$C_{64}H_{132}N_{26}O_{12}S_2$	1,522	2.2
2	16	$C_{144}H_{292}N_{58}O_{28}S_2$	3,348	2.9
3	32	$C_{304}H_{612}N_{122}O_{60}S_2$	7,001	3.6
4	64	$C_{624}H_{1252}N_{250}O_{124}S_2$	14,307	4.5
5	128	$C_{1264}H_{2532}N_{506}O_{252}S_2$	28,918	5.4
6	256	$C_{2544}H_{5092}N_{1018}O_{508}S_2$	58,140	6.7
7	512	$C_{5104}H_{10212}N_{2042}O_{1020}S_2$	116,585	8.1

Z = monomer-shell-saturation level, N_c = core (cystamine) multiplicity, N_b = branch-cell (BC) multiplicity, *G* = generation.

FIGURE 1.8 Mathematical expressions for calculating the theoretical number of surface groups (*Z*), branch cells (BC), and molecular weights (Mw) for [*core*:1, cystamine]; ($G = 0$–7); {*dendri*-poly(amidoamine)-$(NH_2)_Z$} (PAMAM) dendrimers as a function of generation. Approximate hydrodynamic diameters ($G = 0$–7) based on gel electrophoretic comparison for the corresponding PAMAM dendrimers.

Dendrimer mass and size uniformity has been exhaustively demonstrated by ESI/MALDI-TOF mass spectrometry [29], gel electrophoresis [30–32,39], electron microscopy (TEM) [40], (Figure 1.10), atomic force microscopy (AFM) [41] and other methodologies, to mention a few [42]. Within a specific dendrimer family, it is possible to produce a systematic, reproducible continuum of nano-sizes and precise masses as a function of generations (see Figures 1.8–1.10). It should be noted, that dendrimer mass approximately doubles, generation to generation, with a remarkable uniformity of mass distribution over five generations exhibiting polydispersities ranging from 1.005 to 1.130.

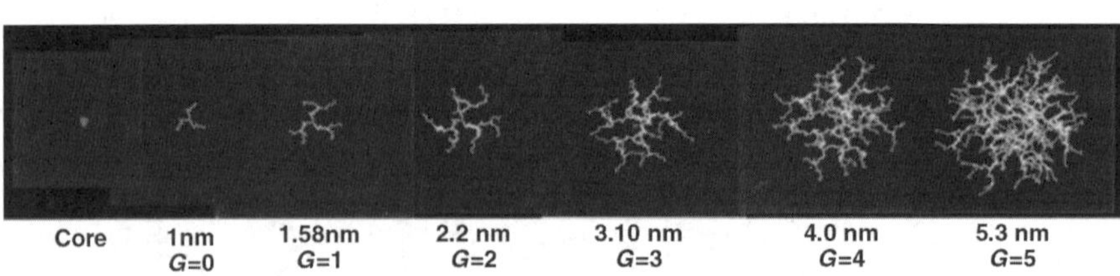

FIGURE 1.9 Molecular simulations for [core:NH_3]; ($G = 0$–5); {*dendri*-poly(amidoamine)-$(NH_2)_n$} (PAMAM) dendrimers and a generational comparison of hydrodynamic diameters. (See the color version of this figure in Color Plates section.)

1.3.2 Quantized Shapes of Dendrimers

Several simple parameters may be used to design and control covalent dendrimer shape. As described in Figures 1.5 and 1.11, the multiplicity and the shape of the core can be selected to produce a desired shape (i.e., spheroidal, ellipsoidal, cylindrical, etc.). For example, covalent 1-D cylindrical dendrimers are readily synthesized with shape control by dendronization or initiating dendron growth from multiple sites along the backbone of a linear polymer [23,43]. Iterative monomer additions to produce generational growth will then produce precise amplifications of these programmed shapes. The rigidity/flexibility of these shapes may be controlled and tuned by appropriate selection of branch-cell monomer symmetry, compositions, and arm lengths (flexibility).

Pioneering work by Rudick and Percec [44], demonstrated that quantized shape design of dendrons as a function of their generation level produced predictable hierarchical patterns of periodic nano-object shapes that could be defined as solid angle projections (α') of the dendron onto a plane (Figure 1.11). The (α') angle is defined as the projection of the solid angle of the dendron onto a plane, and can be

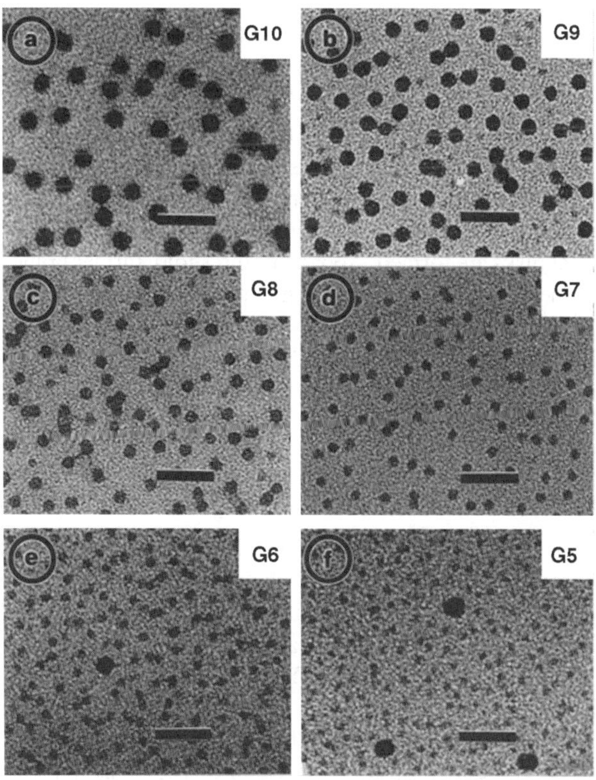

FIGURE 1.10 (a–f) Transmission electron micrographs (TEMs) of $G = 5$–10 PAMAM dendrimers. Sample (f) contains three molecules of $G = 10$ dendrimer for comparison. Bar length = 50 nm [40]. Copyright 1998 American Chemical Society.

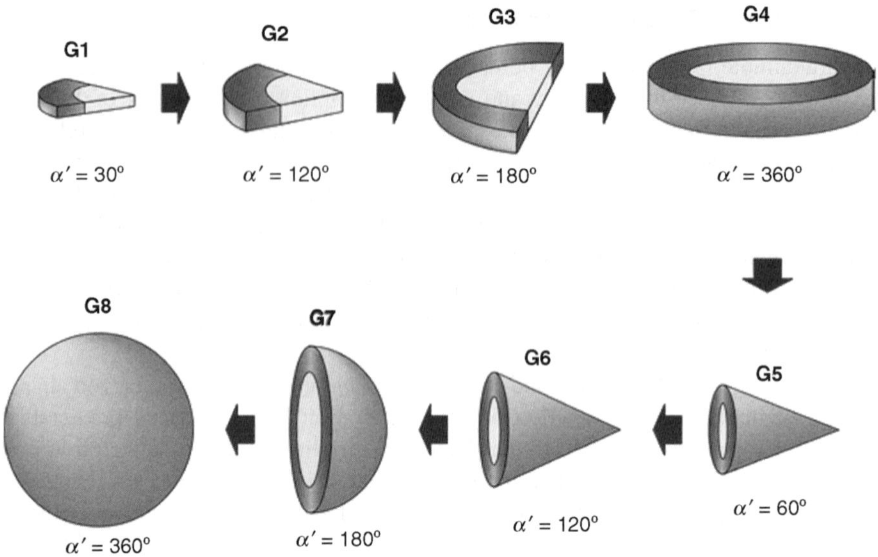

FIGURE 1.11 Hierarchical control of self-assembly *via* molecular solid angle [46].

determined in all cases according to $(\alpha') = 360/u$, where u is the number of dendrons in a column stratum or supramolecular sphere. Increasing the branching *via* a change in sequence or increase in generation number of taper-like dendrons increases the (α') and the fraction of the disk occupied in columnar self-assembly. At a certain threshold only unimolecular disks are formed. Above this threshold, further branching results in deformation of the disk into a conical segment with diminished (α'). Beyond this point, increased branching increases (α') and the fraction of a sphere formed. Ultimately, a unimolecular sphere should result. A wide variety of periodic and quasiperiodic lattices can be formed by simply tuning the dendron shape parameters as described in Figure 1.12 [45].

1.3.3 Quantized Surface Chemistry

More than 1000 different surface reactions have been reported for dendrons/dendrimers. Essentially, all mechanistic reaction types (i.e., covalent, ionic, radical, etc.) have been reported. Several examples of these reactions are illustrated in Figure 1.13. Both subnanoscale, and nanoscale reagents (i.e., proposed nano-elements) have been used in these surface reactions. The surface valency and stoichiometry are mathematically quantified by relationships described in Figure 1.8 (i.e., $Z = N_c N_b{}^G$). Generally, these valencies and stoichiometric binding ratios are operational for all generations preceding serious onset of de Gennes-type dense packing congestion. Surface congested generations (i.e., $>G = 4$ for poly(amidoamine) (PAMAM) dendrimers) usually exhibit lower than theoretical stoichiometries even with small-sized reagents. Large-sized reagents (i.e., >1 nm) may lead to less than ideal stoichiometries

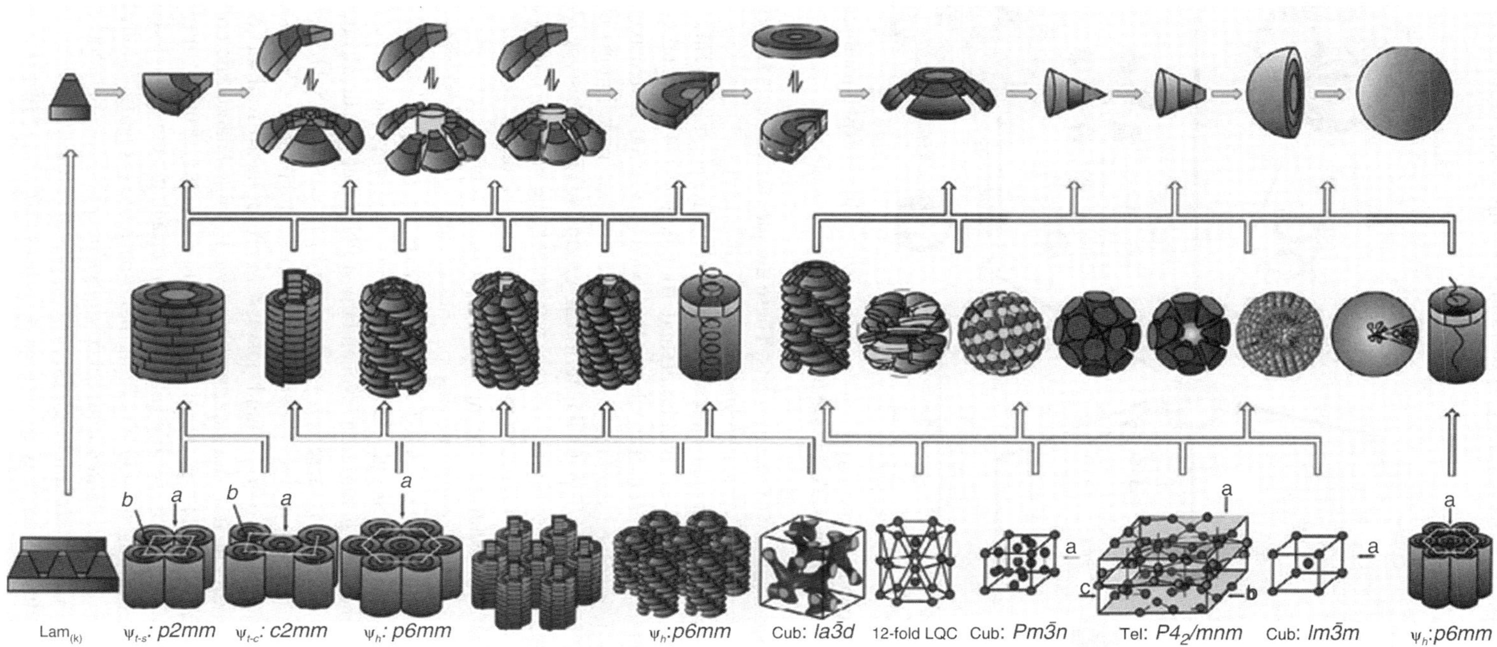

FIGURE 1.12 Periodic and quasiperiodic lattices formed via the self-assembly of Percec-type dendrons [45]. (See the color version of this figure in Color Plates section.)

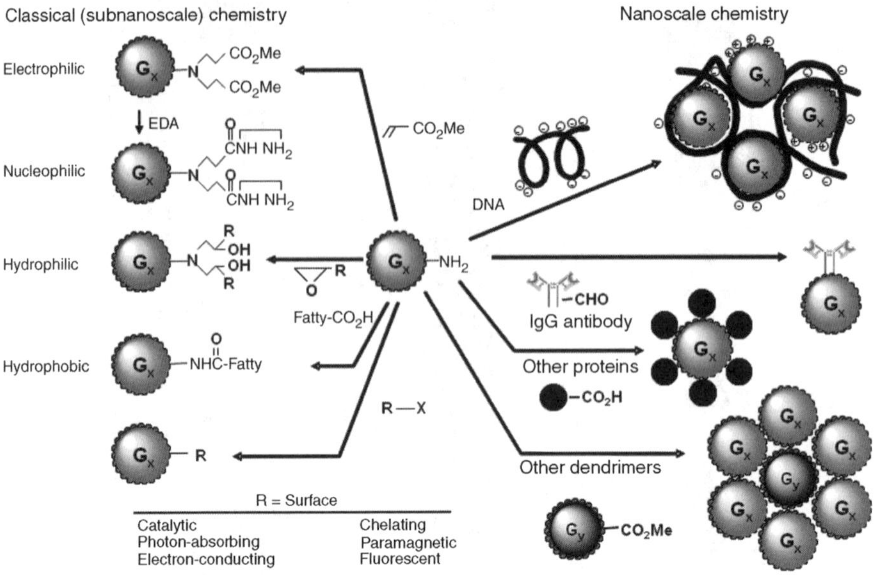

FIGURE 1.13 Options for modifying amine-terminated dendrimer-based nano scaffolding by utilizing classical subnanoscale and nanoscale reagents.

even at low generations due to steric effects. These unique steric effects are referred to as *nanoscale sterically induced stoichiometries* (N-SIS) [33].

The use of dendrimers as nanoscale scaffolding for MRI contrast agents was first reported by Lauterbur, Wiener, Brechbiel, and Tomalia in 1994 [47] and generated tremendous interest and activity which has been reviewed extensively elsewhere [48,49]. It involved conjugating metal-specific (i.e., gadolinium) chelating groups (i.e., DTPA or DOTA) to the terminal groups of Tomalia-type PAMAM dendrimers to produce a continuum of very precisely sized nanoparticles with well-defined nano dimensions (Figure 1.14). These surface saturated, dendrimer–metal chelates were well-characterized examples of: [dendrimer (core)-ligated metal (shell) nano-compounds]. The stoichiometries of these nano-compounds were defined accordingly as: [dendrimer]:[ligated metal]$_Z$, where $Z = N_c N_b{}^G$. These core:shell nano-compounds manifested enhancements in relaxivity properties (i.e., R_1) as a function of the dendrimer generation level and the stoichiometry of dendrimer:metal. Furthermore, the features of these soft particle–hard particle (i.e., [S-1]:[H-1] nano-compounds) fit the proposed criteria for nano-compounds described elsewhere [3,4,66]. These nano-compound possess a core derived from an [S-1] type dendrimer nano-element and shells derived from well-defined nanoscale collections of ligated metals (i.e., [H-1] type metal cluster nano-elements). Finally this series of [core:shell nano-compounds] fulfilled additional criteria that were predicted for these entities [4] by exhibiting unique emerging nano-periodic property patterns [4] as evidenced by their unprecedented high relaxivity values (i.e., R_1) that increased as a function of dendrimer generation level. Empirical formulae for members of this core:shell nano-compound series may be written as follows: $[G = 1–8]@[M]N_c N_b{}^G$.

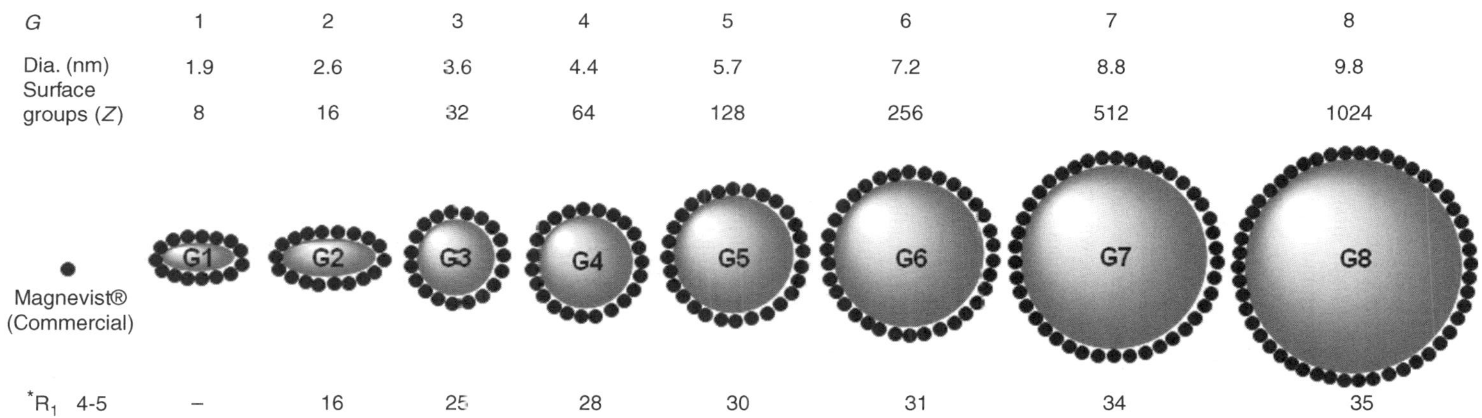

FIGURE 1.14 A Tomalia-type PAMAM dendrimer series (i.e., [*core*:1,2-diaminoethane]; ($G = 1$–8); {*dendri*-poly(amidoamine)-(metal ligation sites)$_Z$} dendrimers) illustrating; the core:shell topology, naked dendrimer scaffolding diameters (nm), number of terminal groups (Z), and relaxivity values (R_1) as a function of generation.

1.3.4 Quantized Interior Chemistry

Metal complexations with dendrimers containing nitrogen-type ligation sites have been known since the phenomenon was first observed in the earliest reported examples of the *dendri*-poly(amidoamine) (PAMAM) series [50]. More recently considerable attention has been focused on the use of certain rigid (i.e., less flexible) Yamamoto-type (i.e., *dendri*-poly(phenylazomethine) (DPA) dendrimers as nanoscale templates for assembling very well-defined metal-based nano-cluster [9] and metal oxide nanoparticles [8]. In contrast to the more flexible/less rigid nitrogen-containing PAMAM or PPI type dendrimers, the Yamamoto-type dendrimers [51] have been shown to complex and assemble a variety of metal salts in a very precise, shell by shell, sequential manner as shown in Figure 1.15. Empirical formulae describing members of this core:shell nano-compound series may be written as follows: $[M]\ N_cN_b^{G-1}$ @ $[G=4]$; where $N_cN_b^{G-1} = 4, 12, 28$, and 60, respectively. The generational shells 1–4 are saturated from the core outwardly in a sequential fashion to produce stoichiometric, dendrimer-ligated metal nano-assemblies. These nano-assemblies exhibited well-defined, mass combining ratios (i.e., dendrimer-ligated metal salt masses) reminiscent of traditional small molecule compounds. It is notable that these nano-assemblies consisted of discrete nano-module components (i.e., dendrimers and ligated metal atom collections), each of which were of nanoscale dimensions. Later it will be shown that these ligated metal domains may be transformed into precise metal nano-clusters or metal oxide nanocrystals, respectively, that remain associated as guests within their dendrimer hosts in stoichiometric proportions to produce unique core:shell nano-compounds.

1.4 HEURISTIC ATOM MIMICRY OF DENDRIMERS: *NANO-LEVEL CORE–SHELL ANALOGS OF ATOMS*

A heuristic comparison of the core–shell architectures that are present in dendrimer-based nanoscale modules and picoscale atoms was made as early as 1990 [5,6,11]. This comparison was used to point out the unique similarities that exist between aufbau components in atoms (i.e., nucleons and electrons) and those that are involved in dendrimer constructions (i.e., cores and branch-cell monomers). Remarkable analogies were also noted between dimensionally different parameters, but recognized components shared by both systems, such as (a) electron shells versus monomer shells (generations), (b) electron shell versus monomer shell aufbau filling patterns (i.e., mathematically defined), (c) electron shell versus monomer shell saturation levels, (d) atomic weights versus dendrimer molecular weights as a function of shell level and saturation level, and (e) atomic (elemental) reactivity versus dendrimer reactivity as a function of shell saturation level. We have referred to these remarkable similarities between picoscale (atomic elements) and nanoscale dendrimers as *atom mimicry*, keeping in mind that picoscale structures are best described by non-Newtonian physics, whereas, the dendrimer structures are expected to adhere to and be described by Newtonian physics. Furthermore, it appears that a very interesting

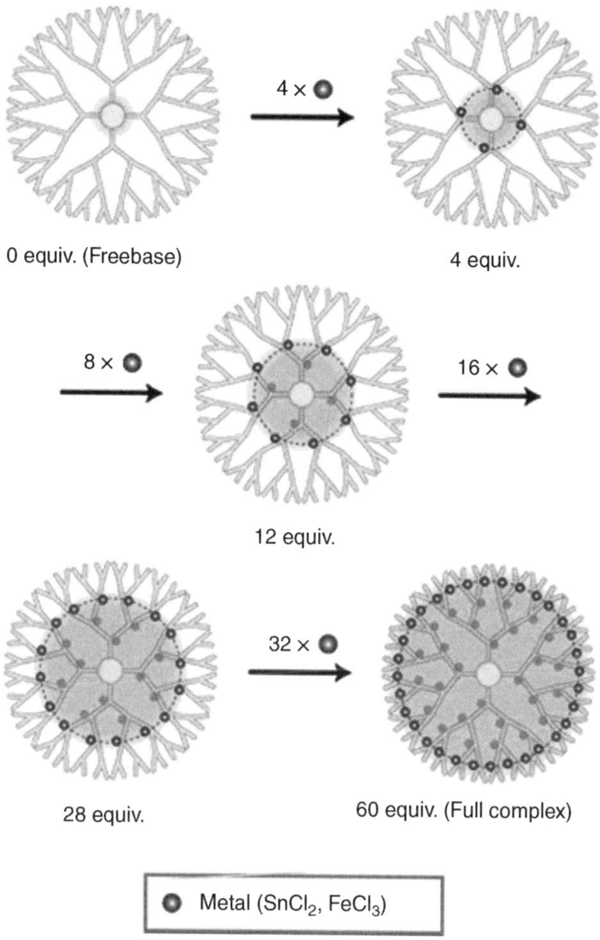

FIGURE 1.15 Schematic representation of the stepwise complexation of $SnCl_2$ or $FeCl_3$ with a Yamamoto-type dendrimer [51].

size continuum exists in the transition from picoscale (atomic) structures to nanoscale dendrimer structures as illustrated in Figure 1.16.

1.5 CHEMICAL BOND FORMATION/VALENCY AND STOICHIOMETRIC BINDING RATIOS WITH DENDRIMERS TO FORM NANO-COMPOUNDS/-ASSEMBLIES

Some of the most compelling experimental examples of dendrimer-based atom mimicry were documented with AFM. These studies clearly demonstrated the remarkably rich array of nano-assembly patterns and related stoichiometric binding

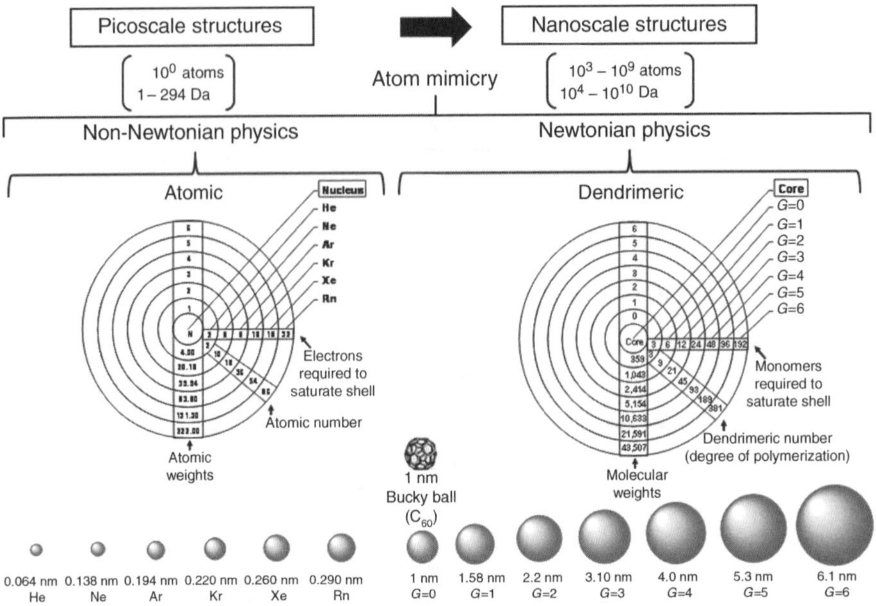

FIGURE 1.16 An example of atom mimicry. A comparison of core–shell structures representing picoscale atoms and nanoscale dendrimers, as well as the continuum of sizes that prevails over the 2-D ranges that are controlled by quantum mechanics and Newtonian physics, respectively.

ratios that were possible by simply spreading dilute solutions of amine terminated, $G = 9$; poly(amidoamine) (PAMAM) dendrimers on a mica surface with a 30° stream of argon. As shown in Figure 1.17, one can readily observe single isolated $G = 9$ modules, dimers, trimers, and a wide variety of oligomeric assemblies (i.e., megamers) that clearly exhibit well-defined 2-D combining ratios on the mica surface.

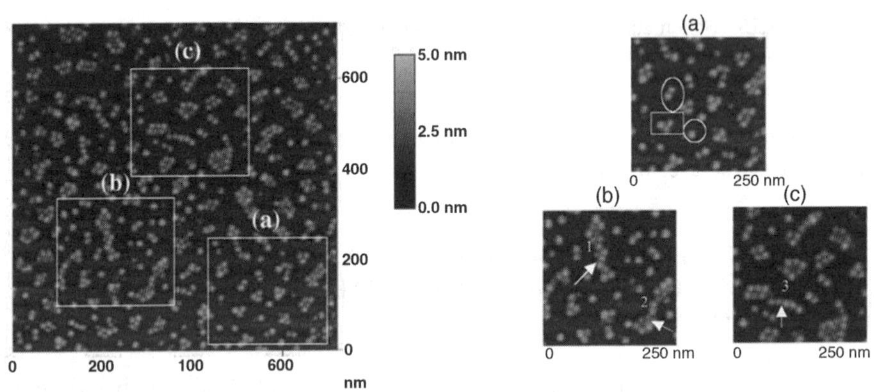

FIGURE 1.17 Tapping mode AFM images of $G = 9$: PAMAM dendrimer molecules on a mica surface [52].

Using first principles and step logic invoked by Dalton (i.e., *Philosophy for a Chemical System*, 1808) [2], as stated earlier (Figure 1.1), it was possible to experimentally demonstrate that certain quantized nano-modules (i.e., dendrimers, fullerenes, metal nano-clusters, or metal oxide nanocrystals) could be chemically combined or assembled to produce stoichiometric nano-compounds/-assemblies possessing well-defined mass combining ratios. Furthermore, both the quantized nano-modules (i.e., nano-elements) and their resulting nano-compounds were found to exhibit new emerging properties and interesting new nano-periodic property patterns [4].

Both soft matter and hard matter categories of these quantized nano-modules (i.e., referred to as nano-elements), have been proposed based on selection criteria described elsewhere [4] (Figure 1.18). Furthermore, these 12 soft and hard nano-elements, designated [S-n] and [H-n], respectively, have been reported to form a wide range of soft particle and soft–hard particle type nano-compounds. Both the nano-elements and their nano-compounds are widely recognized to exhibit new emerging properties and nano-periodic property patterns [4]. Leading references to these literature examples, designated by X in the combinatorial nano-compound library (Figure 1.18), are described in greater detail elsewhere [4]. This account will focus only on four selected examples of nano-compound formation designated by (*) that involve either reactions or self-assembly with dendrimers (Figure 1.18). Combining dendrimers (i.e., [S-1] type) nano-element categories with other dendrimers (i.e., [S-1]:[S-1]$_n$ core–shell

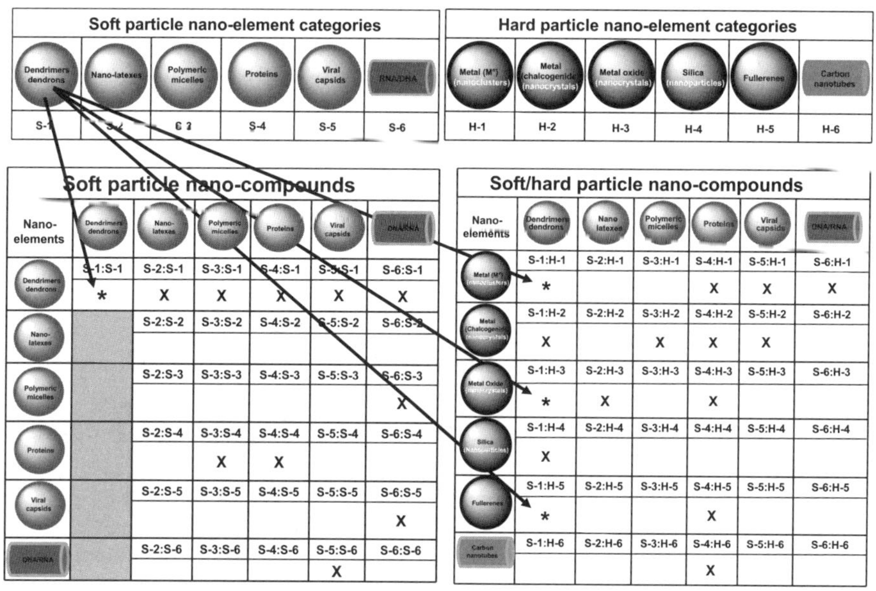

FIGURE 1.18 Proposed hard and soft particle nano-element categories and combinatorial libraries of possible nano-compounds. Nano-compounds designated by [*] are described in the following section. Nano-compounds designated by [X] have been reported in the literature and described elsewhere [4].

types), with fullerenes (i.e., [S-1]:[H-5]$_n$ core–shell type), or with metal nano-clusters (i.e., [H-1]$_n$:[S-1] core–shell type) produced three unique core–shell nano-compound category types designated in Figure 1.18. On the other hand, dendrimers have also been used as templates to produce precisely sized [H-3] type, metal oxide nanocrystals which have exhibited interesting quantum size effects as described in the next section.

1.5.1 Dendrimer–Dendrimer; [S-1]:[S-1]$_n$ Core–Shell Type Nano-Compounds

Saturated shell, nano-compounds (Figure 1.19) are prepared by a two-step approach that involved: first, self-assembly of an excess of carboxylic acid terminated dendrimers (i.e., shell reagent) around a limited amount of amine-terminated dendrimer (i.e., core reagent) in the presence of LiCl. This was followed by covalent amide bond formation between the core and dendrimer shell reagents using a carbodiimide reagent [52–54]. The resulting nano-compounds (i.e., saturated core–shell tecto (dendrimers), referred to as *megamers*), are prime examples of precise poly-dendrimer cluster structures. These structures and their stoichiometries may be mathematically predicted by the Mansfield–Tomalia–Rakesh equation (Figure 1.25) [6,55] and have been unequivocally verified by experimental mass spectrometry, gel electrophoresis, and atomic force field microscopy [6,52,53,56].

1.5.2 Dendrimer–Fullerene; [S-1]:[H-5]$_n$ Core–Shell Type Nano-Compounds

Stoichiometric dendrimer core–fullerene shell nano-compounds were readily formed by allowing a [*core*: 1,2-diaminoethane]; ($G=4$); {*dendri*-poly(amidoamine)-$(NH_2)_{64}$} (PAMAM) dendrimer to react with an excess of buckminsterfullerene (C_{60}) [7]. In the presence of an excess of (C_{60}), only 30 (C_{60}) moieties bonded to the dendrimer surface to produce a well defined, stoichiometric [dendrimer core]:[fullerene shell nano-compound] (i.e., [S-1]:[H-5]$_n$ core–shell type) as shown in Figure 1.20. These structures were extensively characterized by MALDI-TOF, TGA,

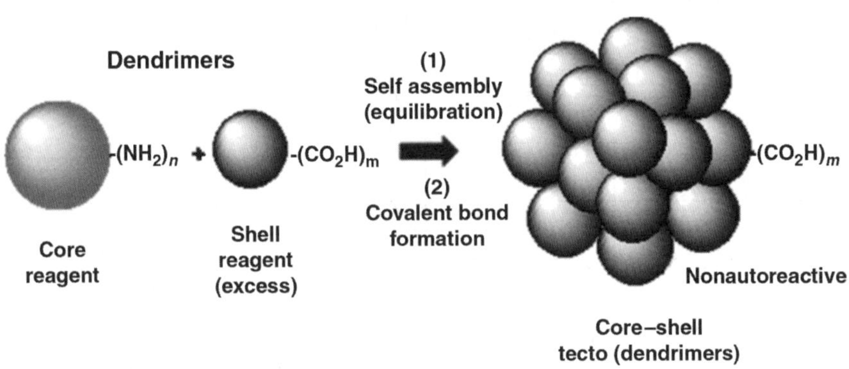

FIGURE 1.19 The saturated-shell-architecture approach to megamer synthesis. All surface dendrimers are carboxylic acid terminated [54].

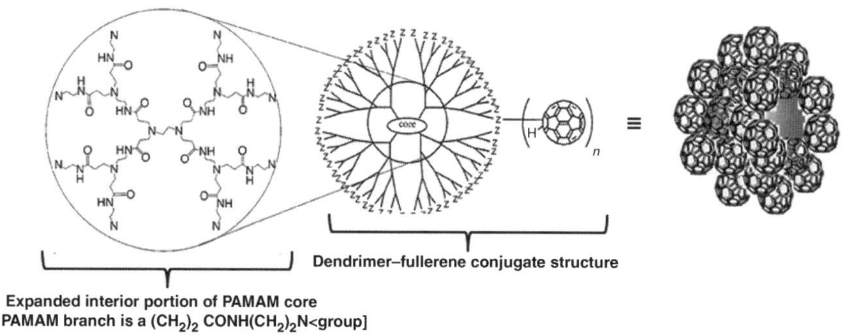

FIGURE 1.20 Core–shell architecture of the PAMAM core:fullerene shell; [S-1]:[H-5] core–shell type nano-compound; where: (Z = terminal $–NH_2$ or–NH–, groups on the PAMAM dendrimer core component of the core–shell nano-compound) [7].

UV-vis, and FTIR. Such nano-compounds exhibited new emerging fullerene-like solubility and photo-chemical properties by readily generating singlet ($^1O_2$) in either aqueous/organic solvents. However, they offered other unique features such as larger size and nano-container type properties that would normally be associated with the dendrimer core interior.

1.5.3 Metal Nano-Cluster–Dendrimer; $[H\text{-}1]_n$:[S-1] Core–Shell Type Nano-Compounds

Using a $G = 4$, phenylazomethine dendrimer as a template, Yamamoto et al. [9] demonstrated that Pt(IV) chloride salts could be sequentially assembled to produce four discretely different, yet precise, collections of ligated Pt salts within the interior of the dendrimer. Subsequent reduction of these ligated Pt(IV) arrays yielded Pt(0) Nan clusters consisting of very *well defined numbers of Pt atoms*. The size of these incarcerated Pt(0) nano-clusters (i.e., by TEM) was in good agreement with the number of Pt atoms present in the ligated metal precursors. For example, the $[H\text{-}1]_n$:[S-1] nano-compound series designated by $[Pt_{12}]$@[G4;dendrimer], $[Pt_{28}]$@[G4;dendrimer], and $[Pt_{60}]$@[G4;dendrimer] core–shell type nomenclature were extensively characterized and determined to exhibit precise stoichiometries as described in Figure 1.21 (Where: n = 12, 28 and 60 respectively). The ability to precisely control these independently sized Pt(0) nano-cluster cores within the G4; dendrimer shells provided discrete nano-cages that prevented metal nano-cluster aggregation. Furthermore, it allowed the evaluation of these four discrete core–shell nano-compounds as catalysts in reactions involving four (4X) electron reductions of oxygen molecules. These catalytic reductions are of high interest in the development of optimized fuel cells. Quite remarkably it was demonstrated that these catalytic reductions are highly dependent on subtle metal nano-cluster size differences.

As such, a significant nano-periodic property pattern was observed within this nano-compound series. It was found that within this Pt(0) nano-cluster size continuum (i.e., sizes ranging from 0.5 to 6.0 nm) it was observed that 3 nm Pt(0) nano-cluster sizes offered optimum catalytic properties [9].

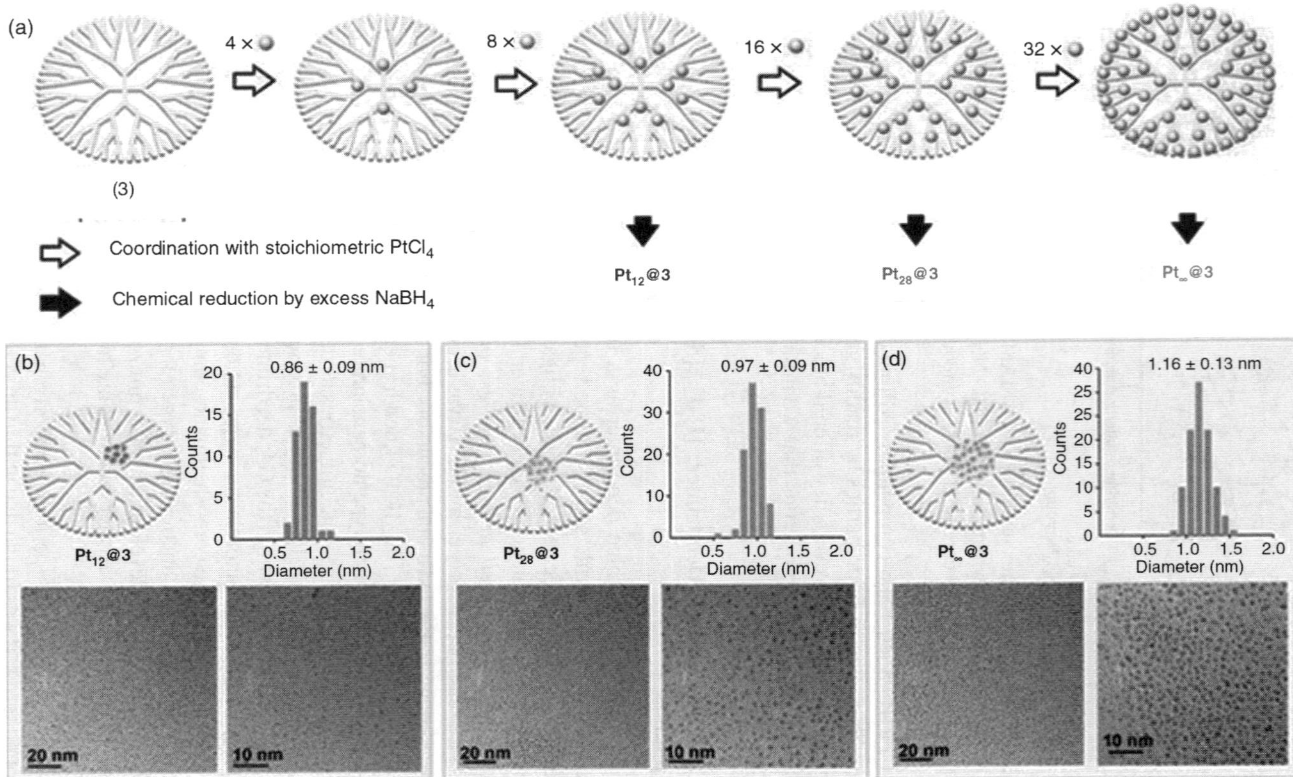

FIGURE 1.21 (a) Sequential complexation of $PtCl_4$ with a Yamamoto-type dendrimer, (3), to produce $Pt_{12}@3$, $Pt_{28}@3$, and $Pt_{60}@3$; $[H\text{-}3]_n$:[S-1] core–shell type nano-compounds, respectively, after reduction with $NaBH_4$. (b–d) Histograms of Pt(0) nano-cluster core size distributions in each nano-compound above with corresponding electromicrograms of these $[H\text{-}3]_n$:[S-1] core–shell type nano-compounds [9]. Reprinted by permission from Macmillan Publishers Ltd., *Nature Chemistry* 2009. (See the color version of this figure in Color Plates section.)

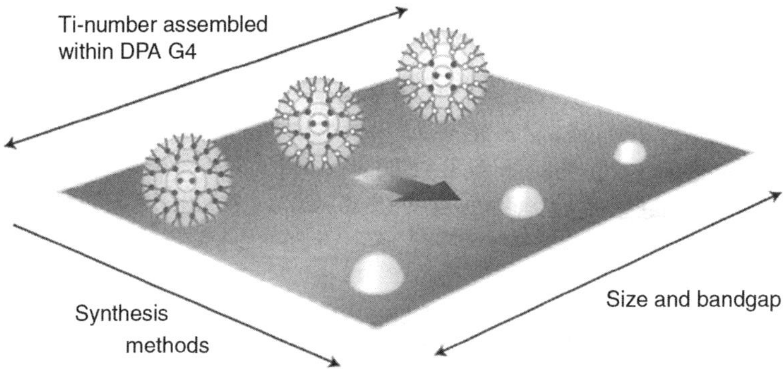

FIGURE 1.22 Illustration of the quantum size effect (Q-size) on semiconductor bandgaps for a series of well-defined $[TiO_2]_n$ nanocrystals. The nanoparticle size can be precisely controlled in a dendrimer template and is influenced by the number of $[TiO_2]_n$ assembled in a dendrimer interior. This is an elegant synthesis strategy for producing well-defined [H-3] type, metal oxide, nano-element particles [8]. Reprinted by permission from Macmillan Publishers Ltd., *Nature Nanotechnology* 2008.

1.5.4 Templated Metal Salt Nanocrystal–Dendrimer; $[H\text{-}3]_n$:[S-1] Core–Shell Nano-Compounds as Precursors to Well-Defined Metal Oxide Nanocrystals (i.e., $[TiO_2]n$, where $n = 6$, 14, and 30)

Much as described above, Yamamoto et al. [9], also demonstrated use of the rigid phenylazomethine dendrimers as templates for ligating discrete titanium salt domains within the dendrimer interior. Based on these precise assemblies, they served as precursors for producing very well-defined [H-3] type metal oxide nanocrystals (i.e., $[TiO_2]_n$, where $n = 6$, 14, and 30). The precursor $[Ti(acac)Cl_3]_n$@[G4;dendrimer] metal ligation complexes were either hydrolyzed or thermolyzed to produce the desired *anatase* or *rutile* forms of TiO_2, with concurrent destruction of the dendrimer template. A unique quantum size effect is observed as one decreases the TiO_2 nanoscale sizes within this series. Such controlled size reductions may be used to raise the conduction band while lowering the valence bands of the metal oxide nanocrystals as illustrated in Figure 1.22. These quantum size (Q-size) effects produce a unique "*nano-periodic property pattern*," which is evidenced by a blue shift as the nanoparticle size decreases.

1.6 NANO-PERIODIC CHEMICAL/PHYSICAL PROPERTY PATTERNS

1.6.1 Periodic Chemical Reactivity/Physical Size Property Patterns

Soft particle, dendrimer-based, [S-1] type nano-elements are unique macromolecules that exhibit completely different physicochemical properties (i.e., nano-periodic property patterns) compared to traditional polymers. This is largely due to congestion properties that emerge as a function of generational growth, as shown in Figure 1.23,

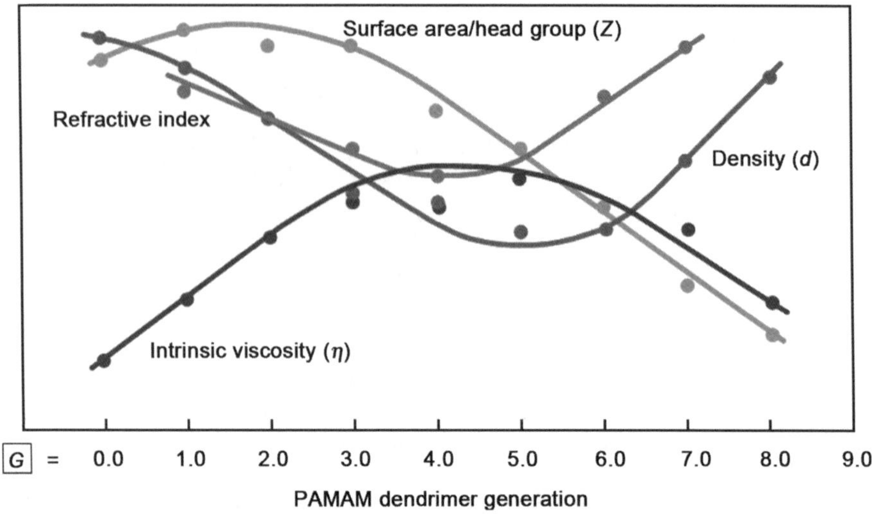

FIGURE 1.23 Comparison of surface area/head group (Z), refractive index, density (d), and viscosity (η) as a function of generation: $G = 1$–9 [11]. Copyright Wiley-VCH Verlag GmbH & Co. KGaA.

to produce unprecedented nano-periodic property patterns that are intrinsic and uniquely characteristic of dendrimers.

Plots of intrinsic viscosity $[\eta]$, density (d), surface area per Z group (A_Z), and refractive index, as a function of generation, clearly show intrinsic maxima or minima at generations = 3–5 for this Tomalia-type PAMAM dendrimer series. These data corroborate computer-assisted molecular-simulation predictions [11,57], and extensive photochemical probe experiments reported by Turro et al. [18,58–62].

Dendrimer-based intrinsic viscosities $[\eta]$ initially increase in a classical fashion as a function of molar mass (generation), but dramatically decline beyond a critical generation due to a congestion-induced shape change. A dendrimer shape change occurs from an extended, compressible, floppy configuration in the early generations (i.e., $G = 0$–3) to more rigid globular shapes in the later generations (i.e., $G = 4$–10). In effect, at critical generations (i.e., $G = 3$–4 and >) the dendrimer acts more like an Einstein spheroid. Intrinsic viscosity is a physical property expressed in dL/g, which in essence is the ratio of volume to mass. As the generation number increases and transition occurs to a spherical shape, the volume of the spherical dendrimer increases in cubic fashion while its mass increases exponentially; hence, the value of $[\eta]$ must decrease once a certain generation is reached. This prediction has now been widely confirmed for many different dendrimer families [11,25,63]. Dendrimer surface congestion may be appraised mathematically as a function of generation according to the following simple relationship:

$$A_Z = \frac{A_D}{N_Z} \alpha \frac{r^2}{N_c N_b^G}$$

where A_Z is the surface area per terminal group Z, A_D the dendrimer surface area, and N_Z the number of surface groups Z per generation. This relationship predicts the surface area per Z group at higher generations (G) and becomes increasingly smaller as it finally approaches the cross-sectional area or van der Waals dimension of the surface groups Z at higher generations. Congestion at these higher generations (G) is referred to as "*de Gennes dense-packing*" [11]. Ideal dendritic growth without branch defects is possible only for those generations preceding this dense-packed state. This critical dendrimer property gives rise to self-limiting dendrimer dimensions, which are a function of the branch-cell segment length (l), the core multiplicity N_c, the branch-cell juncture multiplicity N_b, and the steric dimensions of the terminal group Z. Dendrimer radius r in the above expression is dependent on the branch-cell segment lengths l; wherein, large l values delay congestion. On the other hand, larger N_c, N_b values and larger Z dimensions dramatically enhance congestion. These congestion properties are unique for each dendrimer family; wherein N_c and N_b determine the generation levels within a family that will exhibit nano-encapsulation properties. Higher N_c and N_b values predict that lower generation levels will produce appropriate surface congestion properties to manifest encapsulation features as shown in Figure 1.24.

1.6.2 Spheroidal Valency Defined by Nano-Sterics

Clearly these fundamental dendrimer properties illustrate the unique and intrinsic nano-periodic property patterns manifested by this soft matter, [S-1] type nano-element category. Many other nano-periodic property patterns have been documented for the behavior, assembly, and reactions of dendrimers with other dendrimers and with other well-defined nano-element categories. For example, work on this soft matter, [S-1] type nano-element category [6,52,55] has demonstrated that mathematically defined, periodic size properties of spheroidal dendrimers can determine chemical reactivity patterns with other dendrimers. These reactivity patterns, based on the relative sizes of a targeted dendrimer cores and dendrimer shell components, strongly influence the assembly of precise dendrimer clusters (i.e., core–shell (tecto) dendrimers). Mathematical relationships (i.e., the *Mansfield–Tomalia–Rakesh equation*) predict dendrimer cluster saturation levels (i.e., magic numbers for dendrimer shells) as a function of the core dendrimer size relative to the size of the shell dendrimers that are being used to construct the dendrimer cluster (Figure 1.25) [55,64]. These periodic property patterns and magic shell relationships are reminiscent of those observed for the self assembly of [H-1] type metal nanocrystals; wherein, the predicted number of touching spheroids surrounding a central core metal atom is 12 as shown below (Figure 1.25), when $r_1/r_2 = 1.00$.

1.6.3 Atom Mimicry and Nano-Periodic Property Patterns Observed for Yamamoto-Type Dendrimers

The importance of controlling the CNDP's related to flexibility/rigidity was clearly demonstrated with Yamamoto-type dendrimers. Based on the rigidity features and

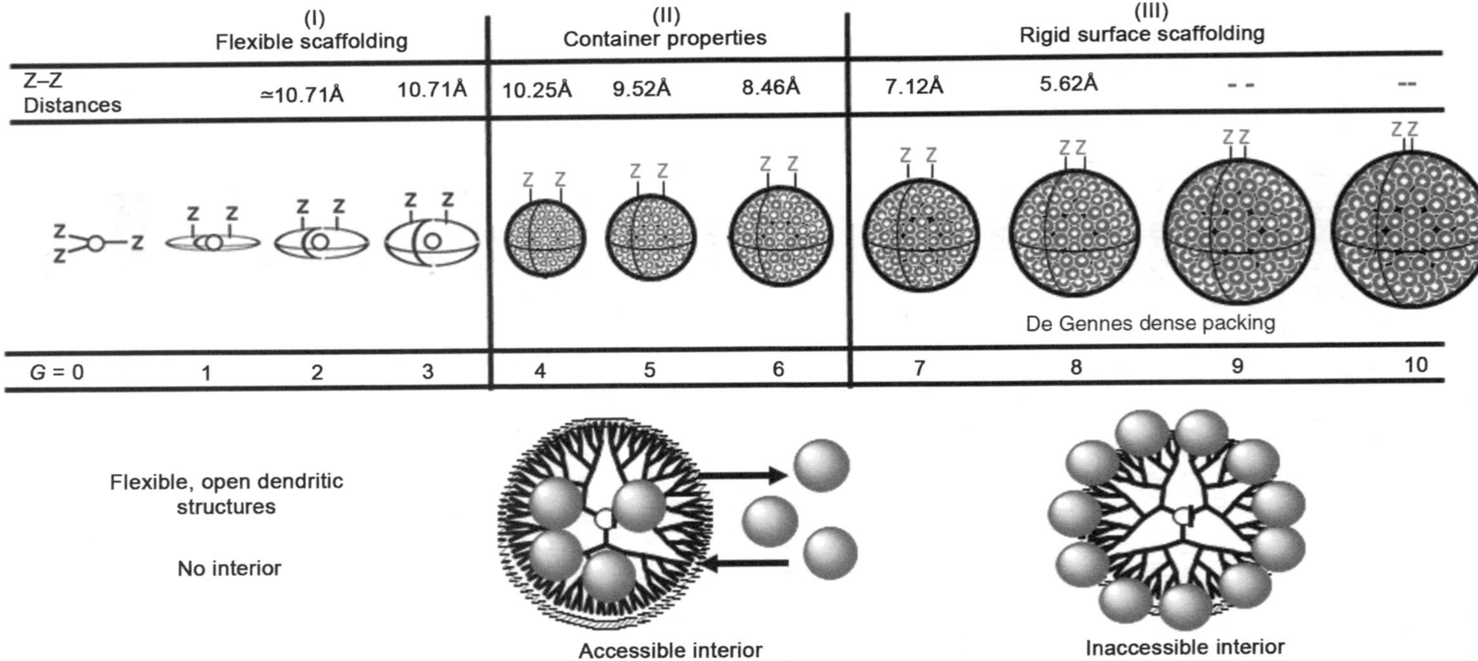

FIGURE 1.24 Congestion-induced dendrimer shape changes (I, II, and III) with development of nanocontainer properties for a family of [*core*:1,2-diaminoethane]; ($G = 0$–4); {*dendri*-poly(amidoamine)-$(NH_2)_Z$}, (PAMAM) dendrimers: $N_c = 4$; $N_b = 2$, where Z–Z = distance between surface groups as a function of generation.

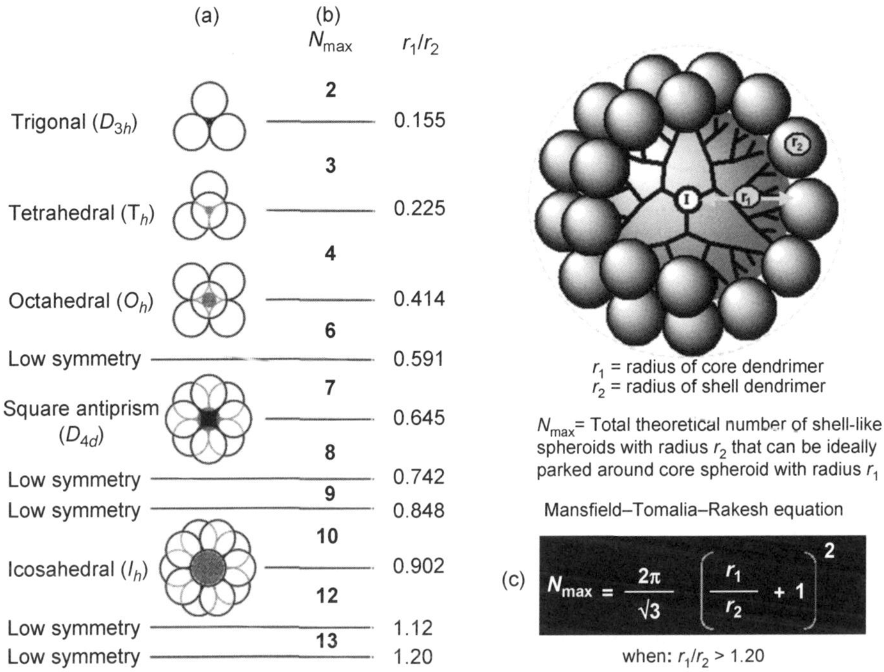

FIGURE 1.25 (a) Symmetry properties of core–shell tecto(dendrimer) structures, when: $r_1/r_2 < 1.20$. (b) Sterically induced stoichiometry (SIS) defined shell capacities (N_{max}), based on the respective core and shell radii, when: $r_1/r_2 < 1.20$. (c) Mansfield–Tomalia–Rakesh equation for calculating the maximum shell filling value (capacity) (N_{max}), when: $r_1/r_2 > 1.20$ [6,55].

mathematically quantized ligation sites present in Yamamoto-type dendrimers, it was possible to quantitatively document an amazing sequence of metal–dendrimer shell filling events. These metal-shell filling events are heuristically reminiscent of traditional electron-shell filling events observed for elemental atoms. It was shown in this present case that quantized amounts of metal salts will first ligate closest to the core at the lowest generation level to produce a saturated generational shell with well-defined and perfect stoichiometry (i.e., $4 \times @G = 1$), based on ligation sites present. Subsequent addition of metal reagent led to a stepwise filling of the next highest generational shell until it reached a shell-saturated state (i.e., $8 \times @G = 2$) and higher generation saturation states as shown in Figure 1.26.

1.7 FIRST STEPS TOWARD A "CENTRAL DOGMA" FOR SYNTHETIC NANO-CHEMISTRY? DENDRIMER-BASED NANO-CHEMISTRY

As stated earlier (Figure 1.1), the "central dogma" for traditional soft matter chemistry emerged from the first initiatives of Lavoisier and Dalton in the early nineteenth

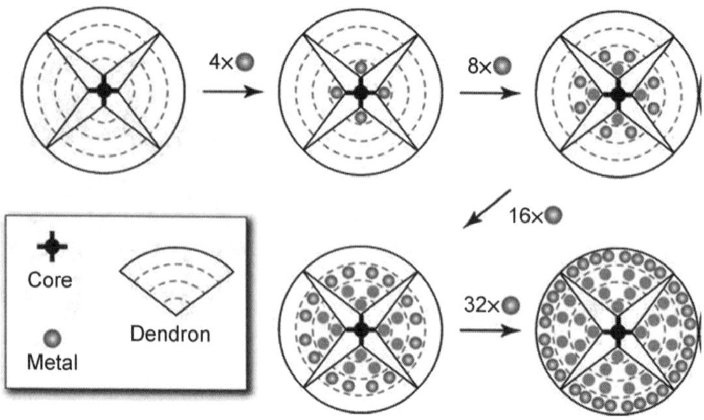

FIGURE 1.26 A Yamamoto-type poly(phenylazomethine) dendrimer series illustrating quantized, sequential dendrimer shell filling with metal salts using; [$G = 4$]; {*dendri*-poly(phenylazomethine)} (DPA) dendrimers [65]. Copyright Wiley-VCH Verlag Gmbh & Co. KGaA.

century and was initially focused on the simple combinatorial bonding of atoms to form small molecules (i.e., monomers, branch-cell monomers), much as illustrated in Figure 1.28. Synthetic soft matter chemistry throughout the nineteenth and twentieth century witnessed steady progress toward more complex molecular structure and architecture, including dendrons and dendrimers. Figure 1.28 illustrates the "aufbau process" for the bottom-up construction of such well-defined nanoscale structures (i.e., dendrons/dendrimers), which one might refer to as soft matter nano-elements [3,4,66]. Essentially all other proposed hard–soft nano-element categories

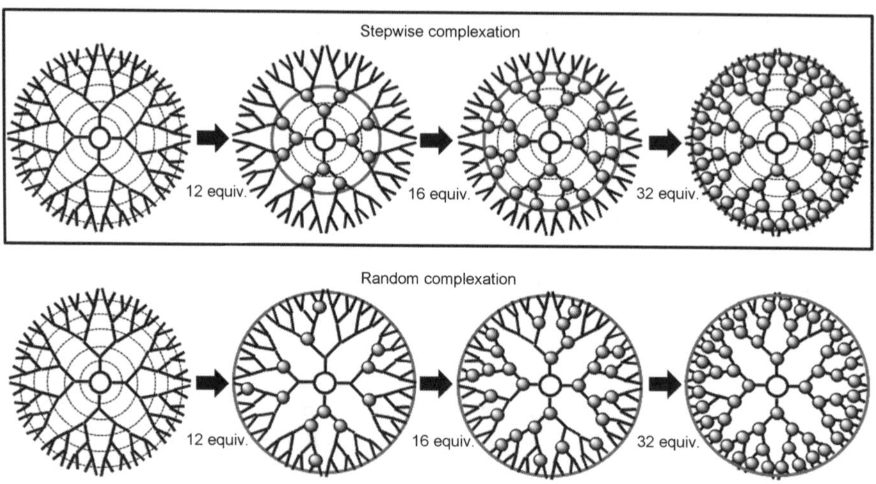

FIGURE 1.27 A comparison of stepwise and random ligation processes that are observed for $PtCl_4$ complexation with a rigid Yamamoto-type poly(phenylazomethine) dendrimer and a flexible Tomalia-type poly(amidoamine) (PAMAM) dendrimer, respectively [8].

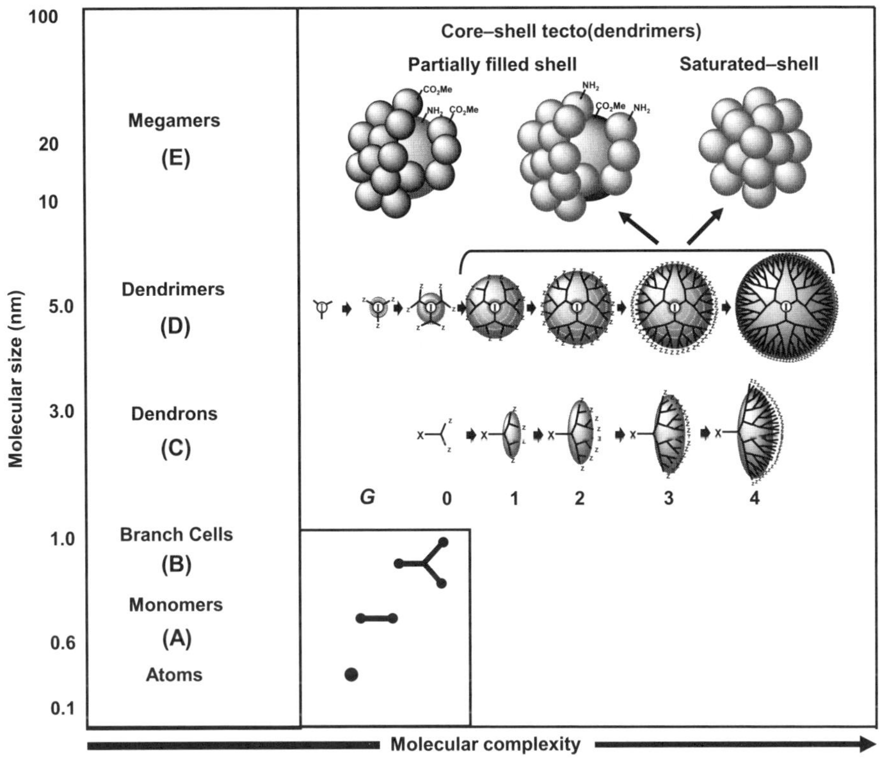

FIGURE 1.28 Approximate nanoscale dimensions as a function of hierarchical building blocks (i.e., atoms, monomers, branch cells, dendrimers, and megamers) and presumed conservation of critical hierarchical design parameters.

(Figure 1.18) evolve from an analogous aufbau strategy that allows the control and conservation of critical hierarchical design parameters from the atom to the nanoscale level (i.e., CADP → CMDP → CNDP). Vivid examples of such aufbau processes for two hard matter nano-element categories (i.e., [H-1] and [H-2] types) were described earlier (see Figures 1.21 and 1.22). In each case, metal atoms were precisely controlled and assembled to produce both [H-1] type metal nano-clusters or [H-3] type metal oxide nanocrystals by using soft matter [S-1] type nano-element (i.e., dendrimer structures) as templates. Nature has already presented very exquisite evolutionary aufbau strategies for synthesizing proposed soft matter nano-element categories such as (a) proteins [S-4], (b) viral capsids [S-5], and (c) DNA/RNA [S-6] (see Figure 1.18).

Clearly a deeper understanding of the critical "aufbau processes" leading to each of these well-defined nano-element categories, and their interconnections within a larger picture (i.e., a nano-periodic system) should be expected to provide important first steps toward the evolution of a central dogma for the new emerging area of *synthetic nano-chemistry.*

1.8 CONCLUSIONS

In this account, we have presented several examples of atom mimicry that have been observed for dendrimers (i.e., quantized nano-element-like behavior). It is apparent that the conservation and control of critical hierarchical design parameters (i.e., size, shape, surface chemistry, flexibility, and architecture) from the atomic (~0.001 nm) to the nanoscale level (i.e., 1–100 nm) are intrinsic and well-defined features (see Figures 1.23–1.25) of dendrimers (Figure 1.28). These structure-controlled CNDPs are important parameters that define chemical bonding, self-assembly (Figure 1.12) [45,46], stoichiometries (Figures 1.21, 1.26, and 1.27), nano-steric effects (Figure 1.25), and architectural patterns that are observed when combining dendrimers with other well-defined nano-modules. These new emerging properties and nano-periodic property patterns associated with dendrimers and their nano-compounds are distinctly different, yet reminiscent of similar features observed for their hierarchical precursors (i.e., atomic elements and small molecules). Many of these early observations on dendrimers inspired our more comprehensive proposal to view and classify all well-defined nanoparticles (i.e., hard and soft matter) that fit similar atom mimicry criteria [3,4,66]. It is from this perspective and based on first principles and rationale invoked for traditional chemistry, that we take initial steps toward a deeper understanding of the implications and possibilities for this new emerging area of *synthetic nano-chemistry.* It goes without saying that this new field

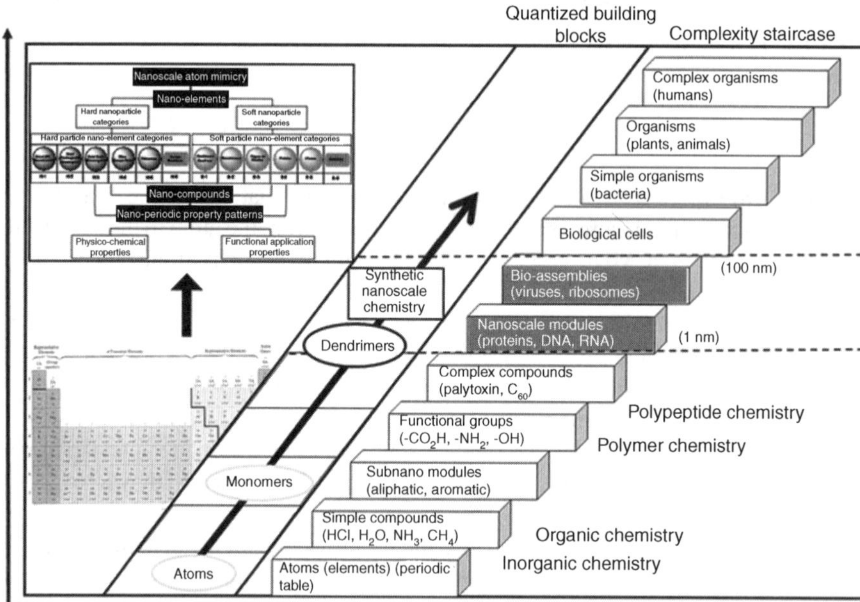

FIGURE 1.29 A chronological overview of traditional disciplines, structural complexity, and quantized building blocks relative to traditional and nano-periodic systems.

will be built on the shoulders of earlier traditional synthesis disciplines as described in Figure 1.29.

> Finally, the spirit of this perspective is not to advocate the disruption/violation of any natural physicochemical laws, but to encourage new and different thinking steeped in historical first principles which may evolve into a comprehensive systematic framework for unifying nanoscience. Much more remains to be done.
>
> Donald A. Tomalia

ACKNOWLEDGMENTS

I gratefully acknowledge the National Science Foundation for financial support of the CMU-NSF Workshop entitled: *Periodic Patterns, Relationships and Categories of Well-Defined Nanoscale Building Blocks*, NSF Award #0707510 the participants and especially the plenary speakers: Bradley D. Fahlman (Central Michigan University), William A. Goddard (Cal. Tech.), Theodore Goodson III (University of Michigan), Port Grodzinski (National Cancer Institute), Donald T. Haynie (Artificial Cell Technologies/Central Michigan), Scott McNeil (Nanotechnology Characterization Laboratory, NCI), Stephen O'Brien (Columbia University), Virgil Percec (University of Pennsylvania), Dmitrii F. Perepichka (McGill University), Mihail C. Roco (National Science Foundation), Robert Rodriquez (Cornell University), Dwight S. Seferos (Northwestern University), and Ulrich Wiesner (Cornell University) for the many stimulating discussions both during and after the workshop. I extend special thanks to Prof. Nicholas Turro (Columbia University) for many in-depth discussions and helpful suggestions in the development of the present concept. Finally, we wish to express our sincere gratitude to Ms. Linda S. Nixon for her invaluable skills in manuscript and graphics preparation.

REFERENCES

[1] R. Hoffmann, *Angew. Chem. Int. Ed.* **1987**, *26*, 846.

[2] E. Heilbronner, J. D. Dunitz, *Reflections on Symmetry*, VCH Publishers, Inc., New York, **1993**.

[3] D. A. Tomalia, *National Science Foundation Final Workshop Report* **2008**, p. 1.

[4] D. A. Tomalia, *J. Nanopart. Res.* **2009**, *11*, 1251.

[5] D. A. Tomalia, *Adv. Mater.* **1994**, *6*, 529.

[6] D. A. Tomalia, *Prog. Polym. Sci.* **2005**, *30*, 294.

[7] A. W. Jensen, B. S. Maru, X. Zhang, D. Mohanty, B. D. Fahlman, D. R. Swanson, D. A. Tomalia, *NanoLetters* **2005**, *5*, 1171.

[8] N. Satoh, T. Nakashima, K. Kamikura, K. Yamamoto, *Nat. Nanotechnol.* **2008**, *3*, 106.

[9] K. Yamamoto, T. Imaoka, W.-J. Chun, O. Enoki, H. Katoh, M. Takenaga, A. Sonoi, *Nat. Chem.* **2009**, *1*, 397.

[10] D. A. Tomalia, J. M. J. Fréchet, *J. Polym. Sci. Part A Polym. Chem.* **2002**, *40*, 2719.

[11] D. A. Tomalia, A. M. Naylor, W. A. Goddard III, *Angew. Chem. Int. Ed. Engl.* **1990**, *29*, 138.

[12] F. Zeng, S. C. Zimmerman, *Chem. Rev.* **1997**, *97*, 1681.

[13] A. Carlmark, C. Hawker, A. Hult, M. Malkoch, *Chem. Soc. Rev.* **2009**, *38*, 352.

[14] Y. Ma, S. V. Kolotuchin, S. C. Zimmerman, *J. Am. Chem. Soc.* **2002**, *124*, 13757.

[15] J.-P. Majoral, A.-M. Caminade, *Chem. Rev.* **1999**, *99*, 845.

[16] P. Antoni, Y. Hed, A. Nordberg, D. Nystrom, H. von Holst, A. Hult, M. Malkoch, *Angew. Chem. Int. Ed.* **2009**, *48*, 2126.

[17] R. Esfand, D. A. Tomalia, in *Dendrimers and Other Dendritic Polymers* (Eds.: J. M. J. Fréchet, D. A. Tomalia), Wiley, Chichester, **2001**, p. 587.

[18] J. M. J. Fréchet, D. A. Tomalia, *Dendrimers and Other Dendritic Polymers*, Wiley, Chichester, **2001**.

[19] M. H. P. Van Genderen, M. H. A. Mak, D. B.-V. D. Berg, E. W. Meijer, in *Dendrimers and Other Dendritic Polymers* (Eds.: J. M. J. Fréchet, D. A. Tomalia), Wiley, Chichester, **2001**, p. 605.

[20] D. A. Tomalia, S. A. Henderson, M. S. Diallo, S. A. Henderson, in *Handbook of Nanoscience, Engineering and Technology* (Eds.: W. A. GoddardIII, D. W. Brenner, S. E. Lyshevski, G. J. Iafrate), CRC Press, Boca Raton, FL, **2007**, p. 24.1.

[21] M. Higuchi, S. Shiki, K. Ariga, K. Yamamoto, *J. Am. Chem. Soc.* **2001**, *123*, 4414.

[22] J. C. Hummelen, J. L. J. van Dongen, E. W. Meijer, *Chem. Eur. J.* **1997**, *3*, 1489.

[23] R. Yin, Y. Zhu, D. A. Tomalia, *J. Am. Chem. Soc.* **1998**, *120*, 2678.

[24] R. Moors, F. Vogtle, *Chem. Ber.* **1993**, *126*, 2133.

[25] J. M. J. Fréchet, *Science* **1994** *263*, **1710**.

[26] G. R. Newkome, C. N. Moorfield, F. Vögtle, *Dendritic Molecules*, VCH, Weinheim, **1996**.

[27] C. J. Hawker, J. M. J. Fréchet, *J. Am. Chem. Soc.* **1990**, *112*, 7638.

[28] P. R. Dvornic, D. A. Tomalia, *Macromol. Symp.* **1995**, *98*, 403.

[29] G. J. Kallos, D. A. Tomalia, D. M. Hedstrand, S. Lewis, J. Zhou, *Rapid Commun. Mass Spectrom.* **1991**, *5*, 383.

[30] H. M. BrothersII, L. T. Piehler, D. A. Tomalia, *J. Chromatogr. A* **1998**, *814*, 233.

[31] C. A. Cason, S. A. Oehrle, T. A. Fabre, C. D. Girten, K. A. Walters, D. A. Tomalia, K. L. Haik, H. A. Bullen, *J. Nanomater.* **2008**, *2008, Article ID 456082*.

[32] A. Sharma, A. Desai, R. Ali, D. A. Tomalia, *J. Chromatogr. A* **2005**, *1081*, 238.

[33] D. R. Swanson, B. Huang, H. G. Abdelbady, D. A. Tomalia, *New J. Chem.* **2007**, *31*, 1368.

[34] P. Singh, *Bioconjugate Chem.* **1998**, *9*, 54.

[35] P. Singh, in *Dendrimers and Dendritic Polymers* (Eds.: J. M. J. Fréchet, D. A. Tomalia), Wiley, Chichester, **2001**, p. 463.

[36] D. A. Tomalia, B. Huang, D. R. Swanson, H. M. BrothersII, J. W. Klimash, *Tetrahedron* **2003**, *59*, 3799.

[37] L. A. Kubasiak, D. A. Tomalia, in *Polymeric Gene Delivery Principles and Applications* (Ed.: M. M. Amiji), CRC Press, Boca Raton, FL, **2005**, p. 133.

[38] U. Boas, J. B. Christensen, P. M. H. Heegaard, *Dendrimers in Medicine and Biotechnology*, The Royal Society of Chemistry, Cambridge, UK, **2006**.

[39] C. Zhang, D. A. Tomalia, in *Dendrimers and Other Dendritic Polymers* (Eds.: J. M. J. Fréchet, D. A. Tomalia), Wiley, Chichester, **2001**, p. 239.

[40] J. L. Jackson, H. D. Chanzy, F. P. Booy, B. J. Drake, D. A. Tomalia, B. J. Bauer, E. J. Amis, *Macromolecules* **1998**, *31*, 6259.

[41] J. Li, D. A. Tomalia, in *Dendrimers and Other Dendritic Polymers* (Eds.: J. M. J. Fréchet, D. A. Tomalia), Wiley, Chichester, **2001**, p. 285.

[42] A. Sharma, M. Rao, R. Miller, A. Desai, *Anal. Biochem.* **2005**, *344*, 70.

[43] A.-D. Schluter, P. J. Rabe, *Angew. Chem. Int. Ed.* **2000**, *39*, 865.

[44] J. G. Rudick, V. Percec, *Accts. Chem. Res.* **2008**, *41*, 1641.

[45] V. Percec, B. C. Won, M. Peterca, P. A. Heiney, *J. Am. Chem. Soc.* **2007**, *129*, 11265.

[46] V. Percec, W.-D. Cho, G. Ungar, D. J. P. Yeardley, *J. Am. Chem. Soc.* **2000**, *122*, 10273.

[47] E. C. Wiener, M. W. Brechbiel, H. M. Brothers II, R. L. Magin, O. A. Gansow, D. A. Tomalia, P. C. Lauterbur, *Magn. Reson. Med.* **1994**, *31*, 1.

[48] W. Andra, H. Nowak, *Magnetism in Medicine: A Handbook*, Wiley-VCH Verlag Gmbh & Co. KGaA, Berlin, **2006**.

[49] R. B. Lauffer, *Chem. Rev.* **1987**, *87*, 901.

[50] D. A. Tomalia, H. Baker, J. Dewald, M. Hall, G. Kallos, S. Martin, J. Roeck, J. Ryder, P. Smith, *Polym. J. (Tokyo)* **1985**, *17*, 117.

[51] T. Imaoka, R. Tanaka, S. Arimoto, M. Sakai, M. Fujii, K. Yamamoto, *J. Am. Chem. Soc.* **2005**, *127*, 13896.

[52] D. A. Tomalia, S. Uppuluri, D. R. Swanson, J. Li, *Pure Appl. Chem.* **2000**, *72*, 2343.

[53] T. A. Betley, J. A. Hessler, A. Mecke, M. M. Banaszak Holl, B. G. Orr, S. Uppuluri, D. A. Tomalia, J. R. Baker Jr., *Langmuir* **2002**, *18*, 3127.

[54] S. Uppuluri, L. T. Piehler, J. Li, D. R. Swanson, G. L. Hagnauer, D. A. Tomalia, *Adv. Mater.* **2000**, *12*(11), 796.

[55] M. L. Mansfield, L. Rakesh, D. A. Tomalia, *J. Chem. Phys.* **1996**, *105*, 3245.

[56] S. Uppuluri, D. R. Swanson, H. M. Brothers II, L. T. Piehler, J. Li, D. J. Meier, G. L. Hagnauer, D. A. Tomalia, *Polym. Mater. Sci. Eng. (ACS)* **1999**, *80*, 55.

[57] I. Tanis, D. Tragoudaras, K. Karatasos, S. H. Anastasiadis, *J. Phys. Chem. B* **2009**, *113*, 5356.

[58] O. Enoki, H. Katoh, K. Yamamoto, *Org. Lett.* **2006**, *8*, 569.

[59] K. R. Gopidas, A. R. Leheny, G. Caminati, N. J. Turro, D. A. Tomalia, *J. Am. Chem. Soc.* **1991**, *113*, 7335.

[60] J. Jockusch, J. Ramirez, K. Sanghvi, R. Nociti, N. J. Turro, D. A. Tomalia, *Macromolecules* **1999**, *32*, 4419.

[61] M. F. Ottaviani, N. J. Turro, S. Jockusch, D. A. Tomalia, *J. Phys. Chem.* **1996**, *100*, 13675.

[62] N. J. Turro, J. K. Barton, D. A. Tomalia, *Acc. Chem. Res.* **1991**, *24*(11), 332.

[63] A.-M. Caminade, R. Laurent, J.-P. Majoral, *Adv. Drug Deliv. Rev.* **2005**, *57*, 2130.

[64] D. A. Tomalia, J. M. Frechet, *Prog. Polym. Sci.* **2005**, *30*, 217.

[65] T. Imaoka, R. Tanaka, K. Yamamoto, *Chem. Eur. J.* **2006**, *12*, 7328.

[66] D. A. Tomalia, *Soft Matter,* **2010**, *6*, 456.

2

NOVEL METHODS FOR DENDRIMER SYNTHESIS

ISAO WASHIO[1] AND MITSURU UEDA[2]

[1]*New Materials Development Center, Mitsui Chemicals, Inc., 580-32 Nagaura, Sodegaura, Chiba 299-0265, Japan*

[2]*Department of Organic & Polymeric Materials, Graduate School of Science & Engineering, Tokyo Institute of Technology, 2-12-1 O-okayama, Meguro, Tokyo 152-8552, Japan*

2.1 GENERAL SYNTHETIC APPROACHES

There are two basic types of polymers that consist entirely of branched repeat units: dendrimers and hyperbranched polymers.

Hyperbranched polymers [1] are usually the product of a one-pot polymerization procedure and therefore exhibit an irregular architecture with incompletely reacted branch points throughout the structure (Figure 2.1), leading to an extremely broad molecular weight distribution and a low degree of branching around 0.5 in a theoretical sense.

A wide variety of hyperbranched polymers such as polyesters [2], polyethers [3], polyphenylenes [4], poly(ether ketone)s [5], and polyamides [6] have been reported in the literatures. However, hyperbranched polymers cannot be expected to possess ultimate properties with respect to solubility, viscosity, encapsulating abilities, and accessibility of terminal groups based on a precise structure.

Dendrimers [7], on the other hand, are synthesized by two complementary general approaches, the divergent [8] and convergent [9] method.

The divergent method results from sequential monomer (AB_n-building block) addition beginning from a core and proceeding outward toward the macromolecular

Designing Dendrimers, Edited by Sebastiano Campagna, Paola Ceroni, Fausto Puntoriero.
© 2012 John Wiley & Sons, Inc. Published 2012 by John Wiley & Sons, Inc.

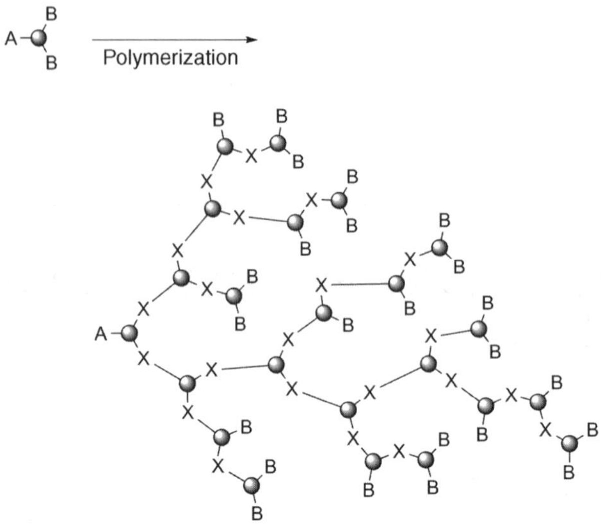

FIGURE 2.1 Synthesis of hyperbranched polymer.

surface. This methodology is illustrated in Figure 2.2. Reaction of the peripheral functional groups of the core with the reactive group of the AB_n-building block gives a new latent branch point at each coupling site, and results in an increase in the number of peripheral functional groups. The peripheral functional groups (protecting group) on each AB_n-building block should be inert to focal functional group, thereby preventing self-condensation of AB_n-building block. After the complete coupling reaction is achieved, these protecting groups can be deprotected to afford a new layer of reactive end groups capable of coupling with additional AB_n-building block.

A characteristic feature of the divergent method is the exponentially increasing number of reactions that require for providing each subsequent generation layer. Therefore, a large excess of reagents is generally required to afford a perfect growth of dendrimer structure. The divergent method is essentially suited for the large-scale synthesis of dendrimers because each generation dendrimer could be isolated from the excess reagents by a simple distillation, precipitation, or ultra filtration due to the large difference in molecular weight. However, to ensure the integrity of the final product, every reaction must be very selective and quantitative. In addition, any defective growth resulting from side or incomplete reaction cannot be easily removed because of similarity of their structures. As the result, the divergently prepared dendrimers frequently contain an appreciable number of structural defects.

The convergent method, on the contrary, initiates from the periphery and growths inward as shown in Figure 2.3. A molecule that eventually becomes the exterior of the dendrimer, terminal unit, first reacts with the functional group B of AB_n-building block. After completion of the coupling reaction, the single functional group A located at the focal point of dendron can be deprotected or changed the functionality for the next coupling reaction. Coupling of this deprotected

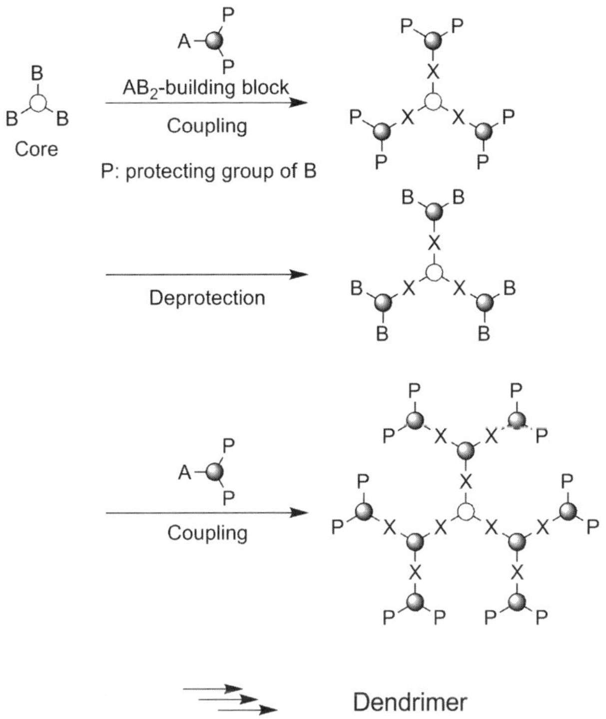

FIGURE 2.2 Divergent synthesis of dendrimer.

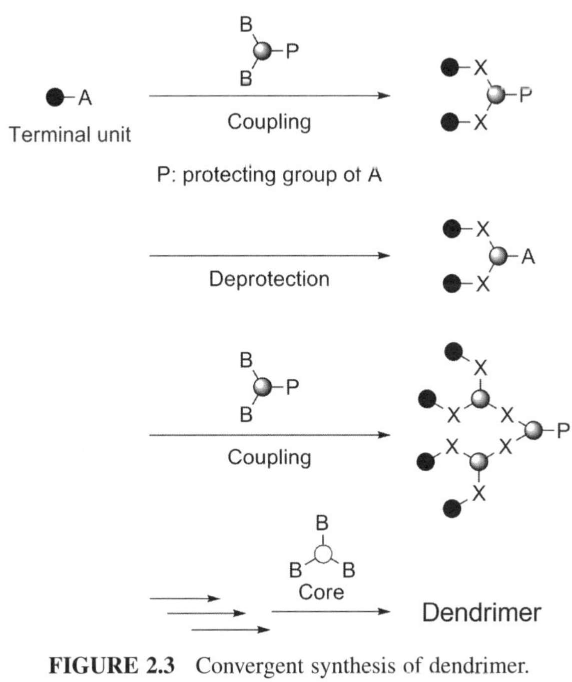

FIGURE 2.3 Convergent synthesis of dendrimer.

dendron with additional AB_n-building block affords a higher generation dendron. Finally, these dendrons can be attached to a core possessing multifunctional groups to form a globular architecture.

A notable advantage of this procedure is the requirement of a small number of reactions per molecule during coupling and deprotection steps. As the result, the reactions can be driven to completion with only a slight excess of reagent, in contrast to the case of divergent method. In addition, product purification after the coupling step can be performed especially by chromatographic purification because of the little dissimilarity between the desired dendron and the defect molecules. Thus, the convergent method can be a defect-free synthetic approach. However, since purification by chromatography results in some losses and difficulty in a large-scale synthesis, the convergent method seems to be difficult to apply industrial use.

In both synthetic methods, a tedious stepwise procedure is required: attaching one generation to the proceeding one, purifying, followed by changing the functional groups for the next-stage reaction. Therefore, the development of precise and facile synthetic methods for dendrimers is becoming one of the significant aspects of current work in this field.

2.2 ACCELERATED APPROACHES

In response to these problems, many researchers have developed accelerated approaches.

Fréchet and coworkers developed the "double-stage" approach, which implies the grafting of dendrons to the surface of small dendrimers called "hypercores" (Figure 2.4) [10]. However, the total number of reactions required for preparing dendron and hypercore is the same as for the conventional synthesis of the final dendrimer.

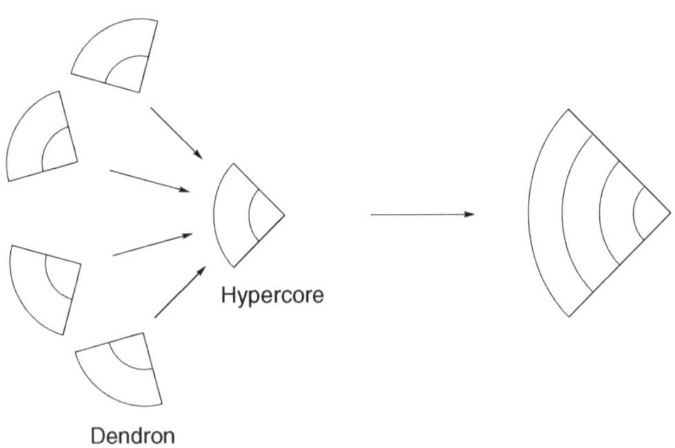

FIGURE 2.4 Double-stage synthesis of dendrimer.

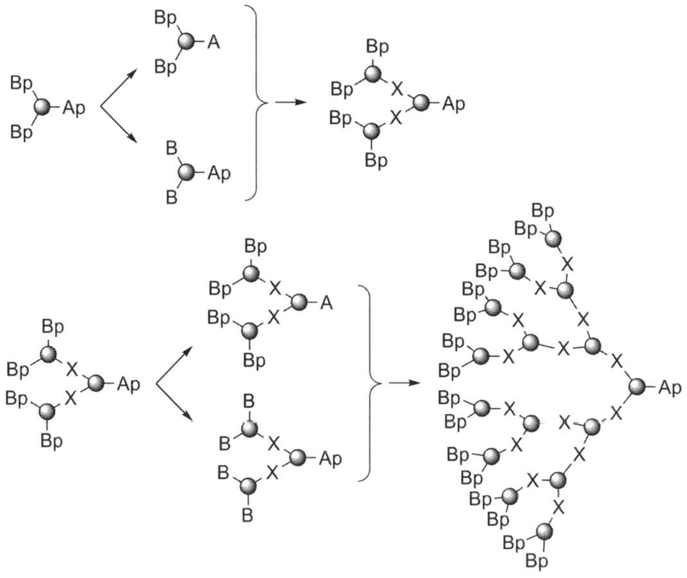

FIGURE 2.5 Double exponential growth synthesis of dendrimer.

The rapid synthesis of dendrimer using "hypermonomer" such as a AB_4- or AB_8-building block instead of a AB_2- or AB_3-building block have been reported [11]. This strategy rapidly increases the number of end groups; however, it does not improve the number of steps needed to obtain one generation because the coupling and deprotection steps may involve reactions identical to those used with a traditional dendrimer synthesis.

The "double exponential growth" approach was proposed by More and coworkers in 1995 and takes direct advantage of both divergent and convergent techniques (Figure 2.5) [12]. This procedure requires an AB_n-building block with different protecting groups for focal and peripheral functionalities. The first-generation protected dendron can be deprotected either at the focal point or at the periphery, resulting in the dendron possessing activated A group and the hypermonomer possessing activated B groups, respectively. Thus, coupling of the activated-dendron and hypermonomer yields a second-generation dendron, which can be deprotected at either the focal point or the periphery again. Coupling of the resultant activated second-generation dendrons to the second-generation hypermonomer provides a fourth-generation dendron. Although this type of strategy could be interesting for the rapid synthesis of high-generation dendrimers, it is not effective for that of middle-sized dendrimers. In the case of the synthesis of fourth-generation dendrons, it requires seven steps while a conventional method requires eight steps.

The other accelerated approach for the dendrimer synthesis is called an "orthogonal coupling strategy," which involves the convergent growth with two different monomers, that is, AB_2- and CD_2-building blocks (Figure 2.6) [13]. These building blocks must be carefully selected such that the focal functional group of each

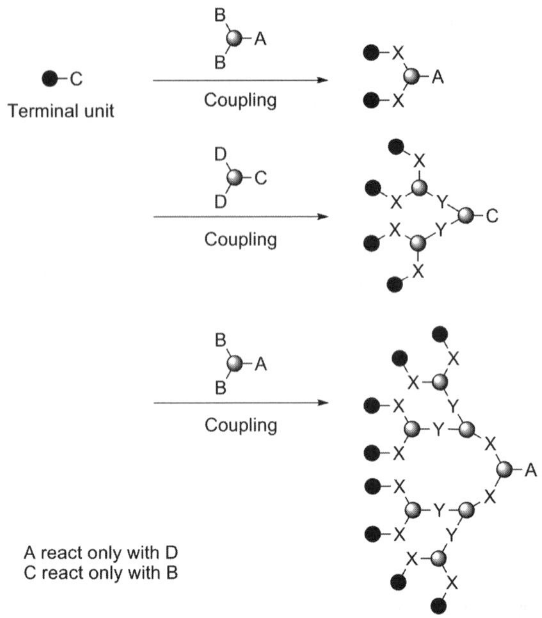

FIGURE 2.6 Orthogonal synthesis of dendrimer.

individual building block will react only with the periphery of the other building block (A reacts only with D, and C only with B). Therefore, the total number of reactions is decreased to half of that for a conventional method. However, this strategy generally requires complicated reaction systems to perform the quantitative coupling in the presence of different types of functional groups, especially for dendrimers having the same repeat unit of the linkage.

2.3 RAPID SYNTHESIS OF DENDRIMERS

There are other several recent reports that appear to be particularly noteworthy in the area of rapid synthesis of dendrimers.

Rannard and Davis [14] reported a one-pot multiple-addition convergent synthesis of polycarbonate dendrimers, in which the second-generation dendrimer was obtained by sequential activation of an alcohol unit with 1,1-carbonyl diimidazole followed by addition of an unprotected AB_2 triol (Scheme 2.1).

Majoral and coworkers [15] reported the one-pot synthesis of a fourth-generation dendrimer via divergent strategy using two different unprotected AB_2-building blocks (Scheme 2.2). In this method, to obtain phosphorus-containing dendrimers, they demonstrated that the condensation reaction between phophorhydrazides and aldehydes on one side, and the Staudinger reaction between phosphines and azides on the other side. This sequence of reactions does not require any isolation, since the only by-products are H_2O and N_2. Although this method has poor versatility because the

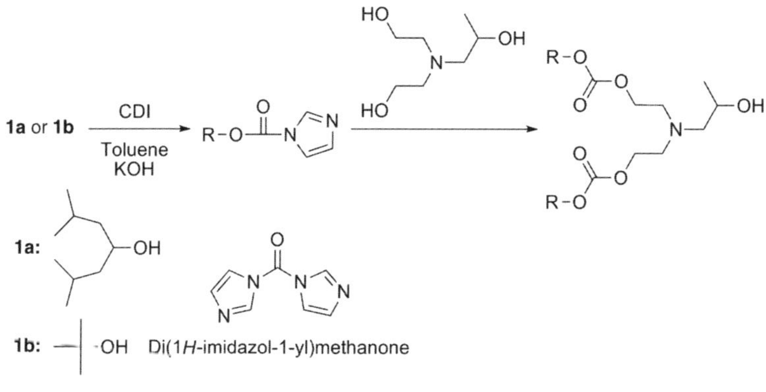

SCHEME 2.1 One-pot convergent synthesis of polycarbonate dendrimer.

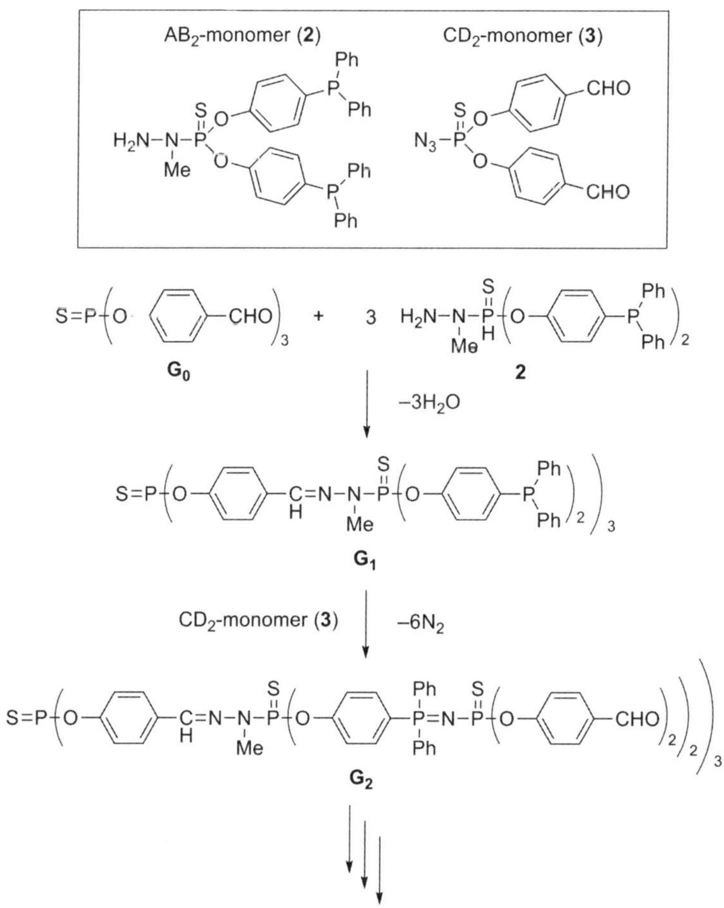

SCHEME 2.2 One-pot divergent synthesis of phosphorous dendrimer.

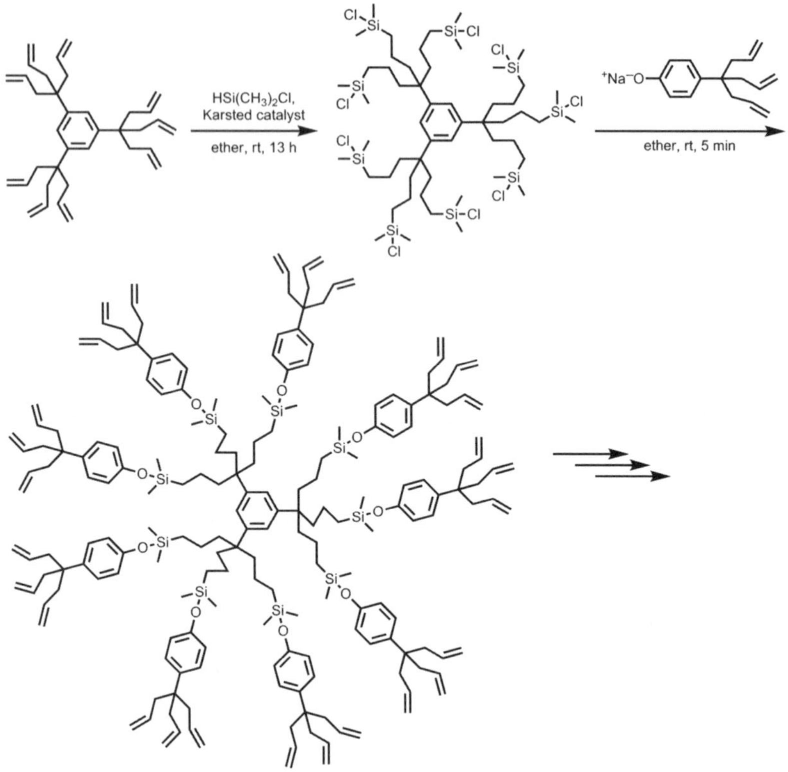

SCHEME 2.3 One-pot allyl-terminated dendrimer synthesis.

reactions used in this system cannot be applied for another structure, it is really attractive for industrial large-scale synthesis. In fact, this type of phosphorus dendrimers is now commercially available.

Eilbracht and Hagg's group has synthesized amino-functionalized dendritic architectures in a one-pot manner by hydroformylation-reductive amination sequences of dendritic polyglycerol with 30–40 amino termini at 75–80°C under 60 atm of CO/H_2 pressure [16].

Also a one-pot synthesis of a 243-allyl-terminated dendrimer has been developed (Scheme 2.3) [17]. Hydrosilylation of a nona-allyl dendritic core using $HSi(Me)_2Cl$ with an AB_3-building block possessing one phenolate and three allyl groups, followed by the repetition of this sequence of reactions twice, gives a 243-allyl-terminated dendrimer under ambient conditions.

All of the reports mentioned above utilize a one-pot synthetic approach to prepare dendrimers, in which one must add very precise amount of reagents at every step for performing the quantitative consumptions of the reactants. Therefore, formation of a small amount of defect molecules in final products is unavoidable.

On the other hand, Hawker and coworkers [18] have reported on an efficient and preparatively simple approach for the generation of diverse triazole-based dendrimers

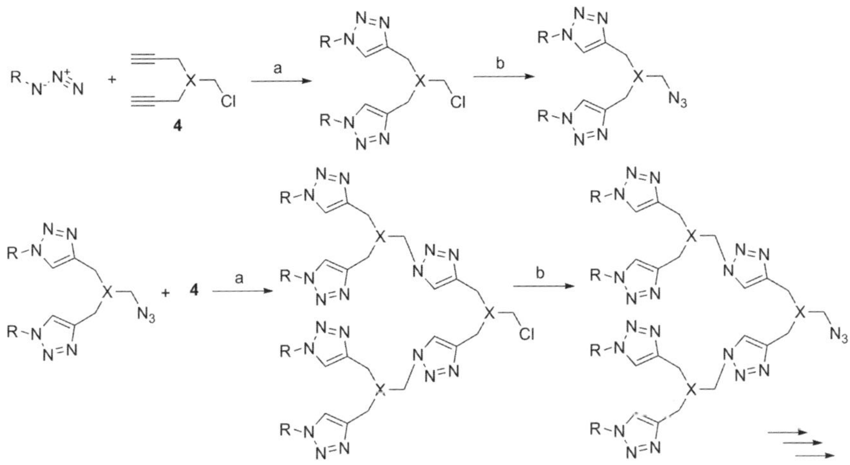

(a) $CuSO_4$ (5 mol%), sodium ascorbate (10 mol%), H_2O/tBuOH (1:1), near-quantitative yield
(b) 1.5 equiv. NaN_3, CH_3COCH_3/H_2O (4:1), 60°C, 1–3 h, >95%

SCHEME 2.4 Effective synthesis of triazole dendrimers using click chemistry.

of high purity and excellent yield (Scheme 2.4). In this approach, they utilized the click-chemistry transformation using copper-catalyzed ligation of azides and alkynes. A variety of functional groups are compatible with this process, and the only major by-products formed in the reaction are NaCl. Although this approach utilizes the conventional stepwise convergent method, it is remarkable that, in some cases, filtration or solvent extraction is the only method required for purification in this highly efficient construction of the triazole units of the dendrimers.

More recently, they have developed a robust, efficient orthogonal synthesis of poly(thio-ether) dendrimers using thiol-ene click chemistry via divergent approach [19]. The overall synthetic strategy is shown in Scheme 2.5, starting from 2,4,6-triallyloxy-1,3,5-triazine **5** as a core and 1-thioglycerol **6** as a AB_2-building block. The thiol-ene reaction between **5** and **6** yields the first-generation, hexahydroxy dendrimer [G1]-OH_6 in quantitative yield under photochemical initiation and subsequent esterification of [G1]-OH_6 with 4-pentenoic anhydride **7** gives the ene-functional dendrimer [G1]-ene_6. The all of the thiol-ene generation growth steps is compatible with water and oxygen and can be performed in the absence of solvent at room temperature, which allows for an environmental-friendly process. Although the purification of ene-functional dendrimers after esterification of hydroxy-functional dendrimers with 4-pentenoic anhydride needs chromatographic separations, the efficient nature of thiol-ene chemistry and the lack of by-products allow all generation hydroxyl-terminated dendrimers to be purified by simple precipitation into diethyl ether in 90 + % yields.

One of the most challenging approaches in this region is the formation of dendrimers by only mixing the building blocks in one pot all at once, without multiple-addition reaction systems. Hirsch and coworkers [20] have demonstrated the

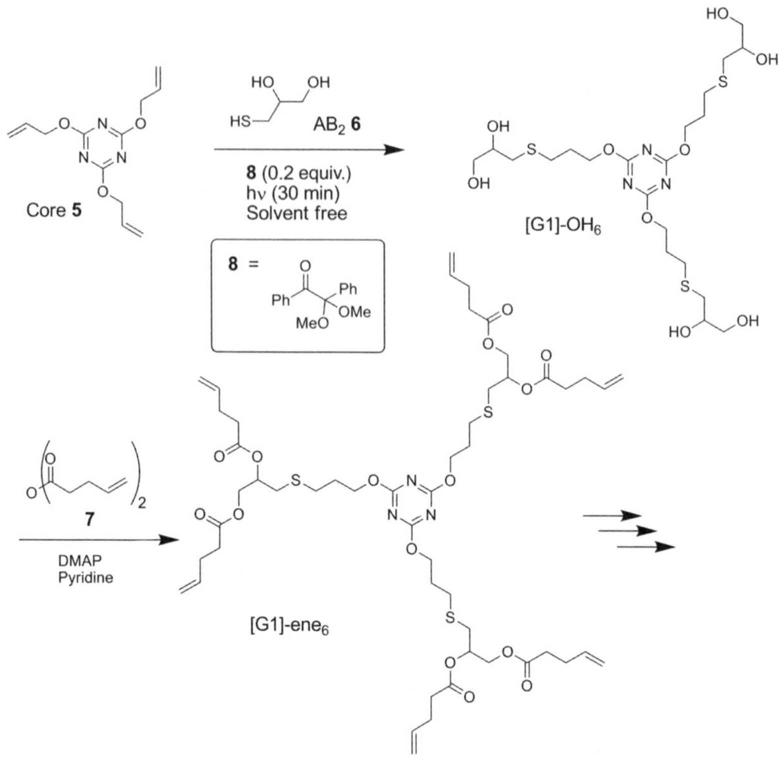

SCHEME 2.5 Efficient synthesis for thio-ether dendrimers using thiol-ene click chemistry.

first approach for the complete self-assembly of discrete supramolecular dendrimers where the repetition reactions are no longer necessary. The building blocks, core, AB_2-building block, and terminal units, prepared in this approach are shown in Scheme 2.6. They employed the hydrogen bonds, which act between Hamilton receptor and cyanuric acid derivatives, as the noncovalent linkages of each building block. When components C **9**, AB_2**10**, and EC **11** are mixed in a $1:(3 \times 2^n - 3):(3 \times 2^n)$ ratio, complete self-assembly of a supramolecular dendrimer of generation *n* can be expected, provided that all possible hydrogen bonds are formed (Figure 2.7). Due to the nature of self-assembly supramolecular system, general analytic method for common dendrimers unable to attempt to prove the formation of desired discrete dendrimers. Nonetheless, they have clarified that discrete supramolecular dendrimers must be formed and present in the solid state, and proved for the first time the programmed self-assembly of discrete supramolecular dendrimers from nondendritic building blocks.

2.3.1 Polyamide Dendrimer

Our research group has focused on the synthesis of dendritic polyamide so far because polyamide is very attractive from the materials point of view. Linear

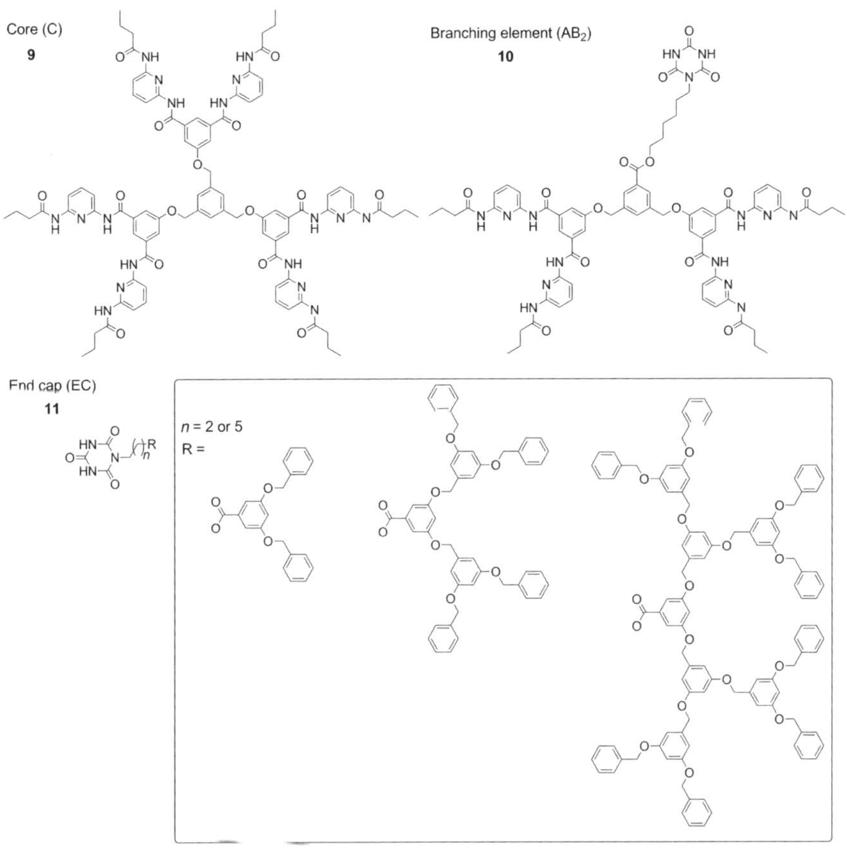

SCHEME 2.6 Building blocks for the discrete supramolecular dendrimers.

polyamides are commonly available as engineering plastic materials with high modulus due to semicrystallinity or high glass transition temperatures. However, the semicrystallinity and the strong tendency to form hydrogen bonding cause low solubility, high melting point, and limiting the processing. A highly branched structure such as dendrimers, which is usually amorphous materials with excellent solubility and low viscosity, might improve the processing of polyamides. Thus, the introduction of dendritic structure into polyamides will result in new applications as an engineering plastic.

Although various synthetic methods have been developed to prepare dendrimers containing amide functions, they have problems such as poor yield and tedious multisynthetic steps [21]. Kakimoto and coworkers [21d] developed an orthogonal synthesis to poly(aryl amide) dendrimers (Scheme 2.7). Condensation of the first AB_2-building block, 3,5-diaminobromobenzene, with a carboxylic acid G1-COOH yielded a dendritic bromide G2-Br. Dendron G2-Br could then be activated by a palladium-catalyzed insertion of carbon monoxide, enabling the reaction with the second AB_2-building block, 3,5-diaminobenzoic acid, to yield a dendritic carboxylic

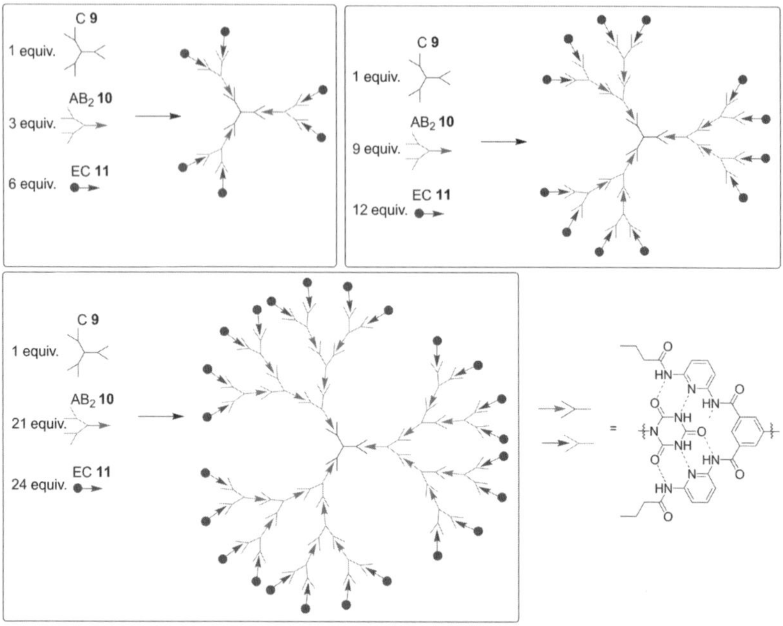

FIGURE 2.7 Supramolecular assembly of discrete dendrimers.

acid, G3-COOH. However, isolation of the third- and fourth-generation dendrons required the use of preparative GPC, resulting in low yields and difficulty in a large-scale synthetic use.

2.3.1.1 Facile Synthetic Route by Convergent Method Recently, we reported rapid synthesis of a perfectly branched third-generation polyamide dendrimer by the convergent method without repetitive protection–deprotection procedures, leading to a decrease of number of steps to half of that required in conventional dendrimer formation (Scheme 2.8) [22]. This synthesis utilizes a two-step method that consists of the activation of carboxylic acids and condensation with an unprotected AB_2-building block having diamine moieties, 3,5-bis(4-aminophenoxy)benzoic acid, using the activating agent diphenyl(2,3-dihydro-2-thioxo-3-benzoxazolyl)phosphonate (DBOP). In this approach, all-generation dendrons and dendrimers are prepared using slightly excess amount of former generation dendrons to achieve the formation of perfectly branched structures. AB_2-building block **14** reacted with acetic acid as the terminal unit in the presence of DBOP and TEA to yield G1 dendron. Both activation and condensation reactions were conducted in one pot for 1 h at 25°C. The reaction mixture was diluted with methanol and poured into water, producing G1 dendron in 90% yield. G2 and G3 dendrons also can be prepared by the same coupling method. Furthermore, all products were purified simply only by reprecipitation technique to remove the impurities including former generation dendrons and the leaving group of the reaction in excellent yields.

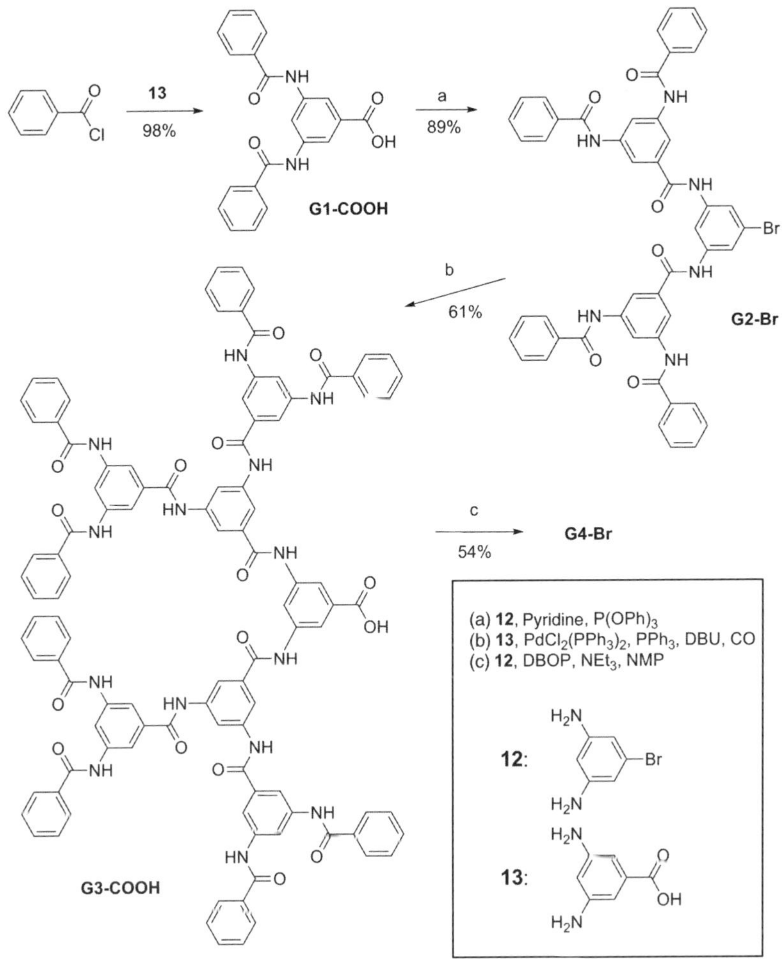

SCHEME 2.7 Orthogonal synthesis of poly(aryl amide) dendrimer.

Later, the same concept was applied to more versatile and inexpensive activating agent, thionyl chloride [23]. The use of thionyl chloride as an activating agent makes the purification process more simple, because the only by-products are gasses such as SO_2 and HCl.

2.3.1.2 Facile Synthetic Route by Divergent Method The divergent approach has several advantages compared to convergent one such as the simplicity of the purification process and accessibility in large-scale industrial synthesis; this is demonstrated by the fact that most commercially available dendrimers are currently prepared using the divergent method.

However, the two-step methods described above are limited to only the convergent approach because of the difficulties in the complete activation of end

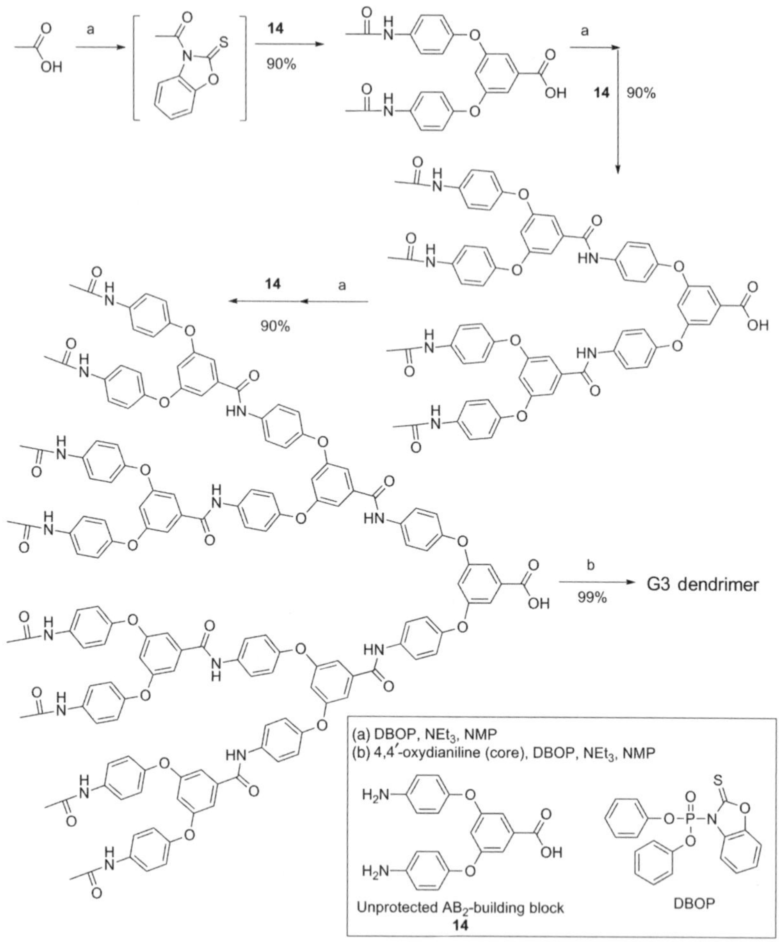

SCHEME 2.8 Rapid synthesis of polyamide dendrimer from unprotected AB_2.

groups at the dendrimer periphery and the quantitative condensation of the resulting active intermediates with AB_2-building blocks without side reactions such as the hydrolysis of active intermediates, which would result in the formation of defects, in a divergent approach.

Authors have developed a two-step method for the facile synthesis of amine-terminated aromatic polyamide dendrimers using 3,5-bis(trifluoroacetamido)benzoyl chloride **15** as an novel AB_2-building block via a divergent approach in which the condensation and deprotection reactions are carried out rapidly in one pot [24].

The G1 dendrimer was prepared using *p*-phenylenediamine (PDA) as a core molecule in one pot (Scheme 2.9). Further, 1.1 equiv. of AB_2-building block **15** was added to each amino group in a solution of PDA in *N*-methyl-2-pyrrolidinone (NMP) at 0°C and stirred for 1 h at room temperature to yield the trifluoroacetyl-terminated

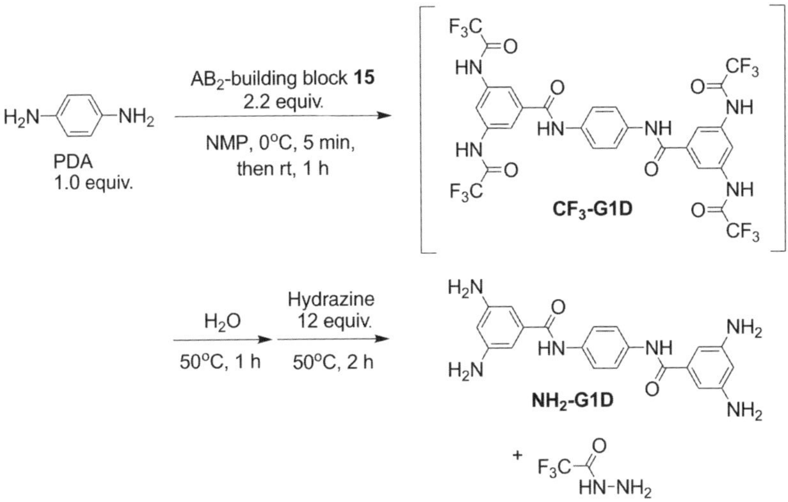

SCHEME 2.9 Synthesis of G1 dendrimer.

G1 dendrimer (CF_3-G1D) *in situ*. After achieving the complete condensation, 5.5 equiv. of hydrazine to **15** was added to form the amine-terminated G1 dendrimer (NH_2-G1D) by a transamidation reaction at 50°C for 2 h. Since the by-products in the resulting final solution are only trifluoroacetyl hydrazide, 3,5-diaminobenzoic acid, HCl, and hydrazine, NH_2-G1D can be purified simply by precipitation in $NaHCO_{3(aq)}$ in a quantitative yield.

The overall synthetic route for the 32-amine-terminated G4 polyamide dendrimer is shown in Scheme 2.10. NH_2-G2D, NH_2-G3D, and NH_2-G4D were synthesized using a similar protocol in the case of NH_2-G1D. The deprotection reaction was completely achieved at 50°C within 3 h even in NH_2-G4D synthesis. The yields of NH_2-G2D, NH_2-G3D, and NH_2-G4D after precipitation in alkaline water were 94, 95, and 95%, respectively.

2.3.1.3 Facile Synthetic Route to Tadpole Dendrimer Recent advances in the area of dendrimer chemistry have demonstrated that properties of dendrimers can be significantly improved if dendritic segments are covalently linked to a different linear polymer block to create a hybrid organic structure. A number of such systems have been explored since they were first introduced in the previous decade [25]. However, almost such diblock structures described in the literature contain flexible linear chain segments such as poly(ethylene oxide) or polystyrene and, therefore, cannot easily retain shape in a solution or in a solid state. From the standpoint of a molecular design, the introduction of a rigid rod moiety to the focal point of dendron is interesting for preparing molecules with a precise characteristic shape that is different from the dendron-coil block copolymers [26,27].

A monodisperse *N*-butylated oligo(*p*-benzamide) was first prepared as described in Scheme 2.11 [27]. In this method, thionyl chloride was used as an activation agent

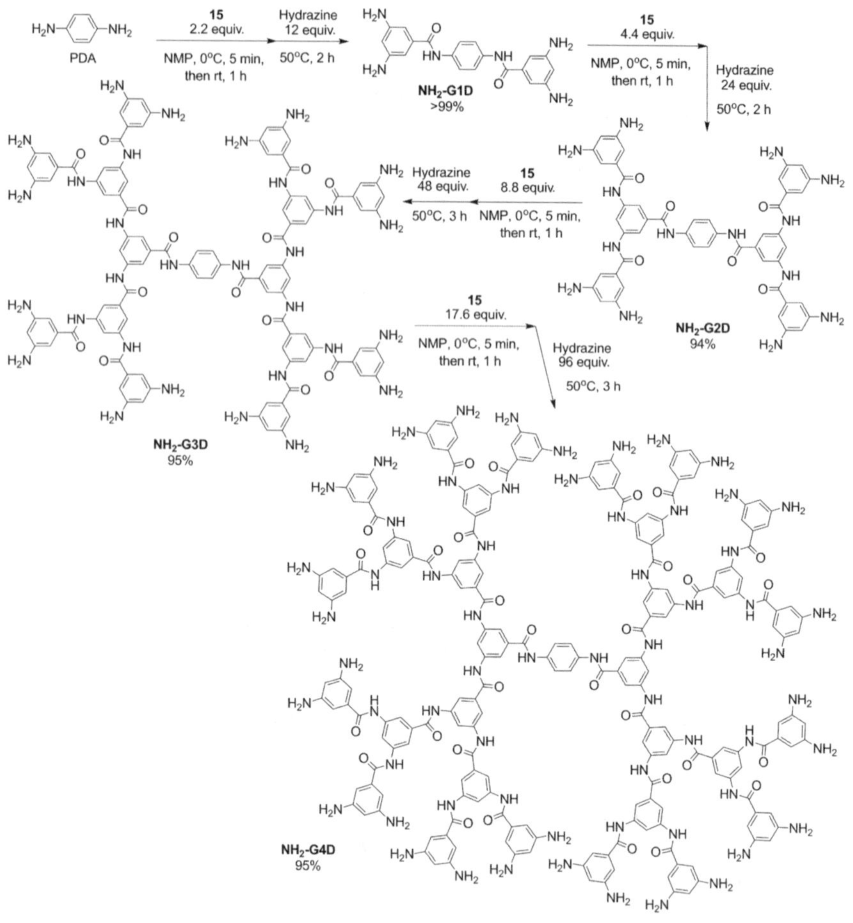

SCHEME 2.10 Overall synthetic route for the 32-amine-terminated G4 polyamide dendrimer.

and trifluoroacetamide as the protecting group for the end amine, in which the number of amide units dramatically propagated only through every deprotection and condensation.

Trifluoroacetamide-protected trimer **19** was synthesized from protected monomer **17** and monomer **16** utilizing a one-pot multiple addition condensation reaction. It is noteworthy that this approach enables four reactions, two times of activation and two times of condensation, to be carried out in one pot. Thus, the nonamer **23** was obtained only through five-step processes. In addition, all the products are purified using only crystallization or precipitation.

The synthesis of dendron-rod molecules were grown by divergent approach from the secondary amine group of *N*-butylated nona(*p*-benzamide) **23** using 3,5-bis

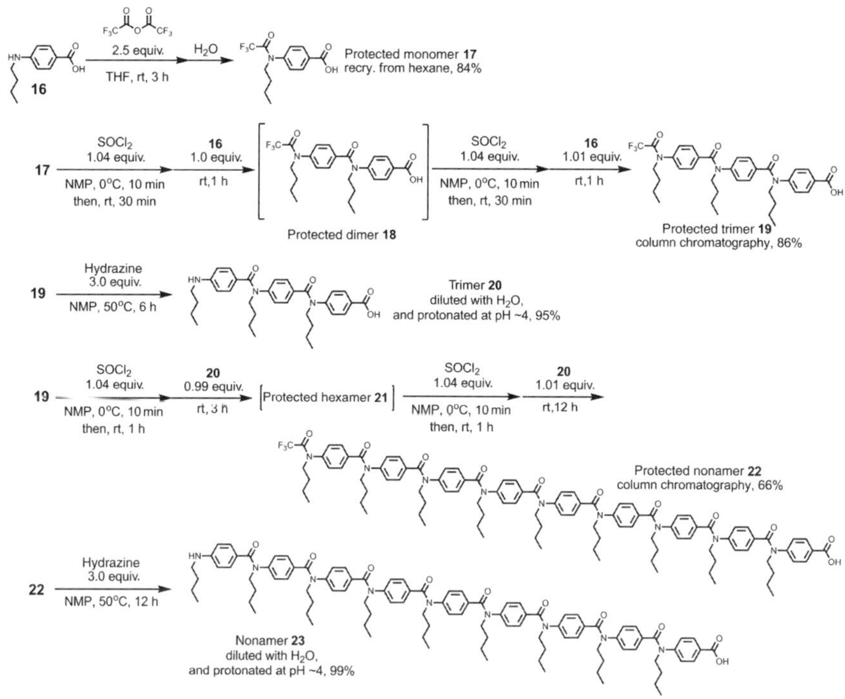

SCHEME 2.11 Synthesis of monodisperse *N*-butylated oligo(*p*-benzamide)s.

(trifluoroacetamido)benzoyl chloride **15** as an AB_2-building block that enable to construct a two-step method as described in the previous section. The reaction of nonamer **23** with small excess amount of AB_2-building block **15** yields the intermediate, trifluoroactamide-terminated G1d-rod (first-generation dendron coupled with rod block) molecule **24**, quantitatively *in situ*. After hydrolyzing the acid chloride group of the unreacted-excess AB_2-building block, the intermediate **24** was converted to amine-terminated G1d-rod **25** by transamidation reaction with excess amount of hydrazine. Since all the by-products in the resulting final solution are soluble in alkaline water, G1d-rod **25** can be purified simply by precipitation in $NaHCO_{3(aq)}$ in a quantitative yield (Scheme 2.12).

The larger amine-terminated rod-dendron molecules were synthesized using a similar protocol, and isolated in excellent yield after precipitation in alkaline water (Scheme 2.13). Finally, G4d-rod **28** was successfully obtained only through four-step processes from nonamer **23**.

This novel structure is particularly attractive for constructing direct applications of theory relating self-assembly system. By modifying the functionalities at the periphery and the focal point of this tadpole dendrimer would lead to its promising applications for producing self-assembled architectures.

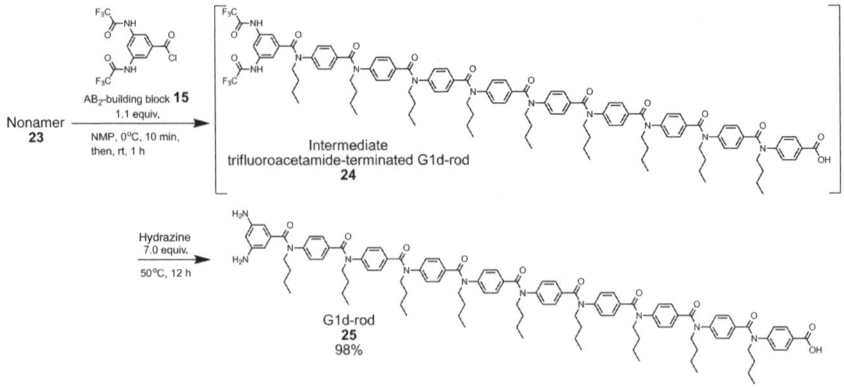

SCHEME 2.12 Synthesis of G1d-rod **25**.

2.3.2 Aryl Ether Dendrimer

The Fréchet-type aryl ether dendrimer [28] is one of the most accepted macromolecules in various fields such as biology, although their synthesis is still difficult [7b]. To avoid the most difficult step of the bromination in the Fréchet process, several methods were reported so far. Utilization of Mitsunobu reaction and of solid support system was developed [29]. They are, however, still tedious to prepare the high-quality dendrimer in a practical scale.

Thionyl chloride is the most attractive reagent for the chlorination of benzyl alcohols due to the short reaction time, low reaction temperature, and low price [30]. Percec et al. [31] have reported utilization thionyl chloride as a chlorination agent of benzyl alcohols and demonstrated the benzyl chloride is sufficient for the synthesis of aryl ether dendrimers. However, they have used 3,5-dihydroxymethylbenzoate as a protected building block instead of 3,5-dihydroxybenzylalcohol that is conventionally used for preparation of aryl ether dendrimers. Therefore, the methyl benzoate group needed two-step reactions, reduction and chlorination, to prepare benzyl chloride of each generation dendron.

Authors developed a successful rapid synthesis of a perfectly branched fourth-generation (G4) Fréchet-type dendrimer from unprotected AB_2-building block 3,5-dihydroxylbenzylalcohol **29** by a convergent method using thionyl chloride as an activating agent (Scheme 2.14) [32]. In this method, each generation dendron and dendrimer can be prepared in one pot, no isolation of intermediate chlorinated dendrons. Furthermore, the purification of every dendritic molecule requires only precipitation, recrystallization, and solvent extraction.

First-generation (G1) dendron **30** was prepared by the direct condensation of benzyl chloride with **29** in the presence of K_2CO_3 in NMP under nitrogen atmosphere. Recrystallization from toluene/hexane gave the G1 dendron **30** in 85%. One-pot synthesis of G2 dendron **32** was then prepared from G1 dendron **30**

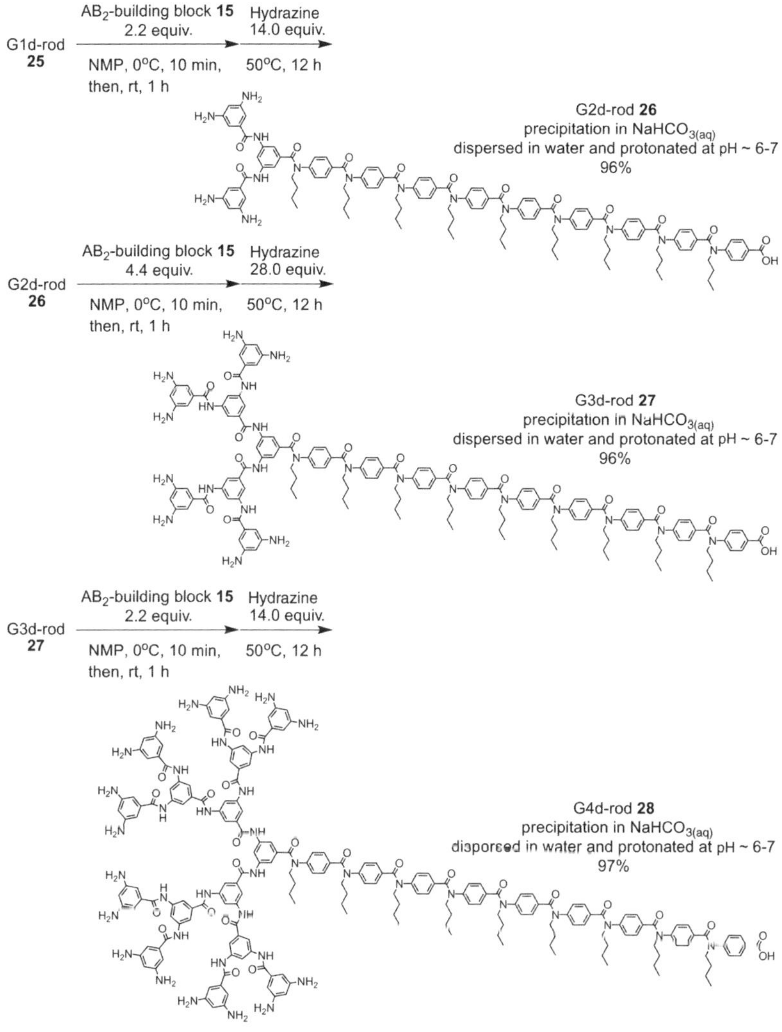

SCHEME 2.13 Synthesis of dendron-rod molecules.

and **29**. G1 dendron **30** was activated with 1.10 equiv. of thionyl chloride in NMP at 0°C for 10 min in NMP to generate the corresponding benzyl chloride **31** *in situ*. Followed by condensation with **29** was conducted at 120°C for 5 h. The desired G2 dendron **32** was purified by reprecipitation to remove G1 dendron **30**, and obtained in 80% yield as a white powder. G3 dendron **33**, G4 dendron **34**, and G4 dendrimer **35** were prepared by the same method and purified by reprecipitation in 82, 73, and 73% yield, respectively. This novel route enables the final G4 dendrimer **35** to be formed only achieving five processes.

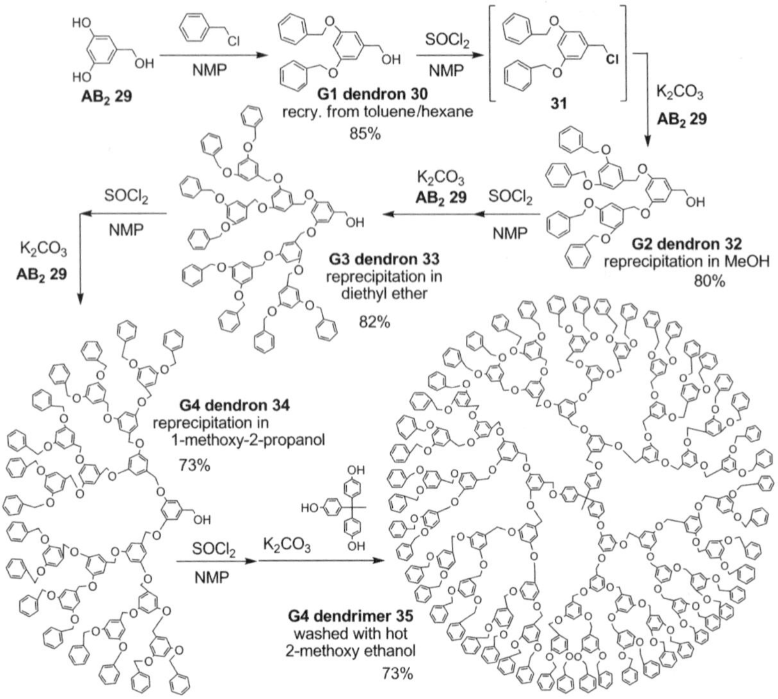

SCHEME 2.14 Synthesis of aryl ether dendrimer.

2.4 SUMMARY

Dendrimers are attractive scaffolds for a variety of high-end applications because of their well-defined, unique macromolecular structures. However, it is practically difficult to apply dendrimers in industrial large-scale applications because of the huge cost of their preparations. Therefore, the development of synthetic methodology as introduced in this chapter will open the new aspect of dendrimer chemistry in industry.

REFERENCES

[1] (a) B. I. Vöit, *Acta Polym.* **1995**, *46*, 87; (b) E. Malmström, A. Hult, *J. Macromol. Sci., Rev. Macromol. Chem. Phys.* **1997**, *C37*, 555; (c) Y. H. Kim, *J. Polym. Sci., Polym. Chem.* **1998**, *36*, 1685.

[2] (a) C. J. Hawker, R. Lee, J. M. J. Fréchet, *J. Am. Chem. Soc.* **1991**, *113*, 4583; (b) S. R. Turner, B. I. Vöit, T. H. Mourey, *Macromolecules* **1993**, *26*, 4617.

[3] K. E. Uhrich, C. J. Hawker, J. M. J. Fréchet, S. R. Turner, *Macromolecules* **1992**, *25*, 4583.

[4] Y. H. Kim, O. W. Webster, *Macromolecules* **1992**, *25*, 5561.

[5] (a) T. M. Miller, T. X. Neenan, E. W. Kwock, S. M. Stein, *J. Am. Chem. Soc.* **1993**, *115*, 356; (b) F. Chu, C. J. Hawker, *J. Polym. Bull.* **1993**, *30*, 265.

[6] Y. H. Kim, *J. Am. Chem. Soc.* **1992**, *114*, 4947.

[7] For reviews, see for example: (a) G. R. Newcome, C. N. Moorefield, F. Vögtle (Eds.), *Dendrimers and Dendrons, Concepts, Syntheses, Applications)*, VCH, Weinheim, **2001**; (b) J. M. J. Fréchet, D. A. Tomalia (Eds.), *Dendrimers and Other Dendritic Polymers*, VCH, Weinheim, **2002**; (c) M. S. Grayson, J. M. J. Fréchet, *Chem. Rev.* **2001**, *101*, 3819; (d) A. W. Bosman, H. M. Janssen, E. W. Meijer, *Chem. Rev.* **1999**, *99*, 1665; (e) K. Inoue, *Prog. Polym. Sci.* **2000**, *25*, 453; (f) F. Vögtle, S. Gestermann, R. Hesse, H. Schwierz, B. Windisch, *Prog. Polym. Sci.* **2000**, *25*, 987.

[8] (a) D. A. Tomalia, J. Dewald, M. Hall, S. Martin, P. Smith, *Prepr. 1st SPSJ Int. Polym. Conf., Soc. Polym. Sci. Jpn. (Kyoto)* **1984**, 65; (b) D. A. Tomalia, H. Baker, J. Dewald, M. Hall, G. Kallos, S. Martin, J. Roeck, J. Ryder, P. Smith, *Polym. J.* **1985**, *17*, 117; (c) D. A. Tomalia, A. N. Naylor, W. A. Goddard, *Angew. Chem. Int. Ed. Engl.* **1990**, *29*, 138.

[9] C. J. Hawker, J. M. J. Fréchet, *J. Am. Chem. Soc.* **1990**, *112*, 7638.

[10] K. L. Wooley, C. J. Hawker, J. M. J. Fréchet, *J. Am. Chem. Soc.* **1991**, *113*, 4252.

[11] K. L. Wooley, C. J. Hawker, J. M. J. Fréchet, *Angew. Chem. Int. Ed. Engl.* **1994**, *33*, 82.

[12] T. Kawaguchi, K. L. Walker, C. L. Wilkins, J. S. More, *J. Am. Chem. Soc.* **1995**, *117*, 2159.

[13] R. Spindler, J. M. J. Fréchet, *J. Chem. Soc., Perkin Trans. 1* **1993**, 913.

[14] S. P. Rannard, N. J. Davis, *J. Am. Chem. Soc.* **2000**, *122*, 11729.

[15] L. Brauge, G. Magro, A.-M. Caminade, J.-P. Majoral, *J. Am. Chem. Soc.* **2001**, *123*, 6698.

[16] F. Koç, M. Wyszogrodzka, P. Eilbracht, R. Haag, *J. Org. Chem.* **2005**, *70*, 2021.

[17] C. Ornelas, J. R. Aranzaes, E. Cloutet, D. Astruc, *Org. Lett.* **2006**, *8*, 2751.

[18] P. Wu, A. K. Feldman, A. K. Nugent, C. J. Hawker, A. Scheel, B. Vöit, J. Pyun, J. M. J. Fréchet, K. B. Sharpless, V. V. Fokin, *Angew. Chem. Int. Ed. Engl.* **2004**, *43*, 3928.

[19] K. L. Kato, L. M. Campos, C. J. Hawker, *J. Am. Chem. Soc.* **2008**, *130*, 5062.

[20] A. Franz, W. Bauer, A. Hirsch, *Angew. Chem. Int. Ed. Engl.* **2005**, *44*, 1564.

[21] (a) T. M. Miller, T. X. Neenan, *Chem. Mater.* **1990**, *2*, 346; (b) P. M. Bayliff, W. J. Feast, D. Parker, *Polym. Bull.* **1992**, *29*, 265; (c) S. C. E. Backson, P. M. Bayliff, W. J. Feast, A. M. Kenwright, D. Parker, R. W. Richards, *Macromol. Symp.* **1994**, *77*, 1; (d) Y. Ishida, M. Jikei, M. Kakimoto, *Macromolecules* **2000**, *33*, 3202; (e) S. Rannard, N. Davis, H. McFarland, *Polym. Int.* **2000**, *49*, 1002; (f) B. Romagnoli, P. R. Ashton, L. M. Harwood, D. Philp, D. W. Price, M. H. Smith, W. Hayes, *Tetrahedron* **2003**, *59*, 3975.

[22] M. Okazaki, I. Washio, Y. Shibasaki, M. Ueda, *J. Am. Chem. Soc.* **2003**, *125*, 8120.

[23] (a) I. Washio, Y. Shibasaki, M. Ueda, *Org. Lett.* **2003**, *5*, 4159; (b) I. Washio, Y. Shibasaki, M. Ueda, *Macromolecules* **2005**, *38*, 2237.

[24] I. Washio, Y. Shibasaki, M. Ueda, *Org. Lett.* **2007**, *9*, 1363.

[25] (a) I. Gitsov, K. L. Wooley, J. M. J. Fréchet, *Angew. Chem. Int. Ed. Engl.* **1992**, *31*, 1200–1202; (b) I. Gitsov, J. M. J. Fréchet, *Macromolecules* **1994**, *27*, 7309–7315; (c) T. M. Chapman, G. L. Hillyer, E. J. Mahan, K. A. Shaffer, *J. Am. Chem. Soc.* **1994**, *116*, 11195–11196; (d) J. C. M. van Hest, D. A. P. Delnoye, M. W. P. L. Baars, M. H. P. van Genderen, E. W. Meijer, *Science* **1995**, *268*, 1592–1595; (e) K. Aoi, A. Motoda, M. Okada, *Macromol. Rapid. Commun.* **1997**, *18*, 945–952; (f) J. Iyer, K. Fleming, P. T. Hammond, *Macromolecules* **1998**, *31*, 8757–8765; (g) I. Gitsov, K. R. Lambrych, V. A. Remnant, R. Pracitto, *J. Polym. Sci. Polym. Chem.* **2000**, *38*, 2711–2727; (h) H. Ihre, O. L.

Padilla De Jesus, J. M. J. Fréchet, *J. Am. Chem. Soc.* **2001**, *123*, 5908–5917; (i) V. Istratov, H. Kautz, Y.-K. Kim, R. Schubert, H. Frey, *Tetrahedron* **2003**, *59*, 4017–4024; (j) E. R. Gillies, J. M. J. Fréchet, *Chem. Commun.* **2003**, *14*, 1640–1641.

[26] (a) Y.-S. Yoo, J.-H. Choi, J.-H. Song, N.-K. Oh, W.-C. Zin, S. Park, T. Chang, M. Lee, *J. Am. Chem. Soc.* **2004**, *126*, 6294–6300; (b) C.-J. Jang, J.-H. Ryu, J.-D. Lee, D. Sohn, M. Lee, *Chem. Mater.* **2004**, *16*, 4226–4231; (c) J. Holzmueller, K. L. Genson, Y. Park, Y.-S. Yoo, M.-H. Park, M. Lee, V. Tsukruk, *Langmuir* **2005**, *21*, 6392–6398; (d) E. R. Zubarev, S. I. Stupp, *J. Am. Chem. Soc.* **2002**, *124*, 5762–5773; (e) B. J. de Gans, S. Wiegand, E. R. Zubarev, S. I. Stupp, *J. Phys. Chem. B* **2002**, *106*, 9730–9736; (f) L. Tian, T. Hammond, *Chem. Mater.* **2006**, *18*, 3976.

[27] Y. Ito, I. Washio, M. Ueda, *Macromolecules* **2008**, *41*, 2778.

[28] C. J. Hawker, J. M. J. Fréchet, *J. Am. Chem. Soc.* **1990**, *112*, 7638.

[29] (a) B. Forier, W. Dehaen, *Tetrahedron* **1999**, *55*, 9829; (b) A. Basso, B. Evans, N. Pegg, M. Bradley, *Chem. Commun.* **2001**, 697.

[30] F. C. Whitmore, F. A. Karnatz, A. H. Popkin, *J. Am. Chem. Soc.* **1938**, *60*, 2540.

[31] (a) V. S. K. Balagurusamy, G. Ungar, V. Percec, G. Johansson, *J. Am. Chem. Soc.* **1997**, *119*, 1539; (b) V. Percec, W.-D. Cho, P. E. Mosier, G. Ungar, D. J. P. Yeardley, *J. Am. Chem. Soc.* **1998**, *120*, 11061; (c) V. Percec, W.-D. Cho, G. Ungar, D. J. P. Yeardley, *J. Am. Chem. Soc.* **2001**, *123*, 1302; (d) V. Percec, C. M. Mitchell, W.-D. Cho, S. Uchida, M. Glodde, G. Ungar, X. Zeng, Y. Liu, V. S. K. Balagurusamy, P. A. Heiney, *J. Am. Chem. Soc.* **2004**, *126*, 6078.

[32] N. Yamazaki, I. Washio, Y. Shibasaki, M. Ueda, *Org. Lett.* **2006**, *8*, 2321.

3

DESIGNER MONOMERS TO TAILORED DENDRIMERS

George R. Newkome[1] and Carol Shreiner[2]

[1]*Departments of Polymer Science and Chemistry, University of Akron, Akron, Ohio 44325-4717, USA*

[2]*Department of Chemistry, Hiram College, Hiram, Ohio 44234, USA*

3.1 INTRODUCTION

On a lonely evening in Baton Rouge, Louisiana in 1982, with my trusty lapboard in place—reading my journals and listening to the TV drone on in the background, I ran across an interesting article in the *American Scientist* [1] by Professor P. B. Tomlinson of the Harvard Forest at Harvard University, entitled "Tree Architecture". How can one pictorially summarize the myriad of "trees" into specific architectural patterns, considering the structural errors caused by the environment? Although this article had little or nothing to do with chemistry—it altered my life and, in essence, started the topic of this chapter. Specifically, it was the last pattern in a figure depicting a family of different idealized tree architectures—the Leeuwenberg model—that suggested to me that Mother Nature's patterns for trees could be transformed to chemical patterns [2,3] and thus, the creation of 1 → 3 and later 1 → (2 + 1) branching motifs for a series of monomers from which molecular trees were conceived, at least for me. Arborols, *neé* dendrimers, were the outcome, if following a perfect pattern; of course, the structural abuses by the environment on the forest—generated trees possessing less-than-idealized architectures—led to structural designs related to today's hyperbranched polymers at the nanoregime.

In general to the academic chemist, organic target molecules, for example, drugs and natural products produced in those rain forests, have a specific structure, whereas

Designing Dendrimers, Edited by Sebastiano Campagna, Paola Ceroni, Fausto Puntoriero.
© 2012 John Wiley & Sons, Inc. Published 2012 by John Wiley & Sons, Inc.

to industrial chemists, organic targets related to polymers or random assemblies of specific monomers result. There are two different approaches to the creation of these new macromolecules possessing a fractal design. One, in which the branching sites are located at the termini, is generally comprised of industrially available monomers; whereas, the other route utilizes a monomer or monomers possessing an internal branching center as part of its composition. We undertook the later—that is, the creation of new monomers that can be assembled to make 1 $\rightarrow$ 3 branching motifs and have an inherent nonreversibility leading to an overall structural stability.

An iterative or repeating assembly is the second part of the molecular construction. If traditional industrial monomers are used, retro-reactions and diverse side reactions can occur resulting in a rather heterogeneous assembly composed of alternating patterns. Easily accessible and readily assembled but uniform final families of products resulted. The vast majority of all reported dendritic studies have been conducted with only a few families of divergently generated dendrimers, of which PAMAMs and PPIs dominate the older literature. In view of Mother Nature's forests, why would only one or two or three tree designs suffice? Thus, in general, convenience or commercial availability leads to fractal-structured families possessing a uniform exterior, which can be uniformly or randomly/combinatorially coated with little or no control of the resultant product.

This chapter addresses the creation of branching monomers that can be assembled in different ways, thus challenging chemists to devise interesting new patterns to address specific objectives, targets, or needs. In our initial example [4] appearing 25 years ago, the core, the first branching layer and the surface—although divergently (inside-to-outside) assembled—were comprised of different branching monomers; this started to address molecular diversity. It was 6 years later, that Fréchet and Hawker reported [5] the convergent (outside-to-inside) approach to molecular growth and introduced the use of the term "dendrons" for these wedges.

Since there is a litany of potential monomers or dendrons that can be devised, each resultant synthetic target should start to address new and potentially utilitarian composition. Herein, we briefly review families of designer monomers, then propose other possibilities and, hopefully, challenge others to be adventuresome and create new predendrons and dendrons to meet tomorrow's needs. Mother Nature has demonstrated that there is a multitude of diverse structural patterns for the trees in her forest; can they be mimicked within the chemist's domain?

Upon construction of the layers—or generations as they are commonly called—the number of surface groups on a tetravalent core for a 1 $\rightarrow$ 3 branched dendrimer increases (4, 12, 36, 108, ...), which is faster than that of a 1 $\rightarrow$ 2 branched dendrimer (4, 8, 16, 32, ...), leading to increasing steric crowding at the periphery. This construction makes use of preferred branched monomers or dendrons, allowing for higher generations with fewer peripheral gaps and defects. Scheme 3.1 shows the different modes of construction with a tetravalent core moiety. Route A illustrates the divergent method of construction, building outward from a tetravalent core, generally using industrial monomers. Transformation of the new termini at each generation allows for further substitution and dendritic growth. Attaching preformed dendrons of specific generation size via the convergent method (Route B) avoids the lack-of-control

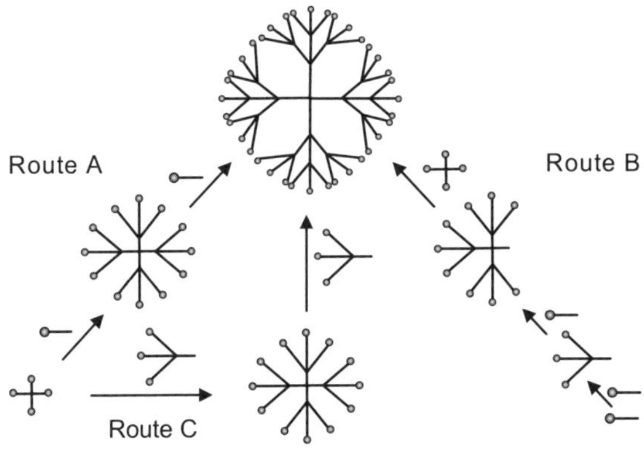

SCHEME 3.1 Generalized protocols for dendrimer construction.

encountered with the divergent procedure and allows for a more uniform macromolecule, especially at higher generations. Route C utilizes branched monomers (minidendrons) via a divergent process. The synthesis of different 1 → 3 branched monomers permits the divergent—and convergent—construction of different dendritic families maximizing on the strengths of both approaches. The following overview presents different families of branched monomers and their synthesis as well as structural modifications.

3.2 1 → 3 *C*-BRANCHED MONOMERS

3.2.1 TRIS-Based

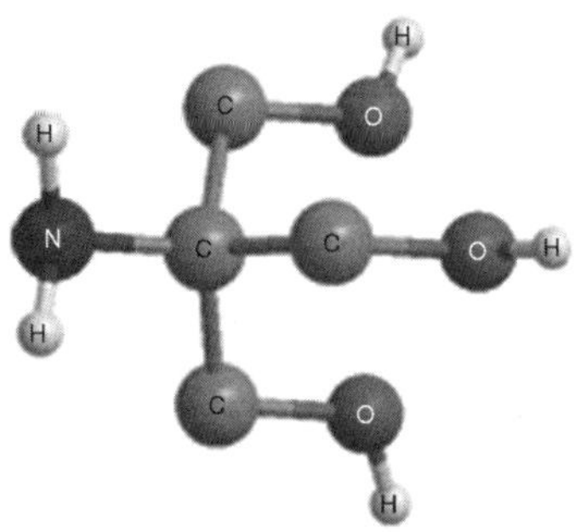

Tris(hydroxymethyl)aminomethane (TRIS) is commercially available and easily transforms most esters in the presence of K_2CO_3 in DMSO to the corresponding amide triol [4] thus enhancing their water-solubility. TRIS can be *N*-protected (Boc_2O) to give (97%) $BocHNC(CH_2OH)_3$ then *O*-benzylation or *O*-allylation affording (63–65%) $BocHNC(CH_2OBn)_3$ or $BocHNC(CH_2OCH_2CH{=}CH_2)_3$; lastly, deprotection (TFA) removes the Boc group to generate $H_2NC(CH_2OBn)_3$ or $H_2NC(CH_2OCH_2CH{=}CH_2)_3$, respectively [6]. The reaction of succinic or glutaric anhydride with TRIS gave $HO_2C(CH_2)_nCONHC(CH_2OH)_3$ ($n = 2$ or 3, respectively) [7].

The predendron $O_2NC(CH_2OH)_3$ can be silylated ($ClSiMe_2$*tert*-Bu/DBU/MeCN) to give $O_2NC(CH_2OTBDMS)_3$, which was reduced (H_2NNH_2/Raney-Ni) to give H_2NC-$(CH_2OTBDMS)_3$, then extension in three steps to generate H_2NCH_2CONHC-$(CH_2OTBDMS)_3$ [8]. TRIS was reacted with $SOCl_2$ in pyridine to give OSNC-$(CH_2Cl)_3$ and $ClH_3NC(CH_2Cl)_3$, which with warm base gave $H_2NC(CH_2Cl)_3$ [9].

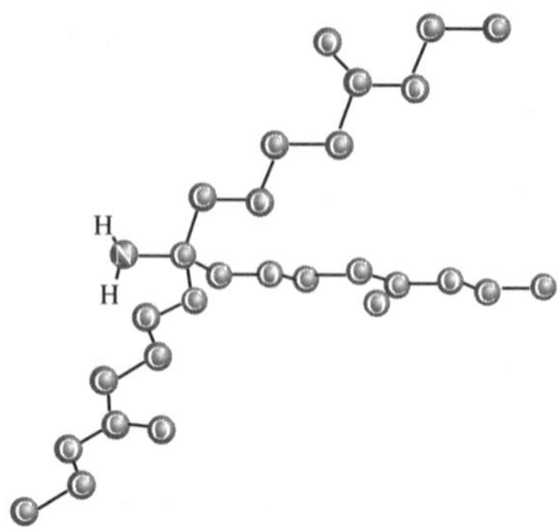

Tris[(carbethoxy)ethoxymethyl]aminomethane (Lin's amine) can be easily prepared [10] in two steps from initially TRIS and acrylonitrile via a Michael-type addition, followed by ethanolysis. Amidation with a carboxylic acid uses standard peptide coupling conditions (DCC/1-HOBT/DMF) [10]. The initial formed H_2NC-$(CH_2OCH_2CH_2CN)_3$ has been transformed in high yield to the protected ZHNC-$[CH_2O(CH_2)_3NH_2]_3$ or $BocHNC(CH_2OCH_2CH_2CO_2H)_3$ so surface functionalization can be easily altered [11]; whereas, treatment with succinic anhydride quantitatively gave $HO_2CCH_2CH_2CONHC(CH_2OCH_2CH_2CN)_3$ [12]. Further modification was conducted in which esterification (MeOH, H^+), followed by reduction ($NaBH_4$, $NiCl_2$, MeOH), Boc–protection [$(Boc)_2O$], and saponification (NaOH, DME) leads to the desired $HO_2CCH_2CH_2CONHC[CH_2O(CH_2)_3NHBoc]_3$ in overall excellent conversion. Further, $H_2NC(CH_2OCH_2CH_2CN)_3$ was transformed [13] into $C_6H_5CH_2O_2CNHC(CH_2OCH_2CH_2CN)_3$, then sequentially reduced (BH_3–THF), Boc-protection of the terminal amino groups, debenzylated (10% Pt/C, H_2; 92%), and treated with $(Boc)_2O$/DMAP [14] to generate (91%) the corresponding isocyanate $O{=}C{=}NC[CH_2O(CH_2)_3NHBoc]_3$. The related monomer, for example, $O{=}C{=}NC[CH_2O(CH_2)_3NHdansyl]_3$, was prepared [15]. The $H_2N(CH_2OCH_2CH_2$-$CO_2Me)_3$ was converted by treatment with either $(Boc)_2O$ [16] or triphosgene [17] to the corresponding isocyanate, $O{=}C{=}NC(CH_2OCH_2CH_2CO_2Me)_3$.

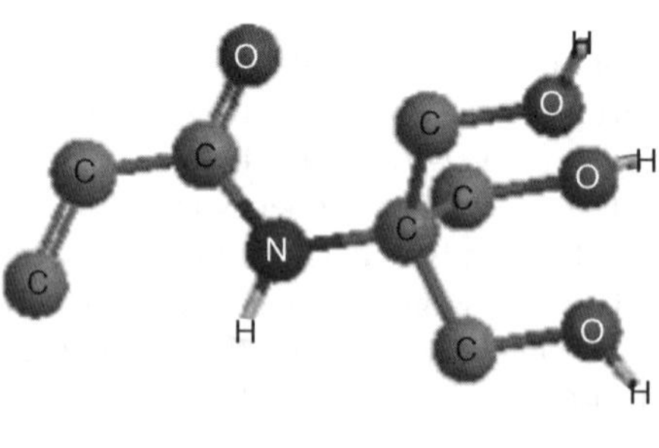

Tris(hydroxymethyl)acrylamidomethane [THAM] [18–20] and its protected acetonide [CH_2=CHCONHC[$(CH_2O)_2CMe_2$](CH_2OH)] [21] have been synthesized and utilized in the creation of hyperbranched materials [22] or glucose-based surfactants [22].

3.2.2 *Methanetricarboxylate*-Based

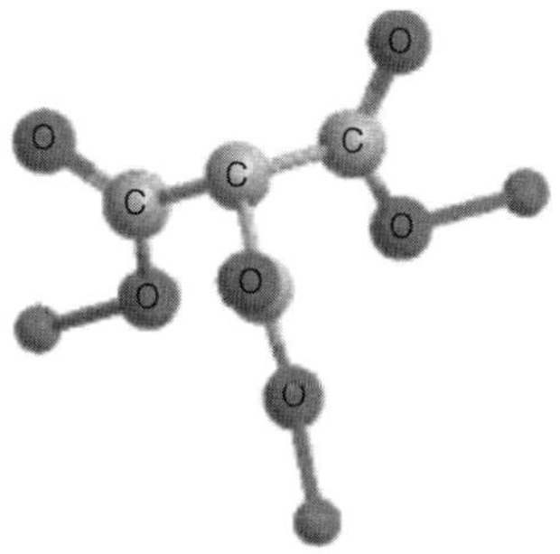

Trialkyl methanetricarboxylates were readily prepared by treating dialkyl malonate with alkyl chloroformate in basic conditions [23–25]; the sodium salt can be isolated and subsequently *C*-alkylated using a primary alkyl halide (R′X) or equivalent leaving group to generate $R'C(CO_2R)_3$. Hydrolysis leads to the initial *tris*-acid, $R'C(CO_2H)_3$, which easily loses carbon dioxide [26] generating the free monoacid, $R'CH_2CO_2H$; whereas, reduction (BH_3–THF) gave the corresponding triol R′C-$(CH_2OH)_3$. Subsequent extension of this triol can be accomplished by the following sequence of simple transformations: $ClCH_2CO_2H$, MeOH/H^+, LAH, TsCl giving $R'C(CH_2OCH_2CH_2OTs)_3$. Whereas, nucleophilic substitution of the terminal leaving group of $R'C(CH_2OCH_2CH_2OTs)_3$ is a facile process. Treatment of the intermediate ester $R'C(CH_2OCH_2CO_2Me)_3$ with TRIS (see above) easily creates a new amide bond and access to $R'C[CH_2OCH_2CONHC(CH_2OH)_3]_3$ via initial transesterification, that is, <$R'C[CH_2OCH_2CO_2CH_2C(NH_2)(CH_2OH)_2]_3$>, followed by a facile intramolecular rearrangement to the desired amide product [4].

3.2.3 *Nitromethane*-Based

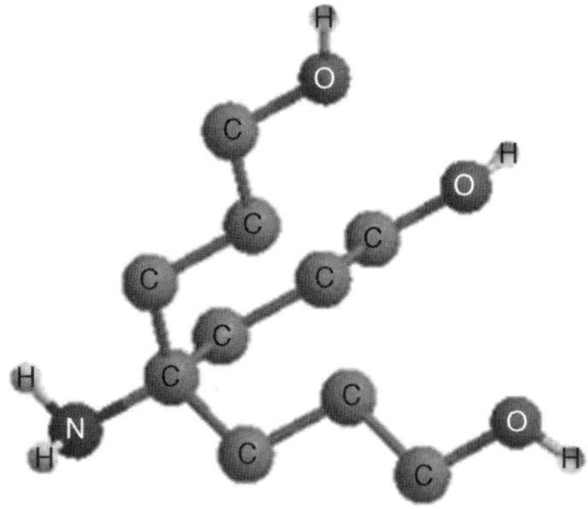

Tris(hydroxypropyl)aminomethane (bis*homo*TRIS) was synthesized by starting with $MeNO_2$ with acrylonitrile to give $O_2NC(CH_2CH_2CN)_3$, which was hydrolyzed to $O_2NC(CH_2CH_2CO_2H)_3$ and reduced (BH_3–THF; 95%) to the desired predendron $O_2NC[(CH_2)_3OH]_3$ [27]. Heterogeneous reduction then afforded the colorless H_2NC-$[(CH_2)_3OH]_3$ [27] that can be either acylated (Ac_2O) affording $H_2NC[(CH_2)_3OAc]_3$ or benzylated ($C_6H_5CH_2Cl$) to create the $H_2NC[(CH_2)_3OBn]_3$ in overall high yield. The treatment of bis*homo*TRIS with $SOCl_2$ gave (84%) the corresponding H_3N^+C-$[(CH_2)_3Cl]_3$ (Cl^-). The related $H_2NC[(CH_2)_3SH]_3$ was prepared from bis*homo*TRIS in four steps [28].

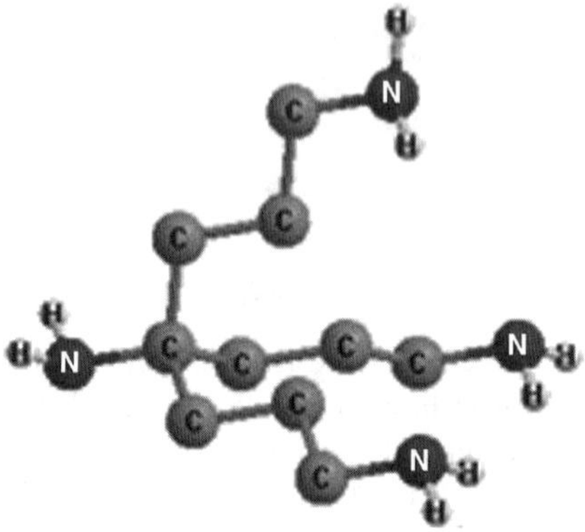

Tris(aminopropyl)aminomethane was prepared by treatment of the above O_2NC $(CH_2CH_2CN)_3$ with (BH_3) to give (~100%) $O_2NC[(CH_2)_3NH_2]_3$, which upon treatment with di-*tert*-butyl dicarbonate [29,30] gave $O_2NC[(CH_2)_3NHBoc]_3$, followed by catalytic reduction (T-1 Raney nickel) [31] to afford $H_2NC[(CH_2)_3NHBoc]_3$ in excellent overall yield.

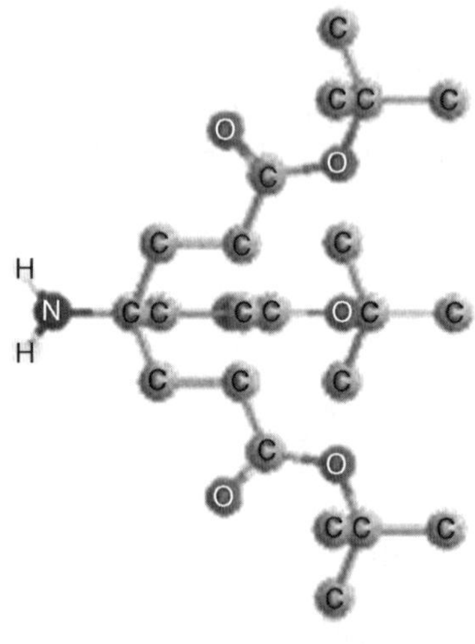

Di-*tert*-butyl 4-[2-*tert*-butoxycarbonyl)ethyl]-4-aminoheptanedioate (Behera's amine) was prepared [32,33] by treatment of $MeNO_2$ with *tert*-butyl acrylate in the presence of base (Michael reaction; [34] Triton-B) to give (ca. 80–95%) O_2NC-$(CH_2CH_2CO_2CMe_3)_3$, which was catalytically reduced [35] to quantitatively afford

$H_2NC(CH_2CH_2CO_2CMe_3)_3$ [33,36,37]. Although the original procedure used Raney nickel in EtOH [32], when *n*-heptane was used for the hydrogenation, the amine was isolated in 98% yield with an increased melting point [38].

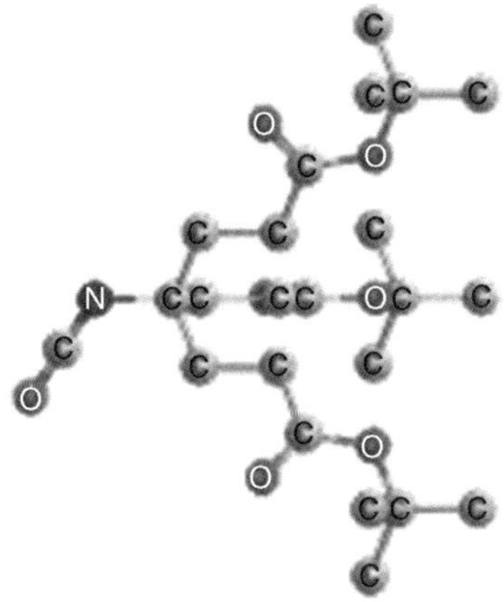

Di-*tert*-butyl 4-[2-(*tert*-butoxycarbonyl)ethyl]-4-isocyanatoheptanedioate, OCNC-$(CH_2CH_2CO_2CMe_3)_3$, was formed by treatment of $H_2NC(CH_2CH_2CO_2CMe_3)_3$ with phosgene or triphosgene [39–43]. The related, divergently generated G2 and G3 dendrons were similarly formed [44]. A convergent procedure [45], which utilized $O_2NC(CH_2CH_2CO_2H)_3$, derived from hydrolysis (HCO_2H) of O_2NC-$(CH_2CH_2CO_2CMe_3)_3$, followed by $H_2NC(CH_2CH_2CO_2CMe_3)_3$ using standard amidation (HOBT and DCC) and then reduction using Raney-nickel generated the related G2 and G3 dendrons. A series of these isocyanates [17,39,42,43] and their use in combinatorial chemistry [40,46] has been reported in which $O_2NC[(CH_2)_3CN]_3$ [47] was reduced (BH_3–THF) to $O_2NC[(CH_2)_3NH_2]_3$, which was Boc-protected giving $O_2NC[(CH_2)_3NHBoc]_3$, then reduction [Ni(R), H_2] to the amine H_2NC-$[(CH_2)_3NHBoc]_3$ and lastly conversion to the isocyanate $O{=}C{=}NC[(CH_2)_3NHBoc]_3$ with triphosgene [48] or DMAP and $(Boc)_2O$ [49]; [R = 2,6-Pyridine [50];

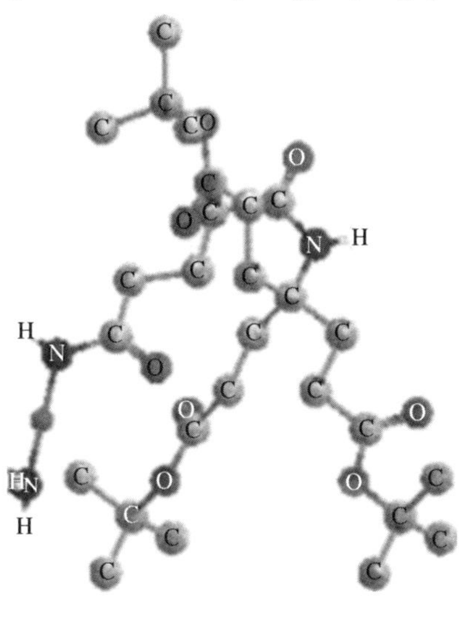

1,4-diaminoanthraquinone [51]; *N,N'*-bis(3-aminopropyl)piperazine; 3,3'- and 5,5'-(2,2'-dipyridino) [51–55]. Synthesis of the internally functionalized monomers was achieved via a three-component, single-pot reaction using glutaryl dichloride, $H_2NC(CH_2CH_2CO_2CMe_3)_3$, and the bis-diamine to give the homologated amino-triester, $H_2NRNHCO(CH_2)_4CONHC(CH_2CH_2CO_2CMe_3)_3$, where R = an appropriate diamine possessing the desired host–guest recognition component.

3.2.4 *Pentaerythritol*-Based

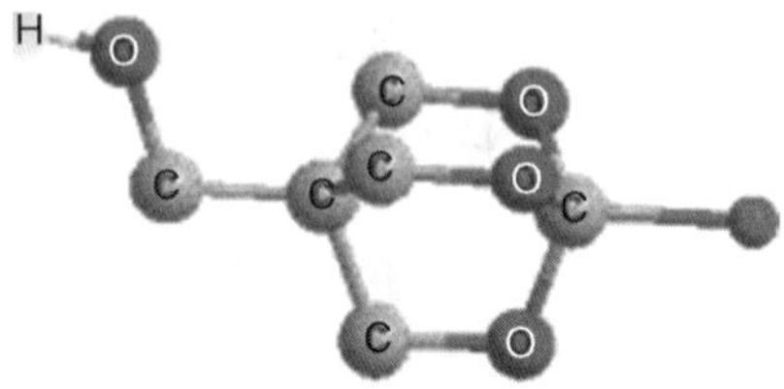

Although 4-(hydroxymethyl)-2,6,7-trioxabicyclo[2.2.2]acetate (R = H) has been constructed (~50%) by treatment of pentaerythritol with $HC(OEt)_3$ [56,57], the orthoacetate was shown to enhance the monomer's versatility [58]. Treatment of $HOCH_2C(CH_2O)_3CH$ with $CH_2{=}CHCOCl$ gave $CH_2{=}CHCO_2CH_2C(CH_2O)_3CH$; whereas, pentaerythritol with HBr in AcOH gave (70%) $BrCH_2C(CH_2OH)_3$, which was transformed (60%) to $BrCH_2C(CH_2O)_3CH$ [59]. Treatment of $C(CH_2OH)_4$ with $MeC(OMe)_3$ gave (85%) 1-methyl-4-(hydroxymethyl)-2,6,7-trioxabicyclo[2.2.2]-acetate (R = Me) [60], $HOCH_2C(CH_2O)_3CMe$, or (80%) $HOCH_2C(CH_2O)_3CEt$ when $EtC(OEt)_3$ was used [61]. The free hydroxy permits selective functionalization at one arm, then deprotection of the orthoester and simple conversion to a series of useful reagents: $ROCH_2C(CH_2OSCN)_3$, $C_6H_5CH_2OCH_2C(CH_2SH)_3$, and $CH_2{=}CHCH_2OCH_2C(CH_2SH)_3$ [62]. Treatment of $HOCH_2C(CH_2O)_3CMe$ with allyl bromide, followed by hydrolysis, gave $CH_2{=}CHCH_2OCH_2C(CH_2OH)_3$, which with $SOCl_2$ (pyr) generated $CH_2{=}CHCH_2OCH_2C(CH_2Cl)_3$, and with $LiPPh_2$ afforded $CH_2{=}CHCH_2OCH_2C(CH_2PPh_2)_3$ [63]. Treatment of $HOCH_2C(CH_2Cl)_3$, prepared directly from pentaerythritol with 3.3 equiv. of $SOCl_2$, with $Me_3SiC{\equiv}CCH_2Br$ (KOH) gave $HC{\equiv}CCH_2OCH_2C(CH_2Cl)_3$, which was converted to $HC{\equiv}CCH_2OCH_2C(CH_2PPh_2)_3$ [63]. Treatment of $C(CH_2OCOCH_2CH_2SH)_4$ with ethyl acrylate underwent a thio-Michael addition reaction in the presence of $HNEt_2$ to give the 1 → 3 monomer, $EtO_2CCH_2CH_2SCH_2CH_2CO_2CH_2C(CH_2$-$OCOCH_2CH_2SH)_3$, along with other addition products [64].

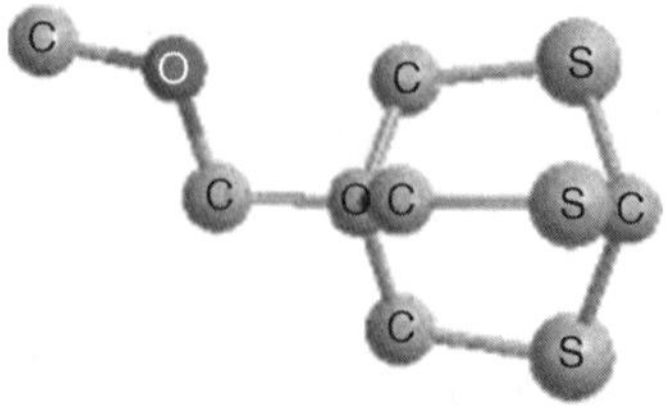

The commercially available $HOCH_2C(CH_2Br)_3$ was initially alkylated via the Williamson synthesis, then converted ($NaSH/S/DMF$, Cu/C_6H_5Me, $Zn/HCl/C_6H_6$) to form the $MeOCH_2C(CH_2SH)_3$ and lastly treated with $HC(OMe)_3$ to generate the 4-methyloxymethyl-2,6,7-trithiabicyclo[2.2.2]octane, $MeOCH_2C(CH_2S)_3CH$ [59].

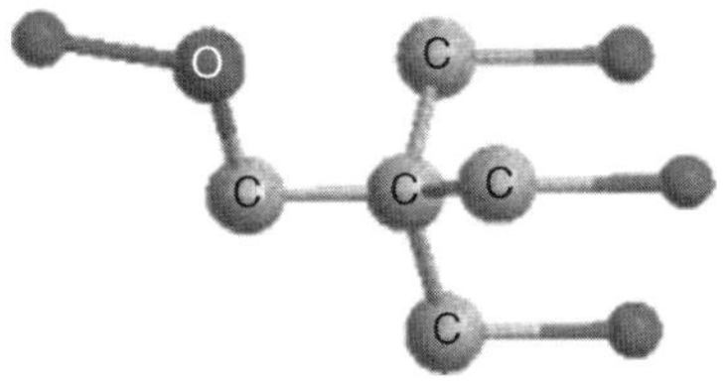

Pentaerythritol with 3.1 equiv. of acrylonitrile in the presence of base and then using Fischer esterification conditions generated $HOCH_2C(CH_2OCH_2CH_2CO_2Me)_3$, which can be *O*-protected, then reduced to the triol $ROCH_2C(CH_2OCH_2CH_2$-$CH_2OH)_3$ [65,66]. Conversely, pentaerythritol was initially monoprotected with *tert*-butylchlorodiphenylsilane ($Z = TBDPS$) then transformed into a family of very useful $1 \rightarrow 3$ branched monomers, for example, $ZO_2CHC(CH_2OH)_3$, ZO_2CHC-$(CH_2OCH_2CH{=}CH_2)_3$, $ZOCHC(CH_2OCH_2CH_2OH)_3$, $ZOCH_2C(CH_2OCH_2CH_2N_3)_3$, $HOCH_2C(CH_2OCH_2CH_2N_3)_3$, $CH_2{=}CHCH_2OCH_2C(CH_2OCH_2CH_2NH_2)_3$, $CH_2{=}CHCH_2OCH_2C(CH_2N_3)_3$, and $HO_2C(CH_2)_6OCH_2C(CH_2N_3)_3$ by a series of simple conversions [67]. The $TBDMSOC_2HC(CH_2OH)_3$ was transformed to $TBDMSOC_2HC$-$(CH_2OCH_2C{\equiv}CH)_3$ upon treatment (NaH/THF; 82%) with $HC{\equiv}CCH_2Br$ [68].

The commercially available pentaerythritol triallyl ether, $HOCH_2C(CH_2OCH_2CH{=}CH_2)_3$, was benzylated (NaH/DMF) with 4-bromobenzyl bromide to give (91%) $BrC_6H_4CH_2OCH_2C(CH_2OCH_2CH{=}CH_2)_3$, which was hydroborated (9-BBN, $NaOH/H_2O_2$; 78%) affording $BrC_6H_4CH_2OCH_2C[CH_2O(CH_2)_3OH]_3$, then transformed ($HC{\equiv}CC(OH)Me_2$/(dppf)$PdCl_2$/CuI/pyrrolidine; 81%, then NaOH/C_6H_5Me; 79%) to $HC{\equiv}CC_6H_4CH_2OCH_2C[CH_2O(CH_2)_3OH]_3$ [69]. Pentaerythritol triallyl ether can be prepared by treatment of pentaerythritol with allyl bromide (NaOH) in 50% yield and subsequently converted (I_2/PPh_3/imidazole) to $ICH_2C(CH_2OCH_2CH{=}CH_2)_3$ [70,71]; whereas Jones oxidation of $HOCH_2C(CH_2$-$OCH_2CH{=}CH_2)_3$ formed the desired focal acid group, which was treated $SOCl_2$, followed by ethylenediamine to give $H_2NCH_2CH_2NHOCCH_2C(CH_2$-$OCH_2CH{=}CH_2)_3$ [72].

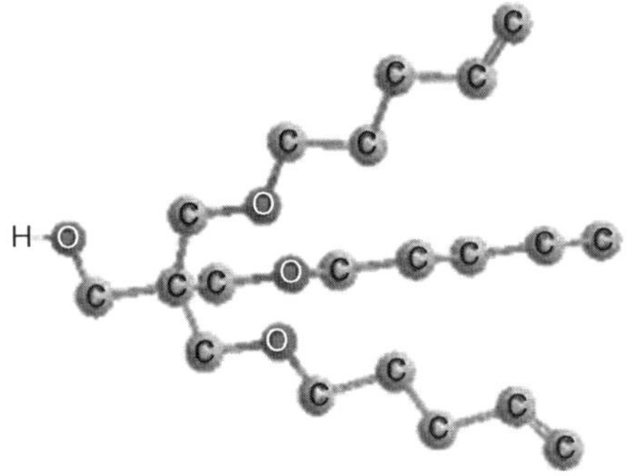

3-(5-Pentenyloxy)-2,2-bis(5-pentenyloxymethyl)propan-1-ol was prepared (26%) by treatment of pentaerythritol with 5-bromopentene [73] also isolated was tetrakis(5-pentenyl)pentaerythritol in 41% yield. Reaction of the free hydroxy group with 5-bromopentanoic acid in the presence of DMAP gave (72%) of the elongated $Br(CH_2)_4CO_2CH_2C[CH_2O(CH_2)_3CH{=}CH_2]_3$. The above $HOCH_2C[CH_2O(CH_2)_3CH{=}CH_2]_3$ was readily converted (PPh_3, I_2, imidazole; 90%) to $ICH_2C[CH_2O\text{-}(CH_2)_3CH{=}CH_2]_3$ [74,75].

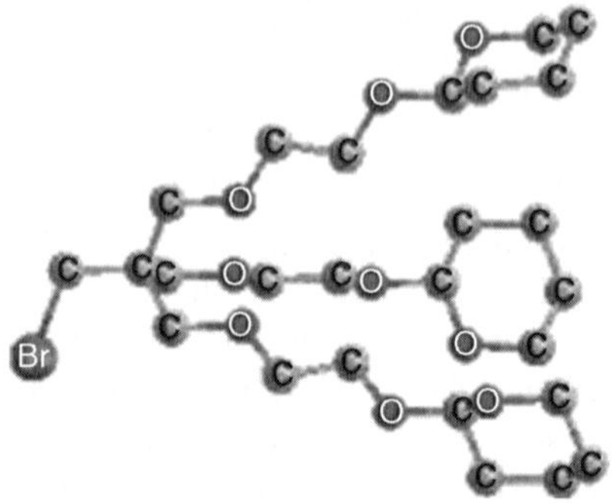

The monoprotection of ethylene glycol with dihydropyran gave (84%) $HOCH_2\text{-}CH_2OTHP$, which with $C(CH_2Br)_4$ in the presence of NaH–diglyme afforded (53%) $BrCH_2C(CH_2OCH_2CH_2OTHP)_3$ then addition of $PMBO(CH_2CH_2O)_2H$ (PMB = *p*-methoxybenzyl) gave (68%) $PMBO(CH_2CH_2O)_2CH_2C(CH_2OCH_2CH_2OTHP)_3$, which can be easily selectively deprotected (DDQ) generating (68%) dendron $HO\text{-}(CH_2CH_2O)_2CH_2C(CH_2OCH_2CH_2OTHP)_3$ [76].

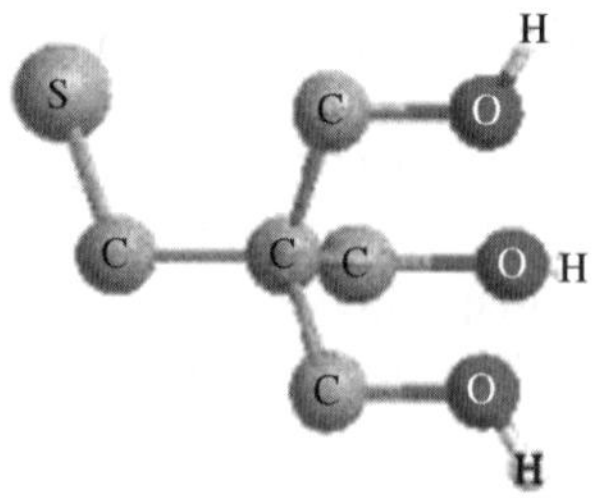

3-Hydroxy-2,2-di(hydroxymethyl)propanethiol [77] was prepared [78] by the treatment of pentaerythritol with HBr in 49–54% yield; similar products via $RSCH_2C(CH_2OCH_2CH_2CN)_3$ and $RSCH_2C(CH_2OCH_2CH_2CO_2R')_3$, where R′ = Me or $-C_6Cl_5$, were reported.

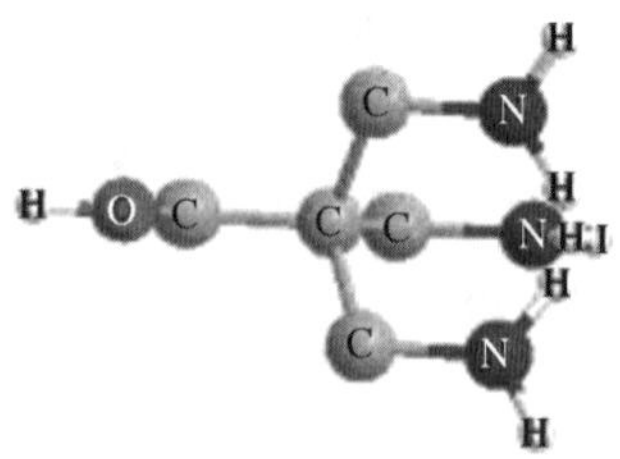

2-Hydroxy-1,1,1-*tris*(aminomethyl)ethane was prepared by the treatment of $HOCH_2C(CH_2O)_3CMe$ with $C_6H_5CH_2Br$, followed by hydrolysis gave $C_6H_5CH_2OCH_2C(CH_2OH)_3$, which was tosylated, subjected to NaN_3, followed by reduction to give $C_6H_5CH_2OCH_2C(CH_2NH_2)_3$ and lastly, deprotected (H_2/Cat) to give $HOCH_2C(CH_2NH_2)_3$ [60]. A novel triphenyl derivative of 1-hydroxymethyl-1,3,5-triazaadamantane, which was prepared from $HOCH_2C(CH_2NH_2)_3$ and an activated benzaldehyde, has been reported [79].

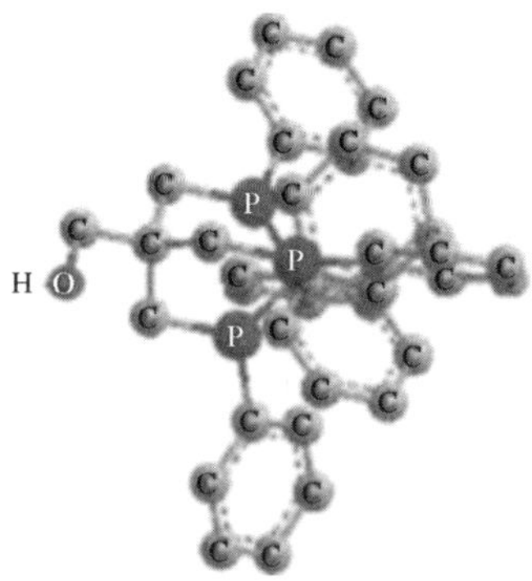

2-Hydroxy-1,1,1-*tris*(diphenylphosphinomethyl)ethane was readily alkylated and extended via, for example, $Cl(CH_2)_6OTs$ to give $Cl(CH_2)_6OCH_2C(CH_2PPh_2)_3$ and related members of this tripodal family were also synthesized: $HOCH_2C$-$(CH_2PPh_2)_{1-3}$ $(CH_2X)_{0-2}$, where $X = PPh_2$, SR [80].

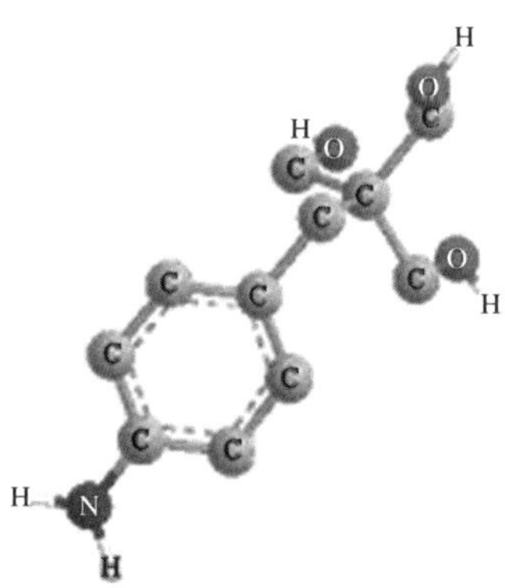

The phenyl-extended 2-(4-aminobenzyl)-2-(hydroxymethyl)propan-1,3-diol, $H_2NC_6H_4CH_2C(CH_2OH)_3$ [81], has been synthesized from 4-nitrobenzyl bromide with ethyl malonate, followed by alkylation with $ClCH_2OCH_2C_6H_5$, reduction ($NaBH_4$; Pd/H_2), and protection and deprotection.

3.2.5 *Neopentane*-Based

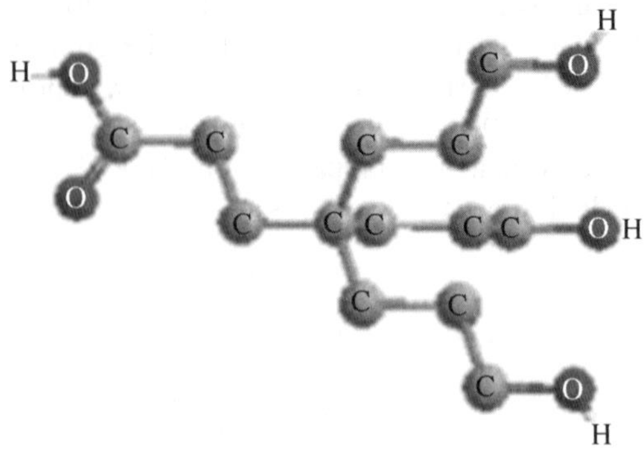

The $O_2NC[(CH_2)_3OH]_3$ was protected ($C_6H_5CH_2Cl$) to give O_2NC-$[(CH_2)_3OCH_2Bn]_3$, which underwent denitration-cyanoethylation [82] upon treatment with *n*-Bu_3SnH and AIBN in the presence of acrylonitrile generating intermediate $NCCH_2CH_2C[(CH_2)_3OH]_3$, which was hydrolyzed to $HO_2CCH_2CH_2C$-$[(CH_2)_3OH]_3$ [83,84].

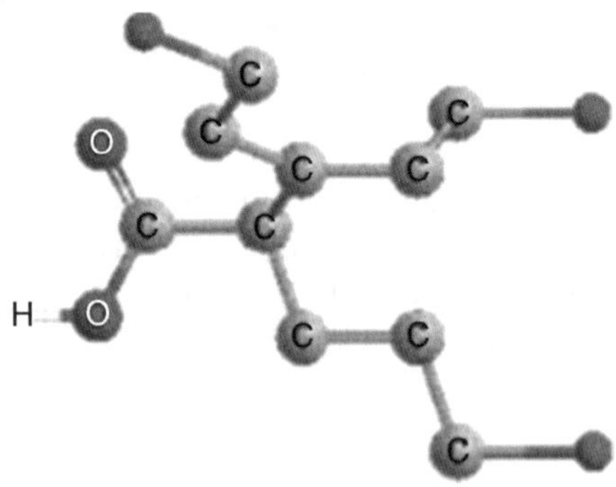

Acetaldehyde with $CH_2{=}CHCN$ was converted to $OHCC(CH_2CH_2CN)_3$, which was reduced and protected giving $TBSOCH_2C(CH_2CH_2CN)_3$; then reduction of the cyano groups (RaNi, H_2NNH_2) gave $TBSOCH_2C[(CH_2)_3NH_2]_3$, which was amidated, deprotected, and oxidized (Jones) forming $HO_2CC[(CH_2)_3NHCOR]_3$ [85,86].

3.2.6 Aryl*C*-Based

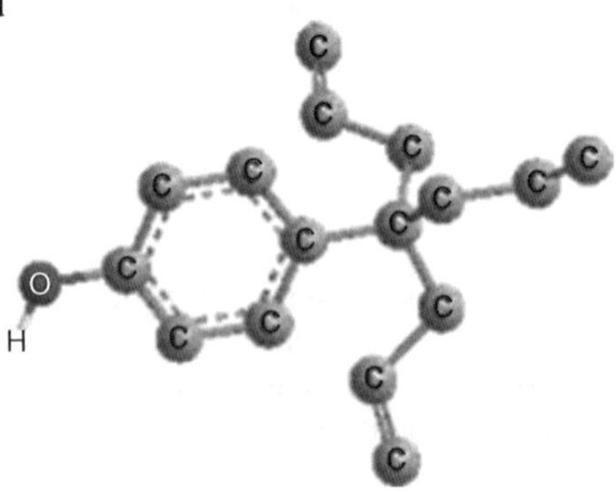

Treatment of <[FeCp(η^6-*p*-EtOC$_6$H$_4$Me)][PF$_6$]> [87] with allyl bromide gave (50%) <[FeCp(η^6-*p*-triallylmethylphenate), which was decomplexed in MeCN using a Hg lamp affording the desired *p*-*tris*(allyl)methylphenol, $HOC_6H_4C(CH_2CH{=}CH_2)_3$ [88,89]. The $HC{\equiv}CCH_2OC_6H_4C(CH_2CH{=}CH_2)_3$ was prepared from $HOC_6H_4C(CH_2CH{=}CH_2)_3$ with $HC{\equiv}CCH_2Br$ with base [90,91]. The $HOC_6H_4C(CH_2CH{=}CH_2)_3$ can be coated with azide moieties by treatment of the polyolefin with $HSiMe_2CH_2Cl$ and Karstedt catalyst, followed by NaN_3 [92], then via a click reaction with $HC{\equiv}CCH_2OC_6H_4C(CH_2CH{=}CH_2)_3$ the next tier can be generated. The reaction of $HOC_6H_4C(CH_2CH{=}CH_2)_3$ with $HSiMe_2CH_2Cl$ generated $HOC_6H_4C[(CH_2)_3SiMe_2Cl]_3$ [93] that was extended (K_2CO_3, DMF) with *p*-iodomethylstyrene affording $CH_2{=}CHC_6H_4CH_2OC_6H_4C[(CH_2)_3SiMe_2Cl]_3$ then subjected to AIBN-induced radical polymerization, followed by conversion to $(CH_2{=}CH)_nC_6H_4CH_2OC_6H_4C[(CH_2)_3SiMe_2CH_2N_3]_3$ [94], which is set up for the click reaction [95,96]. The nucleophilic displacement of chloride from the terminal R-$SiMe_2Cl$ with the triallylmethylphenoxide was catalyzed with NaI in DMF [88,89]; whereas $HOC_6H_4C(CH_2CH{=}CH_2)_3$ with $HSiMe_2CH_2Cl$ was catalyzed to give $HOC_6H_4C[(CH_2)_3SiMe_2CH_2Cl]_3$, which can be converted (NaI, butanone) to the related $HOC_6H_4C[(CH_2)_3SiMe_2CH_2I]_3$. Another monomer in this family is the protected $EtCO_2C_6H_4C[(CH_2)_3I]_3$ [97] and the related sulfide reagents were also derived from $MeSC_6H_4C(CH_2CH{=}CH_2)_3$ [98].

3.2.7 *Tetraphenylmethane*-Based

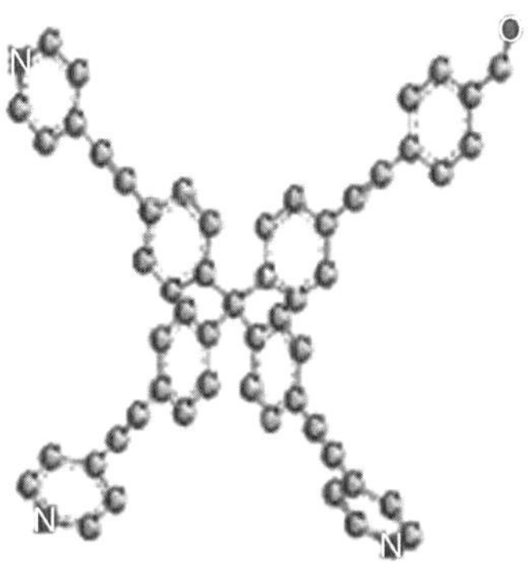

4-[2-(4-Formylphenyl)ethynyl]phenyl-*tris*[4-[2-(4-pyridinyl)ethynyl]phenyl]-methane was prepared by the treatment [$Pd(PPh_3)_2Cl_2$/CuI] of tetra*kis*(4-iodophenyl)methane with 4-ethynylbenzaldehyde to generate (38%) $OHCC_6H_4C{\equiv}CC_6H_4C(C_6H_4I)_3$, which with 4-ethynylpyridine under similar conditions gave (54%) the desired $OHCC_6H_4C{\equiv}CC_6H_4C(C_6H_4C{\equiv}CC_5H_4N)_3$ [99].

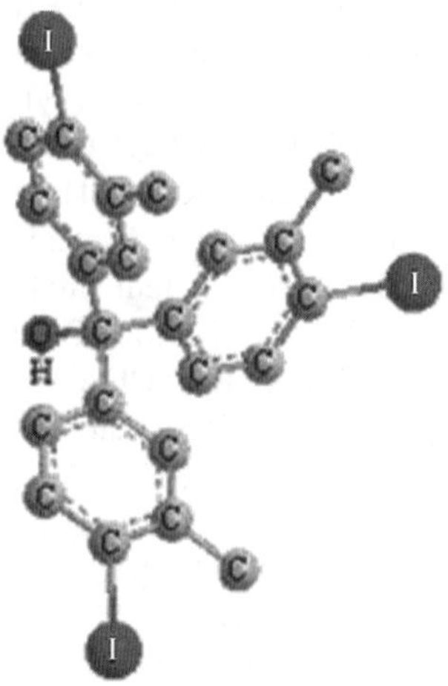

New fuchsin was diazotized via a Sandmeyer reaction ($NaNO_2$, H_2SO_4; KI) giving the *tris*(4-iodo-3-methylphenyl)methanol, which in the presence of phenol and concentrated sulfuric acid afforded the *tris*(4′-iodo-3′-phenyl)-4-methylphenol [100]. Utilizing the Sonogashira coupling [101], this triiodo intermediate was treated with phenylacetylene to generate *tris*(4′-phenylethynyl-3′-methylphenyl)-4-methylphenol.

3.3 1 → 3 *SI*-BRANCHED MONOMERS

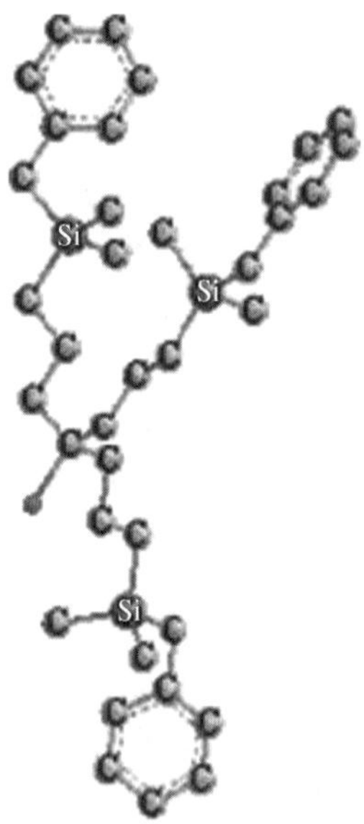

The Si-Ph bond in $PhSi[(CH_2)_3SiMe_2Bn]_3$ was cleaved with triflic acid to give $TfOSi[(CH_2)_3SiMe_2Bn]_3$ that can generate either $ClSi[(CH_2)_3SiMe_2Bn]_3$ or $(C_5H_5)Si$-$[(CH_2)_3SiMe_2Bn]_3$ when treated with Et_3NHCl or potassium cyclopentadienide, respectively; the G2 and G3 members were also prepared [102]. The related $C_6H_5Si[(CH_2)_3Si[(CH_2)_3SiMe_3]_3]_3$ has been reported and shown to undergo acidolysis [103] removing the phenyl group to generate $TfOSi[(CH_2)_3Si[(CH_2)_3SiMe_3]_3]_3$ thus activating the focal position.

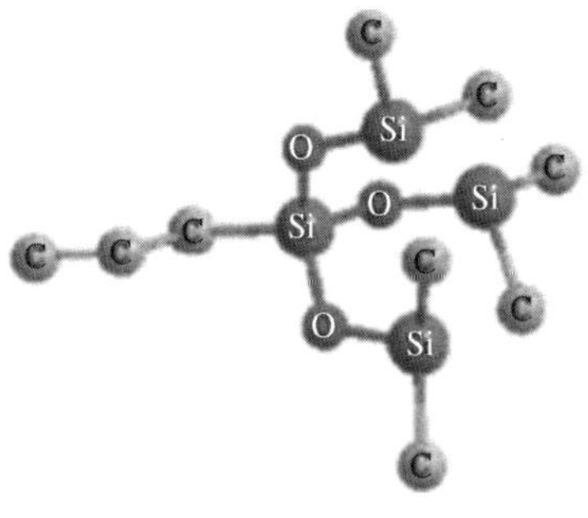

A one-step synthesis of poly(siloxysilanes) was reported [104–106] in which $CH_2{=}CHCH_2Si(OSiMe_2H)_3$ in Et_2O/MeCN (1:1) was treated with $H_2PtCl_6{\cdot}nH_2O$ under nitrogen.

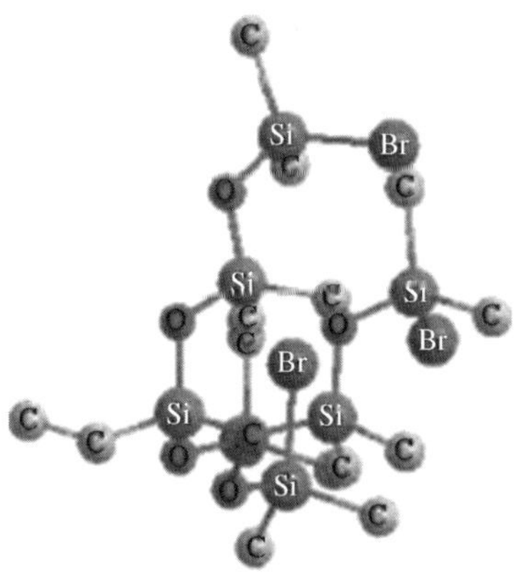

Treatment of $MeSi(OSiMe_2OSiMe_2Ph)_3$ with bromine gave $MeSi(OSiMe_2OSi$-$Me_2Br)_3$, which was transformed to the corresponding diethylamino derivative, $MeSi$-$(OSiMe_2OSiMe_2NEt_2)_3$ [107]. The related $HOSiMe_2OSi(OSiMe_2Ph)_3$, as the building block, should give the related 1 → 3 series. Treatment of $EtOSi[OSi$-$Me(OSiMe_3)_2]_3$ with $C_6H_5Si(OSiMe_2H)_3$ gave $C_6H_5Si[OSiMe_2OSi(OSiMe$-$(OSiMe_3)_2)_3]_3$ [108].

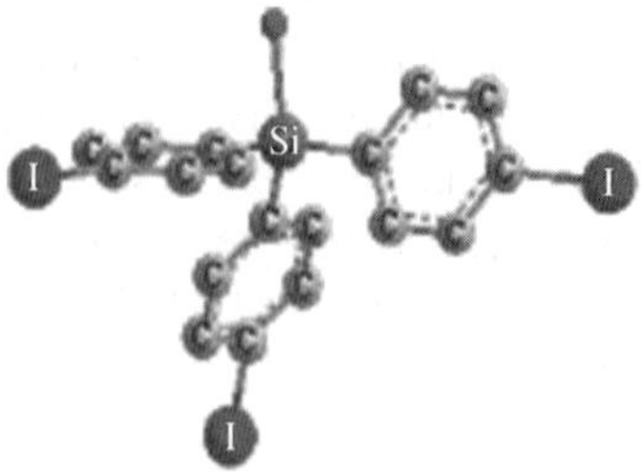

Tris(4-iodophenyl)ethoxysilane, $EtOSi(C_6H_4I)_3$ [109], was reacted with *tert*-butyl (3-ethynylbenzyloxy)dimethylsilane under Sonogashira coupling conditions [Pd $(dba)_2$, PPh_3, R_3N] to give $EtOSi(C_6H_4C{\equiv}CC_6H_4CH_2OTBDMS)_3$, which with 1,4-diiodobenzene and BuLi, then hydrolysis (TBAF, AcOH) generated (90%) $IC_6H_4Si(C_6H_4C{\equiv}CC_6H_4CH_2OH)_3$. This triol was transformed (AcSH, DIAD, PPh_3; 70%) to the corresponding RCH_2SAc, a novel tripod leading to a field-driven molecular motor [110]. Similarly, $HC{\equiv}CC_6H_4Si(C_6H_4C{\equiv}CC_6H_4SAc)_3$ and H-$(C{\equiv}CC_6H_4)_2Si(C_6H_4C{\equiv}CC_6H_4SAc)_3$ were constructed [109].

3.4 1 → 3 *ARYL*-BRANCHED MONOMERS

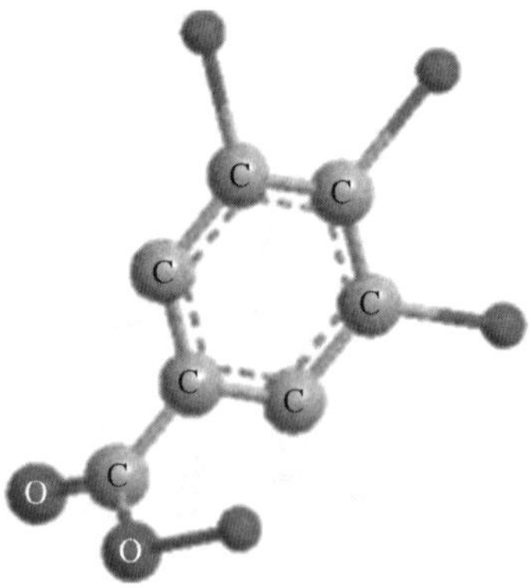

A 1 → 3 aryl branching series [111,112] was readily prepared from methyl gallate (R=Me, R′=H; methyl 3,4,5-trihydroxybenzoate) by either allylation of the free hydroxyl groups or transformation of the focal ester moiety. 3,4,5-Trihydroxybenzoic acid was converted in four steps to 3,4,5-tribenzyloxybenzyl chloride [113]. The treatment of methyl gallate with $HC{\equiv}CCH_2Br$ in acetone gave (K_2CO_3, 93%) $RO_2CC_6H_2(OCH_2C{\equiv}CH)_3$, which was reduced (LAH, 70%) to $HOCH_2C_6H_2(OCH_2C{\equiv}CH)_3$ and then transformed (PBr_3, 77%) to $BrCH_2C_6H_2(OCH_2C{\equiv}CH)_3$ [95]; focal elongation was accomplished (92%) by reaction with hydroquinone in DMF with K_2CO_3 giving $HOC_6H_4OCH_2$-$C_6H_2(OCH_2C{\equiv}CH)_3$, which can be clicked with diverse azides to give easy conversion to macromolecules [95,96]. Tetraethylene glycol was ditosylated (81%) and subsequently transformed (39%) to the monoazide, which was treated with methyl 3,4,5-trihydroxybenzoate to give (68%) $MeO_2CC_6H_2$-$[(OCH_2CH_2)_4N_3]_3$ [114]. This ester was either saponified (100%) to $HO_2CC_6H_2$-$[(OCH_2CH_2)_4N_3]_3$ or reduced (93%) to $MeO_2CC_6H_2[(OCH_2CH_2)_4NH_2]_3$; the combination of these two monomers gave (83%) the G2 PEGed dendron. Similarly, methyl 3,4,5-trihydroxybenzoate with triethylene glycol monotosylate (KI, K_2CO_3), followed by either protection (THP) of the termini and reduction to $HOH_2CC_6H_2[(OCH_2CH_2)_3OTHP]_3$ or tosylation to give $MeO_2CC_6H_2$-$[(OCH_2CH_2)_3OTs]_3$; combination gave (57%) the G2 dendron [115]. $HOCH_2C_6H_2[(OCH_2CH_2)_3OZ]_3$ were subjected to methacroyl chloride to give $CH_2{=}CMeCO_2CH_2C_6H_2[(OCH_2CH_2)_3OZ]_3$, then the termini were deprotected affording $CH_2{=}CMeCO_2CH_2C_6H_2[(OCH_2CH_2)_3OH]_3$. A related series of monomers was devised starting from ethyl 3,4,5-trihydroxybenzoate by treatment with $Br(CH_2)_3X$, where X = NHBoc, NHCbz, or OBn [116]. A click approach to functionalize azide-terminated dendrimers [92] utilized the polyPEGed $HC{\equiv}CCH_2O$-$(CH_2CH_2O)_4CH_2C_6H_2[O(CH_2CH_2O)_3Me]_3$ [117,118].

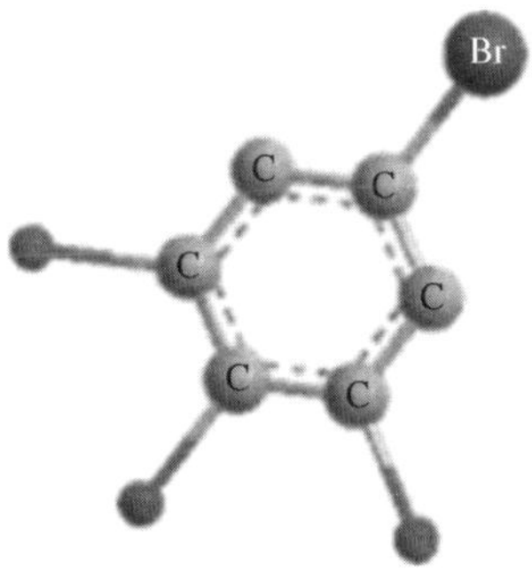

2,6-Dimethoxyphenol was easily transformed [119] to 3,4,5-trialkoxy-1-bromobenzene in three steps and then to the corresponding boronic acid derivative, which underwent a Suzuki-coupling reaction [120,121] with cyanuric acid.

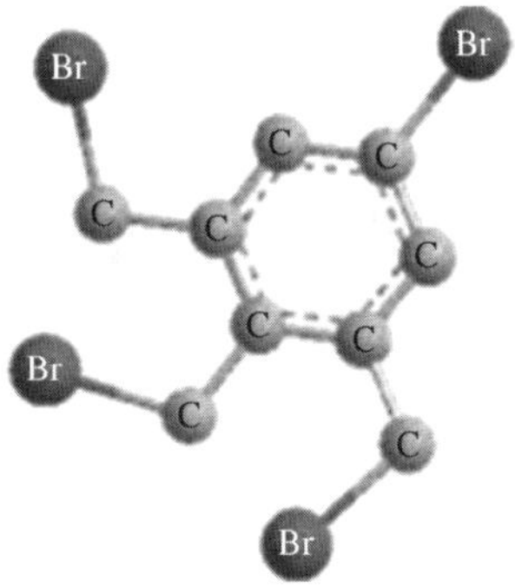

The reaction of 2-bromomestylene ($BrC_6H_2Me_3$) with *N*-bromosuccinimide and dibenzoyl peroxide in HCO_2Me gave (26%) 2,4,6-*tris*(bromomethyl)bromobenzene, which when treated with thiourea generated 2,4,6-*tris*(isothiouroniummethyl)bromobenzene tribromide [122].

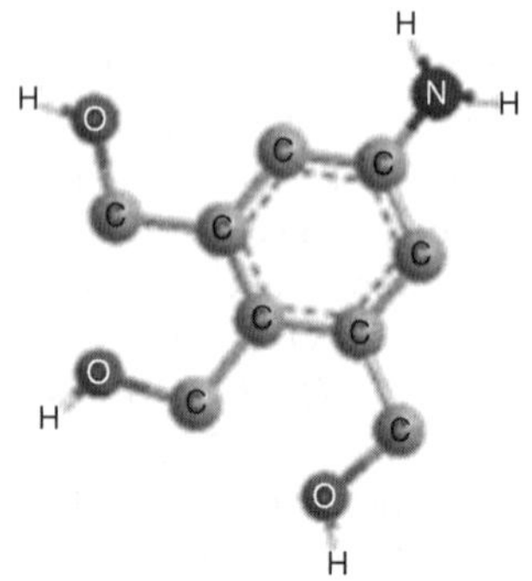

2,4,6-*Tris*(hydroxymethyl)aniline has been shown to possesses a drug [123] or reporter [124,125] properties at the dendron's surface by the incorporation of an enzyme substrate or H_2O_2 trigger at the focal position, respectively. The treatment of 1,3,5-*tris*(hydroxymethyl)benzene with *tert*-butyldimethylsilyl chloride gave $C_6H_3(CH_2OTBS)_3$, which was nitrated and reduced to the *tris-O*-protected aniline, which was transformed to the corresponding isocyanate, $OCNC_6H_2(CH_2OTBS)_3$ [124,126]. This isocyanate was reacted with a substituted phenol and the resultant carbamate was *O*-deprotected free-up the hydroxymethyl moieties for further elaboration.

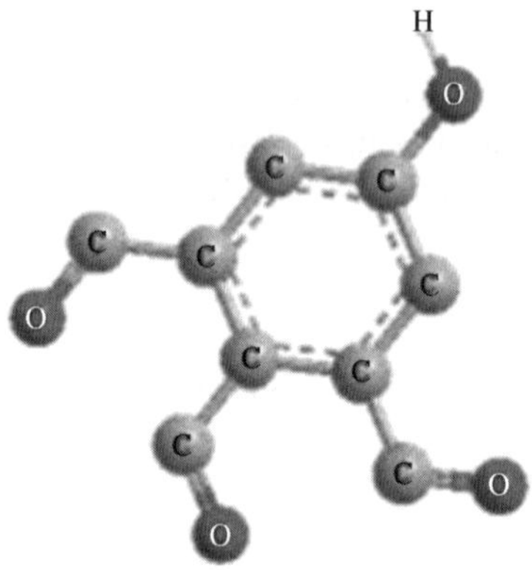

2,4,6-*Tris*(formyl)phenol [127] was prepared (48–54%) or by the reaction of either 4-hydroxybenzaldehyde or phenol with HMTA in trifluoroacetic acid. Treatment of 2,4,6-*tris*(formyl)phenol with the chlorocarbamoyl of *N*,*N*′-dimethylethylenediamine generated $BocNMeCH_2CH_2NMeCO_2C_6H_2(CHO)_3$, then the addition of diethyl malonate, followed by reduction (DIBALH) gave $BocNMeCH_2CH_2NMeCO_2C_6H_2[CH{=}C(CH_2OH)_2]_3$ [128].

3.5 1 → 3 *ADAMANTANE*-BRANCHED MONOMERS

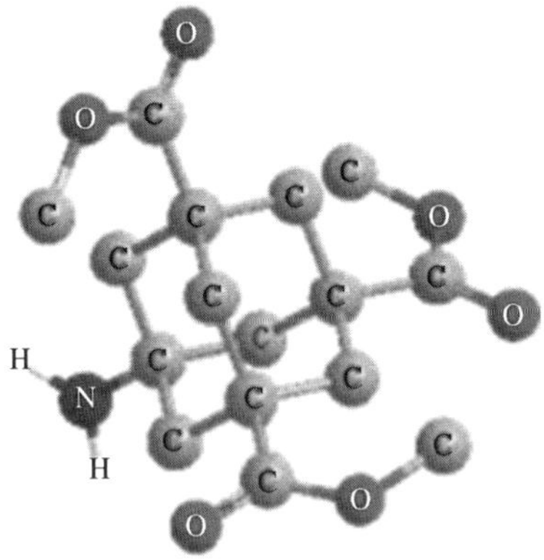

Photolysis of 1-nitroadamantane in $(COCl)_2$, followed by treatment with MeOH afforded (ca. 25%) 1-nitro-3,5,7-*tris*(methoxycarbonyl)adamantane [129], which was catalytically reduced to give the 1-amino-3,5,7-*tris*(methoxycarbonyl)adamantane in low overall yields [130]; notably, this is a rigid analog to the above Behera's amine [32]. Similarly, 1-bromoadamantane gave (24%) 1-bromo-3,5,7-*tris*(methoxycarbonyl)adamantane under similar conditions, then reduction (LAH, 95%) gave the 1-bromo-3,5,7-*tris*(hydroxymethyl)adamantane, which was transformed (Tf_2O/pyr, AcSK/18crown6, KOH) to 1-bromo-3,5,7-*tris*(mercaptomethyl)adamantane [131]. A related series of 1 → 3 monomers was prepared by using 1-(4-bromo/iodophenyl)-3,5,7-*tris*(hydroxymethyl)adamantane [131]. Alternatively, when 1,3,5,7-tetraphenyladamantane was oxidized and esterified 1,3,5,7-tetra*kis*(methoxycarbonyl)adamantane was isolated [132], it was selectively saponified (90%) to the monoacid, then subjected to the Curtius reaction [133] to give 1-amino-3,5,7-*tris*(methoxycarbonyl)adamantane in 83% yield [134]. The synthesis of the related 1-amino-3,5,7-*tris*(3-propionic acid)adamantane, from adamantane has been reported [135]. The related 1-amido-3,5,7-tricarboxyadamantane and 1-carboxy-3,5,7-amidoadamantane have been isolated from a mixture of products when 1,3,5,7-*tris*(chlorocarbonyl)adamantane was treated with aminofluorescein [136]. Tetrakis(4-iodophenyl)adamantane [137], prepared from tetraphenyladamantane with [*bis*(trifluoroacetoxy)iodo]benzene and iodine in CCl_4 or $CHCl_3$ [138], was treated with *t*-BuLi, followed by CO_2, hydrolysis, and CH_2N_2 generated in low yield (6–10%) the 1-(4-iodophenyl)-3,5,7-*tris*(4-methoxycarbonylphenyl)adamantane [139]; whereas, 1-(4-ethynylphenyl)-3,5,7-*tris*(4-cyanophenyl)adamantane was generated from

1-(4-trimethylsilylphenyl)-3,5,7-*tris*-4-iodophenyladamantane in two steps [140]. The related 1-(4-ethynylphenyl)-3,5,7-*tris*(4-methoxycarbonylphenyl)adamantane was prepared [141] in a simple five-step sequence from tetra*kis*(4-iodophenyl)-adamantane. The 1,3,5,7-tetra*kis*(aminomethyl)adamantane [142] was treated with 1 equiv. of CF_3CO_2Et, followed by a slight excess of 2,3-dimethoxybenzoic acid with BOP to generate a mixture of products from which the 1-(trifluoroactamido-methyl)-3,5,7-*tris*(2,3-dimethoxybenzoylamidomethyl)adamantane was isolated in 32% yield [143].

A novel triphenyl derivative of 1-hydroxymethyl-1,3,5-triazaadamantane, which was prepared from $HOCH_2C(CH_2NH_2)_3$ and an activated benzaldehyde has been reported [79]. Recently, an unusual intramolecular cyclization occurred in which *tris*(β-oximinoalkyl) amines, derived from NH_3 and $MeRHCNO_2$, were transformed into 4,6,10-trihydroxy-1,4,6,10-tetraazaadamantanes in 72–100% yield [144].

3.6 1 → 3 *MACROCYCLIC*-BRANCHED MONOMERS

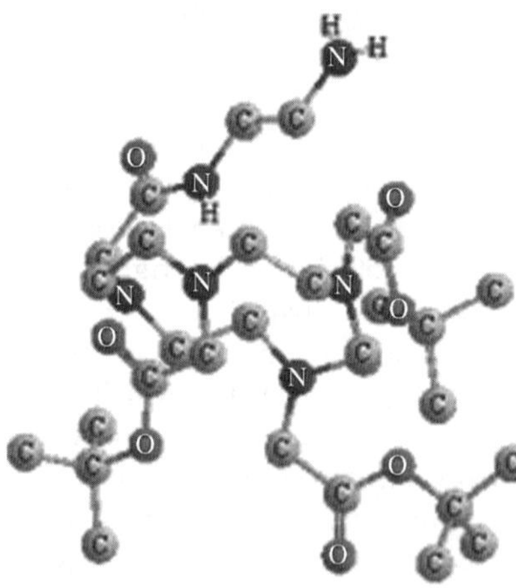

Cyclen (1,4,7,10-tetraazadodecane) was monoalkylated with ethyl bromoacetate and then subjected to 3 equiv. of *tert*-butyl bromoacetate to generate (84%) 1-(ethoxycarbonylmethyl)-4,7,10-*tris*(*tert*-butylcarbonylmethyl)tetraazadodecane, which was treated with neat ethylenediamine to give (68%) 1-(2-aminoethylazacarbonylmethyl)-4,7,10-*tris*(*tert*-butylcarbonylmethyl)tetraazadodecane [145]. This amine was transformed into a 1 → 3 branched dendrimer, then mild hydrolysis removed the *tert*-butyl esters. The 1-(2-aminoethyl)-4,7,10-*tris*(Boc)-tetraazadodacane was prepared by initial selective monoalkylation and separation, followed by exhaustive alkylation to give the 1 → 3 protected monomers [146]. In a reverse manner, cyclen was trialkylated with $BrCH_2CO_2CMe_3$ with DMA, then deesterified (TFA), and treated with 1,2-epoxy-butane or -octane to give access to another 1 → 3 branched monomer [147]. The specific syntheses of the related 1-benzyl-4,8,12-tetraazacyclopentadecane, 1-benzyl-4,8,12-tri-Boc-tetraazacyclopentadecane, 4,8,12-triBoc-1,4,8,12-tetraazacyclopentadecane, (1,4,8,12-tetraazacyclopentadecanyl)acetic acid, and 3-(1,4,8,12-tetraazacyclopentadecanyl)-propionic acid as well as the related 8-isomer, created from known unsymmetrical synthons, have been reported [148].

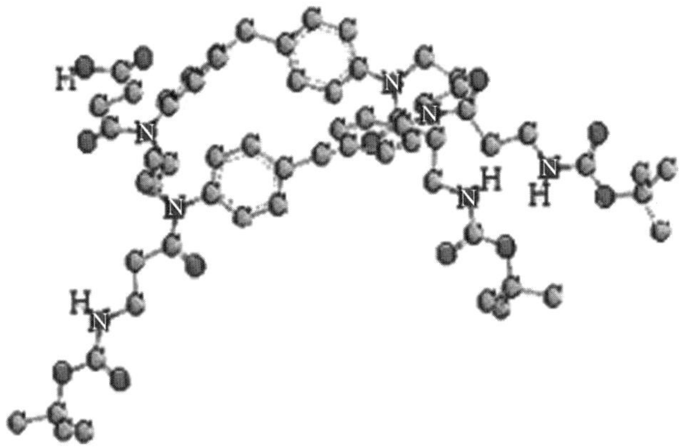

The selected substitution of 1,6,20,25-tetraaza[6.1.6.1]paracyclophane [149] with protected β-alanine (3 equiv.) in the presence of DCC gave (33%) *tris*(Boc-β-alanyl)-1,6,20,25-tetraaza[6.1.6.1]paracyclophane [150,151]. Activation of the remaining free amino site was conducted with succinic anhydride to give acid for future connectivity.

3.7 1 → 3 *CALIXARENE*-BRANCHED MONOMERS

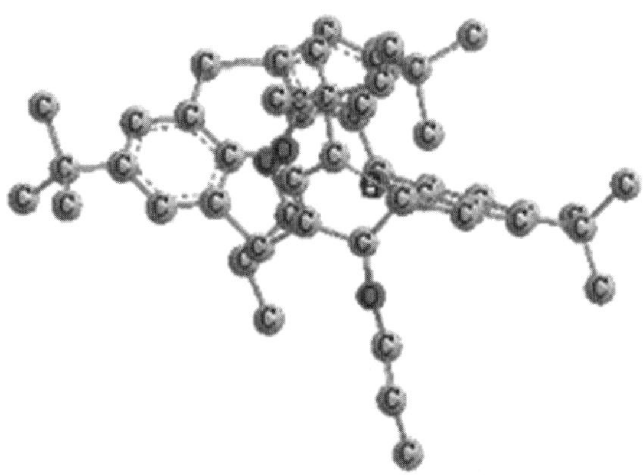

Selective protection of 25,26,27,28-tetrahydroxy-5,11,17,23-tetra(*tert*-butyl)calix-[4]arene by alkylation ($Ba(OH)_2/BaO$) with *n*-PrBr in THF gave 25,26,27-tripropoxy-

28-hydroxy-5,11,17,23-tetra(*tert*-butyl)calix[4]arene, which was subjected to an excess of α,ω-dibromoalkanes in DMF to generate the 25,26,27-tripropoxy-28-bromoalkoxy-5,11,17,23-tetra(*tert*-butyl)calix[4]arene [$(CH_2)_n$, $n=2$, 51%; $n=3$, 81%; $n=6$, 76%] [152]. Since in the initial alkylation step, the mono- and two di-alkylated derivatives were also isolated, thus other interesting polycalixarenes were reported [152] and can be envisioned.

3.8 1 → 3 *PORPHYRIN*-BRANCHED MONOMERS

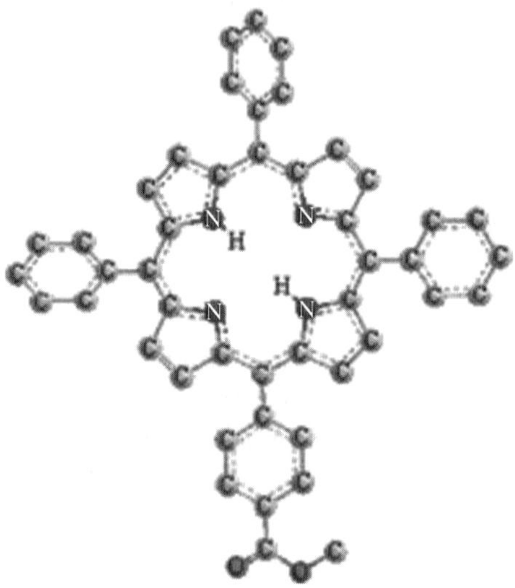

In a one-step procedure [153], from methyl 4-formylbenzoate, benzaldehyde, and pyrrole in a 1:3:4 ratio gave 10,15,20-triphenyl-5-(methoxycarbonylphenyl)porphyrin following a known procedure [154]. The dendronized 5-(4-hydroxyphenyl)-dipyrromethane, 5-(4-methoxycarbonylphenyl)dipyrromethane, and dendronized benzaldehyde in presence of $BF_3{\cdot}OEt_2$ at 25°C, followed by oxidation (DDQ), incorporation of platinum, and lastly hydrolysis gave the desired 10,15,20-*tris*-[(4′-dendronized)phenyl]-5-(methoxycarbonylphenyl)porphyrin [153,155,156]. The 10,15,20-triphenyl-5-(tetramethyldioxaborate)porphyrin as well as 10,15,20-triaryl-5-iodo(or bromo)porphyrin were reported in a route to a molecular "waterwheel"-shaped porphyrin pentamers [157]. In the construction of these amazing structures—porphyrin square planar henicosamer [158] comprised of 21 porphyrins measuring 6.5 nm per side—two different starting porphyrins were prepared from 3,5-di(*iso*-amyloxy)benzaldehyde, pyrrole, and methyl 4-formylbenzoate in propanoic acid, followed by separation (12 and 3%) from which 10,15,20-triaryl-5-(methoxycarbonylphenyl)porphyrin and 15,20-triaryl-5,10-di(methoxycarbonylphenyl)porphyrin were obtained. The monoester was transformed to the corresponding 10,15,20-triaryl-5-(formylphenyl)porphyrin, while the later was reduced, metalated, and selectively oxidized to the 1 → (2 + 1) branching 15,20-triaryl-5-(formylphenyl)-10-(4-ethylcarbonyloxymethyl)porphyrin.

3.9 1 → 3 *PHTHALOCYANINE*-BRANCHED MONOMERS

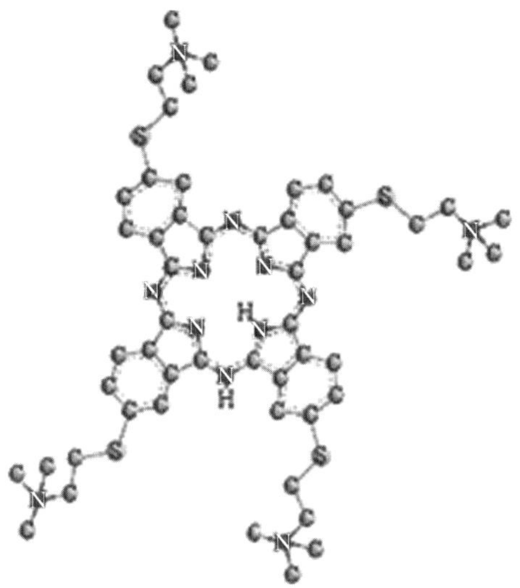

The condensation of 4-nitrophthalonitrile and 4-(2-dimethylaminoethylsulfanyl)-phthalonitrile, then reduction of the nitro group gave the corresponding monoamine, which was quaternized with MeI to give 2-amino[*tris*((2-dimethylaminoethylsulfanyl)phthalocyanine)triiodide] hydrochloride [159].

3.10 1 → (2 + 1) *C*-BRANCHED MONOMERS

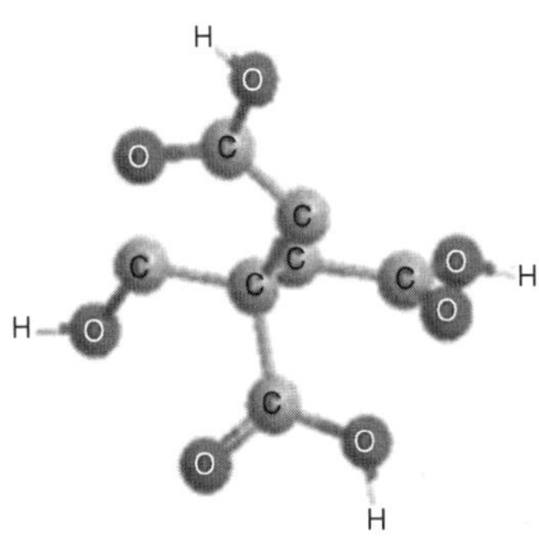

Citric acid can be esterified with an acyl chloride to generate the 3-monoester and leaving a terminal 1 → (2 + 1) tricarboxylic acid surface [160].

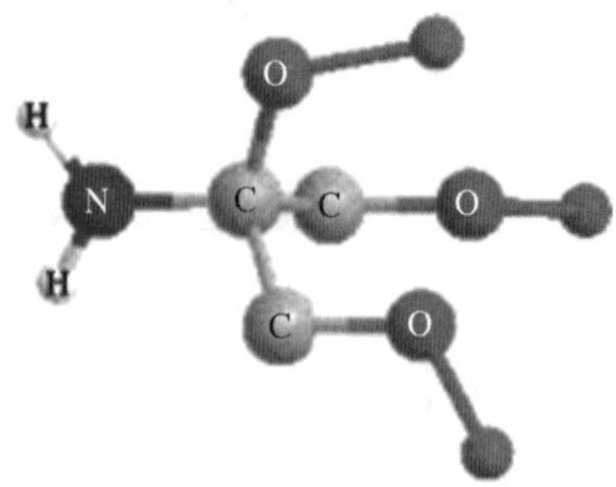

Glycerol (propane-1,2,3-triol) was treated with benzoyl chloride, followed by a Jones oxidation affording the $(C_6H_5CO_2CH_2)_2CO$, which with di(*iso*-propyl)amine generated an azahemiacetal; removal of the protecting moieties gave the free hydroxy groups that are esterified and lastly deprotected to afford $H_2NC[OR(CH_2\text{-}OR)_2]$ [160,161].

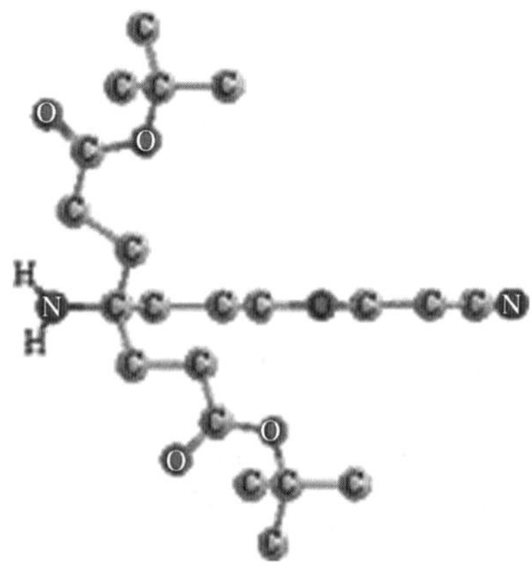

The $1 \rightarrow (2+1)$ monomer, <$H_2NC(CH_2CH_2CO_2CMe_3)_2[(CH_2)_3O\text{-}CH_2CH_2CN]$>, was synthesized from 4-nitrobutanol, which was initially *O*-alkylated with $CH_2{=}CHCN$, then excess $CH_2{=}CHCO_2CMe_3$, followed by reduction (RaNi, H_2) of the nitro group [162]. Amidation, (1) then deprotection (reverse Michael reaction, mild base) gave $RCOHNC(CH_2CH_2CO_2CMe_3)_2[(CH_2)_3OH]$, whereas (2) with acid, deesterification occurred giving $RCOHNC(CH_2CH_2CO_2H)_2\text{-}[(CH_2)_3OCH_2CH_2CN]$, or (3) with catalytic reduction $RCOHNC(CH_2CH_2CO_2\text{-}CMe_3)_2[(CH_2)_3O(CH_2)_3NH_2]$. The monomer $H_2NC(CH_2CH_2CO_2CMe_3)_2[(CH_2)_3\text{-}OCH_2CH_2CN]$ was treated with $HO_2C(CH_2)_3OR$ (R = 4′-terpyridine) [163] affording $RO(CH_2)_3COHNC(CH_2CH_2CO_2CMe_3)_2[(CH_2)_3O(CH_2)_2CN]$, which was reduced (RaNi) to give $RO(CH_2)_3COHNC(CH_2CH_2CO_2CMe_3)_2[(CH_2)_3O(CH_2)_3\text{-}NH_2]$, followed by acylation with 1-chlorocarbonyladamantane [162]. In a related sequence of transformations, $H_2N(CH_2)_5OR$ (R = 4′-terpyridine) [163] was reacted with $CH_2{=}CHCOCl$ to give $CH_2{=}CHCOHN(CH_2)_5OR$, which with $MeNO_2$ then a slight twofold excess of $CH_2{=}CHCO_2CMe_3$, followed by reduction gave the minidendron—$H_2NC(CH_2CH_2CO_2CMe_3)_2[(CH_2)_2CON(CH_2)_5OR]$ [164].

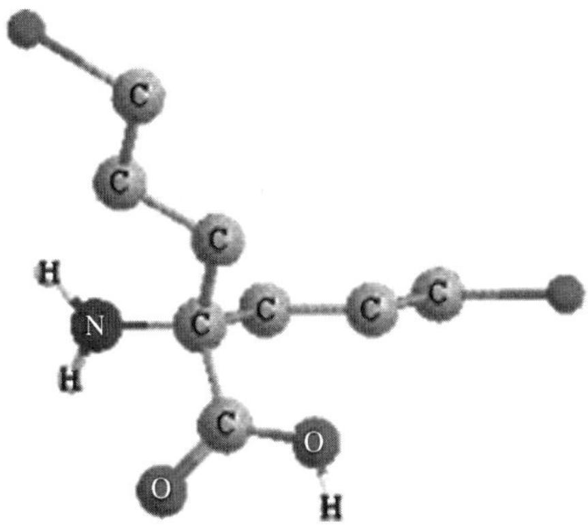

Treatment of methyl *N*-parachlorobenzylidene glycinate ($MeO_2CCH_2N{=}CHC_6H_4Cl$) with a slight excess of $CH_2{=}CHCN$, then hydrolysis gave the interesting methyl *bis*(cyanoethyl)glycinate, $H_2NC(CO_2Me)(CH_2CH_2CN)_2$, which with $(Boc)_2O$ gave the *N*-Boc derivative that was reduced (RaNi, H_2), and terminal *N*-protected to afford $BocHNC(CO_2Me)[(CH_2)_3NHZ]_2$. These derivatives were converted to a "*bis*-ornithine" [2,2-*bis*(aminopropyl)glycine] series [165–167].

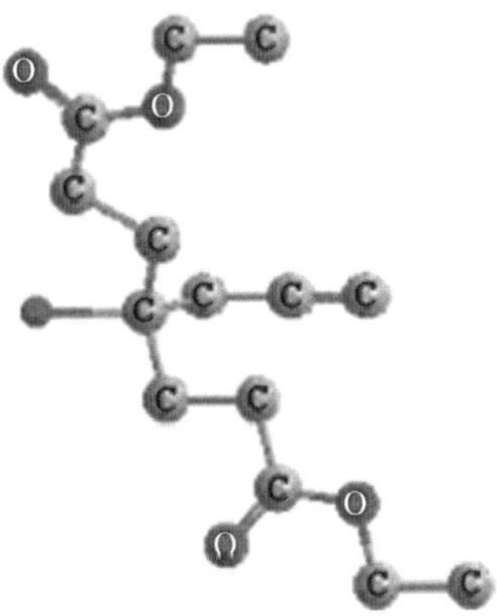

Diethyl acetamidopropargylmalonate was prepared from the salt of diethyl acetamidomalonate with $CH_2{=}C(OTf)CH_2Br$ generating $ZHNC(CO_2Et)_2[CH_2C(OTf){=}CH_2]$, which upon warming [$Pd(PPh_3)_4$,CO, Bu_3N] gave $ZHNC(CO_2Et)_2(CH_2C{\equiv}CH)$ [168].

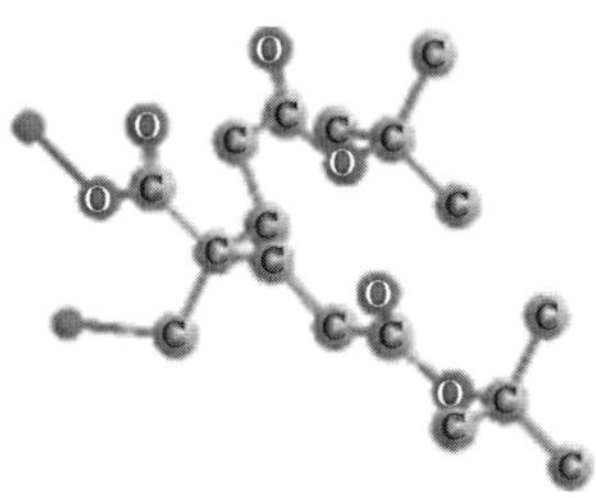

Treatment of benzyl acetoacetate with 2 equiv. of CH_2=$CHCO_2CMe_3$ gave the desired $RO_2CC(COR')(CH_2CH_2CO_2CMe_3)_2$ possessing a unique ketonic moiety [169]; whereas with dialkyl malonate, symmetric products were isolated; however, with unsymmetrical malonates, access to 1 → (2 + 1) *C*-branched monomers should be possible. Other related monomers have also been reported: $(MeO_2C)_2C(CH_2C{\equiv}CH)(CH_2CH_2COMe)_2$, $(MeO_2C)_2C(CH_2C{\equiv}CH)(CH_2CH_2CN)_2$, $(MeO_2C)_2C(CH_2CH{=}CH_2)(CH_2CH_2COMe)_2$, and $(MeO_2C)_2C(CH_2CH{=}CH_2OH)(CH_2CH_2COMe)_2$.

3.11 1 → (2 + 1) *ARYL*-BRANCHED MONOMERS

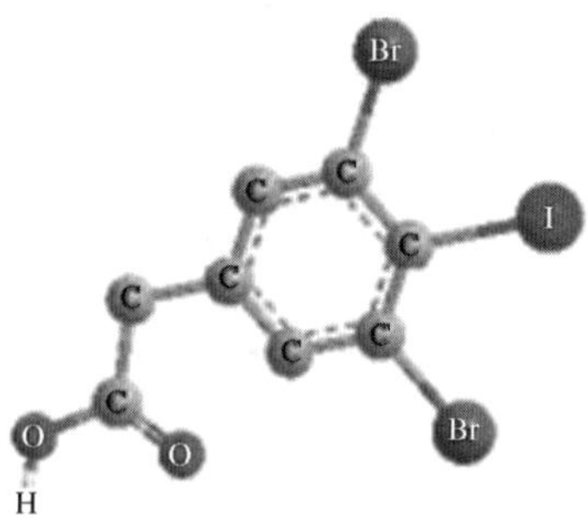

A novel series of dendrons possessing a 1 → (2 + 1)-branching pattern in which 3,5-dibromo-4-iodophenylacetic acid were prepared [170] from ethyl 4-aminophenylacetate via bromination (Br_2/AcOH) to give ethyl 3,5-dibromo-4-aminophenylacetate, then diazotization ($NaNO_2$/HCl, KI) to afford ethyl 3,5-dibromo-4-iodophenylacetate, which was selectively substituted by the use of the Suzuki cross-coupling reaction [171].

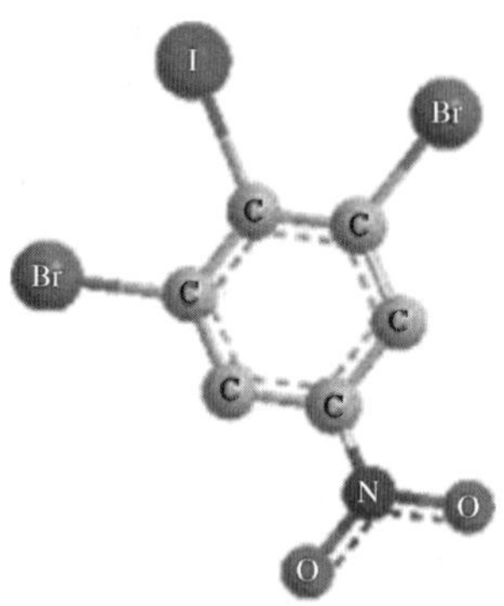

4-Iodo-2,6-dibromonitrobenzene was transformed in three steps to 4-iodo-2,6-dibromodiethyltriazenylbenzene, which was coupled [$Pd(Ph_3P)_2Cl_2/Et_3N$]

with $HC{\equiv}CC_6H_5$ to give 4-phenylethynyl-2,6-dibromodiethyltriazenylbenzene, then further coupling [$Pd(Ph_3)_4/Na_2CO_3$/DME] with $(HO)_2BC_6H_4OC_6H_5$ and deprotected generating 4-phenylethynyl-2,6-di(4-benzyloxyphenyl)iodobenzene [172].

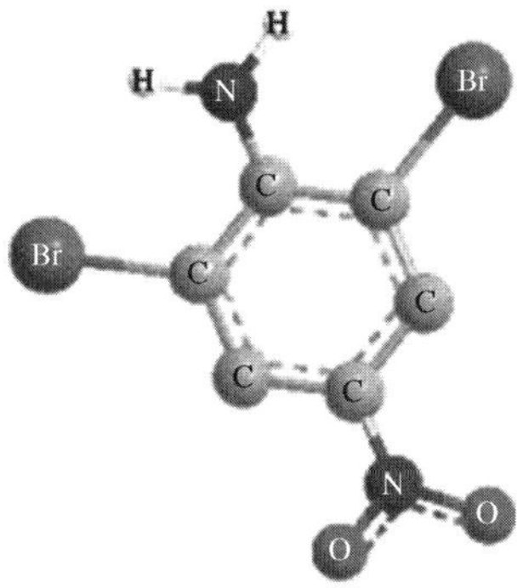

2,6-Dibromo-4-nitroaniline was initially converted into 2,6-di(4′-tolylethynyl)-4-nitroaniline, which *N*-acylated with succinyl, 4-bromobutyryl or 6-bromohexanoyl chloride, and then a capping reagent to form 3,5-di(4′-tolylethynyl)-4-[1-azacarbonylalkyl(aryl substituent)]nitrobenzene [173].

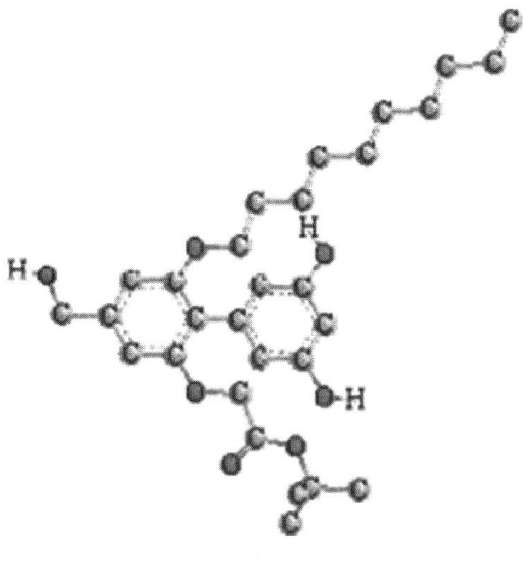

The 1-bromo-3,5-*bis*(*tert*-butoxydimethylsilanyloxy)benzene was transformed (BuLi/Bu_3SnCl) to the tin derivative, which was coupled with 5-decyloxy-3-(*tert*-butoxycarbonylmethoxy)-4-bromobenzaldehyde, derived from ethyl 4-bromo-3,5-dihydroxybenzoate in four steps, to give after reduction and selective hydrolysis the desired 5-decyloxy-3-(*tert*-butoxycarbonylmethoxy)-4-(3′,5′-dihydroxyphenyl)-

benzyl alcohol [174]. The construction details [175–179] of their interesting monomers are given in their supplemental materials.

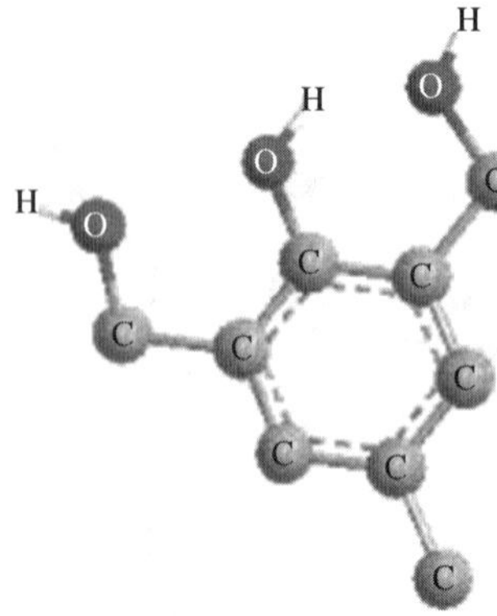

The 3,5-di(hydroxymethyl)-4-hydroxytoluene was treated (K_2CO_3/18-crown-6) with $Br(CH_2CH_2O)_2Me$ to form 3,5-di(hydroxymethyl)-4-(3,6-dioxaheptyloxy)toluene, which was brominated (CBr_4/PPh_3) affording 3,5-di(bromomethyl)-4-(3,6-dioxaheptyloxy)toluene, then after a five-step sequence [$P(OEt)_3$, NBS, HMTA, $CH(OMe)_3$, then NCC_6H_4CHO], 3,5-di(4-cyanophenylethenyl)-4-(3,6-dioxaheptyloxy)benzaldehyde was prepared and subsequently attached to cyclen [180].

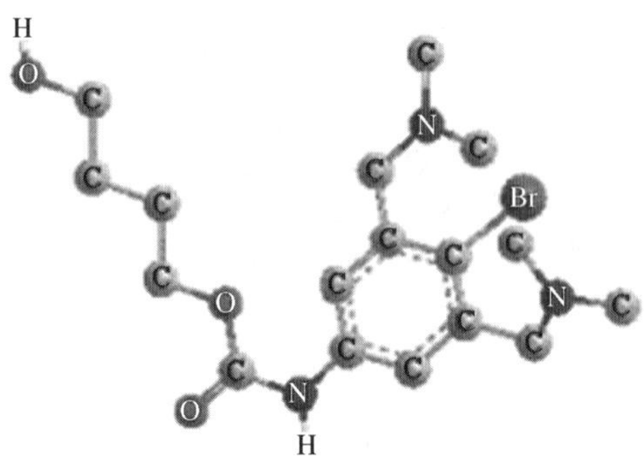

2,6-*Bis*(dimethylaminomethyl)bromobenzene was nitrated to give 4-nitro-3,5-*bis*(dimethylaminomethyl)bromobenzene, which was easily reduced [Pt or Ni(R), N_2H_4] to give 4-bromo-3,5-*bis*(dimethylaminomethyl)aniline [181]. Treatment with triphosgene generated the corresponding 4-bromo-3,5-*bis*(dimethylaminomethyl)benzeneisocyanate that was reacted with butanediol to give the desired carbamate [182]; metal incorporation at the "pincer" site has been demonstrated [183].

3.12 1 → (1 + 1 + 1) BRANCHED MONOMERS

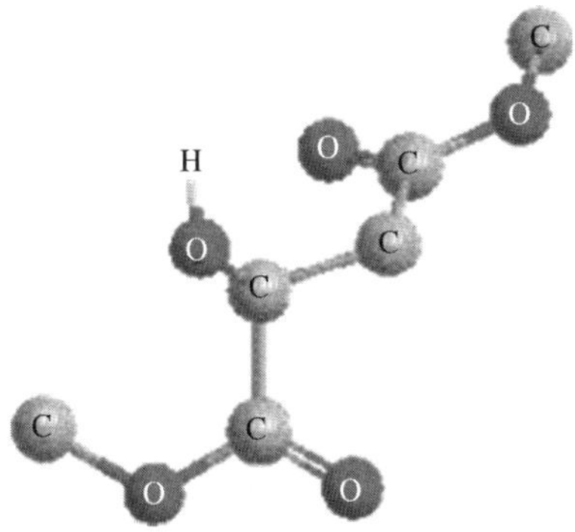

Methyl isocitrate ($MeO_2CCHOHCH_2CO_2Me$) was alkylated with CH_2=$CHCH_2Br$, then MOM-protection to give $MeO_2CCHO(MOM)CH(CH_2CH{=}CH_2)$-$CO_2Me$, subsequent ozonation, oxidation, esterification, and deprotection gave $MeO_2CCOHCH(CO_2Me)(CH_2CO_2R)$; lastly *O*-esterification and selective terminal deesterification generated $MeO_2CCO(OR)CH(CO_2Me)(CH_2CO_2H)$ [184].

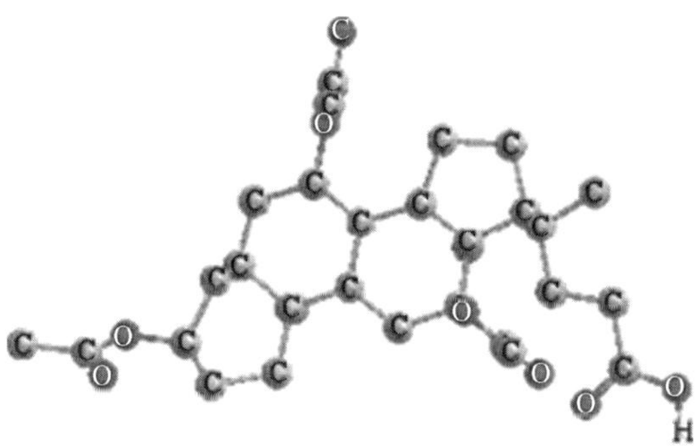

The treatment of 3α,7α,12α-triacetyloxy-5β-cholanic acid with 1-naphthylmethyl-3α,7α,12α-trihydroxy-5β-cholan-24-ate generated the tetramer via its use as a 1 → 3 branched monomer [185,186].

3.13 CONCLUSION

The above examples of 1 → 3 branching monomers are taken from the literature and some have been incorporated into macromolecular frameworks; others have not but possess the attributes to permit their consideration. Almost all dendritic assemblies are based on a single, iterative procedure, thus the introduction of synthetic diversity is still lacking in most cases. It would appear that, in general, the insertion of 1 → (2 + 1) branching monomers possessing a useful attachment point would add a tailored internal component for molecular inclusion/attachment. Literature possesses

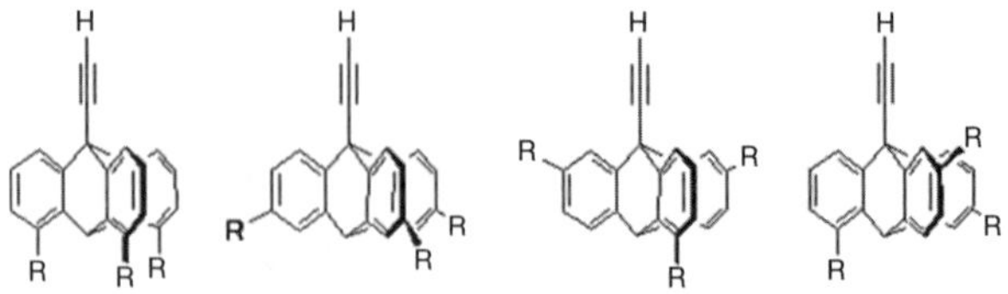

FIGURE 3.1 Functionalized triptycene monomers offer some interesting options for construction.

numerous other potential 1 → 3 branched systems that should/could be introduced into the lexicon of dendritic construction. In the world of molecular motors and rotors [187], a family of 9-triptycylethynes suggests numerous isomeric possibilities. Figure 3.1 provides an interesting approach to rigid dendritic structures and offers a potential to evaluate the effect of symmetrical and unsymmetrical isomers.

Although briefly touched on above, the use of natural products, such as steroid monomers possessing a 1 → (1 + 1 + 1) branching motif, has only scratched the surface of possibilities. The natural products possessing branching motifs derived from the forest or from the sea are endless but highly under utilized. For example, *N*-β-carboxyethyl-L-aspartic acid, ethyl α-cyano-α-ethoxycarbonylpropylglutarate, shikimic and dehydroquinic acid, phloracetophenone, cephalosporin P1, as well as the related helvoic and fusidic acids and from red algae, 4-hydroxy-3,5-dibromobenzyl alcohol and 2,3-dibromo-3,4-dihydroxybenzyl alcohol, and 2,3,5-trihydroxybromobenzene are interesting possibilities. It is amazing how a stroll through the woods or a scuba diving/snorkeling trip can introduce new avenues to synthetic chemistry.

ACKNOWLEDGMENTS

We thank the National Science Foundation (DMR-0705015 and DMR-0812337) for continued support over the years, the Air Force Research Office of Scientific Research, and the Ohio Board of Regents.

REFERENCES

[1] P. B. Tomlinson, *Am. Sci.* **1983**, *71*(March–April), 141–149.

[2] F. Hallé, R. A. A. Oldeman, *Essai sur l'architecture et la dynamique de croissance des arbres tropicaux*, Masson, Paris, **1970**.

[3] F. Hallé, R. A. A. Oldeman, P. B. Tomlinson, *Tropical Trees and Forests: An Architectural Analysis*, Springer, Berlin, **1982**.

[4] G. R. Newkome, Z. Yao, G. R. Baker, V. K. Gupta, *J. Org. Chem.* **1985**, *50*(11), 2003–2004.

[5] C. J. Hawker, J. M. J. Fréchet, *J. Am. Chem. Soc.* **1990**, *112*(21), 7638–7647.

[6] M. Segura, F. Sansone, A. Casnati, R. Ungaro, *Synthesis* **2001**, (14), 2105–2112.

[7] X. Li, J. Zhan, Y. Li, *Macromolecules* **2004**, *37*(20), 7584–7594.

[8] M. H. B. G. Gansey, F. J. Steemers, W. Verboom, D. N. Reinhoudt, *Synthesis* **1997**, (6), 643–648.

[9] T. Martinu, W. P. Dailey, *J. Org. Chem.* **2000**, *65*, 6784–6786.

[10] G. R. Newkome, X. Lin, *Macromolecules* **1991**, *24*(6), 1443–1444.

[11] C. Sun, P. Wirsching, K. D. Janda, *Bioorg. Med. Chem. Lett.* **2002**, *12*(16), 2213–2215.

[12] J. K. Cho, D.-W. Kim, J. Namgung, Y.-S. Lee, *Tetrahedron Lett.* **2001**, *42*(42), 7443–7445.

[13] S. Lebreton, N. Newcombe, M. Bradley, *Tetrahedron Lett.* **2002**, *43*(13), 2479–2482.

[14] H.-J. Knoelker, T. Braxmeier, G. Schlechtingen, *Angew. Chem. Int. Ed. Engl.* **1995**, *34* (22), 2497–2500.

[15] M. Ternon, M. Bradley, *Chem. Commun.* **2003**, (19), 2402–2403.

[16] C. Fromont, M. Bradley, *Chem. Commun.* **2000**, (4), 283–284.

[17] G. R. Newkome, C. D. Weis, C. N. Moorefield, G. R. Baker, B. J. Childs, J. D. Epperson, *Angew. Chem. Int. Ed.* **1998**, *37*(3), 307–310.

[18] C. Breyton, E. Chabaud, Y. Chaudier, B. Pucci, J.-L. Popot, *FEBS Lett.* **2004**, *564*(3), 312–318.

[19] E. Wallin, G. Von-Heijne, *Protein Sci.* **1998**, *7*, 1029–1038.

[20] J. C. Cuggino, C. I. A. Igarzabal, J. C. Rueda, L. M. Quinzani, H. Komber, M. C. Strumia, *Eur. Polym. J.* **2008**, *44*(11), 3548–3555.

[21] A. Polidori, O. Braun, N. Mora, B. Pucci, *Tetrahedron Lett.* **1997**, *38*(14), 2475–2478.

[22] M. Abla, G. Durand, B. Pucci, *J. Org. Chem.* **2008**, *73*(21), 8142–8153.

[23] G. R. Newkome, G. R. Baker, *Org. Prep. Proced. Int.* **1986**, *18*(2), 117–144.

[24] J. Skarzewski, *Tetrahedron* **1989**, *45*(14), 4593–4598.

[25] J. Skarzewski, *Synthesis* **1990**, (12), 1125–1127.

[26] M. B. Smith, J. March, *March's Advanced Organic Chemistry*, 5th edition, John Wiley & Sons, Inc., New York, **2001**, pp. 808–811.

[27] M. Broussard, B. Juma, F. R. Fronczek, S. F. Watkins, G. R. Newkome, C. N. Moorefield, *Acta Cryst. C* **1991**, *C47*, 1245–1247.

[28] J. K. Whitesell, H. K. Chang, *Science* **1993**, *261*(5117), 73–76.

[29] D. S. Tarbell, Y. Yamamoto, B. M. Pope, *Proc. Natl. Acad. Sci. U.S.A.* **1972**, *69*(3), 730–732.

[30] E. Ponnusamy, U. Fotadar, A. Spisni, D. Fiat, *Synthesis* **1986**, (1), 48–49.

[31] X. A. Dominguez, I. C. Lopez, R. Franco, *J. Org. Chem.* **1961**, *26*(5), 1625.

[32] G. R. Newkome, R. K. Behera, C. N. Moorefield, G. R. Baker, *J. Org. Chem.* **1991**, *56* (25), 7162–7167.

[33] G. R. Newkome, C. D. Weis, *Org. Prep. Proced. Int.* **1996**, *28*(4), 485–488.

[34] B. D. Mather, K. Viswanathan, K. M. Miller, T. E. Long, *Prog. Polym. Sci.* **2006**, *31*(5), 487–531.

[35] C. D. Weis, G. R. Newkome, *Synthesis* **1995**, (9), 1053–1065.

[36] Available from Frontier Scientific (www.frontiersci.com).

[37] G. R. Newkome, C. D. Weis, C. N. Moorefield, F. R. Fronczek, *Tetrahedron Lett.* **1997**, *38*(40), 7053–7056.

[38] C. Akpo, E. Weber, J. Reich, *New J. Chem.* **2006**, *30*(12), 1820–1833.

[39] G. R. Newkome, C. D. Weis, B. J. Childs, *Designed Monom. Polym.* **1998**, *1*(1), 3–14.

[40] G. R. Newkome, C. D. Weis, U.S. Pat. 5,703,271, 12-30-1997.

[41] G. R. Newkome, G. R. Baker, C. N. Moorefield, E. He, J. D. Epperson, C. D. Weis, *Polym. Mater. Sci. Eng.* **1997**, *77*, 65–66.

[42] G. R. Newkome, C. N. Moorefield, U.S. Pat. 5,886,126, 3-23-1999.

[43] G. R. Newkome, C. N. Moorefield, U.S. Pat. 5,886,127, 3-23-1999.

[44] G. R. Newkome, K. S. Yoo, C. N. Moorefield, *Designed Monom. Polym.* **2002**, *5*(1), 67–77.

[45] M. Brettreich, A. Hirsch, *Synlett* **1998**, (12), 1396–1398.

[46] G. R. Newkome, C. D. Weis, Eur. Pat. 97917126.1, 3-31-1997.

[47] G. R. Newkome, C. N. Moorefield, K. J. Theriot, *J. Org. Chem.* **1988**, *53*(23), 5552–5554.

[48] G. R. Newkome, A. Mishra, C. N. Moorefield, *Polym. Mater. Sci. Eng.* **2001**, *84*, 1–2.

[49] S. Lebreton, N. Newcombe, M. Bradley, *Tetrahedron Lett.* **2002**, *43*(13), 2475–2478.

[50] M. C. Strumia, A. Halabi, P. A. Pucci, G. R. Newkome, C. N. Moorefield, J. D. Epperson, *J. Polym. Sci. Part A Polym. Chem.* **2000**, *38*(15), 2779–2786.

[51] V. V. Narayanan, G. R. Newkome, L. Echegoyen, E. Pérez-Cordero, *Polym. Prepr.* **1996**, *37*(2), 419–420.

[52] G. R. Newkome, V. V. Narayanan, A. K. Patri, J. Groß, C. N. Moorefield, G. R. Baker, *Polym. Mater. Sci. Eng.* **1995**, *73* 222–223.

[53] G. R. Newkome, R. Güther, F. Cardullo, *Macromol. Symp.* **1995**, *98*(35th IUPAC International Symposium on Macromolecules, 1995), 467–474.

[54] G. R. Newkome, V. V. Narayanan, L. Echegoyen, E. Pèrez-Cordero, H. Luftmann, *Macromolecules* **1997**, *30*(17), 5187–5191.

[55] G. R. Newkome, V. V. Narayanan, L. A. Godínez, *Designed Monom. Polym.* **2000**, *3*(1), 17–24.

[56] A. B. Padias, H. K. Hall, Jr., D. A. Tomalia, J. R. McConnell, *J. Org. Chem.* **1987**, *52* (24), 5305–5312.

[57] A. B. Padias, H. K. Hall, Jr., D. A. Tomalia, *Polym. Prepr.* **1989**, *30*(1), 119–120.

[58] A. A. Shukla, S. S. Bae, J. A. Moore, K. A. Barnthouse, S. M. Cramer, *Ind. Eng. Chem. Res.* **1998**, *37*(10), 4090–4098.

[59] R. A. Valiulin, A. G. Kutateladze, *J. Org. Chem.* **2008**, *73*(1), 335–338.

[60] T. J. Dunn, W. L. Neumann, M. M. Rogic, S. R. Woulfe, *J. Org. Chem.* **1990**, *55*(26), 6368–6373.

[61] T. Endo, M. Okawara, W. J. Bailey, *Polym. J.* **1981**, *13*(7), 715–718.

[62] J. A. Camerano, M. A. Casado, M. A. Ciriano, F. J. Lohoz, L. A. Oro, *Organometallics* **2005**, *24*(21), 5147–5156.

[63] R. A. Findeis, L. H. Gade, *Dalton Trans.* **2003**, (2), 249–254.

[64] L. Kwisnek, S. Nazarenko, C. E. Hoyle, *Macromolecules* **2009**, *42*(18), 7031–7041.

[65] M. S. Shchepinov, E. M. Southern, *Russ. J. Bioorg. Chem.* **1998**, *24*(19), 794–797.

[66] M. S. Shchepinov, I. A. Udalova, A. J. Bridgman, E. M. Southern, *Nucleic Acids Res.* **1997**, *25*(22), 4447–4454.

[67] S. Hanessian, H. Prabhanjan, D. Qiu, S. Nambiar, *Can. J. Chem.* **1996**, *74*(9), 1731–1734.

[68] A. Mollard, I. Zharov, *Inorg. Chem.* **2006**, *45*(25), 10172–10179.

[69] D. Tzalis, Y. Tor, *Tetrahedron Lett.* **1996**, *37*(46), 8293–8296.

[70] A. Lubineau, A. Malleron, C. Le Narvor, *Tetrahedron Lett.* **2000**, *41*(46), 8887–8891.

[71] A. Fishman, M. E. Farrah, J.-H. Zhong, S. Paramananthan, C. Cerrera, E. Lee-Ruff, *J. Org. Chem.* **2003**, *68*(25), 9843–9846.

[72] R. A. Cormier, B. A. Gregg, *Chem. Mater.* **1998**, *10*(5), 1309–1319.

[73] I. M. Saez, J. W. Goodby, *J. Mater. Chem.* **2003**, *13*(11), 2727–2739.

[74] P. L. Nicohls, Jr., E. Yanovsky, *J. Am. Chem. Soc.* **1945**, *67*(1), 46–49.

[75] P. L. Garegg, B. Samuelson, *J. Chem. Soc. Perkin Trans. 1* **1980**, (12), 2866–2869.

[76] K. Li, L. Ran, Y.-H. Yu, Y. Tang, *J. Org. Chem.* **2004**, *69*(11), 3986–3989.

[77] K. Tunaboylu, G. Schwarzenbach, *Helv. Chim. Acta* **1971**, *54*(7), 2166–2184.

[78] S. Wawzonek, A. Matar, C. H. Issidorides, *Org. Synth.* **1963**, *Coll. Vol. 4*, 681–683.

[79] R. E. Kohman, S. C. Zimmerman, *Chem. Commun.* **2009**, (7), 794–796.

[80] P. Schober, R. Soltek, G. Huttner, L. Zsolnai, K. Heinze, *Eur. J. Inorg. Chem.* **1998**, 1407–1415.

[81] E. Dulière, J. Marchand-Brynaert, *Synthesis* **2002**, (1), 39–42.

[82] N. Ono, H. Miyake, A. Kamimura, I. Hamamoto, R. Tamura, A. Kaji, *Tetrahedron* **1985**, *41*(19), 4013–4023.

[83] G. R. Newkome, C. N. Moorefield, G. R. Baker, A. L. Johnson, R. K. Behera, *Angew. Chem. Int. Ed. Engl.* **1991**, *30*(9), 1176–1178.

[84] G. R. Newkome, C. N. Moorefield, G. R. Baker, M. J. Saunders, S. H. Grossman, *Angew. Chem. Int. Ed. Engl.* **1991**, *30*(9), 1178–1180.

[85] S. E. Van der Plas, E. V. Hoeck, F. Lynen, P. Sandra, A. Madder, *Eur. J. Org. Chem.* **2009**, *11*(1796), 1805.

[86] S. E. Van der Plas, A. Gea, S. Figaroli, P. J. De Clercq, A. Madder, *Eur. J. Org. Chem.* **2009**, (9), 1582–1588.

[87] D. Astruc, *Tetrahedron* **1979**, *39*(24), 4027–4095.

[88] V. Sartor, L. Djakovitch, J.-L. Fillaut, F. Moulines, F. Neveu, V. Marvaud, J. Guittard, J.-C. Blais, D. Astruc, *J. Am. Chem. Soc.* **1999**, *121*(12), 2929–2930.

[89] V. Sartor, S. Nlate, J.-L. Fillaut, L. Djakovitch, F. Moulines, V. Marvaud, F. Neveu, J.-C. Blais, J.-F. Létard, D. Astruc, *New J. Chem.* **2000**, *24*(6), 351–370.

[90] C. Ornelas, J. R. Aranzaes, E. Cloutet, S. Alves, D. Astruc, *Angew. Chem. Int. Ed.* **2007**, *46*(6), 872–877.

[91] D. Astruc, C. Ornelas, J. R. Aranzaes, *J. Inorg. Organometal. Polym. Mater.* **2008**, *18*(1), 1–4.

[92] C. Ornelas, J. Ruiz, L. Salmon, D. Astruc, *Adv. Synth. Catal.* **2008**, *350*(6), 837–845.

[93] R. A. Gossage, E. Muñoz-Martínez, G. van Koten, *Tetrahedron Lett.* **1998**, *39*(16), 2397–2400.

[94] E. Boisselier, A. C. K. Shun, J. Ruiz, E. Cloutet, C. Belin, D. Astruc, *New J. Chem.* **2009**, *33*(2), 246–253.

[95] J. Camponovo, J. Ruiz, E. Cloutet, D. Astruc, *Chem. Eur. J.* **2009**, *15*(12), 2990–3002.

[96] J. Camponovo, C. Hadad, J. Ruiz, E. Cloutet, S. Gatard, J. Muzart, S. Bouquillon, D. Astruc, *J. Org. Chem* **2009**, *74*(14), 5071–5074.

[97] D. Astruc, J.-C. Blais, M.-C. Daniel, S. Gatard, S. Nlate, J. Ruiz, *C. R. Chimie* **2008**, *6* (8–10), 1117–1127.

[98] S. Nlate, J.-C. Blais, D. Astruc, *New J. Chem.* **2003**, *27*(1), 178–183.

[99] Y. Kuramochi, A. S. D. Sandanayaka, A. Satake, Y. Araki, K. Ogawa, O. Ito, Y. Kobuke, *Chem. Eur. J.* **2009**, *15*(10), 2317–2327.

[100] S. Sengupta, P. Purkayastha, *Org. Biomol. Chem.* **2003**, *1*(2), 436–440.

[101] R. Chinchilla, C. Nájera, *Chem. Rev.* **2007**, *107*(3), 874–922.

[102] R. Andrés, E. de Jesús, F. J. de la Mata, J. C. Flores, R. Gómes, *Eur. J. Inorg. Chem.* **2005**, (18), 3742–3749.

[103] A. Tuchbreiter, H. Werner, L. H. Gade, *Dalton Trans.* **2005**, (8), 1394–1402.

[104] L. J. Mathias, T. W. Carothers, *J. Am. Chem. Soc.* **1991**, *113*(10), 4043–4044.

[105] L. J. Mathias, T. W. Carothers, R. M. Bozen, *Polym. Prepr.* **1991**, *32*(1), 82–83.

[106] L. J. Mathias, T. W. Carothers, *Polym. Prepr.* **1991**, *32*(3), 633–634.

[107] A. Morikawa, M. Kakimoto, Y. Imai, *Macromolecules* **1991**, *24*(12), 3469–3474.

[108] F. Moingeon, P. Masson, S. Méry, *Macromolecules* **2007**, *40*(1), 55–64.

[109] Y. Yao, J. M. Tour, *J. Org. Chem.* **1999**, *64*(6), 1968–1971.

[110] H. Jian, J. M. Tour, *J. Org. Chem.* **2003**, *68*(13), 5091–5103.

[111] V. S. K. Balagurusamy, G. Ungar, V. Percec, G. Johansson, *J. Am. Chem. Soc.* **1997**, *119* (7), 1539–1555.

[112] V. Percec, W.-D. Cho, P. E. Mosier, G. Ungar, D. J. P. Yeardley, *J. Am. Chem. Soc.* **1998**, *120*(43), 11061–11070.

[113] T. Chen, L. Wang, J. Wang, X. Wang, J. Zhou, W. Wang, *Eur. Polym. J.* **2006**, *42*(3), 687–693.

[114] R. Roy, W. K. C. Park, Q. Wu, S.-N. Wang, *Tetrahedron Lett.* **1995**, *36*(25), 4377–4380.

[115] W. Li, A. Zhang, A. D. Schlüter, *Macromolecules* **2008**, *41*(1), 43–49.

[116] S. Müller, A. D. Schlüter, *Chem. Eur. J.* **2005**, *11*(20), 5589–5610.

[117] E. Boisselier, A. K. Diallo, L. Salmon, J. Ruiz, D. Astruc, *Chem. Commun.* **2008**, (39), 4819–4821.

[118] E. Boisselier, L. Salmon, J. Ruiz, D. Astruc, *Chem. Commun.* **2008**, (44), 5788–5790.

[119] H. Lee, D. Kim, H.-K. Lee, W. Qiu, N.-K. Oh, W.-C. Zin, K. Kim, *Tetrahedron Lett.* **2004**, *45*(5), 1019–1022.

[120] F. Alonso, I. P. Beletskaya, M. Yus, *Tetrahedron* **2008**, *64*(14), 3047–3101.

[121] F. Bellina, A. Carpita, R. Rossi, *Synthesis* **2004**, (15), 2419–2440.

[122] D. M. Tal, S. J. Karlish, *Tetrahedron* **1995**, *51*(13), 3823–3830.

[123] K. Haba, M. Popkov, M. Shamis, R. A. Lerner, C. F. BarbasIII, D. Shabat, *Angew. Chem. Int. Ed.* **2005**, *44*(5), 716–720.

[124] E. Sella, D. Shabat, *Chem. Commun.* **2008**, (44), 5701–5703.

[125] M. Shamis, D. Shabat, *Chem. Eur. J.* **2007**, *13*(16), 4523–4528.

[126] A. Sagi, E. Segal, R. Satchi-Fainaro, D. Shabat, *Bioorg. Med. Chem.* **2007**, *15*(7), 3720–3727.

[127] A. A. Anderson, T. Goetzen, S. A. Shackelford, S. Tsank, D. Alanex, *Syn. Commun.* **2000**, *30*(17), 3227–3232.

[128] M. Shamis, D. Shabat, *Chem. Eur. J.* **2007**, *13*(16), 4523–4528.

[129] A. Bashir-Hashemi, *Unpublished data* **1996**.

[130] G. R. Newkome, *Unpublished data* **1996**.

[131] T. Kitagawa, Y. Idomoto, H. Matsubara, D. Hobara, T. Kakiuchi, T. Okazaki, K. Komatsu, *J. Org. Chem.* **2006**, *71*(4), 1362–1369.

[132] A. Bashir-Hashemi, J. Li, N. Gelber, *Tetrahedron Lett.* **1995**, *36*(8), 1233–1236.

[133] P. A. S. Smith, *Org. React.* **1946**, *3*, 337–449.

[134] K. Nasr, N. Pannier, J. V. Frangioni, W. Maison, *J. Org. Chem.* **2008**, *73*(3), 1056–1060.

[135] V. Humblet, P. Misra, K. R. Bushan, Y.-S. Ko, T. Tsukamoto, N. Pannier, J. V. Frangioni, W. Maison, *J. Med. Chem.* **2009**, *52*(2), 544–550.

[136] V. V. Martin, I. S. Alferiev, A. L. Weis, *Tetrahedron Lett.* **1999**, *40*(2), 223–226.

[137] O. Mongin, A. Gossauer, *Tetrahedron* **1997**, *53*(20), 6835–6846.

[138] J. K. Young, G. R. Baker, G. R. Newkome, K. F. Morris, C. S. Johnson, Jr., *Macromolecules* **1994**, *27*(13), 3464–3471.

[139] M. Lamberto, C. Pagba, P. Piotrowiak, E. Galoppini, *Tetrahedron Lett.* **2005**, *46*(29), 4895–4899.

[140] S. Zarwell, S. Dietrich, C. Schulz, P. Dietrich, F. Michalik, K. Rück-Braun, *Eur. J. Org. Chem.* **2009**, (13), 2088–2095.

[141] W. Guo, E. Galloppini, G. Rydja, G. Pardi, *Tetrahedron Lett.* **2000**, *41*(39), 7419–7421.

[142] G. S. Lee, J. N. Bashara, G. Sabih, A. Oganesyan, G. Godjoian, H. M. Duong, E. R. Marinez, C. G. Gutiérrez, *Org. Lett.* **2004**, *6*(11), 1705–1707.

[143] A. Oganesyan, I. A. Cruz, R. B. Amador, N. A. Sorto, J. Lozano, C. E. Godinez, J. Anguiano, H. Pace, G. Sabih, C. G. Gutierrez, *Org. Lett.* **2007**, *9*(24), 4967–4970

[144] A. N. Semakin, A. Y. Sukhorukov, A. V. Lesiv, S. L. Ioffe, K. A. Lyssenko, Y. V. Nelyubina, V. A. Tartakovsky, *Org. Lett.* **2009**, *11*(18), 4072–4075.

[145] J. P. André, C. F. G. C. Geraldes, J. A. Martins, A. E. Merbach, M. I. M. Prata, A. C. Santos, J. J. P. de Lima, É. Tóth, *Chem. Eur. J.* **2004**, *10*(22), 5804–5816.

[146] D. A. Jose, S. Stadbauer, B. König, *Chem. Eur. J.* **2009**, *15*(30), 7404–7412.

[147] C. Gløgård, R. Hovland, S. L. Fossheim, A. J. Aasen, J. Klaveness, *J. Chem. Soc. Perkin Trans. 2* **2000**, (5), 1047–1052.

[148] C. Granier, R. Guilard, *Tetrahedron* **1995**, *51*(4), 1197–1208.

[149] O. Hayashida, Y. Takaoka, I. Hamachi, *Tetrahedron Lett.* **2005**, *46*(38), 6589–6592.

[150] O. Hayashida, *J. Synth. Org. Chem. Jpn.* **2006**, *64*(10), 1041–1048.

[151] O. Hayashida, A. Kitaura, *Chem. Lett.* **2006**, *35*(7), 808–809.

[152] P. Lhotak, S. Shinkai, *Tetrahedron* **1995**, *51*(28), 7681–7696.

[153] J. B. Oh, K. L. Paik, J.-W. Ka, S.-G. Roh, M. K. Nah, H. K. Kim, *Mater. Sci. Eng. C* **2004**, *24*(1–2), 257–260.

[154] C.-H. Lee, *Tetarhedron* **2009**, *50*(39), 11427–11440.

[155] J. B. Oh, M.-K. Nah, Y. H. Kim, M. S. Kang, J.-W. Ka, H. K. Kim, *Adv. Funct. Mater.* **2007**, *17*(3), 413–424.

[156] J. B. Oh, Y. H. Kim, M. K. Nah, H. K. Kim, *J. Lumin.* **2005**, *111*(4), 255–264.

[157] C.-A. Wu, C.-L. Chiu, C.-L. Mai, Y.-S. Lin, C. Yeh, *Chem. Eur. J.* **2009**, *15*(18), 4534–4537.

[158] K. Sugiura, H. Tanaka, T. Matsumoto, T. Kawai, Y. Sakata, *Chem. Lett.* **1999**, *28*(11), 1193–1194.

[159] M. Sülü, A. Altindal, Ö. Bekaroglu, *Synth. Met.* **2005**, *155*(1), 211–221.

[160] H. Namazi, M. Adeli, *Eur. Polym. J.* **2003**, *39*(7), 1491–1500.

[161] D. Patel, B. D. McKinley, T. P. Davis, F. Porreca, H. I. Yamamura, V. J. Hruby, *Bioconj. Chem.* **1997**, *8*(3), 434–441.

[162] G. R. Newkome, H. J. Kim, K. H. Choi, C. N. Moorefield, *Macromolecules* **2004**, *37* (17), 6268–6274.

[163] G. R. Newkome, E. He, L. A. Godínez, G. R. Baker, *J. Am. Chem. Soc.* **2000**, *122* 9993–10006.

[164] G. R. Newkome, K. S. Yoo, C. N. Moorefield, *Chem. Commun.* **2002**, (18), 2164–2165.

[165] B. Aussedat, C. Chassaing, S. Lavielle, F. Burlina, *Tetrahedron Lett.* **2006**, *47*(22), 3723–3726.

[166] B. Aussedat, S. Sagan, C. Chassaing, G. Bolbach, F. Burlina, *Biochem. Biophys. Acta* **2006**, *1758*(3), 375–383.

[167] B. Aussedat, E. Dupont, S. Sagan, A. Joliot, S. Lavielle, G. Chassaing, F. Burlina, *Chem. Commun.* **2008**, (12), 1398–1400.

[168] G. T. Crisp, A. G. Meyer, *Tetrahedron* **1995**, *51*(19), 5585–5596.

[169] M. Iwamura, Y. Gotoh, T. Hashimoto, R. Sakurai, *Tetrahedron Lett.* **2005**, *46*, 6275–6277.

[170] C. Modrakowski, S. C. Flores, M. Beinhoff, A.-D. Schlüter, *Synthesis* **2001**, (14), 2143–2155.

[171] A. Suzuki, in *Modern Arene Chemistry* (Ed.: D. Astruc), Wiley-VCH Verlag Gmbh & Co. KGaA, Weinheim, **2002**, pp. 53–106.

[172] M. Kozaki, K. Okada, *Org. Lett.* **2004**, *6*(4), 485–488.

[173] A. Khatyr, H. Maas, G. Calzaferri, *J. Org. Chem.* **2002**, *67*(19), 6705–6710.

[174] D. R. Vutukuri, S. Basu, S. Thayumanavan, *J. Am. Chem. Soc.* **2004**, *126*(48), 15636–15637.

[175] A. V. Ambade, S. V. Aathimanikandan, D. van der Poll, S. Thayumanavan, *J. Org. Chem.* **2007**, *72*(22), 8167–8174.

[176] A. V. Ambade, Y. Chen, S. Thayumanavan, *New J. Chem.* **2007**, *31*(7), 1052–1063.

[177] A. V. Ambade, A. V. Sivakumar, S. Thayumanavan, *Polym. Prepr.* **2005**, *46*(2), 1180–1181.

[178] K. N. Jayakumar, P. Bharathi, S. Thayumanavan, *Org. Lett.* **2004**, *6*(15), 2547–2550.

[179] K. Sivanandan, S. V. Aathimanikandan, C. G. Arges, C. J. Bardeen, S. Thayumanavan, *J. Am. Chem. Soc.* **2005**, *127*(7), 2020–2021.

[180] J. Weng, Q. Zhang, *J. Polym. Sci. Part A Polym. Chem.* **2005**, *43*(22), 5414–5428.

[181] L. A. van de Kuil, H. Luitjes, D. M. Grove, J. W. Zwikker, J. G. M. van der Linden, A. M. Roelofsen, L. W. Jenneskens, W. Drenth, G. van Koten, *Organometallics* **2002**, *13*(2), 468–477.

[182] J. W. J. Knapen, A. W. van der Made, J. C. de Wilde, P. W. N. M. van Leeuwen, P. Wijkens, D. M. Grove, G. van Koten, *Nature* **1994**, *372*(12/15), 659–663.

[183] R. A. Gossage, L. A. van de Kuil, G. van Koten, *Acc. Chem. Res.* **1998**, *31*(7), 423–431.

[184] M. Tori, Y. Sono, Y. Nakashiba, N. Hamada, M. Sono, Y. Asakawa, M. Suganuma, S. Okabe, H. Fujiki, *Tetrahedron Lett.* **2000**, *41*, 3095–3098.

[185] R. Balasubramanian, U. Maitra, *J. Org. Chem.* **2001**, *66*(9), 3035–3040.

[186] S. Balasubramanian, P. Rao, U. Maitra, *Chem. Commun.* **1999**, (23), 2353–2354.

[187] G. S. Kottas, L. I. Clarke, D. Horinek, J. Michl, *Chem. Rev.* **2005**, *105*(4), 1281–1376.

4

DENDRONIZED POLYMERS: STATE OF THE ART IN ZURICH

AFANG ZHANG AND A. DIETER SCHLÜTER

Laboratory of Polymer Chemistry, Department of Materials, ETH Zurich, Wolfgang-Pauli Strasse 10, HCI J 541, 8093 Zurich, Switzerland

4.1 INTRODUCTION

Thinking in terms of linear macromolecules with large cross-sectional diameters and a high number of terminal units per segment length, dendronized polymers come to mind. They have backbones with dendritically branched side chains at each repeat unit and are therefore not only unprecedentedly "thick" but also carry at the same time thousands of end groups per molecule. With these rather unusual characteristics, which set them apart from linear, branched, hyperbranched, and bottlebrush polymers, dendronized polymers can be regarded molecular objects rather than flexible chain macromolecules. From the early days in 1994 on [1] the unique structure of dendronized polymers has fascinated numerous researchers and led to contributions ranging from how to make and characterize representatives of these intriguing entities all the way to where they could have interesting applications [2]. The most recent, rather comprehensive accounts date back to the year 2005 [3]. In view of the considerable activity since then it is appropriate to highlight at least a few important developments that took place after 2005. For space reasons, the present chapter is restricted to findings from the authors' own laboratory but many of those from other laboratories should at least be mentioned [4]. It covers rather different aspects ranging from the g-scale synthesis of high-generation dendronized polymers with long main chains and the stimuli-responsive behavior of certain representatives all the way to extraordinarily stable helical dendronized polymers, the handedness of which can be

Designing Dendrimers, Edited by Sebastiano Campagna, Paola Ceroni, Fausto Puntoriero.
© 2012 John Wiley & Sons, Inc. Published 2012 by John Wiley & Sons, Inc.

predetermined by choice of monomer. The chapter concludes with a discovery trip into the world of single molecule chemistry. The symbiosis of the thick dendronized polymers and the atomic force microscope with its resolution in the nanometer range allows to covalently connecting individual macromolecules at predetermined positions. These four cases together with the beautiful work of other laboratories [4] show the richness and diversity of both basic research and application aspects of macromolecules with regularly branched lateral substituents. The nomenclature used in this chapter is to be understood as follows: M and P stand for macromonomer and polymer, respectively, and G for generation of the appending dendrons. Please note that for simplicity a polymer of the same name, for example, **PG2**, may refer to different chemical structures in different sections of this chapter.

4.2 SYSTEMATIC THICKENING OF POLYMER CHAINS: THE ATTACH-TO ROUTE TO DENDRONIZED POLYMERS

The large majority of dendronized polymers are synthesized according to the macromonomer route. This route uses monomers that already carry the dendron of the final size desired as the lateral substituent of the polymer's main chain repeat units. It is attractive because of synthetic ease and the high level of perfection achievable for the dendritic branchwork in the polymer. Ease of synthesis is of course a relative assessment. Obviously, the effort to synthesize a highly branched macromonomer is substantially higher than that for simple monomers like methylmethacrylate, which are often even commercially available. Compared to the attach-to approach to dendronized polymers (see below), however, in which the generations are built up by applying synthetic steps repeatedly to polymers, the macromonomer route is in fact much simpler. All operations are restricted to the relatively low molar mass—monomeric—regime. This facilitates purification and analytics. The corresponding dendronized polymer is then obtained in just a single step by polymerizing the macromonomer. Ideally the dendrons stay untouched during this polymerization.[1] Thus, the macromonomer route directly translates the level of structural perfection contained in the macromonomers into the product, a substantial benefit, which in the attach-to alternative has to be heavily fought for. These two reasons made the macromonomer route what it is today, the prime tool for the synthesis of dendronized polymers specifically if low-generation representatives are concerned. As far as higher generations are concerned (beyond approximately G3) a feature becomes visible that can be disadvantageous and should therefore not be hidden: The main chain length decreases with increasing monomer generation, sometimes even drastically (Figure 4.1). This certainly applies within the same series of polymers. A recent example concerning the monomers **M1(G4)** and **M2(G3)** [5] (Figure 4.2) illustrates, however, that the term "generation" should be used with care when representatives of structurally different series are compared. What actually has an impact on chain

[1] The dendrons should be designed such that chain transfer to monomer and polymer is negligible.

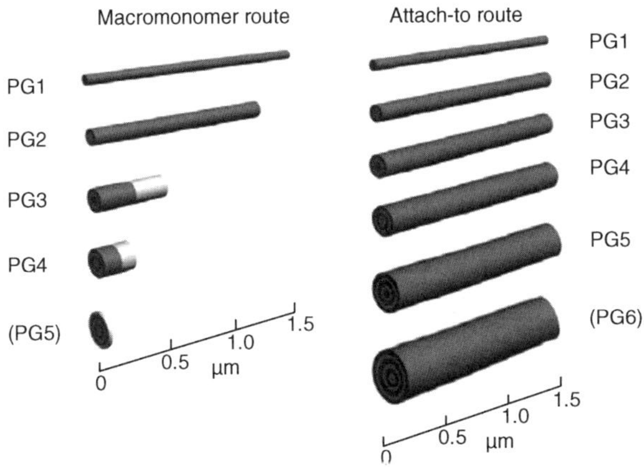

FIGURE 4.1 Comparison of achievable main chain length in dependence of generation for the macromonomer (left) and attach-to route (right) to dendronized polymers. For the former the main chain lengths decrease with increasing generation (dark gray cylinders) and as a result there is no report on a fifth-generation polymer by this method. The light gray extensions indicate that under supercritical carbon dioxide conditions longer chains may be achievable. Accordingly, at least an oligomeric **PG5** could recently be obtained for the first time. For the latter method the main chain length and its distribution may stay more or less constant if appropriate synthetic measures are taken. This is the key advantage of the attach-to route. Preliminary studies show that **PG6** may be within reach.

length is not the generation of the dendrons as such but rather their steric demand that is reflected to some degree by molar mass [6]. This shows nicely when comparing monomer **M1(G4)** that has a higher generation than **M2(G3)** and should therefore give the shorter chains at first glance. As expected for the macromonomer route, both monomers gave relatively short chains with number average degrees of polymerization P_n of 30–40 for **P1(G4)** and 10–15 for **P2(G3)**. Thus, despite its higher generation, **P1(G4)** has the longer chains than **P2(G3)**.[2] This can qualitatively be understood in terms of the macromonomers' molar masses. Whereas **M1(G4)** has ~4900 g/mol, **M2(G3)** has ~7700 g/mol.[3] The latter monomer's steric demand is therefore more pronounced, which results in shorter chains.

This disadvantage of the macromonomer route is known since the early days of dendronized polymers and is of very similar nature to what had priory been observed with linear macromonomers [7]. It has thus always been of some concern to develop

[2] It is delicate to compare main chain lengths obtained with different monomers in different experimental settings. Though the authors' laboratory tries to meet certain standards as to monomer quality (e.g., purity), it cannot be excluded that both monomers have unknown impurities in unknown (low) molar ratios. Given the substantial experience in macromonomer polymerizations, the authors nevertheless consider the described difference relevant. Note that the above argumentation implies that the elution behavior of the two different dendronized polymers on the GPC column is the same that is unproven.

[3] The dendron of the lower generation **M2G3** has a higher branching multiplicity (3) than **M1G4** (2) and also longer segments between the branch points.

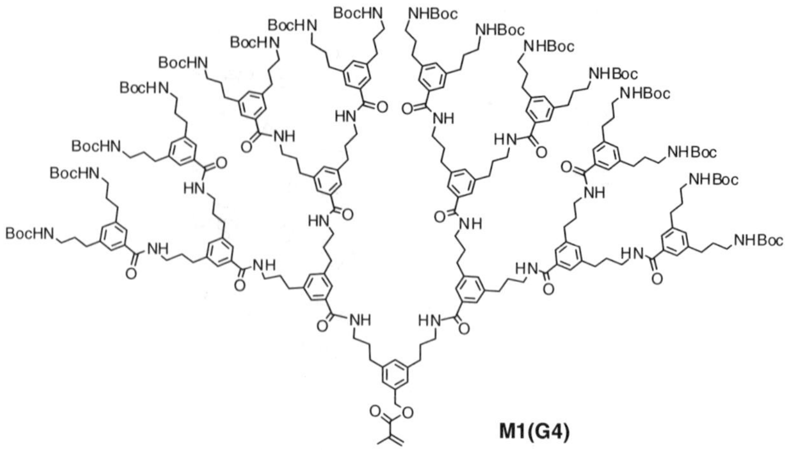

M1(G4)

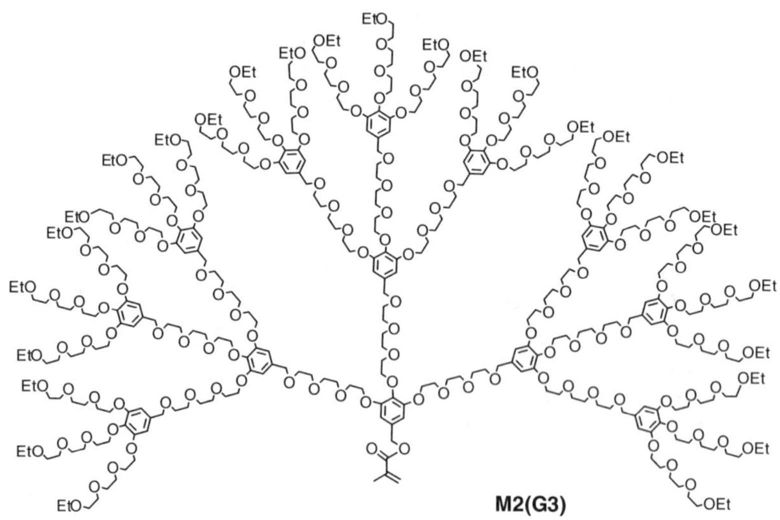

M2(G3)

FIGURE 4.2 Chemical structures of **M1(G4)** and **M2(G3)** with branching multiplicities of 2 and 3, respectively. Note that the molar mass of **M2(G3)** is significantly higher than **M1(G4)** despite its lower generation.

alternative routes that would help circumventing it. As a matter of fact, the first mentioning of a polymer, which according to nowadays terminology would be referred to as dendronized, was synthesized in a different way. In a patent report of 1987, Tomalia et al. described the grafting of a linear polyethyleneimine layer by layer, generation by generation with dendrons [8]. In the first step, the dendrons were bonded to the main chain amines and in all subsequent steps to the attached dendrons' peripheral functional groups. In principle, the length of the main chain stays the same through this built-up of the generations.[4] In light of what has just been discussed above as the disadvantage of the macromonomer route, this sounds attractive. There is a price to be paid for this attractiveness though, which is the lack of perfection of the dendritic branchwork. Considering a polyethyleneimine with an average of 500 repeat units [8b] in the first growth step 500 reactions have to be performed per macromolecule, followed by thousands of such reactions for each of the following steps. Since (i) there is no chemical reaction proceeding with 100% conversion and (ii) the growth steps become increasingly more critical as the generation increases because of growing congestion among the peripheral functional groups, it is evident that such an approach unavoidably leads to structural imperfection. Experimentally this showed in a couple of cases in the open literature published from 1994 onwards, in which the degree of coverage for such attach-to approaches was estimated and found to be far away from the theoretically possible [9]. Despite quite some effort in different research groups [10], no synthetically convincing solution was provided until recently. The main deficiencies are (i) the practically exclusive use of very short **PG1** main chains (e.g., $P_n = 30$–300) for which it is intrinsically easier to reach higher coverage; (ii) the lack of a proper determination of the coverage achieved. Often the coverage is just believed to be virtually quantitative; (iii) the occasional failures to provide essential experimental conditions like reaction scale. This unsatisfactory situation has nourished the perception that based on synthetic effort and achievable structure perfection the attach-to protocol is premature and cannot compete with the macromonomer route, though it is highly attractive on paper and seemingly the only method capable of furnishing dendronized polymers with long main chains and at the same time high-generation dendrons. Thus, there was a demand for synthetic development.

Two key observations made this need even more obvious so that the authors undertook a serious effort to improve on this situation by developing a reliable and robust attach-to route. Both these observations relate to charged dendronized polymers, the hundreds of peripheral charges of which render them polyelectrolytes with an unusually high linear charge density. First, it was found that charged third and fourth-generation dendronized polymers at a certain concentration in aqueous medium self-organize into a three dimensional network, each segment of which consists of exactly two tightly connected strands [11]. Second, it was discovered that in a lower concentration range a second and a third-generation charged dendronized polymer folds back on itself to give a tight duplex [12]. At first glance both findings look counterintuitive. Why should chains with like charges approach each other to give a network or a back-folded duplex? A coarse-grained theoretical treatment of this matter showed that dendronized polymers have a responsive thickness. There is enough free volume that allows the branches to change their conformations if an external stimulus makes this attractive [13]. Let's consider the back-folding process

[4] Typically the dendrons are used in considerable excess that can result in losses of polymer during work-up. These losses can be associated with inadvertent fractionations leading to a preferred depletion of the long chains. Nevertheless, even in complicated cases truly long chains (P_n's of several thousands) of high generation polymers were obtained according to the attach-to protocol in the authors' laboratory.

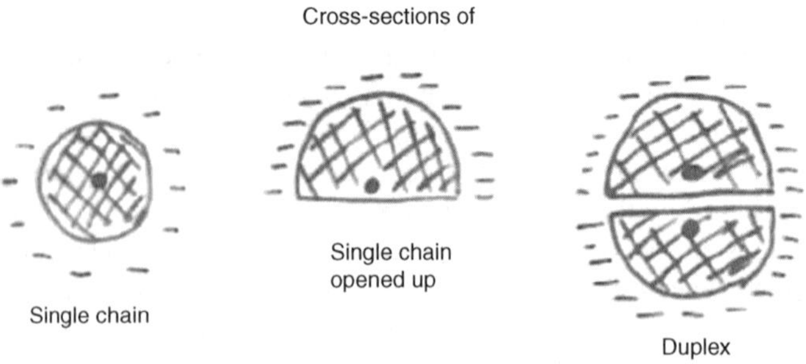

FIGURE 4.3 Hand drawn sketches of the cross-sectional geometries of a charged dendronized polymer as single chain (left), hypothetic lengthwise opened-up single chain (center), and duplex structure (right). The hydrophobic contact area in the duplex provides the driving force for the duplex formation and is correlated with "thickness" of the individual chain.

as an example. Upon bending, the charges in the bend's inner part can be pushed toward the periphery away from the area with the highest curvature. This increases the interaction between the hydrophobic interiors of the two polymer segments next to the bend. The thus increased hydrophobic interaction upon bending is a key factor making it attractive for the system to continue the back-folding process until a complete duplex is formed. It also compensates the energy needed for the bending. In addition, the back-folding leads to an improved protection of the hydrophobic parts against the surrounding water by the increased charge density at the surface of the developing duplex. Figure 4.3 shows a sketch illustrating this responsiveness that is based on what one could call a responsive or "dendritic" thickness of linear polymer chains.

Neither for bottle brushes nor for conventional polyelectrolytes such effects have been reported, which supports the view that it is the unprecedentedly large overlap area of the hydrophobic parts and thus the unique thickness of the dendronized polymers that makes this unusual behavior possible. Similar considerations apply to explain the network formation at higher concentrations. If such and other thickness effects should be more systematically explored, it would be highly desirable to have homologous series of dendronized polymers at hand all of which have the same main chain length but differ by the dendron generation and, thus, their cross-sectional diameter. In order to allow for quantitative structure/property relations, the quantitative knowledge of the branch work's structural fidelity is essential and any synthetic method must account for this.

Figure 4.4 shows the core steps of the synthetic sequence developed for this purpose. It is a powerful, gram-scale attach-to protocol that in principle can be applied to all polymers with primary amines in the side chain [14]. The advantages of this strategy are manifold. Monomer **MG1** and the two polymers **PG1** and **PG2** are known from earlier work that facilitates their characterization [15]. Also the deprotection chemistry, using neat trifluoroacetic acid, has been employed numerous times and found to be highly reliable [16]. The key building block for dendronization, compound **1** (Scheme 4.1), is new. However, its synthesis rests upon commercially

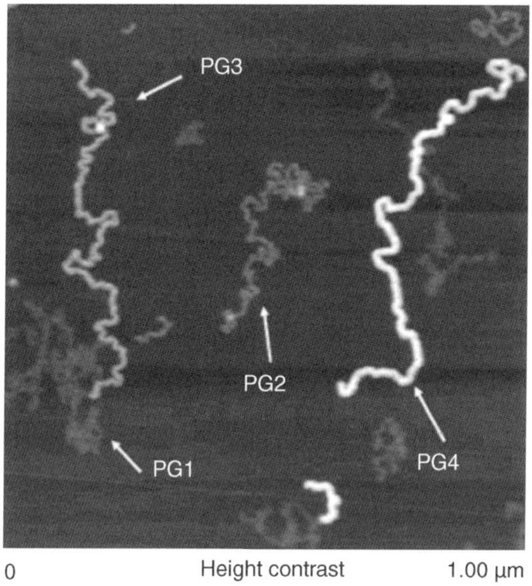

FIGURE 4.4 AFM height image of coprepared **PG1–PG4** dendronized polymers on mica. Each chain can unambiguously be assigned to a generation. Apparent thicknesses and widths are quantized by generation. There is no intermediate situation. Of hundreds of objects each can be assigned.

available dendrons, **2a** and **2b**. Furthermore, the way it bonds to peripheral amines during the dendronization event, succinidyl active ester amidation, has been quantified in a few similar cases. Under carefully optimized conditions, coverages of up to 99.8% were reproducibly achieved [17] that obviously suggested this building block for the present purpose. Finally, monomer **MG1** can be polymerized by controlled or, alternatively, free radical polymerization. This allowed obtaining starting polymers **PG1** with different chain lengths (L) in the range of several hundred nanometers up to more than a micrometer.

Considering the enormous increase in molar mass during the homologization, which may also be referred to as consecutive dendronization or thickening, the molar mass of the starting polymer **PG1** had to be carefully chosen. High molar masses may result in handling and characterization problems. The initial focus was therefore placed on a "short" **PG1** with $P_n \approx 1000$ and $L \approx 250\,\text{nm}$. Assuming complete coverage throughout, its corresponding **PG4** had approximately a molar mass of 5.4×10^6. With our experience, such a **PG4** was expected to be soluble in organic solvents at room temperature and still exhibit sufficiently resolved NMR spectra. It was therefore considered a "safe" series. A longer and potentially more interesting series starting from a **PG1** with $L > 1\,\mu\text{m}$ was also planned. This of course involved the risk of lower solubility and complicated structural analysis especially for higher generation polymers. In order to synthesize these two **PG1**'s of different length both *reversible addition fragmentation chain transfer(RAFT) polymerization* [18] and free radical polymerization of **MG1** suggested themselves.

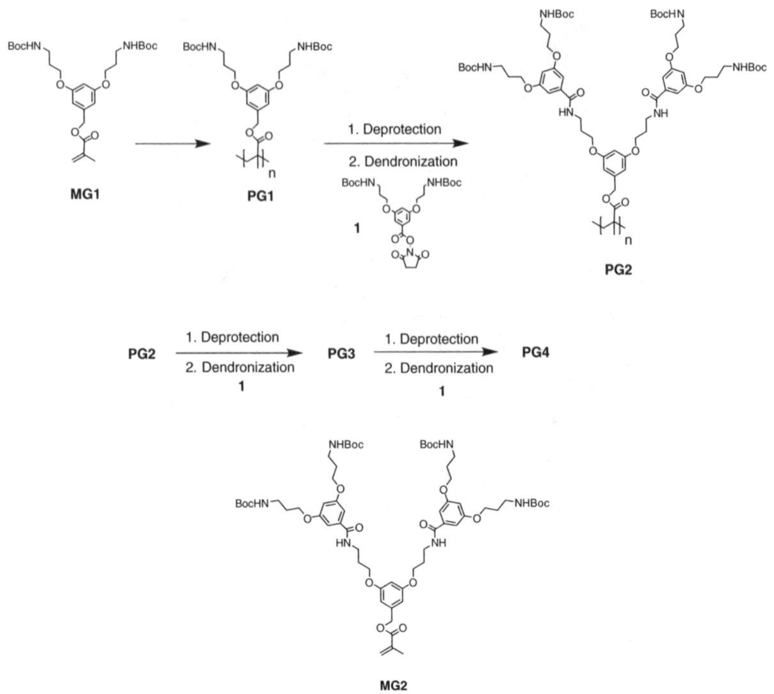

SCHEME 4.1 Synthesis of a homologous series of dendronized polymers (**PG1–PG4**) according to a powerful attach-to protocol and structure of macromonomer **MG2** used to compare the two **PG2**s synthesized according to the attach-to and macromonomer protocol, respectively. Compound **1** is the key building block used for dendronization, which is based on active ester amidation.

The main characterization effort involved methods providing *qualitative* insights concerning the growth processes. These comprised gel permeation chromatography (GPC), ^{13}C NMR spectroscopy, and atomic force microscopy (AFM) apparent height measurements. To be on the safe side all the samples were also checked by a *quantitative* method based on UV absorption. In them the unreacted terminal amines of the dendronized polymers were converted into UV labels with high extinction coefficient by reacting them with an excess of Sanger reagent [17]. Finally, the NMR spectra of the second-generation polymers obtained by the attach-to route were compared to those obtained by the macromonomer route starting from **MG2** (Scheme 4.1). For details the reader is referred to Ref. [14].

After having discussed so much about thickening and thickness of polymer chains it is time to provide some experimental insight into this matter. Figure 4.4 shows the AFM height image recorded in tapping mode of a sample containing four dendronized polymers **PG1–PG4** (structures derived from M1 in Scheme 4.1) at the same time. As can be seen, each object can unambiguously be assigned to a particular generation. Since absolute heights are difficult to determine by the AFM, TEM images of low angle shadowed samples were also recorded [19].

4.3 STIMULI-RESPONSIVE DENDRONIZED HOMO- AND DIBLOCK COPOLYMERS

"Thickness" has already been referred to in Chapters 1 and 2 as an unique feature of dendronized polymers. The authors of this book chapter tend to believe that it represents a new variable in polymer chemistry in that the properties of such polymers should not only depend on chain length and persistence length [20] anymore but now also on thickness. With this view in mind areas were explored, in which thickness either could directly result in new properties or at least provide additional insight into known behavior of linear macromolecules under certain conditions. The area of thickness-induced aggregation behavior has already been mentioned above. Another such area was in rather general terms "responsiveness to stimuli such as temperature change or addition of complexing salts." In the following a few examples will be discussed that have the potential to shed more light on the impact thickness of polymer chains may exhibit also in such an area.

Oligoethylene glycols (OEGs) and their derivatives are thermoresponsive with lower critical solution temperatures (LCSTs) normally above 100°C. These LCSTs, which are too high for many practical purposes, can be tuned by the overall hydrophilicity/hydrophobicity ratio commonly referred to as structure effect [21] and by the way in which the hydrophobic and hydrophilic units are arranged in the macromolecule. This latter aspect is referred to as architecture effect and has attracted considerable attention recently [22]. Given the rather unusual structure of dendronized polymers it was perhaps a not far fetching idea to investigate those representatives that carry thermoresponsive OEG-based dendrons. They should not only allow exploring the architecture effect very systematically generation by generation but also gaining new insights into how the collapse of such macromolecules takes place on the molecular level in dependence of the main chain thickness. Stimulated by these perspectives a series of first- to third-generation OEG-based dendronized polymers (Figure 4.5) were synthesized according to the macromonomer route (not shown) [5,23]. Because of the impact this synthesis route has on main chain length, care was taken that the collapse temperature was not affected it.

Three structure parameters determine these polymers overall hydrophilicity [24] and, thus, their thermoresponsiveness:

(1) Nature of peripheral groups [hydroxyl (OH), methoxy (M), ethoxy (E)].
(2) Length of OEG segment [triethylene (T), diethylene glycol (D)].
(3) Dendron generation [first (**PG1**), second (**PG2**), third (**PG3**)].

The first two parameters are more or less conventional and apply to numerous OEG-based thermoresponsive polymers. The last parameter however is associated with thickness and therefore new. Except for **PG1(OH)** and **PG2(OH)**, which are water soluble at all temperature explored, all other polymers collapse from water. The aqueous solutions turn turbid once heated above the respective LCST. This behavior was investigated, for example, by proton NMR spectroscopy and turbidity

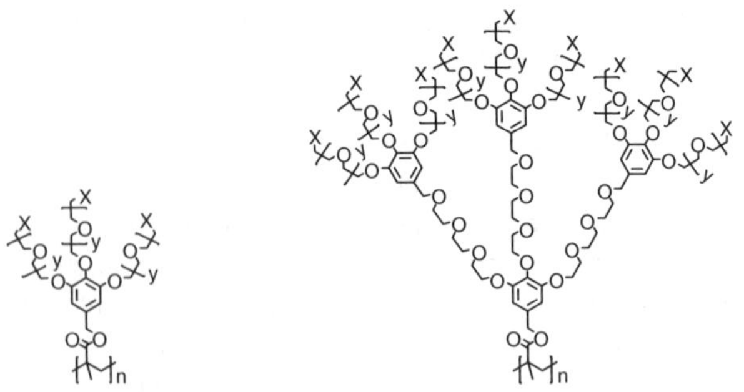

X=OH, Y=3: **PG1(OH)**
X=OMe, Y=2: **PG1(MT)**, log P = 0.35
X=OMe, Y=1: **PG1(MD)**, log P = 0.61
X=OEt, Y=2: **PG1(ET)**, log P = 1.38

X=OH, Y=2: **PG2(OH)**
X=OMe, Y=2: **PG2(MT)**, log P = -1.04
X=OMe, Y=1: **PG2(MD)**, log P = -0.27
X=OEt, Y=2: **PG2(ET)**, log P = 2.04

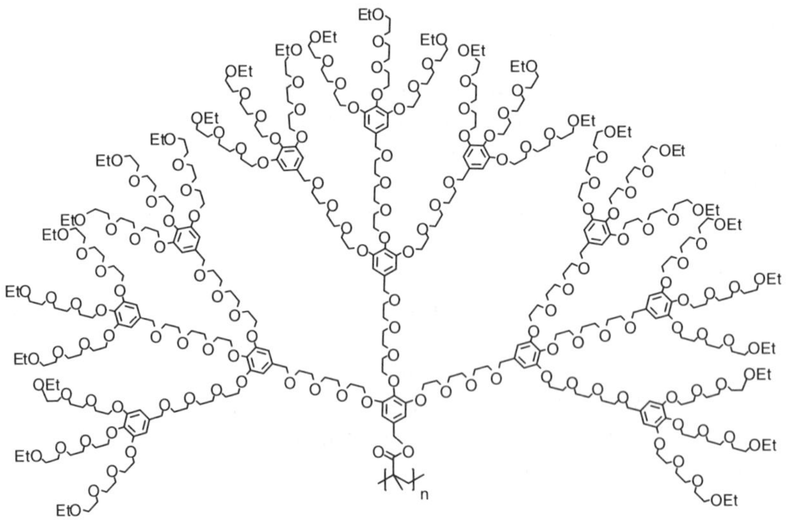

PG3(ET), log P = 3.40

FIGURE 4.5 Chemical structures of OEG-based dendronized polymers **PG1–PG3**. The "atomic hydrophobicity parameters" (log P) calculated with the method by Ghose [24] are given.

measurements. As expected, the proton signal intensities decrease with increasing temperature around the critical points due to dehydration (Figure 4.6), while turbidity measurements revealed very fast and sharp transitions in both heating and cooling direction. Typical transition curves are shown in Figure 4.7. The hystereses are extremely small ($\leq$0.6°C) and among the smallest ever observed. The apparent LCSTs for **PG1(ET)**, **PG2(ET)**, **PG3(ET)**, **PG1(MD)**, **PG2(MD)**, **PG1(MT)**, and **PG2(MT)** are 33, 36, 34, 43, 48, 63, and 64°C, respectively. These values nicely

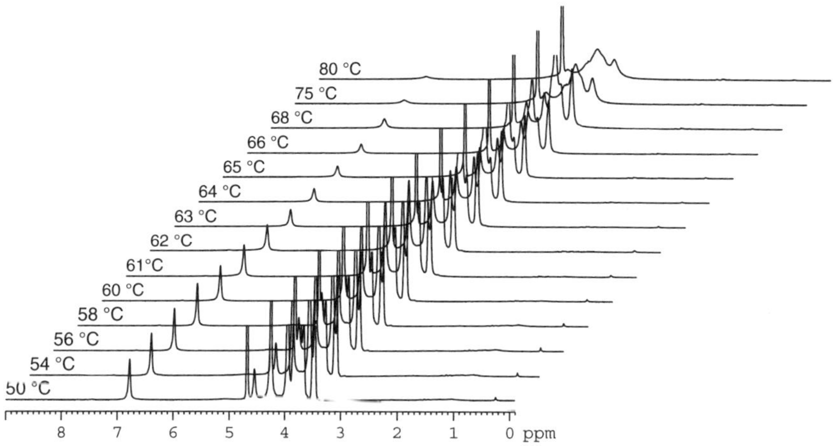

FIGURE 4.6 Fully reversible temperature-dependent proton NMR spectra of **PG2(MT)** (2.0 wt% in D_2O) in the range from 50 to 80°C.

reflect the expectations based on hydrophilicity/hydrophobicity considerations and show that even without any additions (e.g., of salts) a broad and highly relevant temperature window can be covered with slightly modified polymers. Interestingly, it can also be concluded that there does not seem to be a significant generation dependence. This is perhaps the most important finding as it may point toward mechanisms in which the collapse of the outermost units determines the LCST to a large degree and the internal dendritic units play less of a role [25]. This proposal is

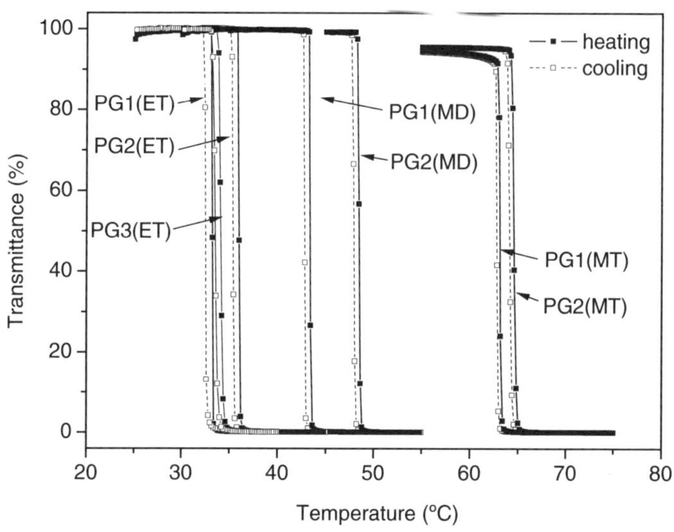

FIGURE 4.7 Transmittance versus temperature for the various OEG-based dendronized polymers synthesized, indicating unprecedentedly sharp and fully reversible transitions.

supported by the fact that the hydrophilicity of the various polymers changes with generation (see log *P* values in Figure 4.5). These changes however are not mirrored by the LCST values. If this assumption is correct, **PG3** could possibly serve as an interesting container that in its interior does not collapse and, thus, is capable of holding its load while the periphery is closed-up above LCST. Further work along these lines is in progress and should eventually furnish the desired molecular scale picture of the collapse [26].

The collapse of **PG2(ET)** was further investigated by static and dynamic light scattering, and its aggregation by *cryo*-TEM. Interestingly, a coil-to-mesoglobule transition was observed that results in stable and nearly monodisperse mesoglobules (Figure 4.8). Possible reasons why the aggregation of collapsed chains stops automatically when the mesoglobules have reached a certain size have been discussed [27].

This rather successful journey into thermoresponsive dendronized polymers triggered further investigations along the lines indicated but at the same time also on related systems. Dendronized block copolymers (DBCP) combine the structure characteristics of both dendronized polymers and conventional block copolymers. They are therefore obvious candidates for stimuli-responsive materials and also, for example, for bioapplications and supramolecular constructions. poly(*N*-isopropylacrylamide) (P*N*iPAM) is a prominent thermoresponsive polymer [28]. It shows a LCST around body temperature, is biocompatible, and has therefore been widely used also in thermoresponsive block copolymers. Comb-like polymethacrylates with OEG side chains are another type of attractive thermoresponsive polymer [29]. They show smaller hystereses than P*N*iPAM due to their much weak ability to form hydrogen bonds and are also biocompatible. Two different types of DBCPs were prepared namely those with a P*N*iPAM and a dendronized block [30] and those with comb-like OEG units and a dendronized block [31].

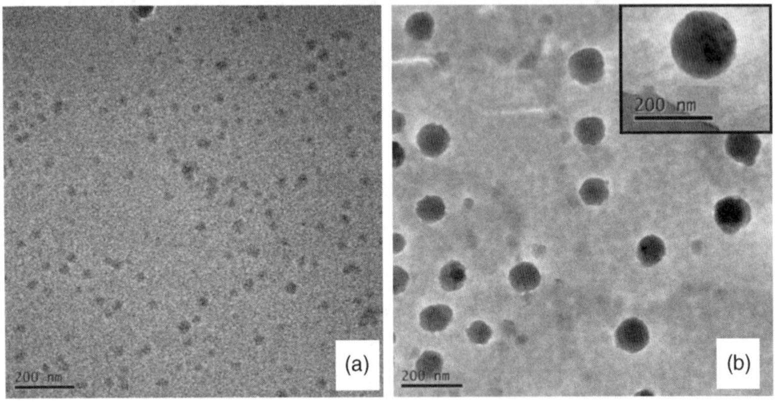

FIGURE 4.8 *Cryo*-TEM images of the dendronized polymer **PG2(ET)** at different temperatures. (a) Image taken at a solution temperature of 25°C. The individual dendronized polymers are visible. (b) At 60°C, the chains aggregate and form stable spherical mesoglobules. The inset shows an individual single mesoglobule. Reprinted from Ref. [27] with permission of the publisher.

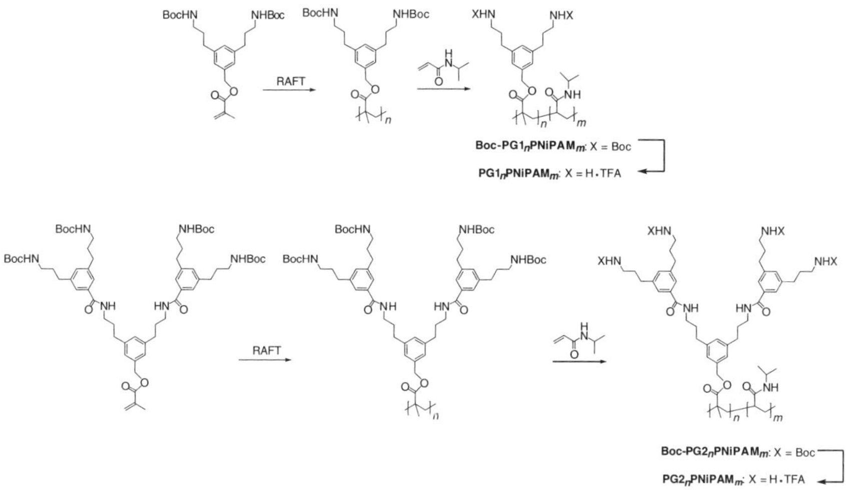

SCHEME 4.2 Synthetic sequences to *tert*-butyloxycarbonyl (Boc) protected DBCPs with P*Ni*PAM block and their deprotected counterparts.

DBCPs with P*Ni*PAM block were prepared by RAFT polymerization technique (Scheme 4.2) and the block lengths and ratios varied. After deprotection of the dendronized block, these copolymers are water soluble at room temperatures, but start aggregating at their LCSTs. The aggregate morphologies were examined by TEM and found to depend not only on the block ratios and block lengths but also on the dendron generations [32]. For DBCPs with short **PG1** block uniform spherical objects with sizes in the range of 40–120 nm were observed. With increasing chain length of the **PG1** block, the aggregates become irregular. For DBCPs with **PG2** blocks so called "large compound micelles" were observed [33]. The reversibility of DBCPs aggregates with **PG1** block was investigated to some detail for $\mathbf{PG1_{45}PNiPAM_{263}}$. During three heating and cooling cycles, samples for TEM measurements were taken at 50°C. In each cycle very similar spherical objects were observed (Figure 4.9) that indicates that these objects represent a thermodynamic minimum under the applied conditions.

First- and second-generation representatives of the second class of DBCPs were synthesized by atom transfer radical polymerization (ATRP) (Figure 4.10) [31]. They showed a better phase separation in bulk as compared to the first system, a property that could be exploited in terms of three-dimensional bulk structure design. For this purpose the deprotected and, thus, charged dendronized blocks were decorated with differently long simple surfactant molecules and the resulting bulk structures investigated by X-ray diffraction and TEM.

Finally, ion-responsive cases were studied for which the first syntheses and orienting charging experiments were finished very recently [34]. For this purpose crown ether units were incorporated into the dendron side chains and their capacity to load and unload metal ions investigated. The chemical structures are shown in Figure 4.11. Complexation with potassium ions proved the coil conformation of these polymers to be manipulable. For loadings of one potassium ion per crown ether

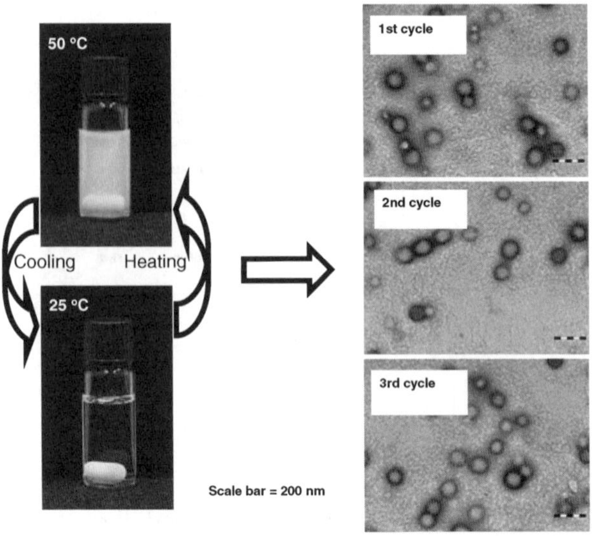

FIGURE 4.9 Reversible aggregate formation: aqueous buffered solution of **$PG1_{45}PNiPAM_{263}$** above and below the LCST (left) and the corresponding TEM images taken above the LCST (scale bars: 200 nm) (right).

unit the chain dimensions increased by up to 80% and at the same time the chain stiffness also increased significantly. For example, for a **PG1** with $DP_w \sim 4360$, the Kuhn length l_k increased from 8 to 19 nm, and for a **PG2** with $DP_w \sim 2730$, l_k increased from 19 to 45 nm. It would be interesting to explore whether charging and

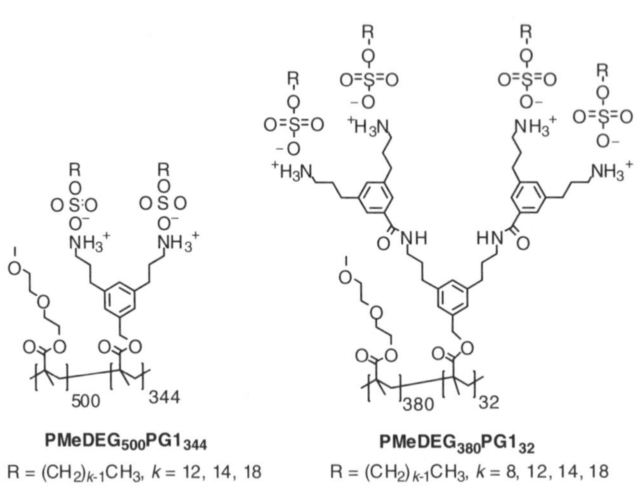

FIGURE 4.10 Chemical structures of DBCPs with comb-like OEG blocks and surfactant molecules complexed to the positively charged "surface" groups for bulk structure design.

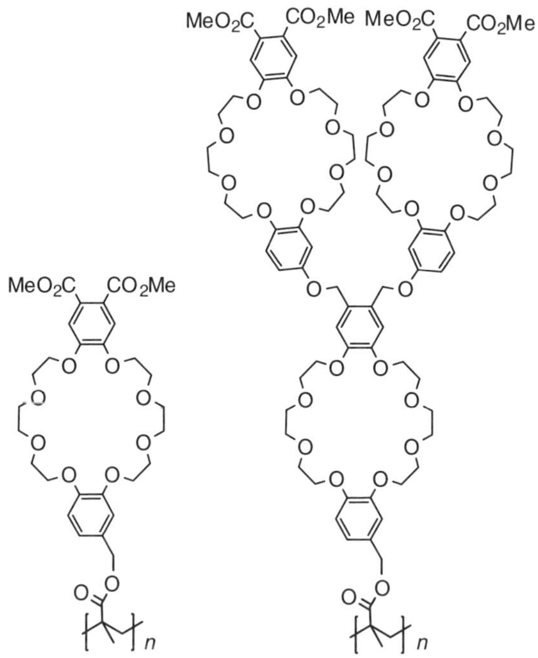

FIGURE 4.11 Chemical structures of crown ether-based dendronized polymers **PG1** and **PG2** from macromonomer route.

decharging of the polymers with ions can be exploited for inducing shape changes of macroscopic materials like films prepared from these polymers [35]. Whether there are any thickness effects is too early to speculate about.

4.4 HELICAL DENDRONIZED POLYMERS

Having spatially rather demanding substituents bonded in tight distance to a linear polymer backbone raises the question whether this unique situation could be used to create dendronized polymers with a predetermined helical screw sense. The dense packing of the (dendritic) substituents may result in novel helical polymers exhibiting unprecedentedly stable helical conformations over a wide temperature and solvent polarity range. This way dendronized polymers with their congested structures could contribute to the important field of helical polymers, which has been explored over the years, for example, by Okamoto [36], Percec [37], Nolte [38], and Yashima [39]. In order to achieve such a goal chiral and very compact dendrons had to be developed and incorporated into dendronized polymers such that their "cross-talking" results in stable either right- or left-handed helices. 4-Aminoproline moieties were selected as the branching units and highly efficient synthetic ways developed how to construct first- and second-generation macromonomers from them as diastereomerically pure isomers [40]. Ball-and-stick model considerations suggested that certain

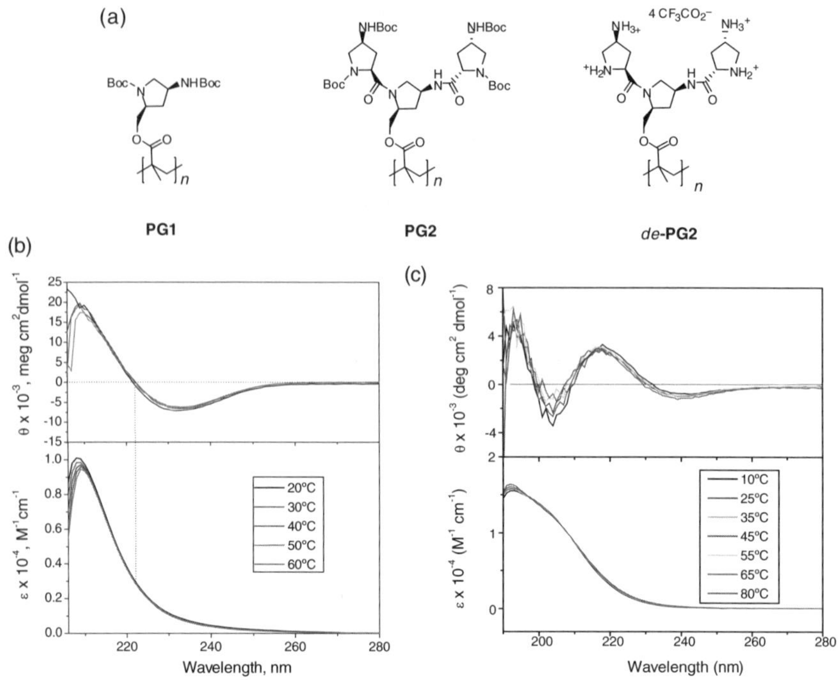

FIGURE 4.12 Polymer structures **PG1**, **PG2**, and *de*-**PG2** (a), and the normalized CD spectra of **PG2** in THF (b) and its deprotected congener *de*-**PG2** in water (c).

diastereomers should give right-handed helices and others the corresponding left-handed ones. The first two examples (**PG1** and **PG2**) are based on (2*S*,4*S*)-4-aminoproline, and their chemical structures are shown in Figure 4.12. These polymers were synthesized with molar masses above 1 MDa by spontaneous polymerization of the corresponding macromonomers (not shown). Both polymers are very rigid as evidenced by high glass transition temperatures and extremely broad NMR signals. This was further substantiated by molecular dynamic (MD) simulations by C. Aleman, indicating no change in end-to-end distance with time. Their conformation was investigated with optical rotation (OR) and circular dichroism (CD). The results prove **PG2** to adopt an ordered conformation, while **PG1** does not. Temperature-varied CD measurements reveal that the chiral conformation of **PG2** is stable in protic and aprotic solvents up to temperature of 80°C. Even after deprotection with a mass loss of around 50%, the chiral conformation of the polymer still remains! MD simulations using initially 20 and later 50 repeat units of **PG2** provided a right-handed helix model irrespective of whether the backbone is isotactic, syndiotactic, or heterotactic [41]. The details of the model of course depended on the tacticity. It was concluded that the helical conformation of the polymer formed during polymerization is mainly driven by dendron/dendron interaction rather than the backbone tacticity. The corresponding results of the left-handed helix have not yet been reported.

4.5 DENDRONIZED POLYMERS FOR SINGLE MOLECULE CHEMISTRY

Synthetic chemists in academia are used doing their experiments all the way from the 500 g to 1 kg scale in a small reactor to a few milligrams in an NMR tube or nanograms in biochemical assays. All these scales have their own specifics, challenges, and fascinations. An ultimate goal of synthetic chemistry would, of course, be to work on the smallest possible scale that is using individual molecules and do specific reactions with them. Obviously this poses localization, handling, and analytical problems. Among the pioneers in this field are Hla and Rieder, who back in 2000 reported a covalent reaction between individual small molecules on a solid substrate. They combined two iodo benzene molecules to biphenyl on a Cu surface by using the interactions between the scanning tunneling microscope (STM) tip and the molecules [42]. Other bimolecular reactions [43] and even polymerizations [44] initiated by tip molecule interactions followed. Related are experiments in which chemical reactions on solid substrates were initiated by other means (e.g., electrochemical or photochemical ones) but the successful course of which was monitored by the STM [45]. This is the point where dendronized polymers and the AFM come into play. This fruitful symbiosis of a specific chemical system and a specific manipulation and analysis tool laid the foundation for a series of interesting experiments on the single macromolecule level and brought them to ripeness. Nowadays this technology can in principle be used by everybody. What are the key factors of both object and tool that made this symbiosis a success story? The first to mention regarding object certainly is thickness. Compared to other linear polymers, dendronized polymers have a considerable thickness even when in the collapsed state adsorbed on a substrate. Qualitatively this can nicely be seen from Figure 4.4 (Section 4.2). This allows the AFM tip to actually grab the chains and drag them across when scanning near to the substrate surface. The tip neither snaps over the object nor scratches into the substrate. Thus, if the adsorption forces are properly tuned (see below), two entities can be moved together to create a situation in which an intermolecular bond can be formed at a predetermined position. In this context it is another important feature of dendronized polymers that they can easily be equipped with groups such as azides, capable of initiating covalent cross-linking when properly triggered, for example, by photochemical means. Many of these azide groups will be located near or at the object's "surface" and, thus, ideally positioned to bring about interchain cross-linking events if the other chain is in sufficiently tight contact. Finally, for dendronized polymers prepared on highly oriented pyrolytic graphite (HOPG) or carpet-modified HOPG [46] the adsorption forces are in a very attractive intermediate range. Neither do the chains adhere too strongly (e.g., as on mica) that the tip would tear them apart, nor do they adhere too weakly to complicate imaging before and after manipulation due to an uncontrolled lateral diffusion.

The key factor that comes to mind regarding the tool AFM are the ambient conditions under which it can be operated. No ultrahigh vacuum conditions are required as they are typically applied for the STM. It is fully sufficient to spin-cast a polymer solution onto the substrate, and the AFM experiment can be started. Though

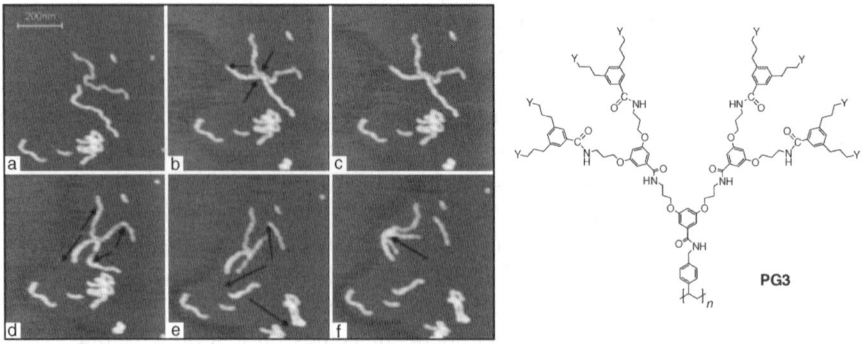

FIGURE 4.13 Tapping-mode AFM images of two individual dendronized polymers **PG3**, moved toward each other ("move"; a → b), irradiated by UV light ("connect"; b → c), and challenged mechanically ("prove"; d, e, f). The arrows indicate the movement of the AFM tip during manipulation. Reprinted from Ref. [47] with permission of Wiley-VCH.

one admittedly never knows the exact state of the substrate when working under ambient conditions, the ease of operation has its attraction. A sample is ready for manipulation and analysis after a few minutes time! Additionally, dendronized polymers are such huge objects that adsorbed molecules, for example, water, will alter the adsorption forces but cannot render unreliable the interpretation of the experiments dealing with covalent bond formation between these objects.

In 2003, the first experiment was reported in which two dendronized polymers were prepared as single entities on a solid substrate, moved together by the AFM and connected by photochemical treatment. The connection was then proven again with the help of the AFM, by dragging the combined macromolecules across the surface [47].[5] These experiments, which were coined *move-connect-prove* sequence, were done with a third-generation dendronized polymer **PG3** [Y mostly NH_3^+ $CF_3CO_2^-$ but also $NHCO(CH_2)_5NHC_6H_3(NO_2)(N_3)_3$] in which some of the peripheral ammonium groups had been converted into azide derivatives (Figure 4.13). The actual "gluing" process was achieved through these azide groups, which are known to decompose to the highly reactive nitrene intermediates upon irradiation. At the site where the two chains were in tight contact, these nitrenes caused a structurally ill-defined intermolecular cross-linking.

Because of the potential importance of such protocols for the bottom-up approach to the nanosciences, in the following years this experiment was developed to maturity [48]. The key steps were (i) the introduction of a dense monolayer ("carpet") of long-chain fatty acids between the solid substrate (HOPG) and the deposited macromolecules to be subjected to such a "move-connect-prove" sequence and (ii) the use of the structurally precisely defined dendronized polymer **PG3A**

[5] This initial experiment was carried out with a not fully characterized azidified polymer that had been prepared on site and not purified. It is therefore likely that the substrate was covered with small molecules of different nature. They of course cannot have altered the key conclusion namely the successful execution of a move-connect-prove sequence but will have influenced the lateral mobility of all components involved.

(A = acidified, structure not shown). This way any irrationalities stemming from unknown adsorbents on HOPG and the not fully known chemical structure of the polymer were prevented. The carpet technology [46] also improved the dragging and imaging of the deposited dendronized polymers.

On this basis it was an obvious next step not only to try to connect two neutral macromolecules with one another but rather to expand the range of employable components to charged ones. This would open access for this technology to the repertoire of polyelectrolytes the bio world offers [49]. An issue that had to be considered in this regard is the need for copreparation of neutral and charged species on the same substrate. In light of the present interest in graphenes [50] it was an additional challenge to explore whether the edges of these unique sheet-like entities could be attacked and thus modified at predetermined sites by the gluing protocol. Two examples may suffice here to illustrate the potential.

Figure 4.14 shows a *move-connect-prove* sequence involving the neutral **PG3A** with *ds*DNA, which is negatively charged, when coprepared on HOPG modified by a carpet of the long-chain alkyl amine $C_{12}H_{25}NH_2$. The linear **PG3A** can be readily

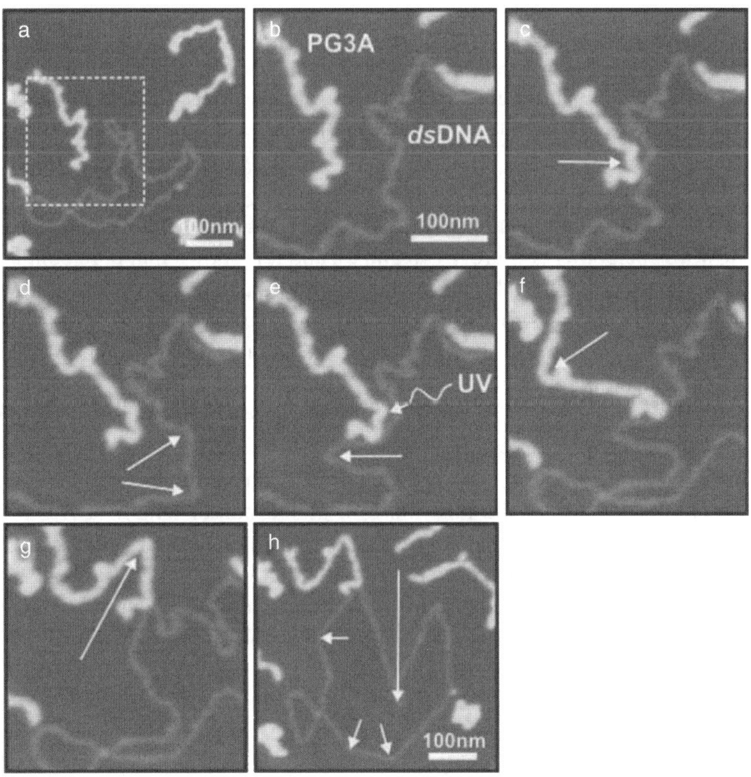

FIGURE 4.14 "Move-prove-connect" sequence with the neutral and azide-carrying dendronized polymer **PG3A** and plasmid DNA **puC19** on HOPG precoated with a monolayer of $C_{12}H_{25}NH_2$. Reprinted from Ref. [49] with permission of Wiley-VCH.

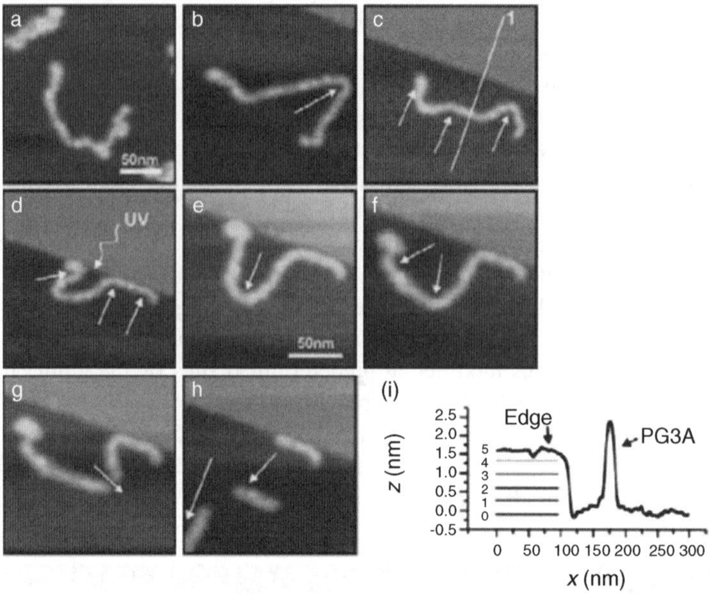

FIGURE 4.15 "Move-prove-connect" sequence with the neutral and azide-carrying dendronized polymer **PG3A** and the edge of the topmost graphite plane of HOPG precoated with a monolayer of $C_{12}H_{25}NH_2$. Reprinted from Ref. [49] with permission of Wiley-VCH.

distinguished from the circular DNA **puC19** by its larger thickness and its topology since **puC19** is a circular plasmid (Figure 4.14a). First **PG3A** and the DNA have been moved toward each other (Figure 4.14b and c), and it is shown that pure contact does not link the chains such that the junction would hold as one moves one chain apart from the other (Figure 4.14d). After pushing the chains toward each other in tight contact again (Figure 4.14e) and illuminating the sample for 5 min with UV-light at 254 nm the induced junction was mechanically tested by pulling first on the **PG3A** (Figure 4.14f and g) and then on the DNA (Figure 4.14h), showing that the junction withstands the employed forces when being dragged across the surface and the chains come straight off the junction. The DNA chain was furthermore over-stretched by a factor of 1.3 during testing the junction to a global contor-length of $L = 1136$ nm ($L_0 = 878$ nm). Finally, Figure 4.15 shows a case in which the same polymer, **PG3A**, on $C_{12}H_{25}NH_2$ precoated HOPG is welded to the edges of a topmost graphene plane of HOPG. These rather different cases illustrate the considerable potential evolving from the symbiosis of a powerful physical method, the AFM, with a powerful chemical system, the dendronized polymers.

4.6 SUMMARY AND OUTLOOK

The four examples illuminated in this chapter mirror the width of basic and applied research presently going on in the field of dendronized polymers. Now that finally also a robust attach-to protocol is available, it seems that the main synthetic challenges have been solved. Clearly more work is needed, for example, in the direction of higher generations and less perfect structures but this is considered more as a fine-tuning. The third section together with many aspects that are being researched in other groups [4] shows that the whole field is presently moving toward materials applications. Surprising aspects have been discovered and it seems that at least for some of them the "thickness" of dendronized polymers is a key ingredient. Whether thickness as such will develop into a recognized variable of polymer chemistry only the future can tell. The fourth section also points toward a possible future direction of research. It is this tight register in which consecutive dendrons are aligned along the backbone, which causes unprecedented helix stabilities. We foresee that this will find its use in connection with novel supramolecular structures, for example, in coassemblies of dendronized polymers with DNA or proteins. The final section may look a bit esoteric at first glance. Why caring about developing chemistry between two individual macromolecules on a substrate at all? Compared to the situation at the beginning of this research, the methodological development was steep and nowadays in principle everybody, who has an AFM and the proper software can do similar experiments. We do expect this field to soon have the ripeness to allow for systematic constructions on the molecular level. This for the first time would open access to designed complex and possibly functional molecular constructs that otherwise would not be available. Dendronized polymers are an exciting field of research and we wish to thank not only our coworkers for their invaluable input over the years but also all those in other laboratories, who have helped us with their ideas to develop it to the state of harvesting in which it presently is.

ACKNOWLEDGMENTS

The authors are indebted to their coworkers that substantially and creatively contributed to the results described here. There names can be found in the references. The research was supported over the years by the German Science Foundation (DFG), the Swiss National Science Foundation (SNF), The Chinese Natural Science Foundation, the Freie Universität Berlin, and the ETH Zurich, which is gratefully acknowledged. We thank Dr. Zhang Baozhong for critically reading the manuscript.

REFERENCES

[1] R. Freudenberger, W. Claussen, A. D. Schlüter, H. Wallmeier, *Polymer* **1994**, *35*, 4496–4501.

[2] (a) A. D. Schlüter, J. P. Rabe, *Angew. Chem. Int. Ed.* **2000**, *39*, 864–883; (b) A. Zhang, L. Shu, Z. Bo, A. D. Schlüter, *Macromol. Chem. Phys.* **2003**, *204*, 328–339.

[3] (a) A. Zhang, *Prog. Chem.* **2005**, *17*, 157–171. (b) H. Frauenrath, *Prog. Polym. Sci.* **2005**, *30*, 325–384; (c) A. D. Schlüter, *Top. Curr. Chem.* **2005**, *245*, 151–191.

[4] (a) For a related review, see: M. Marcos, R. Martin-Rapun, A. Omenat, J. L. Serrano, *Chem. Soc. Rev.* **2007**, *36*, 1889–1901. For synthesis of dendronized homopolymers via attach-to route, see for example: (b) M. L. Hassan, C. N. Moorefield, G. R. Newkome, *Macromol. Rapid Commun.* **2004**, *25*, 1999–2002. (c) M. L. Hassan, C. N. Moorefield, K. Kotta, G. R. Newkome, *Polymer* **2005**, *46*, 8947–8955. (d) M. Yoshida, Z. M. Fresco, S. Ohnishi, J. M. J. Fréchet, *Macromolecules* **2005**, *38*, 334–344. (e) J. L. Mynar, T.-L. Choi, M. Yoshida, V. Kim, C. J. Hawker, J. M. J. Fréchet, *Chem. Commun.* **2005**, 5169–5171. (f) C. Zhang, L. M. Price, W. H. Daly, *Biomacromolecules* **2006**, *7*, 139–145. (g) F. Z. Khan, M. Shiotsuki, Y. Nishio, T. Masuda, *Macromolecules* **2007**, *40*, 9293–9303. (h) E. Oestmark, J. Lindqvist, D. Nystroem, E. Malmstroem, *Biomacromolecules* **2007**, *8*, 3815–3822. (i) T. Heinze, M. Schoebitz, M. Pohl, F. Meister, *J. Polym. Sci., Part A: Polym. Chem.* **2008**, *46*, 3853–3859. (j) M. Pohl, J. Schaller, F. Meister, T. Heinze, *Macromol. Rapid Commun.* **2008**, *29*, 142–148. (k) M. Pohl, J. Schaller, F. Meister, T. Heinze, *Macromol. Symposia* **2008**, *262*, 119–128. (l) M. Pohl, N. Michaelis, F. Meister, T. Heinze, *Biomacromolecules* **2009**, *10*, 382–389. (m) M. Pohl, G. A. Morris, S. E. Harding, T. Heinze, *Eur. Polym. J.* **2009**, *45*, 1098–1110. (n) W.-H. Ting, C.-C. Chen, S. A. Dai, S.-Y. Suen, I.-K. Yang, Y.-L. Liu, F. M. C. Chen, R.-J. Jeng, *J. Mater. Chem.* **2009**, *19*, 4819–4828. (o) E. Boisselier, A. C. K. Shun, J. Ruiz, E. Cloutet, C. Belin, D. Astruc, *New J. Chem.* **2009**, *33*, 246–253. For synthesis of dendronized homopolymers via macromonomer route, see for example: (p) E. Kasemi, W. Zhuang, J. P. Rabe, K. Fischer, M. Schmidt, M. Colussi, H. Keul, D. Yi, H. Cölfen, A. D. Schlüter, *J. Am. Chem. Soc.* **2006**, *128*, 5091–5099. (q) A. Nystroem, M. Malkoch, I. Furo, D. Nystroem, K. Unal, P. Antoni, G. Vamvounis, C. Hawker, K. Wooley, E. Malmstroem, A. Hult, *Macromolecules* **2006**, *39*, 7241–7249. (r) K. T. Kim, J. Han, C. Y. Ryu, F. C. Sun, S. S. Sheiko, M. A. Winnik, I. Manners, *Macromolecules* **2006**, *39*, 7922–7930. (s) M. Liu, W. J. Xu, X. Xu, Y. B. Lu, R. R. Miao, *Chin. Chem. Lett.* **2006**, *17*, 1507–1510. (t) A. Luebbert, T. Q. Nguyen, F. Sun, S. S. Sheiko, H.-A. Klok, *Macromolecules* **2005**, *38*, 2064–2071. (u) M. Liu, W.-J. Xu, X. Xu, Y. Lu, *J. Appl. Polym. Sci.* **2007**, *105*, 3087–3096. (v) S. R. Benhabbour, M. C. Parrott, S. E. A. Gratton, A. Adronov, *Macromolecules* **2007**, *40*, 5678–5688. (w) E. Kasemi, A. D. Schlüter, *New J. Chem.* **2007**, *31*, 1313–1320. (x) P. Sonar, H. Benmansour, T. Geiger, A. D. Schlüter, *Polymer* **2007**, *48*, 4996–5004. (y) L. I. Costa, E. Kasemi, G. Storti, M. Morbidelli, P. Walde, A. D. Schlüter, *Macromol. Rapid Commun.* **2008**, *29*, 1609–1613. (z) A. J. Boydston, T. W. Holcombe, D. A. Unruh, J. M. J. Fréchet, R. H. Grubbs, *J. Am. Chem. Soc.* **2009**, *131*, 5388–5389. For helical dendronized polymers, see: (aa) Y. Kamikawa, T. Kato, H. Onouchi, D. Kashiwagi, K. Maeda, E. Yashima, *J. Polym. Sci., Part A: Polym. Chem.* **2004**, *42*, 4580–4586. (ab) H.-J. Kim, W.-C. Zin, M. Lee, *J. Am. Chem. Soc.* **2004**, *126*, 7009–7014. (ac) J.-H. Ryu, J. Bae, M. Lee, *Macromolecules*, **2005**, *38*, 2050–2052. (ad) H. Zhao, F. Sanda, T. Masuda, *Macromol. Chem. Phys.* **2006**, *207*, 1921–1926. (ae) H.-J. Kim, E. Lee, H.-S. Park, M. Lee, *J. Am. Chem. Soc.* **2007**, *129*, 10994–10995. (af) Z.-T. Liu, Y.-M. He, Z.-J. Wang, Y. Feng, Q.-H. Fan, *J. Polym. Sci., Part A: Polym. Chem.* **2008**, *46*, 886–896. (ag) J. G. Rudick, V. Percec, *New J. Chem.* **2007**, *31*, 1083–1096. (ah) V. Percec, J. G. Rudick, M. Peterca, P. A. Heiney, *J. Am. Chem. Soc.* **2008**, *130*, 7503–7508. (ai) H.-J. Kim, E. Lee, M. G. Kim, M.-C. Kim, M. Lee, E. Sim, *Chem. Eur. J.* **2008**, *14*, 3883–3888. (aj) A. Zhang, *Macromol. Rapid Commun.* **2008**, 839–845. For stimuli-responsive dendronized polymers, see: (ak) M. Gao, X. Jia, G. Kuang, Y. Li, D. Liang, Y. Wei, *Macromolecules* **2009**, *42*, 4273–4281. (al) H.-J. Kim, E. Lee, H.-S. Park, M. Lee, *J. Am. Chem. Soc.* **2007**, *129*, 10994–10995. For dendronized (block) copolymers, see: (am) C. X. Cheng, R. P. Tang, Y. L. Zhao, F. Xi,

J. Appl. Polym. Sci. **2004**, *91*, 2733–2737. (an) C. Cheng, R. Tang, F. Xi, *J. Polym. Sci., Part A: Polym. Chem.* **2005**, *43*, 2291–2297. (ao) C. Cheng, Y. Tian, Y. Shi, R. Tang, F. Xi, *Langmuir* **2005**, *21*, 6576–6581. (ap) C. Cheng, R. Tang, F. Xi, *Macromol. Rapid Commun.* **2005**, *26*, 744–749. (aq) C. Cheng, Y. Tian, Y. Shi, R. Tang, F. Xi, *Macromol. Rapid Commun.* **2005**, *26*, 1266–1272. (ar) Y. Zhang, J. Huang, Y. Chen, *Macromolecules* **2005**, *38*, 5069–5077. (as) C.-X. Cheng, T.-F. Jiao, R.-P. Tang, E.-Q. Chen, M.-H. Liu, F. Xi, *Macromolecules* **2006**, *39*, 6327–6330. (at) Y. Zhang, X. Li, G. Deng, Y. Chen, *Macromol. Chem. Phys.* **2006**, *207*, 1394–1403. (au) Q. Peng, M. Li, S. Lu, X. Tang, *Macromol. Rapid Commun.* **2007**, *28*, 785–791. (av) S. Rajaram, T.-L. Choi, M. Rolandi, J. M. J. Fréchet, *J. Am. Chem. Soc.* **2007**, *129*, 9619–9621. (aw) Y. Zhang, Z. Xu, X. Li, Y. Chen, *J. Polym. Sci., Part A: Polym. Chem.* **2007**, *45*, 3303–3310. (ax) Y. Zhang, Z. Xu, X. Li, Y. Chen, *J. Polym. Sci., Part A: Polym. Chem.* **2007**, *45*, 3994–4001. (ay) X. Xiong, Y. Chen, S. Feng, W. Wang, *Macromolecules* **2007**, *40*, 9084–9093. (az) Z. Yi, Y. Zhang, Y. Chen, F. Xi, *Macromol. Rapid Commun.* **2008**, *29*, 757–762. (aaa) Z. Yi, X. Liu, Q. Jiao, E. Chen, Y. Chen, F. Xi, *J. Polym. Sci., Part A: Polym. Chem.* **2008**, *46*, 4205–4217. For supramolecular dendronized polymers, see:. (aab) T. Tozawa, *Chem. Commun.* **2004**, 1904–1905. (aac) K. C.-F. Leung, P. M. Mendes, S. N. Magonov, B. H. Northrop, S. Kim, K. Patel, A. H. Flood, H.-R. Tseng, J. F. Stoddart, *J. Am. Chem. Soc.*, **2006**, *128*, 10707–10715. (aad) F. Wuerthner, V. Stepanenko, A. Sautter, *Angew. Chem., Int. Ed.* **2006**, *45*, 1939–1942. (aae) D. Onoshima, T. Imae, *Soft Matter* **2006**, *2*, 141–148. (aaf) D. Xie, M. Jiang, G. Zhang, D. Chen, *Chem. Eur. J.* **2007**, *13*, 3346–3353. (aag) H.-J. Kim, E.-Y. Jung, L. Y. Jin, M. Lee, *Macromolecules* **2008**, *41*, 6066–6072. (aah) P.-J. Yang, C.-W. Wu, D. Sahu, H.-C. Lin, *Macromolecules* **2008**, *41*, 9692–9703. (aai) Z. Cheng, B. Ren, H. Shan, X. Liu, Z. Tong, *Macromolecules* **2008**, *41*, 2656–2662. (aaj) X. Zhang, Y. Wang, W. Wang, S. Bolisetty, Y. Lu, M. Ballauff, *Langmuir* **2009**, *25*, 2075–2080. (aak) Z. Zhang, M. Yang, X. Zhang, L. Zhang, B. Liu, P. Zheng, W. Wang, *Chem. Eur. J.* **2009**, *15*, 2352–2361. For dendronized polymers as templates for nanoparticles formation, see:. (aal) Y. Zhang, Y. Chen, H. Niu, M. Gao, *Small* **2006**, *2*, 1314–1319. (aam) H. Ozawa, M. Kawao, H. Tanaka, T. Ogawa, *Langmuir*, **2007**, *23*, 6365–6371. For single molecular force microscopy, see:. (aan) W. Shi, Z. Wang, S. Cui, X. Zhang, Z. Bo, *Macromolecules* **2005**, *38*, 861–866. (aao) W. Shi, Y. Zhang, C. Liu, Z. Wang, X. Zhang, Y. Zhang, Y. Chen, *Polymer* **2006**, *47*, 2499–2504. For catalysis, see:. (aap) G.-J. Deng, B. Yi, Y.-Y. Huang, W.-J. Tang, Y.-M. He, Q.-H. Fan, *Adv. Synth. Catal.* **2004**, *346*, 1440–1444. (aaq) B. Helms, C. O. Liang, C. J. Hawker, J. M. J. Fréchet, *Macromolecules* **2005**, *38*, 5411–5415. (aar) T. Kehat, M. Portnoy, *Chem. Commun.* **2007**, 2823–2825. (aas) A. Dahan, M. Portnoy, *J. Am. Chem. Soc.* **2007**, *129*, 5860–5869. (aat) A. Mansour, T. Kehat, M. Portnoy, *Org. Biomol. Chem.* **2008**, *6*, 3382–3387. For solution and bulk assembly, see for example:. (aau) C. Park, K. S. Choi, Y. Song, H.-J. Jeon, H. H. Song, J. Y. Chang, C. Kim, *Langmuir* **2006**, *22*, 3812–3817. (aav) N. Canilho, E. Kasemi, R. Mezzenga, A. D. Schlüter, *J. Am. Chem. Soc.* **2006**, *128*, 13998–13999. (aaw) N. Canilho, E. Kasemi, A. D. Schlüter, J. Ruokolainen, R. Mezzenga, *Macromolecules* **2007**, *40*, 7609–7616. (aax) K.-N. Lau, H.-F. Chow, M.-C. Chan, K.-W. Wong, *Angew. Chem., Int. Ed.* **2008**, *47*, 6912–6916. (aay) S. Feng, X. Xiong, G. Zhang, N. Xia, Y. Chen, W. Wang, *Macromolecules* **2009**, *42*, 281–287. For biorelated applications, see:. (aaz) C. C. Lee, M. Yoshida, J. M. J. Fréchet, E. E. Dy, F. C. Szoka, *Bioconjugate Chem.* **2005**, *16*, 535–541. (aaaa) C. C. Lee, S. M. Grayson, J. M. J. Fréchet, *J. Polym. Sci., Part A: Polym. Chem.* **2004**, *42*, 3563–3578. For optical-electro applications, see:. (aaab) C.-W. Wu, C.-M. Tsai, H.-C. Lin, *Macromolecules* **2006**, *39*, 4298–4305. (aaac) B. Li, Y. Fu, Y. Han, Z. Bo,

Macromol. Rapid Commun. **2006**, *27*, 1355–1361. (aaad) B. Zhu, Y. Han, M. Sun, Z. Bo, *Macromolecules* **2007**, *40*, 4494–4500. (aaae) Y.-S. Wu, J. Li, X.-C. Ai, L.-M. Fu, J.-P. Zhang, Y.-Q. Fu, J.-J. Zhou, L. Li, Z.-S. Bo, *J. Phys. Chem. A* **2007**, *111*, 11473–11479. (aaaf) X. Xiao, Y. Wu, M. Sun, J. Zhou, Z. Bo, L. Li, C. Chan, *J. Polym. Sci., Part A: Polym. Chem.* **2008**, *46*, 574–584. (aaag) Z. Fei, Y. Han, Z. Bo, *J. Polym. Sci., Part A: Polym. Chem.* **2008**, *46*, 4030–4037.

[5] W. Li, D. Wu, A. D. Schlüter, A. Zhang, *J. Polym. Sci., Part A: Polym. Chem.* **2009**, *41*, 6630–6640.

[6] There are additional factors impacting the growth of dendronized polymers. Unfortunately not much is known about the molecular scale growth mechanism with all its thermodynamic and kinetic features. For a study of the polymerization of pre-organized mesogenic macromonomers, see: V. Percec, C. H. Ahn, B. Barboiu, *J. Am. Chem. Soc.* **1997**, *119*, 12978–12979.

[7] We are not aware of systematic studies on this matter in the field of bottle brush polymers. The required low concentrations of initiator tend to render the polymerizations somewhat irreproducible. For a publication in which this trend can nevertheless be seen for macromonomers with methylmethacrylate polymerizable unit and linear polystyrene side chains with molar masses from 800–5500 g/mol, see: M. Wintermantel, M. Gerle, K. Fischer, M. Schmidt, I. Wataoka, H. Urakawa, K. Kajiwara, Y. Tsukahara, *Macromolecules* **1996**, *29*, 978–983. For an oligomerization of a linear macromonomer with a side chain of 12000 g/mol, see: M. Gehrle, Diploma Thesis, University of Mainz, Germany, **1995**.

[8] (a) D. A. Tomalia, P. M. Kirchhoff, Dow Chemical, Pat. USA 4,694,064, **1987**; (b) R. Yin, Y. Zhu, D. A. Tomalia, H. Ibuki, *J. Am. Chem. Soc.* **1998**, *120*, 2678–2679.

[9] For example: B. Karakaya, W. Claussen, K. Gessler, W. Saenger, A. D. Schlüter, *J. Am. Chem. Soc.* **1997**, *119*, 3296–3301;A. Desai, N. Atkinson, F. Rivera, Jr., W. Devonport, I. Rees, S. E. Branz, C. J. Hawker, *J. Polym. Sci., Part A: Polym. Chem.* **2000**, *38*, 1033–1044.

[10] (a) For recent examples, see: L. Shu, A. Schäfer, A. D. Schlüter, *Macromolecules* **2000**, *33*, 4321–4328; (b) S. M. Grayson, J. M. J. Fréchet, *Macromolecules* **2001**, *34*, 6542–6544; (c) A. Zhang, S. Vetter, A. D. Schlüter, *Macromol. Chem. Phys.* **2001**, *202*, 3301–3315; (d) L. Shu, I. Gössl, J. P. Rabe, A. D. Schlüter, *Macromol. Chem. Phys.* **2002**, *203*, 2540–2550; (e) B. Helms, J. L. Mynar, C. J. Hawker, J. M. J. Fréchet, *J. Am. Chem. Soc.* **2004**, *126*, 15020–15021; (f) M. Yoshida, Z. M. Fresco, S. Ohnishi, J. M. J. Fréchet, *Macromolecules* **2005**, *38*, 334–344; (g) R. Al-Hellani, A. D. Schlüter, *Helv. Chim. Acta* **2006**, *89*, 2745–2763; (h) R. Al-Hellani, J. Barner, J. P. Rabe, A. D. Schlüter, *Chem. Eur. J.* **2006**, *12*, 6542–6551; (i) C. C. Lee, J. M. J. Fréchet, *Macromolecules* **2006**, *39*, 476–481; (j) S. R. Benhabbour, M. C. Parrott, S. E. A. Gratton, A. Adronov, *Macromolecules* **2007**, *40*, 5678–5688; (k) Z.-T. Liu, Y.-M. He, B.-L. Li, J. Liu, Q.-H. Fan, *Macromol. Rapid Commun.* **2007**, *28*, 2249–2255.

[11] C. Böttcher, B. Schade, C. Ecker, J. P. Rabe, L. Shu, A. D. Schlüter, *Chem. Eur. J.* **2005**, *11*, 2923–2928.

[12] W. Zhuang, E. Kasemi, Y. Ding, M. Kröger, A. D. Schlüter, J. P. Rabe, *Adv. Mater.* **2008**, *20*, 3204–3210.

[13] Y. Ding, H. C. Öttinger, A. D. Schlüter, M. Kröger, *J. Chem. Phys.* **2007**, *127*, 094904. It should be noted that bottlebrushes do not seem to show such effects. It is the dendritic

compactness of the structure of dendronized polymers which provides sufficient hydrophobic interactions upon back-folding.
[14] Y. Guo, J. D. van Beek, B. Zhang, M. Colussi, P. Walde, A. Zhang, M. Kröger, A. Halperin, A. D. Schlüter, *J. Am. Chem. Soc.* **2009**, *131*, 11841–11854.
[15] N. Canilho, E. Kasemi, R. Mezzenga, A. D. Schlüter, *J. Am. Chem. Soc.* **2006**, *128*, 13998–13999.
[16] For example, see: L. Shu, A. D. Schlüter, C. Ecker, N. Severin, J. P. Rabe, *Angew. Chem. Int. Ed.* **2001**, *40*, 4666–4669.
[17] L. Shu, I. Gössl, J. P. Rabe, A. D. Schlüter, *Macromol. Chem. Phys.* **2002**, *203*, 2540–2550.
[18] J. T. Lai, D. Filla, R. Shea, *Macromolecules* **2002**, *35*, 6754–6756.
[19] B. Zhang, R. Wepf, M. Kröger, A. Halperin, A. D. Schlüter, in preparation.
[20] P. J. Flory, *Principle of Polymer Chemistry*, Cornell University, **1953**.
[21] H. G. Schild, *Prog. Polym. Sci.* **1992**, *17*, 163–249.
[22] (a) Y. Z. You, C. Y. Hong, C. Y. Pan, P. H. Wang, *Adv. Mater.* **2004**, *16*, 1953–1957. (b) Z. Jia, H. Chen, X. Zhu, D. Yan, *J. Am. Chem. Soc.* **2006**, *128*, 8144–8145. (c) Y. J. Jun, U. S. Toti, H. Y. Kim, J. Y. Yu, B. Jeong, M. J. Jun, Y. S. Sohn, *Angew. Chem., Int. Ed.* **2006**, *45*, 6173–6176. (d) K. Van Durme, G. Van Assche, V. Aseyev, J. Raula, H. Tenhu, B. Van Mele, *Macromolecules* **2007**, *40*, 3765–3772.
[23] (a) W. Li, A. Zhang, K. Feldman, P. Walde, A. D. Schlüter, *Macromolecules* **2008**, *41*, 3659–3667. (b) W. Li, A. Zhang, A. D. Schlüter, *Chem. Commun.* **2008**, 5523–5525.
[24] V. N. Viswanadhan, A. K. Ghose, G. R. Revankar, R. K. Robins, *J. Chem. Inf. Comput. Sci.* **1989**, *29*, 163–172.
[25] M. J. N. Junk, U. Jonas, D. Hinderberger, *Small* **2008**, *4*, 1485–1493.
[26] M. J. N. Junk, W. Li, A. D. Schlüter, G. Wegner, H. W. Spiess, A. Zhang, D. Hinderberger, *Angew. Chem. Int. Ed.* **2010**, *212*, 5818–5823.
[27] S. Bolisetty, C. Schneider, F. Polzer, M. Ballauff, W. Li, A. Zhang, A. D. Schlüter, *Macromolecules* **2009**, *42*, 7122–7128.
[28] J. H. Chang, C. H. Shim, B. J. Kim, Y. Shin, G. J. Exarhos, K. Kim, *J. Adv. Mater.* **2005**, *17*, 634–637.
[29] (a) S. Han, M. Hagiwara, T. Ishizone, *Macromolecules* **2003**, *36*, 8312–8319. (b) F. Hua, X. Jiang, D. Li, B. Zhao, *J. Polym. Sci., Part A: Polym. Chem.* **2006**, *44*, 2454–2467. (c) J.-F. Lutz, O. Akdemir, A. Hoth, *J. Am. Chem. Soc.* **2006**, *128*, 13046–13047.
[30] C. Cheng, M. Schmidt, A. Zhang, A. D. Schlüter, *Macromolecules* **2007**, *40*, 220–227.
[31] C. Li, A. D. Schlüter, A. Zhang, R. Mezzenga, *Adv. Mater.* **2008**, *20*, 4530–4534.
[32] C. Cheng, A. D. Schlüter, A. Zhang, *Israel J. Chem.* **2009**, *49*, 49–53.
[33] K. Yu, L. F. Zhang, A. Eisenberg, *Langmuir* **1996**, *12*, 5980–5984.
[34] A. Ossenbach, H. Rüegger, A. Zhang, K. Fischer, A. D. Schlüter, M. Schmidt, *Macromolecules* **2009**, *42*, 8781–8793.
[35] T. Ikeda, O. Tsutsmi, *Science* **1995**, *268*, 1873–1875.
[36] T. Nakano, Y. Okamoto, *Chem. Rev.* **2001**, *101*, 4013–4038.
[37] J. G. Rudick, V. Percec, *Acc. Chem. Res.* **2008**, *41*, 1641–1652.
[38] J. J. L. M. Cornelissen, A. E. Rowan, R. J. M. Nolte, N. A. J. M. Sommerdijk, *Chem. Rev.* **2001**, *101*, 4039–4070.

[39] E. Yashima, K. Maeda, *Macromolecules* **2008**, *41*, 3–12.

[40] A. Zhang, A. D. Schlüter, *Chem. Asian J.* **2007**, *2*, 1540–1548.

[41] A. Zhang, F. Rodríguez-Ropero, D. Zanuy, C. Alemán, E. W. Meijer, A. D. Schlüter, *Chem. Eur. J.* **2008**, *14*, 6924–6934.

[42] S.-W. Hla, L. Bartels, G. Meyer, K.-H. Rieder, *Phys. Rev. Lett.* **2000**, *85*, 2777–2780.

[43] L. J. Lauhon, W. Ho, *Faraday Discuss.* **2000**, *117*, 249–255.

[44] (a) Y. Okawa, M. Aono, *Nature* **2001**, *409*, 683. (b) A. Miura, S. De Fyter, M. M. S. Abdel-Mottaleb, A. Gesquire, P. C. M. Grim, G. Moessner, M. Sieffert, M. Klapper, K. Müllen, F. C. De Schryver, *Langmuir* **2003**, *19*, 6474–6482.

[45] (a) M. M. S. Abdel-Mottaleb, S. de Feyter, A. Gesquière, M. Sieffert, M. Klapper, K. Müllen, F. C. De Schryver, *Nano Lett.* **2001**, *1*, 353–359. (b) H. Sakaguchi, H. Matsumura, H. Gong, A. M. Abouelwafa, *Science* **2005**, *310*, 1002–1006. (c) L. Grill, M. Dyer, L. Lafferentz, M. Persson, M. V. Peters, S. Hecht, *Nat. Nanotechnol.* **2007**, *2*, 687–691.

[46] N. Severin, J. Barner, A. Kalachev, J. P. Rabe, *Nano Lett.* **2004**, *4*, 577–579.

[47] J. Barner, F. Mallwitz, L. Shu, A. D. Schlüter, J. P. Rabe, *Angew. Chem. Int. Ed.* **2003**, *42*, 1932–1935.

[48] R. Al-Hellani, J. Barner, J. P. Rabe, A. D. Schlüter, *Chem. Eur. J.* **2006**, *12*, 6542–6551.

[49] J. Barner, R. Al-Hellani, A. D. Schlüter, J. P. Rabe, *Macromol. Rapid Commun.* **2009**, *31*, 362–367.

[50] A. K. Geim, K. S. Novoselov, *Nat. Mater.* **2007**, *6*, 183–191.

5

SHAPE PERSISTENT POLYPHENYLENE-BASED DENDRIMERS

MARTIN BAUMGARTEN, TIANSHI QIN, AND KLAUS MÜLLEN
Max-Planck-Institut für Polymerforschung, Ackermannweg 10, D-55128 Mainz, Germany

5.1 INTRODUCTION

Functional organic and inorganic particles are an essential element of today's materials science [1–3]. Properties such as the surface-to-volume ratio of polymer lattices or the color control of quantum dots as a result of quantum confinement provide convincing evidence that "size matters." There are various ways to fabricate functional nanoparticles whereby the controls of size, shape, and the chemical functionalization, both, inside and outside the particles are the key issues. A special challenge is the design and synthesis of core–shell systems since this approach allows for the combination of different behaviors such as charge function and charge carrier mobility [4–6]. Dendrimers play an outstanding role among organic functional nanoparticles because they can be made molecularly defined [7–12], and polyphenylene dendrimers (PPDs) are a very special family of dendrimers. For instance, the PPDs offer the highest possible rigidity among all types of dendrimers and possess high thermal stability, solubility, shape persistence, and structural perfectness [13,14]. Based on the synthetic protocols of the dendritic generations, a large variation of these molecules has been made available as a result of using different cores, branching units, and outer phenylene shells (endcapping) [15–18].

Besides additional functional groups to enable further reactions even chromophores and electrophores have been included at will in the cores, the scaffold, and on

Designing Dendrimers, Edited by Sebastiano Campagna, Paola Ceroni, Fausto Puntoriero.
© 2012 John Wiley & Sons, Inc. Published 2012 by John Wiley & Sons, Inc.

the surface leading to a number of new properties, that also give rise to new applications in materials science and biology-related applications. The fascination of polyphenylene dendrimers is thus not only based on their structure and dynamics but also on their synthetic variability.

The three-dimensional (3-D) structures of dendrimers have gained tremendous interest, since the early work of de Gennes and Hervet [19], who developed a model of increasing density at the surface, caused by the exponential growth of the dendrimers with the number of generations. However, such ideal structures with functional groups remaining on the surface when introduced on the last sphere could not be found at the desired location in flexible dendritic structures such as Fréchet-type polyaryl-ether dendrimers [20]. Surprisingly, in such flexible cases end groups have been found throughout the entire dendrimer volume, leading to homogenous density distribution [9]. This raised tremendous interest in stiffening the dendrimer generations, in order to prevent backfolding to the interior.

Prior to the introduction of a new concept of polyphenylene dendrimer synthesis, which allowed easy extension to higher generations in a divergent method, only a few examples of relatively stable shape persistent dendrimers are known in the literature. Hart introduced nanometer-sized dendrimers in which benzene units were bound to each other via two σ bonds (**1**, Scheme 5.1) [21]. These dendrimers, based on extended iptycenes, although extremely stiff and shape persistent, did not allow any rotational movement, but could not be applied for attachment of additional functional groups at desired positions by demand. Another example of rigid and shape-persistent dendrimers based on branched triangulenes, such as **2**, was published by DeMeijere and coworkers [22], but further applications were hampered since these dendrimers neither could be extended to higher generations nor could other functional groups be introduced. Moore et al. [23] presented dendrimers such as **3** constructed from phenylacetylene units (Scheme 5.1). Miller and coworkers [24] contributed to some polyphenylene dendrimers as **4**. Both these kinds of dendrimers **3** and **4** were synthesized by the convergent method, because the metal catalyzed coupling reactions led to side products and incomplete reactions rendering a divergent synthesis not practicable. Therefore, a new strategy based on noncatalytic Diels–Alder cycloaddition reaction was developed [25], allowing the divergent growth and synthesis of higher generations where one example **5** is depicted in Scheme 5.1. Furthermore, it is immediately clear from the structures in Scheme 5.1, that many more phenylene units can be incorporated in **5** compared to **3** and **4**. Additionally, it turned out that dendrimers **3** and **4**, based on 1,3,5-substituted benzene rings, led to conformational isomers.

In general, the synthesis of dendrimers is performed by two routes either (i) the expanding divergent synthesis with stepwise growth of generations starting from a core or (ii) the convergent synthesis of large dendrons finally coupled to a core [13]. The latter approach is, however, often limited to lower generations of dendrimers such as G2 and G3 caused by the steric hindrance of accessing the core properly. For **5**, it was shown nicely that both approaches toward G2 yielded the same products in slightly different yields, but due to the many phenylene units and steric hindrance it was not possible to go beyond the second generation via the convergent

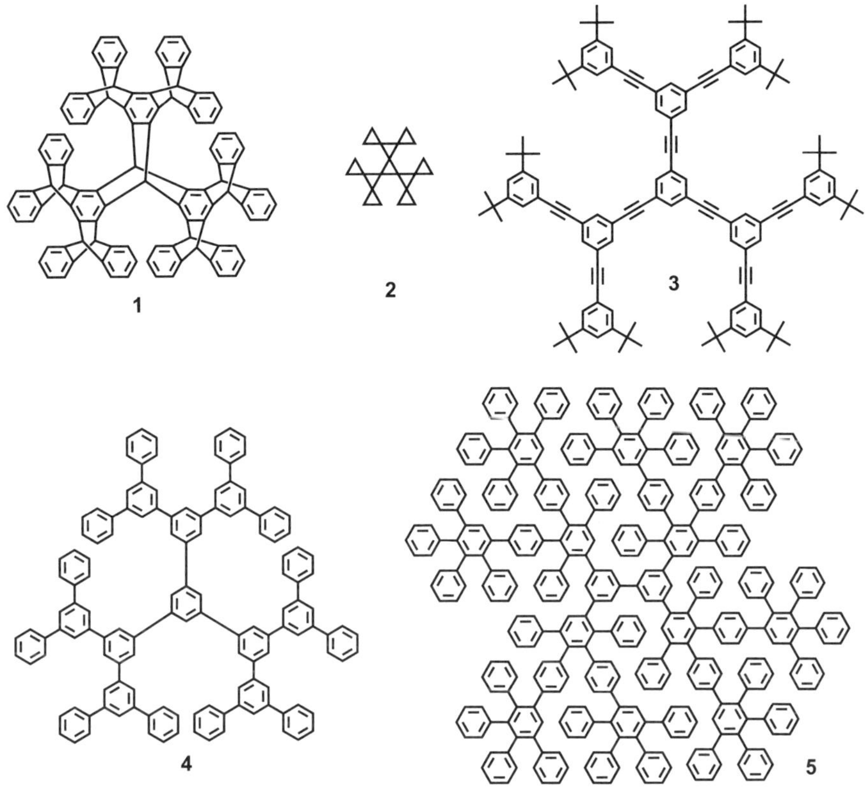

SCHEME 5.1 Different stiff and rigid dendrimers **1–5**.

pathway. For other dendrimers with extended cores or more flexible dendrons, a combination of both approaches could be applied to achieve faster growth of higher generation dendrimers, for example, G2–G4–G6 with a second-generation end functionalized dendron.

Since some reviews on functional polyphenylene dendrimers have been published [13–17,26], the aim here is to focus on more recent achievements, while briefly introducing the whole concept to begin with.

5.2 SYNTHESIS

The synthesis of polyphenylene dendrimers is mainly based on two reactions with nearly no side products and close to quantitative yields as the (i) growth step via Diels–Alder cycloaddition between cyclopentadienones and ethynylenes and the (ii) deprotection step via desilylation of triisopropyl-substituted alkynes [13,25]. The functionalized tetraphenyl-cyclopentadienones **6**, **7** containing two (A_2B) or four triisopropylsilylethenyl (A_4B) groups were chosen as building blocks for the

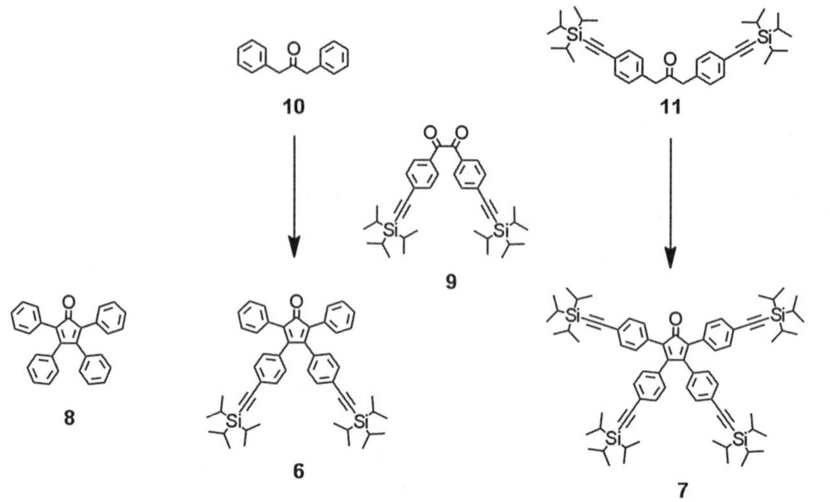

SCHEME 5.2 Cyclopentadienones **6** and **7** as A_2B- and A_4B-building blocks, respectively, together with their synthesis from benzil **9** and acetone derivatives **10** and **11**, and a nonfunctionalized endcapping unit **8**.

enhancement of generations after deprotection and a standard tetraphenylcyclopentadienone (**8**) unit for endcapping of the generations. The cyclopentadienones (CP)s building blocks are typically available in high yields via double Knoevenagel condensation of the bis(triisopropylsilylethynyl)-benzil (**9**) with either 1,3-diphenyl-2-propanone (**10**) or bis(triisopropylsilylethynylbenzene)-2-propanone (**11**) (Scheme 5.2). The bulky triisopropylsilyl (TIPS) substituents were selected as protecting groups after having proved that they prevent the self-cycloaddition, which would have resulted in hyperbranched polymers [27].

The above-mentioned cyclopentadienones **6–8** were then applied to Diels–Alder cycloaddition reaction with various cores carrying multiple ethynyl subunits **12–15** (Scheme 5.3).

The additional labeling of the cores (**B** for 1,3,5-trisethynylbenzene, **Biph** for 3,3′,5,5′-tetraethynylbiphenyl, **TPM** for tetrakis(4-ethynylphenyl)methane, and

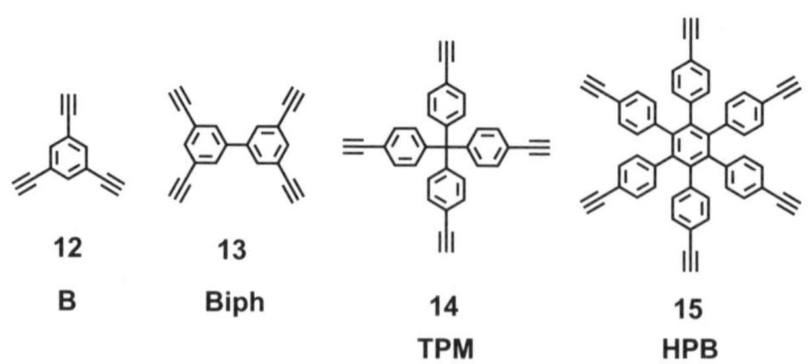

SCHEME 5.3 Different ethynyl-carrying cores **12–15** and the abbreviations used for the corresponding dendrimers (G1–G5) based on this core.

HPB for 1,2,3,4,5,6-hexakis(4-ethynylphenyl)benzene) was necessary to, later on, distinguish the polyphenylene dendrimer generations (G1–G7) on different cores. The typical divergent reaction sequence is shown for 1,3,5-triethynylbenzene **12** yielding the first-generation dendrimer **B-G1** (**16**) upon coupling with **8** or the protected triisopropylethynyl functionalized dendrimer **17** upon reaction with **6**. All TIPS groups could be removed with fluoride anions yielding the first-generation functionalized dendrimer **18**, which was then endcapped with **8** resulting in a second-generation dendrimer (B-G2) **19** (Scheme 5.4). Alternatively **18** could be further extended with **6** for growth reactions via **20** up to the fourth generation.

In addition the convergent route has been shown to be practicable for achieving these polyphenylene dendrimers by extending the cyclopentadienones to the next higher generation dendrons. However, to prevent self-coupling, this approach started from the 4,4′-diethynylbenzil **21** that became available after deprotection of **9** and contained the two dienophile groups for Diels–Alder reaction with **8** yielding the dendronized benzil **22** for Knoevenagel condensation. Thereby the second-generation dendron **23** was available for coupling with one of the aforementioned cores **12–15** (Scheme 5.5).

The divergent and the convergent approach resulted in the same monodisperse products in comparable yields. The convergent approach provided a fast access to second-generation dendrimers, but could not be used to synthesize higher generation PPDs with the relatively small cores considered here so far. A functionalized second-generation dendron is an extremely powerful tool for desymmetrization, as it is shown later [28–30].

Also the other cores **13–15** were transformed into the first-, second-, and higher-generation dendrimers, in order to study their influence on the shape and rigidity of the resulting dendrimers. Luckily, for three polyphenylene dendrimers of the first generation with different cores, single crystals could also be grown large enough for X-ray structural investigation [31] (Figure 5.1).

For the first-generation dendrimers geometry optimizations, even on a semi-empirical level, were well able to reproduce the shape and distances compared to their X-ray structure. The calculated distances (radii) derived from semiempirical AM1 calculations (which are well-suited for analyzing large and twisted π-systems) between the center of the core and the outside protons, for instance were within 0.02 nm, and only slightly worse for the molecular mechanics calculations.

As nicely seen in Figure 5.2, the variation of the cores causes the dendritic branches to grow in certain directions as a dumbbell shape for **13** based **Biph-G2** and more flat globular shape for **14** based **TPM-G2**.

The shape persistence of the newly derived polyphenylene dendrimers has further been investigated for different generations by various techniques. Advanced solid-state nuclear magnetic resonance (NMR) experiments with rapid magic angle spinning were applied for gaining more insight into the dynamics of these molecules. Fast processes in the megahertz regime were shown to be restricted to fast vibrations of terminal phenyl rings, while the phenyl rings in the core and scaffolds displayed very limited dynamics with restricted movements [32,33] Comparing **Biph-G2** and

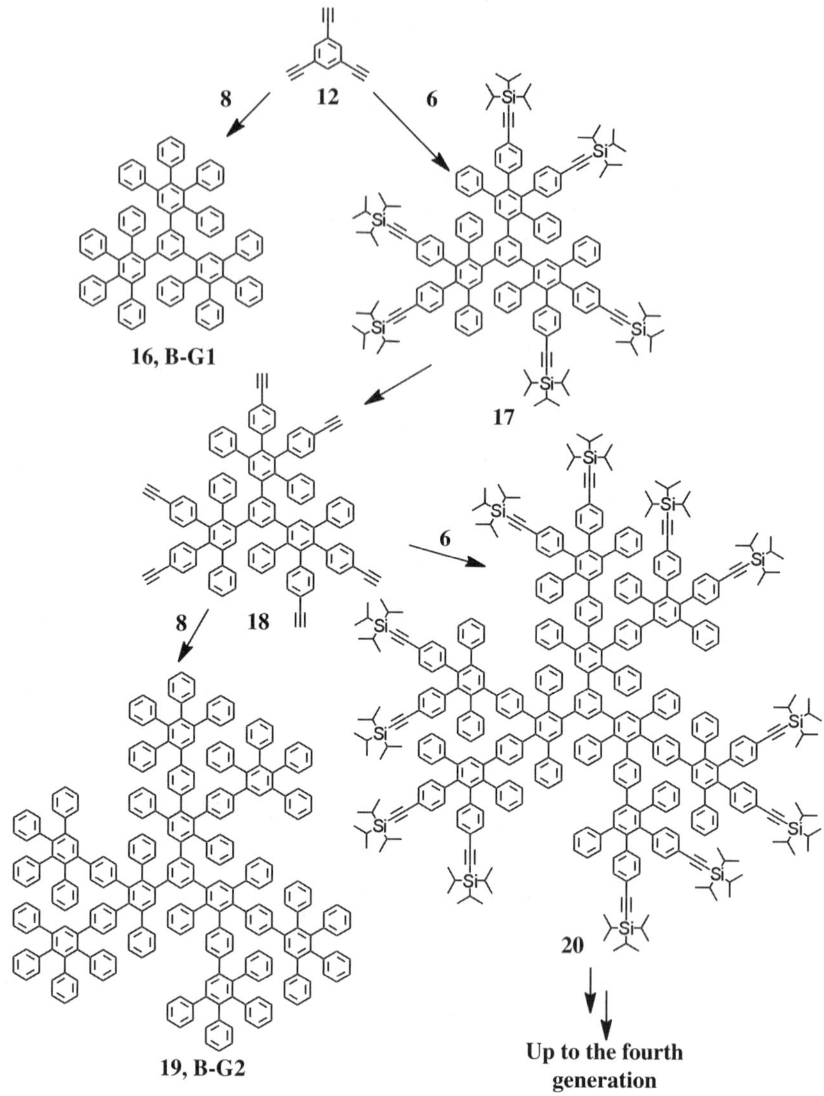

SCHEME 5.4 Divergent synthesis of polyphenylene dendrimers starting from the 1,3,5-trisethynylbenzene core **12** using cyclopentadienones **6** or **8** for endcapping and growth, respectively.

TPM-G2 by these NMR experiments, it turned out that the biphenyl core causes a slightly higher flexibility, since an additional freedom exists in the central aryl–aryl bond, while the tetrahedral core in **14** is more rigid and directs the branches in tetrahedral alignment toward a more space-filling object. Thus, the choice of the central cores turned out to have a major influence on the overall shape, rigidity, and

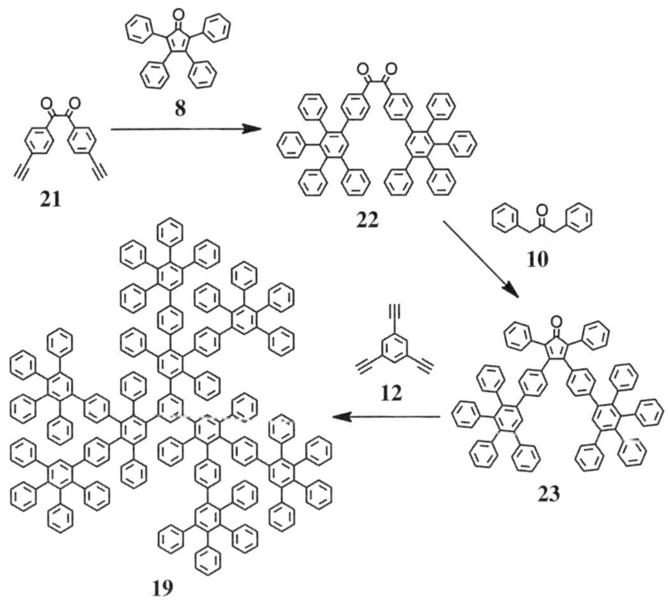

SCHEME 5.5 Convergent synthesis of second-generation dendrimer **19** (**B-G2**).

space filling of these dendrimers. It was found that **19** (**B-G2**) adopts a false propeller structure while **HPB-G2** a true propeller shape.

Furthermore, these NMR studies have proven only slow angular motion for the terminal groups, while the motion of whole dendrons was excluded even at higher

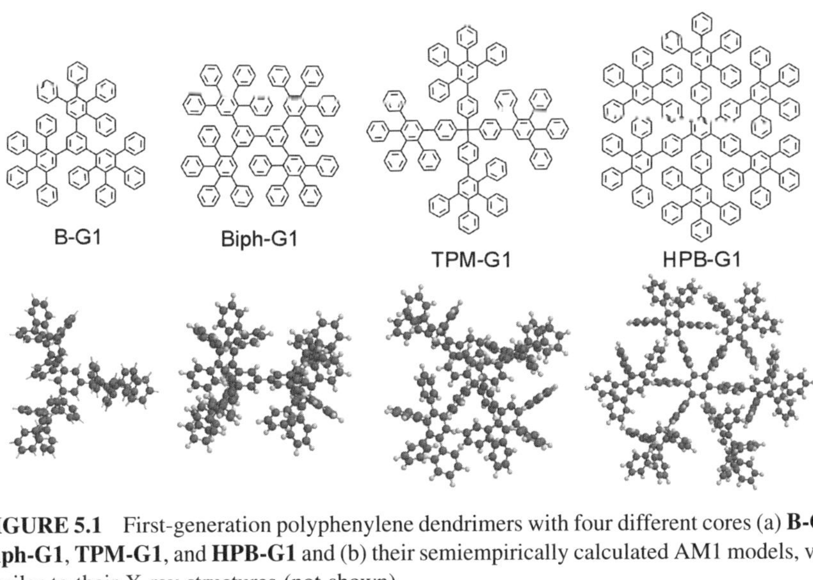

FIGURE 5.1 First-generation polyphenylene dendrimers with four different cores (a) **B-G1**, **Biph-G1**, **TPM-G1**, and **HPB-G1** and (b) their semiempirically calculated AM1 models, very similar to their X-ray structures (not shown).

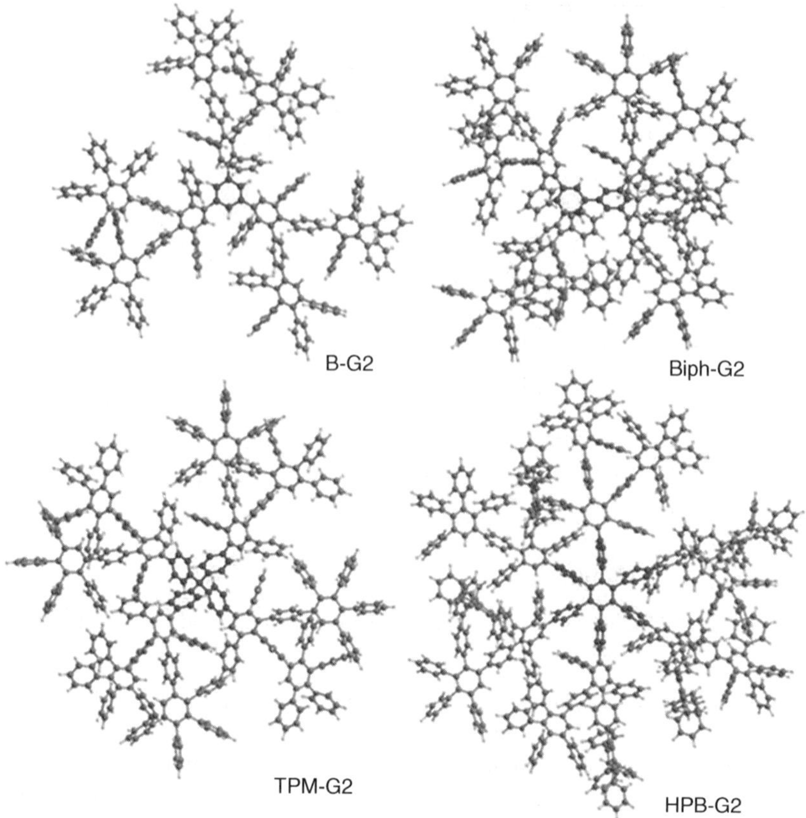

FIGURE 5.2 Second-generation dendrimers **B-G2**, **Biph-G2**, **TPM-G2**, and **HPB-G2** as 3-D model.

temperatures [32]. However, the larger flexibility was obtained with increasing dendrimer size, which is considered to be in agreement with increasing solubility. Atomic force microscopy (AFM) experiments of such dendritic molecules absorbed on mica gave similar height values as those from the modeling-derived radial dimensions [34]. Small-angle neutron scattering (SANS) techniques were used to determine the structure of dendrimers in solution [35]. For polyphenylene dendrimers, the segments of different generations were found to be in a well-defined distance to the center of the dendrimer, proving the stiff and shape persistent nature of the backbone [36,37].

Molecular modeling demonstrated some possibility of collapsing dendrons for the second-generation B-G2 at higher temperatures, while at lower temperatures the rigidity prevailed [37]. The dendrimers built up from the tetraphenylmethane core, however, possessed much higher shape persistence than those built from 1,3,5-trisubstituted benzene or a biphenyl core.

The density of the dendrimers can be increased even further upon using the A_4B branching unit **7** doubling the number of functionalized end groups. Such extremely

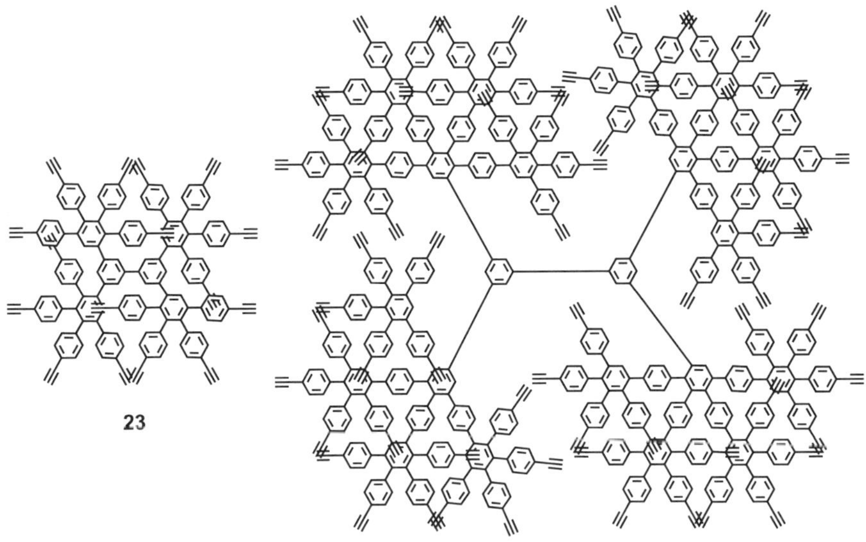

23

24

SCHEME 5.6 The extension of core **13** with A_4B CP unit **7** to dendrimer **23** (G1) and **24** (G2).

dense dendrimers (Scheme 5.6), however, were not extended beyond G2, even for the highly efficient divergent growth pathway.

5.3 HIGHER GENERATIONS TPM-G4–5 AND EXTENDED ARMS TPM-leG1–7

In the case of polyphenylene dendrimers with a tetrahedral core, so far it was shown that divergent synthetic growth led to generations up to G4. The TPM-G4 thereby yielded a nanoparticle with a radius of about 6.3 nm, where 304 benzene rings were included resulting in a molecular weight of 23,150 g/mol (Figure 5.1) [25]. Here the dense shell seemed to hinder further well-defined extension to higher generations as G5 and G6. However, recently we were also successful in synthesizing the fifth generation (TPM-G5), which demanded very special care and extended reaction times. Then the molecular weight already reached 47,501 g/mol and included 624 benzene rings. It turned out that matrix-assisted laser desorption ionization—time of flight (MALDI-TOF) mass spectrometry was probably the most efficient and powerful method for proving the structural perfection of PPDs with high molecular weights (Figure 5.3). In addition, standard NMR spectroscopy proved to be extremely valuable for characterizing the single proton left on each branching unit upon dendrimer growth [38].

In order to go beyond the limits of the hitherto described polyphenylene dendrimer generations, more elongated CP units were introduced since linear enlargement of the arms of the general dendritic repeat unit should directly translate into increased

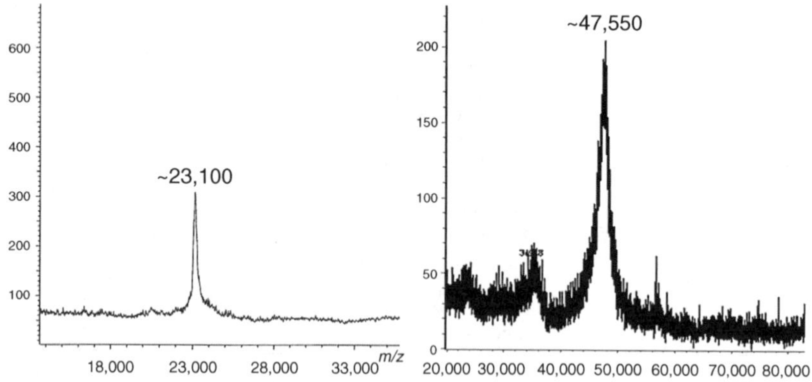

FIGURE 5.3 MALDI-TOF mass spectra of TPM-G4 and TPM-G5 proving the monodisperse nature.

dendrimer radii leading to decreased congestions at the chain ends without requiring a larger core to start from.

This could be achieved by coupling an end protected bromo-terphenyl **25** to the bis-dioxoborolane derivative **26** achieving the terphenyl extended CP unit **27** for the Diels–Alder induced growth reaction of the dendrimers (Scheme 5.7) [39]. Two methyl groups were introduced as side chains from the beginning in order to increase the solubility of the building blocks and to induce torsion between the phenylenes preventing them from π-stacking.

These terphenyl extended CP units **27** enabled the stepwise synthesis of well-defined fifth to seventh-generation dendrimers with amazing molecular weights of 135 kg/mol, 271 kg/mol, and even 542 kg/mol, respectively. Thus, huge nanoobjects as structurally perfect particles with radii between 10 and 16 nm were obtained (Table 5.1, Figure 5.4), which exceeded by far the molecular weight of standard conjugated polymers and even the dimensions of nanoparticles synthesized by self-assembly or multiphase approaches. Such nanoparticles can at best be prepared narrowly disperse but never monodisperse [2,40,41].

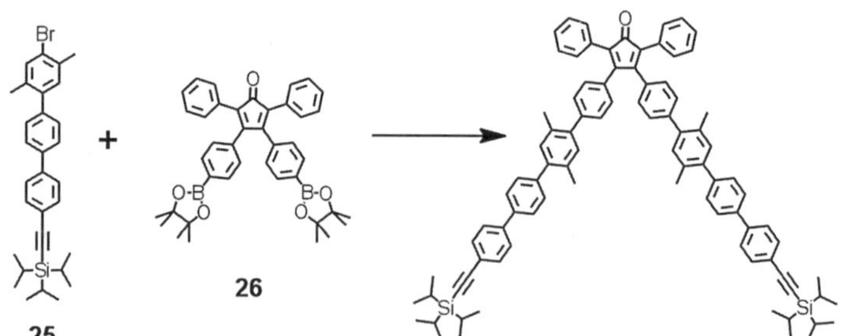

SCHEME 5.7 Linear extended cyclopentadienone (leCP) branching unit **27** from Suzuki coupling of terphenyl **25** with diboronicester derivative of CP **26**.

TABLE 5.1 Comparison of Radii of Endcapped TPM-G and TPM-leG from the Center to the Outer Hydrogen from Geometry Optimizations by Semiempirical AM1 Calculations

Dendrimers TPM-G	Radius (nm)	Dendrimers TPM-leG	Radius (nm)
G1	1.40	leG1	2.1
G2	2.20	leG2	4.7
G3	3.15	leG3	6.9
G4	3.99	leG4	9.1
G5	4.85	leG5	11.2
		leG6	13.4
		leG7	15.5

Some additional shoulders occurred in the size exclusion chromatography (SEC) as demonstrated in Figure 5.5, which indicated even higher molecular weight fractions, and which were been further examined. It was concluded that they must stem from interlocked dimers, covalently linked moieties, or higher aggregates. These intermolecular connected molecules could be assumed by some possible cross-linking of the reactive triple bonds upon intra- and intermolecular couplings. The higher aggregates, nonetheless, also occurred even after repeated synthesis under higher dilutions, such that entanglements of side arms before the growth reaction could also be responsible. Light scattering experiments of the partially separated higher molecular weight fractions provided further proof of their dimer nature [42,43].

Concerning the shape persistence of those extended PPDs, these results confirmed that the flexibility of the scaffold is strongly enhanced with increasing length of the branching units. The fact that the four dendrons clearly reached and finally exceeded the persistence length of a polyphenylene chain supported the idea of their loss of rigidity. The applied atomic force microscopy to the dendrimers, **TPM-leG5** and **TPM-leG6**, demonstrated their partial flattening on surfaces. Two effects responsible for this flattening were discussed and have to be considered (i) deformation from strict linearity with increasing number of phenylene units and (ii) some collapse against the surface used for AFM, which does not exist in solution. Both factors seemed to contribute to such flattening effects on surfaces and in addition it should be considered that size extension beyond the persistence length of conjugated dendrons finally break the rigidity of any conjugated branch of a dendrimer or a conjugated polymer. Nevertheless, it should be highlighted in this context that discrete monodisperse macromolecules became available exceeding molecular weights of 500 kg/mol still being soluble in common organic solvents such as THF and chloroform.

For the introduction of further functionalities, however, the standard PPDs of generations 2–4 as **19** or **20** are perfectly suited, since they prevent attached or incorporated functional units from aggregation. This is important since aggregation causes a red shift of absorption with strong quenching of the fluorescence and excimer formation.

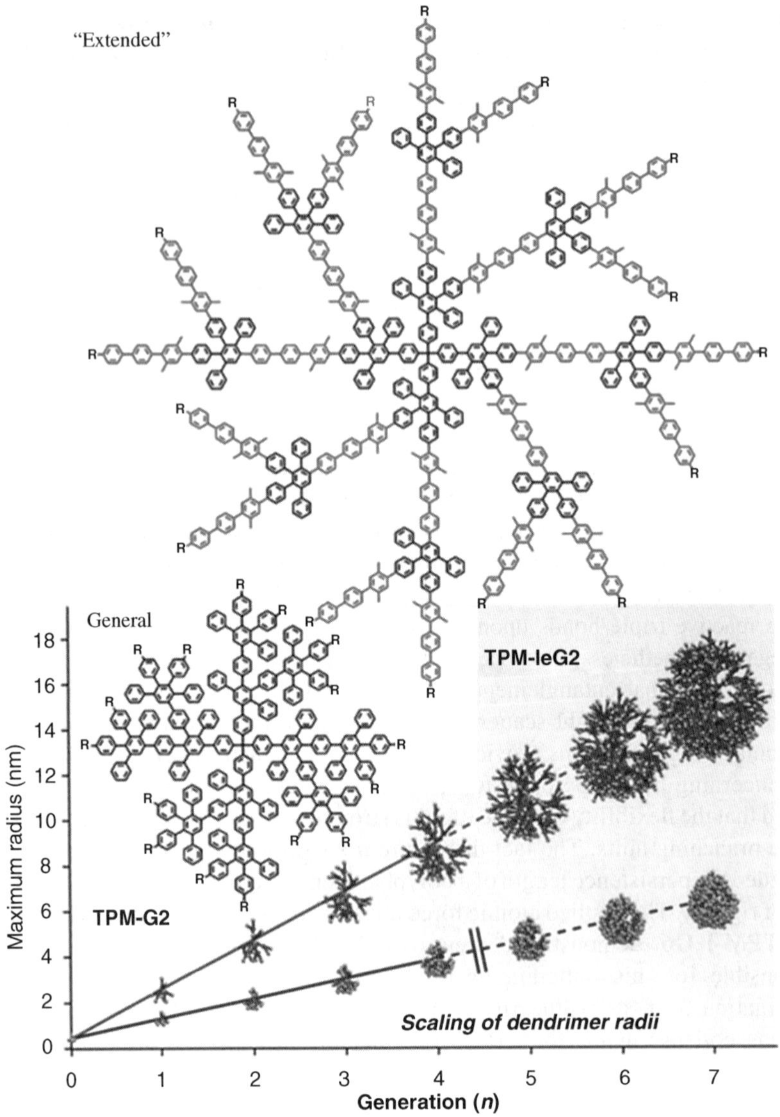

FIGURE 5.4 The generation-induced changes of the maximum radius of *standard* polyphenylenes TPM-G1–5 and those with linear triphenylene extended dendrons TPM-leG1–7. (See the color version of this figure in Color Plates section.)

Before discussing additional functions, it should be pointed out here that some of the dendrimer structures have been applied for fusion of their phenyl rings via dehydrogenation into new grapheme molecules as semiconductors [44,45]. For the oxidative cyclodehydrogenation via iron(III)chloride especially those dendrimers with fully aromatic cores as **B-G1** and **HPB-G1** (Figure 5.1) were perfectly suited,

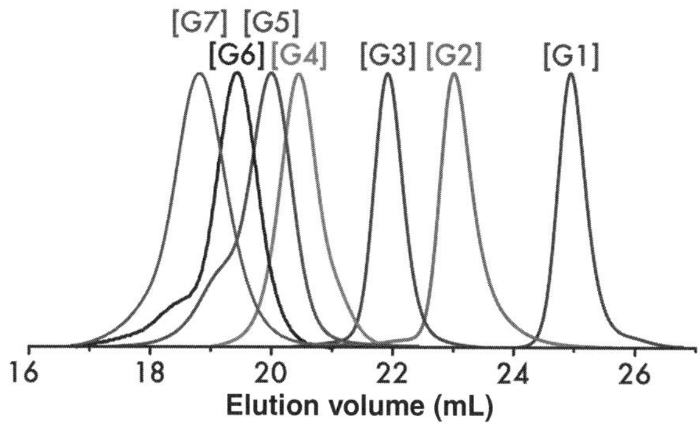

FIGURE 5.5 Traces of size exclusion chromatography for the dendrimers **TPM-leG1–7**.

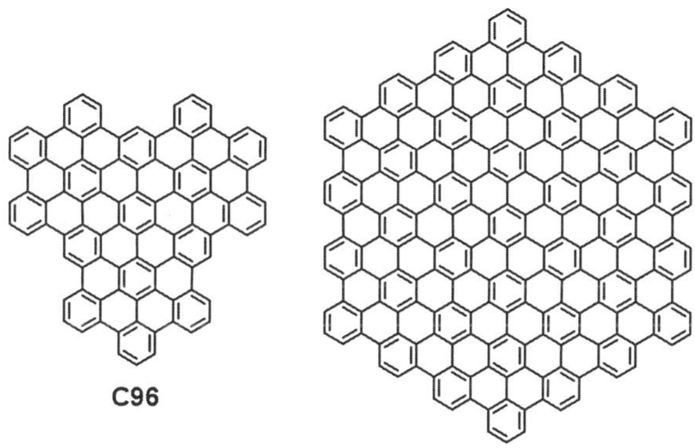

SCHEME 5.8 Graphene molecules **C96** and **C222**.

leading to planar extended disk-like molecules with 16 and 36 benzene rings (Scheme 5.8), to **C96** [46] and **C222** [47], respectively. These graphene molecules have been reviewed many times and will not be further discussed here [48].

5.4 POLYPHENYLENE DENDRIMERS WITH FUNCTIONAL CORES

It should be noted that essentially any functional group, which can withstand the Diels–Alder cycloaddition conditions (high temperatures up to 170°C) and can be substituted with ethynyl units may be used as a core for a polyphenylene dendrimer growth. So far we demonstrated that a wide variety of hydrocarbons can be used as cores for dendrimers; however, their molecular geometry only influenced the overall shape and rigidity of the resulting dendrimers [49]. Another core function could be a carrier of a special optical or electronic property, where in such a dendrimer the dendrons act as a shield, protecting the core from interacting with other cores or with

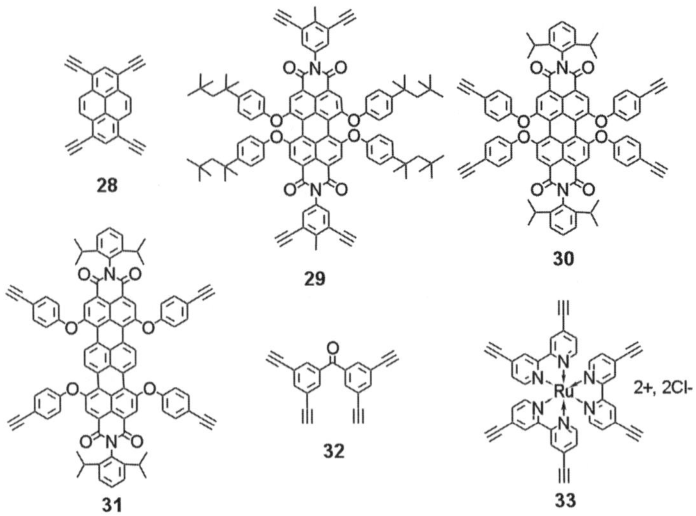

SCHEME 5.9 Functional cores for extension of dendrimers.

other species, which is particularly important for controlling the optical and electrical properties, for example, photoluminescence (PL) or electroluminescence (EL). Therefore, such new core building blocks were synthesized, possessing several ethynyl functions for dendritic growths (Scheme 5.9). This was extensively explored for dye molecules such as pyrenes (**28**) [50], perylenes (**29,30**) [51,52], terrylenes (**31**) [53,54], the functional precursor benzophenone (**32**) [38], and the charge-carrying complex ruthenium(II)tris-bipyridyl (**33**) as functional cores [55] for polyphenylene dendrimer generations (Scheme 5.9). In general, single pyrene and perylene dyes show a strong aggregation tendency, which induces a luminescence quenching and leads to pronounced bathochromic shifts in the solid-state absorption spectra. Upon multiple covalent dendritic encapsulation of such a chromophore in a polyphenylene shell this aggregation could be avoided. Furthermore, the polyphenylene dendrons also served as solubilizing groups enabling solution processing of the resulting compounds.

For the perylenetetracarboxydiimides two different functionalizations were considered as (i) **29** with ethynyl groups attached to the imide moieties [56] and (ii) **30**, where the ethynyl groups are located in the bay positions of the dye [52]. The shielding has been demonstrated to be more efficient when the dendrons grow from the bay position. These perylene dyes have been regarded perfectly eligible for single molecule spectroscopy (SMS), which enable us to study the interaction of the fluorophores with the surrounding matrix. These results, however, will be summarized in a following chapter by the group from Leuven.

For the pyrene **28** based dendrimers dilution and temperature-dependent fluorescence spectroscopy studies were performed to investigate the site isolation of the pyrene core. It was found that a second-generation dendrimer layer is needed to efficiently shield the encapsulated pyrene and to prevent aggregate formation.

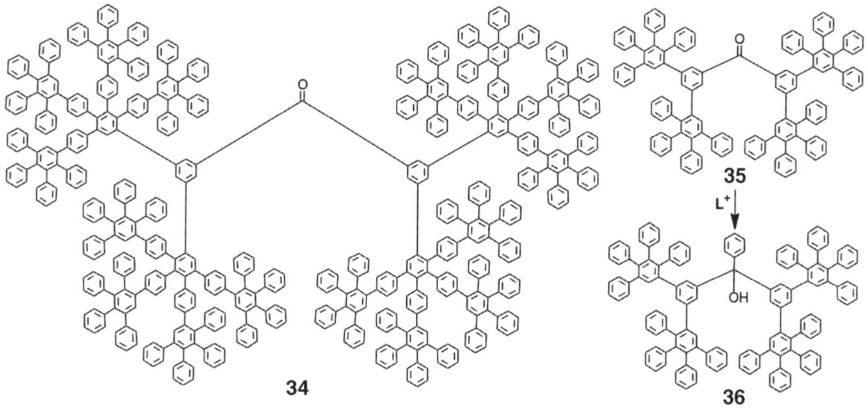

SCHEME 5.10 First- and second-generation polyphenylene dendrimers with benzophenone cores.

Alkaline metal reduction of the encapsulated pyrene core was carried out to afford the corresponding pyrene radical anions, for which hampered electron transfer to the core was observed with increasing dendrimer generation. This served as further proof of the site isolation due to the polyphenylene shell. But these charging experiments also demonstrated that for device applications such as organic light-emitting diodes (OLEDs), one should not go to higher dendrimer generations beyond G3, since the charge recombination may be hindered. High quantum efficiency yield in solution (Qf > 0.92) was already achieved by applying a second-generation dendrimer shell. Even a gram-scale synthesis of this new materials based on second-generation dendrimers for device applications became possible.

Only a very few reports have tackled the question of synthetic transformation of the dendrimer core. For testing the core accessibilities toward charging activities and postsynthetic transformations, first- and second-generation PPDs were synthesized starting from benzophenone [57]. In order to obtain high density and thereby a strong shielding by polyphenylenes around the benzophenone core (Scheme 5.10), a large number of starting points for the dendrimer growth was necessary, which was achieved applying the tetraethynylated building block **32**.

Variation of the size and the density of the surrounding dendrimer shell resulted in different isolation of the cores. The dense-shielded dendrimer **34** was obtained from the core **32** using the A_4B CP unit **7** for second-generation dendrimer **34**. Upon modeling, it was directly shown how well the keto function is protected by the surrounding phenylene units.

To investigate the reactivity of the core upon shielding, chemical reaction of the core with aryl-/alkyllithium and Grignard reagents, as well as the alkali-metal reduction of the encapsulated benzophenone core were performed. The phenyllithium reaction took place easily with the first-generation dendrimer **35** to achieve a functional phenylated hydroxyl methyl group **36**. For the second-generation dendrimer **34** very prolonged reaction times (3 days of reflux) were needed to find some incomplete hydroxylation, while larger nucleophiles as biphenyl lithium did not react

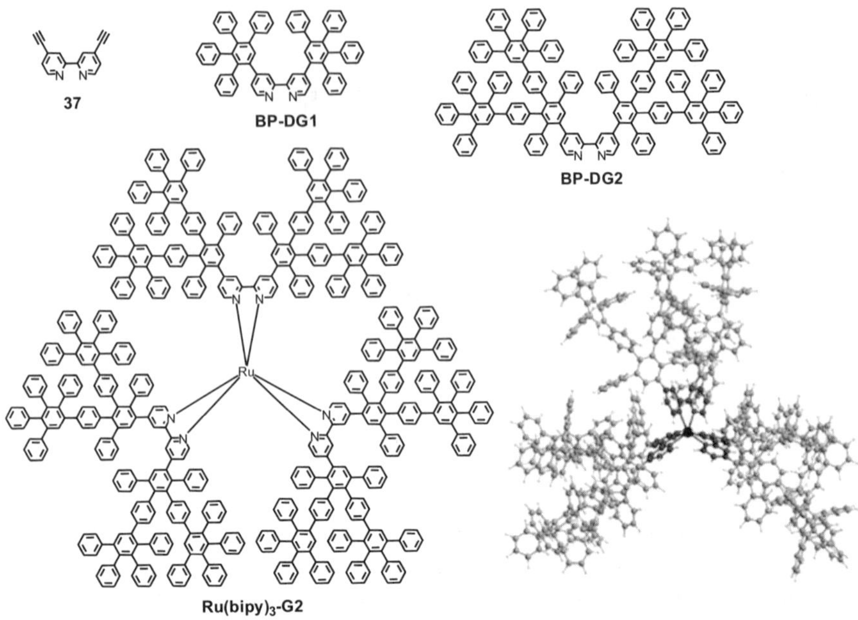

SCHEME 5.11 Bipyridine-containing dendrons **BP-DG** and the **Ru(bipy)$_3$-G2** dendrimer.

at all. From the charging experiments with alkaline metal, potassium-bridged benzophenone anions could only be detected for the first-generation dendrimer. The larger and denser dendrimer shell of the second-generation dendrimer, led to an isolation of the radical anion species in the core. This synthetic concept opened the way to spatially well-defined spherical nanoparticles bearing a single isolated function in the center.

The highest core symmetry that approached spherical polyphenylene dendrimers (TPM-G4) has, however, been limited to a four-armed, tetrahedral tetraphenylmethane core **14** (Scheme 5.3), since higher symmetries, such as octahedral, are challenging to achieve in organic chemistry. A powerful tool for building octahedral structures with controlled symmetry is instead provided by the use of organometallic complexes as core units, such as the well-known tris(2,2′-bipyridyl)ruthenium(II) complex Ru(bipy)$_3$ (Scheme 5.11). This metal complex possesses an almost perfect octahedral coordination geometry [58], is shape persistent itself, and thus can serve as the desired core when dendritic wedges are attached to the six positions para to its nitrogen atoms. The synthesis could be achieved through the functionalized bipyridyl moiety **37** carrying two ethynyl groups for growing the polyphenylene dendrons. The complexation with ruthenium dichloride salt was performed at the end in a convergent manner after first to fourth generations of bipyridine dendrons (**BP-DG1**, **BP-DG2**) were synthesized. The largest complex **Ru(bipy)$_3$-G4** failed, however, while first- to third-generation dendrimers with an integrated Ru^{2+}-core became available [55].

Later it was shown that also the divergent synthetic approach is applicable. Even a mixture of both the convergent and divergent route could be applied, where the ruthenium core was coupled to one extended dendron and protected smaller bipyridine units leading to desymmetrized dendrimers, as demonstrated in Chapter 7.

The size, shape, and polyphenylene density of the G3 Ru(bipy)$_3$-cored dendrimer (**Ru(bpy)$_3$-G3**) were compared with those of the related G3 species with a TPM core (**TPM-G3**) [13]. Diffusion NMR measurements revealed a hydrodynamic radius of 2.5 nm for dendrimer **Ru(bpy)$_3$-G3**, which is slightly larger than that determined for the **TPM-G3** species (2.2 nm). This finding indicated a decrease in the large voids between branches, as the physical radii were almost identical. Whereas the tetrahedral dendrimer **TPM-G3** distributed 144 phenylene rings within the hydrodynamic volume, the density of the **Ru(bipy)$_3$-G3** was increased to 216 aromatic rings.

Another intriguing aspect, which is still under consideration, is the interaction of the core with the counter ions, their distance, and the influence of the counter ions on the structure of the Ru-dendrimers or other charge-carrying cores.

5.5 FUNCTIONALIZED SCAFFOLD OF POLYPHENYLENE DENDRIMERS

Functionalization of the scaffold can serve several purposes, for example, multiplying the number of dyes in a defined environment and distances, or fine-tuning of charge or energy transfer to a functional core. As opposed to core functionalities, functionalization of the scaffold needs to be introduced by the CP unit, which has been established for the growth reaction and needs to be verified further to carry the desired function, and also possess the TIPS-ethynyl groups for growth reactions. Sketching the accessibility of such CP units the Knoevenagel condensation of difunctionalized benzils **38** with the disubstituted acetones **39** results in product **40** (Scheme 5.12). The most common way is the functionalization that has been introduced either in the α or in the β position of the CP units, as presented by the two guiding structures dibromotetraphenyl-CPs **41** (β-substituted) and **42**, (α-substituted), which can also easily be converted to their diborolanes, as **26** (Scheme 5.7).

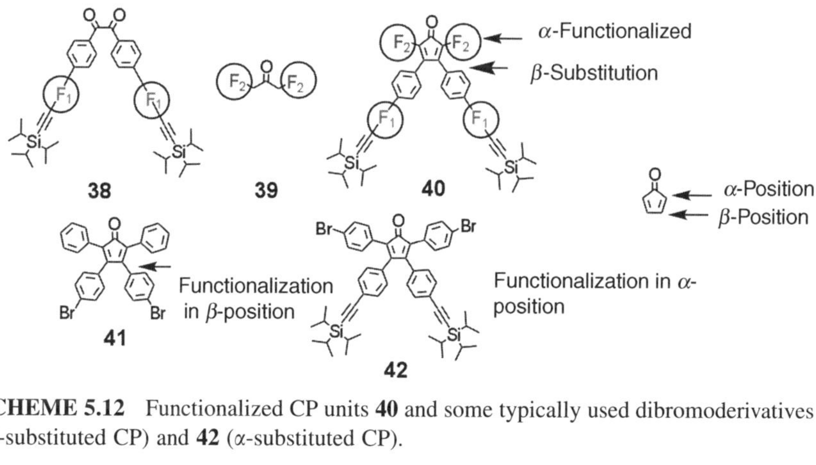

SCHEME 5.12 Functionalized CP units **40** and some typically used dibromoderivatives **41** (β-substituted CP) and **42** (α-substituted CP).

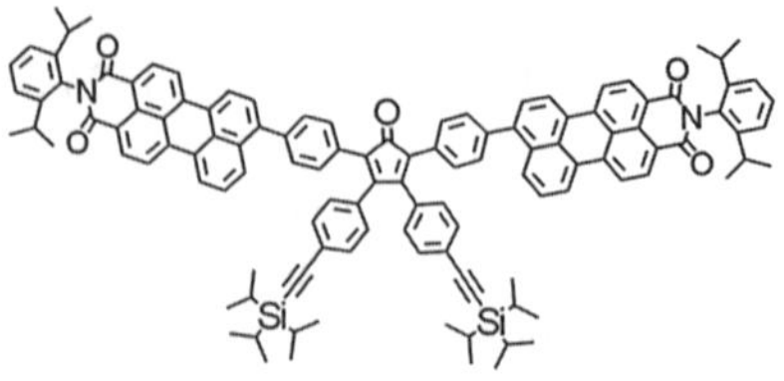

43

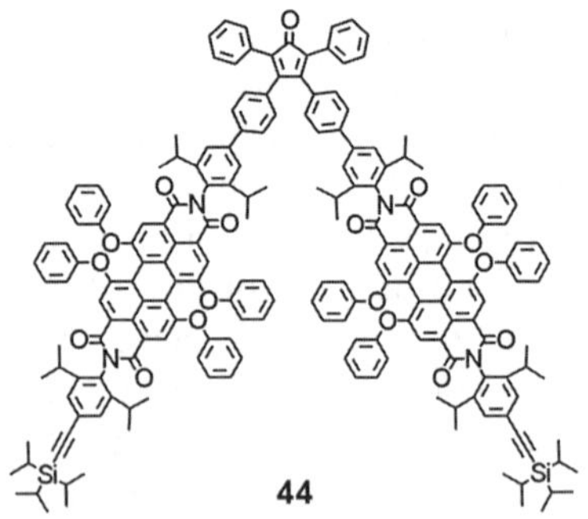

44

SCHEME 5.13 CP units with perylene dicarboxymonoimide **43** and perylentetracarboxydiimide **44**.

These two routes outlined above were compared for the case of incorporation of highly fluorescent perylene mono- and diimide chromophores into the scaffold of polyphenylene dendrimers via iterative Diels–Alder cycloadditions. The key intermediates for the divergent dendrimer build-up were two cyclopentadienone branching units carrying two perylene imides, and two masked terminal alkynes as **43** and **44**, respectively (Scheme 5.13). The major difference between the two CP units is the mode of incorporation of the chromophores, either in the α-position of the CP (**43**) or in the β-position (**44**), where in the latter case the perylene dyes can be considered as functional linear extension, similar to the terphenyl extended case **TPM-leG** discussed before, but even more lengthy.

The use of the β-substituted CP units **44** caused an extension of molecular diameters for the second-generation dendrimer with up to 12 nm, and inclusion of 24 chromophores within their scaffold. Absorption and emission characteristics of the new multichromophoric dendrimers (**45** and **46**) were investigated and compared to those of the parent dyes, indicating that the self-quenching among chromophores was controllable by the shape-persistent architectures. As was mentioned before, dyes approaching each other induce bathochromically shifted absorption bands and a quenching of their photoluminescence. For the dendrimer with the CP unit functionalized in α-position (**45**) a very dense packed dendrimer resulted in which the dyes surrounding the core, however, were too close and interacted with each other. For **46** this was prevented by the linear elongation away from the core (Scheme 5.14).

Dyes like **46** have intensively been studied by single molecule fluorescence spectroscopy, demonstrating their uniqueness as single-photon source due to the singlet–singlet annihilation [59–61]. They are perfectly suited for this kind of experiment, since they possess large molar extinction coefficients in the visible region, a high quantum yield of fluorescence and excellent photostability [62].

The CP units can be modified with other functions at will, keeping in mind that they should be thermally stable. The β-substituted CPs (**47**, **48**) were mainly synthesized from the above-mentioned dibromo derivative **41** or the corresponding bis-dioxoborolane derivative **26** [38,63], while for the synthesis of α-substituted CP units different

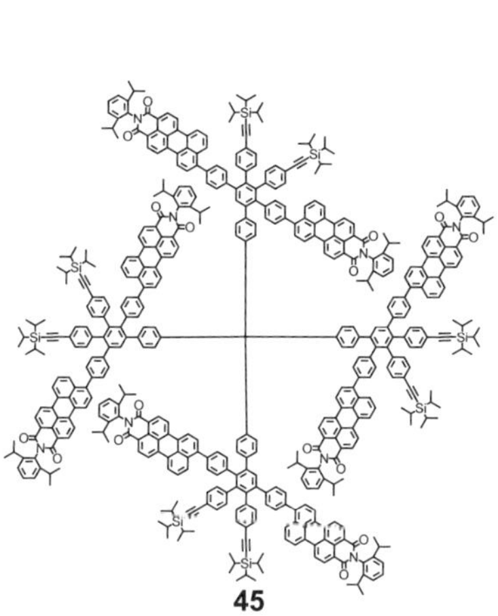

45

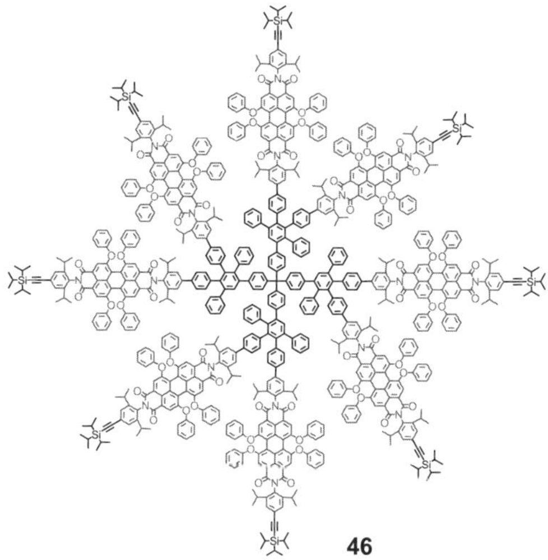

46

SCHEME 5.14 Perylene dicarboxymonoimide and perylene tetracarboxydiimide chromophores containing dendrimers **45** and **46**.

acetone derivatives were applied for the Knoevenagel condensation achieving the new CPs **49, 50** (Scheme 5.15) [64].

In the following part, two representative examples from the polyphenylene dendrimer chemistry will be discussed keeping the focus on the postsynthetic functionalization of the backbone or scaffold.

The insight into the shielding of a single core by polyphenylene dendrons allowed us to extend the postsynthetic concept to multiple benzophenones incorporated in the dendritic scaffold. Thus, a nonfunctional tetraphenylmethane core was used, where the first dendritic growth step introduced the benzophenone units, which were then shielded by the subsequent polyphenylene generations. Accordingly, a benzophenone-substituted tetraphenylcyclopentadienone branching unit **51** was applied to synthesize monodisperse polyphenylene dendrimers, bearing a defined number of keto groups in their backbone (Figure 5.6).

β-Functionalized

α-Functionalized

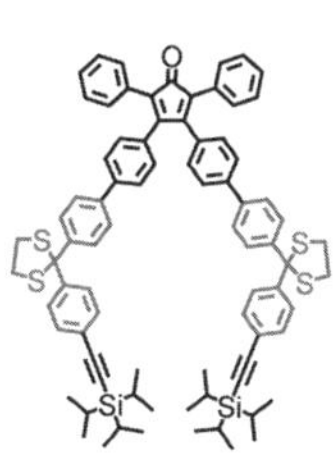

47

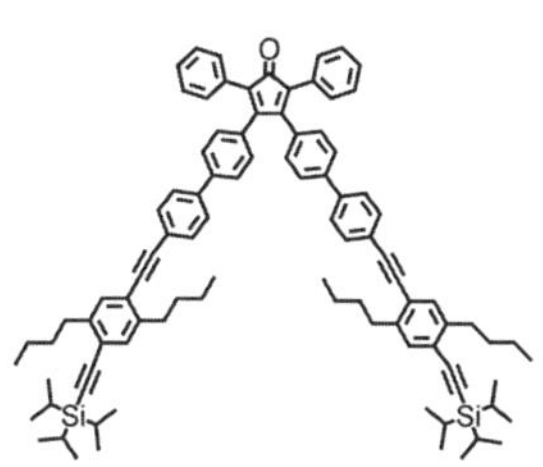

48

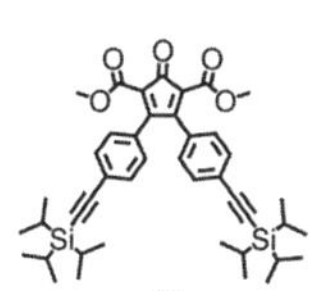

49

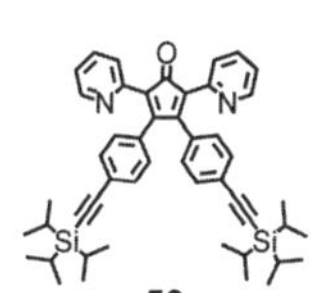

50

SCHEME 5.15 Further substituted CP units with extra functions in the *β*- (**47**, **48**) and in the *α*- (**49**, **50**) positions for scaffold functionalization.

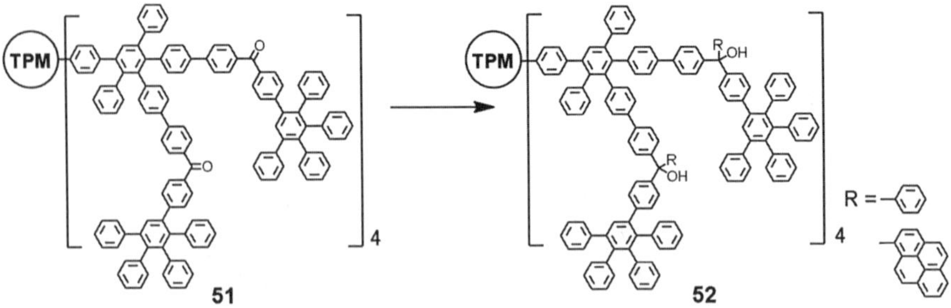

FIGURE 5.6 Second-generation polyphenylene dendrimer **51** carrying eight benzophenones in its scaffold and its postsynthetic hydroxyl functionalization **52** via organolithium reagents.

The second-generation dendrimer **51** turned out to be a suitable substrate for a perfect and versatile postsynthetic functionalization of the dendritic scaffold using organolithium reagents [38]. Even large reaction partners like pyrene **52** could be introduced quantitatively allowing an easy and versatile modification of the scaffold. Alkali metal reduction of the inner benzophenones quantitatively resulted in the octa-radical anion species **51**, where potassium-bridged anions could be detected, stemming from interpenetrating dendritic arms. These facts were pointing to the formation of some kind of three-dimensional network of radical anions bearing dendrimer **51**.

Another representative example for the postsynthetic variation of the molecular structure was obtained with regard to the embedding of acetylene groups into the first-generation layer of the polyphenylene dendrimers. The synthesis of the required CP-building blocks followed a similar protocol in which the benzophenones were replaced by two diphenylacetylene moieties. The use of this building block for the synthesis of the first-generation layer led to a radial size extension by the length of a diphenyl acetylene unit in the new dendrimers **53** (Figure 5.7) [38,63].

The heterogeneous hydrogenation of the embedded triple bonds by Pd on carbon was probed in order to permit a direct assessment of the shielding effect of the outer polyphenylene shell. Moreover, a softening effect of the dendritic structure **54** in

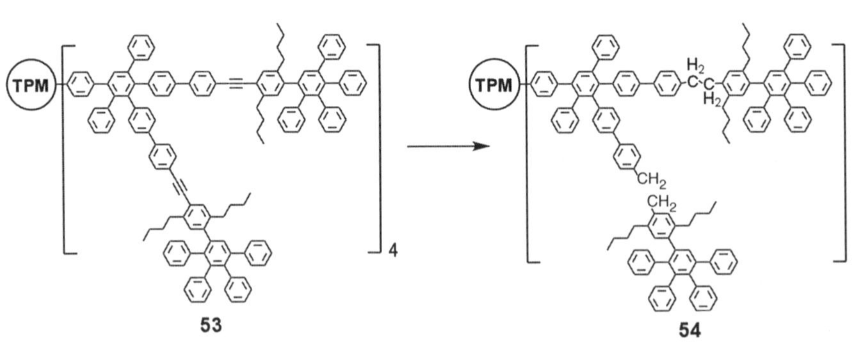

FIGURE 5.7 Second generation of diphenylacetylene extended dendrimer **53** transformed to **54**.

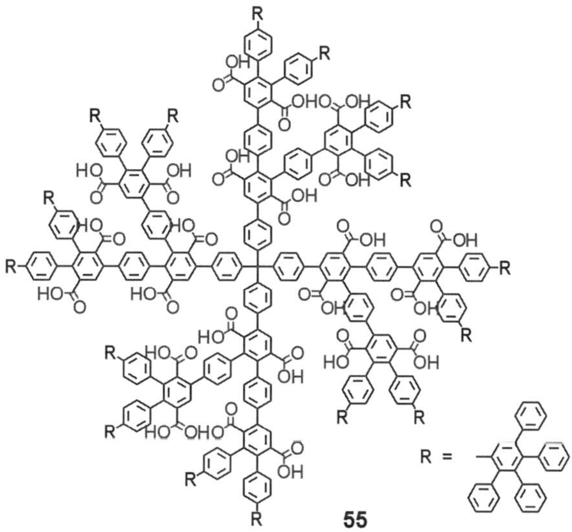

SCHEME 5.16 Dendrimer **55** with carboxylic functions in the scaffold.

consequence of the hydrogenation was expected. For the first time ever the investigation of this effect upon size, shape, and intramolecular voids in the case of similar dendrimer pairs could be accomplished. Quartz microbalance studies revealed that upon hydrogenation the capacity in host uptake is decreased, allowing the incorporation of a lower number of guest molecules compared to the parent material **53** [65].

The effect of the substituents in the α-position is to introduce functionality within the scaffold influencing the internal density and polarity, but not affecting the number of branches or the lipophilic surface. Let us now discuss two examples explaining their access and function.

For the synthesis of polyphenylene dendrimers with backbone-pendant internal carboxylic acids and nonpolar tetraphenylbenzene chain ends, the carboxylic ester carrying CP unit **49** was used. This way up to 24 ester groups in the second generation were incorporated **55** (Scheme 5.16). For a more effective isolation of the inner microenvironment, the final generational layer was formed by reaction with the commercially available tetraphenylcyclopentadienone shielding the polyphenylenes dendrimers outside (Scheme 5.16). The ester groups could be completely hydrolyzed to carboxylates, further increasing the hydrophilic character of the interior. Then the host–guest and transport properties of these dendrimers with proflavine, an acridine dye, were investigated [64]. The host–guest chemistry of such acridine dyes based on ammonium carboxylates is well established, and they find wide applications as spectroscopic probes. Ammonium carboxylate formation (guest uptake) with the polyphenylene dendrimer hosts (Figure 5.1) was observed by monitoring spectrophotometric properties of proflavine hydrochloride. These properties were found not to depend strongly on the number of carboxylic acids, but rather on the structural

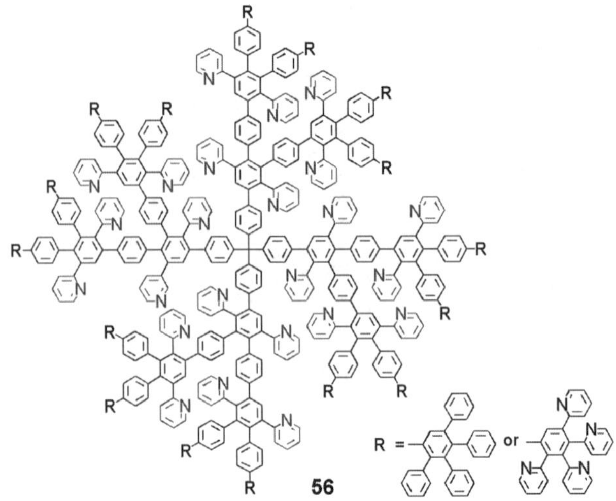

SCHEME 5.17 Polypyridyl dendrimer **56**.

properties of the stiff dendrimers, binding only 3–4 dyes each, for the first- and second-generation dendrimers.

The other example for α-functionalized CP units is a dendrimer with pyridine in the scaffold based on **50** with protected dienophile functions. By varying the building blocks at the last step of dendrimer construction, one could design molecules **56** possessing either phenyl- or pyridyl-decorated outer shells (Scheme 5.17) [66,67]. The pyridine then could serve as ligation site for metal ions. Therefore, these stiff polyphenylene dendrimers with embedded pyridyls were used as a matrix for encapsulation of Pd nanoparticles [67]. The fourth-generation poly(phenylenepyridyl) dendrimer with a phenyl-decorated periphery and pyridine-containing interior was found to serve as a powerful template for metal-nanoparticle formation, resulting in excellent nanoparticle stability with catalytic function. Further, these special functionalized dendrimers exhibited very strong sensitivity toward the explosive triacetonetriperoxide (TATP), which will shortly be outlined in Chapter 8 [68].

5.6 FUNCTIONALIZED SURFACES

Functionalization at the surface is the most simple and flexible route to obtain functionalized dendrimers and it offers the highest flexibility to control the type, number, and position of the substituents.

There are two possibilities to achieve functionalized surfaces of polyphenylene dendrimers. The first and most straightforward one is to introduce any functional cyclopentadienone unit "a priory" as endcapping reagent in the dendrimer synthesis of desired generation just as discussed above for pyridyls. Since the Diels–Alder reaction takes place at high temperatures only, an important requirement of the functional groups is their thermal stability, which in some cases necessitates

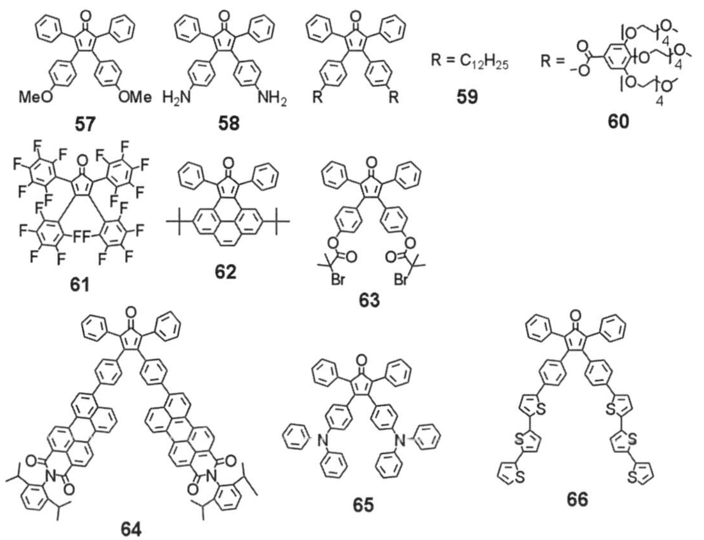

SCHEME 5.18 Functional CPs for surface activation.

protection. The second possibility is a reactive group transformation after dendrimer synthesis in an "a posteriori" way. Both approaches have been considered.

By the "a priori" method, a decoration of the polyphenylene dendrimers with numerous functional groups affects their solubilities, tendencies to intermolecular interactions, their self-assembly behavior, and surface affinity or just polarity (Scheme 5.18, **57–66**). Furthermore photonic and electronic properties can be adjusted by dye groups as perylenmonoimide (PMI, **64**) [56] and conducting properties upon using terthiophenes **66** [69] or the hole conductor triphenylamine as in **65** [70].

At this point, we will not discuss such extended surface functionalizations since they have been reviewed earlier [14].

An example is the core–shell dendrimer using **66** as endcapping reagent. The resulting core–shell dendrimer possesses oligothiophene chains on its surface [69]. Its chemical or electrochemical oxidative coupling yielded a 3-D network in which dendrimer spheres were connected by sexithiophene spacers. The electrical conductivity of this 3-D structure provided insight into the mechanism of charge carrier transport. Interestingly, the electrical conductivity was further increased when the 3-D polyphenylene dendrimer cores were oxidatively dehydrogenated to planarized graphitic propellers [69]. One thus obtained a unique hybrid network in which graphene structures where embedded in an oligothiophene matrix.

The conversion of an already existing group A on the dendrimer surface according to the "a posteriori" method into a substituent B via subsequent addition of an electrophile is presented in Scheme 5.19 [71]. In the ideal case, this reaction should be quantitative, but the possibility of incomplete conversions or side products must always be considered.

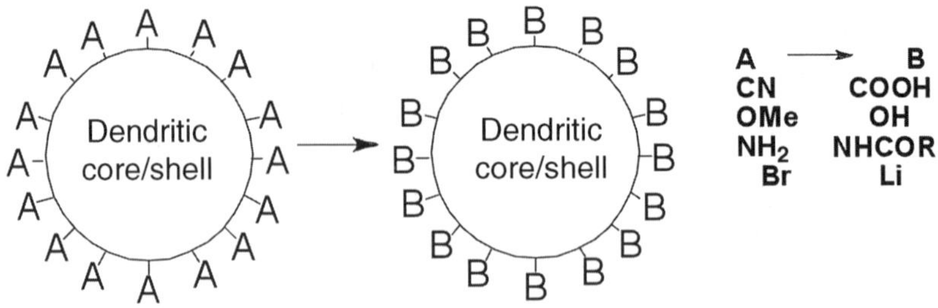

SCHEME 5.19 Reactive group transformation.

We were able to functionalize our dendrimers via this method with carboxyl, hydroxyl, amide, and lithium groups at the periphery (Scheme 5.19).

Except for the conversion of an amino to an amide group, all the other groups given in the scheme were obtained quantitatively. For example, the hydrolysis of a cyano group on a second-generation polyphenylene dendrimer with a tetrahedral core led to a dendritic molecule **TPM-G2** bearing 16 carboxylic acid functions (**67**) at the periphery (Scheme 5.20). Even other polar groups such as nitriles, amines, and thiols have been introduced [13]. Thiomethyl substituents have shown to be valuable for binding to gold surfaces and also for the formation of gold nanoparticles [72].

It is this multiple surface functionalization which qualified these dendrimers for supramolecular structures even allowing us to combine such rigid surface functionalized dendrimers with ring opening and radical polymerizations which led to novel

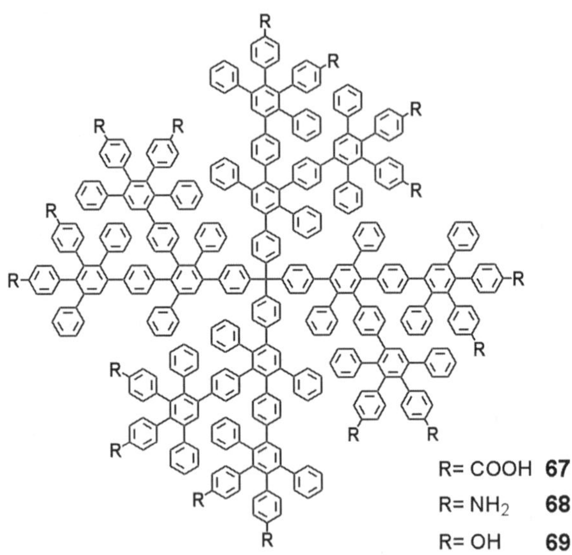

SCHEME 5.20 TPM-G2 with 16 polar end groups, R=COOH (**67**), R=NH_2 (**68**), R=OH (**69**).

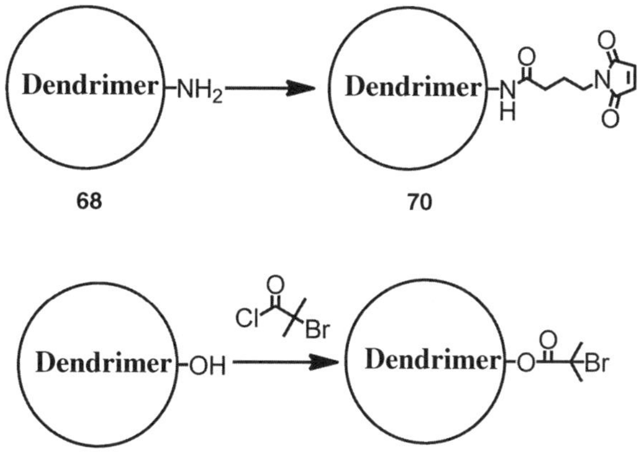

SCHEME 5.21 TPM-G2 and B-G2 dendrimers **68–71**.

core–shell polymers [73–76]. For instance, the amine groups in dendrimer **68** (Scheme 5.21) could be further linked to the maleimide groups (**70**), which are the precursors for the reaction with the thiol groups of peptides [77–80]. The hydroxyl groups in dendrimer **69** can also be converted into 2-bromo-2-methylpropionic ester groups (**71**), which are the macroinitiators for atom transfer radical polymerization (ATRP) of *tert*-butyl acrylate [73,74,76]. Although the rigidity of phenylene dendrimers is lost upon extended outer shell polymerizations, these water-soluble core–shell structures derived from the dendrimer surface functionalization are highly interesting for biological applications [65,73]. The outer shell containing functional groups were shown to possess efficient interaction with a diverse set of DNA fragments and stable complexes were formed.

The observed tight binding of the negatively charged DNA to the positively charged dendritic star polymers, which was found in the subnanomolar range, was mainly attributed to strong electrostatic interactions. DNA was released from the complexes after treatment with high concentrations of sodium chloride in aqueous solution. Such core–shell systems based on polyphenylene dendrimers, which allow the binding and release of DNA, represent attractive candidates for the development of novel ion-exchange resins for DNA purification or as nonviral vector systems for gene delivery [74]. In addition fluorophores could be centered in the core in order to use them for cell nucleus staining [73,76]. These core–shell nanoparticles consisted of a central PDI chromophore with a first- and second-generation polyphenylene scaffold for suppressing aggregation of the central chromophore in aqueous media and a shell with polycarboxylic acid groups (Scheme 5.22). Staining was found exclusively in the cell nucleus. A tight binding between the carboxylated PDI derivatives **72** and positively charged lysine-rich histones was found.

Moreover, the ATRP polymerized core–shell macromolecules based on polyphenylene dendrimers **73** were applied as templates to control the structure of TiO_2

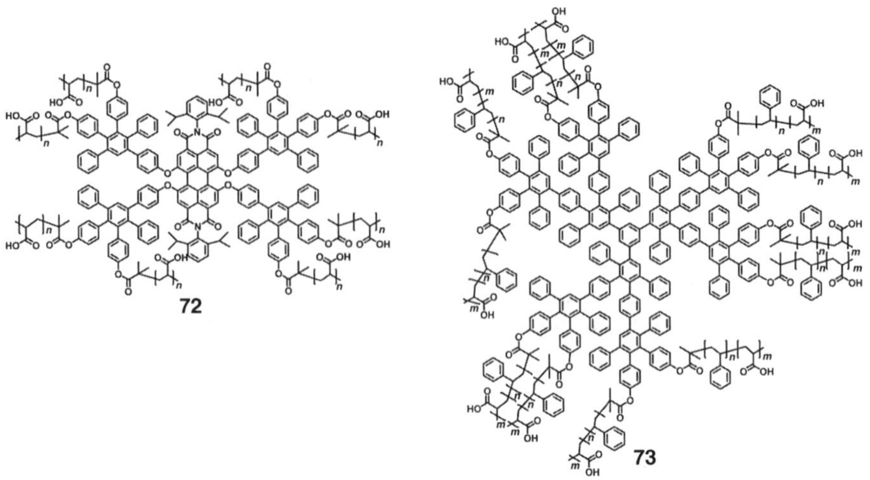

SCHEME 5.22 Polyphenylene dendrimer-based core–shell nanoparticles **72** and **73**.

particles due to their dendritic shape-persistent polyphenylene cores [75]. The phenylene cores were surrounded by flexible amphibilic double polymer shell with 12 arms, where a polystyrene block served as hydrophobic segment and the outer poly (acrylic acid) provided water solubility and served as a binding site for titanium

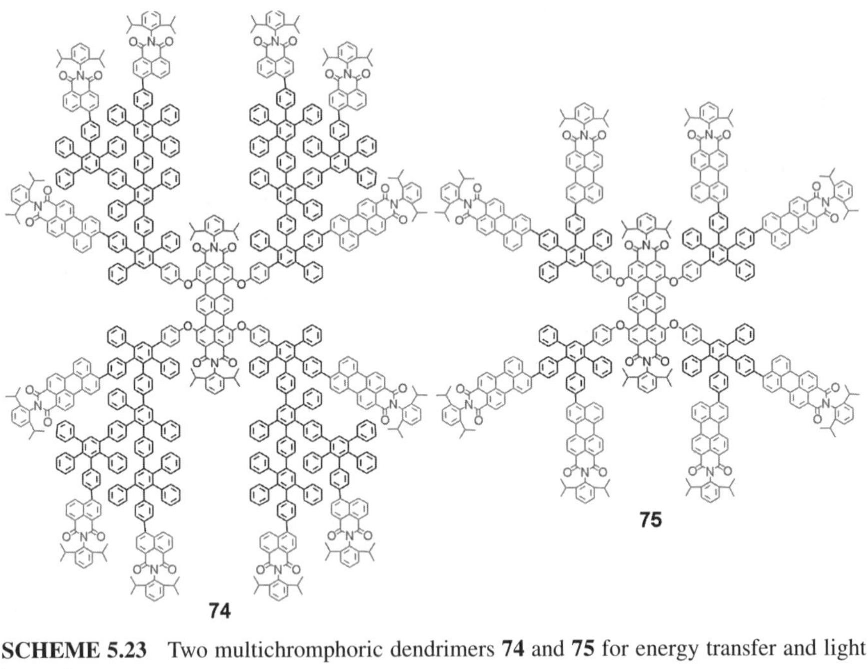

SCHEME 5.23 Two multichromphoric dendrimers **74** and **75** for energy transfer and light harvesting.

tetraisopropoxide. After sol gel synthesis of TiO_2 in the outer shell and removal of the organic dendritic template by calcination at 500°C hollow nanospheres of TiO_2 could be obtained. The controllable TiO_2 morphology provides tremendous applications in many different fields such as photocatalysis, gas sensors, lithium-ion batteries, and dye-sensitized solar cells [81–83].

At the end of this chapter on surface functionalization, we should mention that combinations of functions at different locations have also been considered, for example, from the surface, through the scaffold to the core as shown by some multichromophoric dendrimers (Scheme 5.23) [53,54,84]. For example, the dendrimer **74** consists of a single terrylenetetracarboxydiimide (**TDI**) core and multiple **PMI** groups in the periphery, while on the surface naphthalenedicarboxymonoimids (**NMI**)s are located [83]. This is a nice model for light-harvesting systems, in which energy transfer processes from **NMI** as donors to **TDI** as an acceptor were studied. It was further shown that the number of perylenes can easily be increased by using a CP-building block with two PMI units yielding **75** or the perylendiimide **44** mentioned earlier for further extension.

5.7 DESYMMETRIZATION APPROACHES

The cyclopentadienone chemistry also represented a powerful method for the synthesis of dendrimers bearing two different types of functional groups at the surface of a single dendrimer. In the beginning, the only known synthetic route to desymmetrized PPDs proceeded via several protection and deprotection steps of the ethynyl groups of tetrakis(4-ethynylphenyl)methane and subsequent stepwise Diels–Alder cycloaddition as schematically depicted in Scheme 5.24 [56]. This synthesis started from a partially protected tetrahedral core (**76**, Scheme 5.24) and a cyclopentadienone building block containing a functional group of type "A" **77**. Deprotection of the acetylene units of the core and Diels–Alder reaction with a cyclopentadienone coated with a functionality of type "B" **79** led to a first-generation dendrimer **80** with two kinds of functionalities (type A and type B) in a well-defined spatial arrangement at the rim. This pathway enabled the synthesis of a second-generation bifunctional dendrimer **81** using the convergent synthetic approach. It had the disadvantages of low yields and time-consuming purifications.

Thus, a new synthetic approach for the desymmetrization of PPDs was developed and described. Tetrakis(4-ethynylphenyl)methane **14** undergoes facile Diels–Alder cycloaddition with substoichiometric quantities of tetraphenylcyclopentadienones bearing one polar functional group **83** (Scheme 5.25). This method allowed the preparation of dendrimers possessing two types of functionalities in only two steps. A single ethynyl group was thereby converted to a rigid, selectively functionalized polyphenylene moiety, which serves as a focal point for further transformations [28,85–87].

By using the above-described core molecule **82a**, it was possible to build up the first-generation dendrimer **84**, bearing six PDI chromophores [51] and one aliphatic ester group for anchoring (Scheme 5.26).

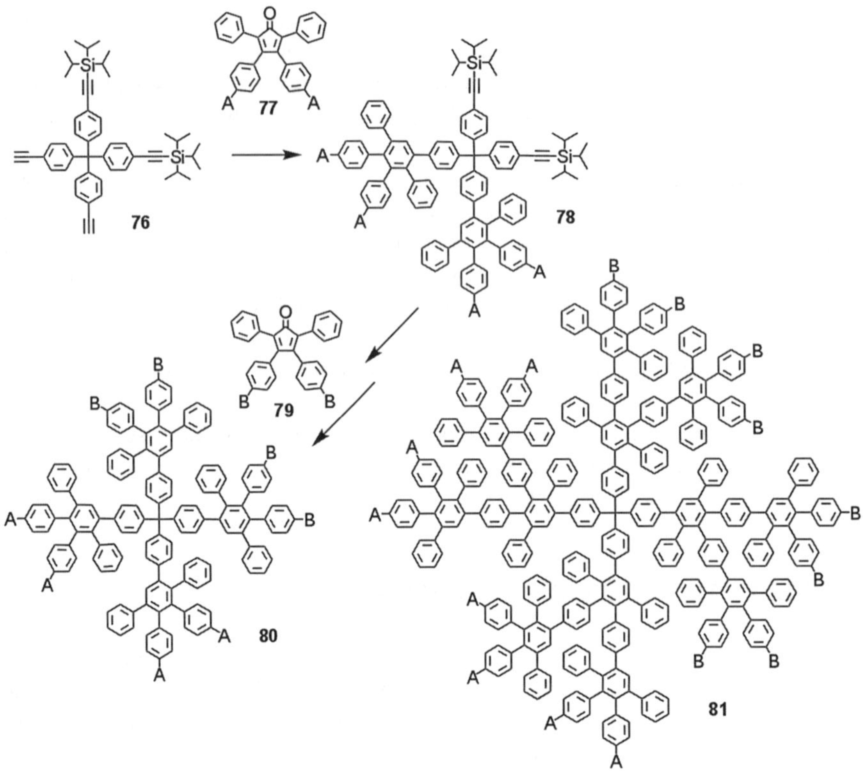

SCHEME 5.24 Time-consuming desymmetrization leading to **80** and **81**.

The ester group enables interfacial anchoring on functionalized surfaces, particles, or biologically active moieties with the perylenetetracarboxydiimides possessing fluorescence quantum yields near unity serving as highly emitting sensors.

Another asymmetric dendrimer with a Ru-core was achieved starting from the dichlororuthenium derivative **85** (Scheme 5.27). It was synthesized and coupled with the second-generation bipyridine dendron **85** in order to establish the macromolecule **86** [55].

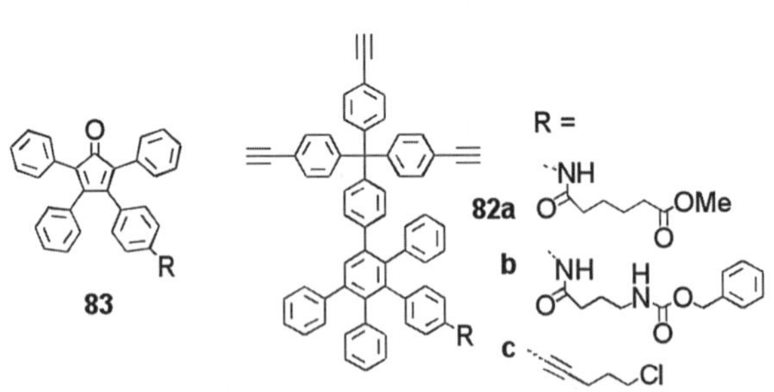

SCHEME 5.25 One-step synthesis of desymmetrized core **82**.

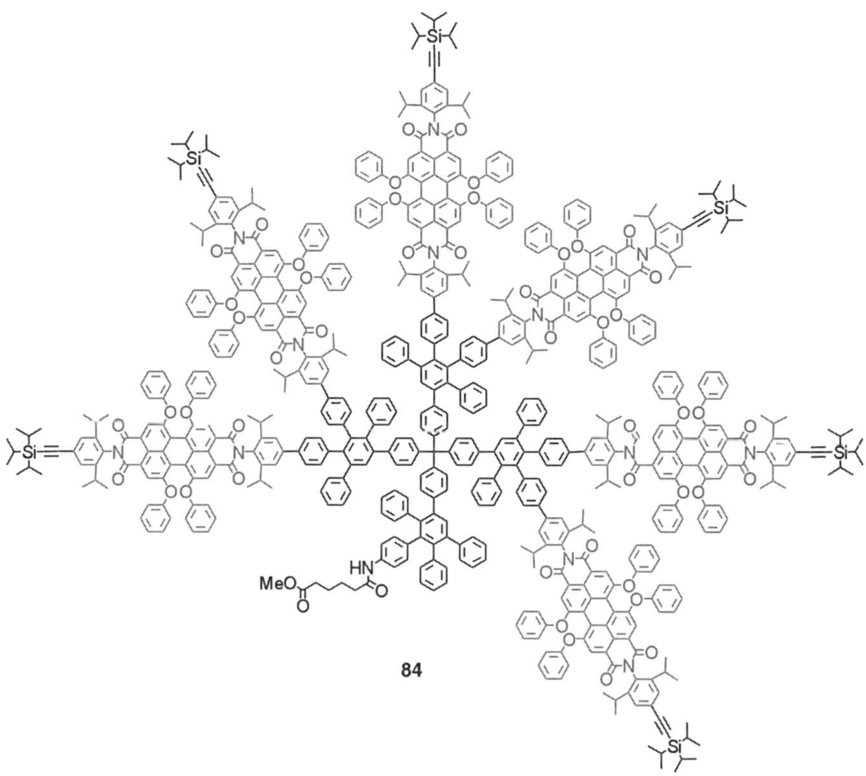

SCHEME 5.26 Asymmetric perylenetetracarboxydiimides containing dendrimer **84**.

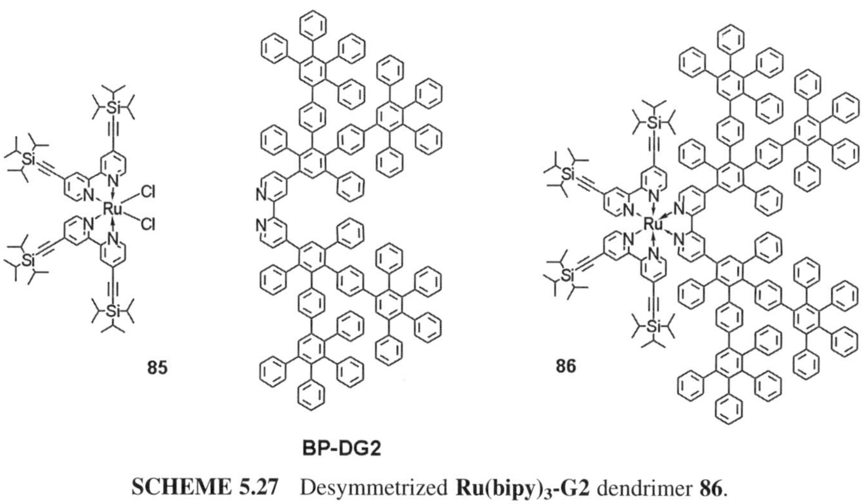

SCHEME 5.27 Desymmetrized **Ru(bipy)$_3$-G2** dendrimer **86**.

This approach demonstrated the ease of desymmetrization opening new access to multifunctional and amphiphilic structures. Ru–bipy complexes are important chromophores and sensitizers in solar cells, which underlines the coupling with other chemical entities such as organic dyes and the anchoring on surfaces.

5.8 SENSING OF VOLATILE ORGANIC COMPOUNDS

Voids of interior are perfectly suited for the uptake of volatile solvents/gases. The suitability of polyphenylene dendrimers as sensor layers for monitoring volatile organic compounds (VOCs) in the gas phase was demonstrated. The coating of the quartz micro balance (QMB) with second-generation dendrimers possessing different cores (**B-G2, Biph-G2, TPM-G2, HPB-G2**) was performed by using a procedure from mass spectrometry known as electrospray. The dendrimers were dissolved in THF and accelerated by a high DC voltage through a thin capillary and sprayed onto the top electrode of the QMB. All investigated dendrimers showed similar behavior; all compounds responded very selectively to polar aromatic VOCs, such as acetophenone, aniline, benzaldehyde, benzonitrile, fluorobenzene, nitrobenzene, and 2-methyl-benzonitrile. Remarkably neither chlorinated nor unsubstituted aliphatic hydrocarbons, alcohols, amines, aldehydes, and carbonyl compounds became included into these hosts. The number of included guest molecules depends on both the structure and the size of the dendrimers. A structure–property relationship may be feasible. Incorporating functional groups enables fine tuning of the sensor properties, which makes it possible to detect even aliphatic solvents. The sensitivity of dendrimer-coated QMB sensors was remarkably high, for example, for aniline and acetophenone around 5 ppm and it has now been further optimized in the group of Waldvogel [68].

Upon such optimizations, it was found that with a combination of three different matrices triacetonetriperoxide can be distinguished from water, hydrogen peroxide, acetone, and other omnipresent molecules. Specific shape-persistent phenylene dendrimers with pyridine in the scaffold (**56**) were found to have an extremely high affinity toward TATP, while rejecting polar compounds. The additional internal pyridyl moieties turned out to be very beneficial for the selectivity, while other internal functionalities are under investigation. The crucial role of the voids was revealed by testing different dendrimer generations G1–G3 and, in particular, G4 are unique affinity materials since they provide cavities, which are complementary to the shape of the TATP molecule. The limit of detection for TATP could be optimized far below 0.1 ppm. Thereby these dendritic sensors can be applied for many volatile compounds. It is considered that the design of new functionalized scaffolds is the key element for further promising sensors with high affinity for such analytes.

In another attempt to use the hydrophobic polyphenylene dendrimers for sensing of volatile organic compounds they were functionalized with disulfide and attached to gold nanoparticles. The disulfide functionalization was achieved via coupling of thioacetic acid (5-[1,2]dithiolan-3-ylpentanoicacid) to polyphenylene branches via esterification resulting in **87** (Scheme 5.28) [88].

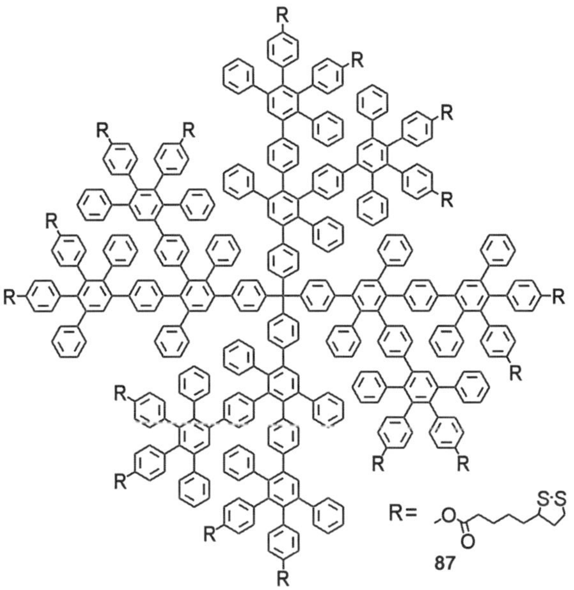

SCHEME 5.28 Disulfide functionalized dendrimer **87**.

The composite films were assembled layer-by-layer via alternated and repeated exposure of amino-functionalized substrates to solutions of the gold nanoparticles and dendrimers [88–90].

During film deposition, the dendrimers served to cross-link the nanoparticles, and thereby enabled the stepwise assembly of homogeneous and mechanically reinforced films with precise control over the film thickness. When using a film as a chemical resistor sensor device, high sensitivity to vapors of toluene and tetrachlorethane was observed, but the undesired sensitivity to humidity was negligible. The short response times, the high sensitivity to VOCs, and the low sensitivity to humidity made the materials promising candidates for sensor applications.

5.9 OLED APPLICATIONS

In order to make PPDs promising targets for organic light-emitting diode applications, the building blocks should be easy to make and should cover the desired color range. Usually the dendritic polyphenylenes themselves are not active fluorophores in the visible due to their inherent steric hindrance between the phenylene units. As demonstrated before, however, chromophores as perylene tetracarboxydiimide or pyrene can be built into the core [50,70] or the scaffold [51,56], where the polyphenylene dendron shell is nonemissive and dilutes the dyes to considerable amount, but helps to avoid aggregation and excimer formation.

Considering that the blue light-emitting diodes are still the most challengeable ones in red green blue (RGB) light-emitting diodes (LEDs) displays [91–95], only the

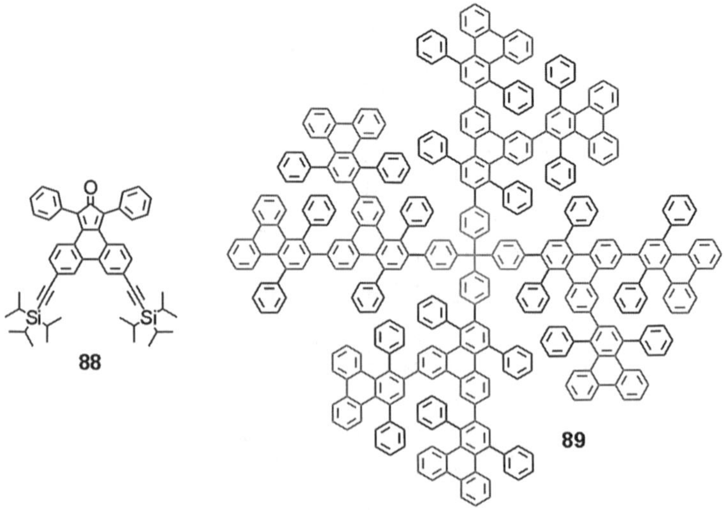

SCHEME 5.29 Diphenylcyclopentaphenanthrenone **88** and polytriphenylene dendrimer **89**.

pyrene would follow that demand, while the perylene tetracarboxydiimides usually exhibit red emission. Stable blue light-emitting polymers for commercial use are still under development [96].

Therefore, new conjugated dendrons were considered which are emissive themselves, but still do not undergo quenching effects following a new design principle. Accordingly, blue light emission should be caused by the presence of electronically decoupled polycyclic aromatic hydrocarbon (PAH) units, which should adopt sterically defined positions and disallow intra-dendrimer chromophore–chromophore interactions, thus forming amorphous films. Finally the amount of "useless" substituents and coupling units should be kept at a minimum. This led to the design of diphenylcyclopentaphenanthrenone **88**, which is the precursor of a fluorescent triphenylene. With this building block the polytriphenylene dendrimer **89** up to third generation became available (Scheme 5.29) [97]. Most importantly, the twisted triphenylene units not only function as chromophores but also effectively prevent the inter- or intramolecular fluorescence quenching, as proved by increased PL quantum yield with increasing generation. These dendrimers exhibited stable and pure-blue emission in both PL and EL spectra. While the triphenylene itself has a low PL quantum yield ($\sim$4%) the quantum yield increased considerably for the emission of higher generation up to 35% for G3. These dendrimers could provide an avenue for dendritic emitters with the optimized EL efficiency/color purity trade-offs needed for pure blue light emission.

The fluorescent polymers and dendrimers are limited to 25% internal quantum yield due to the singlet:triplet ratio [98]. Therefore, the phosphorescent dopants recently raised a lot of interest because of their quantum yield up to unity [98–100]. In this way, iridium(III) complexes are particularly promising phosphorescent materials because they can emit light from both singlet and triplet excitons with extremely high

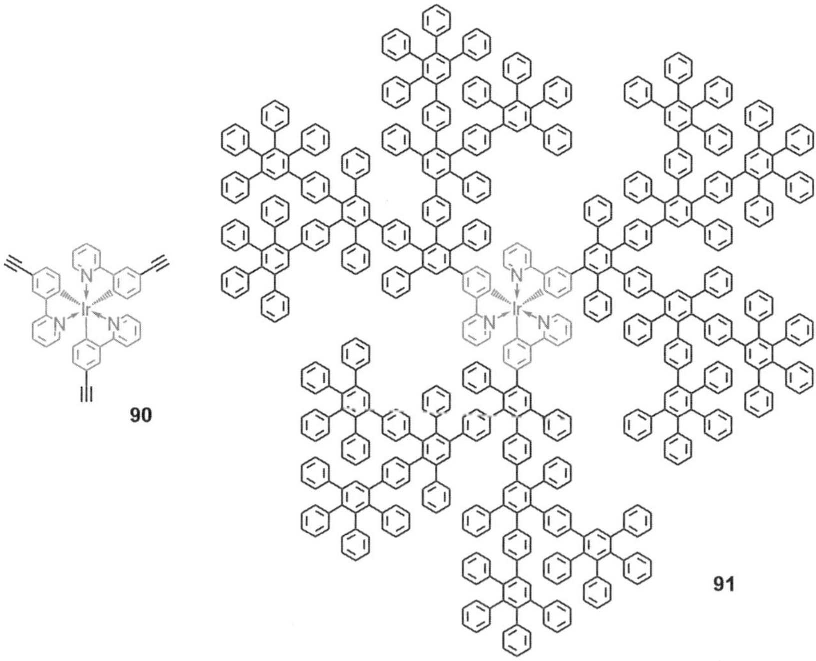

SCHEME 5.30 Iridium(III) complex **90** and iridium(III) dendrimer **91**.

emission quantum yield [101–105]. They often have to be blended into host materials to avoid self-quenching [106–108]. Thus, a new divergent synthesis from tris(2-(3-ethynylphenyl)-pyridyl)-iridium(III) core **90** was developed, where the polyphenylene dendrons were shielding the emissive core **91** [109]. These dendrimers were grown up to the fourth generation, and fully characterized by analytical methods before their phosphorescent and device properties were studied (Scheme 5.30). Thereby it turned out that the phosphorescence quantum yield still increased up to the fourth generation (36%), however, OLED devices showed optimum efficiencies for the third generation, which already has a radius of about 3 nm. The third-generation PPDs may therefore be the limit for good charge injection and still protect the core against triplet–triplet annihilation.

5.10 CONCLUSION AND OUTLOOK

It should have become clear by now that polyphenylene dendrimers are outstanding for quite different reasons:

(i) Making dendrimers only from benzene rings brings about an outstanding chemical and thermal stability.

(ii) The build-up from twisted, tightly packed, interlocked benzene rings while not being perfectly rigid, reduces the mobility of the dendron arms and excludes the back bending of arms. This creates a high degree of shape persistence with, however, sufficient solubility in organic solvents.

(iii) The method of synthesis via repetitive Diels–Alder cycloaddition ensures a high structural perfection, which is superior to that in other dendrimers. The most convincing proof for this unique character is the synthesis of monodisperse defect-free, three-dimensional polymers with molecular weights in excess of half a million Dalton.

(iv) It is a characteristic feature of the polyphenylene dendrimers that in the course of their synthesis a large variety of functional groups can be site-selectively placed in the core, in the scaffold, or on the surface of the dendrimers. It is this property that qualifies the PPDs as unique functional nanoparticles with perfect nanosite definition of their functional groups.

(v) An illustrating example of complex functional nanoparticles is a whole family of functional multichromophores, which can serve for example in light-harvesting systems, single photon emitters, and bioassays.

(vi) Polyphenylene dendrimers, due to their molecular design create a lot of empty space in their interior, giving rise to unique host–guest interactions, which can be best utilized in sensor fabrication.

(vii) The ability of carrying a payload of organic molecules is tightly connected with solubility and mobility of dendrimers in organic or aqueous media or across a membrane. There, the controlled surface modification, for example, by grafting reactions under formation of polyelectrolyte shells is highly relevant. Here again, the structural perfection and functional versatility of PPDs are of utmost importance.

These few points outlined above do not define the end of the polyphenylene dendrimer story, but rather open numerous opportunities for further work, some of which are be mentioned below:

(i) While it is possible to build electrolyte functions into the interior or onto the surface of a strongly hydrophobic polyphenylene dendrimer, it is challenging to have one single cationic or anionic function in the dendrimer. This is already the case in the Ru^{2+}-centered PPDs. The idea is to surround a single charge by rigid, unpolar encapsulation, which would create totally new questions for the chemistry and physics of ion pairs.

(ii) Decorating PPDs with multiple cationic charges leads to opportunities of making polyelectrolyte–polyelectrolyte complexes for example with DNA molecules. The resulting question of DNA transfection into cells strongly suggests a comparison between surface functionalized dendrimers and other viral and nonviral DNA carriers.

(iii) The endcapping of surface functionalized dendrimers allows the simultaneous introduction of functional groups which would otherwise avoid each other, such as polar and unpolar or alkyl and perfluoro alkyl moieties. This will again create exciting questions regarding solute–solvent interactions and transport properties.

(iv) Polyphenylene dendrimers are carbon-rich macromolecules, although the benzene repeat units are not in conjugative interaction. In the case of properly chosen topologies, the dendrimers can be transformed into giant graphene molecules by an oxidative cyclodehydrogenation. This leads to a whole new family of graphene molecules with outstanding electronic and self-assembly properties.

These few examples are far from complete. They are meant to encourage the reader to follow the current literature since there is more to come.

ACKNOWLEDGMENT

The authors gratefully acknowledge support from the SFB 625 and the interuniversity attraction poole (IUAP) Belgium.

REFERENCES

[1] A. N. Shipway, E. Katz, I. Willner, *Chemphyschem* **2000**, *1*, 18.
[2] J. H. Fendler, *Chem. Mater.* **1996**, *8*, 1616.
[3] Y. Lin, A. Boker, J. B. He, K. Sill, H. Q. Xiang, C. Abetz, X. F. Li, J. Wang, T. Emrick, S. Long, Q. Wang, A. Balazs, T. P. Russell, *Nature* **2005**, *434*, 55.
[4] A. R. Clapp, I. L. Medintz, J. M. Mauro, B. R. Fisher, M. G. Bawendi, H. Mattoussi, *J. Am. Chem. Soc.* **2004**, *126*, 301.
[5] A. D. Yoffe, *Adv. Phys.* **2001**, *50*, 1.
[6] Y. Kakizawa, K. Kataoka, *Adv. Drug. Deliv. Rev.* **2002**, *54*, 203.
[7] R. M. Crooks, M. Q. Zhao, L. Sun, V. Chechik, L. K. Yeung, *Accounts. Chem. Res.* **2001**, *34*, 181.
[8] D. Luo, W. M. Saltzman, *Nat. Biotechnol.* **2000**, *18*, 33.
[9] S. Hecht, J. M. J. Frechet, *Angew. Chem. Int. Ed.* **2001**, *40*, 74.
[10] A. Roucoux, J. Schulz, H. Patin, *Chem. Rev.* **2002**, *102*, 3757.
[11] R. Scherrenberg, B. Coussens, P. van Vliet, G. Edouard, J. Brackman, E. de Brabander, K. Mortensen, *Macromolecules* **1998**, *31*, 456.
[12] R. C. Triulzi, M. Micic, S. Giordani, M. Serry, W. A. Chiou, R. M. Leblanc, *Chem. Commun.* **2006**, 5068.
[13] U. M. Wiesler, T. Weil, K. Müllen, *Dendrimers III: Design, Dimension, Function*, Vol. 212, Springer, Berlin, **2001**, p. 1.
[14] R. E. Bauer, A. C. Grimsdale, K. Müllen, *Top. Curr. Chem.* **2005**, *245*, 253.
[15] A. J. Berresheim, M. Müller, K. Müllen, *Chem. Rev.* **1999**, *99*, 1747.
[16] A. C. Grimsdale, K. Müllen, *Angew. Chem. Int. Ed.* **2005**, *44*, 5592.
[17] M. Fischer, F. Vogtle, *Angew. Chem. Int. Ed.* **1999**, *38*, 885.

[18] F. Morgenroth, E. Reuther, K. Müllen, *Angew. Chem. Int. Ed.* **1997**, *36*, 631.
[19] P. G. Degennes, H. Hervet, *J. Phys. Lett.* **1983**, *44*, L351.
[20] T. H. Mourey, S. R. Turner, M. Rubinstein, J. M. J. Frechet, C. J. Hawker, K. L. Wooley, *Macromolecules* **1992**, *25*, 2401.
[21] H. Hart, *Pure Appl. Chem.* **1993**, *65*, 27.
[22] S. I. Kozhushkov, T. Haumann, R. Boese, A. Demeijere, *Angew. Chem. Int. Ed. Engl.* **1993**, *32*, 401.
[23] J. S. Moore, *Acc. Chem. Res.* **1997**, *30*, 402–413.
[24] T. M. Miller, T. X. Neenan, R. Zayas, H. E. Bair, *J. Am. Chem. Soc.* **1992**, *114*, 1018.
[25] U. M. Wiesler, A. J. Berresheim, F. Morgenroth, G. Lieser, K. Müllen, *Macromolecules* **2001**, *34*, 187.
[26] M. Müller, C. Kubel, K. Müllen, *Chem. Eur. J.* **1998**, *4*, 2099.
[27] F. Morgenroth, K. Müllen, *Tetrahedron* **1997**, *53*, 15349.
[28] G. Mihov, I. Scheppelmann, K. Müllen, *J. Org. Chem.* **2004**, *69*, 8029.
[29] A. C. Grimsdale, R. Bauer, T. Weil, N. Tchebotareva, J. S. Wu, M. Watson, K. Müllen, *Synthesis* **2002**, 1229.
[30] M. Schlupp, T. Weil, A. J. Berresheim, U. M. Wiesler, J. Bargon, K. Müllen, *Angew. Chem. Int. Ed.* **2001**, *40*, 4011.
[31] R. E. Bauer, V. Enkelmann, U. M. Wiesler, A. J. Berresheim, K. Müllen, *Chem. Eur. J.* **2002**, *8*, 3858.
[32] M. Wind, K. Saalwachter, U. M. Wiesler, K. Müllen, H. W. Spiess, *Macromolecules* **2002**, *35*, 10071.
[33] M. Wind, U. M. Wiesler, K. Saalwachter, K. Müllen, H. W. Spiess, *Adv. Mater.* **2001**, *13*, 752.
[34] H. Zhang, P. C. M. Grim, T. Vosch, U. M. Wiesler, A. J. Berresheim, K. Mullen, F. C. De Schryver, *Langmuir* **2000**, *16*, 9294.
[35] S. Rosenfeldt, N. Dingenouts, D. Potschke, M. Ballauff, A. J. Berresheim, K. Müllen, P. Lindner, K. Saalwachter, *J. Lumin.* **2005**, *111*, 225.
[36] P. Brocorens, E. Zojer, J. Cornil, Z. Shuai, G. Leising, K. Müllen, J. L. Bredas, *Synth. Met.* **1999**, *100*, 141.
[37] P. Carbone, A. Calabretta, M. Di Stefano, F. Negri, K. Müllen, *J. Phys. Chem. A* **2006**, *110*, 2214.
[38] S. Bernhardt, M. Baumgarten, M. Wagner, K. Müllen, *J. Am. Chem. Soc.* **2005**, *127*, 12392.
[39] E. V. Andreitchenko, C. G. Clark, R. E. Bauer, G. Lieser, K. Müllen, *Angew. Chem. Int. Ed.* **2005**, *44*, 6348.
[40] M. Li, H. Schnablegger, S. Mann, *Nature* **1999**, *402*, 393.
[41] H. Colfen, S. Mann, *Angew. Chem. Int. Ed.* **2003**, *42*, 2350.
[42] C. G. Clark, R. J. Wenzel, E. V. Andreitchenko, W. Steffen, R. Zenobi, K. Müllen, *J. Am. Chem. Soc.* **2007**, *129*, 3292.
[43] C. G. Clark, R. J. Wenzel, E. V. Andreitchenko, W. Steffen, R. Zenobi, K. Müllen, *New J. Chem.* **2007**, *31*, 1300.
[44] S. Müller, K. Müllen, *Philos. Trans. R. Soc. A Math. Phys. Eng. Sci.* **2007**, *365*, 1453.
[45] M. D. Watson, A. Fechtenkotter, K. Müllen, *Chem. Rev.* **2001**, *101*, 1267.

[46] W. Pisula, Z. Tomovic, C. Simpson, M. Kastler, T. Pakula, K. Müllen, *Chem. Mater.* **2005**, *17*, 4296.

[47] C. D. Simpson, J. D. Brand, A. J. Berresheim, L. Przybilla, H. J. Rader, K. Müllen, *Chem. Eur. J.* **2002**, *8*, 1424.

[48] L. J. Zhi, K. Müllen, *J. Mater. Chem.* **2008**, *18*, 1472.

[49] R. E. Bauer, A. C. Grimsdale, K. Müllen, *Top. Curr. Chem.* **2005**, *245*, 253.

[50] S. Bernhardt, M. Kastler, V. Enkelmann, M. Baumgarten, K. Müllen, *Chem. Eur. J.* **2006**, *12*, 6117.

[51] I. Oesterling, K. Müllen, *J. Am. Chem. Soc.* **2007**, *129*, 4595.

[52] C. Kohl, T. Weil, J. Q. Qu, K. Müllen, *Chem. Eur. J.* **2004**, *10*, 5297.

[53] S. Jordens, G. De Belder, M. Lor, G. Schweitzer, M. Van der Auweraer, T. Weil, E. Reuther, L. Müllen, F. C. De Schryver, *Photochem. Photobiol. Sci.* **2003**, *2*, 177.

[54] G. Schweitzer, R. Gronheid, S. Jordens, M. Lor, G. De Belder, T. Weil, E. Reuther, M. Müllen, F. C. De Schryver, *J. Phys. Chem. A* **2003**, *107*, 3199.

[55] M. C. Haberecht, J. M. Schnorr, E. V. Andreitchenko, C. G. Christopher, M. Wagner, K. Müllen, *Angew. Chem. Int. Ed.* **2008**, *47*, 1662.

[56] T. Weil, U. M. Wiesler, A. Herrmann, R. Bauer, J. Hofkens, F. C. De Schryver, K. Müllen, *J. Am. Chem. Soc.* **2001**, *123*, 8101.

[57] S. Bernhardt, M. Baumgarten, K. Müllen, *Eur. J. Org. Chem.* **2006**, 2523.

[58] D. P. Rillema, D. S. Jones, H. A. Levy, *J. Chem. Soc. Chem. Commun.* **1979**, 849.

[59] M. Sliwa, C. Flors, I. Oesterling, J. Hotta, K. Müllen, F. C. De Schryver, J. Hofkens, *J. Phys. Condens. Matter* **2007**, 445004 (14 pp.).

[60] G. De Belder, G. Schweitzer, S. Jordens, M. Lor, S. Mitra, J. Hofkens, S. De Feyter, M. Van der Auweraer, A. Herrmann, T. Weil, K. Müllen, F. C. De Schryver, *Chemphyschem* **2001**, *2*, 49.

[61] J. Hofkens, M. Maus, T. Gensch, T. Vosch, M. Cotlet, F. Kohn, A. Herrmann, K. Mullen, F. De Schryver, *J. Am. Chem. Soc.* **2000**, *122*, 9278.

[62] H. Zollinger, *Color Chemistry*, Wiley-VCH, Weinheim, **2003**.

[63] E. V. Andreitchenko, R. E. Bauer, C. Kreutz, M. Baumgarten, J. Bargon, K. Müllen, *Macromolecules* **2008**, *41*, 548.

[64] R. E. Bauer, C. G. Clark, K. Müllen, *New J. Chem.* **2007**, *31*, 1275.

[65] M. Z. Yin, K. Ding, R. A. Gropeanu, J. Shen, R. Berger, T. Weil, K. Müllen, *Biomacromolecules* **2008**, *9*, 3231.

[66] Z. B. Shifrina, M. S. Averina, N. V. Firsova, A. L. Rusanov, K. Muellen, *Doklady Chem.* **2005**, *400*, 34.

[67] Z. B. Shifrina, M. S. Rajadurai, N. V. Firsova, L. M. Bronstein, X. L. Huang, A. L. Rusanov, K. Muellen, *Macromolecules* **2005**, *38*, 9920.

[68] J. Loergen, D. Lubczyk, K. Muellen, R. Bauer, S. R. Waldvogel, Eur. Pat. DE102008008660, WO2009101118.

[69] H. John, R. Bauer, P. Espindola, P. Sonar, J. Heinze, K. Müllen, *Angew. Chem. Int. Ed.* **2005**, *44*, 2447.

[70] J. Q. Qu, N. G. Pschirer, D. J. Liu, A. Stefan, F. C. De Schryver, K. Müllen, *Chem. Eur. J.* **2004**, *10*, 528.

[71] M. P. Stevens, *Polymer Chemistry*, Oxford University Press, New York, **1999**.

[72] A. Taubert, U. M. Wiesler, K. Mullen, *J. Mater. Chem.* **2003**, *13*, 1090.

[73] M. Yin, C. R. W. Kuhlmann, K. Sorokina, C. Li, G. Mihov, E. Pietrowski, K. Koynov, M. Klapper, H. J. Luhmann, K. Müllen, T. Weil, *Biomacromolecules* **2008**, *9*, 1381.

[74] M. Yin, J. Shen, G. O. Pflugfelder, K. Müllen, *J. Am. Chem. Soc.* **2008**, *130*, 7806.

[75] M. Z. Yin, Y. J. Cheng, M. Y. Liu, J. S. Gutmann, K. Müllen, *Angew. Chem. Int. Ed.* **2008**, *47*, 8400.

[76] M. Z. Yin, J. Shen, R. Gropeanu, G. O. Pflugfelder, T. Weil, K. Müllen, *Small* **2008**, *4*, 894.

[77] M. Mondeshki, G. Mihov, R. Graf, H. W. Spiess, K. Müllen, P. Papadopoulos, A. Gitsas, G. Floudas, *Macromolecules* **2006**, *39*, 9605.

[78] D. H. Kim, J. L. Hernandez-Lopez, J. Y. Liu, G. Mihov, L. J. Zhi, R. E. Bauer, D. Grebel-Kohler, M. Klapper, T. Weil, K. Müllen, S. Mittler, W. Knoll, *Macromol. Chem. Phys.* **2005**, *206*, 52.

[79] G. Mihov, D. Grebel-Koehler, A. Lubbert, G. W. M. Vandermeulen, A. Herrmann, H. A. Klok, K. Müllen, *Bioconjugate Chem.* **2005**, *16*, 283.

[80] A. Herrmann, G. Mihov, G. W. M. Vandermeulen, H. A. Klok, K. Müllen, *Tetrahedron* **2003**, *59*, 3925.

[81] A. L. Linsebigler, G. Q. Lu, J. T. Yates, *Chem. Rev.* **1995**, *95*, 735.

[82] R. Wang, K. Hashimoto, A. Fujishima, M. Chikuni, E. Kojima, A. Kitamura, M. Shimohigoshi, T. Watanabe, *Nature* **1997**, *388*, 431.

[83] H. J. Snaith, L. Schmidt-Mende, *Adv. Mater.* **2007**, *19*, 3187.

[84] T. Weil, E. Reuther, K. Müllen, *Angew. Chem. Int. Ed.* **2002**, *41*, 1900.

[85] C. J. Hawker, K. L. Wooley, *Science* **2005**, *309*, 1200.

[86] R. Bauer, D. Liu, A. Ver Heyen, F. C. De Schryver, S. De Feyter, K. Müllen, *Macromolecules* **2007**, *40*, 4753.

[87] B. Voit, *New J. Chem.* **2007**, *31, 1139.*

[88] T. Vossmeyer, B. Guse, I. Besnard, R. E. Bauer, K. Mullen, A. Yasuda, *Adv. Mater.* **2002**, *14*, 238.

[89] N. Krasteva, Y. Fogel, R. E. Bauer, K. Mullen, Y. Joseph, N. Matsuzawa, A. Yasuda, T. Vossmeyer, *Adv. Funct. Mater.* **2007**, *17*, 881.

[90] N. Krasteva, I. Besnard, B. Guse, R. E. Bauer, K. Mullen, A. Yasuda, T. Vossmeyer, *Nano Lett.* **2002**, *2*, 551.

[91] A. Kraft, A. C. Grimsdale, A. B. Holmes, *Angew. Chem. Int. Ed.* **1998**, *37*, 402.

[92] U. Mitschke, P. Bauerle, *J. Mater. Chem.* **2000**, *10*, 1471.

[93] S. Nakamura, *Science* **1998**, *281*, 956.

[94] F. A. Ponce, D. P. Bour, *Nature* **1997**, *386*, 351.

[95] U. Scherf, E. J. W. List, *Adv. Mater.* **2002**, *14*, 477.

[96] K. Müllen, U. Scherf, *Organic Light Emitting Devices: Synthesis, Properties and Applications*, Wiley-VCH, Weinheim, **2006**.

[97] T. S. Qin, G. Zhou, H. Scheiber, R. E. Bauer, M. Baunigarten, C. E. Anson, E. J. W. List, K. Müllen, *Angew. Chem. Int. Ed.* **2008**, *47*, 8292.

[98] Y. R. Sun, N. C. Giebink, H. Kanno, B. W. Ma, M. E. Thompson, S. R. Forrest, *Nature* **2006**, *440*, 908.

[99] M. A. Baldo, D. F. O'Brien, Y. You, A. Shoustikov, S. Sibley, M. E. Thompson, S. R. Forrest, *Nature* **1998**, *395*, 151.

[100] M. A. Baldo, M. E. Thompson, S. R. Forrest, *Nature* **2000**, *403*, 750.

[101] S. Welter, K. Brunner, J. W. Hofstraat, L. De Cola, *Nature* **2003**, *421*, 54.

[102] M. A. Baldo, S. Lamansky, P. E. Burrows, M. E. Thompson, S. R. Forrest, *Appl. Phys. Lett.* **1999**, *75*, 4.

[103] S. Lamansky, P. Djurovich, D. Murphy, F. Abdel-Razzaq, H. E. Lee, C. Adachi, P. E. Burrows, S. R. Forrest, M. E. Thompson, *J. Am. Chem. Soc.* **2001**, *123*, 4304.

[104] S. C. Lo, N. A. H. Male, J. P. J. Markham, S. W. Magennis, P. L. Burn, O. V. Salata, I. D. W. Samuel, *Adv. Mater.* **2002**, *14*, 975.

[105] J. Q. Ding, J. Gao, Y. X. Cheng, Z. Y. Xie, L. X. Wang, D. G. Ma, X. B. Jing, F. S. Wang, *Adv. Funct. Mater.* **2006**, *16*, 575.

[106] E. Holder, B. M. W. Langeveld, U. S. Schubert, *Adv. Mater.* **2005**, *17*, 1109.

[107] M. K. Nazeeruddin, R. Humphry-Baker, D. Berner, S. Rivier, L. Zuppiroli, M. Graetzel, *J. Am. Chem. Soc.* **2003**, *125*, 8790.

[108] S. J. Yeh, M. F. Wu, C. T. Chen, Y. H. Song, Y. Chi, M. H. Ho, S. F. Hsu, C. H. Chen, *Adv. Mater.* **2005**, *17*, 285.

[109] T. S. Qin, J. Q. Ding, L. X. Wang, M. Baumgarten, G. Zhou, K. Müllen, *J. Am. Chem. Soc.* **2009**, *131*, 14329.

6

DENDRIMER CHEMISTRY WITH FULLERENES

Jean-François Nierengarten

Laboratoire de Chimie des Matériaux Moléculaires, Université de Strasbourg et CNRS, Ecole Européenne de Chimie, Polymères et Matériaux (ECPM), 25 rue Becquerel, 67087 Strasbourg Cedex 2, France

6.1 INTRODUCTION

In light of their unique molecular structure, dendrimers have attracted increasing attention in the past years and the design of functional dendrimers is an area with unlimited possibilities for fundamental new discoveries and practical applications [1]. Of the various electro- and photoactive chromophores used in dendrimer chemistry, C_{60} appears to be a versatile building block and at present a growing interest is developing in fullerene-functionalized dendrimers, that is, *fullerodendrimers* [2,3]. In particular, the unusual chemical and physical properties of fullerene derivatives [4] make fullerodendrimers attractive candidates for a variety of interesting features in supramolecular chemistry and materials science. Dendritic structures have been used as supramolecular receptors for fullerenes or as templates for the self-assembly of fullerene-rich nanostructures [5]. C_{60} itself is also a convenient core for covalent dendrimer chemistry [6,7]. Furthermore, examples of dendritic structures with fullerene units at their surface or with C_{60} spheres in the dendritic branches have been reported [7]. These different aspects of dendrimer chemistry with fullerenes will be presented therein. The aim of this chapter is not to show an exhaustive review on such systems but to present significant examples to illustrate the current state-of-the-art of fullerene chemistry for the development of new functional dendrimers. In particular, specific features resulting from the dendritic structures will be highlighted.

Designing Dendrimers, Edited by Sebastiano Campagna, Paola Ceroni, Fausto Puntoriero.
© 2012 John Wiley & Sons, Inc. Published 2012 by John Wiley & Sons, Inc.

6.2 SUPRAMOLECULAR CHEMISTRY WITH FULLERENES AND DENDRIMERS

In light of their multifunctionality and specific shape, dendrimers are attractive candidates for applications in fullerene supramolecular chemistry [8]. Several approaches are possible when using such architectures for the complexation of fullerene derivatives. Indeed, all three topologically distinct regions (core, branching shell, and surface) of dendrimers can be used for association with fullerene substrates.

6.2.1 Host–Guest Chemistry with Fullerenes and Dendrimers

The formation of host–guest complexes between C_{60} and various dendritic receptors has been reported [9,10]. For example, Shinkai and coworkers have shown that the branching shell of polybenzyl ether dendrimers is capable of providing the space size for the inclusion of C_{60} [9]. In particular, polybenzyl ether dendrimers with a phloroglucinol central core (Figure 6.1) have been investigated in organic solvents, by means of ^{13}C-NMR and UV-vis spectroscopy.

The formation of complexes between [60]fullerene and **1**–**3** in toluene solution was followed by the continuous changes observed in the UV-vis spectra upon successive additions of dendrimers to solutions of fullerene. Assuming a 1:1 stoichiometry, the association constants (K_a) for the binding of C_{60} to **1**, **2**, and **3** were estimated to be 5 ± 2, 12 ± 1, and $68 \pm 4\,M^{-1}$, respectively [10]. The increase of the K_a values in parallel with the generation number of the dendrimers, has been attributed to the increasing number of aromatic units, thus allowing a higher number of π–π interactions with the fullerene sphere. However, the authors also believed that along the series the interior space of the dendrimers tends to approach the appropriate size for the inclusion of C_{60}. Therefore, they assumed that the internal cavity within the dendrimers becomes more suitably defined as the dendrimer becomes larger. In order to provide more accurate evidence, ^{13}C-NMR studies were also carried out. In toluene-d_8 at 25°C, the ^{13}C-enriched (10–15%) C_{60} peak was observed at 143.200 ppm, which was shifted to 143.197 ppm in the presence of an equimolar amount of **3**. Whereas the change in chemical shift was not very informative at 25°C, it appeared significant at −60°C when the exchange became slower (from 143.232 to 142.942 ppm in the presence of **3**). Several peaks in **3** were also shifted but the largest shift was observed for the phloroglucinol carbons. These results suggest that the C_{60} guest mainly resides in the space present around the phloroglucinol moiety. A similar trend in the chemical shift change was also observed for the complexation with **1** and **2**, but the magnitude was smaller than that observed for **3**. These important observations were in full agreement with their hypothesis that the interior space of the branching shell located close to the central core of the polyaryl ether dendrimers is capable of providing the space size necessary for the inclusion of [60]fullerene.

Several macrocyclic host systems have been used as supramolecular receptors for fullerenes, among them cyclotriveratrylene (CTV) derivatives [11]. Even if the shape and the dimension of the CTV macrocycle appears well predetermined for the complexation of the fullerene sphere, the binding constants for C_{60} found in organic

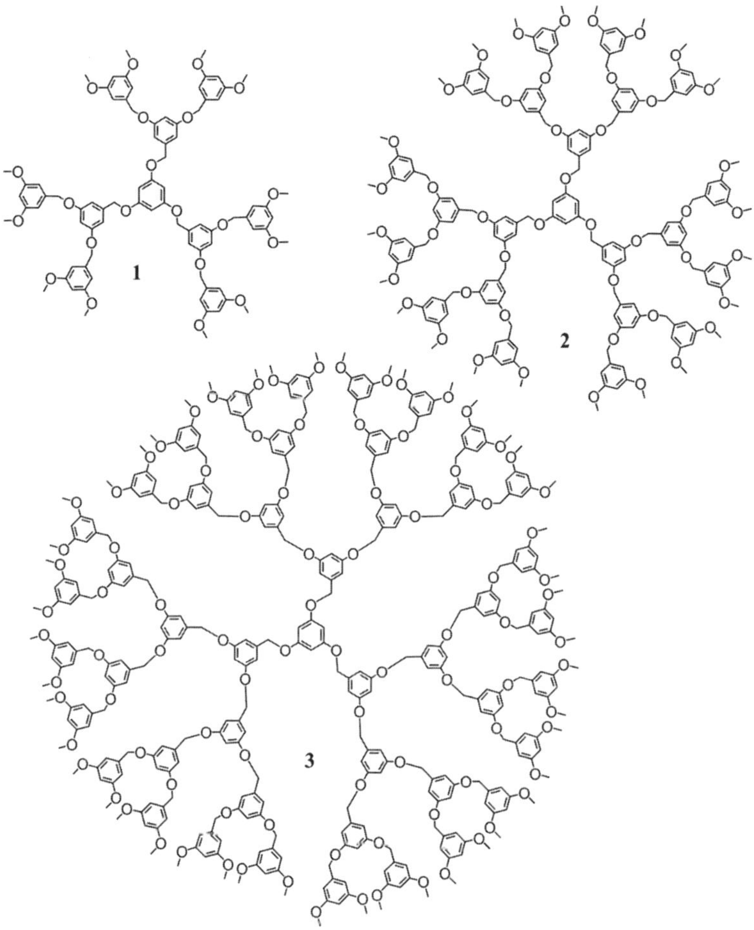

FIGURE 6.1 Dendrimers **1–3** used for the complexation of C_{60}.

solvents with CTV derivatives are typically quite low [10]. This may be related to the fact that only a small part of the C_{60} surface is actually in contact with the cavity of the macrocyclic CTV receptor. Based on the results obtained with dendrimers **1**–**3**, CTV derivatives bearing polybenzyl ether dendrons should actually provide an internal cavity capable of interacting with the main part of the fullerene's surface. Therefore, one can expect that the surrounding dendritic branches will be able to increase the inclusion abilities of the CTV central core for fullerenes. Dendritic CTV derivatives **4**–**7** (Figure 6.2) have been prepared and their inclusion abilities for C_{60} systematically investigated [12].

The formation of host–guest complexes between C_{60} and **4**–**7** was shown by the continuous changes observed in the UV-vis spectra upon successive additions of the host to C_{60} solutions [10]. Treatment of the titration data with the Benesi-Hildebrand equation gave the association constants values reported in Table 6.1. For all the

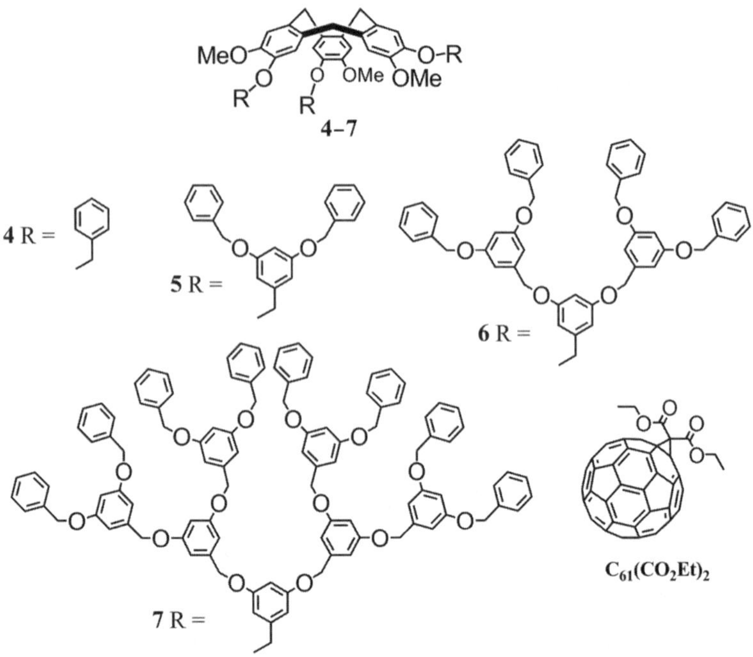

FIGURE 6.2 Dendritic receptors **4–7** with a CTV core and guest $\mathbf{C_{61}(CO_2Et)_2}$.

dendritic CTV derivatives, Job's plot provided good evidence for 1:1 complex formation in solution. The formation of host–guest complexes in CH_2Cl_2 or C_6H_6 solutions between methanofullerene $\mathbf{C_{61}(CO_2Et)_2}$ (Figure 6.2) and the dendritic CTV derivatives was also investigated. The association constant values, determined by standard UV-vis titrations, are summarized in Table 6.1.

Interestingly, in both solvents and for both substrates, the K_a values are significantly increased as the generation number of the dendritic substituents is increased. Therefore, the surrounding dendritic branches appear to be able to increase the

TABLE 6.1 Association Constants K_a (M^{-1}) from UV-vis Binding Titrations for 1:1 Complexes of C_{60} and $C_{61}(CO_2Et)_2$ to 4–7 in CH_2Cl_2 and C_6H_6 at 298K[a]

	C_{60}		$C_{61}(CO_2Et)_2$	
	CH_2Cl_2	C_6H_6	CH_2Cl_2	C_6H_6
4	85 ± 20	%[b]	90 ± 30	%[b]
5	120 ± 20	115 ± 15	130 ± 20	110 ± 30
6	200 ± 10	190 ± 20	190 ± 20	180 ± 20
7	340 ± 20	345 ± 20	300 ± 30	290 ± 30

[a] The association constants K_a have been determined by monitoring the variations of absorbance at different wavelengths in the 430–445 nm region where the strongest spectral changes are observed. Identical results within the error range in duplicate and triplicate runs were obtained, and average values are reported.
[b] Not determined due to the poor solubility of **4** in C_6H_6.

inclusion abilities of the CTV central core for fullerenes. This effect could be attributed to additional electronic donor–acceptor π–π interactions between the polyaryl ether dendrons and the fullerene guest. However, it is also believed that the additional phenyl subunits are not the sole explanation for the observed effect; the dendritic structure itself must play an important role for the binding of C_{60} as shown with **1–3**. The CTV core itself is a poor receptor for the complexation of fullerenes in solution and its functionalization with polybenzyl ether dendrons is capable of generating an internal cavity with more appropriate shape and dimension for interactions with the fullerene guest. This effect is more pronounced as the size of the dendritic substituents is increased. As a result, the K_a values increases when the surrounding dendrons become larger. Based on this observation, Nierengarten and coworkers have shown that the properties of CTV–C_{60} supramolecular complexes can also be modulated by the nature of the surrounding dendrons (Figure 6.3). On the one hand, CTV derivatives with peripheral triethyleneglycol chains have been used for the preparation of CTV–C_{60} complexes soluble in water thus making them interesting candidates for biological applications [13]. On the other hand, by using a CTV derivative surrounded with long alkyl chains, a liquid-crystalline CTV–C_{60} host–guest complex was obtained thus showing that the supramolecular chemistry of fullerenes can be used for the preparation of ordered fullerene assemblies easy to process for applications in materials science [14,15].

6.2.2 Self-Assembly of Fullerenes on Dendritic Templates

C_{60} itself and simple fullerene derivatives have been assembled on dendritic templates by using purely electrostatic interactions. For example, Astruc and coworkers have taken advantage of the electrochemical properties of C_{60} to assemble a fullerene-rich supramolecular dendritic structure [16]. When a toluene solution of C_{60} (64 equiv.) was added to an acetonitrile solution of the 64-Fe(I) dendrimer **10** (1 equiv.), a black precipitate of **11** was obtained (Figure 6.4). Tentative extraction of this precipitate with toluene yielded a colorless solution, thus indicating that no C_{60} was present. The Mössbauer spectrum of this black solid **11** is a clean quadrupole doublet whose parameters at 77K are consistent with the presence of an Fe(II) sandwich complex. Its EPR spectrum recorded at 298K shows the characteristic feature observed for a model compound obtained from the reaction of C_{60} with the 19-electron complex [Fe(I)Cp (η^6-C_6Me_6)] [17]. It can thus be concluded that C_{60} has been reduced to its monoanion, as designed for a process that is exergonic by 0.9 eV. The peripheral cationic Fe(II) units with their C_{60}^- counteranion being very large, they are most likely located at the dendrimer periphery, presumably with rather tight ion pairs although the number of fullerene layers and overall molecular size are unknown.

Another system based on electrostatic interactions has been reported by van Koten and coworkers [18]. A core-shell dendrimer with a cationic tetra[bis(benzylammonium)aryl] silane core has been used as a template for the assembly of fullerene-carboxylate derivatives via a straightforward anion exchange reaction of [**12**]Br_8 with Na[**13**] (Figure 6.5). In contrast to dendrimer **40**, the supramolecular fullerene-rich assembly [**12**][**13**]$_8$ thus prepared is soluble in common organic solvents and its

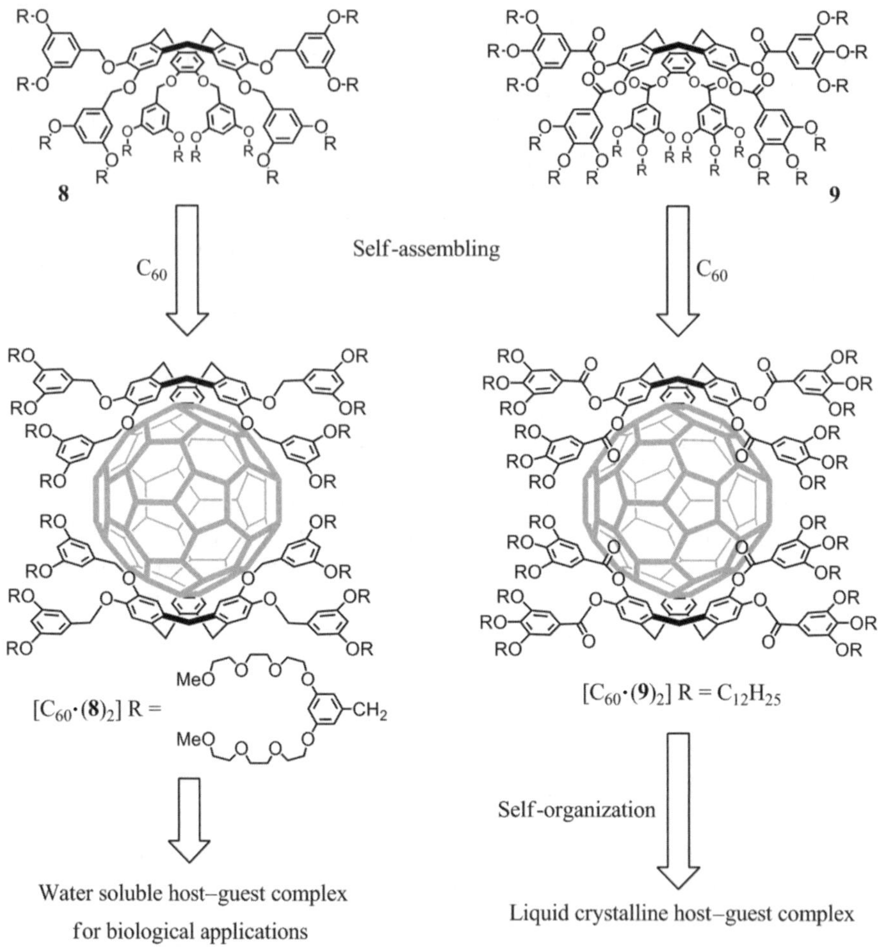

FIGURE 6.3 CTV receptors with different peripheral groups leading to supramolecular systems with new properties upon complexation of C_{60}.

spectroscopic characterization was easily achieved. In the ^1H NMR spectrum of a solution of [**12**][**13**]$_8$ in $CDCl_3$, the diagnostic signals of both [**12**]$^{8+}$ and [**13**]$^-$ can clearly be observed. Furthermore, specific peak integrals show that the octa-cationic dendritic moiety [**12**]$^{8+}$ and the anions [**13**]$^-$ are present in a 1 to 8 molar ratio.

The UV-absorption spectrum of a solution of [**12**][**13**]$_8$ in CH_2Cl_2 shows characteristic absorption features that can be ascribed to the fullerene units as well as a band diagnostic for the cationic dendrimer. Actually, the UV-absorption spectrum of [**12**][**13**]$_8$ matches the spectral profile of a 1 to 8 molar mixture of authentic [**12**]Br_8 and the *t*-butyl ester of the [**13**]$^-$ carboxylate, which is unable to self-assemble. The similarity of these spectra indicates that there are no significant ground state interactions between the respective chromophores in the supramolecular complex [**12**][**13**]$_8$. The luminescence properties of solutions of [**12**][**13**]$_8$ in CH_2Cl_2 were also

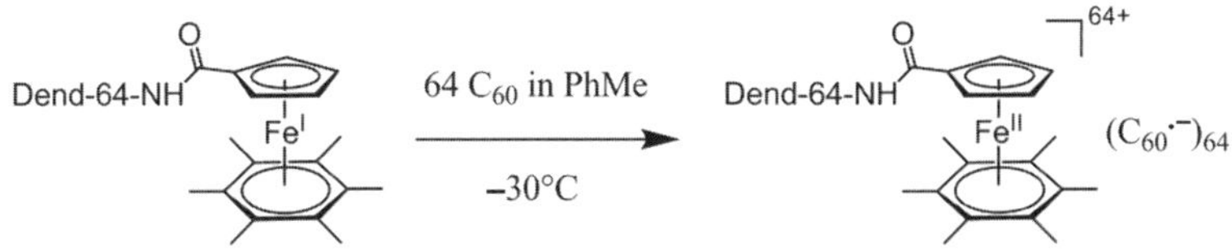

10 in MeCN **11** (black precipitate)

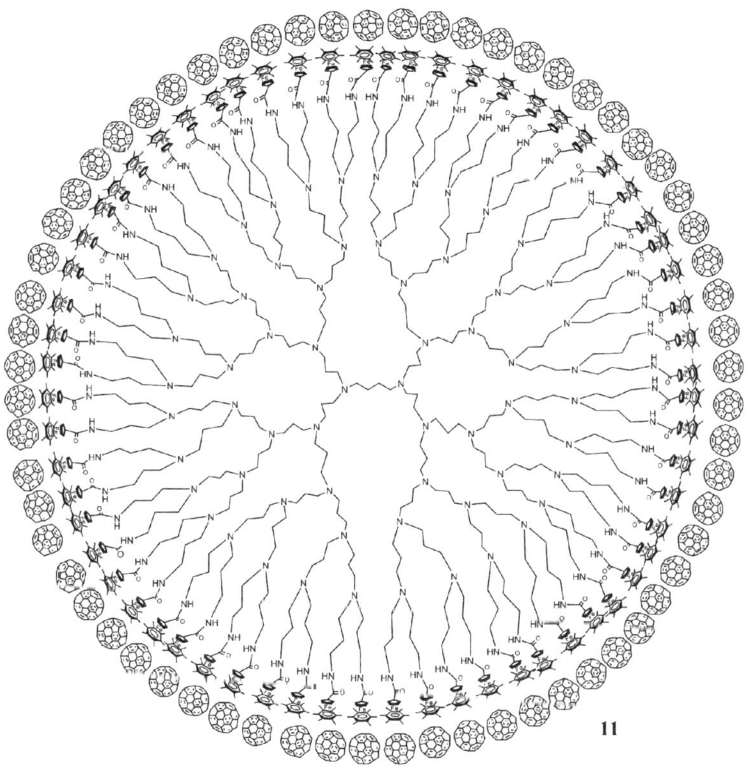

11

FIGURE 6.4 Dendrimer **11** resulting from the reaction of **10** (1 equiv.) with C_{60} (64 equiv.) in MeCN/PhMe at −30°C.

investigated and compared to the behavior of an isoabsorbing model mixture of authentic [**12**]Br_8 and the *t*-butyl ester derived from the [**13**]$^-$ carboxylate. Upon excitation of the latter model mixture, in the UV-region where the main part of the light is absorbed by the dendritic wedges of the octa-cationic moiety [**12**]$^{8+}$, the characteristic emission of the polybenzyl aryl ether dendritic wedges is observed. In contrast, complete quenching of this emission of [**12**]$^{8+}$ is observed in [**12**][**13**]$_8$, which suggests that upon excitation intramolecular energy transfer occurs from the lowest singlet excited state of the Fréchet-type dendritic branch to the low lying fullerene singlet excited state.

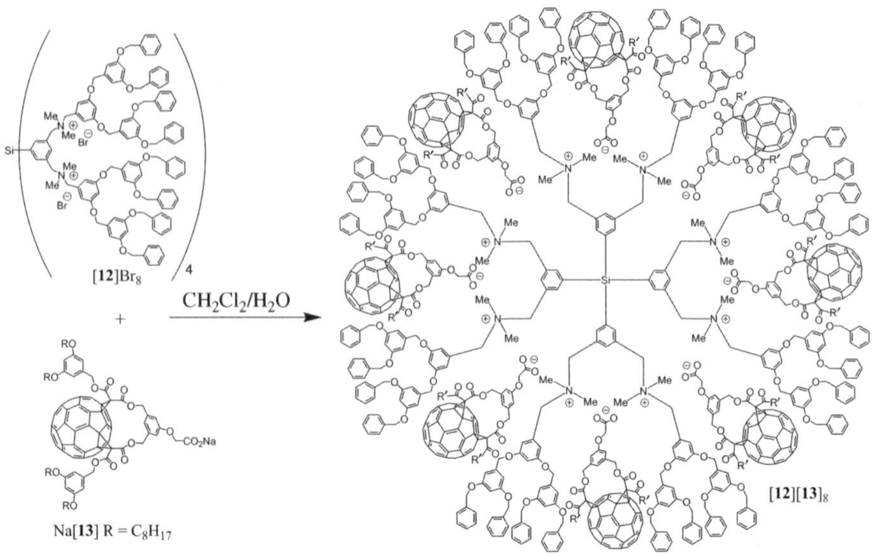

FIGURE 6.5 Anion exchange reaction of [**12**]Br_8 with Na[**13**] leading to the fullerene-rich dendrimer [**12**][**13**]$_8$.

The attractive van der Waals interaction of the fullerene sphere with the planar π-surface of porphyrins has also been exploited to produce dendritic supramolecular fullerene–porphyrin conjugates from C_{60} and multiporphyrins **14**–**16** (Figure 6.6) [19].

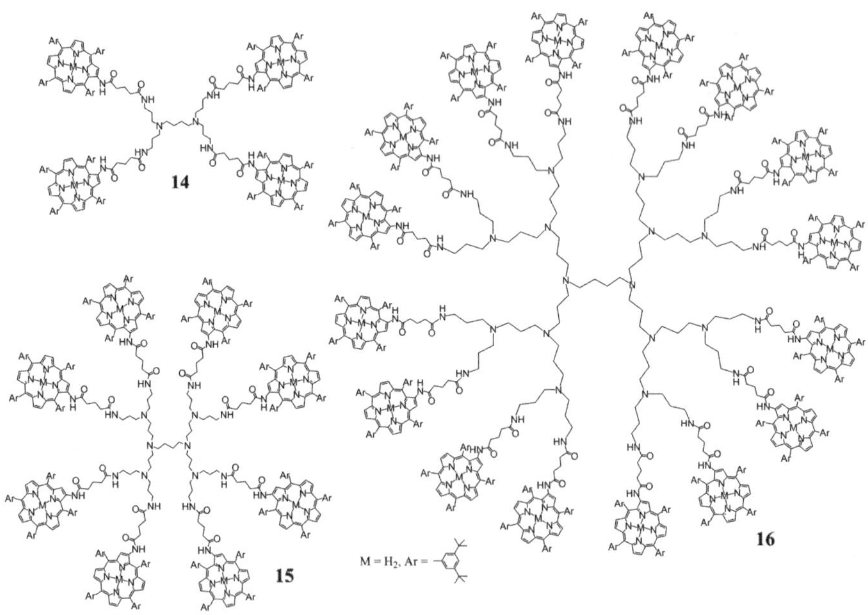

FIGURE 6.6 Dendrimers **14**–**16** with peripheral porphyrin subunits.

The supermolecules obtained from **14–16** and C_{60} have not been investigated in solution but used to successfully construct light energy conversion systems according to Kamat's method [20]. For this purpose, nanoclusters were first prepared by injecting a mixed toluene solution of **14–16** and C_{60} into an acetonitrile/toluene mixed solvent system. The electronic absorption spectra of the colloidal suspensions thus obtained are much broader than those of the starting toluene solutions, indicating the formation of clusters in the mixed solvent. Importantly, the broad long-wavelength absorption in the 700–800 nm region is diagnostic of the charge-transfer absorption band due to the complex formed between the peripheral porphyrin subunits of **14–16** and C_{60}. Upon subjecting the resulting cluster suspension to a high electric dc field, mixed **14–16**/C_{60} clusters were deposited onto an ITO electrode covered with SnO_2 nanoparticles (ITO/SnO_2) to give modified electrodes (denoted as ITO/SnO_2/(**14–16** + C_{60})$_n$). Photoelectrochemical measurements were performed in acetonitrile containing 0.5 M LiI and 0.01 M I_2 with ITO/SnO_2/(**14–16** + C_{60})$_n$ as a working electrode and a platinum wire as a counter electrode. Photocurrent generation in the present device is initiated by photo-induced charge separation from the porphyrin excited singlet state (P^*) in the dendrimer to C_{60} in the porphyrin dendrimer–C_{60} complex rather than by direct electron injection into the conduction band (CB) of the SnO_2 system, as illustrated in Figure 6.7. The reduced C_{60} ($C_{60}^{\cdot-}$) injects electrons into the SnO_2 nanocrystallites, whereas the oxidized porphyrin ($P^{\cdot+}$) undergoes an electron-transfer reduction with the iodide in the electrolyte system.

Incident photon-to-photocurrent efficiency (IPCE) values of ITO/SnO_2/(**14–15** + C_{60})$_n$/NaI + I_2/Pt devices exhibit a remarkable increase as compared with those of a reference device prepared from a simple porphyrin derivative and C_{60}. In particular, the ITO/SnO_2/(**14** + C_{60})$_n$/NaI + I_2/Pt device has a maximum IPCE value of 15% as well as an enhanced photoelectrochemical response in the UV-vis and near-IR regions (400–1000 nm). It should be however noted that the IPCE value of the ITO/SnO_2/(**14** + C_{60})$_n$/NaI + I_2/Pt device becomes smaller than the value of the reference

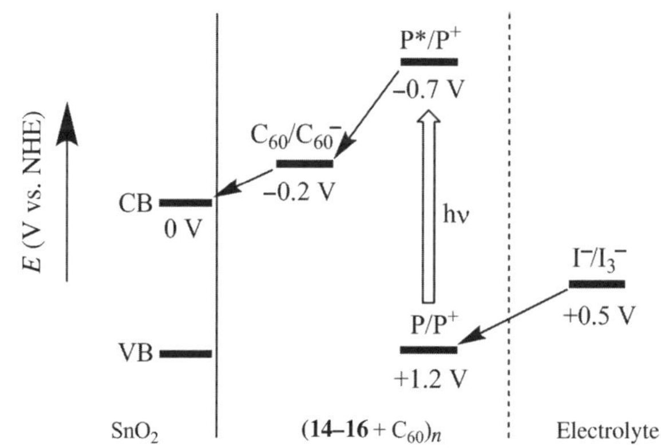

FIGURE 6.7 Energy-level diagram of light energy conversion systems using clusters of supramolecular complexes of porphyrin dendrimers **14–16** with fullerene on nanostructured SnO_2 electrodes.

device irrespective of the high generation. This has been explained by negative steric effects in the higher generation, which partially inhibits the π–π interactions with C_{60}.

Aida and coworkers [21] have reported the preparation of fullerene-rich dendritic structures resulting from the apical coordination of C_{60} derivatives bearing pyridyl moieties to dendritic molecules appended with multiple Zn(II) porphyrin units. For example, compound **17** bound **18** strongly to form stable [(**17**)·(**18**)$_{12}$] (Figure 6.8). Upon titration with **18** in $CHCl_3$ at 25°C, **17** displayed a large spectral change in the Soret and Q-bands, characteristic of the axial coordination of zinc porphyrins, with a clear saturation profile at a molar ratio **18**/**17** exceeding 12. The average binding affinity (K), as estimated by simply assuming a one-to-one coordination between the individual zinc porphyrin and pyridine units, is $1.2 \times 10^6\ M^{-1}$. This value is more than 2 orders of magnitude greater than association constants reported for monodentate coordination between zinc porphyrins and pyridine derivatives [22]. The sizeable increase of stability can be ascribed to the simultaneous coordination of two Zn centers of **17** by the two pyridine moieties of **18**. Similar increases in the

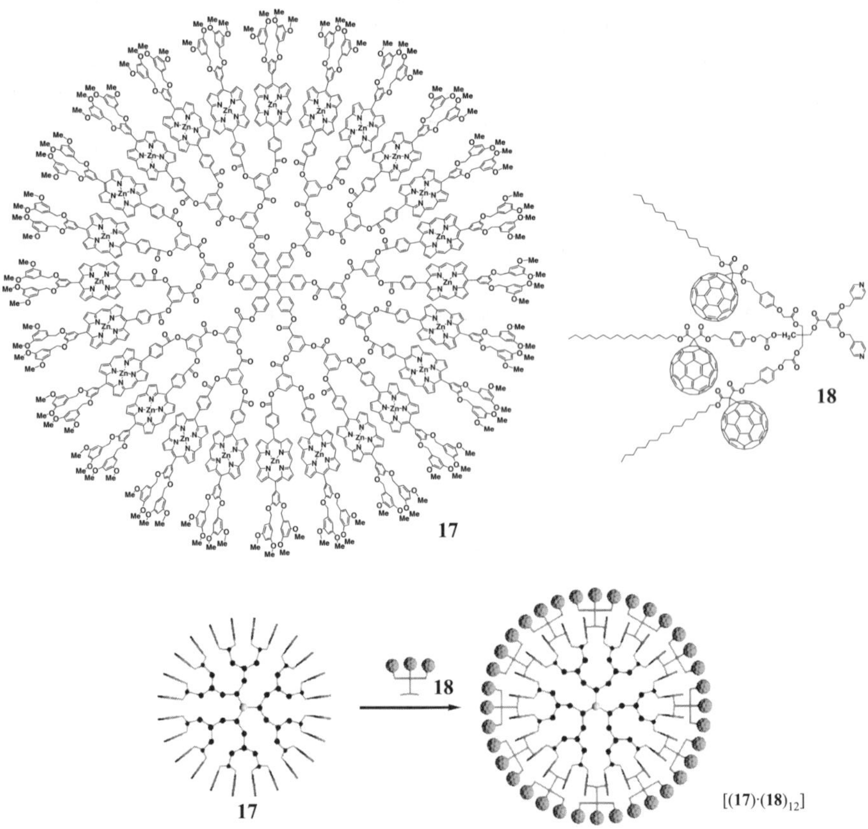

FIGURE 6.8 Compounds **17** and **18** and schematic representation of the self-assembly process leading to the fullerene-rich dendrimer [(**17**)·(**18**)$_{12}$].

association constants have been reported for supramolecular systems resulting from the axial coordination of a bis-Zn(II)-porphyrinic receptor to substrates bearing two pyridine subunits [23].

Supramolecular assembly [(**17**)·(**18**)$_{12}$] combining C_{60} units and porphyrin moieties [21] is also a photochemical molecular device. Indeed, the photophysical properties of this system have been studied in detail and an almost quantitative intramolecular photo-induced electron transfer from the photoexcited porphyrins to the C_{60} units evidenced by means of steady-state emission spectroscopy and nanosecond flash photolysis measurements. Excited-state dynamic studies have been carried out to investigate both charge-separation and charge-recombination events in [(**17**)·(**18**)$_{12}$]. The charge-separation rate constants (k_{CS}) and the charge-recombination rate constants (k_{CR}) have been thus deduced. Importantly, the k_{CS}/k_{CR} ratio for [(**17**)·(**18**)$_{12}$] is more than an order of magnitude greater than those reported for precedent porphyrin-fullerene supramolecular dyads and triads [24]. It is obvious that a larger number of the fullerene units in [(**17**)·(**18**)$_{12}$] can enhance the probability of the electron transfer from the zinc porphyrin units. However, in addition to this, one can also presume that an efficient energy migration along the densely packed Zn(II) porphyrin array [21] may enhance the opportunity for this electron transfer.

6.3 COVALENT FULLERENE-CONTAINING DENDRIMERS

6.3.1 Dendrimers with a Fullerene Core

C_{60} itself is a convenient core for dendrimer chemistry [6] and the functionalization of C_{60} with a controlled number of dendrons dramatically improves the solubility of the fullerenes [25–31]. Furthermore, variable degrees of addition within the fullerene core are possible and its almost spherical shape leads to globular systems even with low-generation dendrons [32]. In this respect, fullerene hexakis-adducts with a T_h-symmetrical octahedral addition pattern are of particular interest [33]. Whereas such compounds can be prepared in good yields from relatively simple malonates [33], structurally more complicated systems are generally obtained in rather low yields. In order to overcome this problem, Nierengarten and coworkers have recently developed a simple C_{60} hexakis-adduct derivative bearing 12 alkyne subunits and shown that the copper-mediated Huisgen 1,3-dipolar cycloaddition of azides and alkynes resulting in 1,2,3-triazoles is an ideal tool to efficiently produce dendritic hexa-substituted fullerenes **19** and **20** (Figure 6.9) [34,35].

The 1,2,3-triazole-linked fullerodendrimers **19** and **20** were thus obtained in good yields. It is worth noting that a first-generation dendritic C_{60} hexa-adduct similar to **19** was already reported by Hirsch and coworkers [32]. The latter was obtained in a low yield (5%) from C_{60} and the corresponding dendritic malonate, furthermore preparative HPLC was required for its purification. In contrast, compound **20** was prepared in a good yield (84%) and its purification was easily achieved on a silica-gel column. Furthermore, the second-generation derivative (**20**) was also obtained in a good yield (67%) under the copper-mediated Huisgen reaction

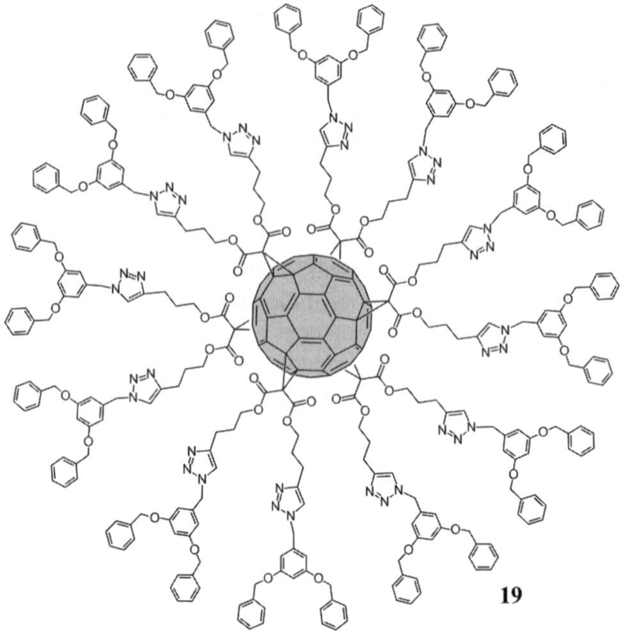

19

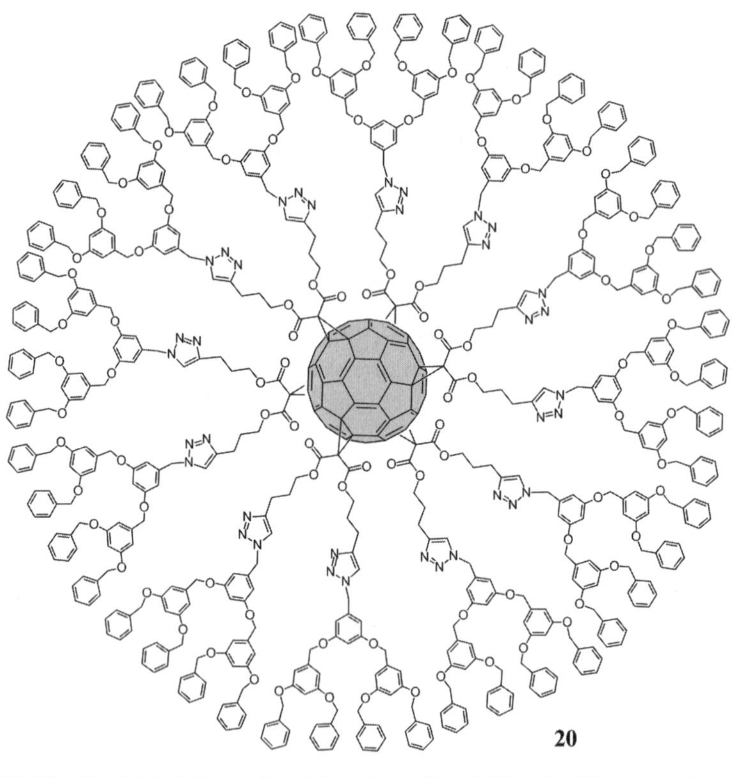

20

FIGURE 6.9 Dendritic fullerene hexakis-adducts **19** and **20** with a T_h-symmetrical octahedral addition pattern.

conditions. Actually, the preparation of such a second-generation fullerodendrimer would be nearly impossible from the corresponding malonate and C_{60}.

Specific advantages are brought about by the encapsulation of a fullerene moiety in the middle of a dendritic structure [36]. For example, the shielding effect resulting from the presence of the surrounding shell has been found useful to obtain amphiphilic derivatives with good spreading characteristics [37–39]. Effectively, the dendrons provides a compact insulating layer around the carbon sphere capable of preventing contact between neighboring fullerenes when the film is compressed, thus the irreversible aggregation usually observed for amphiphilic fullerene derivatives cannot occur. As part of this research, Nierengarten and coworkers have reported the synthesis of diblock globular fullerodendrimers **21–23** with hydrophobic chains on one hemisphere and hydrophilic groups on the other one [40] (Figure 6.10). Actually, the hydrophobic/hydrophilic balance of these dendrimers has been systematically modified by changing the size of the polar head group in order to investigate the role

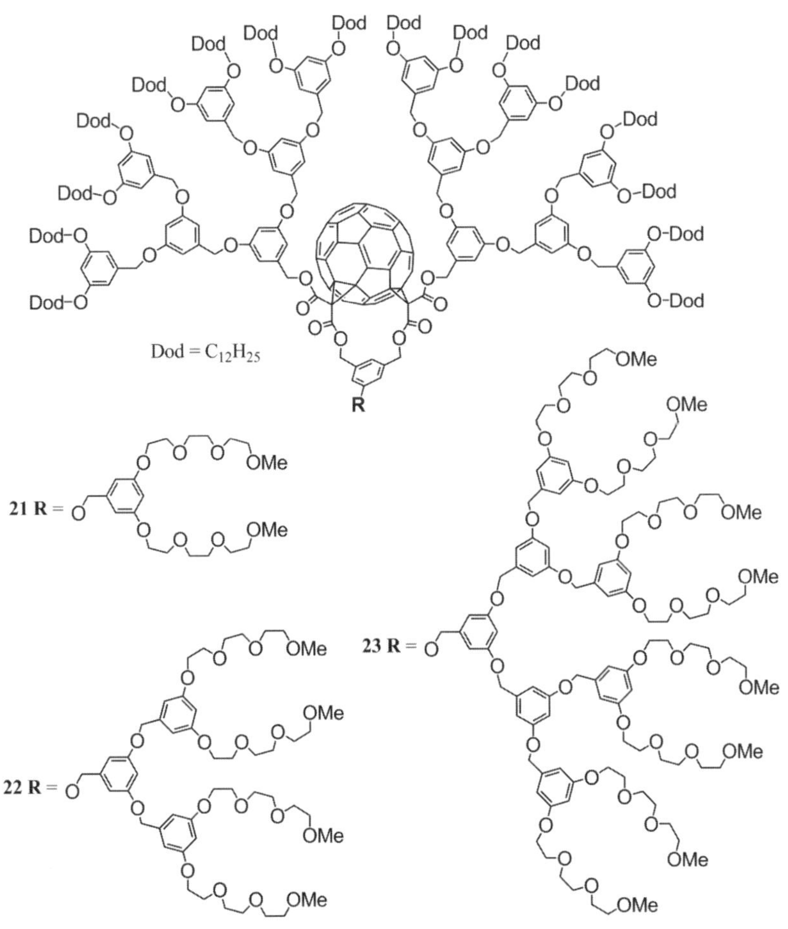

FIGURE 6.10 Dibloc fullerodendrimers **21–23**.

of the amphiphilicity at the air–water interface and during the deposition onto solid substrates.

Langmuir films have been obtained with dendrimers **21–23**. Regardless of its size, the polar-head group of these compounds is responsible for an attractive interaction with the aqueous subphase thus forcing the molecules toward the water surface in a two-dimensional arrangement. The pressure/area (Π/A) isotherms of fullerodendrimers **21–23** are depicted in Figure 6.11. The four compounds exhibit a similar behavior: the surface pressure Π increases smoothly at molecular areas between 400 and 500 Å^2 before taking a sharper rise between 250 and 350 Å^2, depending on the compound. The general shape of the isotherms indicates that the films are at first in a liquid-condensed phase. Final molecular areas A_0 extrapolated at zero surface pressure are 320 ± 3 (**21**), 295 ± 3 (**22**), and 280 ± 3 Å^2 (**23**) in good agreement with the value that can be estimated by molecular modeling. Interestingly, the A_0 values decrease when the size of the polar-head group increases. The latter observation could be ascribed to a conformational change in the dendritic structure when the anchoring on the water surface is stronger. The repulsion of the $C_{12}H_{25}$ terminated dendrons from the water surface must be more effective in the case of **22** and **23** and, as a result, the molecules adopt a more compact structure. In other words, the two $C_{12}H_{25}$ terminated dendritic branches are forced to wrap the fullerene core. In contrast, for compound **21**, the dendritic structure may be less densely packed around the central fullerene core due to the weaker anchoring on the water surface. This model is further supported by observations done on the LB films prepared from the Langmuir films of compounds **21–23** (see below).

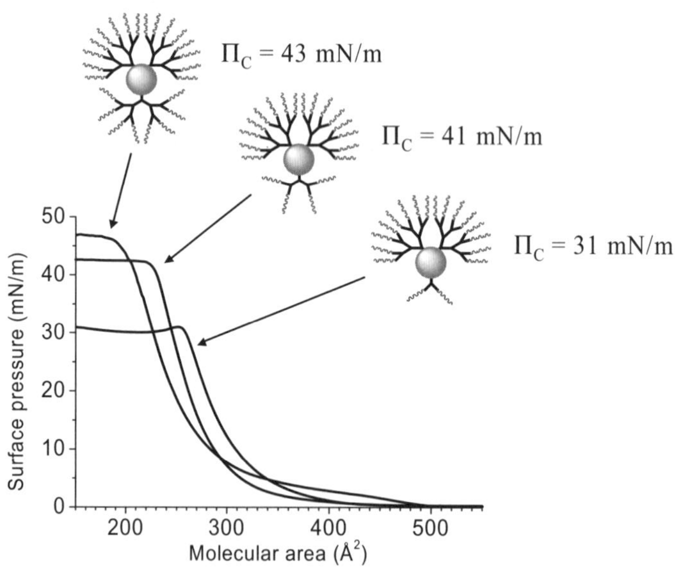

FIGURE 6.11 Pressure/area isotherms of compounds **21–23** taken on pure water at 20°C. The collapse pressure increases with the size of the polar-head group.

Using the LB technique, it has been possible to transfer the Langmuir films made from **21**–**23** and to obtain multilayers on solid substrates. In accordance with previous observations, transfer is more efficient upon increasing the size of the polar-head group. The excellent quality of the LB films prepared from these amphiphilic dendrimers is deduced from the plot of their UV-vis absorbance as a function of the layer number that results in straight lines, indicating that the films obtained from **21**–**23** grow regularly. The main feature of the UV-vis spectrum of the LB film of **21** is the broadening of the absorption in the film when compared to the solution. The latter observation is indicative of fullerene–fullerene interactions resulting from contacts of carbon spheres among neighboring layers. Remarkably, a clear evolution can be seen by going from **21** to **23**: the broadening of the absorption spectrum seen for **21** is almost vanished for **22** and **23**. Actually, the UV-vis spectra of LB films of **22** and **23** are close to the ones recorded in CH_2Cl_2 solutions suggesting limited fullerene--fullerene interactions within the LB film. This observation is in full agreement with the more compact structures proposed for **22** and **23** at the air–water interface when compared to **21**. In summary, the Langmuir studies of fullerodendrimers **21**–**23** have revealed a conformational change in the dendritic structure with the size of the polar-head group. Due to a better anchoring onto the water surface, the compounds with the largest polar-head groups adopt a more compact structure and the dendritic branches are forced to wrap the fullerene core. This model is nicely confirmed by the amount of fullerene–fullerene interactions within the LB films deduced from their absorption spectra. On the one hand, the results obtained in this systematic study show some of the fundamental architectural requirements for obtaining stable Langmuir films with amphiphilic dendrimers. On the other hand, it is worth noting that the fullerene chromophores are almost completely isolated from external contacts by the dendritic structure thus paving the way toward ordered thin film of isolated functional molecular units. This appears to be an important finding for future nanotechnological applications, in particular for data storage at a molecular level.

Although the incorporation of fullerenes in thin ordered films has been extensively studied during the past few years, liquid crystal ordering of such materials has been probed to a much lesser degree. Indeed, the fullerene sphere is a quite large molecular unit and the preparation of fullerene-containing liquid crystals appears to be difficult. Deschenaux and coworkers [41] have shown that the functionalization of C_{60} with a malonic ester bearing two mesogenic cholesterol subunits resulted in a fullerene derivative with liquid-crystalline properties. However, the mesomorphic behavior of this fullerene derivative is limited in comparison with that of the corresponding malonic ester precursor due to the presence of the C_{60} core that acts as a bulky spacer between the two mesogenic units. The same group has shown that a dendritic addend exhibits similar mesogenic properties as those of the corresponding fullerene-functionalized dendrimer **24** [42] (Figure 6.12). In this case, the C_{60} core of **24** is buried in the middle of the dendritic structure and thus prevents unfavorable effects of the C_{60} unit such as aggregation or steric hindering. Therefore, incorporation of fullerenes into well ordered structures can be easily achieved.

The concept developed to elaborate fullerene-containing thermotropic liquid crystals, that is, grafting dendritic mesogenic subunits onto the C_{60} core, was

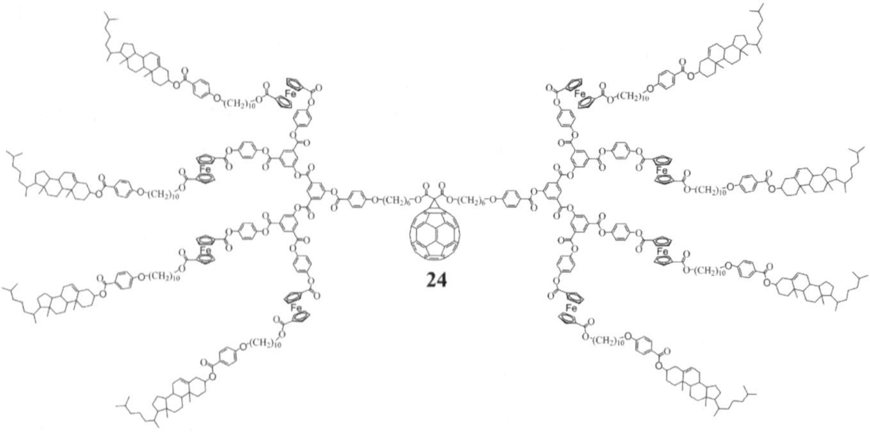

FIGURE 6.12 Liquid-crystalline fullerodendrimer **24**.

successfully extended to produce several other series of fullerene-containing derivatives with mesomorphic properties [43,44]. For example, dendritic liquid-crystalline fulleropyrrolidines **25–28** were synthesized from the corresponding aldehyde-based dendrimers and sarcosine by applying a 1,3-dipolar cycloaddition reaction [45] (Figure 6.13). With the exception of the first-generation dendrimer (**25**), which was found to be nonmesomorphic, all fullerene-based dendrimers gave rise to a smectic A phase. Indeed, the mesogenic moiety in **25** is not large enough to prevent the steric effects generated by the bulky fulleropyrrolidine framework thus showing the importance of the dendritic structure to obtain liquid-crystalline fullerene-based materials.

The versatility of the 1,3-dipolar cycloaddition reaction [46] is also an ideal platform for the introduction of various functional groups onto mesomorphic fullerenes. For example, this strategy was applied to the design of liquid-crystalline fullerenes bearing oligophenylenevinylene (OPV) subunits **29** and **30** (Figure 6.14) [47]. Fullerene derivatives **29** and **30** gave rise to smectic A phases. Compounds **29** and **30** exhibit similar results indicating that the overall behavior is mainly governed by the second-generation dendritic framework in agreement with results obtained for the above-described fullerene-containing liquid-crystalline dendrimers. The functionalization of C_{60}–OPV conjugates with a dendritic mesogenic group allows the liquid-crystalline ordering of such donor–acceptor systems that present all the characteristics required for photovoltaic applications [48,49]. Indeed, the use of liquid-crystalline C_{60}–OPV conjugates could be of particular interest since such materials would spontaneously form ordered assemblies that could then be oriented and lead to high-performance thin films.

Dendrimers with a fullerene core appear also to be appealing candidates to demonstrate the shielding effects resulting from the presence of the surrounding dendritic shell [50]. Effectively, the lifetime of the first triplet excited state of fullerene derivatives is sensitive to the solvent [50]. Therefore, lifetime measurements in different solvents can be used to evaluate the degree of isolation of the central

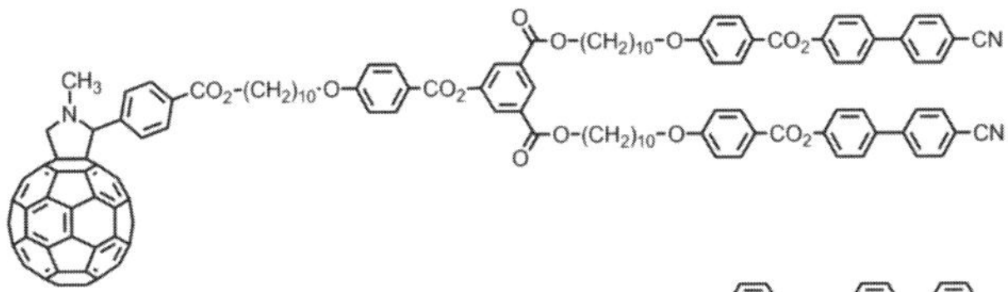

25

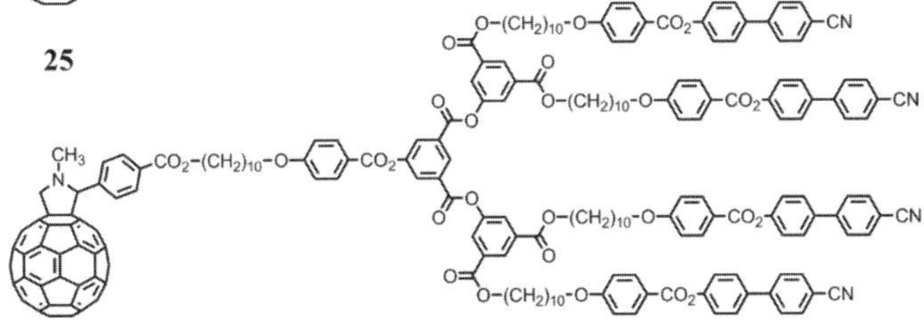

26

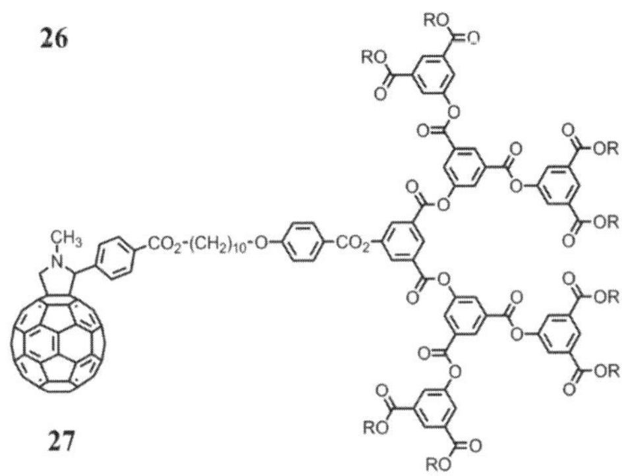

27

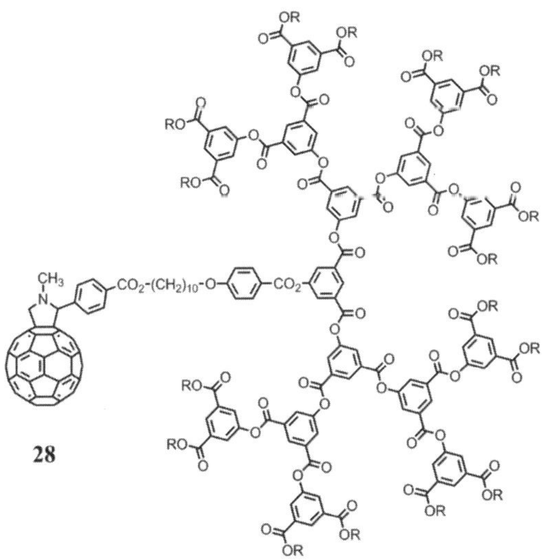

28

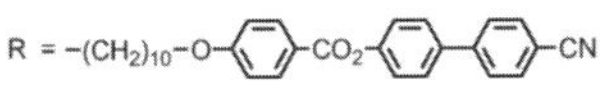

FIGURE 6.13 Dendritic liquid-crystalline fulleropyrrolidines **25–28**.

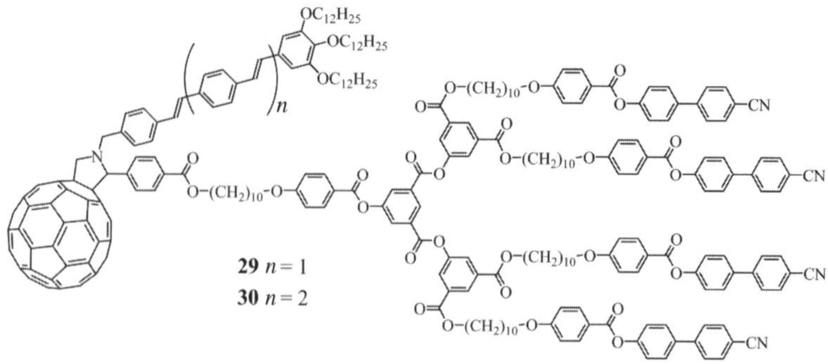

FIGURE 6.14 Liquid-crystalline fullerene–OPV conjugates.

C_{60} moiety from external contacts. With this idea in mind, two series of fullerodendrimers have been prepared (Figure 6.15) [24,51,52]. In the design of these compounds, it was decided to attach poly(aryl ether) dendritic branches terminated with peripheral triethyleneglycol chains to obtain derivatives soluble in a wide range of solvents [24,51,52]. The photophysical properties of **31–38** have been studied in different solvents (PhMe, CH_2Cl_2, and CH_3CN). The lifetimes of the lowest triplet excited states are summarized in Table 6.2.

For both series of dendrimers interesting trends can be obtained from the analysis of triplet lifetimes in air-equilibrated solutions [51,52]. A steady increase of lifetimes is found by increasing the dendrimers size in all solvents, suggesting that the dendritic wedges are able to shield, at least partially, the fullerene core from external contacts

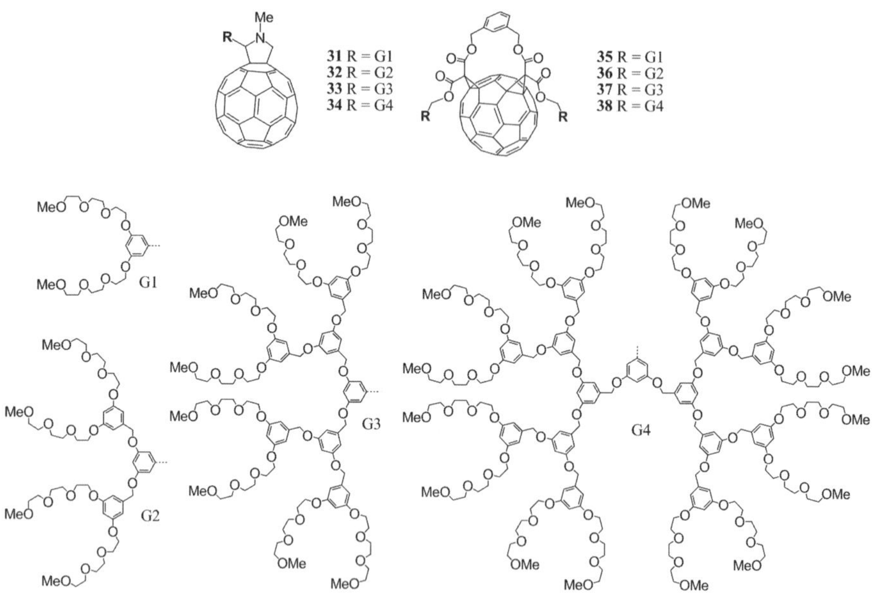

FIGURE 6.15 Fullerodendrimers **31–38**.

TABLE 6.2 Lifetime of the First Triplet Excited State of 31–38 in Air Equilibrated Solutions Determined by Transient Absorption at Room Temperature

Compound	τ (ns) in PhMe	τ (ns) in CH_2Cl_2	τ (ns) in CH_3CN
31	279	598	%[a]
32	304	643	330
33	318	732	412
34	374	827	605
35	288	611	314
36	317	742	380
37	448	873	581
38	877	1103	1068

[a] Not soluble in this solvent.

with the solvent and from quenchers such as molecular oxygen. It must be however emphasized that the triplet lifetimes of **34** in the three solvents are rather different from each other, likely reflecting specific solvent–fullerene interactions that affect excited state deactivation rates. This suggests that, albeit a dendritic effect is evidenced, even the largest wedge is not able to provide a complete shielding of the central fulleropyrrolidine core in **34**. The latter hypothesis was confirmed by computational studies. The calculated structure of **34** reveals that the dendritic shell is unable to completely cover the fullerene core. In contrast, the triplet lifetimes of **38** [52] in the three solvents lead toward a similar value suggesting that the fullerene core is in a similar environment whatever the nature of the solvent is. In other words, the C_{60} unit is, to a large extent, not surrounded by solvent molecules but substantially buried in the middle of the dendritic structure that is capable of creating a specific site-isolated microenvironment around the fullerene moiety. The latter hypothesis is quite reasonable based on the calculated structure of **38** showing that the dendritic branches are able to fully cover the central fullerene core. The dendritic effect evidenced for **31–38** was found to be useful to optimize the optical limiting properties characteristic of fullerene derivatives. Effectively, the intensity dependant absorption of fullerenes originates from larger absorption cross sections of excited states compared to that of the ground state [53], therefore the increased triplet lifetime observed for the largest fullerodendrimers may allow for an effective limitation on a longer time scale. For practical applications, the use of solid devices is largely preferred to solutions and inclusion of fullerene derivatives in sol–gel glasses has shown interesting perspectives [53]. However, faster de-excitation dynamics and reduced triplet yields are typically observed for fullerene-doped sol–gel glasses when compared to solutions [53]. The latter observations are mainly explained by two factors: (i) perturbation of the molecular energy levels due to the interactions with the sol–gel matrix and (ii) interactions between neighboring fullerene spheres due to aggregation [53]. Therefore, the encapsulation of the C_{60} core evidenced by the photophysical studies for both series of fullerodendrimers might also be useful to prevent such undesirable effects. The incorporation of fullerodendrimers **31–34** in sol–gel glasses has been easily achieved by soaking mesoporous silica glasses with a solution of

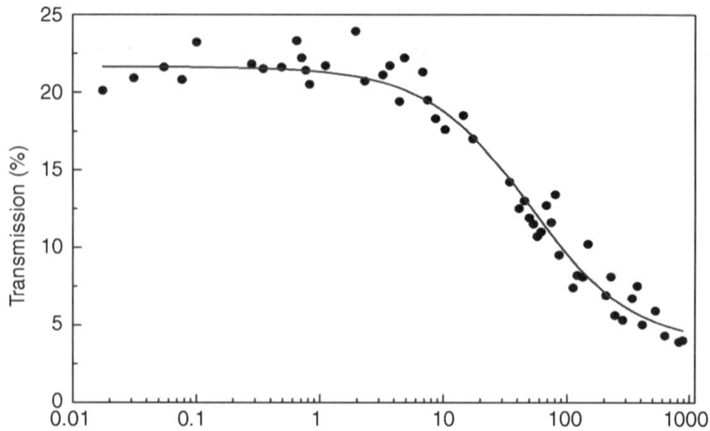

FIGURE 6.16 Transmission versus incident fluence at 532 nm of a sol–gel sample containing compound **34**.

31–34 [51,53]. For the largest compounds, the resulting samples only contain well-dispersed fullerodendrimer molecules. Measurements on the resulting doped samples have revealed efficient optical limiting properties [51]. The transmission as a function of the fluence of the laser pulses remains nearly constant for fluences lower than 5 mJ/cm^2 (Figure 6.16). When the intensity increases above this threshold, the effect of induced absorption appears, and the transmission diminishes rapidly, thus showing the potential of these materials for optical limiting applications.

C_{60} is also an attractive functional core for the preparation of light harvesting dendrimers [54–60]. Effectively, its first singlet and triplet excited states are relatively low in energy and photo-induced energy transfer events have been evidenced in some fullerene-based dyads [61]. In particular, photophysical investigations of some fulleropyrrolidine derivatives substituted with OPV moieties revealed a very efficient singlet–singlet OPV $\rightarrow$ C_{60} photo-induced energy transfer [54,62]. Based on this observation, dendrimers **39–41** with a fullerene core and peripheral OPV subunits (Figure 6.17) have been prepared [63].

The photophysical properties of fullerodendrimers **39–41** have been first investigated in CH_2Cl_2 solutions. Upon excitation at the OPV band maximum, dramatic quenching of OPV fluorescence is observed for all fullerodendrimers. At 394 nm (corresponding to OPV band maxima) the molar absorptivities (ε) of these fullerodendrimers are 134,800 $M^{-1}cm^{-1}$ for **39**, 255,100 $M^{-1}cm^{-1}$ for **40**, and 730,400 $M^{-1}cm^{-1}$ for **41**. Since the ε of the ubiquitous *N*-methyl-fulleropyrrolidine at 394 nm is only 7600 a remarkable light harvesting capability of the peripheral units relative to the central core is evidenced along the series. UV-vis-NIR luminescence and transient absorption spectroscopy have been used to elucidate in more details the photo-induced processes in fullerodendrimers **39–41** as a function of the dendritic generation and of the solvent polarity (toluene, CH_2Cl_2, benzonitrile), taking into account that the free energy change for electron transfer is the same along the series due to invariability of the donor–acceptor couple. In any solvents, all of the fullerodendrimers exhibit ultrafast OPV $\rightarrow$ C_{60} singlet energy transfer (k_{EnT} ca.

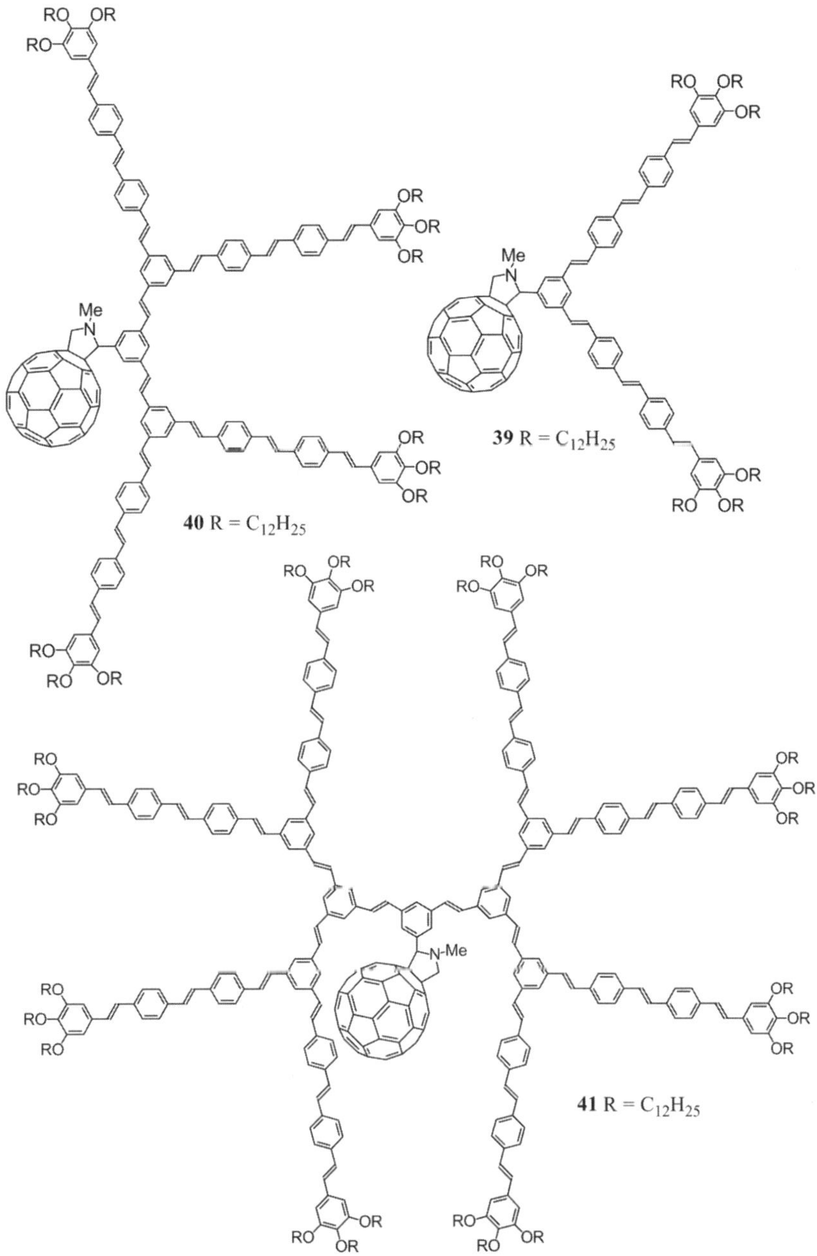

FIGURE 6.17 Fullerodendrimers with peripheral OPV units.

10^{10}–10^{12} s^{-1}). In CH_2Cl_2, a slightly exergonic OPV $\rightarrow$ C_{60} electron transfer from the lowest fullerene singlet level ($^1C_{60}{}^*$) is made possible ($\Delta G_{CS} \approx 0.07$ eV), but it is observed, to an increasing extent, only in the largest systems **40** and **41** characterized by a lower activation barrier for electron transfer. This effect has been related to a decrease of the reorganization energy upon enlargement of the molecular architecture. Structural factors are also at the origin of an unprecedented OPV → C_{60} electron transfer observed for **40** and **41** in apolar toluene, whereas in benzonitrile electron transfer occurs in all cases.

6.3.2 Fullerene-Rich Dendrimers

Whereas the main part of the fullerene-containing dendrimers reported so far have been prepared with a C_{60} core, dendritic structures with fullerene units at their surface or with C_{60} spheres in the dendritic branches have been much scarcely considered. This is mainly associated with the difficulties related to the synthesis of fullerene-rich molecules [7]. Indeed the two major problems for the preparation of such dendrimers are the low solubility of C_{60} and its chemical reactivity limiting the range of reactions that can be used for the synthesis of branched structures bearing multiple C_{60} units. Over the past years, several examples of fullerene-rich dendrimers have been reported. These results will be summarized in the following sections.

6.3.2.1 Divergent Synthesis The divergent preparation of fullerodendrimers by grafting C_{60} units on the peripheral reactive groups of commercially available dendrimers is the easiest way to produce large dendritic structures. This synthetic approach was applied to prepare compounds **42–44** (Figure 6.18) from fullerene derivative **13** and polypropyleneimine (PPI) dendrimers [64].

The choice of the appropriate activating group for the carboxylic acid function of the C_{60} building block **13** was the key to this synthesis. Effectively, the reaction conditions for the activation step may not be strongly acid or basic to preserve the ester functions. Furthermore, the grafting onto the dendritic polyamines requires an extremely efficient reaction to obtain the corresponding functionalized derivatives with good yields and to avoid the formation of defected dendrimers. The preparation of the pentafluorothiophenolester derived from compound **13** and pentafluorothiophenol under *N,N′*-dicyclohexylcarbodiimide (DCC)-mediated esterification conditions appeared as a good choice. Indeed, the reaction conditions for the preparation of this activated acid are mild and the efficient grafting of pentafluorothiophenolesters onto PPI dendrimers has already been reported [65]. The activated acid was obtained in nearly quantitative yields by reaction of carboxylic acid **13** with pentathiofluorothiophenol in the presence of DCC and a catalytic amount of 4-dimethylaminopyridine (DMAP). Subsequent reaction of the resulting pentafluorothiophenolester with the PPI dendrimers of first-, second- and third generation in the presence of triethylamine provided the corresponding dendritic derivatives **42–44** in good yields. Owing to the presence of four pendant alkyl chains per fullerene moiety, compounds **42–44** are all well soluble in common organic solvents such as CH_2Cl_2, $CHCl_3$, THF, or toluene, and spectroscopic characterization was easily achieved. The ^{1}H-NMR spectra of dendrimers **42–44** show the typical pattern of the fullerene *cis*-2 bis-adduct with the expected additional signals arising from the PPI skeleton. The integration ratio are also consistent with the proposed molecular structures. The structure of **42** and **43** was further confirmed by MALDI-TOF mass spectrometry. The expected

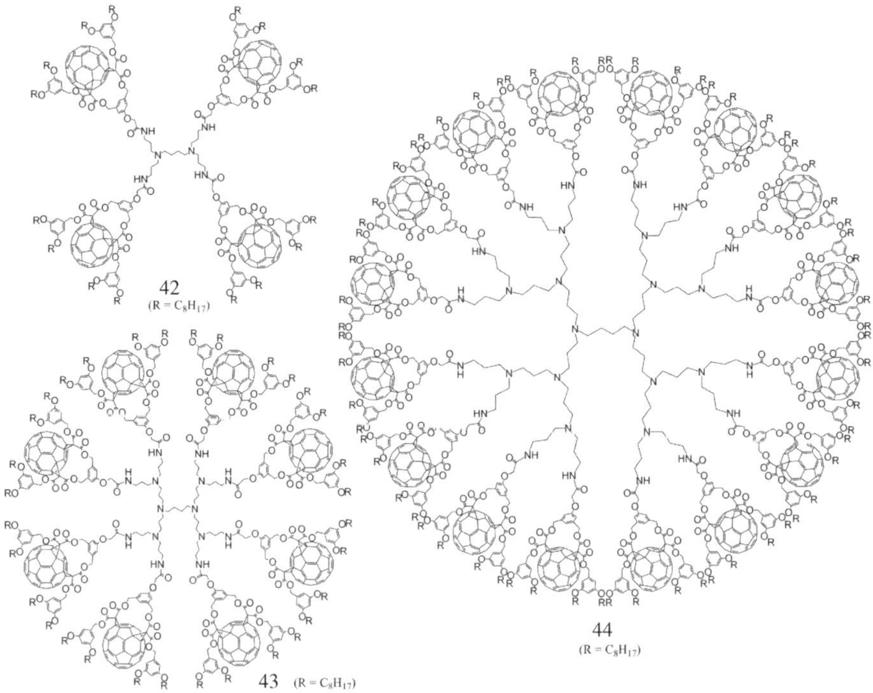

FIGURE 6.18 Fullerene-rich dendrimers **42–44**.

molecular ion peaks were clearly observed for both compounds. It should be pointed out that no peaks corresponding to defected dendrimers were observed in the mass spectra of **42** and **43**, thus providing clear evidence for their monodispersity. For **44**, a high level of fragmentation prevented the observation of the expected molecular ion peak and its monodispersity could not be unambiguously demonstrated.

6.3.2.2 Convergent Synthesis In the divergent strategy, dendrimers are constructed from the central core to the periphery [1]. In each repeat cycle, the *n* reactive groups on the dendrimer periphery react with *n* monomer units to add a new generation to the dendrimer. However, due to the increasing number of reactive terminal units, defects in the structures appear rapidly. In the convergent strategy that has been introduced by Hawker and Fréchet [66], dendritic branches are first built up, then attached to the central core in the final step. In this case, the number of coupling reactions needed to add each new generation, usually 2 or 3, is the same throughout the synthesis making defective products easier to separate. Whereas exactly the same number of steps is required by using a divergent or a convergent approach for the preparation of a given dendrimer, the convergent one appears to be more efficient for the construction of monodisperse dendrimers. The first example of convergent preparation of dendritic branches with fullerene subunits at the periphery has been reported by Nierengarten and coworkers [67]. The synthesis of these fullerodendrons

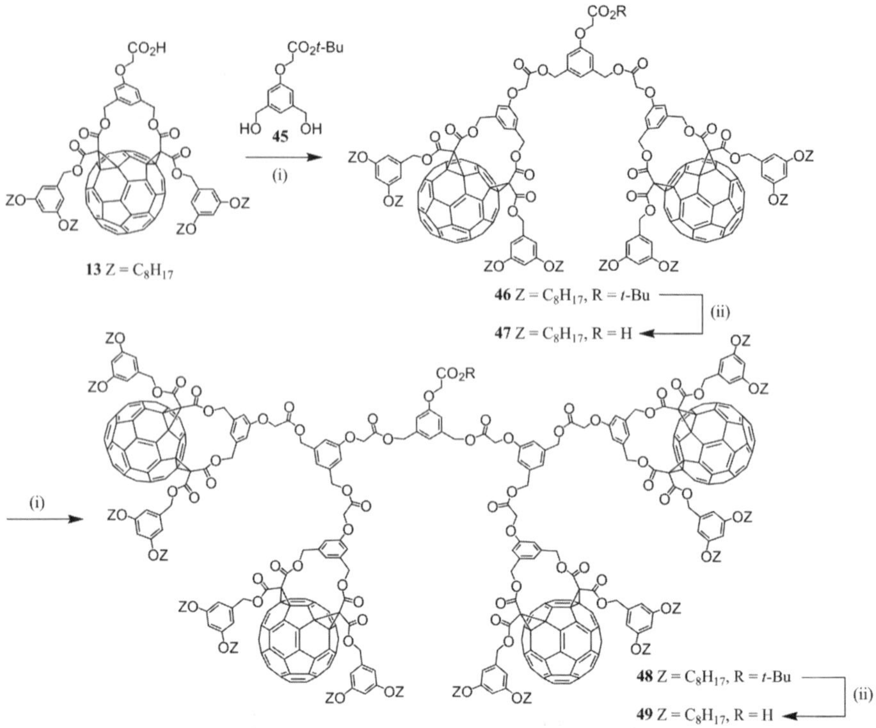

FIGURE 6.19 Preparation of fullerodendrons **46**–**49**. Reagents and conditions: (i) **45**, DCC, DMAP, HOBt, CH_2Cl_2; (ii) TFA, CH_2Cl_2.

is depicted in Figure 6.19. The starting fullerene derivative **13** is easily obtained on a multigram scale and is highly soluble in common organic solvents owing to the presence of the four long alkyl chains. The iterative reaction sequence used for the preparation of the subsequent dendrimer generations relies upon successive DCC-mediated esterification reaction with the A_2B-building block **45** possessing two benzylic alcohol functions and a protected carboxylic acid group followed by cleavage of a *t*-butyl ester moiety under acidic conditions [67].

Reaction of diol **45** with carboxylic acid **13** under esterification conditions using DCC, DMAP, and 1-hydroxybenzotriazole (HOBt) in CH_2Cl_2 gave the protected dendron of second generation **46** in 90% yield. Hydrolysis of the *t*-butyl ester moiety under acidic conditions then afforded the corresponding carboxylic acid **47** in a quantitative yield. Esterification of **47** with diol **45** (DCC, HOBt, DMAP) afforded the *t*-butyl-protected fullerodendron **48** in 95% yield. Selective hydrolysis of the *t*-butyl ester under acidic conditions afforded acid **7** in 97% yield. Dendrons **13**, **47**, and **49** are easily prepared on a multigram scale and are highly soluble in common organic solvents.

Following this first successful preparation of dendritic branches with peripheral C_{60} subunits, Nierengarten and coworkers have reported several synthesis of fullerodendrons having either peripheral C_{60} subunits (**50**) or containing C_{60} groups at each branching units (**51**) [68–70]. A few examples are shown in Figure 6.20. In all the cases, the synthesis of the dendrons relies upon iterative the same iterative reaction sequence (esterification/cleavage of a *t*-butyl ester) as the one used for the preparation of **47** and **49**.

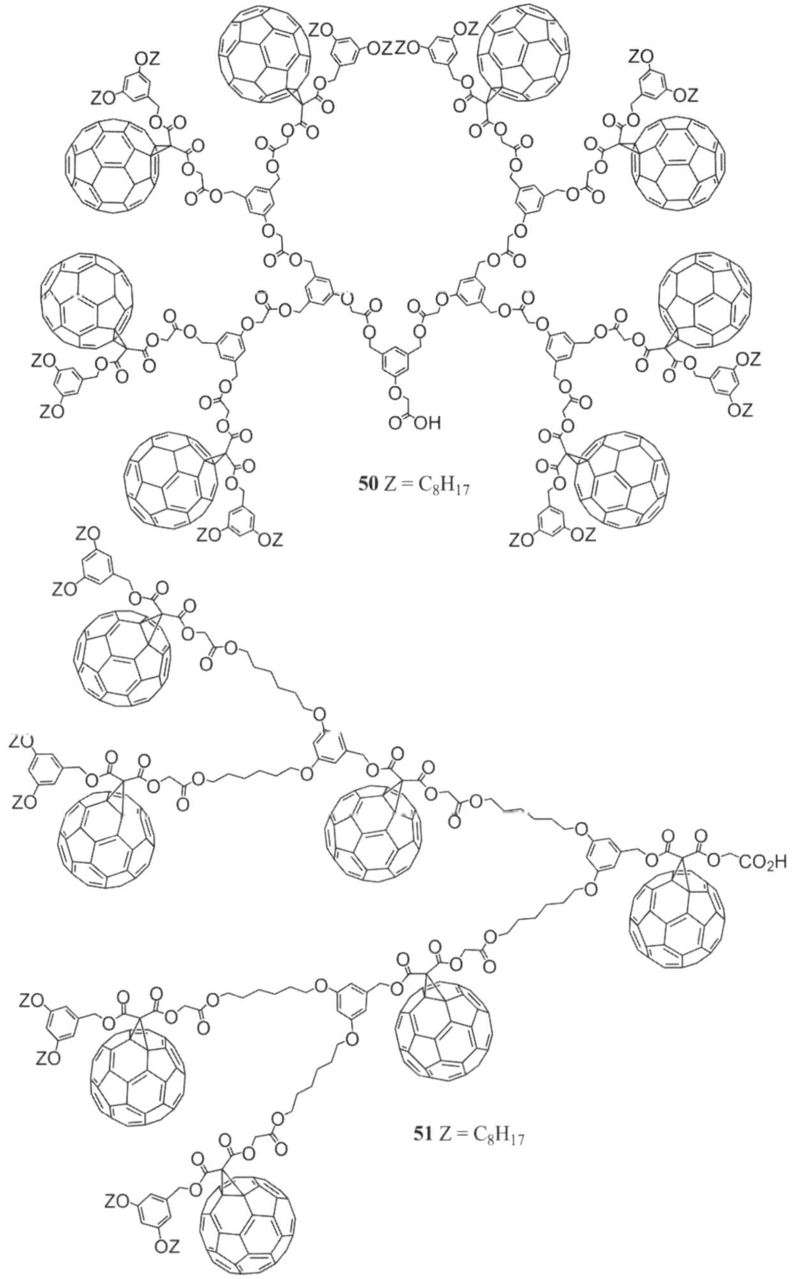

50 Z = C_8H_{17}

51 Z = C_8H_{17}

FIGURE 6.20 Fullerodendrons **50** and **51**.

These fullerodendrons are interesting building blocks for the preparation of monodisperse fullerene-rich macromolecules. For example, Nierengarten and coworkers have prepared alkyne **53–55** by reaction of diol **52** with fullerodendrons **13**, **47**, and **49**, respectively, under esterification conditions using DCC, HOBt, and DMAP (Figure 6.21). The cyclotrimerization of the resulting dendronized bis-arylalkynes was a perfect tool for the synthesis of fullerene-rich dendrimers **56–58** [71].

The reaction conditions for the cyclotrimerization were first optimized for the first-generation compound. The choice of the appropriate catalyst was the key to this synthesis. Indeed, $Co_2(CO)_8$ appeared as a good candidate. It is a known catalyst for the cyclotrimerization of alkynes [72] and Martín and coworkers have shown that C_{60} is only reactive in the presence of $Co_2(CO)_8$ under Pauson–Khand conditions when a very specific stereochemical orientation is possible [73]. It was effectively found that $Co_2(CO)_8$ is an efficient catalyst for the preparation of compound **56** from alkyne **53**. Under optimized conditions, treatment of **53** with a catalytic amount of $Co_2(CO)_8$ in dioxane at room temperature for 24 h afforded **56** in 93% yield. The same conditions were used for the preparation of the highest generation compounds. The reaction of the second-generation derivative **54** was finished after one day and compound **57** was isolated in 62% yield. In contrast, the reaction of the highest generation precursor was very slow, most probably as a result of steric effects. After 5 days, the starting material was not completely consumed but the reaction was stopped since significant

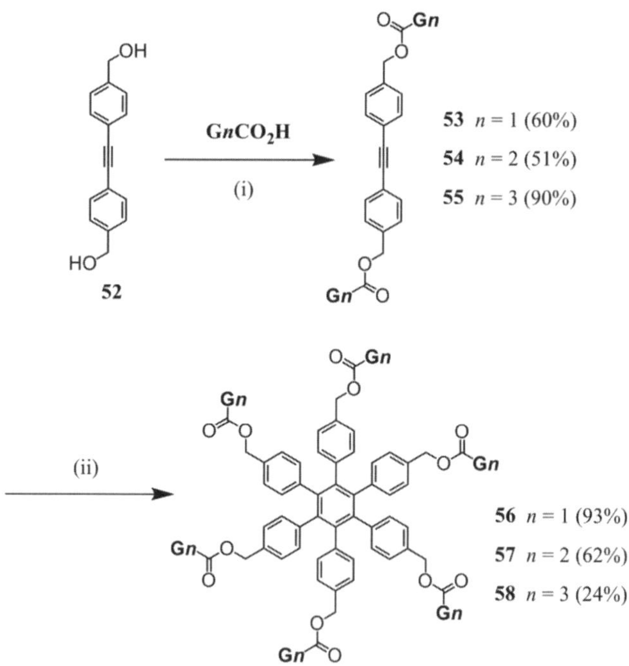

FIGURE 6.21 Preparation of fullerene-rich dendrimers **56–58**. Reagents and conditions: (i) **13** (Gn = 1), **47** (Gn = 2), or **49** (Gn = 3), DCC, DMAP, HOBt, CH_2Cl_2; (ii) $Co_2(CO)_8$, dioxane.

degradation was evidenced. After purification by column chromatography on SiO_2 followed by gel permeation chromatography, compound **58** (Figure 6.22) was isolated in 24% yield.

Having shown that the construction of dendrimers containing multiple C_{60} subunits is successful, the next challenge was to prepare fullerene-rich nanostructures with new properties. With this idea in mind, dendrimer **59** with a bis(1,10-phenanthroline)copper(I) core has been prepared from fullerodendron **49** (Figure 6.23) [74].

The central core in **59** appears inaccessible to external contacts as shown by molecular modeling and confirmed by electrochemical investigations [75]. Effectively, the bulky fullerodendrons around the Cu center prevent its approach on the electrode surface and its oxidation could no longer be observed. Furthermore, due to the high number of fullerene subunits in **59**, a strong shielding effect is observed and only a small part of the incident light is available to the central core relative to the periphery. A photophysical study revealed that the small portion of light energy able to reach the central Cu(I) complex is returned to the external fullerenes by energy

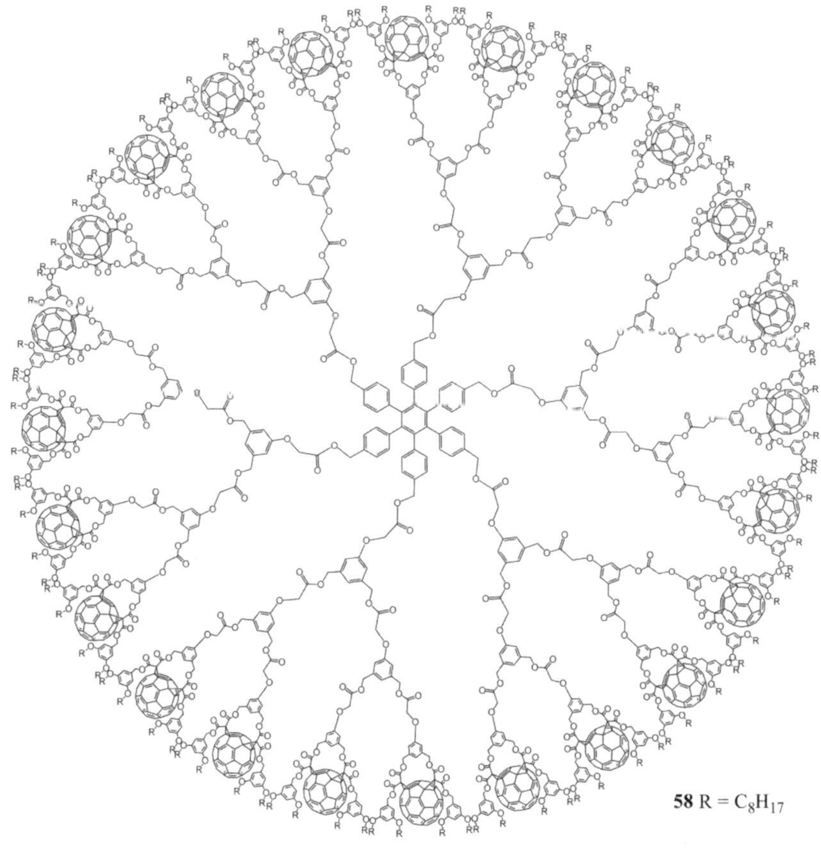

58 R = C_8H_{17}

FIGURE 6.22 Fullerene-rich dendrimer **58**.

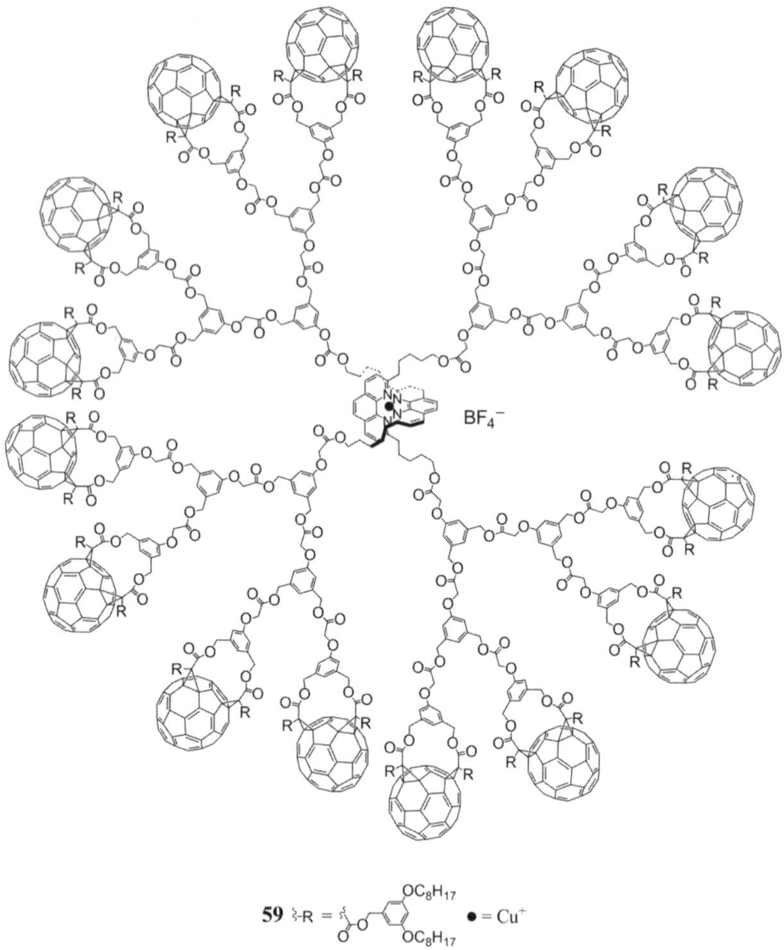

FIGURE 6.23 Fullerene-rich dendrimer **59**.

transfer [75]. Therefore, one can conclude that the central core is buried in a dendritic black box.

The dodecyloxy analogs of dendritic branches **13**, **47**, and **49** have also been attached to an OPV core bearing two alcohol functions to yield dendrimers **60**, **61**, and **62** with two, four, or eight peripheral C_{60} groups, respectively (Figure 6.24) [76,77].

The photophysical properties of **60–62** have been systematically investigated in solvents of increasing polarity, that is, toluene, dichloromethane, and benzonitrile. Ultrafast OPV → C_{60} singlet energy transfer takes place upon photoexcitation of the OPV core for the whole series of dendrimers, whatever the solvent is. Electron transfer from the fullerene singlet is thermodynamically allowed in CH_2Cl_2 and benzonitrile, but not in apolar toluene. For a given solvent, the extent of electron transfer, signaled by the quenching of the fullerene fluorescence, is not the same along the series, despite the fact that identical electron transfer partners are present. By increasing the

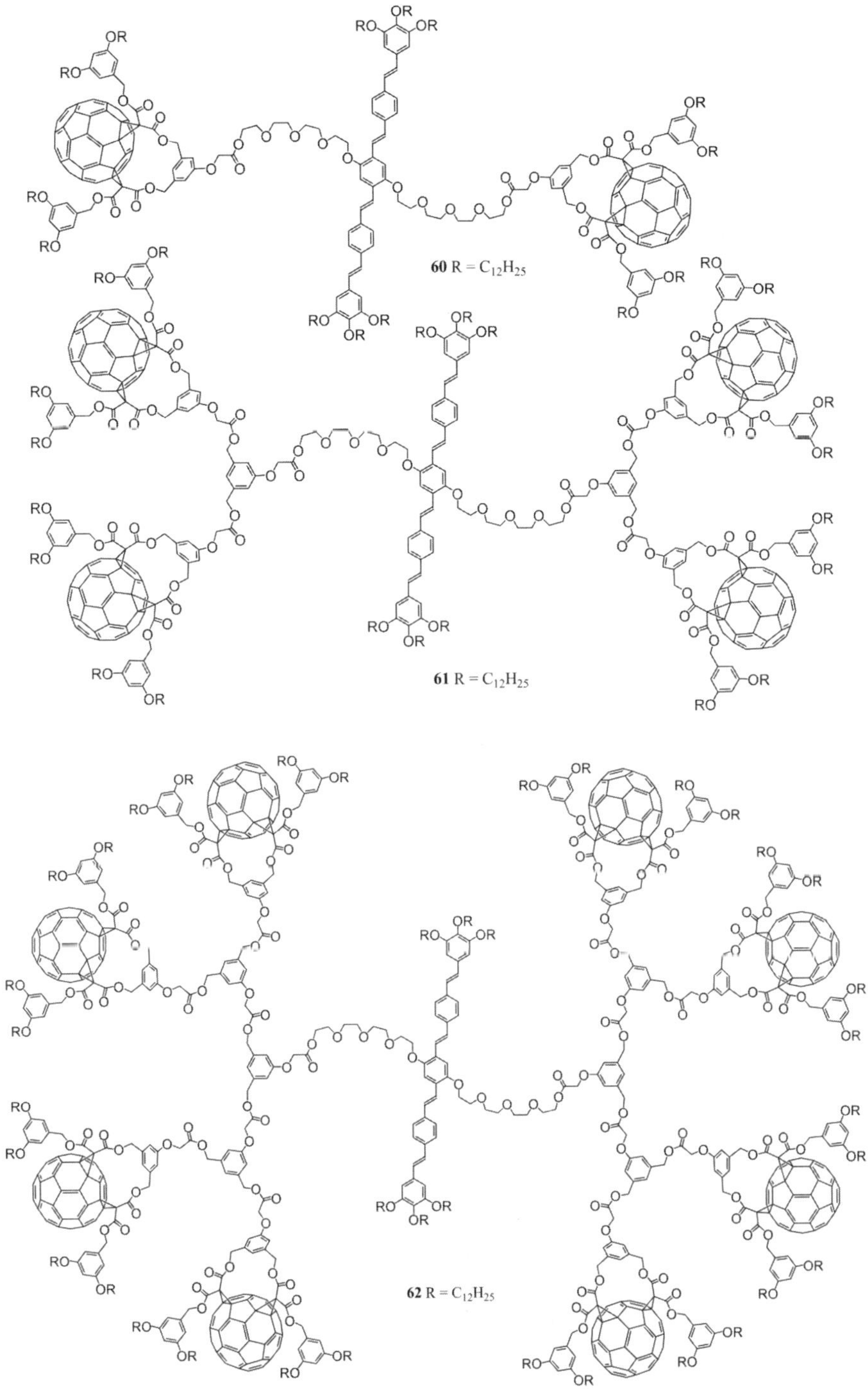

FIGURE 6.24 Fullerodendrimers **60–62**.

dendrimer size, electron transfer is progressively more difficult. Practically no electron transfer from the fullerene singlet occurs for **62** in CH_2Cl_2, whereas some of it is still detected in the more polar PhCN. These trends can be rationalized by considering increasingly compact dendrimer structures in more polar solvents [76]. This implies that the actual polarity experienced by the involved electron transfer partners, particularly the central OPV, is no longer that of the bulk solvent. This strongly affects electron transfer thermodynamics that, being reasonably located in the normal region of the Marcus parabola, becomes less exergonic and thus slower and less competitive toward intrinsic deactivation of the fullerene singlet state. This dendritic effect is in line with the molecular dynamics studies that suggest that the central OPV unit is more and more protected by the dendritic branches when the generation number is increased. Actually, the calculated structure of **62** shows that the two dendrons of third generation are able to fully cover the central OPV core.

6.4 CONCLUSIONS

Dendrimer chemistry and fullerenes crossed each other to give rise to a new interdisciplinary field in which the imagination of chemists has allowed to design and construct unprecedented molecular and supramolecular fullerene-based nanoarchitectures. Substantial research efforts have been carried out to organize such compounds onto surface as well as in liquid crystals. Owing to their special photophysical properties, fullerene-based dendrimers are also good candidates for evidencing dendritic effects or to produce original photoactive molecular devices. Despite some remarkable recent achievements, it is clear that the examples discussed herein represent only the first steps toward the design of fullerene-containing macromolecules that can display functionality at the macroscopic level. More research in this area is clearly needed to fully explore the possibilities offered by these materials, for example, in nanotechnology or in photovoltaics.

ACKOWLEDGMENTS

This research was supported by the CNRS. All my coworkers and collaborators are acknowledged for their outstanding contributions, their names are cited in the references. I further thank Profs. D. Astruc, T. Aida, and R. Deschenaux for having kindly provided the original files of Figures 6.4, 6.8 and 6.13, respectively.

REFERENCES

[1] G. R. Newkome, C. N. Moorefield, F. Vögtle, *Dendrimers and Dendrons: Concepts, Syntheses, Applications*, VCH, Weinheim, **2001**.

[2] J.-F. Nierengarten, *Chem. Eur. J.* **2000**, *6*, 3667–3670.

[3] M. Holler, J.-F. Nierengarten, *Aust. J. Chem.* **2009**, *62*, 605–623.

[4] F. Langa, J.-F. Nierengarten (Eds.), *Fullerenes: Principles and Applications*, RSC Nanoscience and Nanotechnology Series, **2007**.

[5] U. Hahn, F. Cardinali, J.-F. Nierengarten, *New J. Chem.* **2007**, *31*, 1128–1138.

[6] A. Hirsch, O. Vostrowsky, *Top. Curr. Chem.* **2001**, *217*, 51–93.

[7] J.-F. Nierengarten, *Top. Curr. Chem.* **2003**, *228*, 87–110.

[8] D. K. Smith, F. Diederich, *Chem. Eur. J.* **1998**, *4*, 1353–1361.

[9] M. Munenori, A. Ikeda, C. Fukuhara, S. Shinkai, *Tetrahedron Lett.* **1999**, *40*, 6945–6948.

[10] J.-F. Eckert, D. Byrne, J.-F. Nicoud, L. Oswald, J.-F. Nierengarten, M. Numata, A. Ikeda, S. Shinkai, N. Armaroli, *New J. Chem.* **2000**, *24*, 749–758.

[11] J.-F. Nierengarten, *Fullerenes, Nanotubes and Carbon Nanostructures* **2005**, *13*, 229–242.

[12] J.-F. Nierengarten, L. Oswald, J.-F. Eckert, J.-F. Nicoud, N. Armaroli, *Tetrahedron Lett.* **1999**, *40*, 5681–5684.

[13] Y. Rio, J.-F. Nierengarten, *Tetrahedron Lett.* **2002**, *43*, 4321–4324.

[14] D. Felder, B. Heinrich, D. Guillon, J.-F. Nicoud, J.-F. Nierengarten, *Chem. Eur. J.* **2000**, *6*, 3501–3507.

[15] D. Felder, D. Guillon, J.-F. Nierengarten, *Mater. Sci. Eng. C* **2001**, *C18*, 161–164.

[16] J. Ruiz, C. Pradet, F. Varret, D. Astruc, *Chem. Commun.* **2002**, 1108–1109.

[17] C. Bossard, S. Rigaut, D. Astruc, M.-H. Delville, G. Félix, A. Février-Bouvier, J. Amiell, S. Flandrois, P. Delhaès, *J. Chem. Soc., Chem. Commun.* **1993**, 333–334.

[18] R. van de Coevering, R. Kreiter, F. Cardinali, G. van Koten, J.-F. Nierengarten, R. J. M. Klein Gebbink, *Tetrahedron Lett.* **2005**, *46*, 3353–3356.

[19] T. Hasobe, Y. Kashiwagi, M. A. Absalom, J. Sly, K. Hosomizu, M. J. Crossley, H. Imahori, P. V. Kamat, S. Fukuzumi, *Adv. Mater.* **2004**, *16*, 975–979.

[20] P. V. Kamat, S. Barazzouk, K. G. Thomas, S. Hötchandani, *J. Phys. Chem. B* **2000**, *104*, 4014–4024.

[21] W.-S. Li, K. S. Kim, D.-L. Jiang, H. Tanaka, T. Kawai, J. H. Kwon, D. Kim, T. Aida, *J. Am. Chem. Soc.* **2006**, *128*, 10527–10532.

[22] N. Armaroli, F. Diederich, L. Echegoyen, T. Habicher, L. Flamigni, G. Marconi, J.-F. Nierengarten, *New J. Chem.* **1999**, 77–83.

[23] A. Trabolsi, M. Elhabiri, M. Urbani, J. L. Delgado de la Cruz, F. Ajamaa, N. Solladié, A.-M. Albrecht-Gary, J.-F. Nierengarten, *Chem. Commun.* **2005**, 5736–5738.

[24] Y. Rio, J.-F. Nicoud, J.-L. Rehspringer, J.-F. Nierengarten, *Tetrahedron Lett.* **2000**, *41*, 10207–10210.

[25] K. L. Wooley, C. J. Hawker, J. M. J. Fréchet, F. Wudl, G. Srdanov, S. Shi, Q. Li, M. Kao, *J. Am. Chem. Soc.* **1993**, *115*, 9836–9337.

[26] C. J. Hawker, K. L. Wooley, J. M. J. Fréchet, *J. Chem. Soc., Chem. Commun.* **1994**, 925–926.

[27] J.-F. Nierengarten, T. Habicher, R. Kessinger, F. Cardullo, F. Diederich, V. Gramlich, J.-P. Gisselbrecht, C. Boudon, M. Gross, *Helv. Chim. Acta* **1997**, *80*, 2238–2276.

[28] M. Brettreich, A. Hirsch, *Tetrahedron Lett.* **1998**, *39*, 2731–2734.

[29] R. Kunieda, M. Fujitsuka, O. Ito, M. Ito, Y. Murata, K. Komatsu, *J. Phys. Chem. B* **2002**, *106*, 7193–7199.

[30] Y. Murata, M. Ito, K. Komatsu, *J. Mater. Chem.* **2002**, *12*, 2009–2020.

[31] I. M. Mahmud, N. Zhou, L. Wang, Y. Zhao, *Tetrahedron* **2008**, *64*, 11420–11432.

[32] X. Camps, H. Schönberger, A. Hirsch, *Chem. Eur. J.* **1997**, *3*, 561–567.

[33] A. Herzog, A. Hirsch, O. Vostrowsky, *Eur. J. Org. Chem.* **2000**, 171–180.

[34] J. Iehl, R. Pereira de Freitas, B. Delavaux-Nicot, J.-F. Nierengarten, *Chem. Commun.* **2008**, 2450–2452.

[35] J. Iehl, J.-F. Nierengarten, *Chem. Eur. J.* **2009**, *15*, 7306–7309.

[36] J.-F. Nierengarten, *Comptes Rendus Chimie* **2003**, *6*, 725–733.

[37] F. Cardullo, F. Diederich, E. Echegoyen, T. Habicher, N. Jayaraman, R. M. Leblanc, J. F. Stoddart, S. Wang, *Langmuir* **1998**, *14*, 1955–1959.

[38] J.-F. Nierengarten, J.-F. Eckert, Y. Rio, M. P. Carreon, J.-L. Gallani, D. Guillon, *J. Am. Chem. Soc.* **2001**, *123*, 9743–9748.

[39] D. Felder, M. Gutiérrez Nava, M. del Pilar Carreon, J.-F. Eckert, M. Luccisano, C. Schall, P. Masson, J.-L. Gallani, B. Heinrich, D. Guillon, J.-F. Nierengarten, *Helv. Chim. Acta* **2002**, *85*, 288–319.

[40] S. Zhang, Y. Rio, F. Cardinali, C. Bourgogne, J.-L. Gallani, J.-F. Nierengarten, *J. Org. Chem.* **2003**, *68*, 9787–9797.

[41] T. Chuard, R. Deschenaux, *Helv. Chim. Acta* **1996**, *79*, 736–741.

[42] B. Dardel, R. Deschenaux, M. Even, E. Serrano, *Macromolecules* **1999**, *32*, 5193–5198.

[43] T. Chuard, R. Deschenaux, *J. Mater. Chem.* **2002**, *12*, 1944–1951.

[44] R. Deschenaux, B. Donnio, D. Guillon, *New J. Chem.* **2007**, *31*, 1064–1073.

[45] S. Campidelli, J. Lenoble, J. Barbera, F. Paolucci, M. Marcaccio, D. Paolucci, R. Deschenaux, *Macromolecules* **2005**, *38*, 7915–7925.

[46] M. Prato, M. Maggini, *Acc. Chem. Res.* **1998**, *31*, 519–529.

[47] S. Campidelli, R. Deschenaux, J.-F. Eckert, D. Guillon, J.-F. Nierengarten, *Chem. Commun.* **2002**, 656–657.

[48] J.-F. Nierengarten, *New J. Chem.* **2004**, *28*, 1177–1191.

[49] J.-F. Nierengarten, *Solar Energy Materials and Solar Cells*, **2004**, *83*, 187–199.

[50] J.-F. Nierengarten, N. Armaroli, G. Accorsi, Y. Rio, J.-F. Eckert, *Chem. Eur. J.* **2003**, *9*, 36–41.

[51] Y. Rio, G. Accorsi, H. Nierengarten, J.-L. Rehspringer, B. Hönerlage, G. Kopitkovas, A. Chugreev, A. Van Dorsselaer, N. Armaroli, J.-F. Nierengarten, *New J. Chem.* **2002**, *26*, 1146–1154.

[52] Y. Rio, G. Accorsi, H. Nierengarten, C. Bourgogne, J.-M. Strub, A. Van Dorsselaer, N. Armaroli, J.-F. Nierengarten, *Tetrahedron* **2003**, *59*, 3833–3844.

[53] G. Kopitkovas, A. Chugreev, J.-F. Nierengarten, Y. Rio, J. L. Rehspringer, B. Hönerlage, *Opt. Mater.* **2004**, *27*, 285–291.

[54] N. Armaroli, F. Barigelletti, P. Ceroni, J.-F. Eckert, J.-F. Nicoud, J.-F. Nierengarten, *Chem. Commun.* **2000**, 599–600.

[55] G. Accorsi, N. Armaroli, J.-F. Eckert, J.-F. Nierengarten, *Tetrahedron Lett.* **2002**, *43*, 65–68.

[56] J. L. Segura, R. Gomez, N. Martin, C. P. Luo, A. Swartz, D. M. Guldi, *Chem. Commun.* **2001**, 707–708.

[57] D. M. Guldi, A. Swartz, C. P. Luo, R. Gomez, J. L. Segura, N. Martin, *J. Am. Chem. Soc.* **2002**, *124*, 10875–10886.

[58] F. Langa, M. J. Gomez-Escalonilla, E. Diez-Barra, J. C. Garcia-Martinez, A. de la Hoz, J. Rodriguez-Lopez, A. Gonzalez-Cortes, V. Lopez-Arza, *Tetrahedron Lett.* **2001**, *42*, 3435–3438.

[59] J. N. Clifford, A. Gégout, S. Zhang, R. Pereira de Freitas, M. Urbani, M. Holler, P. Ceroni, J.-F. Nierengarten, N. Armaroli, *Eur. J. Org. Chem.* **2007**, 5899–5908.

[60] M. Holler, M. Urbani, A. Gégout, S. Zhang, J.-F. Nierengarten, *Comptes Rendus Chimie* **2009**, *12*, 479–488.

[61] D. M. Guldi, M. Prato, *Acc. Chem. Res.* **2000**, *33*, 695–703.

[62] J.-F. Eckert, J.-F. Nicoud, J.-F. Nierengarten, S.-G. Liu, L. Echegoyen, F. Barigelletti, N. Armaroli, L. Ouali, V. Krasnikov, G. Hadziioannou, *J. Am. Chem. Soc.* **2000**, *122*, 7467–7479.

[63] N. Armaroli, G. Accorsi, J. N. Clifford, J.-F. Eckert, J.-F. Nierengarten, *Chem. Asian J.* **2006**, *1*, 564–574.

[64] U. Hahn, J. F. Nierengarten, F. Vögtle, A. Listorti, F. Monti, N. Armaroli, *New J. Chem.* **2009**, *33*, 337–344.

[65] A. P. H. J. Schenning, C. Elissen-Román, J.-W. Weener, M. W. P. L. Baars, S. J. van der Gaast, E. W. Meijer, *J. Am. Chem. Soc.* **1998**, *120*, 8199–8208.

[66] C. Hawker, J. M. J. Fréchet, *J. Chem. Soc., Chem. Commun.* **1990**, 1010–1013.

[67] J.-F. Nierengarten, D. Felder, J.-F. Nicoud, *Tetrahedron Lett.* **1999**, *40*, 269–272.

[68] J.-F. Nierengarten, D. Felder, J.-F. Nicoud, *Tetrahedron Lett.* **2000**, *41*, 41–44.

[69] D. Felder, H. Nierengarten, J.-P. Gisselbrecht, C. Boudon, E. Leize, J.-F. Nicoud, M. Gross, A. Van Dorsselaer, J.-F. Nierengarten, *New J. Chem.* **2000**, *24*, 687–695.

[70] U. Hahn, K. Hosomizu, H. Imahori, J.-F. Nierengarten, *Eur. J. Org. Chem.* **2006**, 85–91.

[71] U. Hahn, E. Maisonhaute, C. Amatore, J.-F. Nierengarten, *Angew. Chem. Int. Ed.* **2007**, *46*, 951–954.

[72] S. Kotha, E. Brahmachary, K. Lahiri, *Eur. J. Org. Chem.* **2005**, 4741–4767.

[73] N. Martín, M. Altable, S. Filippone, Á. Martín-Domenech, A. Poater, M. Solà, *Chem. Eur. J.* **2005**, *11*, 2716–2729.

[74] J.-F. Nierengarten, D. Felder, J.-F. Nicoud, *Tetrahedron Lett.* **1999**, *40*, 273–276.

[75] N. Armaroli, C. Boudon, D. Felder, J. P. Gisselbrecht, M. Gross, G. Marconi, J.-F. Nicoud, J.-F. Nierengarten, V. Vicinelli, *Angew. Chem. Int. Ed. Engl.* **1999**, *38*, 3730–3733.

[76] M. Gutierrez-Nava, G. Accorsi, P. Masson, N. Armaroli, J.-F. Nierengarten, *Chem. Eur. J.* **2004**, *10*, 5076–5086.

[77] T. M. Figueira-Duarte, A. Gégout, J.-F. Nierengarten, *Chem. Commun.* **2007**, 109–119.

7

REDOX AND FLUORESCENT OPEN CORE DENDRIMERS

ANGEL E. KAIFER
Center for Supramolecular Science and Department of Chemistry, University of Miami, Coral Gables, FL 33124-0431, USA

7.1 INTRODUCTION

Dendrimers offer a large number of molecular design possibilities. In many cases their molecular frameworks can be described using a relatively small branching unit that contains perhaps one or two simple functional groups. For instance, one can consider that the commercially available poly(amidoamine) (PAMAM) dendrimers [1] are composed of many repeating units of the type $-CH_2CH_2N(CH_2CH_2CONH-)_2$, in which a branching bifurcation exists at the tertiary nitrogen atom in the middle. The corresponding functional groups, tertiary amine and amide, are integral constituents of the molecular framework in this class of dendrimers. The internal structure of the macromolecule becomes thus rather repetitive, although this does not detract from some of their many fascinating properties. However, one of the most interesting aspects of dendrimer chemistry is the possibility of placing additional functional groups in well-defined locations within the dendrimer structure [2–7]. Of particular concern in this chapter are functional groups with redox active or fluorescent properties.

The covalent attachment of redox active or fluorescent residues is in fact one of the many possibilities involved in dendrimer functionalization. The most commonly used methods involve either the functionalization of the dendrimer periphery with multiple copies of the same residue or the functionalization of the core, from which

Designing Dendrimers, Edited by Sebastiano Campagna, Paola Ceroni, Fausto Puntoriero.
© 2012 John Wiley & Sons, Inc. Published 2012 by John Wiley & Sons, Inc.

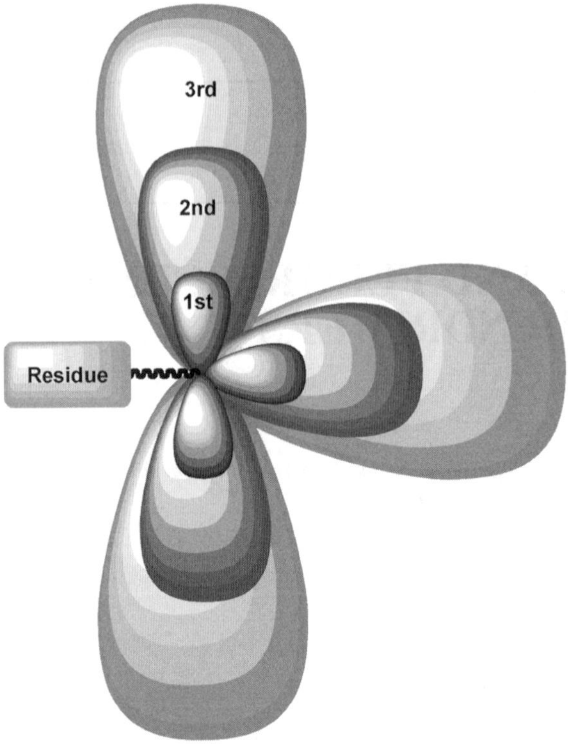

FIGURE 7.1 Schematic representation of an "open core" dendrimer composed of a redox or fluorescent residue and a tri-lobed dendron of variable size (first to third generation).

the dendrimer structure is grown in all directions. In the latter case the growth of the dendrimer framework will eventually insulate the functional core, leading to its relative protection from other solutes or solvent molecules, which leads to the concept of *site isolation*. While we have done some work with dendrimers functionalized on their peripheries or at their cores, most of our dendrimer research has evolved in a slightly different direction [8]. Our basic dendrimer design starts from a functional group of interest (redox, fluorescent) to which a dendron of variable size is covalently attached (Figure 7.1). The resulting macromolecule is a functionalized dendrimer in which the size of the attached dendron affects to some extent the microenvironment around the functional group and, thus, some of its properties. As the dendron grows and starts to fold back, it is reasonable to visualize situations in which the functional group is partially or completely encapsulated by the dendritic framework. It is also true that the functional group may have access to reacting partners in the solution, that is, site isolation may or may not be complete (Figure 7.1). The direction of approach of a reacting partner may also have a profound influence on the kinetics of the resulting reaction. As an example, if we consider a redox active substrate connected to a medium size or large dendron, the effective rate of any electron-transfer reactions will depend to a large extent on the direction of approach of the redox partner.

We have referred to this class of macromolecules as *open core dendrimers* [8] in order to emphasize their similarities to core functionalized dendrimers, while also highlighting that the functional residue on the focal point may remain accessible to interactions or reactions with other solute molecules. It is also equally acceptable to view these macromolecules as "dendronized" residues, to underline the fact that the dendron component can be seen as a chemical substituent on the residue.

This chapter is organized to first summarize our research work on redox active or fluorescent dendrimers based on this molecular design and highly related structures. No attempt will be done here to provide an exhaustive review of the already extensive literature on dendrimers functionalized with redox active or fluorescent groups. The progress of the review will eventually lead us to the discussion of dendrimers, which allow an interesting comparison of two main classes of dendrons that are frequently used in dendrimer research.

7.2 OPEN CORE DENDRIMERS BASED ON NEWKOME DENDRONS

The general schematic structure presented in Figure 7.1 is interesting for a number of reasons. In our own view, the most interesting aspect is that it combines the possibility that the dendron may affect the microenvironment around the focal residue and the possibility that this residue may still be accessible and reactive to substrates in the solution. Furthermore, when considering reactive collisions between a macromolecule of this kind and a smaller reactant, we must recognize that a wide range of possibilities exists. Given the large size of the macromolecule and the fact that the focal residue occupies only a relatively small fraction of its surface, we anticipated that many collisions will be unfruitful, as the reactant and the focal residue will not approach enough to yield a reactive collision. This is particularly interesting as it quickly brings to mind the behavior of many biological macromolecules. For instance, redox proteins usually have the redox active group partially buried inside the protein framework. The redox site is normally not centrally located in the protein structure (see Figure 7.2). In

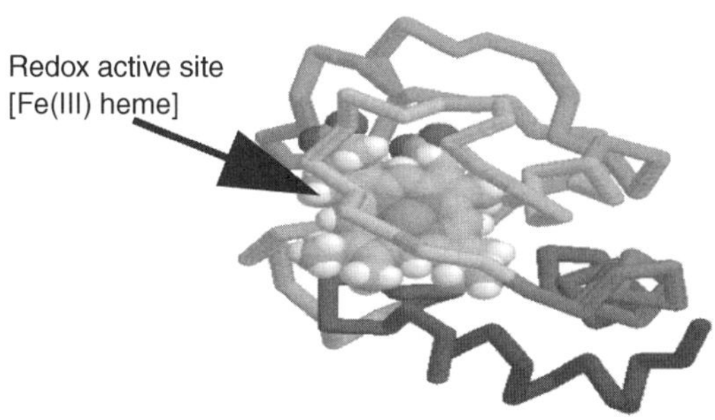

FIGURE 7.2 A schematic representation of horse heart cytochrome *c* showing the location of its redox active site, partially buried inside the protein framework. (See the color version of this figure in Color Plates section.)

this way, the electrostatic character of the protein surface can be used to attract or repel approaching electron-transfer partners either to the proximity or far away from the protein's redox active site, thus favoring or disfavoring the corresponding electron-transfer reactions [9,10]. This approach is one of Nature's preferred ways to selectively enhance or decrease rates of electron transfer, ensuring that electrons will move along intended pathways. We believed that proper design with open core dendrimers would allow the preparation of dendrimers with similar properties.

A key molecular design issue was the selection of dendrimer structure to functionalize the focal residue in our open core dendrimers. Important considerations for dendron selection are (i) ease of synthesis, (ii) space-filling ability, (iii) surface composition, (iv) molecular rigidity, (v) polarity, and (vi) solubility, among others. When we started this work, the number of dendron possibilities was not very large and we quickly selected amino triester building blocks, originally developed by Newkome and coworkers [11,12] (see Scheme 7.1). These dendritic building blocks offered some important advantages for our purposes. First, the synthesis seemed accessible and readily adaptable to divergent and convergent schemes. The AB_3 nature of the building block and the lack of aromatic residues suggested good space-filling ability and reasonable molecular flexibility. A number of possibilities were available regarding the surface composition. These dendrimers are initially prepared as *tert*-butyl esters, giving rise to hydrophobic surfaces. However, the ester functional groups can be easily hydrolyzed to yield more polar carboxylic acid groups. One could anticipate that water solubility might be achieved in neutral or basic solutions. Furthermore, the reactivity of the terminal carboxylic acid groups can be used to continue the dendrimer growth.

We introduced some modifications in the synthetic procedures developed by Newkome's group and prepared second- and third-generation analogs [13] of Behera's amine, and used them to prepare several series of dendrimers by attaching the same functional group to the focal point of these amines. Thus, each series of dendrimers consists of three macromolecules containing the same functional residue

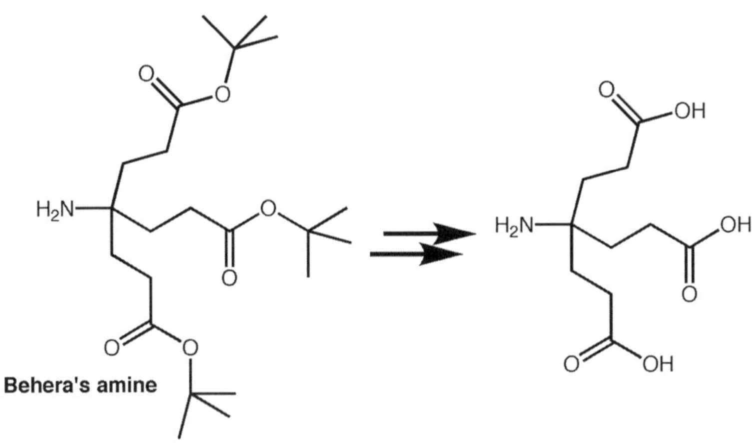

SCHEME 7.1 Structure of the selected dendrimer building block.

connected to Behera's amine, its second- and third-generation analogs. An important question that we must address is: Why stop at the third generation? Our answer is based on two considerations. First, we wanted to work with well-defined molecules. The third-generation dendrimer has 27 esters (or carboxylic acids) on the surface and the fourth-generation analog would have 81 groups on the surface. In the latter case, it becomes statistically much more difficult to guarantee that we are dealing with a single molecular species, without any significant dispersion in the molecular weight. Our mass spectroscopic data indicates that it is possible to prepare third-generation dendrimers of this kind without detectable molecular weight distribution. Furthermore, we have observed in all cases very clear trends of molecular properties from the first to the third generation, which has made it essentially unnecessary to reach the fourth-generation dendrimers.

Figure 7.3 shows all the dendrimers that we have prepared with this general structure. Dendrimers **1–3** and **4–6** are fluorescent as a result of the covalent incorporation of dansyl and pyrene groups at their focal points. Dendrimers **7–24** are electroactive, that is, they undergo electrochemical reactions at accessible

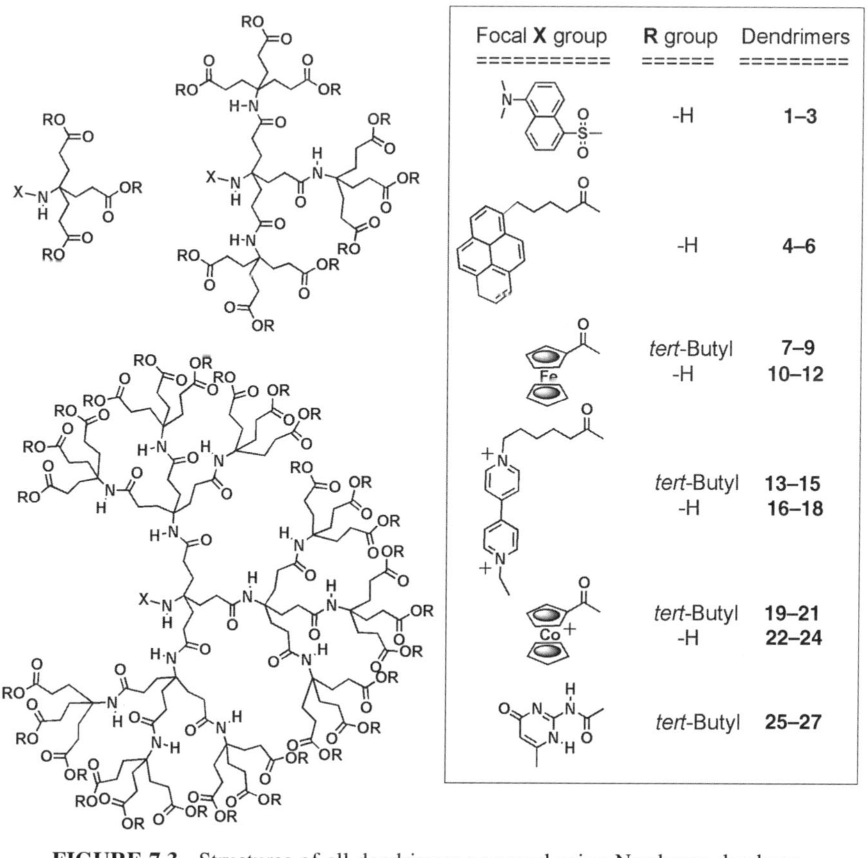

FIGURE 7.3 Structures of all dendrimers prepared using Newkome dendrons.

potentials owing to the incorporation of redox active residues, such as ferrocene, 4,4′-bipyridinium (viologen) or cobaltocenium. Finally, dendrimers **25–27** contain an ureidopyrimidine residue, introduced to express hydrogen bonding recognition properties in these macromolecules. The fluorescent dendrimers were prepared for investigation in aqueous solution, and thus, only the series of dendrimers containing carboxylic acid groups on the periphery were of interest. In contrast to this, we prepared both hydrophobic (*tert*-butyl esters on the surface) and hydrophilic (carboxylic acids on the surface) series of the redox dendrimers for investigation in aqueous and nonaqueous solvents. The ureidopyrimidine dendrimers (**25–27**) were only prepared as *tert*-butyl esters, since the investigation of hydrogen bonding interactions is best carried out in low polarity solvents, such as chloroform or toluene. All these dendrimers were fully characterized by ^{1}H- and ^{13}C-NMR spectroscopies, MALDI-TOF mass spectrometry, and other more specialized techniques such as FTIR spectroscopy, voltammetric techniques, vapor pressure osmometry, and so on.

7.2.1 Fluorescent Open Core Dendrimers

Dendrimers **1–3** were prepared by reaction of dansyl chloride with the focal amine of first- to third-generation Newkome dendrons [14]. Initially prepared with *tert*-butyl ester groups on their surfaces, hydrolysis in formic acid produced the corresponding dendrimers with terminal carboxylic acids. MALDI-TOF mass spectroscopic data were extremely important for the characterization of these dendrimers, providing confirmation of the purity of the amine building blocks and verifying the completeness of the final hydrolysis process. The photophysical behavior of these macromolecules was investigated in joint work with the group of Professor Frank Bright, at the University of Buffalo. Fluorescent measurements on the dendrimers (about 1 μM) dissolved in aqueous buffer (pH 7.0) demonstrated that the dansyl residue is progressively shielded from solvent molecules as the dendrimer generation increases [14]. This is evidenced by the marked changes in spectral parameters (Table 7.1), such as fluorescence quantum yields, excited-state fluorescence lifetimes, radiative and nonradiative decay rates, and rotational correlation times. To provide the 'reader with an approximate measure of the hydrophobic character of the

TABLE 7.1 Selected Photophysical Parameters for Dansylamine (DA) and Dansyl-Containing Dendrimers 1–3 in Aqueous Buffer (pH 7.0)

Compound	λ_{abs} (nm)^a	λ_{fluor} (nm)^a	Φ_f^b	τ_f (ns)^c
DA	324	548	0.028	3.17
1	329	547	0.044	4.90
2	329	542	0.068	6.32
3	324	535	0.086	7.94

[a] Wavelengths of maximum absorption and emission.
[b] Fluorescence quantum yield.
[c] Excited-state fluorescence lifetime.

microenvironment surrounding the dansyl residue in the third-generation dendrimer **3**, we note that its fluorescence quantum yield in aqueous buffer is statistically identical to that exhibited by the model compound dansyl-lysine dissolved in 80:20 (v:v) H_2O:dioxane mixtures.

The excited-state intensity decay kinetics for **1–3** are well described by single exponentials in all cases [14]. Contrary to the common belief that lower generation dendrimers are "floppy" species in solution, the molecular motions of these dendrimers are described by a single rotational reorientation time. This is probably the result of the ionization of the peripheral carboxylic acids in the neutral pH solutions used in these experiments. The ionized, negatively charged carboxylates give rise to internal electrostatic repulsions, which are likely to be responsible for the molecular rigidity exhibited by these dendrimers.

The pyrenyl dendrimers **4–6** were prepared by reaction of the acid chloride of pyrenebutyric acid with the same amine Newkome dendrons, followed by hydrolysis of the peripheral *tert*-butyl ester groups [15]. The photophysical behavior was also investigated in pH 7.0 phosphate buffer solution. In analogy to the behavior of the dansyl dendrimers, fluorescence quantum yields and excited-state fluorescence lifetimes were observed to increase with dendrimer growth, again indicating that the fluorescent pyrenyl residue is gradually shielded from solvent molecules as the attached Newkome dendron increases in size. This work was extended by investigation of the fluorescence quenching of the pyrenyl dendrimers by a wide variety of molecular (nitromethane, acrylamide, *N,N′*-dimethylaniline, and methyl iodide) and ionic (iodide and cupric ions) quenchers to assess the relative structural permeabilities of the macromolecules **4–6**. With the exception of quenching by Cu^{2+}, we observed classic Stern–Volmer behavior in all fluorescence quenching experiments [15]. The recovered Stern–Volmer quenching constants and bimolecular quenching rate constants generally decrease with dendrimer size. This is rationalized by blocking of the pyrenyl residue by the growing dendrimer framework. The decrease is particularly dramatic for the anionic quencher iodide (I^-), which is attributed to the electrostatic repulsions between the quencher and the terminal COO^- groups on the dendrimer surface. Quenching by Cu^{2+} is inconsistent with diffusion control. In fact, we demonstrated that the Cu^{2+} ions bind to the anionic dendrimer surface.

This research work with the two series of fluorescent dendrimers shown in Figure 7.3 shows that growth of the Newkome dendron, even in the limited range from first to third generation, is enough to affect the microenvironment around the fluorophore, slightly increasing its hydrophobic character in aqueous solution. These findings provide some initial validation of our molecular design, in the sense that dendrimer growth leads to some extent of back folding and measurable effects on the photophysical properties of these macromolecules.

7.2.2 Electroactive Open Core Dendrimers

From the beginning of our dendrimer work, we had considerable interest on their functionalization with redox active residues to make possible the investigation of heterogeneous electron-transfer processes involving these macromolecules. Some of our reasons have been described before in this chapter, especially those related to biological mimicry. We were also interested on the idea of detecting dendron back folding and its effect on the microenvironment of the redox active residue through electrochemical potential measurements. Finally, a powerful motivator for the preparation of these dendrimers was the investigation of their heterogeneous rates of electron transfer.

Our first series of electroactive dendrimers was prepared by reaction of chlorocarbonylferrocene with the amine Newkome dendrons, giving rise to the hydrophobic series of macromolecules **7–9** [16]. These hydrophobic dendrimers were investigated in dichloromethane solutions (about 1 mM) also containing 0.1 M tetrabutylammonium hexafluorophosphate as the supporting electrolyte. Our data showed a very clear effect of dendrimer generation on the half-wave potential ($E_{1/2}$) for one-electron oxidation of the ferrocenyl residue. As the dendron grows, the $E_{1/2}$ value shifts to less positive values, suggesting that the generation of positive charge is favored by the dendritic mass [16]. This finding was rationalized by considering that the dendrimer inner phase contains amide groups and its polarity is, thus, higher than that of the bulk dichloromethane solution. As the dendron grows, it folds back and starts to affect the microenvironment around the ferrocenyl residue, increasing its relative polarity and leading to a greater thermodynamic ease for conversion from ferrocene (neutral) to ferrocenium ($+1$ charge).

Voltammetric experiments were also performed to measure the standard rate constants for electron transfer (k^o) between the ferrocenyl residues in the dendrimer and the electrode surface [13,16]. Our data show a clear tendency to lower k^o values with dendrimer growth. This trend has been commonly observed in redox core dendrimers, in which dendrimer growth generally attenuates the electrochemical rate constant [17]. Our hydrophobic ferrocenyl dendrimers (**7–9**) are no exception. It is widely accepted that the decreased rate of heterogeneous electron transfer is due to the longer average distance of maximum approach between the redox center and the electrode surface, which is anticipated as the molecular weight of the macromolecule increases.

We also investigated dendrimers **10–12** in aqueous solution. Here the pH of the solution becomes extremely important as the dendrimers go from bearing no charge at acidic pH values to bearing a relatively large negative charge upon ionization, that is, when the solution pH is neutral or basic. We were very interested on the possibility of using this anionic charge development to orient the dendrimers on the electrode surface. Experiments with bare gold electrodes showed that the voltammetric parameters are rather insensitive to the solution pH. In order to emphasize the electrostatic effects, we decided to experiment with gold working electrodes derivatized with a monolayer of cystamine. These electrodes maintain a positively charged surface throughout the pH range of our experiments (roughly from pH 3 to 8), as the terminal amine group of cystamine remains fully protonated. Under these conditions, we observed that the voltammetric behavior of the second-generation dendrimer, compound **11**, is strongly pH dependent [18]. At low pH, when the dendrimer is fully protonated and uncharged, the one-electron oxidation of

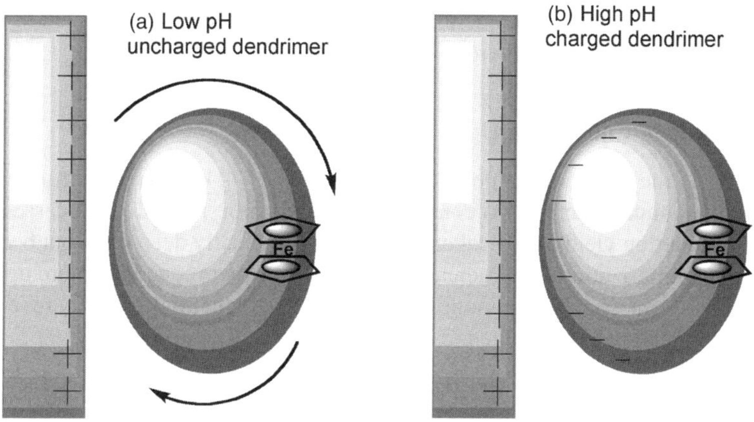

SCHEME 7.2 Cartoon representations of (a) unrestricted interfacial orientation of dendrimer **10** at low pH, and (b) electrostatically restricted orientation of dendrimer **10** at high pH, on positively charged gold electrodes.

the ferrocenyl residue is clearly observed in square wave voltammetric (SWV) experiments. This SWV peak is progressively broadened and eventually disappears as the pH increases. At pH > 7 the voltammetric peak is no longer observed in the surveyed potential range, but it fully develops again if the solution pH is decreased again. This pH dependence was shown to be completely reversible in several cycles, suggesting that decomposition of the ferrocene residue is not behind the loss of voltammetric response at high pH values. After additional experimentation we concluded that the best rationalization of the data relies on the pH-dependent electrostatic orientation of the dendrimer at the electrode solution interface [18] (Scheme 7.2). At low pH, the dendrimer has no charge and can adopt a range of orientations at the interface. In many of these, the ferrocenyl group faces the electrode, which gives rise to relatively fast electron transfer. At high pH, the dendrimer is negatively charged and electrostatic attraction between the positively charged electrode surface and the negatively charged carboxylates on the dendrimer surface orients the dendrimer in such a way that the ferrocenyl group is kept far away from the electrode surface. As a result, the electron-transfer rate is much slower in this situation. This rationalization of the data is further supported by the fact that the third-generation dendrimer, compound **12**, shows similar behavior, while the first-generation dendrimer, **10**, is too small to show these pronounced orientation effects [18].

The demonstration that open core ferrocenyl-containing dendrimers show a pronounced pH controlled, interfacial orientation dependence on their rates of heterogeneous electron transfer was a strong validation of our dendrimer design, as it shows that these macromolecules exhibit properties similar to those of redox proteins. We also studied the electrochemical behavior of hydrophobic (**19–21**) and hydrophilic (**22–24**) cobaltocenium-containing dendrimers and the results were similar to those obtained with the ferrocenyl systems [19]. The viologen-containing dendrimers deserve a separate section as we investigated them in more detail, which eventually led us in some new research directions.

7.2.3 Viologen-Containing Open Core Dendrimers

Viologens is the common name given to the salts of 4,4′-bipyridinium dications. These compounds undergo two consecutive one-electron reduction processes [20]. The first reduction yields a radical cation, which is typically deep blue, but becomes violet upon dimerization. The radical cation can be further reduced at more negative potentials to form a neutral, quinonoid species. These reductions are fully reversible, from a chemical and electrochemical standpoint.

Our general interest in viologen derivatives led Carlos Peinador, a visiting professor from Universidad de A Coruña, Spain, to prepare a series of viologen core dendrimers (compounds **28–30**), using Fréchet dendrons [21] to surround the viologen nucleus [22]. The structures of these macromolecules are shown in Figure 7.4. Investigation of their voltammetric behavior in acetonitrile solution produced surprising results, as their electrochemical behavior was found to be fully reversible (fast in the voltammetric time scale) as the size increases from first to third generation. This was an unusual result, since electroactive core dendrimers had been found to suffer a pronounced attenuation of electrochemical kinetics with dendrimer growth [17]. Even considering that these dendrimers are not very large, their fast electrochemistry was surprising in light of the existing body of literature data. However, our results were confirmed by the simultaneous publication by Balzani and coworkers of similar data on the same structures [23].

Intrigued by the voltammetric findings with compounds **28–30**, we decided to prepare viologen dendrimers using our open core design with Newkome dendrons. We encountered problems with compound stability until we settled on a five-methylene spacer between the viologen nucleus and the amide connection to the focal point of the dendrimer (see structures **13–15** and **16–18** in Figure 7.3). This spacer was not required in the case of ferrocene or cobaltocenium redox residues, and we initially worried that it might diminish the effects of dendron growth on the microenvironment of the viologen nucleus. However, our voltammetric data quickly dispelled this notion, because the half-wave potentials showed a clear dependence on dendron size [24]. In fact, the voltammetric behavior of compounds **13–15**, as their hexafluorophosphate salts, could be investigated in various solvents with a substantial range of polarities, which was not possible with the ferrocenyl dendrimers due to solubility limitations.

Figure 7.5 shows voltammetric data obtained from these solvent studies [24,25]. In dichloromethane solution (Figure 7.5a), we observed that the first reduction potential shifts to more negative values as the dendrimer grows. This clearly suggests that dendritic growth differentially favors the dicationic versus the cationic form of the viologen nucleus. This is the same trend recorded with the ferrocenyl dendrimers **7–9** in dichloromethane solution and suggests that, in both cases, the inner phase of the dendrimer is more polar than the bulk dichloromethane solution. Thus, when the dendrimer component grows the oxidation state with the highest positive charge is

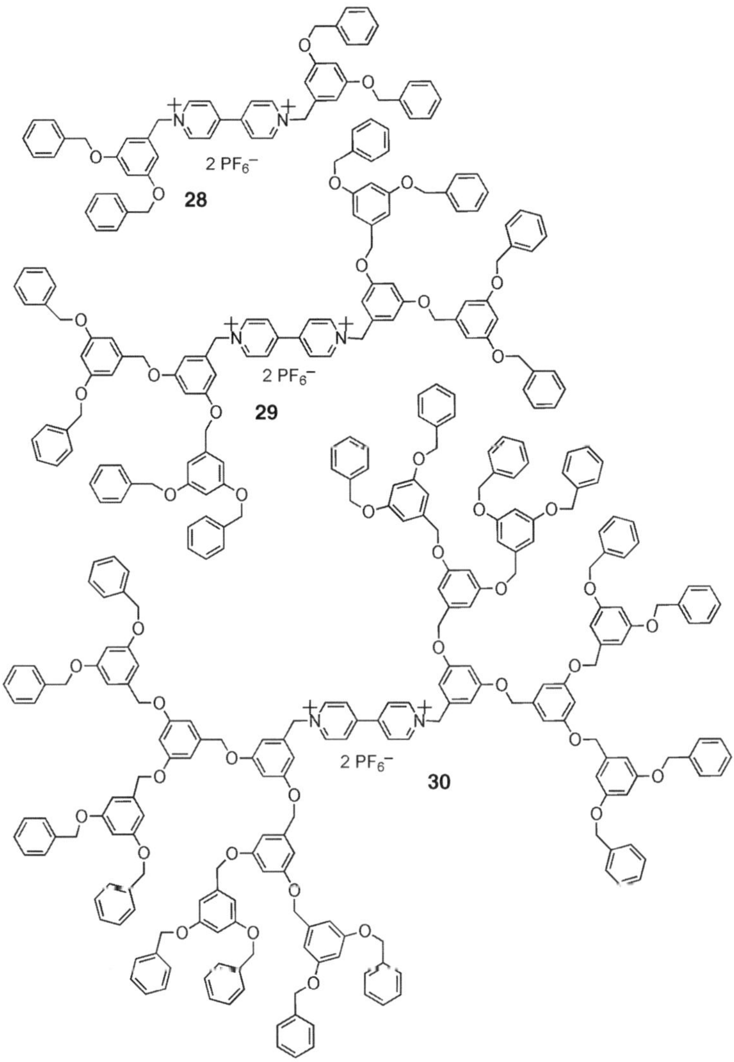

FIGURE 7.4 Structures of viologen-core Fréchet dendrimers.

differentially stabilized. The same trend is observed in THF (Figure 7.5b) and, barely, in acetonitrile (Figure 7.5c). However, this is no longer true in DMSO solution, where the half-wave potentials show a modest trend in the opposite direction. From this set of data, we concluded that the microenvironment around the viologen nucleus is affected by the growing Newkome dendron. The direction of the half-wave potential shift observed as the dendron grows depends on the polarity balance between the inner dendron phase and the bulk solution. Not surprisingly, we observed similar trends in the half-wave potential data for the second one-electron reduction of the viologen nucleus in dendrimers **13–15** [24,25].

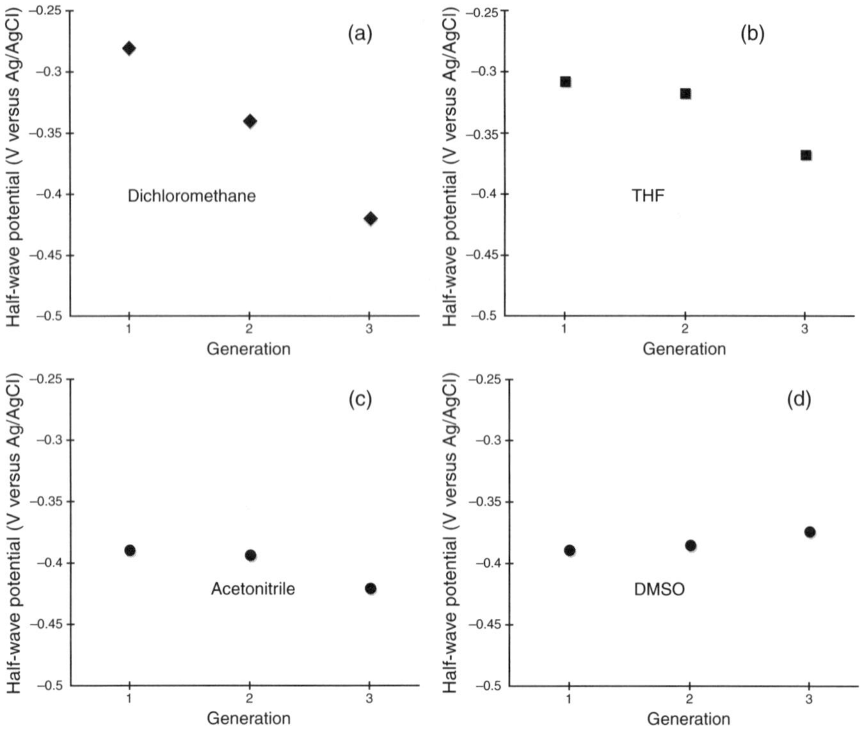

FIGURE 7.5 Half-wave potentials for the first one-electron reduction of dendrimers **13–15** as a function of generation in (a) dichloromethane, (bB) THF, (c) acetonitrile, and (d) DMSO also containing 0.2 M tetrabutylammonium hexafluorophosphate as the supporting electrolyte.

The voltammetric behavior of these dendrimers was also found to maintain complete reversibility from the first to the third generation of growth at scan rates as fast as 2.0 V/s [24]. Again, viologen dendrimers appear to resist the general trend to exhibit slower electrochemical kinetics upon growth of the dendritic mass [17]. Due to these findings we decided to prepare viologen-containing open core dendrimers with Fréchet dendrons [21] in order to compare their behavior to those prepared with Newkome dendrons. Thus, we prepared two new series of viologen-containing dendrimers [25] (Figure 7.6).

Not surprisingly, since these dendrimers can be considered to be approximately half the size of dendrimers **28–30**, their voltammetric behavior also remains perfectly reversible at scan rates up to 2.0 V/s. The fact that viologen-core Fréchet dendrimers **28–30**, open core Newkome dendrimers **13–15,** and open core Fréchet dendrimers **31–33** and **34–36** all show reversible voltammetric behavior, without any detectable attenuation of electrochemical kinetics, constitutes a very interesting finding, as it establishes a clear exception to the otherwise general attenuation of electrochemical kinetics observed with core dendrimers containing other redox active groups [17]. Given that, at the potential values required to cause viologen reduction, the electrode

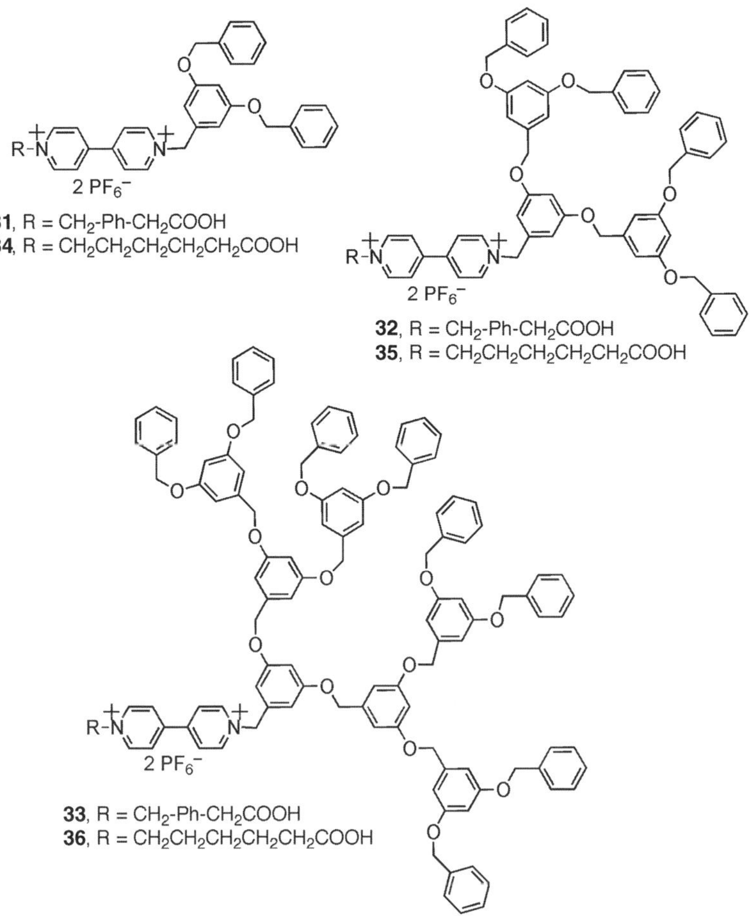

FIGURE 7.6 Viologen-containing, open core Fréchet dendrimers.

surface exhibits a negative charge density, we have speculated that the dicationic viologen residue might be electrostatically attracted to the electrode surface. This attraction may enforce interfacial dendrimer conformations that negate or minimize the effect of dendrimer size, allowing the heterogeneous electron-transfer process to remain fast enough so that the voltammetric behavior remains reversible.

An important feature of the molecular design of open core dendrimers is the possibility of interactions between the functional residue at the macromolecule's focal point and reacting partners in the same solution. We have exploited this feature to investigate host–guest binding interactions between the focal functional groups (as dendronized guests) and freely diffusing hosts. For instance, we have investigated in detail the binding interactions in aqueous solution between the β-cyclodextrin host and the water-soluble dendrimers **1–3** [14] and **10–12** [13], as well as those between the cucurbit[7]uril hosts and dendrimers **10–12** [26], **16–18** [27], and **22–24** [19] also

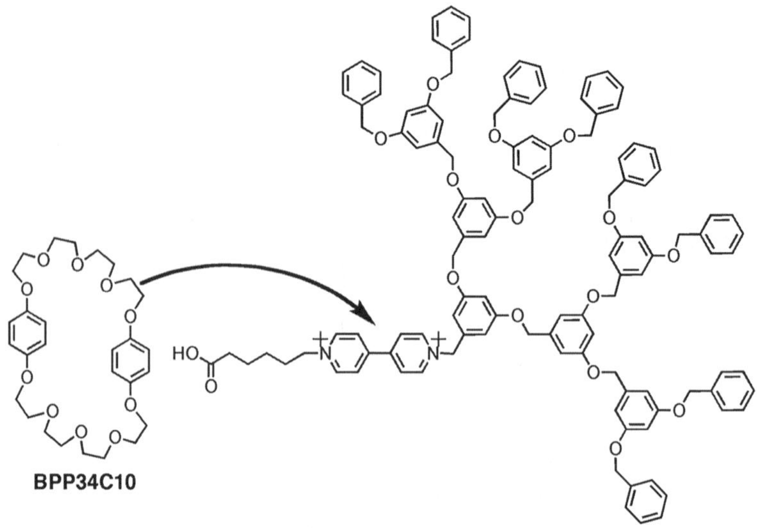

SCHEME 7.3 Binding of dendrimer **36** by BPP34C10 crown ether host.

in aqueous media. The binding interactions of the viologen open core dendrimers **16–18** with the cucurbit[8]uril host have allowed us to report interesting redox-based methods to control the self-assembly of these dendrimers [27–29]. This body of work has been the subject of other reviews [30,31], so we will not devote a lot of space to these issues in this chapter. However, the well-known host–guest interactions between dibenzocrown ether hosts and viologen guests in solvents of intermediate polarity, such as acetonitrile, appeared as an interesting way to compare the properties of viologen open core dendrimers based on Newkome (**16–18**) and Fréchet (**31–33** and **34–36**) dendrons. Therefore, we decided to determine the corresponding equilibrium association constants (K) between these dendronized viologens and bis-p-phenylene-24-crown-8 (BPP34C10, see structure in Scheme 7.3) in acetone and acetonitrile solutions. The K values (Table 7.2) were determined by recording the absorbance of the charge transfer band of the complex in titrations in which the concentration of crown host was varied while maintaining the concentration of dendrimer fixed. The measured absorbance values were fitted to a 1:1 binding isotherms using standard methods [25].

Several trends are clearly evident in the data. First, all complexes show higher K values in acetone than in acetonitrile solution. This finding reflects the ion-dipole nature of the main intermolecular forces holding the complexes together. The more polar solvent, acetonitrile, solvates more effectively the positively charged nitrogen atoms on the viologen and the oxygen atoms on the crown, decreasing the complex stability. The stability of the BPP34C10 complexes with the open core Newkome dendrimers (**13–15**) decreases with dendrimer growth. At the limit of this behavior, no complex formation was observed between the third-generation dendrimer **15** and the crown host in acetonitrile solution [25]. A five-methylene tether connects the viologen

TABLE 7.2 Equilibrium Association Constants (M^{-1}) for Complex Formation between the Host BPP34C10 and Viologen-Containing Open Core Dendrimers at 25°C

Guest	K (in acetone)[a]	K (in acetonitrile)[a]
Methyl viologen	560	200
13	370	93
14	210	85
15	92	NB[b]
31	480	100
32	630	100
33	670	85
34	420	70
35	560	100
36	740	75

[a] Error margin no larger than ±12%.
[b] No binding detected.

group to the focal point of the Newkome dendrons in **13–15**. No such tether exists in the Fréchet dendrimers **31–36**. Despite the closer proximity of the viologen group to the dendrons in this case and in marked contrast to the behavior observed with **13–15**, the K values measured with the two series of Fréchet dendrimers do not show any decreasing trend as the dendron grows. If anything, dendron growth results in small gains in complex stability with these two series of macromolecules. No significant differences in the binding affinities were observed between the two series of Fréchet dendrimers, which differ in the structure of the substituents connected to the viologen residue on the opposite side to the dendron.

The structure of the complex between BPP34C10 and methyl viologen is well known [32,33]. Briefly, the crown's oxygen atoms interact with the two positive charges on the viologen while the two π-donor hydroquinol units on the crown develop charge-transfer interactions with the π-acceptor bipyridinium residue. Overall the crown encircles the viologen groups and the stability of the resulting complex is rather modest, which facilitates the interference on the complexation process by steric hindrance from the growing dendritic mass. Clearly, this is the case with the Newkome dendrons, but Fréchet dendrons fail to bring about similar effects. The more rigid character of the polyaryl ether (Fréchet) dendrons, built around the 1,3,5-substitution pattern on the phenyl ring, may lead to wedge-like dendron shapes, which afford less effective encapsulation of the viologen residues as compared to the more flexible Newkome dendrons. This interpretation is also supported by the very small dendron size effects on the half-wave potentials for viologen reduction that were recorded with dendrimer series **31–33** and **34–36** [25].

We conclude that the investigation of these host–guest binding interactions leads to useful insights into the space filling and encapsulating properties of these two very different, albeit popular, types of dendron that are often used in dendrimer design.

7.3 HYBRID FRÉCHET–NEWKOME DENDRIMERS WITH A VIOLOGEN CORE

Our interest in viologen-containing dendrimers and the differences noted above on the space-filling properties of Newkome and Fréchet dendrons led us to synthesize a new series of dendrimers containing a dendron of each kind covalently attached to the viologen nucleus [34] (see Figure 7.7). The synthesis of these dendrimers relied on procedures previously reported by our group. Briefly, we first monoquaternized 4,4′-bipyridine with 6-bromohexanoic acid. The resulting intermediate still contains a free nitrogen on the bipyridine nucleus to which Fréchet dendrons (as bromides) can be easily attached. Finally, the terminal carboxylic acid on the other end of the molecule can be reacted with the amine Newkome dendrons, to give rise to the final

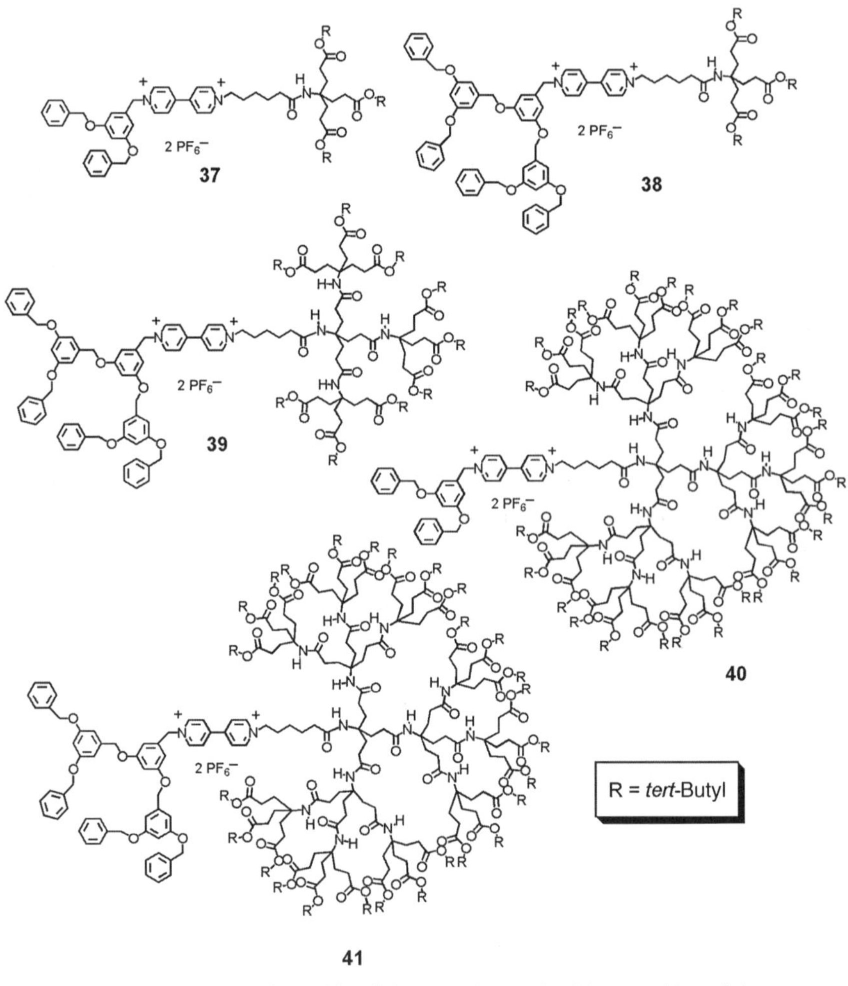

41

FIGURE 7.7 Structures of hybrid Fréchet–Newkome dendrimers with a viologen core.

TABLE 7.3 Half-Wave Potentials[a] ($E_{1/2}{}^1$ and $E_{1/2}{}^2$ in V versus Ag/AgCl) for dendrimers 37–41 Measured at 23°C in Organic Solutions Also Containing 0.1 M Tetrabutylammonium Hexafluorophosphate as Supporting Electrolyte

Solvent		37	38	39	40	41
DCM	$E_{1/2}{}^1$	− 0.240	− 0.240	− 0.277	− 0.351	− 0.292
DCM	$E_{1/2}{}^2$	− 0.755	− 0.764	− 0.722	− 0.821	− 0.784
THF	$E_{1/2}{}^1$	− 0.276	− 0.271	− 0.285	− 0.320	− 0.300
THF	$E_{1/2}{}^2$	− 0.676	− 0.676	− 0.685	− 0.706	− 0.708
ACN	$E_{1/2}{}^1$	− 0.325	− 0.322	− 0.345	− 0.369	− 0.358
ACN	$E_{1/2}{}^2$	− 0.754	− 0.756	− 0.770	− 0.776	− 0.772

[a] Estimated error margin for all potential values is ±0.005 V.

hybrid dendrimer. We elected to designate these dendrimers as "hybrid" to emphasize the fact that their structures combine two entirely different types of dendritic components. Using first- to third-generation dendrons, we could prepare up to nine unsymmetric dendrimers in this series, but we decided to prepare only the five macromolecules shown in Figure 7.7, not only for reasons of economy but also because we were able to answer a good number of questions related to the interplay between the two types of dendrons with this selected set of dendrimers [34].

As anticipated the cathodic voltammetric behavior of dendrimers **37–41** is characterized by the stepwise reduction of the viologen nucleus, which takes place as two consecutive one-electron reduction processes. The half-wave potentials measured in dichloromethane (DCM), tetrahydrofuran (THF), and acetonitrile (ACN) are given in Table 7.3.

The potential values are clearly affected by dendron growth. The first reduction half-wave potentials ($E_{1/2}{}^1$) spread over a range of 110 mV in DCM solution, while the covered range is narrower in THF and ACN [34]. The reason behind this is probably related to the low polarity of DCM, which gives rise to the most pronounced polarity contrast with the inner phase of the dendrons, particularly with the amide-rich and polar Newkome dendrons. Therefore, the careful analysis of the potential values recorded for these dendrimers in DCM solution is of substantial interest. First, we should note that the $E_{1/2}{}^1$ value is almost invariant in going from **37** to **38**, where the only change is the size of the Fréchet dendron (first to second generation). Conversely, from **38** to **39**, there is a measurable $\Delta E_{1/2}{}^1$ of − 37 mV, which is obviously due to the growth of the Newkome dendron from first to second generation, while the Fréchet dendron is identical (second generation). Clearly the reduction potential is more sensitive to Newkome dendron growth than to Fréchet dendron growth. A more pronounced size differential in the Newkome dendron, from dendrimer **37** to **40** leads to a more acute potential shift of − 111 mV. It is instructive to compare this $\Delta E_{1/2}{}^1$ value with that observed in going from **38** to **41** (− 52 mV). In the latter case, the potential shift results from replacing the first generation Newkome dendron by its third-generation analogue while keeping the second generation Fréchet dendron unchanged. However, the magnitude of the potential shift is decreased (compared to the $\Delta E_{1/2}{}^1$ value **37** to **40**). This means that the growth of the Fréchet dendron modulates down the effects of Newkome dendron growth on the other side of the viologen nucleus. Consistent with this conclusion, the $\Delta E_{1/2}{}^{1}$ value measured with the open core viologen dendrimers **13–16** was -140 mV in DCM solution. In other words, our data reveals that the first to third generation dendron-induced $\Delta E_{1/2}{}^{1}$ value is tempered by the steric bulk of the other viologen substituent. These findings indicate that, in spite of its five-methylene tether to the viologen residue, Newkome dendrons are effective at modifying the core's microenvironment. The sign of the recorded $\Delta E_{1/2}{}^{1}$ values in DCM solution is explained by the greater polarity of the Newkome dendron inner phase, which differentially stabilizes the oxidation states with the larger positive charges ($+2$ versus $+1$ in the first-reduction process and $+1$ versus 0 in the second). The potential values recorded in THF and ACN follow similar trends, although the magnitude of the potential shifts is generally smaller [34].

We also assessed the standard rate constants for heterogeneous electron transfer with these viologen core dendrimers. While the smaller dendrimers, **37–39,** exhibit essentially reversible voltammetric behavior, the two largest dendrimers **40** and **41** show measurably slower (quasireversible) voltammetric behavior [34]. These two compounds are the largest molecular weight viologen core dendrimers ever studied by our group and, thus, the loss of voltammetric reversibility is anything but unexpected.

The voltammetric behavior of these hybrid dendrimers is interesting because it allows us to compare the behavior of two different types of dendrons present in the same macromolecule. The results are consistent with previous data obtained by our group and suggest important differences in the rates of space filling and relative inner polarities of Fréchet and Newkome dendrons.

7.4 HYBRID FRÉCHET–NEWKOME DENDRIMERS WITH A TRIAZINE CORE

Our interest on the preparation of dendrimer systems containing both Fréchet and Newkome dendrons eventually led us to consider triazine-based dendrimers of the kind developed by Steffensen and Simanek [35], taking advantage of the differential reactivity of trichlorotriazine (cyanuric chloride). This group and others [36] have demonstrated that it is possible to sequentially replace one, two, or three of the chloro substituents by controlling the temperature of the reaction. In our synthetic procedures, we started by reacting cyanuric chloride with a Fréchet dendron (in its –OH terminated form) at 0°C in THF. In a second step, we reacted the resulting Fréchet dendronized dichlorotriazine with an amine Newkome dendron in acetone solution (at room temperature). This leads to the nucleophilic substitution of the second chloro substituent and the attachment of the Newkome dendron to the triazine nucleus. Finally, the last chloro substituent is replaced by treatment with aminoferrocene in dry refluxing THF, yielding a triazine-based dendrimer containing three different substituents: a Fréchet dendron, a Newkome dendron, and aminoferrocene [37]. The resulting dendrimers are redox active due to the presence of the ferrocenyl residue.

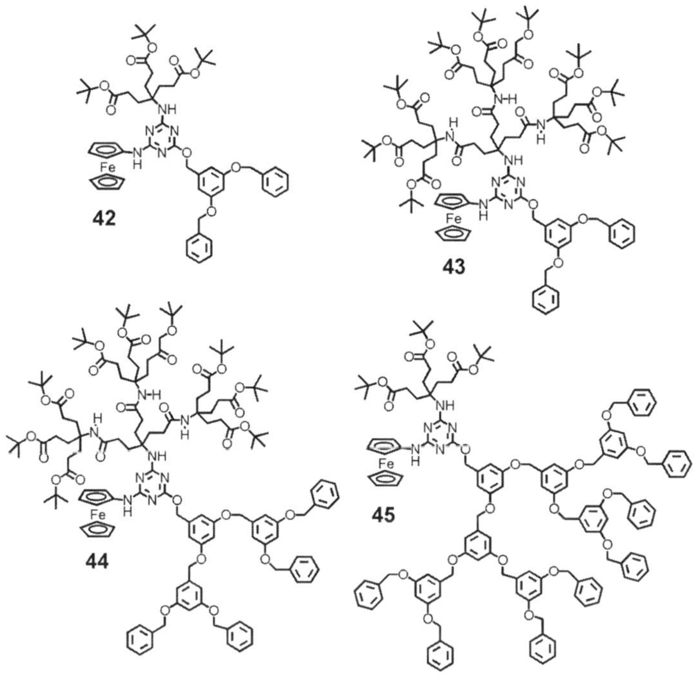

FIGURE 7.8 Structure of the hybrid Fréchet–Newkome dendrimers with a triazine core.

Nine different dendrimers are possible in this series, if we keep the size of the dendrons within the first to third generation. However, as before, we elected to prepare only a set of four dendrimers (Figure 7.8), which were sufficient to understand their properties in considerable detail.

We took advantage of the redox activity of the ferrocenyl group to investigate the voltammetric behavior of these dendrimers in DCM and ACN solutions. The half-wave potential values for one-electron oxidation of the ferrocene center are given in Table 7.4.

In dichloromethane solution, growth of the Newkome dendron has a pronounced effect on the half-wave potentials, which shift to less positive values as the dendron

TABLE 7.4 Half-Wave Potentials ($E_{1/2}$, V versus Ag/AgCl) and Standard Rate Constants (k^o, cm/s) at 25°C for the Oxidation of the Ferrocene Center in Hybrid Dendrimers 42–45

	42	43	44	45
$E_{1/2}$ (CH_2Cl_2)[a]	0.372 ± 0.002	0.318 ± 0.002	0.335 ± 0.001	0.376 ± 0.03
$E_{1/2}$ (CH_3CN)[a]	0.317 ± 0.003	0.309 ± 0.003	0.307 ± 0.001	0.318 ± 0.001
k^o (CH_2Cl_2)[a]	0.086 ± 0.015	0.046 ± 0.008	0.018 ± 0.002	0.026 ± 0.005

[a] Data obtained in the indicated solvent also containing 0.2 M tetrabutylammonium hexafluorophosphate.

grows. For instance, replacement of the first generation (compound **42**) by a second-generation Newkome dendron (compound **43**) causes an anodic shift of 54 mV in the $E_{1/2}$ value [37]. In strong contrast to this, replacement of a first generation (compound **42**) by a third-generation Fréchet dendron (compound **45**) has no measurable effect on the half-wave potential. In acetonitrile solution, the effect of dendron replacements on the observed $E_{1/2}$ values is much less pronounced, because the polarity of the bulk solution is closer to that of the inner phase of the Newkome dendrons [37]. These data are in agreement with our previous observations and suggest that Fréchet dendrons expand as wedges away from the core and have little effect on the microenvironment around the ferrocene residue. Overall, the Newkome dendrons appear to be more effective at changing the microenvironment around the ferrocene center in these hybrid dendrimers. This is probably a consequence of their AB_3 architectural framework and greater flexibility.

In the range of scan rates surveyed (0.1–2.0 V/s) the voltammetric behavior of these dendrimers falls in the reversible to quasireversible regimes [37]. The standard rate constants (k^o) for the heterogeneous electron-transfer processes were determined in DCM solution and are also given in Table 7.4. In general terms, as expected, the k^o values tend to decrease as the overall size of the dendrimer increases, as observed before in most dendrimers containing redox active centers at their cores [17].

7.5 HYDROGEN BONDING OPEN CORE DENDRIMERS

Our general interest on hydrogen bonding systems led us to prepare a series of open core Newkome dendrimers with self-recognition abilities [38]. Therefore, dendrimers **25–27** (Figure 7.3) were prepared with a single ureidopyrimidine residue in their focal points. According to the design of Meijer and coworkers [39], this residue self-recognizes by formation of four parallel hydrogen bonds as shown in Figure 7.9 for the second-generation dendrimer (**26**).

We used NMR spectroscopy, mass spectrometry, and vapor-pressure osmometry in nonpolar solvents (chloroform and toluene) to investigate the dimerization of these dendrimers by quadruple hydrogen bonding [38]. Our data indicate that dendrimers **25** and **26** dimerize extensively at concentrations above the millimolar level. In sharp contrast to these results, the level of dimerization observed with the third-generation dendrimer **27** was very limited. At millimolar concentrations of **27**, a 2% dimerization level was measured by NMR spectroscopy. This is a very surprising result based on the essentially quantitative dimerization observed with the smaller dendrimers in this series. Steric hindrance exerted by the larger dendron in **27** may certainly play a significant role. However, our data with open core Newkome dendrimers suggest a complementary rationalization relying on the increased microenvironment polarity created by the Newkome dendron. Growth of the dendron from the second to the third generation may give rise to a microenvironment polar enough to weaken hydrogen bonding between the ureidopyrimidine residues of the two dendrimers.

We did not prepare an equivalent series of open core hydrogen bonding dendrimers based on Fréchet dendrons. However, a few months after our report, Chow and

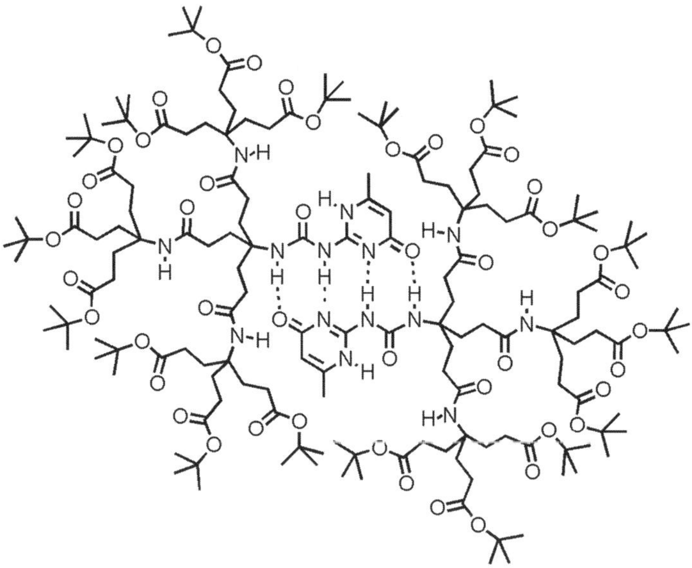

FIGURE 7.9 Dimerization of dendrimer **26** via quadruple hydrogen bonding.

coworkers synthesized a series of Fréchet-dendronized ureidopyrimidine macromolecules [40] and investigated their degree of dimerization via quadrupole hydrogen bond formation. Their data indicate that these open core Fréchet dendrimers dimerize extensively from first to third generation of dendron growth. Ostensibly, we must conclude again that the Fréchet dendrons tend to expand away from the core in wedge-like structures, exerting only minor influences on the polarity of the microenvironment around the core, while Newkome dendrons are more effective at filling space and affecting the core microenvironment. The contrast between our Newkome-based dendrimers and Chow's Fréchet-based dendrimers is completely consistent again with all our data comparing these two types of dendritic architectures.

7.6 CONCLUSIONS AND OUTLOOK

The research work summarized in this chapter demonstrates that it is not truly necessary to use very large dendrons to affect the microenvironment around a functional group connected to the dendron's focal point. Our series of dendrimers, using first to third-generation dendrons, afford many examples with fluorescent or redox active functional groups in which modest dendron growth leads to easily measurable changes attributable to dendron back-folding, which modulates the polarity of the microenvironment around the functional group. Furthermore, our data clearly and consistently demonstrate that Newkome dendrons are more effective than Fréchet dendrons at folding back and affecting the polarity of the functional group surroundings.

Some of the work discussed here includes dendrimers containing Fréchet and Newkome dendrons in the same structure. It is of substantial interest to investigate the assembly of these dendrimer structures, taking advantage of the very different space-filling abilities, flexibilities, and polarities of these two dendron types. We are currently completing an investigation on the monolayer assembly of Fréchet–Newkome hybrid dendrimers at the air–water interface and strongly believe that these hybrid macromolecules may give rise to a variety of supramolecular assemblies with unique properties.

ACKNOWLEDGMENTS

The author is grateful to the National Science Foundation for the long-standing support of this research. This work would have been impossible without the effort and dedication of all the coworkers named in the references.

REFERENCES

[1] D. A. Tomalia, A. M. Naylor, W. A. Goddard III, *Angew. Chem. Int. Ed. Engl.* **1990**, *29*, 138–175.
[2] S. M. Grayson, J. M. J. Fréchet, *Chem. Rev.* **2001**, *101*, 3819–3868.
[3] D. K. Smith, *Chem. Commun.* **2006**, 34–44.
[4] A.-M. Caminade, A. Maraval, J.-P. Majoral, *Eur. J. Inorg. Chem.* **2006**, 887–901.
[5] U. Boas, J. B. Christensen, P. M. H. Heegaard, *J. Mater. Chem.* **2006**, *16*, 3785–3798.
[6] S.-H. Hwang, C. D. Shreiner, C. N. Moorefield, G. R. Newkome, *New J. Chem.* **2007**, *31*, 1192–1217.
[7] S.-C. Lo, P. L. Burn, *Chem. Rev.* **2007**, *107*, 1097–1116.
[8] A. E. Kaifer, *Eur. J. Inorg. Chem.* **2007**, 5015–5027.
[9] V. L. Davidson, *Acc. Chem. Res.* **2000**, *33*, 87–93.
[10] V. L. Davidson, *Acc. Chem. Res.* **2008**, *41*, 730–738.
[11] G. R. Newkome, R. K. Behera, C. N. Moorefield, G. R. Baker, *J. Org. Chem.* **1991**, *56*, 7162–7167.
[12] G. R. Newkome, K. K. Kotta, C. N. Moorefield, *J. Org. Chem.* **2005**, *70*, 4893–4896.
[13] C. M. Cardona, T. D. McCarley, A. E. Kaifer, *J. Org. Chem.* **2000**, *65*, 1857–1864.
[14] C. M. Cardona, J, Alvarez, A. E. Kaifer, T. D. McCarley, S. Pandey, G. A. Baker, N. J. Bonzagni, F. V. Bright, *J. Am. Chem. Soc.* **2000**, *122*, 6139–6144.
[15] C. M. Cardona, T. Wilkes, W. Ong, A. E. Kaifer, T. D. McCarley, S. Pandey, G. A. Baker, M. N. Kane, S. N. Baker, F. V. Bright, *J. Phys. Chem. B.* **2002**, *106*, 8649–8656.
[16] C. M. Cardona, A. E. Kaifer, *J. Am. Chem. Soc.* **1998**, *120*, 4023–4024.
[17] C. S. Cameron, C. B. Gorman, *Adv. Funct. Mater.* **2002**, *12*, 17–20.
[18] Y. Wang, C. M. Cardona, A. E. Kaifer, *J. Am. Chem. Soc.* **1999**, *121*, 9756–9757.
[19] D. Sobransingh, A. E. Kaifer, *Langmuir* **2006**, *22*, 10540–10544.

[20] P. Monk, *The Viologens: Physicochemical Properties, Synthesis and Applications of the Salts of 4,4′-Bipyridine*, Wiley, New York, **1998**, Chapter 9.

[21] C. J. Hawker, J. M. J. Fréchet, *J. Am. Chem. Soc.* **1990**, *112*, 7638–7647.

[22] R. Toba, J. M. Quintela, C. Peinador, E. Roman, A. E. Kaifer, *Chem. Commun.* **2001**, 857–858.

[23] V. Vicinelli, M. Maestri, V. Balzani, W. M. Muller, U. Muller, U. Hahn, F. Osswald, F. Vogtle, *New J. Chem.* **2001**, *25*, 989–993.

[24] W. Ong, A. E. Kaifer, *J. Am. Chem. Soc.* **2002**, *124*, 9358–9359.

[25] W. Ong, J. Grindstaff, D. Sobransingh, R. Toba, J. M. Quintela, C. Peinador, A. E. Kaifer, *J. Am. Chem. Soc.* **2005**, *127*, 3353–3361.

[26] D. Sobransingh, A. E. Kaifer, *Chem. Commun.* **2005**, 5071–5073.

[27] W. Ong, A. E. Kaifer, *Angew. Chem. Int. Ed.* **2003**, *42*, 2164–2167.

[28] K. Moon, J. Grindstaff, D. Sobransingh, A. E. Kaifer, *Angew. Chem. Int. Ed.* **2004**, *43*, 5496–5499.

[29] W. Wang, A. E. Kaifer, *Angew. Chem. Int. Ed.* **2006**, *45*, 7042–7046.

[30] W. Ong, M. Gómez-Kaifer, A. E. Kaifer, *Chem. Commun.* **2004**, 1677–1683.

[31] W. Wang, A. E. Kaifer, *Adv. Polym. Sci.* **2009**, *222*, 205–235.

[32] B. L. Allwood, N. Spencer, H. Shahriari-Zavarech, J. F. Stoddart, D. J. Williams, *J. Chem. Soc. Chem. Commun.* **1987**, 1064–1065.

[33] B. L. Allwood, N. Spencer, H. Shahriari-Zavarech, J. F. Stoddart, D. J. Williams, *J. Chem. Soc. Chem. Commun.* **1987**, 1066–1067.

[34] P. Bhattacharya, A. E. Kaifer, *J. Org. Chem.* **2008**, *73*, 5693–5698.

[35] M. B. Steffensen, E. E. Simanek, *Angew. Chem. Int. Ed.* **2004**, *43*, 5178–5180.

[36] M.-X. Wang, H.-B. Yang, *J. Am. Chem. Soc.* **2004**, *126*, 15412–15422.

[37] W. Wang, H. Sun, A. E. Kaifer, *Org. Lett.* **2007**, *9*, 2657–2660.

[38] H. Sun, A. E. Kaifer, *Org. Lett.* **2005**, *7*, 3845–3848.

[39] A. T. Cate, H. Kkjiman, A. L. Spek, R. P. Sijbesma, E. W. Meijer, *J. Am. Chem. Soc.* **2004**, *126*, 3801–3808.

[40] C.-H. Wong, H.-F. Chow, S.-K. Hui, K. H. Sze, *Org. Lett.* **2006**, *8*, 1811–1814.

8

REDOX-ACTIVE ORGANOMETALLIC DENDRIMERS AS ELECTROCHEMICAL SENSORS

CARMEN M. CASADO[1], BEATRIZ ALONSO[1], JOSÉ LOSADA[2], AND MARÍA PILAR GARCÍA-ARMADA[2]

[1]*Departamento de Química Inorgánica, Universidad Autónoma de Madrid, Madrid, Spain*

[2]*Departamento de Ingeniería Química Industrial, Universidad Politécnica de Madrid Madrid, Spain*

8.1 INTRODUCTION

Dendrimers are nanosized, highly branched molecules that are constructed from a multifunctional central core and expand to a periphery that becomes more dense with increasing generation number, and which exhibit a unique combination of chemical and physical properties [1–5]. Since the pioneering work by Vögtle and coworkers [6] (cascade synthesis), Newkome et al. [7] (arborols), and Tomalia et al. [8] (starburst dendrimers), the synthetic research on dendritic macromolecules has been increasing at an amazing rate. In the beginning, the research on dendrimers focused on the synthesis, characterization, and properties of perfect dendrimers of higher generations, but efforts are now directed toward the study of properties, applications, and uses of dendrimers. The wide range of possibilities offered by dendritic molecular systems has generated enthusiastic studies in many different fields, such as chemistry, physics, engineering, biology, and medicine [1–5,9].

From an application point of view, the incorporation of a range of metals into the dendritic framework is of particular interest. Balzani's group [10] and Newkome et al. [11] independently initiated such studies by using polypyridine Ru complexes in

Designing Dendrimers, Edited by Sebastiano Campagna, Paola Ceroni, Fausto Puntoriero.
© 2012 John Wiley & Sons, Inc. Published 2012 by John Wiley & Sons, Inc.

the early 1990s. The interest in developing metallodendrimers [12–18] resulted from the fact that the introduction of metals into dendritic structures allows access to highly ordered materials with new magnetic, catalytic, optical, electro- and photochemical, and biomedical properties as well as reactivity.

In particular, the incorporation of organometallic entities on the surface or within the dendritic structures represents a stimulating, challenging target in both organometallics and dendrimers research because it provides a unique opportunity for tailoring organometallic dendrimers to achieve desirable properties for well-defined applications such as dendritic catalysts, in multielectron redox and photocatalytic processes, as molecular sensors, and others [19–22].

Dendrimers containing redox-active units are of great importance because the number of electrochemically active functional groups can be precisely controlled. Thus, redox-active dendrimers are excellent candidates to play a key role in understanding biological electron transfer and for use in practical applications such as catalyst, electron transfer mediators, energy conversion, ion sensors, and in electronic devices [23–26].

Since 1994, Morán and his group, pioneering in the dendrimer research in Spain, have reported the synthesis of several families of silicon- and nitrogen-based organometallic dendritic molecules decorated, in most cases, with either ferrocene or cobaltocenium redox-active moieties as surface functional groups [27–40]. Electrochemical properties of both metallocenes are dominated by a reversible monoelectronic oxidation (ferrocene) or reduction (cobaltocenium), which can be easily obtained at very accessible potentials. Some of these dendritic molecules have been used successfully in the modification of electrode surfaces [41], as electron transfer mediators in amperometric biosensors [42–44], as exo-receptors for sensing anions [27,30,36,38,45], and also, as effective guests for the formation of supramolecular complexes with β-cyclodextrin [37,46].

In this chapter, we mainly focus on redox-active organometallic dendrimers, as well as their properties and electrochemical applications, especially in the field of molecular recognition and in amperometric biosensors. This chapter will provide selected references to the work of other groups who are actively working on metallodendrimers in which reversible redox centers have been attached in any way, allowing applications to processes that involve the use of the redox functions, especially in the field of molecular recognition and in amperometric biosensors, but will primarily focus on our own work in this area.

8.2 ORGANOMETALLIC DENDRIMERS AS ION SENSORS

In the last decades the tailored-design and synthesis of host materials for the selective recognition and sensing of anions, cations, as well as neutral molecules has become a topic of increasing interest [47]. In particular, the development of molecule-based anion selective hosts has been a pivotal issue currently receiving considerable attention and the recognition and sensing of environmentally important anions such as nitrate and phosphate, as well as the sensing of analytes in biological systems have emerged as key research themes within the generalized area of supramolecular chemistry [48–51]. Among numerous methodologies, the exploitation of organometallic fragments incorporated with redox transductions as a reporter seems to be a very promising one. At this respect, ferrocene above all other redox-active moieties has been utilized as an electrochemical reporter in anion receptors [48–52].

8.2.1 Ferrocenyl Dendrimers

The electrochemical behavior of dendrimers containing topologically equivalent ferrocene units is dominated by the reversible one-electron oxidation of the ferrocenyl groups. This finding reveals that in one dendrimer there is no electronic communication between the multiple ferrocenyl centers, which are seemingly identical and sufficiently remote from one another to render the electrostatic factor negligible [53]. We have investigated ferrocenyl-terminated dendrimers as anion receptors to selectively bind and recognize electrochemically guest species using the variation of Fe^{II}/Fe^{III} redox potentials provoked by the host–guest interactions.

In the wide and attracting field of dendrimer chemistry, silicon-based ferrocenyl dendrimers represented our first target electroactive dendritic molecules. We exploited the reactivity of organosilicon dendrimers functionalized at their peripheries with Si–Cl and Si–H sites toward suitable reactive ferrocenyl monomers. Thus, the high reactivity of the Si–Cl bonds toward monolithioferrocene (η^5-C_5H_4Li)Fe(η^5-C_5H_5) on one hand, and the amine group in (β-aminoethyl)ferrocene (η^5-$C_5H_4CH_2CH_2NH_2$)Fe(η^5-C_5H_5) on the other hand, allowed facile access to dendrimers **1–6** (Figure 8.1) [28]. We also explored the incorporation of ferrocenyl moieties onto dendrimers via hydrosilylation reactions of vinylferrocene (η^5-$C_5H_4CH{=}CH_2$)Fe(η^5-C_5H_5) with hydrido-terminated organosilicon dendritic cores affording dendrimers **7–9** (Figure 8.1) [26,35].

Redox-active ferrocenyl dendrimers **4–6** containing multiple NH linkages capable of participating in H-bonding, as well as characteristic internal cavities, have been shown suitable hosts for electrochemical recognition of guest anions [30,35] Electrochemical investigations showed that these silicon-based ferrocenyl dendrimer receptors can coordinate anionic guest species via the cooperative forces of favorable hydrogen-bonding interactions in the neutral state and electrostatic attractions with the positively charged ferricinium centers, after the electrochemical oxidation of the ferrocenyl moieties in the receptor (Figure 8.2).

These dendrimers exhibit HSO_4^- and $H_2PO_4^-$ selective preferences. Their cyclic voltammograms (CVs) in CH_2Cl_2/0.1 M TBAH solution, showed significant anion-induced cathodic perturbations. Thus, the addition of increasing amounts of [Bu_4N][H_2PO_4] or [Bu_4N][HSO_4] causes a decrease in the intensity of the redox wave, along with the progressive appearance of a new redox wave at a less positive potential (Figure 8.3). The current associated with the new redox couple increases linearly with the concentration of anions until the original wave disappears and the new redox couple reaches full development, when 8 equiv. of anion have been added [54]. It is noteworthy that the largest magnitude of cathodic shift was observed with the $H_2PO_4^-$ anion. The addition of Cl^- and Br^- salts to the ferrocenyl dendrimer does not give rise

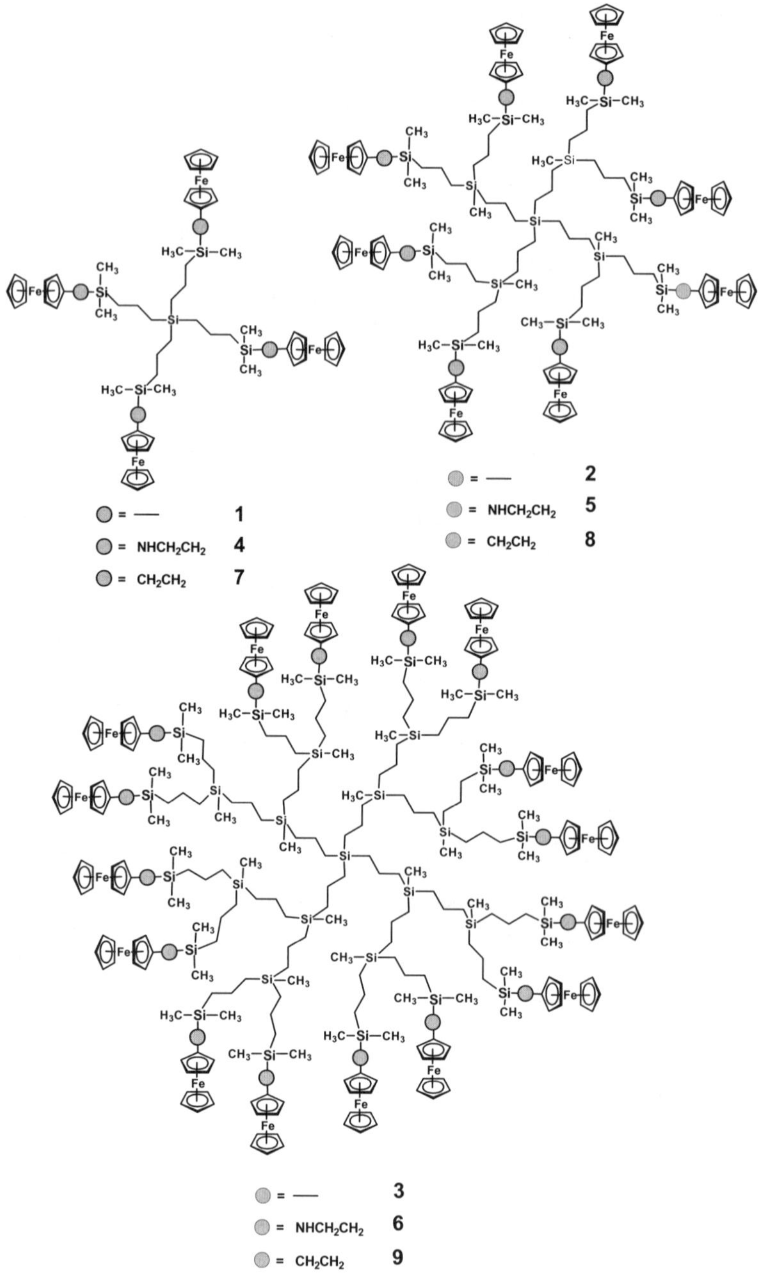

FIGURE 8.1 Carbosilane containing dendrimers with ferrocenyl units located on the external surface.

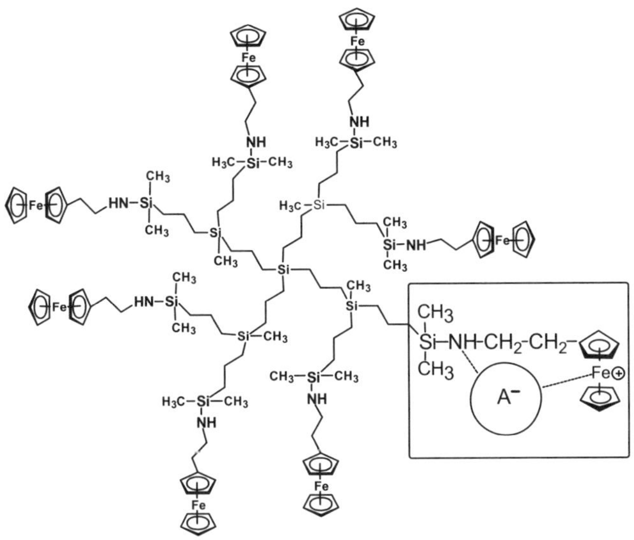

FIGURE 8.2 Schematic representation of dihydrogenphosphate recognition by ferrocenyl dendrimer **5** receptor, bearing SiNH peripheral groups.

to a new wave but only a progressive cathodic shift of the initial wave is observed. The fact that the most striking changes of the CVs, caused by strong interactions [54], are obtained for large and tetrahedral anions such as $H_2PO_4^-$ and HSO_4^- are attributed to the presence of cavities in the receptor with complementary dimensions to these anions. The selectivity of the receptor in the electrochemical recognition was established by competition experiments. The CV recorded when $H_2PO_4^-$ is added

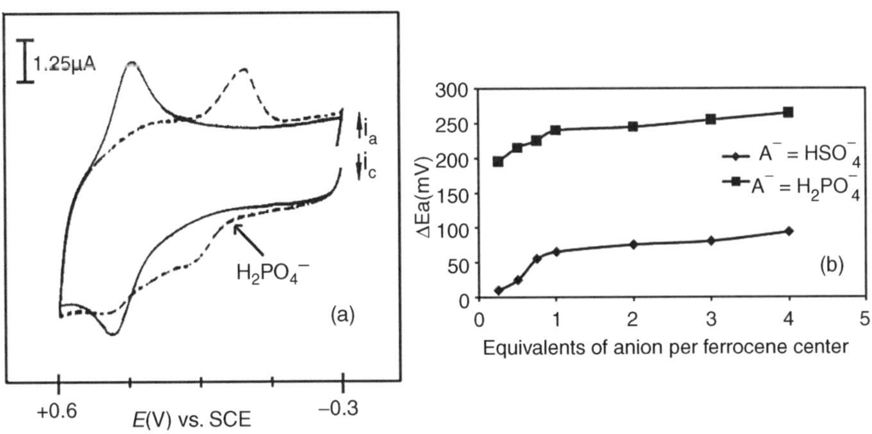

FIGURE 8.3 (a) CV of dendrimer **5** (1.25×10^{-5} M) in the absence (—), and presence (- - -) of 0.5 equiv. of $[n\text{-}Bu_4N][H_2PO_4]$ per ferrocene center. (b) Oxidation potential shift of the ferrocene redox couple versus equivalents of anion added.

to a CH_2Cl_2/TBAH solution of dendrimer **5** in the presence of 10-fold excess of HSO_4^-, Cl^-, and Br^- results to be closely similar to that recorded for **5** in the presence of $H_2PO_4^-$ alone. In a like manner, HSO_4^- can be detected, unambiguously, in the presence of Cl^- and Br^-. The results obtained indicated that the dendrimer displays the selectivity trend $H_2PO_4^- > HSO_4^- > Cl^- > Br^-$. The presence of Si–NH groups in the dendrimers plays a significant role in the electrochemical recognition of anions that is highlighted by the behavior observed for dendrimers **2** and **8**, which do not contain Si–NH groups (see Figure 8.1) [28]. In fact, when 2 equiv. of $H_2PO_4^-$ per branch are added to CH_2Cl_2/0.1 M TBAH solutions of dendrimers **2** and **8**, no splitting is observed and the cathodic shifts of the Fc oxidation wave are smaller. Moreover, dendrimer **2** displayed no change in the electrochemical response upon the addition of 2 equiv. of Cl^- per dendrimer branch, ruling out the possibility of the cathodic perturbations observed for **5** being caused only by ion-pairing interactions.

An especially interesting approach for the preparation of electrochemical sensory devices is the immobilization of redox-responsive receptor systems on electrode surfaces [55,56]. Electrodes modified by electrooxidation of aminoferrocenyl dendrimer **5** resulted sensitive to the presence and concentration of anions and constituted the first example of molecular recognition with dendritic receptors confined on an electrode surface [30,35]. The electroactive dendrimer film behaved almost ideally and showed rapid charge-transfer kinetics. In the presence of $H_2PO_4^-$ a new peak appears at a more negative potential, while the original wave is observed only as a reminiscent weak shoulder (Figure 8.4). Increasing values of ΔE can be measured in the $H_2PO_4^-$ concentration range from 10^{-4} to 2×10^{-3} M.

Neutral amide-substituted ferrocene receptors have been widely used to bind and electrochemically recognize guest anions combining hydrogen-bonding effects and electrostatic effects [48–52,57,58]. Amidoferrocenyl dendrimers have also been used as redox sensors as the acidic amide hydrogen atom can form a hydrogen bond with an oxygen atom of oxoanions [34,59–61]. Astruc et al. have investigated the behavior of the trimetallic ferrocenyl compound **10**, that of the monometallic ferrocene (η^5-C_5H_5)Fe(η^5-$C_5H_4CONHCH_2CH_2OPh$) and those of dendrimers **11** and **12** (Figure 8.5), in the electrochemical recognition of small inorganic anions ($H_2PO_4^-$, HSO_4^-, Cl^-, Br^-, and NO_3^-) and reported a strong positive dendritic effect (i.e., the sensing and recognition ability of the ferrocenyl host increases as the dendrimer generation is higher) [61]. The dendritic core topology as well as the nature of the solvent are critical on the recognition by the amidoferrocenyl groups.

Astruc and coworkers have also undertaken anion recognition studies with diaminobutane-based (DAB) poly(propyleneimine) amidoferrocenyl [34] and pentamethylamidoferrocenyl dendrimers [59–61] (Figure 8.6) for comparison. The electron-releasing methyl groups at the dendrimer periphery stabilize the oxidized ferrocenium 17-electron form and decrease the hydrophilicity and therefore, the pentamethylated series shows much less adsorption and no chemical irreversibility that prevent a clean recognition in the nonmethylated series. In addition, recognition and titration with the pentamethylated series are subjected to a modest dendritic effect in DMF whereas no dendritic effect is noted in CH_2Cl_2, which contrasts with the large dendritic effect found with dendrimers **11** and **12**.

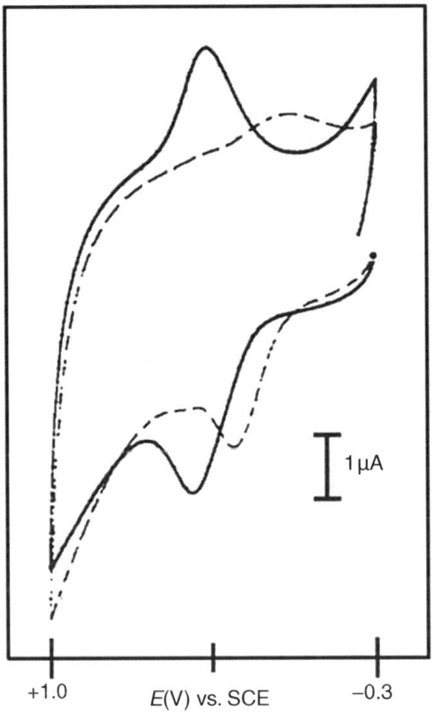

FIGURE 8.4 CV of a glassy-carbon electrode modified with a film of dendrimer **5** in the absence (—), and presence (- - -) of [n-Bu_4N][H_2PO_4] 5×10^{-4} M.

Our interest in extending recognition studies with dendritic macromolecules to more polar solvents prompted us to search for new ion receptors to be incorporated to ferrocenyl dendrimers. Another important type of anion receptors that have been studied extensively are those incorporating the urea group because this unit contributes two relatively strong hydrogen-bonding sites [62]. Vögtle et al. have reported urea-functionalized dendrimers as efficient carriers for oxyanions [63]. There are, however, few examples of ferrocene–urea redox-active ionophores [62,64–68]. In collaboration with Kaifer, we have synthesized DAB-based poly(propyleneimine) dendrimers **13–16** containing 4, 8, 16, and 32 ferrocenyl–urea peripheral groups, by reaction of the DAB dendrimers with isocyanatoferrocene (Scheme 8.1) [27].

Electrochemical anion sensing experiments were carried out in a relatively polar solvent (DMSO), where hydrogen-bonding interactions between the urea functional groups and the anion are usually weakened by competing solvent molecules. Nevertheless, the obtained results clearly show anion sensing at relatively low (submillimolar) anion and dendrimer concentrations, which reflects the considerable stability of the anion complexes formed by these dendrimers. These ferrocenyl–urea dendrimers exhibit a high selectivity for the dihydrogenphosphate anion as shown in the Osteryoung square-wave voltammogram (OSWV) response of dendrimer **13** in Figure 8.7.

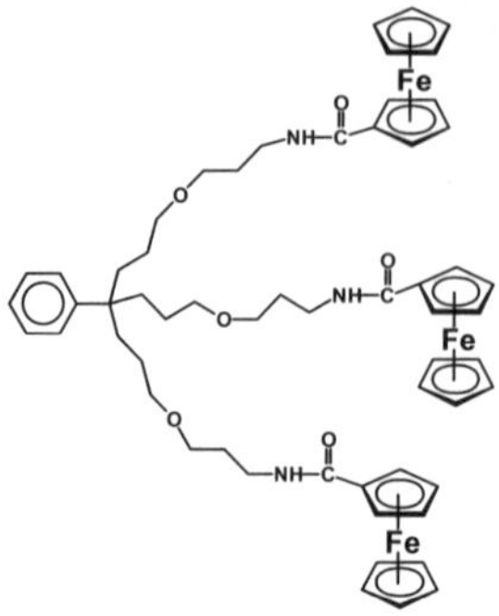

10

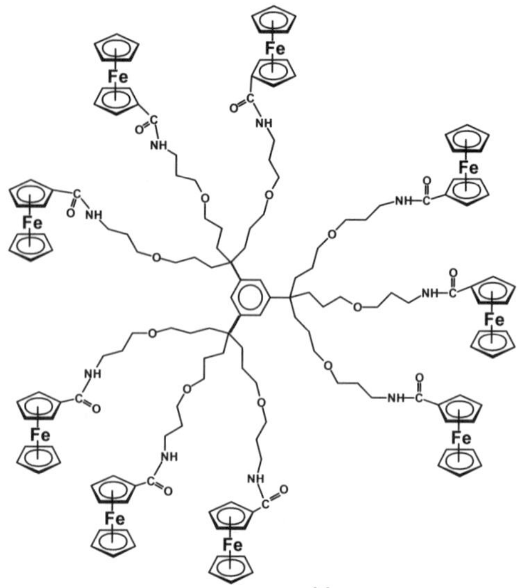

11

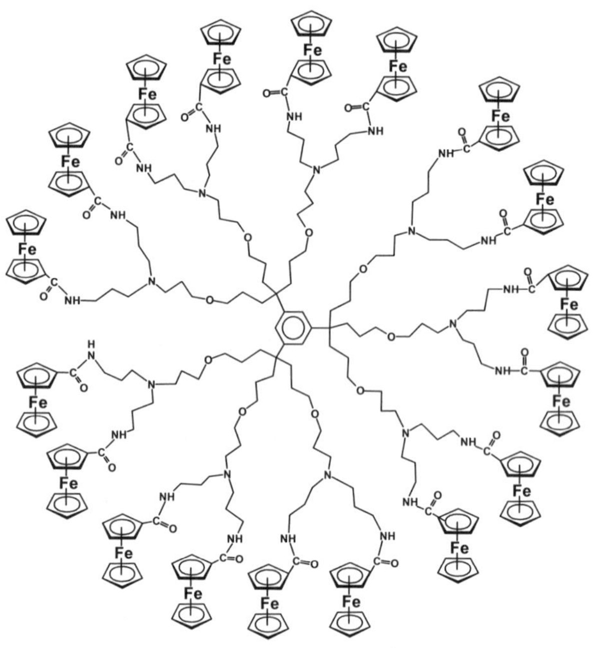

12

FIGURE 8.5 Astruc's benzylated-cored ferrocenyl dendrimers.

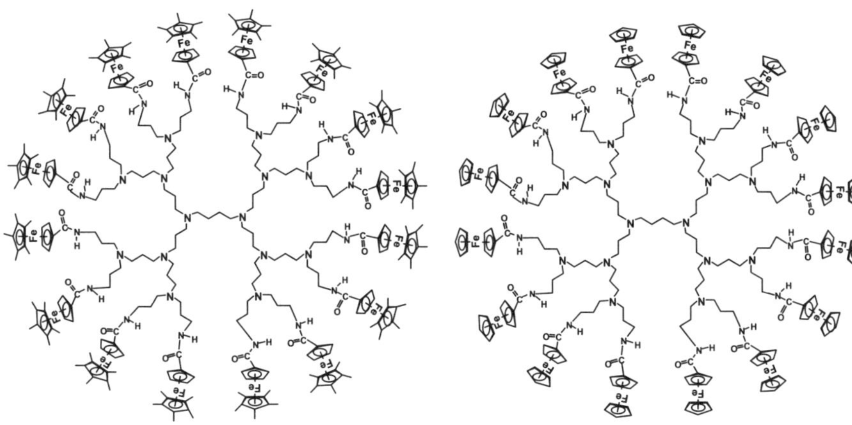

FIGURE 8.6 DAB-based dendrimers containing ferrocenyl and pentamethylferrocenyl units.

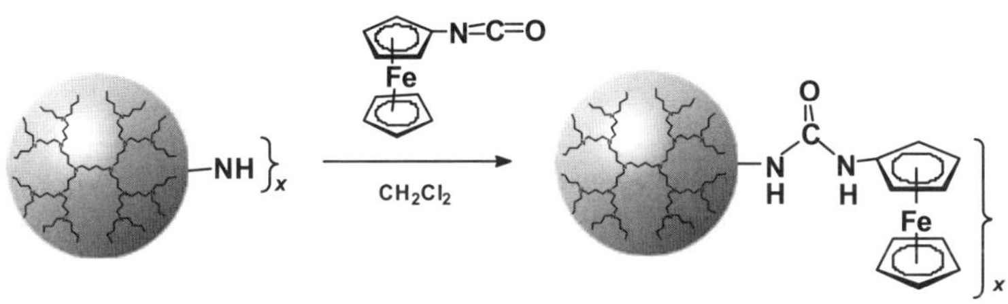

13 : $x = 4$ 15 : $x = 16$
14 : $x = 8$ 16 : $x = 32$

SCHEME 8.1 Synthesis of the ferrocenyl–urea-terminated DAB-based dendrimers **13–16**.

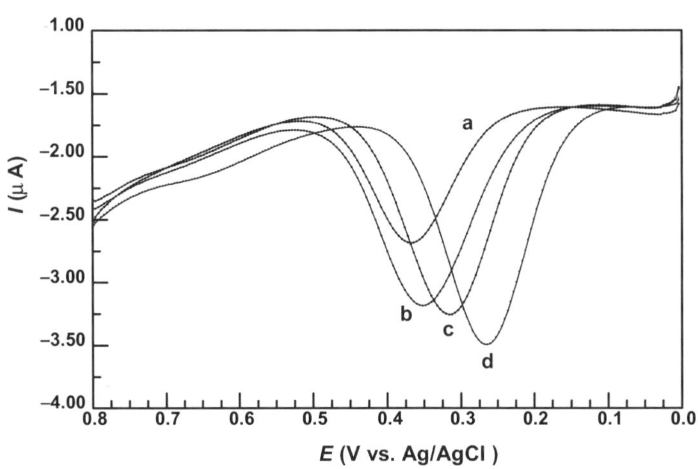

FIGURE 8.7 OSWV of **13** (5×10^{-5} M) in DMSO/TBAH in the absence (a), and presence of 200 μM of [n-Bu_4N][Cl] (b), 200 μM of [n-Bu_4N][Cl] + 100 μM of [n-Bu_4N][HSO_4] (c), 200 μM of [n-Bu_4N][Cl] + 100 μM of [n-Bu_4N][HSO_4] + 100 μM of [n-Bu_4N][H_2PO_4] (d).

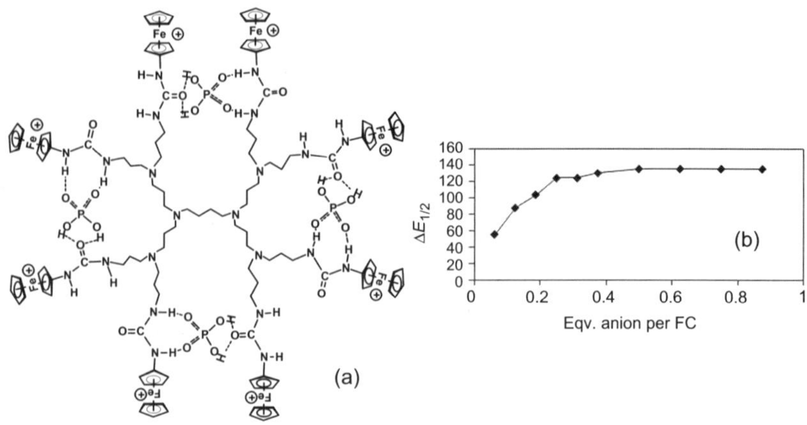

FIGURE 8.8 (a) Schematic representation of the double hydrogen-bonding interaction between two ferrocenium–urea branches of dendrimer **14**. (b) Potential shift of the ferrocene redox couple versus equivalents of $H_2PO_4^-$ added. Concentration of **14** 2.5×10^{-5} M.

Even in the presence of Cl^-, HSO_4^-, and $H_2PO_4^-$ (4 equiv. of each anion) the cathodic shift of the observed half-wave potential is within a few mV of the potential shift observed with only 4 equiv. of $H_2PO_4^-$ (in the absence of Cl^- and HSO_4^-). We also explored the dependence of the half-wave potential of each dendrimer on the concentration of $H_2PO_4^-$. Interestingly, the apparent half-wave potential value levels off at 0.5 equiv. of anion, which strongly suggests that each $H_2PO_4^-$ anion is bound by two ferrocenyl–urea branches (Figure 8.8), in agreement with the fact that the two N–H groups can bind with two adjacent oxygen atoms in an oxyanion to yield an eight-membered chelate ring and the C=O group from the adjacent dendritic arm can bind with the OH groups in $H_2PO_4^-$. These macromolecules also exhibit a dendritic effect in going from the first to the second and third generations, but the degree of steric congestion in the fourth generation seems to deteriorate the binding ability toward hydrogenphosphate.

A simple and efficient synthetic route has been used to incorporate poly(propyleneimine) dendrimers containing 4, 8, and 64 amidoferrocenyl moieties into MCM-41, a mesoporous silicate with a highly regular structure composed of channels in a hexagonal arrangement, thus giving rise to a novel type of redox-active materials that constituted the first examples of mesoporous silica hosts containing dendritic guest molecules within their channels (Figure 8.9) [69]. One significant feature of these composite materials is that the guest dendrimers, containing a controlled number of ferrocenyl units, are easily accessible to electrochemical oxidation. Platinum disk electrodes coated with layers of dend-n-MCM-41/graphite powder/polystyrene composites have been studied by cyclic voltammetry and differential pulse voltammetry showing a well-defined reversible redox process. These modified electrodes are very stable and multiple successive cyclic voltammetric scans can be carried out with no loss of electroactivity. Likewise, after standing in air for several days, the redox response was practically unchanged. These mixed materials can be

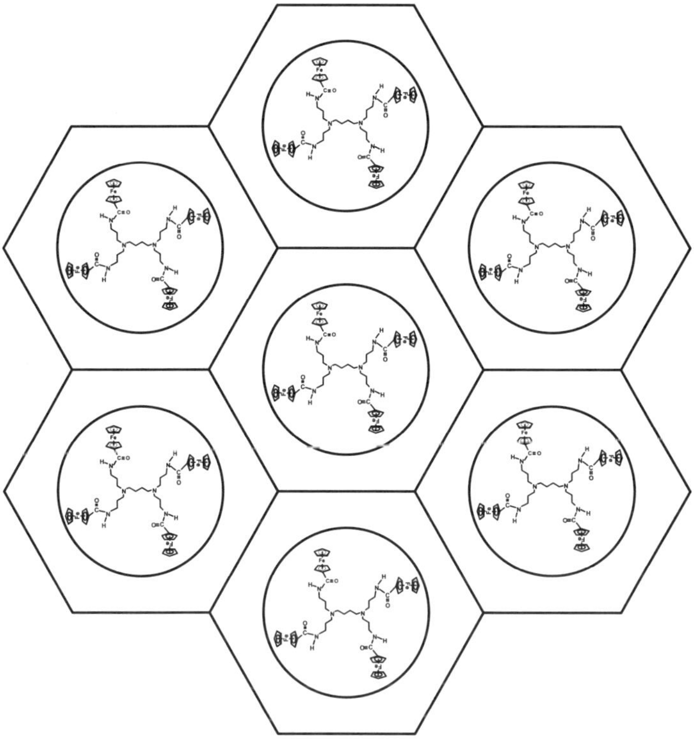

FIGURE 8.9 Schematic representation of the encapsulation of first-generation DAB-based amidoferrocenyl dendrimer within MCM-41.

used as redox-responsive receptor systems for anion recognition. The modified electrodes are sensitive to the presence of $H_2PO_4^-$ anion since a shift of the anodic peak to less positive potentials is observed (ca. 60 mV in the presence of $[n\text{-}Bu_4N][H_2PO_4^-]$ 5×10^{-3} M).

The design and synthesis of heteroditopic receptors capable of simultaneous binding of cationic and anionic guests by multisite receptors has received a great deal of attention, including heteroditopic ferrocenyl derivatives [65,70–74]. In this sense we presented a synthetic strategy aimed to obtain dendritic macromolecules containing redox centers in close proximity to two different host binding sites for the simultaneous complexation of cationic and anionic guest species [36]. The heteroditopic dendrimers **17–20** (Scheme 8.2) were obtained by condensation reactions of the highly reactive organometallic derivative $\{[\eta^5\text{-}C_5H_4COCl]Fe[\eta^5\text{-}C_5H_4CONCH_2\text{-}CH_2(OCH_2CH_2)_5]\}$ with 1,4-diaminobutane and DAB-based poly(propyleneimine) dendrimers. NMR studies confirmed that the dendrimers are capable of simultaneous coordination to cations and anions thanks to the presence of aza-crown ether and amide binding sites.

The ^1H NMR spectrum of the diferrocenyl complex **17**, which could be considered as the generation zero of this family of dendrimers, shows the signal for the amide

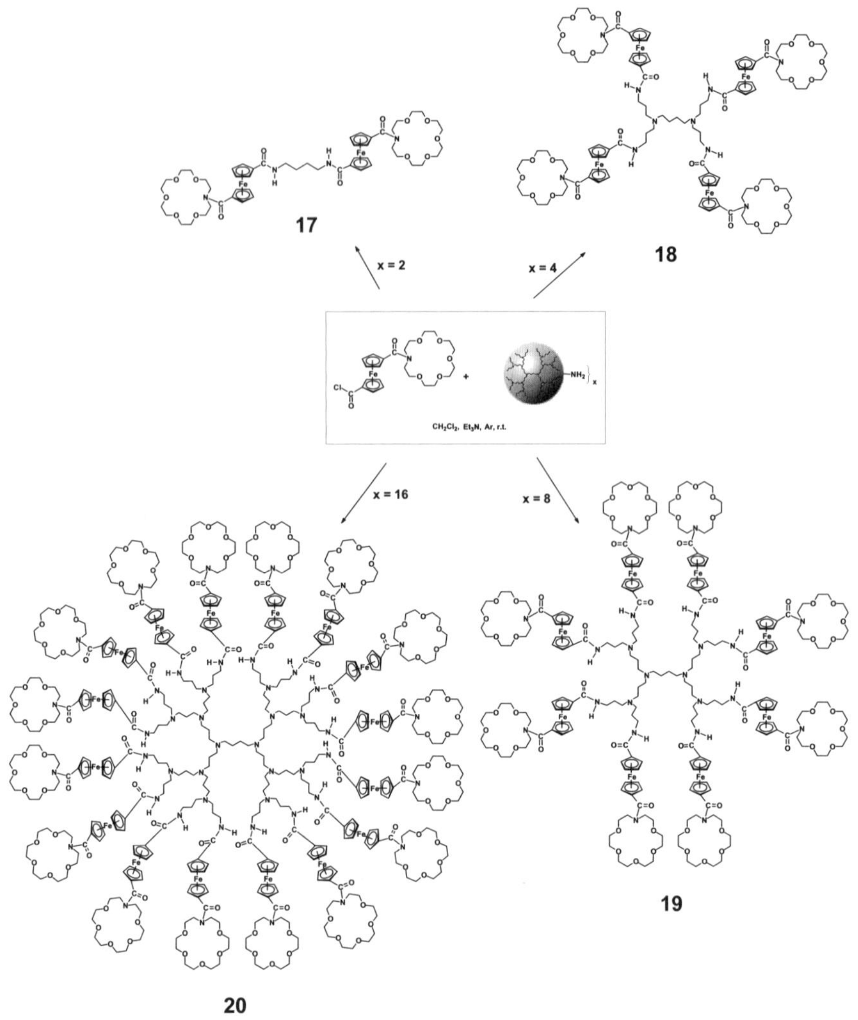

SCHEME 8.2 Functionalization of DAB-based dendrimers with amidoferrocenyl aza-crown ethers.

proton at 8.43 ppm (Figure 8.10a). In the presence of 2 equiv. of fluoride anion a significant downfield shift ($\Delta\delta = 0.59$) was observed for the amide NH protons of **17** (Figure 8.10b) that is consistent with the hydrogen-bonding interaction between the fluoride anion and the amide group. On the other hand, in the presence of 2 equiv. of KPF_6, the NH resonance of **17** is slightly upfield shifted to 8.31 ppm (Figure 8.10c) because cation complexation prevents hydrogen bonding between the amide proton and the oxygen atoms of the aza-crown ether. Finally, upon addition of 2 equiv. of TBAF to this solution, the NH resonance is very slightly downfield shifted again to 8.36 ppm (Figure 8.10d). From these results it was concluded that the presence of K^+

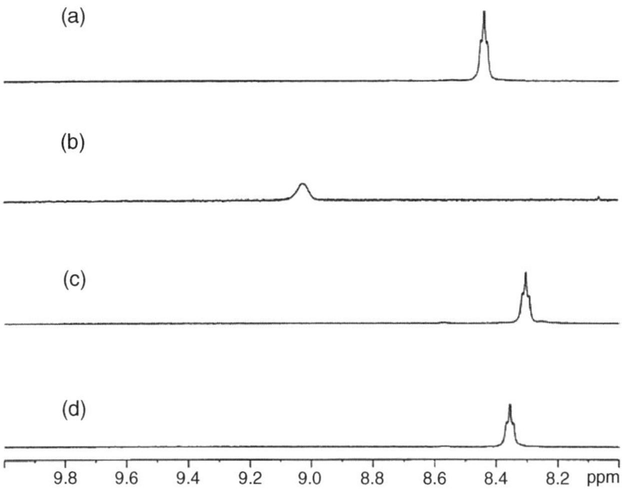

FIGURE 8.10 ^{1}H NMR changes in the chemical shifts of the amide protons of compound **17** in acetone-d_6 (a) alone, (b) in the presence of 2 equiv. (1 equiv. per aza-crown ether) of TBAF, (c) in the presence of 2 equiv. of KPF_6, and (d) in the presence of 2 equiv. of KPF_6 and 2 equiv. of TBAF. Adapted from González et al. [36].

weakens the hydrogen bonding between the fluoride anion and the amide groups in **17** probably due to the ion-pairing interaction K^+-F^-.

The condensation reaction of octamethylferrocenyl aldehyde with DAB-based poly(propyleneimine) dendrimers and further reduction of the imine groups affords up to the fourth generation of the octamethylferrocenylmethylamine dendrimers **21–24** (Scheme 8.3) [40]. These dendrimers represent another type of receptor systems capable of electrochemically recognizing anionic guest species. Solution voltammetric studies are characteristic of reversible oxidation processes with formal potentials considerably more negative than those corresponding to nonmethylated analogs as the result of the strong electron donating effect of the eight methyl groups on the ferrocene rings. The dendrimers exhibit a tendency to adsorb on electrode surfaces. Pt or GC electrodes modified with these dendritic molecules containing redox-active octamethylferrocenyl moieties and amine NH-CH_2 hydrogen-bonding functionalities exhibit a well-defined and persistent electrochemical response that is sensitive to the presence of $H_2PO_4^-$ anion. A shift of the anodic peak to less positive potentials of about 195 mV for the fourth generation **24** (Scheme 8.3) in the presence of [n-Bu_4N][$H_2PO_4^-$] 2×10^{-5} M is reported.

The development of surface immobilized conducting polymers bearing functionalities capable of recognizing and detecting a particular target analyte, constitutes a very active field of research in continuous growth. At this respect, the anodic electropolymerization of pyrrole-substituted ferrocenyl derivatives is a well-established route to fabricate conducting polymer-coated electrodes [57,75–79] with a high potential for the electrochemical sensing of analytes. Likewise, in the field of molecular recognition, designing receptors that can bind and sense anions in water

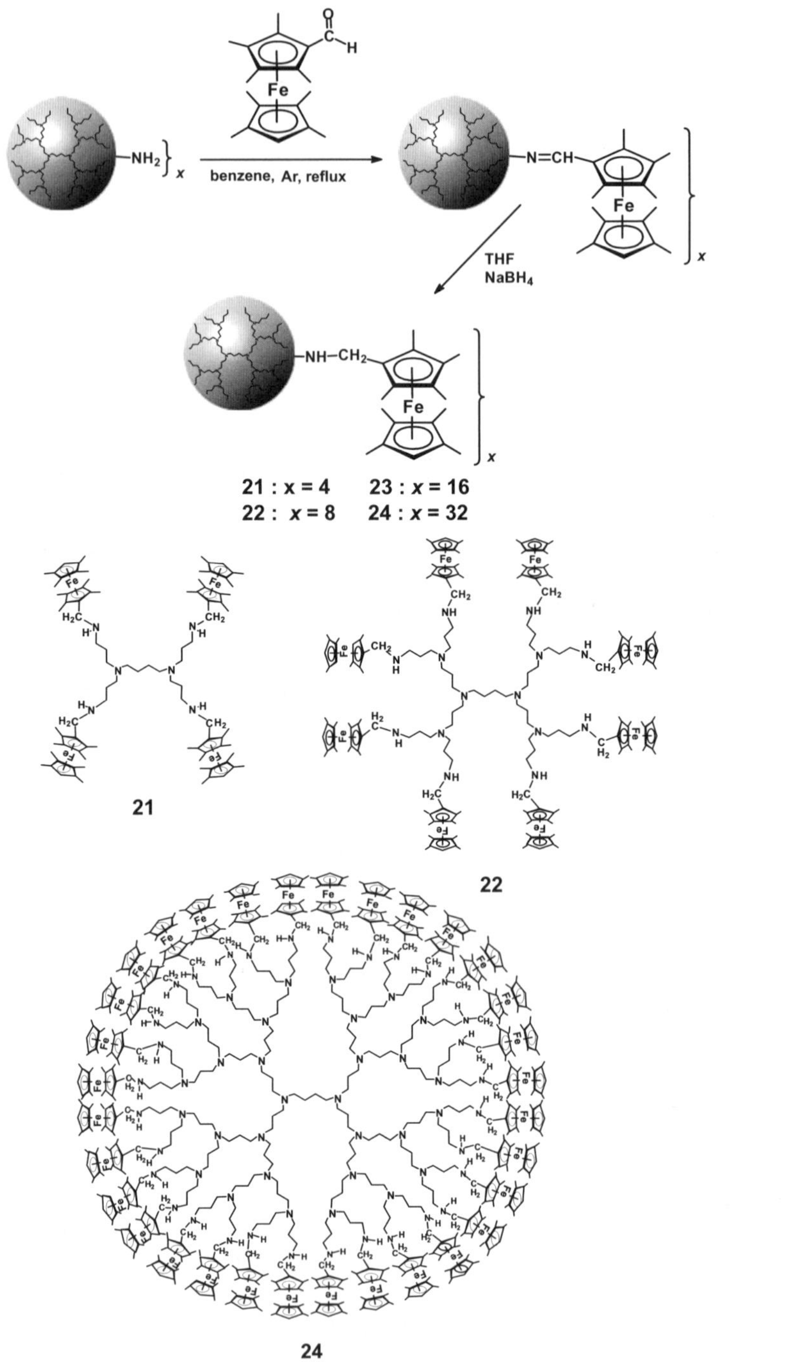

SCHEME 8.3 Synthesis of octamethylferrocenylimine dendrimers **21–24**.

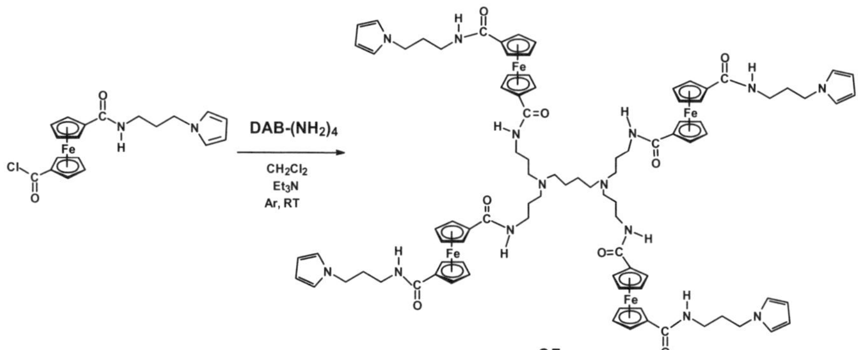

SCHEME 8.4 Synthesis of pyrrole-functionalized ferrocenyl dendrimer **25**.

is a particularly challenging task as anion hydration energies are large [50,80,81]. In this context, we have reported on the synthesis and electrochemical investigations of a DAB-based ferrocenyl dendrimer functionalized with pyrrole substituents **25** (Scheme 8.4), focusing upon its electropolymerization as films onto an electrode surface, and the subsequent use of these films in the electrochemical sensing of the biologically and environmentally important dihydrogenphosphate anion in water [45].

Visually detectable golden electroactive ferrocenyl dendritic polymer films persistently attached to the electrode surfaces were obtained from CH_2Cl_2/CH_3CN 1/1 solutions of **25** onto platinum or glassy-carbon electrode surfaces by repeated cycling of the potential over the range − 1.0 to + 1.3 V versus SCE or by controlled potential electrolysis at + 1.3 V. Anion recognition studies are precluded in CH_3CN solution as well as in aqueous solution using $LiClO_4$ as supporting electrolyte, due to a decrease in the current response with cycle number. However, when using a supporting electrolyte containing a large weakly coordinating anion, tetrakis(pentafluorophenyl)borate, $[B(C_6F_5)_4]^-$ in water, electrodes modified with electropolymerized films of **25** resulted robust and reproducible. Electrochemical $H_2PO_4^-$ sensing experiments were carried out in this media using OSWV, by exploring the dependence of the half-wave potential of the ferrocene/ferrocenium couple on the concentration of the anion (Figure 8.11). Competitive experiments show that when the modified electrode is exposed to a solution containing $H_2PO_4^-$ 10^{-4} M and HSO_4^-, Cl^-, or Br^- all at concentrations of 10^{-4} M, the cathodic perturbation observed in the half-wave potential is approximately the same as that induced by the $H_2PO_4^-$ anion alone.

As a continuation of our search for new macromolecules that could be used to modify electrodes with films within which the amidoferrocenyl units are accessible hosts to interact with anionic guests, we envisaged a family of dendrimers with large and flexible ferrocenylamidoalkyl chains at the periphery. The ferrocenylamidoalkyl dendrimers **26–29** (Scheme 8.5) have been obtained by treatment of the first four generations of DAB-based poly(propyleneimine) dendrimers with an excess of the organometallic fragment $[\eta^5\text{-}C_5H_5]Fe[\eta^5\text{-}C_5H_4CONH(CH_2)_{10}COF]$ [38]. Electrochemical

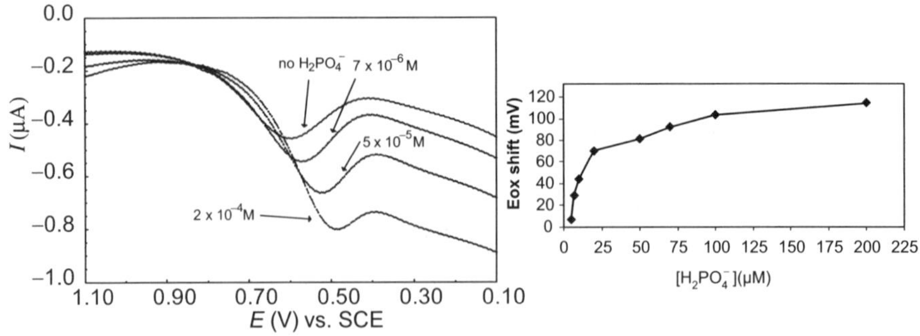

FIGURE 8.11 OSWV curves for a GC electrode modified with poly-**25** measured in H_2O/Li $[B(C_6F_5)_4]$ in the absence and presence of different concentrations of $H_2PO_4^-$. Shift of the oxidation peak potential. Adapted from Martínez et al. [45].

investigations of the ferrocenylamidoalkyl dendrimers show their ability to recognize and sense anionic species. Robust modified electrodes can be successfully prepared with the dendrimers by different procedures and their voltammetric response is also sensitive to the presence and concentration of anions in both organic and aqueous media. The electrochemical behavior of this dendritic family has been investigated by cyclic voltammetry, normal pulse voltammetry, and Osteryoung square-wave voltammetry of the materials in homogeneous solution, as well as confined to electrode surfaces. For all generations a single redox process is observed in CH_2Cl_2 indicating that the multiple ferrocenyl units are seemingly identical and sufficiently remote from one another to render the electrostatic factor negligible [53].

The electrochemical behavior of the dendrimers is strongly affected by the addition of increasing amounts of $H_2PO_4^-$ anion. A "two-wave" behavior is observed both in solution as well as in the modified electrodes (Figure 8.12A) [54]. This is attributed to a large increase in the association constants on oxidation, due to the effect of the synergy between the H-bonding and strong electrostatic interactions between $H_2PO_4^-$ and the oxidized ferricinium, enhanced by the dendritic structure that improve the electrochemical sensing properties toward the dihydrogenphosphate anion. A progressive decrease of the initial Fc/Fc^+ wave is observed along with the growth at less positive potentials of a new peaks system (Figure 8.12A curves b–d). The binding of anions effectively stabilizes the positive charge of the oxidized film, shifting the redox potential (E_p) of the Fc/Fc^+ system to less positive values until the potential peak reaches a constant value. The intensity of the peaks is depressed in the presence of $H_2PO_4^-$ ions, even at low concentrations. The loss of electroactivity has been attributed to the doping of the films that is responsible for a restricted transport of counter ions due to the formation of strong ion-pairs. This leads to a decrease in the rate of charge propagation that is controlled by counter ion diffusivity in the films and ion-trapping effects. Calibration graphs have been established upon the analysis of the OSWV curves for electrodes modified with the films of dendrimers as function of the $[n\text{-}Bu_4N][H_2PO_4]$ concentration and it has been concluded that the strength of the hydrogen bonding and electrostatic interactions between the amidoferrocene groups

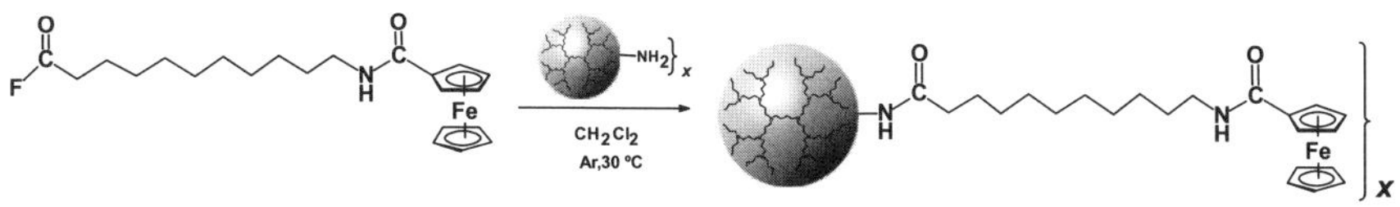

26: $x = 4$ **28**: $x = 16$
27: $x = 8$ **29**: $x = 32$

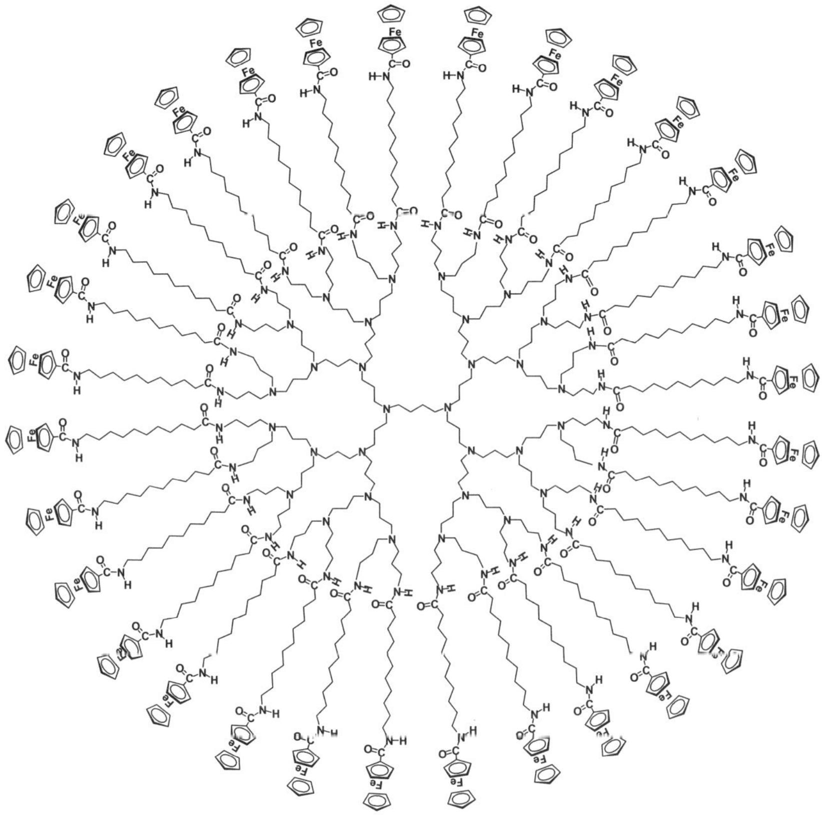

29

SCHEME 8.5 Synthesis of dendrimers **26–29** with ferrocenylamidoalkyl terminal groups.

and the anions depends on the topological characteristics and concentration effects. Likely, in the layers of the dendrimers deposited onto the electrode surface the alkyl organometallic units form cavities and channels that are responsible of the establishment of stronger interactions as the dendrimer generation increases. The sensing properties of the dendrimers toward other anions are also reported. When the modified electrodes are exposed to increasing amounts of [n-Bu_4N][HSO_4] only a progressive shift of the potential of the Fc/Fc^+ peak without the growth of a new redox wave is observed (Figure 8.12B). This behavior is linked to a weaker anion–ligand interaction

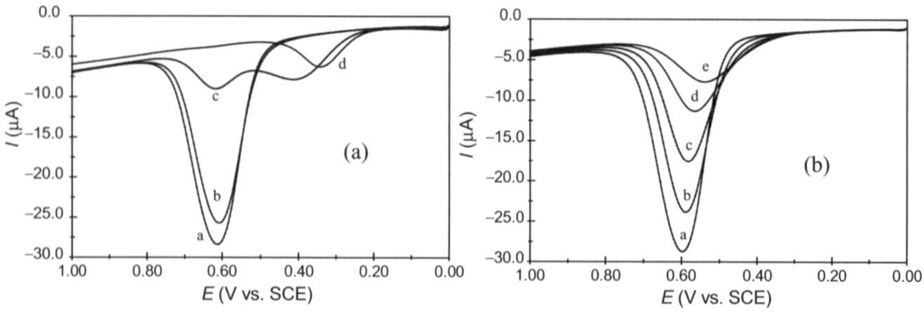

FIGURE 8.12 OSWV in CH_2Cl_2/TBAH, of a GC electrode modified with a film of dendrimer **29** ($\Gamma = 2.3 \times 10^{-10}$ mol Fc/cm^2). (A) In the absence (a) and presence of [n-Bu_4N][H_2PO_4]: (b) 2×10^{-5} M, (c) 3×10^{-5} M, and (d) 4×10^{-5} M. (B) In the absence (a) and presence of [n-Bu_4N][HSO_4]: (b) 5×10^{-6} M, (c) 10^{-5} M, (d) 5×10^{-5} M, and (e) 10^{-4} M. Adapted from Villoslada et al. [38].

following the oxidation of the ferrocene groups [57,82]. The electrochemical response of electrodes modified with the surveyed dendrimers was also tested in water. Decreasing values of the intensity of the Fc/Fc^+ peak were measured with increasing concentrations of KH_2PO_4 or Na_2ATP.

Another family of giant alkylferrocenyl dendrimers with a different dendritic core has been reported by Astruc and coworkers [83,84]. Eight generations (from G_0 with nine branches to G_7 with a theoretical total number of termini of 19,683) of polyferrocenyl dendrimers were synthesized by sequences involving Williamson reaction followed by hydrosilylation and terminal Williamson functionalization with an alkylferrocene derivative. These dendrimers have been used for the redox recognition of the oxoanions $H_2PO_4^-$ and ATP^{2-} by CV, even though there is no interaction site near the ferrocenyl group on the dendrimers tethers to perturb the ferrocenyl group and modify its redox potential (Figure 8.13a). It has been shown that an ion-pairing effect in synergy with the dendritic encapsulation effect is sufficient for the redox recognition of oxoanions such as $H_2PO_4^-$ and ATP^{2-}. Recognition of these oxoanions is also possible with Pt electrodes modified with larger dendrimers.

Supramolecular dendrimers assembled by hydrogen bonding between an amidoferrocenyl dendronic phenol and the primary amines of DAB-based dendrimers (Figure 8.13b) show a positive dendritic effect in the recognition of $H_2PO_4^-$ and a slightly negative effect for ATP^{2-} [85,86].

Gold-nanoparticle-cored silylferrocenyl and amidoferrocenyl dendrimers have also been used as sensors for the selective recognition and titration of $H_2PO_4^-$ and ATP^{2-} [87–91]. Moreover, modified electrodes fabricated with the dendronized nanoparticles serve as reusable sensors.

The ability of the 1,2,3-triazole rings located inside click ferrocenyl dendrimers (Figure 8.14) to recognize, bind, and sense oxoanions such as $H_2PO_4^-$ and ATP^{2-} and metal cations such as Pd^{II}, Pt^{II}, Cu^{I}, and Cu^{II}, using the ferrocenyl termini as a redox monitor directly attached to the triazole ring has been investigated by Astruc and coworkers [92–94]. The electron-withdrawing properties of the metal cations decreased the electron density on the ferrocenyl center,

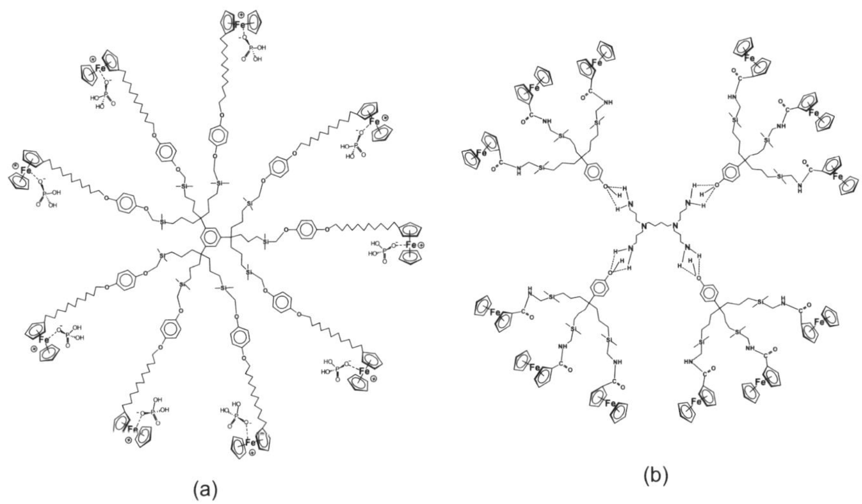

FIGURE 8.13 (a) Schematic representation of an alkylferrocenyl dendrimer and the electrostatic attraction between the ferricinium form and the oxoanion $H_2PO_4^-$. (b) Structure of Astruc's H-bonded dendrimer.

which shifted the CV wave to more positive potential values. On the other hand, the electron-releasing properties of the anion shifted the ferrocenyl CV wave to less positive potential values.

Stone and Smith [95] have reported an example of a dendritic molecule containing a single ferrocene unit as a well-defined anion binding recognition site at the core. The anion binding ability of two ferrocenoyl amide dendritic derivatives is described. OSWV was used to report on halide anion binding via a decrease in the peak current and a cathodic shift in the apparent redox potential of the receptors. The magnitude of this shift decreases with increasing generation indicating that the dendritic branching hinders the penetration of guest anions.

8.2.2 Hyperbranched Ferrocenyl Polymers

Hyperbranched ferrocenyl polymers containing cyclic siloxanes and silsesquioxanes as frameworks have been prepared by Pt-catalyzed hydrosilylation of 1,1′-divinylferrocene or 1,1′-divinyl(octamethyl)ferrocene with 1,3,5,7-tetramethylcyclotetrasiloxane [96] and octakis(hydrodimethylsiloxy)-octasilsesquioxane [97] (Figure 8.15). Solution electrochemical behavior of the ferrocenyl polymers in CH_2Cl_2/TBAH shows that all the ferrocenyl centers of the polymer are independent and electrochemically equivalent. The most notably aspect of the redox behavior of these hyperbranched polymers is their ability to modify electrode surfaces. The electrochemical properties of the modified electrodes are sensitive to the anion composition of the solutes. When a glassy-carbon electrode modified with oxidized films of the polymers are immersed in a 0.1 M $H_2O/LiClO_4$ solution containing thiocyanate, anion exchange takes place and the SCN^- anions are incorporated in the structure of the film

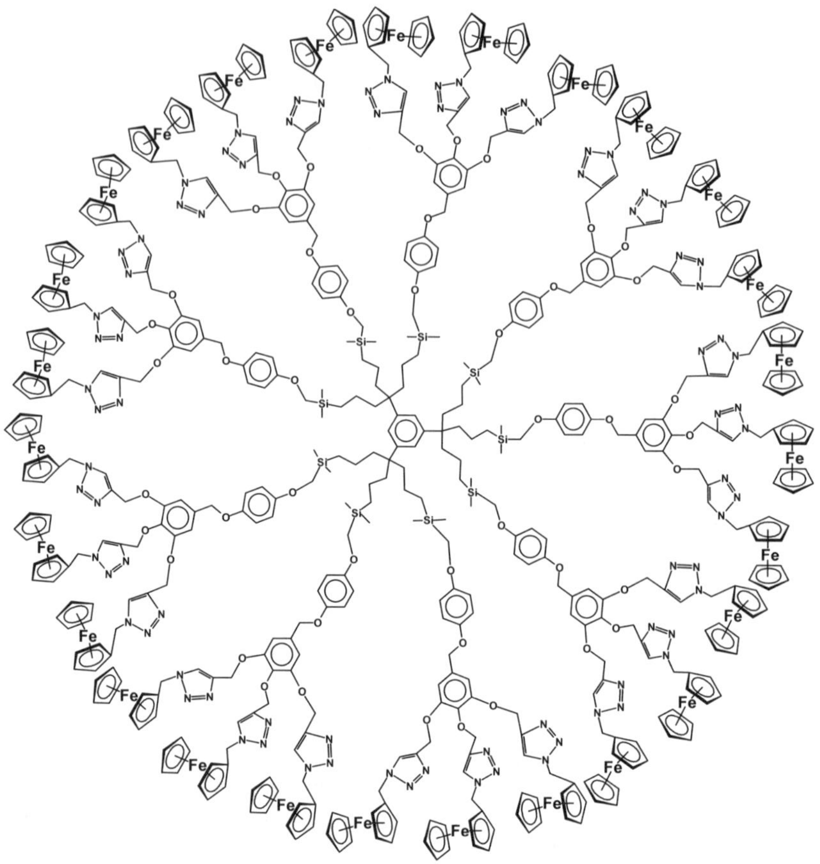

FIGURE 8.14 Schematic representation of a 1,2,3-triazole containing dendrimer reported by Astruc's group.

instead of the ClO_4^- anion. The CV of a film of polymer $\mathbf{32^+ClO_4^-}$ after immersion in 10^{-3} M thiocyanate solution for 15 min showed an oxidation peak at +0.95 V, corresponding to the thiocyanate [31].

Wang and coworkers [98] have investigated the electrochemical anion recognition of ferrocenyl-modified hyperbranched polyethers, through the peak currents decrease rate. The anion binding capacity order depends on the ferrocene content on the polymer.

8.2.3 Other Metallocenyl Dendrimers

As seen above, electrochemistry is in most cases the method of choice for studying anions using ferrocene as reporter, because the anodically generated cationic ferricinium form strongly interacts electrostatically with the anions and this interaction operates in synergy with $X^- \cdots HN$ bonding. In contrast, the anion sensing abilities

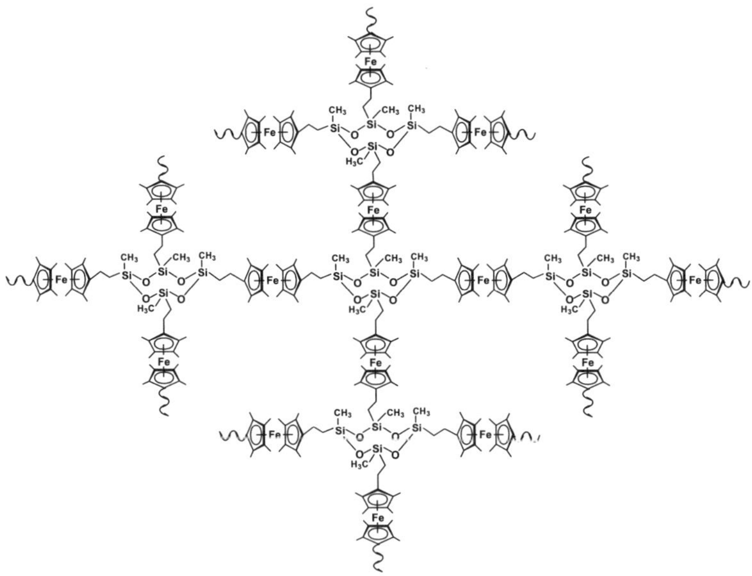

— = H **30**
— = CH_3 **31**

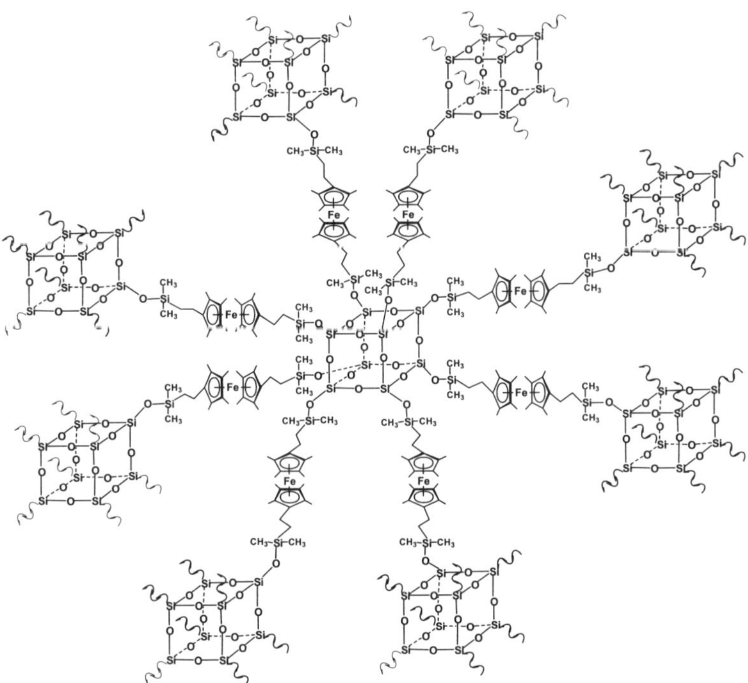

— = H **32**
— = CH_3 **33**

FIGURE 8.15 Ferrocenyl and octamethylferrocenyl cyclosiloxane and silsesquioxane-based hyperbranched polymers. From Losada et al. [118].

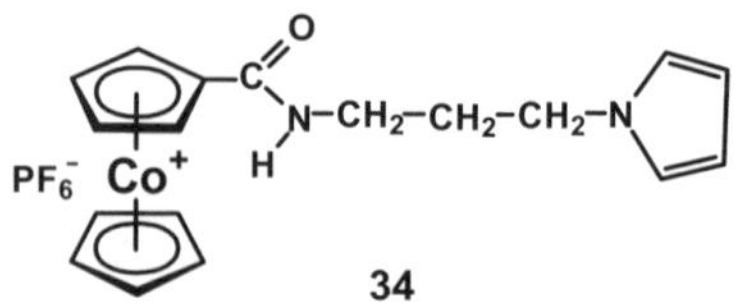

34

FIGURE 8.16 A polymerizable pyrrole-cobaltocenium receptor.

of cobaltocenyl and CpFearene$^+$ units can be also evaluated by ^{1}H NMR as the anionic guests can be attracted by the charged metallocenyl units.

Beer and coworkers [99–102] and Astruc and coworkers [103] have shown that positively charged cobaltocenium-based receptors also containing amide C(O)N-H groups prove particularly effective in anion complexation. Compound **34** (Figure 8.16) [104] in which the cobaltocenium moiety is covalently attached to a pyrrole derivative by an amide group can be successfully electropolymerized resulting in electrode surfaces coated by nonpassivating polymeric films [105].

^{1}H NMR spectroscopy and CV were used to investigate the ability of this monomer as well as polymeric poly-**34** cobaltocenium-containing pyrrole systems to act as selective amperometric sensors for anions. The addition of the tetrabutylammonium dihydrogenphosphate salt ([n-Bu_4N][H_2PO_4]) to dimethyl sulfoxide solutions of **34** causes a substantial downfield shift of the amide proton ($\Delta\delta = 1.62$ ppm). The electrochemical response of **34** in CH_3CN/TBAP is also strongly affected by the addition of increasing amounts of $H_2PO_4^-$ anion, with the CV curves exhibiting a significant negative potential shift ($\Delta E_{pc} = 40$ mV at 1 equiv.). This behavior is in accordance with the increase of the effective electron density on the redox center due to the binding of the anion, which makes the reduction more difficult. Low concentrations of the $H_2PO_4^-$ anion have a significant effect on the voltammetric response of electrodes modified with films of poly-**34** (Figure 8.17). Thus, the presence of

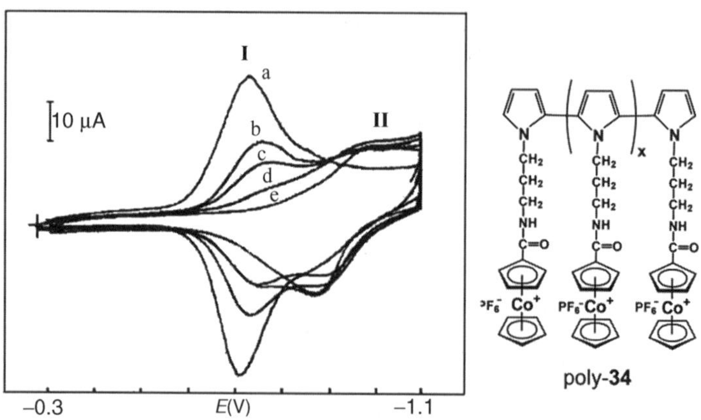

poly-**34**

FIGURE 8.17 CVs in CH_3CN/TBAP of a Pt electrode modified with a film of poly-**34** ($\Gamma = 1.8 \times 10^{-9}$ mol/cm^2), in the absence (a) and presence of [n-Bu_4N][H_2PO_4]: (b) 10^{-5} M, (c) 2×10^{-5} M, (d) 5×10^{-5} M, (e) 10^{-4} M. Adapted from del Peso et al. [104].

$H_2PO_4^-$ anion causes a cathodic shift and a decrease in the intensity of the initial redox wave (I) along with the appearance of a new redox system at more negative potentials (II). The presence of the sulfate anion in the absence of $H_2PO_4^-$ causes a small potential shift, however it perturbs significantly the dihydrogenphosphate anion determination. On the other hand, $H_2PO_4^-$ can be unambiguously detected in the presence of a 10-fold excess of Cl^- and/or Br^- salts.

Astruc and coworkers have reported the specificity of dendritic cationic aminoarene–iron sandwich complexes for halides [106]. Whereas the amidoferrocenyl dendrimers recognize the oxoanions but not the halides (see above), dendrimers terminated with $[Fe(\eta^5\text{-}C_5Me_5)(\eta^6\text{-}C_6H_5NH\text{-dendr})]^+$ recognize chloride and bromide but not the oxoanions. The polycationic metallodendrimers show large dendritic effects for recognition of Cl^- and Br^- according to ^1H NMR spectroscopy.

Dendrimers functionalized with $[CpFe(\mu_3\text{-}CO)]_4$ clusters at the periphery have also been applied as selective redox sensors for oxoanions including ATP^{2-}, by monitoring the shifts of their single reversible wave for the redox change $Fe_4 \rightarrow Fe_4^+$. In contrast to ferrocenyl dendrimers, the largest cathodic shift is observed in the presence of ATP^{2-}. This result is explained on the basis of a more facile interaction between the Fe_4 clusters and ATP^{2-}, which are more similar in size, than between ferrocene and ATP^{2-} [107].

8.3 ORGANOMETALLIC DENDRIMERS IN THE CONSTRUCTION OF BIOSENSORS

In a biosensor a bioactive molecule is used in close contact with a transducing element. The bioactive molecule recognizes the presence of a substrate and the biochemical signal obtained from the sensing device, which reflects the concentration of substrate, is converted into an electronic signal by the transducer. The specificity of the enzymes enables sensors incorporating enzymes to achieve much greater selectivity. Enzyme electrodes result from the combination of any type of electrochemical sensor with a layer of immobilized enzyme in close proximity to the active surface of the transducer [108,109]. The analyte can be either the substrate or a product of the enzyme reaction. The redox enzymes have great interest especially in amperometric biosensors in which an externally applied potential is employed to drive the electrode reaction. When an oxidoreductase is used, the reaction implies oxygen consumption according to Scheme 8.6. Since oxygen reoxidizes the enzyme and produces hydrogen peroxide, the reaction may be monitorized either through the oxygen uptake or through the formation of hydrogen peroxide.

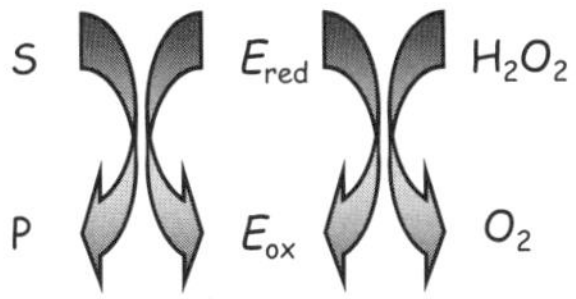

SCHEME 8.6

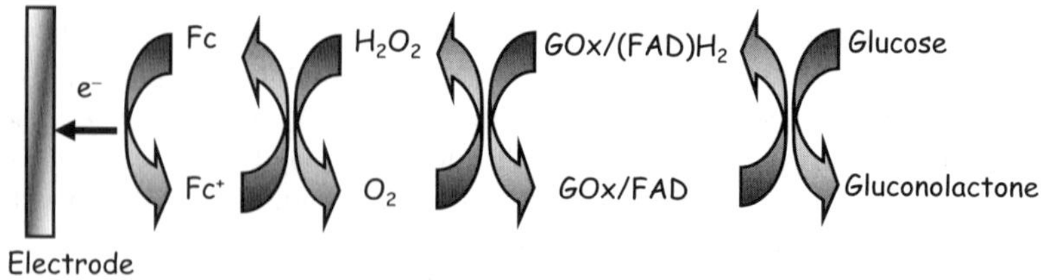

SCHEME 8.7

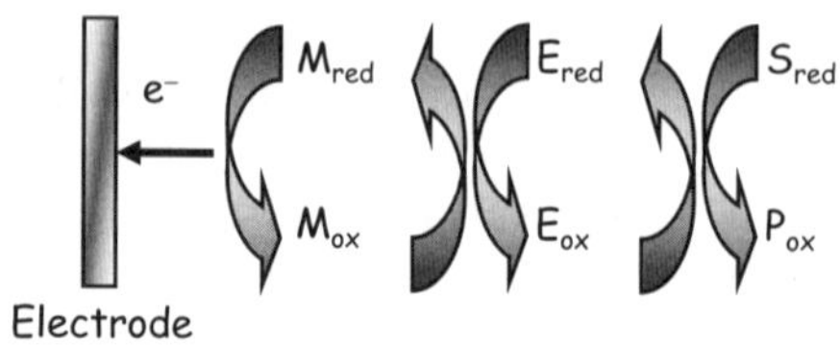

SCHEME 8.8

An important limitation of the amperometric sensors based on the electrochemical determination of H_2O_2 is the high operating potential required for its measurement. One of the methods aimed to minimize interferences and improve the sensitivity is the modification of the electrode surface with catalysts that are able to decrease the overvoltage (Scheme 8.7).

The restricted solubility of oxygen may limit the enzyme reaction and the high potential necessary causes interfering reactions. One way to circumvent these effects is to reach direct electron transfer reactions between enzymes and electrodes in anaerobic conditions using artificial mediators whose role is to shuttle electrons efficiently between the electrode and enzyme (Scheme 8.8).

Many amperometric enzyme electrodes with electroactive species acting as mediators replacing the natural electron acceptor, dissolved oxygen, have been widely investigated. Ferrocene derivatives represent one of the major groups of mediators owing to their attractive redox properties and fast reaction with the reduced oxidases and peroxidases [110,111].

8.3.1 Ferrocenyl Silicon-Based Dendrimers

The construction of carbon paste electrodes prepared doping carbon paste with enzyme and the mediator is simple and in addition, the enzymatic mediator and the sensing sites are located in close proximity. Monomeric mediators as ferrocene were initially used as electron-shuttling redox couples but the solubility of the oxidized form causes instability due to the loss of the mediator from surface electrode [110,112]. Increase of the molecular weight of the mediator should significantly prevent this loss and ferrocenyl silicon-based dendrimers with relatively high molecular weights in which the ferrocene/ferrocenyl units are at the end of long

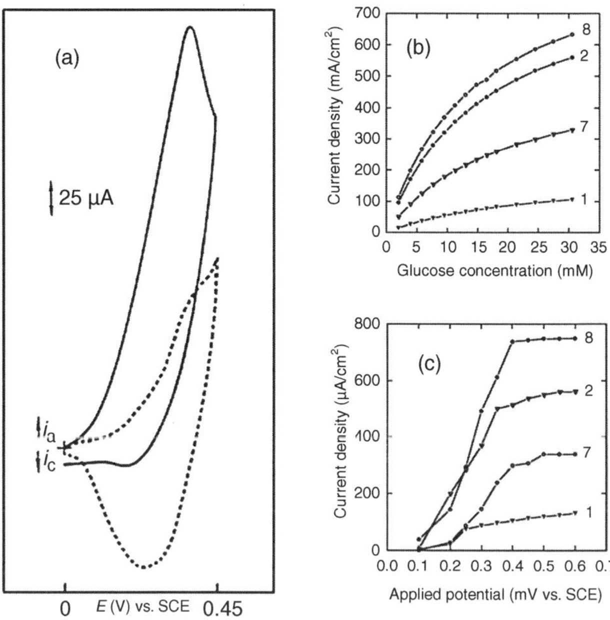

FIGURE 8.18 (a) CVs for dendrimer **8**/glucose-oxidase/carbon paste electrode, in pH 7.0 sodium phosphate buffer (with 0.1 M KCl) solution, with no glucose present (dotted line), and in presence of 0.1 M glucose (solid line). (b) Steady-state polarization curves of the dendrimers/glucose oxidase/carbon paste electrodes in the presence of 25 μM glucose. (c) Variation of the steady-state current of the dendrimers/glucose-oxidase/carbon paste electrodes with glucose concentration (at +350 mV vs. SCE). Adapted from Losada et al. [42].

flexible silicon-containing branches (Figure 8.1) can serve to wire electrically the enzyme [26,35,42].

In order to test the capability of ferrocenyl dendrimers **1**, **2**, **7**, and **8** to act as electron-mediating species, a study of the efficiency of dendrimer/glucose-oxidase/carbon paste electrodes was carried out. The most significant results are illustrated in Figure 8.18. Cyclic voltammograms of these carbon paste electrodes show an electrochemical behavior indicative of enzyme-dependent catalytic reduction of the ferricinium cations. The electrodes are clearly sensitive to small changes in glucose concentration and display a good response over long periods of time. Dependence of glucose response on dendrimer structure was examined. It has been found, from the calibration curves and the results of Michaelis–Menten analysis that for equimolecular amounts of ferrocene moieties, octanuclear dendrimers **2** and **8**, possessing the longest organsilicon branches, mediate electron transfer more efficiently than the relay systems based on tetranuclear dendrimers **1** and **7**. In addition, it is clear that dendrimers **7** and **8** in which the ferrocenyl units are attached to the dendritic framework through a two methylene flexible spacer are more effective at mediating electron transfer between reduced glucose oxidase (GOx) and the carbon paste electrode than the corresponding dendrimers of same nuclearity **1** and **2**. These results indicate that the differences observed in the response to glucose can be related to the length of the branches and to the presence of flexible ethylene groups in the branches. Therefore, the flexibility of the dendrimeric mediators is an important factor in the ability to facilitate the interaction between the mediating species and the flavin adenine dinucleotide (FAD) redox centers of GOx. Ferrocenyl dendrimers-based sensors exhibit a higher sensitivity than ferrocene-modified polymers sensors and a better stability than monomeric-mediated electrodes [42].

8.3.2 DAB-Based Metallocenyl Dendrimers

8.3.2.1 Mixed Ferrocene-Cobaltocenium Dendrimers Dendritic molecules containing various organometallic moieties can combine the properties and advantages of these different functional groups. We have reported the synthesis and studies of metallodendrimers containing ferrocene and cobaltocenium units [32,43]. That was the first time that multiple neutral and cationic redox-active organometallic moieties were present in a dendritic structure. These mixed metal dendrimers were obtained by treatment of the first four generations of DAB-based poly(propyleneimine) dendrimers with a mixture of freshly prepared chlorocarbonylferrocene and the PF_6^- salt of chlorocarbonylcobaltocenium in CH_2Cl_2/CH_3CN (Scheme 8.9).

Dendrimers with higher relative loading of cobaltocenium moieties are soluble in H_2O and insoluble in CH_2Cl_2 and they become increasingly soluble in CH_2Cl_2 as the relative loading of ferrocene increases. The electrochemical properties of the heterometallic dendrimers **35–38** were investigated by cyclic voltammetry in both homogeneous solution and confined onto glassy carbon electrodes. For all generations cyclic voltammograms demonstrate the presence of peripheral amidoferrocene units and multiple noninteracting amidocobaltocenium moieties. One remarkable feature of these dendrimers is the ability to modify electrode surfaces, as a consequence of the solubility change arisen upon reduction of cobaltocenium groups. These modified electrodes are extremely durable and reproducible. Their voltammetric features unequivocally indicate surface-confined nature of the electroactive dendrimers films (see Figure 8.19 as an example). They were shown to exhibit a double function: while the ferrocene units act as mediators in enzymatic processes under anaerobic conditions the cobaltocenium moieties take part in an electrocatalytic process in reactions with presence of oxygen. GOx was immobilized into the conducting film by electrostatic interaction. An advantage of these electrodes is the large amount of enzyme immobilized due to electrostatic interactions between ferrocenium and cobaltocenium groups and the negatively charged enzyme. In addition, the presence of the cobaltocenium groups prevents loss of GOx when the ferrocenium groups are reduced, thus increasing the long-term stability of the sensor. Glucose was determined measuring the amperometric response due to the mediation of the enzymatic reaction by electrooxidation of ferrocene. The dendrimer generation used shows significant effects in the behavior of the sensors. As the generation increases the sensitivity increases and the dynamic range decreases. From the calibration plots and the results of the Michaelis–Menten analysis is derived that the higher generation metallodendrimers are more efficient electron transfer mediators. Dendrimers with relatively

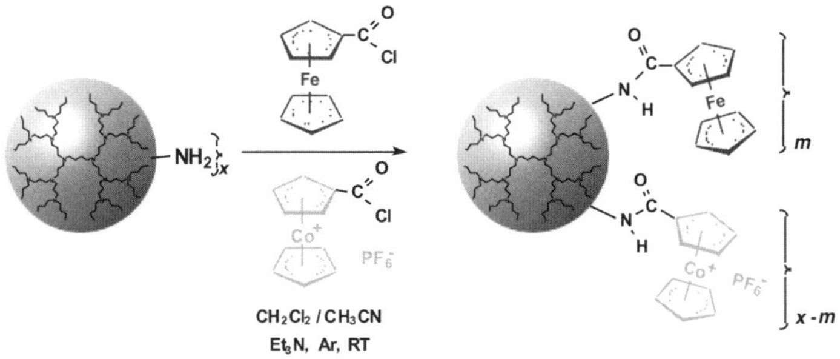

35 : $n=1$, $x=4$ **37** : $n=3$, $x=16$
36 : $n=2$, $x=8$ **38** : $n=4$, $x=32$

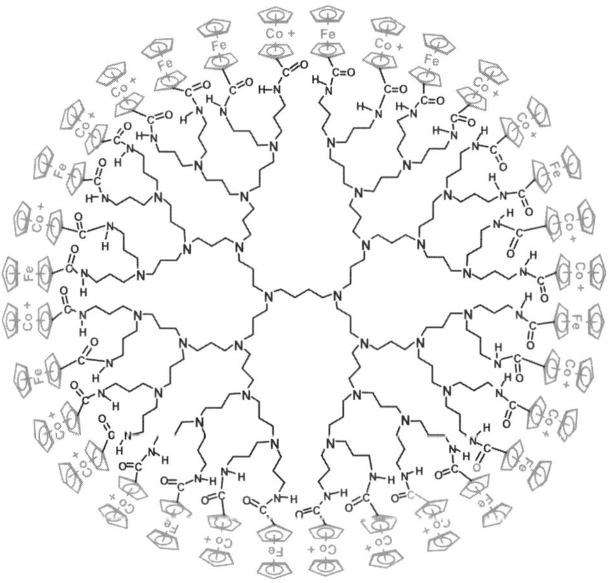

38

SCHEME 8.9 Synthesis of mixed ferrocene–cobaltocenium dendrimers **35–38**.

smaller size give rise to better packed films [113,114] more rigid and likely less permeable, thus making the diffusion of glucose into the film more difficult. The relays contained in more rigid films cannot achieve a close contact with the enzyme redox centers, and therefore they are less efficient mediators. It was found that the electrodeposited films of the smaller dendrimers exhibit a much higher electrocatalytic activity in the reduction of the dissolved oxygen. This can be due to the fact that the distance between organometallic centers becomes small enough that cooperative effects can play a major role. The application of dendrimer–GOx-modified electrodes allows the reduction of oxygen to hydrogen peroxide to occur

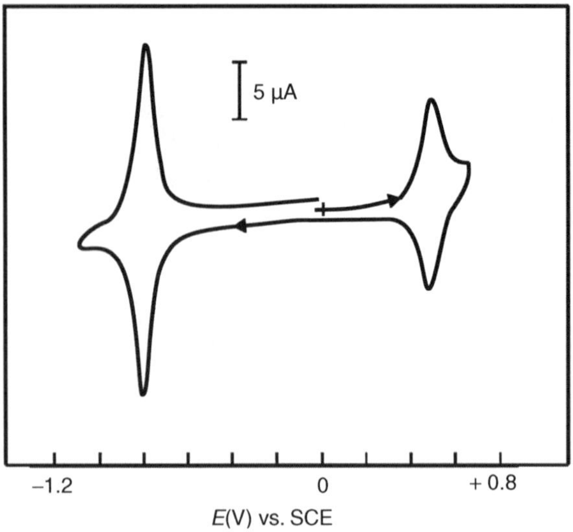

FIGURE 8.19 CV of a glassy carbon disk electrode modified with a film of **37** in $H_2O/LiClO_4$.

at a less negative potential, diminishing interferences, and with higher sensitivity than bare electrodes. The steady-state current due to oxygen reduction decreases as the glucose concentration increases. The sensitivities obtained are in agreement with the variation of the electrocatalytic efficiency of the respective electrodes. On other hand, it was observed that the addition, under aerobic conditions, of the same amounts of glucose produces similar variations in the oxygen concentrations in all sensors prepared with similar coverages of the different dendrimers. This fact demonstrates that the amount of active enzyme immobilized is nearly the same in these sensors.

8.3.2.2 Polymethylferrocenyl Dendrimers We have reported the use of DAB-based dendrimers as core to prepare a polymethylferrocenyl dendrimers family (Scheme 8.3) [40]. While macromolecules containing ferrocenyl moieties are numerous, dendrimers and polymers constructed with permethylferrocenyl units have been relatively less explored. As a result of the enhanced electron donating ability of permethylated cyclopentadienyl rings, polymethylferrocenyl derivatives have lower redox potentials. It is also known that modification of the cyclopentadienyl ligand with methyl substituents causes changes in the rate of the electron transfer [115,116]. In addition, it has been reported that 1,1′dimethylferrocene reoxidises GOx at a rate nearly three times higher than the unsubstituted ferrocene. This improved ability is maintained after covalent attaching to a siloxane polymer. These are important factors when these permethylferrocenyl compounds are used as mediators for biosensors to increase their efficiency and minimize interferences. The electrocatalytical properties of dendrimers **21, 22,** and **24** deposited onto platinum electrodes examined by CV show that the modified electrodes act as electrocatalysts in either oxidation or reduction of H_2O_2. Amperometric enzyme electrodes based on GOx were employed for the determination of glucose under aerobic conditions. The enzyme was

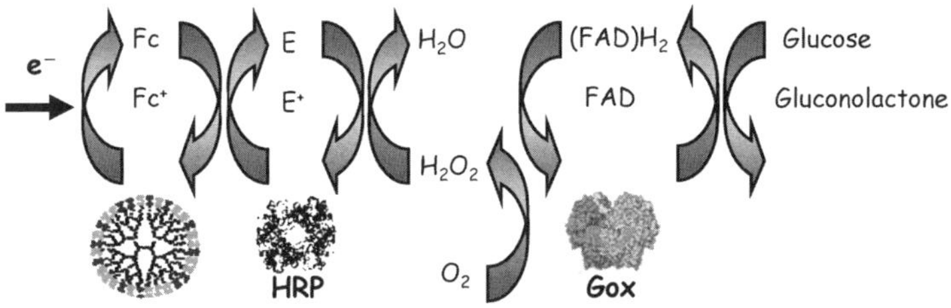

SCHEME 8.10

immobilized by cross-linking using bovine serum albumin (BSA) and glutaraldehyde into the organometallic dendrimer films obtained by evaporation of dichlomethane solution of the corresponding compound on a Pt electrode surface. The results indicate that it is possible to perform both anodic and cathodic determination of the hydrogen peroxide arisen in the enzyme-catalyzed reaction [44].

Several peroxidases have been used for the determination of H_2O_2 and small organic peroxides. The use of peroxidases in combination with hydrogen peroxide producing oxidases and electrochemical detection is well established in analytical chemistry. However, direct electron transfer between common electrode materials and immobilized peroxidase is a slow process and an electron mediator can be used to improve the rate of electron transfer (Scheme 8.10).

Based on the same octamethylferrocenyl dendrimers family, bienzyme electrode sensors were developed. The amperometric sensors were prepared coimmobilizing GOx and Horseradish peroxidase (HRP) on platinum electrodes modified with films of dendrimers **21**, **22**, and **24** (Scheme 8.11).

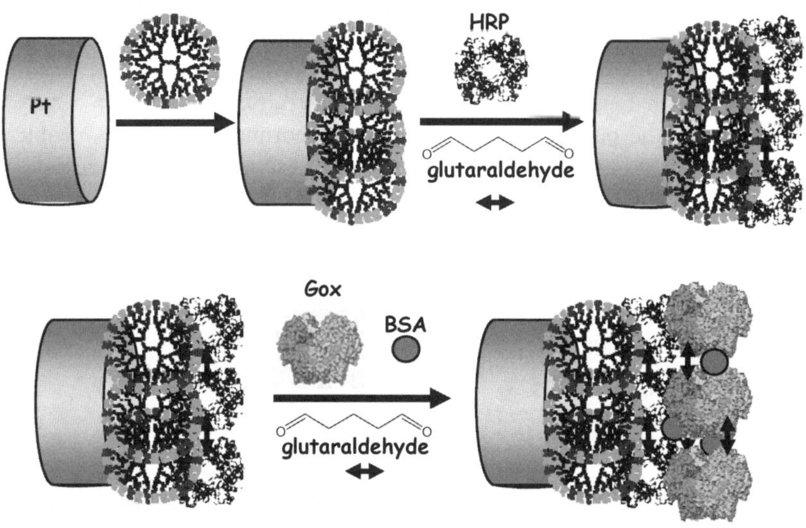

SCHEME 8.11 Schematic representation of the bienzymatic GOx/HRP/dendrimer layers/ network preparation on the platinum electrode surface.

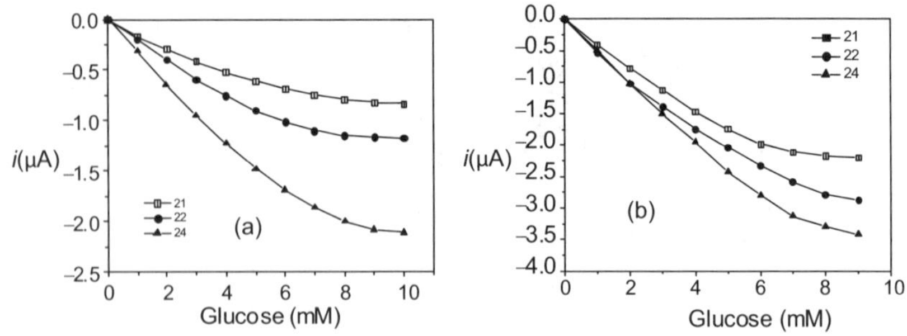

FIGURE 8.20 (a) Glucose calibration plots of monoenzymatic sensors of dendrimers **21**, **22**, and **24** ($\Gamma \approx 3 \times 10^{-9}$ mol ferrocene/cm^2 thickness films). Steady-state anodic currents measured at 400 mV (vs. SCE) applied potential in O_2 saturated phosphate buffer (pH = 7.0). (b) Glucose calibration plots of bienzymatic sensors of dendrimers **21**, **22**, and **24** ($\Gamma \approx 2 \times 10^{-9}$ mol ferrocene/cm^2 thickness films). Steady-state current measured at a 100 mV applied potential in O_2 saturated phosphate buffer (pH 7.0). Adapted from Losada et al. [117].

These sensors were applied to the determination of glucose under aerobic conditions [117]. The results obtained (Figure 8.20) indicate that the sensitivity of both mono and bienzymatic biosensors based on octamethylferrocenyl dendrimers increases with the increasing dendrimer generation. Thus, the dendrimer generation demonstrates to play a significant role in the electrocatalytic activity.

The increase of flexibility of the dendrimer backbone, which is a function of the length of the dendrimer branches, together with the shorter separation between ferrocenyl neighbors, increase the electron transfer efficiency between ferrocene moieties and cause a higher catalytic response of the sensors based on the higher generation dendrimers. On other hand, the analysis of SEM micrographs (Figure 8.21) shows that films of the higher generation dendrimers have a more permeable structure that facilitates diffusion of glucose, minimizing the barrier to transport, and also contributes to signal enhancement.

As expected, the octamethylated ferrocenyl dendrimeric mediators enable the use of lower working potentials for glucose measurements than sensors with

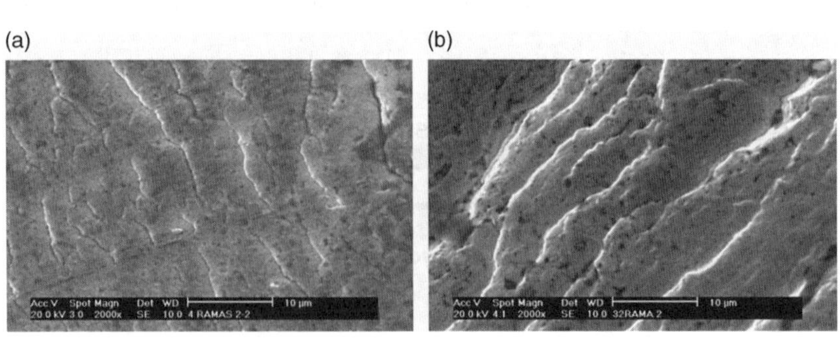

FIGURE 8.21 SEM micrographs of a platinum wire electrode modified with films of (a) dendrimer **21** and (b) dendrimer **24**. From Losada et al. [117].

nonmethylated ferrocenyl compounds. Also the sensitivity, detection limits, and reproducibility were comparable with, or even better than other reported mono- and bienzymatic sensors based on other different mediators.

8.3.3 Dendritic Hyperbranched Macromolecules

The ferrocene dendrimers described provide excellent structural and electrochemical properties. In order to progress in the knowledge of these ferrocenyl materials we prepared macromolecular dendritic hyperbranched materials containing ferrocenyl moieties together with cyclic siloxanes and cubic silsesquioxanes as frameworks (Figure 8.15) [31,118] These dendritic macromolecules have a three-dimensional hyperbranched structure and due to their stereochemical properties both cyclosiloxanes and polyhedral silsesquioxanes are good frames for sterically controlled catalysts systems. Enzyme electrodes were constructed by electrostatic immobilization of GOx onto Pt electrodes modified by electrodeposition or dipping with ferrocenyl and permethylferrocenyl polymers, which act as mediators in the determination of glucose. Thus, it was possible to establish the dependence of the sensors response on the structure of the siloxane framework and on the presence of methyl substituents on the ferrocenyl units. The sensors can be used in different operational modes for glucose determination. In anaerobic conditions these polymer-modified electrodes can act as electrocatalysts in the oxidation of the enzyme. In this case, the sensors based on polymers with cyclotetrasiloxane frameworks display higher electrocatalytic signals and sensitivity than silsesquioxanes. Films of the cyclosiloxane-based polymers exhibit a more permeable structure that leads to facilitate the diffusion of the substrate minimizing the transport barrier and enhancing the bioelectrocatalytic signal. Another reason for the better response has been reported to be the higher flexibility of the cyclotetrasiloxane. The polymethylferrocenylcyclosiloxane polymer has demonstrated to be an electrocalyst for the reduction of H_2O_2, which has allowed the successful determination of glucose, in aerobic conditions and at low potentials, through the electrochemical reduction of hydrogen peroxide produced in the enzymatic reaction. Due to their characteristic catalytic activity the silsesquioxane polymer–GOx electrodes allow the glucose determination in aerobic conditions, by measuring the oxygen spent in the enzyme-catalyzed reaction, at a less negative potential and with a higher sensitivity than a bare electrode. In general, it is found that the permethylated polymers are more effective mediators and allow using lower working potentials in the glucose measurement.

8.3.4 Dendronized Polymers

8.3.4.1 Homopolymers and Copolymers In the last years, we have prepared new redox-active macromolecular materials containing dendritic building blocks with interacting ferrocenyl moieties grafted to multifunctional, flexible poly(methylsiloxane) backbones (see, e.g., polymers **39** and **40** in Figure 8.22). The combination of the exceptional architectural features of dendritic molecules together with the electronic communication between metal centers and the well known, remarkable features and

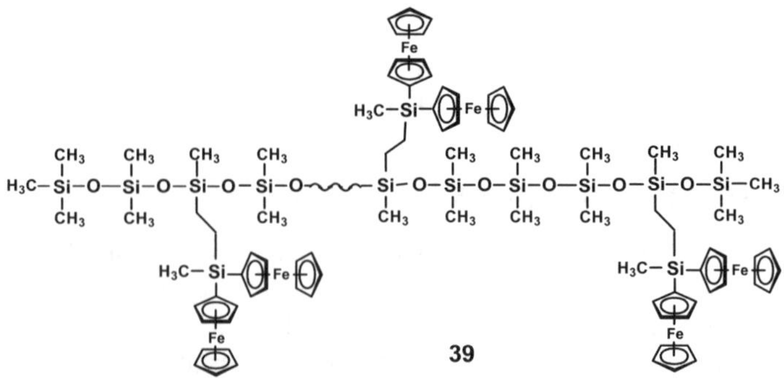

39

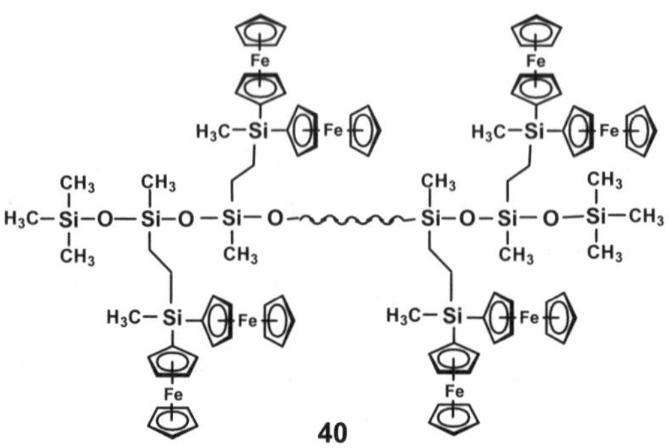

40

FIGURE 8.22 Siloxane-based copolymer **39** and homopolymer **40** containing interacting ferrocenes.

properties of poly(siloxanes) (such as chemical stability, high permeability to gases and low toxicity) is an interesting and useful strategy for controlling the physical and redox properties of the resulting hybrid macromolecules. The polymers were synthesized employing Karstedt-catalyzed hydrosilylation reactions of the poly (methylhydrosiloxane)–poly(dimethylsiloxane) copolymer, $(Me_3SiO)(Me_2SiO)_m(MeSiHO)_n(Me_3Si)$ ($m = 70$–75%, $n = 25$–30%), and poly(methylhydrosiloxane), $(Me_3SiO)(MeSiHO)_n(Me_3Si)$ ($n \approx 35$) with a wedge-shaped dendritic fragment such as the silicon-bridged biferrocene, $CH_2{=}CHSiMe[Fe(\eta^5\text{-}C_5H_4)(\eta^5\text{-}C_5H_5)]_2$ that contains a reactive vinyl group at the focal point [29,119–121].

A CV of a film of homopolymer **40** electrodeposited on a Pt electrode is shown in Figure 8.23. In solvents of low polarity such as CH_2Cl_2, the polymer film displays two reversible and stable redox reactions. This response is due to the existence of appreciable interactions between the two iron centers, which are linked together by a bridging silicon atom. The initial oxidation takes place at nonadjacent ferrocene sites, which makes the subsequent removal of electrons from the remaining ferrocenyl centers, adjacent to those already oxidized more difficult. Pt electrodes modified by electrodeposition with the homopolymer **40** and copolymer **39** have shown to promote the redox reactions of hydrogen peroxide. This electrocatalytic activity may be favored by the capability of the mediators to accomplish bielectronic transfers. Based on this property it has been possible to carry out the amperometric determination of

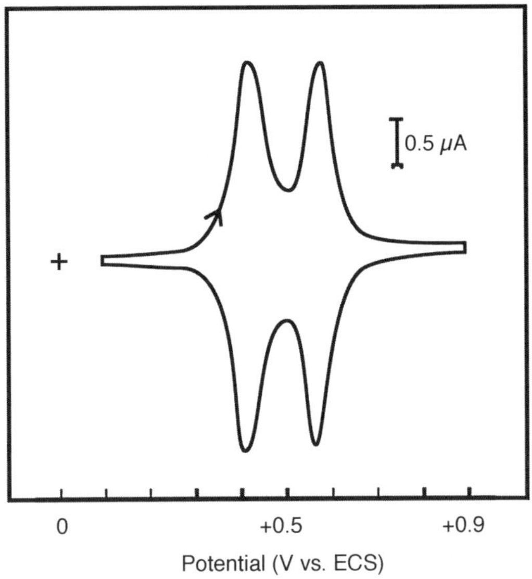

FIGURE 8.23 CV of a Pt disk electrode modified with a film of homopolymer **40**, measured in CH_2Cl_2/TBAH.

H_2O_2. In addition, the redox homopolymer **40** coimmobilized with GOx or Lactate Oxidase (LOx) on Pt electrodes using glutaraldehyde and BSA has been used as amperometric sensors for glucose and lactate respectively, which were measured by H_2O_2 detection at +250 mV (vs. SCE). Figure 8.24 shows the electrocatalytic

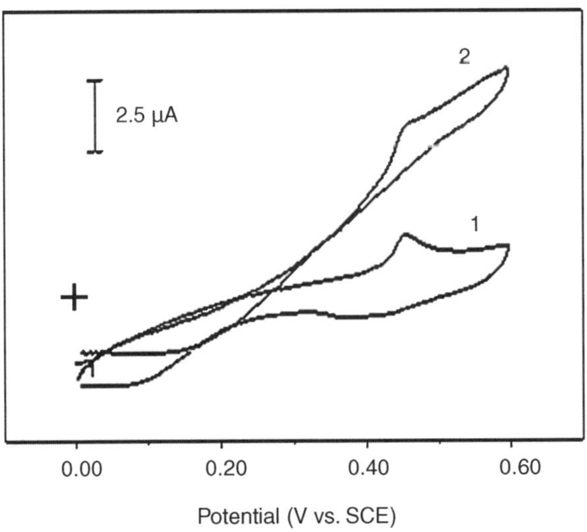

FIGURE 8.24 CVs of lactate oxidase sensor (Pt electrode area = 0.07 cm^2 with 1.1×10^{-9} mol/cm^2 thickness homopolymer **40** film) in absence (1) and presence (2) of 5 mM lactate in air saturated 0.1 M phosphate buffer (pH 7.0). From García-Armada et al. [119].

behavior of one of these modified electrodes. In turn, the copolymer **39** coimmobilized with GOx on a Pt electrode, has been also used as amperometric sensor for glucose determination. The results obtained indicate that both anodic and cathodic operation mode may be used with the copolymer, which acts as electrocatalyst in either oxidation or reduction of the hydrogen peroxide arisen in the enzyme-catalyzed reaction. The sensitivities and detection limits found with both polymers were comparable or even better than other ferrocene-modified polymers-mediated electrodes. On other hand, the homopolymer **40** was used to prepare an enzyme electrode incorporating HRP. The redox polymer was coimmobilized with HRP on Pt electrodes using glutaraldehyde and BSA and applied as amperometric biosensor for the determination of hydrogen peroxide and organic peroxides. It was possible to operate at a low potential (− 50 mV vs. SCE) and a good linear response and protection from interferences was found.

8.3.4.2 ***Block Copolymers*** We have applied the electrocatalytic properties of diferrocenylsilane units attached to other different type of supramolecular structures. Three different block copolymers containing interacting ferrocenyl moieties were synthesized (Scheme 8.12). The anionically prepared diblock copolymers **41–43** consist of a polystyrene (PS) block and a polybutadiene (PB) block functionalized with diferrocenylsilane by Karstedt-catalyzed hydrosilylation [122]. It has been found that the presence of polystyrene blocks confers long-term stability and reproducibility to the electrodes modified with block-copolymer films, which is attributable to the excellent film forming properties of the polystyrene blocks. NAD/NADH-dependent dehydrogenases need the freely diffusing cofactor NAD^+ to catalyze the oxidation of the substrates. The major problem in practical applications of dehydrogenases is the difficulty in the regeneration of the nicotinamide coenzyme since the electrooxidation of NADH on solid electrodes usually requires high potentials.

Electrodes modified with appropriate mediators can be used to shuttle the electrons from NADH to the electrode in order to reduce the overpotential. Another

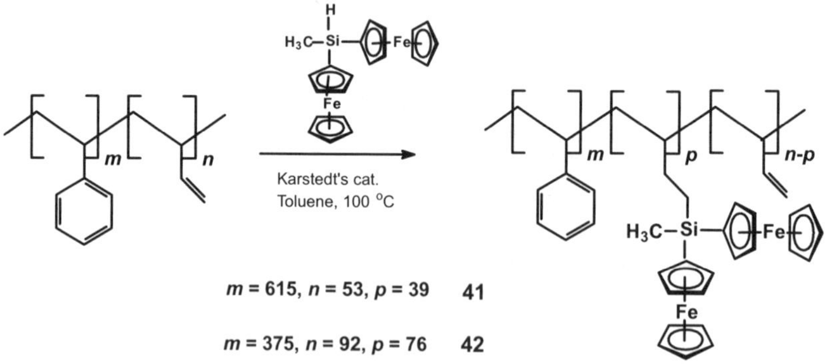

m = 615, *n* = 53, *p* = 39 **41**

m = 375, *n* = 92, *p* = 76 **42**

m = 455, *n* = 204, *p* = 170 **43**

SCHEME 8.12 Synthesis of block copolymers **41–43**.

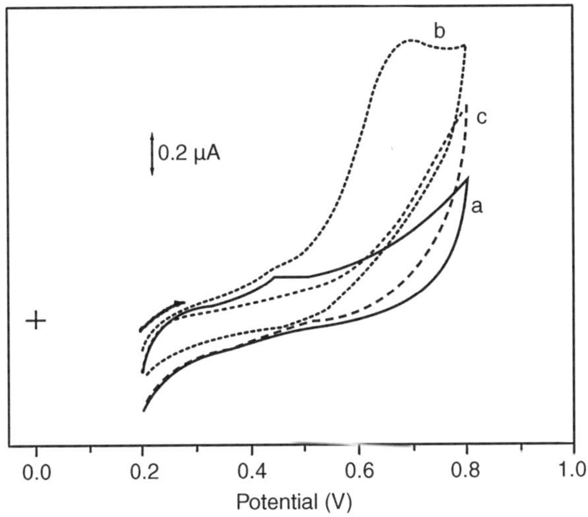

FIGURE 8.25 CVs of copolymer **39** ($\Gamma = 1.0 \times 10^{-10}$ mol ferrocene/cm^2 thickness film) coated platinum electrode in 0.1 M deaerated phosphate buffer (pH = 7.0) solution in absence (a) and presence (b) of NADH 0.5 mM. (c) A platinum bare electrode in presence of NADH 0.5 mM. Scan rate 5 mV/s. From García-Armada et al. [122].

approach is the use of the diaphorase reaction coupled with an electron mediator as ferrocene and its derivatives. The electrodes modified with the copolymers are efficient redox mediators for the electrocatalytic oxidation of both reduced nicotinamide dinucleotide (NADH) cofactor (Figure 8.25) and glucose oxidase. It was shown that the reaction with NADH proceeds via formation of a charge-transfer intermediate. This has been a novel example of electrodes modified with ferrocene derivatives that can be applied to the determination of NADH but in this case it is unnecessary the use of diaphorase. The redox copolymers coimmobilized with GOx have also been successfully used as amperometric biosensors for glucose determinations at lower working potentials (0.25 V vs. SCE) that contribute to reduce the effect of the interfering electrochemical compounds with adequate sensitivities and detection limits.

8.3.5 Ferrocenyl-Tethered Poly(AmidoAmine) Dendrimers

Kim and coworkers [123,124] have reported the application of dendrimers as bioconjugating reagents for preparing multilayered enzyme nanostructures. Primary amines of NH_2-terminated fourth-generation poly(amidoamine) (PAMAM) dendrimers were partially modified with ferrocenyl groups through reaction with ferrocenemonocarboxaldehyde. The resulting partial ferrocenyl-tethered PAMAM dendrimers (Fc-Ds) were effectively employed for the construction of biosensors by simple and direct layer-by-layer depositions of dendrimers and periodate-oxidized GOx on Au electrodes. They demonstrated that the multilayer is constructed in a spatially

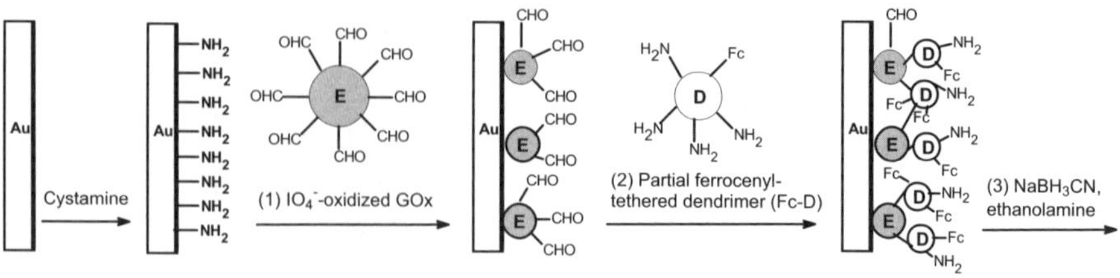

SCHEME 8.13 Schematic organization of the multilayered GOx–Fc-D network on the Au electrodes.

ordered way. The bioelectrocatalytic signals were shown to be amplified with multilayer growth. This fact indicates that the electrode sensitivity is controllable and the method would be applicable to the construction of sensitivity adjustable biosensing devices. The stability of these sensors seems to be due to the integrity of multilayered network with multiple covalent linkages and the rigidity of the dendrimer building blocks (Scheme 8.13).

Using the same Fc-Ds dendrimers as building blocks of a sensing monolayer Yoon et al. [125] have developed an affinity biosensor system based on avidin–biotin interaction. On Au electrodes an amine self-assembled monolayer (SAM) was prepared (3,3-dithiopropionic acid, *N*-hydroxysuccinimide ester), then amine groups of dendrimers are used for immobilization in the SAM monolayer and functionalization with biotinyl groups. An electrochemical signal was generated by free glucose oxidase, depending on the degree of coverage of the sensing surface with avidin. The sensor signal decreased with increasing avidin concentration. From a kinetic analysis the formation of a spatially ordered and compact protein layer on the electrode surface was demonstrated.

A dendrimer Fc-D was also used as an electrocatalyst to enhance the DNA detection as well as a building block to immobilize capture probes. Thus, a sandwich-type enzyme-linked DNA sensor was developed by Kim and coworkers [126]. Fc-Ds have been immobilized on a carboxylic acid-terminated SAM. The target DNA was hybridized with the capture probe and an extension in the DNA was then hybridized with a biotinylated detection probe. The bioterminated layer is detected using avidin-conjugated alkaline phosphatase (Av-ALP), which generates *p*-aminophenol (pAP) from *p*-aminophenyl phosphate (pAPP); pAP is electrochemically oxidized with the mediation of Fc in the layer. The combination of electrocatalysis using Fc-Ds and biochemical amplification enabled the sequence-selective detection of various targets, including single-based mismatched oligonucleotide allowing highly sensitive amperometric measurements.

It has been reported [127] a highly sensitive electrochemical DNA sensor that uses Au nanoparticles (AuNPs), magnetic beads (MBs), and an ITO electrode modified with partially functionalized Fc dendrimers (Fc-Ds) [123,124]. AuNPs are employed as catalytic labels, MBs to permit low nonspecific binding of DNA-conjugated AuNPs and to take advantage of easy magnetic separation. Fc-D-modified ITO electrodes are used to obtain a low background current and easy electron oxidation of pAP.

In a sandwich-type electrochemical sensor; capture probe-conjugated MBs act as a target-binding surface and the modified ITO electrode acts as the signal-generating surface. Biotinylated capture probes were biospecifically immobilized onto streptavidin-coated MBs. Target DNA were hybridized to capture probe-conjugated MBs and detection probe-conjugated AuNPs were also hybridized. AuNPs assemblies were attracted to a working electrode by an external magnet. pAP generated from *p*-nitrophenol in the presence of $NaBH_4$ is electrooxidized giving rise to a redox cycling that considerably increases the electrochemical signals.

Carbon nanotubes (CNTs) have been employed for the development of glucose enzymatic electrodes, because these nanomaterials can provide a suitable microenvironment. Deriu et al. have used multiwalled carbon nanotubes (MWCNTs) and amine-terminated G4-PAMAM dendrimers to build a reagentless glucose biosensor [128]. The G4-PAMAM dendrimers were previously functionalized with ω-alkenyl tethers and then the ferrocenyl units were linked to the tethers. The tether anchoring was accomplished through the reaction of the amino groups of G4-PAMAM dendrimer with undec-10-enoyl chloride and vinylferrocene. The functionalized dendrimers and GOx were immobilized by means of polyazetidine (PAP) on the activated surface of a MWCNTs screen printed electrode (SPE) through the carboxylic groups. PAP acts as cross-linking agent being able to react with several different organic moieties (thiolic, oxydril, amino groups) increasing the possibility to create chemical bonds with the enzyme. The glucose biosensor displays good repeatability, sensitivity and precision.

Closely related dendrimer sensors have been developed taking advantage of the high affinity of the amine groups of PAMAM frameworks for nano-Au. A penicillamine (PCA) biosensor based on tyrosinase immobilized on a nano-Au PAMAM dendrimer-modified gold electrode has been thus designed [129]. Au electrodes were modified with submonolayers of 3-mercaptopropionic acid and further reacted with PAMAM and after contact with gold colloid solution were incubated with tyrosinase. These electrodes catalyze the oxidation of *o*-diphenols to *o*-quinones. In presence of PCA the reduction peak current decreases due to the reaction between PCA and *o*-quinones. This reaction allows to carry out determinations of PCA.

8.3.6 Immunosensors

Immuno- or affinity sensing techniques to detect biospecific interactions such as antibody–antigen, ligand–receptor, and protein–protein interactions have developed into an important area in molecular device technologies. The development of the immunosensors is in continuous focus of many researchers [130]. In the case of the electrochemical immunosensors, the biospecific interaction is electrochemically transformed into an electrical signal. Specifically, the interaction signal can be amplified by an enzyme that continuously generates electroactive products. The electrode surface is very important in electrochemical immunosensors. The modification of electrode surfaces allows immobilization of biomolecules, long-term stability, low nonspecific binding and proper biomolecular orientation.

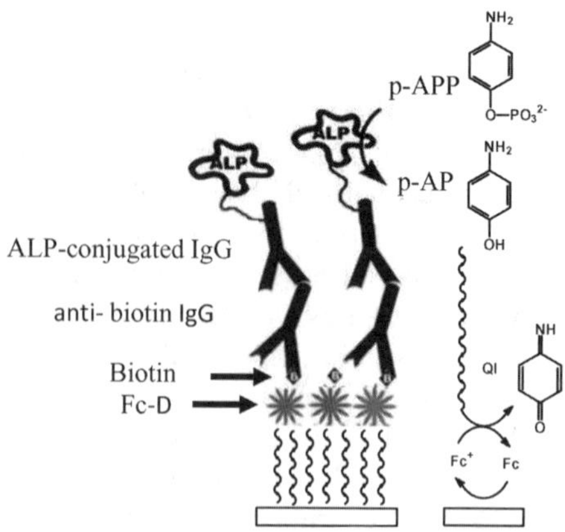

SCHEME 8.14 Schematic illustration of an enzyme-amplified immunosensor using redox mediation of Fc-D.

Yang and coworkers [131–133] reported an enzyme-amplified electrochemical immunosensor applying a ferrocenyl-tethered dendrimer Fc-D [123] immobilized on gold electrodes that acts as a mediator of electron transfer and allows the attachment of biomolecules. Fc-D was immobilized to the electrode surface by covalent binding between the dendrimer amines and the carboxylic acid of a SAM monolayer and the unreacted amines of dendrimers were modified with biotin groups.

As a model system they measured the biospecific interaction between biotin and antibiotin IgG by a sandwich-type immunosensor using alkaline phosphatase as the enzyme for generating electroactive products (Scheme 8.14) [131].

In order to obtain a better sensitivity Yang and coworkers [132] have developed a sandwich-type heterogeneous electrochemical immunosensor (SHEI) for mouse IgG in which the signal amplification is achieved using pAP redox cycling by hydrazine and the noise level is reduced using an ITO electrode modified with a Fc-D dendrimer. This research group has reported an ultrasensitive and simple electrochemical method for signal amplification that is obtained by catalytic reduction of *p*-nitrophenol to pAP using gold-nanocatalysts labels. The electrochemical signal is amplified by chemical reduction of *p*-quinone imine to aminophenol by $NaBH_4$. For the preparation of a SHEI an IgG layer was formed on an ITO electrode in which Fc-D was also immobilized by bonding to carboxylic groups of a phosphonate SAM. The unreacted amines of Fc-D were modified with biotin groups and biotinylated antibodies were immobilized to the streptavidin-modified ITO [133].

A similar amplification strategy is employed in the preparation of a nanocatalyst-based electrochemical immunoassay developed using MBs and gold nanocatalysts. The formed immunosensing complex is strongly attracted to an ITO electrode modified with Fc-Ds by an external magnet [134].

It has been performed a new pAp redox cycling scheme using NADH, which is applied to Au electrocatalytic electrodes. These are modified with a mixed SAM of mercaptododecanoic acid and mercaptoundecanol, and a Fc-D dendrimer layer. In a sandwich-type electrochemical immunosensor for mouse immunoglobulin G (IgG) 1 pg/mL detection limit for mouse IgG was achieved [135].

ACKNOWLEDGMENTS

This work has been supported by the Spanish Dirección General de Investigación (CTQ2004-07381-C02 and CTQ2009-12332-C02) and the Consejería de Educación, Comunidad de Madrid (S-0505/PPQ-0328). The authors are greatly indebted to their colleagues and students who contributed to the work described in this chapter (see references). This chapter is dedicated to the Memory of Professor Moisés Morán our distinguished friend and colleague from the Universidad Autónoma of Madrid who passed away in December 2002.

REFERENCES

[1] G. R. Newkome, C. N. Moorefield, F. Vögtle, *Dendrimers and Dendrons. Concept, Synthesis and Applications*, Wiley-VCH, Weinheim, **2001**.

[2] D. A. Tomalia, J. M. Frechet, *Prog. Polym. Sci.* **2005**, *30*, 217–219.

[3] F. Vögtle, S. Gestermann, R. Hesse, H. Schwierz, B. Windisch, *Prog. Polym. Sci.* **2000**, *25*, 987–1041.

[4] J. Frechet, D. Tomalia, *Dendrimers and Other Dendritic Polymers*, John Wiley & Sons, Chichester, **2001**.

[5] G. R. Newkome, *Advances in Dendritic Macromolecules*, Vols. *1–5*, JAI Press, Greenwich, **1994**, **1995**, **1996**, **1999**, **2002**.

[6] E. Buhleier, W. Wehner, F. Vögtle, *Synthesis* **1978**, 155–158.

[7] G. R. Newkome, Z. Yao, G. R. Baker, V. K. Gupta, *J. Org. Chem.* **1985**, *50*, 2003–2004.

[8] D. A. Tomalia, H. Baker, J. Dewald, M. Hall, G. Kallos, S. Martin, J. Roeck, J. Ryder, P. Smith, *Polym. J. (Tokyo)* **1985**, *17*, 117–132.

[9] U. Boas, J. B. Christensen, P. M. H. Heegaard, *Dendrimers in Medicine and Biotechnology: New Molecular Tools*, Royal Society of Chemistry, Cambridge, **2006**.

[10] G. Denti, S. Campagna, S. Serroni, M. Ciano, V. Balzani, *J. Am. Chem. Soc.* **1992**, *114*, 7326.

[11] G. R. Newkome, F. Cardullo, E. C. Constable, C. N. Moorefield, A. M. W. C. Thompson, *J. Chem. Soc., Chem. Commun.* **1993**, 925–927.

[12] G. R. Newkome, E. He, C. N. Moorefield, *Chem. Rev.* **1999**, *99*, 1689–1746.

[13] S.-H. Hwang, C. D. Shreiner, C. N. Moorefield, G. R. Newkome, *New J. Chem.* **2007**, *31*, 1192–1217.

[14] C. Gorman, *Adv. Mater.* **1998**, *10*, 295–309.

[15] J.-P. Majoral, A.-M. Caminade, *Chem. Rev.* **1999**, *99*, 845–880.

[16] D. Mery, D. Astruc, *Coord. Chem. Rev.* **2006**, *250*, 1965–1979.

[17] A. Berger, R. J. M. K. Gebbink, G. van Koten, *Top. Organomet. Chem.* **2006**, *20*, 1–38.

[18] A.-M. Caminade, J.-P. Majoral, *Coord. Chem. Rev.* **2005**, *249*, 1917–1926.

[19] I. Cuadrado, M. Morán, C. M. Casado, B. Alonso, J. Losada, *Coord. Chem. Rev.* **1999**, *193–195*, 395–445.

[20] M. A. Hearshaw, J. R. Moss, *Chem. Commun.* **1999**, 1–8.

[21] P. A. Chase, R. J. M. K. Gebbink, G. van Koten, *J. Organomet. Chem.* **2004**, *689*, 4016–4054.

[22] D. Astruc, C. Ornelas, J. Ruiz, *Acc. Chem. Res.* **2008**, *41*, 841–856.

[23] A. E. Kaifer, M. Gomez-Kaifer, *Supramolecular Electrochemistry*, Wiley-VCH, Weinheim, **1999**.

[24] P. Zanello, *Inorganic Electrochemistry: Theory, Practice and Application*, Royal Society of Chemistry, Cambridge, **2003**.

[25] D. Astruc, M.-C. Daniel, S. Nlate, J. Ruiz, in *Trends in Molecular Electrochemistry*, (Eds.: A. J. L. Pombeiro, C. Amatore), Marcel Dekker/FontisMedia, New York, **2004**, Ch. 9, pp. 283–310.

[26] C. M. Casado, I. Cuadrado, M. Morán, B. Alonso, B. García, B. González, J. Losada, *Coord. Chem. Rev.* **1999**, *185–186*, 53–79.

[27] B. Alonso, C. M. Casado, I. Cuadrado, M. Morán, A. E. Kaifer, *Chem. Commun.* **2002**, 1778–1779.

[28] B. Alonso, I. Cuadrado, M. Morán, J. Losada, *J. Chem. Soc., Chem. Commun.* **1994**, 2575–2576.

[29] B. Alonso, B. González, B. García, E. Ramírez-Oliva, M. Zamora, C. M. Casado, I. Cuadrado, *J. Organomet. Chem.* **2001**, *637–639*, 642–652.

[30] C. M. Casado, I. Cuadrado, B. Alonso, M. Morán, J. Losada, *J. Electroanal. Chem.* **1999**, *463*, 87–92.

[31] C. M. Casado, I. Cuadrado, M. Morán, B. Alonso, M. Barranco, J. Losada, *Appl. Organomet. Chem.* **1999**, *13*, 245–259.

[32] C. M. Casado, B. González, I. Cuadrado, B. Alonso, M. Morán, J. Losada, *Angew. Chem., Int. Ed.* **2000**, *39*, 2135–2138.

[33] I. Cuadrado, C. M. Casado, B. Alonso, M. Morán, J. Losada, V. Belsky, *J. Am. Chem. Soc.* **1997**, *119*, 7613–7614.

[34] I. Cuadrado, M. Morán, C. M. Casado, B. Alonso, F. Lobete, B. García, M. Ibisate, J. Losada, *Organometallics* **1996**, *15*, 5278–5280.

[35] I. Cuadrado, M. Morán, J. Losada, C. M. Casado, C. Pascual, B. Alonso, F. Lobete, in *Advances in Dendritic Macromolecules*, Vol. *3* (Ed.: G. Newkome), JAI Press, Greenwich, **1996**, pp. 151–195.

[36] B. González, B. Alonso, J. Losada, M. P. García-Armada, C. M. Casado, *Organometallics* **2006**, *25*, 3558–3561.

[37] B. González, C. M. Casado, B. Alonso, I. Cuadrado, M. Morán, Y. Wang, A. E. Kaifer, *Chem. Commun.* **1998**, 2569–2570.

[38] R. Villoslada, B. Alonso, C. M. Casado, P. García-Armada, J. Losada, *Organometallics* **2009**, *28*, 727–733.

[39] M. Zamora, B. Alonso, C. Pastor, I. Cuadrado, *Organometallics* **2007**, *26*, 5153–5164.

[40] M. Zamora, S. Herrero, J. Losada, I. Cuadrado, C. M. Casado, B. Alonso, *Organometallics* **2007**, *26*, 2688–2693.

[41] B. Alonso, M. Morán, C. M. Casado, F. Lobete, J. Losada, I. Cuadrado, *Chem. Mater.* **1995**, *7*, 1440–1442.

[42] J. Losada, I. Cuadrado, M. Morán, C. M. Casado, B. Alonso, M. Barranco, *Anal. Chim. Acta* **1997**, *338*, 191–198.

[43] B. Alonso, P. G. Armada, J. Losada, I. Cuadrado, B. González, C. M. Casado, *Biosens. Bioelectron.* **2004**, *19*, 1617–1625.

[44] M. P. G. Armada, J. Losada, M. Zamora, B. Alonso, I. Cuadrado, C. M. Casado, *Bioelectrochemistry* **2006**, *69*, 65–73.

[45] F. J. Martínez, B. González, B. Alonso, J. Losada, M. P. García-Armada, C. M. Casado, *J. Inorg. Organomet. Polym. Mater.* **2008**, *18*, 51–58.

[46] R. Castro, I. Cuadrado, B. Alonso, C. M. Casado, M. Morán, A. E. Kaifer, *J. Am. Chem. Soc.* **1997**, *119*, 5760–5761.

[47] P. D. Beer, *Adv. Inorg. Chem.* **1992**, *39*, 79–157.

[48] P. D. Beer, S. R. Bayly, *Top. Curr. Chem.* **2005**, *255*, 125–162.

[49] P. D. Beer, E. J. Hayes, *Coord. Chem. Rev.* **2003**, *240*, 167–189.

[50] P. D. Beer, P. A. Gale, *Angew. Chem., Int. Ed.* **2001**, *40*, 486–516.

[51] J. H. R. Tucker, S. R. Collinson, *Chem. Soc. Rev.* **2002**, *31*, 147–156.

[52] S. R. Bayly, P. D. Beer, G. Z. Chen, in *Ferrocenes* (Ed.: P. Stepnicka), John Wiley & Sons, Chichester, **2008**, pp. 281–318.

[53] J. B. Flanagan, S. Margel, A. J. Bard, F. C. Anson, *J. Am. Chem. Soc.* **1978**, *100*, 4248–4253.

[54] S. R. Miller, D. A. Gustowski, Z. H. Chen, G. W. Gokel, L. Echegoyen, A. E. Kaifer, *Anal. Chem.* **1988**, *60*, 2021–2024.

[55] H. Abruña, in *Electroresponsive Molecular and Polymeric Systems* (Ed.: T. A. Stokheim), Marcel Dekker, New York, **1988**.

[56] R. W. Murray, in *Molecular Design of Electrode Surfaces, Techniques of Chemistry*, Vol. *XXII* (Ed.: R. W. Murray), Wiley, New York, **1992**, p. 1.

[57] O. Reynes, T. Gulon, J.-C. Moutet, G. Royal, E. Saint-Aman, *J. Organomet. Chem.* **2002**, *656*, 116–119.

[58] O. Reynes, F. Maillard, J.-C. Moutet, G. Royal, E. Saint-Aman, G. Stanciu, J.-P. Dutasta, I. Gosse, J.-C. Mulatier, *J. Organomet. Chem.* **2001**, *637–639*, 356–363.

[59] M.-C. Daniel, J. Ruiz, J.-C. Blais, N. Daro, D. Astruc, *Chem. Eur. J.* **2003**, *9*, 4371–4379.

[60] J. Ruiz, M. J. R. Medel, M.-C. Daniel, J.-C. Blais, D. Astruc, *Chem. Commun.* **2003**, 464–465.

[61] C. Valerio, J.-L. Fillaut, J. Ruiz, J. Guittard, J.-C. Blais, D. Astruc, *J. Am. Chem. Soc.* **1997**, *119*, 2588–2589.

[62] B. P. Hay, T. K. Firman, B. A. Moyer, *J. Am. Chem. Soc.* **2005**, *127*, 1810–1819.

[63] H. Stephan, H. Spies, B. Johannsen, L. Klein, F. Vögtle, *Chem. Commun.* **1999**, 1875–1876.

[64] A. J. Evans, S. E. Matthews, A. R. Cowley, P. D. Beer, *Dalton Trans.* **2003**, 4644–4650.

[65] H. Miyaji, S. R. Collinson, I. Prokes, J. H. R. Tucker, *Chem. Commun.* **2003**, 64–65.

[66] F. Oton, A. Tárraga, A. Espinosa, M. D. Velasco, P. Molina, *Dalton Trans.* **2006**, 3685–3692.

[67] F. Oton, A. Tárraga, A. Espinosa, M. D. Velasco, P. Molina, *J. Org. Chem.* **2006**, *71*, 4590–4598.

[68] M. D. Pratt, P. D. Beer, *Polyhedron* **2003**, *22*, 649–653.

[69] I. Díaz, B. García, B. Alonso, C. M. Casado, M. Morán, J. Losada, J. Perez-Pariente, *Chem. Mater.* **2003**, *15*, 1073–1079.

[70] J. B. Cooper, M. G. B. Drew, P. D. Beer, *Dalton Trans.* **2000**, 2721–2728.

[71] A. J. Evans, P. D. Beer, *Dalton Trans.* **2003**, 4451–4456.

[72] F. Oton, A. Tárraga, M. D. Velasco, P. Molina, *Dalton Trans.* **2005**, 1159–1161.

[73] C. Suksai, P. Leeladee, D. Jainuknan, T. Tuntulani, N. Muangsin, O. Chailapakul, P. Kongsaeree, C. Pakavatchai, *Tetrahedron Lett.* **2005**, *46*, 2765–2769.

[74] W. W. H. Wong, D. Curiel, S.-W. Lai, M. G. B. Drew, P. D. Beer, *Dalton Trans.* **2005**, 774–781.

[75] J. G. Eaves, R. Mirrazaei, D. Parker, H. S. Munro, *J. Chem. Soc., Perkin Trans. 2* **1989**, 373–376.

[76] J. G. Eaves, H. S. Munro, D. Parker, *Synth. Met.* **1986**, *16*, 123–125.

[77] J.-C. Moutet, E. Saint-Aman, M. Ungureanu, T. Visan, *J. Electroanal. Chem.* **1996**, *410*, 79–85.

[78] A. Naji, M. Cretin, M. Persin, J. Sarrazin, *J. Appl. Polym. Sci.* **2004**, *91*, 3947–3958.

[79] E. Saint-Aman, M. Ungureanu, T. Visan, J. C. Moutet, *Electrochim. Acta* **1997**, *42*, 1829–1837.

[80] P. D. Beer, J. Cadman, *Coord. Chem. Rev.* **2000**, *205*, 131–155.

[81] I. Szymanska, H. Radecka, J. Radecki, P. A. Gale, C. N. Warriner, *J. Electroanal. Chem.* **2006**, *591*, 223–228.

[82] O. Reynes, J.-C. Moutet, G. Royal, E. Saint-Aman, *Electrochim. Acta* **2004**, *49*, 3727–3735.

[83] C. Ornelas, J. Ruiz, D. Astruc, *Organometallics* **2009**, *28*, 4431–4437.

[84] C. Ornelas, J. Ruiz, C. Belin, D. Astruc, *J. Am. Chem. Soc.* **2009**, *131*, 590–601.

[85] M.-C. Daniel, F. Ba, J. R. Aranzaes, D. Astruc, *Inorg. Chem.* **2004**, *43*, 8649–8657.

[86] M.-C. Daniel, J. Ruiz, D. Astruc, *J. Am. Chem. Soc.* **2003**, *125*, 1150–1151.

[87] D. Astruc, M.-C. Daniel, J. Ruiz, *Chem. Commun.* **2004**, 2637–2649.

[88] M.-C. Daniel, J. R. Aranzaes, S. Nlate, D. Astruc, *J. Inorg. Organomet. Polym. Mater.* **2005**, *15*, 107–119.

[89] M.-C. Daniel, D. Astruc, *Chem. Rev.* **2004**, *104*, 293–346.

[90] M.-C. Daniel, J. Ruiz, S. Nlate, J.-C. Blais, D. Astruc, *J. Am. Chem. Soc.* **2003**, *125*, 2617–2628.

[91] M.-C. Daniel, J. Ruiz, S. Nlate, J. Palumbo, J.-C. Blais, D. Astruc, *Chem. Commun.* **2001**, 2000–2001.

[92] J. Camponovo, J. Ruiz, E. Cloutet, D. Astruc, *Chem. Eur. J.* **2009**, *15*, 2990–3002.

[93] C. Ornelas, R. Aranzaes Jaime, L. Salmon, D. Astruc, *Chem. Eur. J.* **2008**, *14*, 50–64.

[94] C. Ornelas, J. Ruiz Aranzaes, E. Cloutet, S. Alves, D. Astruc, *Angew. Chem. Int. Ed.* **2007**, *46*, 872–877.

[95] D. L. Stone, D. K. Smith, *Polyhedron* **2003**, *22*, 763–768.

[96] C. M. Casado, I. Cuadrado, M. Morán, B. Alonso, F. Lobete, J. Losada, *Organometallics* **1995**, *14*, 2618–2620.

[97] M. Morán, C. M. Casado, I. Cuadrado, J. Losada, *Organometallics* **1993**, *12*, 4327–4333.

[98] Q. Tan, L. Wang, L. Ma, H. Yu, Q. Liu, A. Xiao, *Macromolecules* **2009**, *42*, 4500–4510.

[99] P. D. Beer, *Chem. Commun.* **1996**, 689–696.

[100] P. D. Beer, M. G. B. Drew, A. R. Graydon, *J. Chem. Soc., Dalton Trans.* **1996**, 4129–4134.

[101] P. D. Beer, A. R. Graydon, *J. Organomet. Chem.* **1994**, *466*, 241–247.

[102] P. D. Beer, S. E. Stokes, *Polyhedron* **1995**, *14*, 873–879.

[103] C. Valerio, J. Ruiz, J.-L. Fillaut, D. Astruc, *C. R. Acad. Sci., Ser. IIc: Chim.* **1999**, *2*, 79–83.

[104] I. del Peso, B. Alonso, F. Lobete, C. M. Casado, I. Cuadrado, J. Losada del Barrio, *Inorg. Chem. Commun.* **2002**, *5*, 288–291.

[105] I. Cuadrado, C. M. Casado, F. Lobete, B. Alonso, B. González, J. Losada, U. Amador, *Organometallics* **1999**, *18*, 4960–4969.

[106] C. Valerio, E. Alonso, J. Ruiz, J.-C. Blais, D. Astruc, *Angew. Chem., Int. Ed.* **1999**, *38*, 1747–1751.

[107] J. R. Aranzaes, C. Belin, D. Astruc, *Angew. Chem., Int. Ed.* **2006**, *45*, 132–136.

[108] I. Willner, E. Katz, *Angew. Chem., Int. Ed.* **2000**, *39*, 1181–1218.

[109] E. Bakker, Y. Qin, *Anal. Chem.* **2006**, *78*, 3965–3983.

[110] S. P. Hendry, M. F. Cardosi, A. P. F. Turner, E. W. Neuse, *Anal. Chim. Acta* **1993**, *281*, 453–459.

[111] J. M. Dicks, W. J. Aston, G. Davis, A. P. F. Turner, *Anal. Chim. Acta* **1986**, *182*, 103–112.

[112] M. J. Green, H. A. O. Hill, *J. Chem. Soc., Faraday Trans.* **1986**, *82*, 1237–1243.

[113] K. Takada, D. J. Díaz, H. D. Abruña, I. Cuadrado, C. Casado, B. Alonso, M. Morán, J. Losada, *J. Am. Chem. Soc.* **1997**, *119*, 10763–10773.

[114] K. Takada, D. J. Díaz, H. D. Abruña, I. Cuadrado, B. González, C. M. Casado, B. Alonso, M. Morán, J. Losada, *Chem. Eur. J.* **2001**, *7*, 1109–1117.

[115] A. E. G. Cass, G. Davis, G. D. Francis, H. A. O. Hill, W. J. Aston, I. J. Higgins, E. V. Plotkin, L. D. L. Scott, A. P. F. Turner, *Anal. Chem.* **1984**, *56*, 667–671.

[116] P. D. Hale, L. I. Boguslavsky, T. Inagaki, H. I. Karan, H. S. Lee, T. A. Skotheim, Y. Okamoto, *Anal. Chem.* **1991**, *63*, 677–682.

[117] J. Losada, M. Zamora, M. P. García-Armada, I. Cuadrado, B. Alonso, C. M. Casado, *Anal. Bioanal. Chem.* **2006**, *385*, 1209–1217.

[118] J. Losada, M. P. García-Armada, I. Cuadrado, B. Alonso, B. González, C. M. Casado, J. Zhang, *J. Organomet. Chem.* **2004**, *689*, 2799–2807.

[119] M. P. García-Armada, J. Losada, I. Cuadrado, B. Alonso, B. González, C. M. Casado, *Electroanalysis* **2003**, *15*, 1109–1114.

[120] M. P. García-Armada, J. Losada, I. Cuadrado, B. Alonso, B. González, C. M. Casado, J. Zhang, *Sens. Actuators, B* **2004**, *B101*, 143–149.

[121] M. P. García-Armada, J. Losada, I. Cuadrado, B. Alonso, B. González, E. Ramírez-Oliva, C. M. Casado, *Sens. Actuators, B* **2003**, *B88*, 190–197.

[122] M. P. García-Armada, J. Losada, F. J. Lopez-Villanueva, H. Frey, B. Alonso, C. M. Casado, *J. Organomet. Chem.* **2008**, *693*, 2803–2811.

[123] H. C. Yoon, M.-Y. Hong, H.-S. Kim, *Anal. Chem.* **2000**, *72*, 4420–4427.

[124] H. C. Yoon, H.-S. Kim, *Anal. Chem.* **2000**, *72*, 922–926.

[125] H. C. Yoon, M. Y. Hong, H. S. Kim, *Anal. Biochem.* **2000**, *282*, 121–128.

[126] E. Kim, K. Kim, H. Yang, T. Kim Youn, J. Kwak, *Anal. Chem.* **2003**, *75*, 5665–5672.

[127] T. Selvaraju, J. Das, K. Jo, K. Kwon, C.-H. Huh, T. K. Kim, H. Yang, *Langmuir* **2008**, *24*, 9883–9888.

[128] D. Deriu, G. Favero, A. D'Annibale, F. Mazzei, *ECS Trans.* **2008**, *16*, 105–113.

[129] N. B. Li, J. Kwak, *Electroanalysis* **2007**, *19*, 2428–2436.

[130] P. Skladal, *Electroanalysis* **1997**, *9*, 737–745.

[131] S. J. Kwon, E. Kim, H. Yang, J. Kwak, *Analyst* **2006**, *131*, 402–406.

[132] J. Das, K. Jo, W. Lee Jae, H. Yang, *Anal. Chem.* **2007**, *79*, 2790–2796.

[133] J. Das, M. A. Aziz, H. Yang, *J. Am. Chem. Soc.* **2006**, *128*, 16022–16023.

[134] T. Selvaraju, J. Das, S. W. Han, H. Yang, *Biosens. Bioelectron.* **2008**, *23*, 932–938.

[135] S. J. Kwon, H. Yang, K. Jo, J. Kwak, *Analyst* **2008**, *133*, 1599–1604.

9

SHAPE-PERSISTENT CONJUGATED DENDRIMERS FOR ORGANIC ELECTRONICS

JING YAN, FAN GAO, JIAN PEI, AND YUGUO MA

Beijing National Laboratory for Molecular Sciences (BNLMS), Key Laboratories of Polymer Chemistry and Physics and of Bioorganic Chemistry and Molecular Engineering of Ministry of Education, College of Chemistry and Molecular Engineering, Peking University, Beijing 100871, China

9.1 INTRODUCTION

Organic electronics has come a long way since its humble start [1]. Various devices, especially organic light-emitting diodes (OLEDs), are beginning to make significant progress into the commercial world [2]. Others, such as organic field-effect transistors (OFETs) [3], and organic solar cells (OSCs) [4], are also near or at the stage of commercialization. The reason for the interest in organic optoelectronic materials lies in the fact that they offer many properties that cannot be realized in traditional inorganic devices. Compared to their inorganic counterparts, organic active materials have the advantage of low cost, easy solution-processability, and an almost unlimited possibility of molecular designs.

Traditionally, researchers have mainly focused on two kinds of organic semiconductors, namely small molecules and polymers. Small molecules can easily be made in a pure form, so structure–property relationships can often be established [5]. The possibility to obtain single crystals from small molecular semiconductors has further aided detailed investigations and theoretical treatments of the device

Designing Dendrimers, Edited by Sebastiano Campagna, Paola Ceroni, Fausto Puntoriero.
© 2012 John Wiley & Sons, Inc. Published 2012 by John Wiley & Sons, Inc.

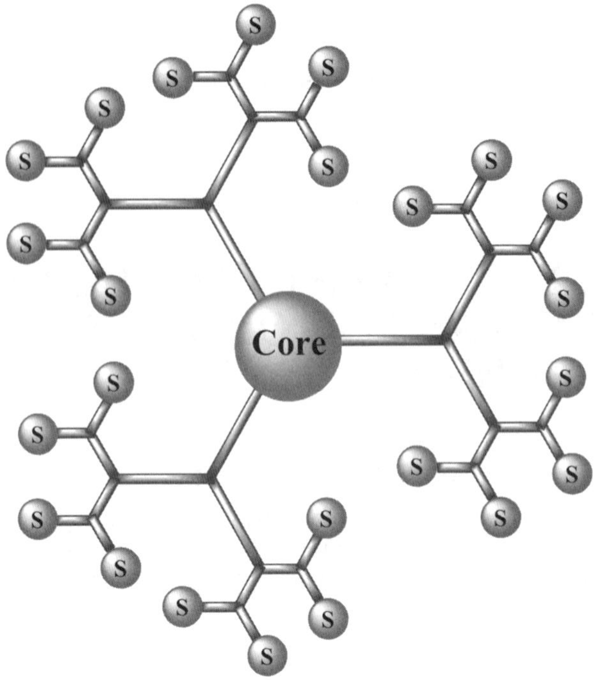

FIGURE 9.1 Schematic representation of a dendrimer's structure.

physics [6]. One of the main drawbacks for small molecules is that they are mostly processed through vapor deposition techniques, which requires high-cost and cumbersome high-vacuum facilities. Meanwhile, a large percentage of materials are wasted during the vacuum deposition process due to the necessity of patterning [7]. These drawbacks post great obstacles for small molecules for truly realizing low-cost manufacturing. As an alternative, conjugated polymers can be solution-processed by spin-coating or inkjet printing techniques, due to their favorable rheological properties [8]. However, their performances in devices are often insufficiently reproducible, owing to difficulties in controlling the polydispersity, molecular weight, backbone defects, and/or end groups [5]. The complexity and heterogeneity of a polymer system also hamper a simple theoretical description of their charge transport properties [9].

Dendrimers present a relatively new type of organic semiconductors for optoelectronic applications. They are hyperbranched, structurally well-defined, monodisperse molecules [10]. Their tree-like structure is shown schematically in Figure 9.1. A dendrimer usually consists of a central core, several layers of branching units, and a large number of surface groups. The syntheses and characterizations of dendrimers have made tremendously progress in the past 30 years, as summarized in excellent reviews and books cited above. Increasing importance and attention are now placed on their functions and applications [11]. Regarding electronic applications, dendrimers beautifully combine the advantages of both small molecules and polymers. On one hand, they can be solution-processed like polymers; on the other hand, their polydispersity can be reduced to close to unity by using highly efficient transformations and careful purification protocols. Structure–property relationship may thus be deduced, just as for small molecules. In addition, the unique structure of dendrimers offers new possibilities for molecular designs and for device configurations. For example, independently tuning the properties of the core, the branches, and the surface groups allows one to create a library of structurally related materials for screening. In addition, employing dendrimers of different generations can be used as a new approach to control intermolecular interactions [5].

Conjugation plays a central role in semiconducting dendrimers [12]. For example, in light-emitting and photovoltaic materials, the π-delocalized structural units absorb or emit light of various wavelengths depending on their conjugation lengths. In dendrimers, chromophores can be linked by either saturated aliphatic chains or stiff conjugate units. The former choice bestows some flexibility in the dendrimer structure, hence improving its solubility. In addition, ground-state electronic coupling among subunits is minimized. However, the introduction of insulating (nonconjugated) moieties has been shown to be detrimental to device performance [13]. Moreover, uncertainty in the distance between chromophores sometimes makes the interpretation of photophysical processes difficult. In some cases, chromophores may come into contact via folding of the flexible connecting chains, resulting in undesirable chromophore–chromophore interactions, such as aggregation, excimer formation, and/or self-quenching [14]. The use of rigid, conjugated linkages can prevent these problems. More importantly, the stiffness of all building blocks in such dendrimers results in a unique property: shape-persistency.

The Moore group pioneered the study of shape-persistent dendrimers [15]. Since their first publication in 1993, many rigid, conjugated dendrimers have been reported [16]. The use of stiff units ensures that the dimension of the entire dendrimer is almost invariant of environmental conditions [17]. As we will see later, the ability to retain a stiff conformation in the solid state results in many favorable properties, such as reduced interchromophore interactions. Additionally, different functional groups can be positioned with exact distances as designed. Note that the conjugation length of a shape-persistent dendrimer does not necessarily extend through the entire molecule. For example, the commonly used *meta*-phenylene linkage disrupts conjugation to large extent. In addition, nonplanarity among different aromatic rings (due to torsional angle) also reduces coupling of π-electrons. Also, certain chromophores such as pyrene [18] and perylene diimide [19] are known to exhibit nearly the same photophysical properties regardless of their being incorporated into a dendrimer or not. Therefore, it is usually appropriate to treat a shape-persistent dendrimer as an ensemble of many chromophores. This fact simplifies the interpretation of the structure–property relationship, and allows independent tuning of different segments of a dendrimer.

These unique properties of shape-persistent dendrimers have brought about their increasing applications in organic electronics. In this chapter, we present an overview of recent researches in this area, including our own contributions.

9.2 SHAPE-PERSISTENT CONJUGATED DENDRIMERS AS LIGHT-EMITTING MATERIALS

9.2.1 Introduction to OLEDs

OLEDs represent a relatively mature area of organic electronics. Since their discovery [20], OLEDs have attracted enormous attention due to their applications in full color flat-panel displays, solid-state lighting, and as backlights for liquid-crystal displays [21]. Before outlining the general requirements in the molecular design of light-emitting dendrimers, it may be necessary to briefly introduce how an OLED works. In a simplest one-layer OLED device, active materials are sandwiched between two electrodes [22]. Electrons are injected into the lowest unoccupied molecular orbital (LUMO) of the active material at the cathode. At the anode, holes are injected into the material's highest occupied molecular orbital (HOMO). These positive and negative charge carriers then move in opposite directions within the active layer under an applied voltage. When coming across, they combine to generate excitons that emit light with wavelength determined by the energy gap between the HOMO and LUMO. However, in a single material, usually either hole transport or electron transport dominates, resulting in a large waste of the excessive charge and reduced device efficiency. Therefore, to balance the charge transport, one or more layers are commonly incorporated between the cathode and anode (Figure 9.2); each layer is optimized for hole transport, electron transport, or light emitting [23]. The overall efficiency of an OLED device can be described in terms of the external quantum efficiency (EQE), which is the number of photons emitted by the device divided by the number of charge pairs injected. However, other parameters, such as lifetime at working conditions, maximum brightness, and color purity, are equally important in evaluating the performance of an OLED device.

The operation mechanism of an OLED device imposes several requirements on the active materials. First of all, the dendrimer should have high luminescent quantum yield (LQY) in the solid state. This requirement has several implications: first, conjugated units with inherent high LQY should be incorporated into the dendritic scaffold. Typical building blocks include 9,10-substituted anthracene [24] and oligofluorene [25] among others. Second, interchromophore interactions should be

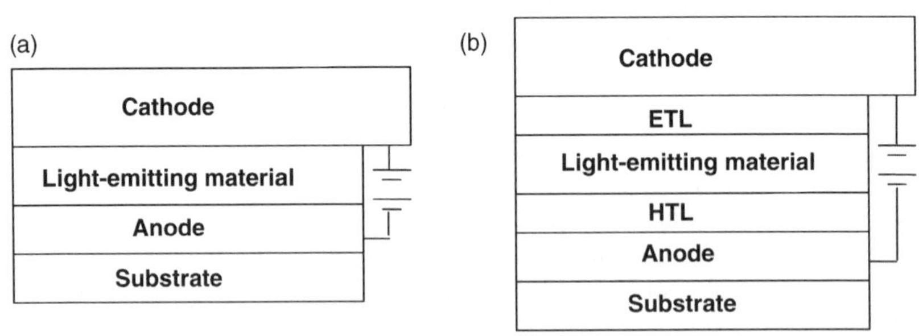

FIGURE 9.2 Illustration of the configuration of an OLED device: (a) a single layer device. (b) A multilayer device with electron- and hole-transporting layers separated from the light-emitting layer.

minimized in the solid state. Such interactions, whether in the ground state or excited state, generally decrease the LQY and affect the color purity of the emitted light [26]. Consequently, it is desirable to introduce steric hindrance to prevent such interactions. Examples include 9,9-substituted fluorene unit [27] and *t*-butyl group decorated periphery of a conjugated dendrimer [28]. It is noteworthy that the branching architecture of dendrimers has the inherent ability to prevent aggregation in the solid state [29]. In the case of shape-persistent dendrimer, the rigidity of the scaffold renders both intra- and interdendrimer chromophore–chromophore interactions difficult. These considerations explain why shape-persistent dendrimers are especially promising for construction of light-emitting devices.

The second requirement for an OLED material is an adequate charge transport ability [30]. Yet, it is a dilemma that decreased intermolecular interaction is always accompanied by decreased charge transport ability. Fortunately for an OLED device the charge carrier mobility does not need to be as high as that of an OFET device. Rather, it is more important to achieve a balance in the transport of both types of charges [23]. The advantage of a dendrimer scaffold is that one can introduce chemical moieties that promote electron/hole transport into different positions of the dendrimer, without interfering with the emission property.

The final remark concerns the synthesis of dendrimers. For commercialization, cheap starting materials, high-yield, efficient, scalable reactions, and easy purification procedures are desired. For use in OLED, a dendrimer does not need to be of a high generation as long as all functionalities are present. As we will see, most dendrimer-based light-emitting diodes (DLEDs) discussed herein are made of dendrimers with a generation number lower than three.

9.2.2 Dendrimers as Light-Emitting Materials in OLED Devices

9.2.2.1 Phenylacetylene Dendrimers In 1996, Moore and coworkers took the first step in applying shape-persistent dendrimers as the active material in an OLED [31]. By introducing a highly fluorescent anthracene core to their phenylacetylene (PA) dendrimer, they constructed up to third-generation blue-emitting dendrimers (Figure 9.3). To improve the hole-transporting ability of these dendrimers, triphenylamine (TPA) units were appended to the periphery. It was expected that wherever excitons were generated within the PA dendron, they could be transferred to the blue-emitting substituted anthracene core. Indeed, electroluminescence (EL) was observed for devices using **1** and **2**, as well as dendrimers of lower generations. However, the EL and photoluminescence (PL) spectra in thin film were significantly different from those in dilute solution, that is, featureless and red-shifted, indicating that the PA dendritic scaffold was unable to completely prevent aggregation in the solid state.

Two additional observations in this report are worth mentioning. First, compared to the parent dendrimers, dendrimers with peripheral TPA units gave increased current densities and therefore much lower turn-on voltages, due to more efficient hole injection. Second, higher generation dendrimers generally gave better device performance, because of their better resistance to crystallization.

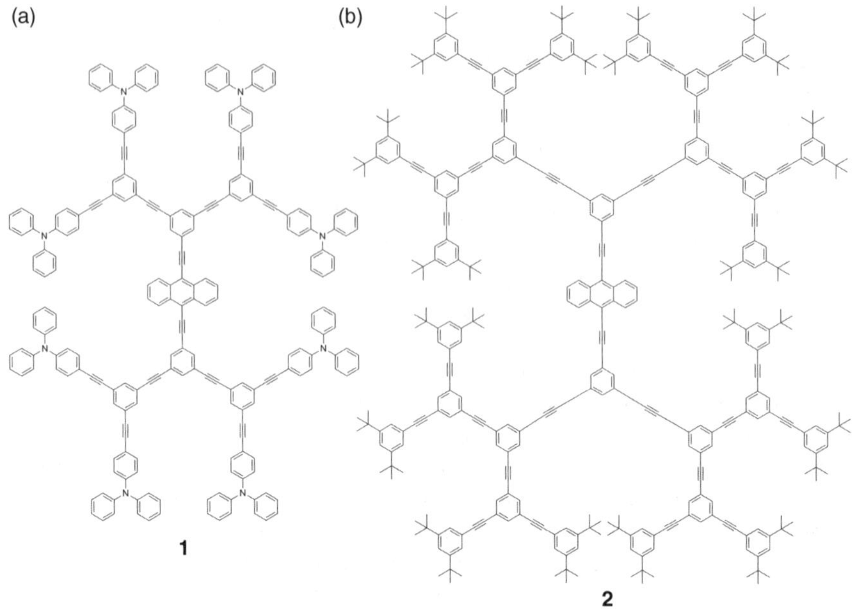

FIGURE 9.3 Chemical structures of the first series of DLED materials. (a) Second-generation dendrimer with peripheral triphenylamine groups. (b) Third-generation PA dendrimer with an anthracene core [31]. Dendrimers of lower generations are not shown.

Further reports on PA dendrimers are rare, and device performances are not particularly remarkable [32]. One possible reason is the strong tendency of the PA units to flatten in the solid state, causing aggregation and excimer formation [31].

9.2.2.2 E-Stilbenyl-Based Dendrimers An important development in DLED is the introduction of *E*-stilbenyl moieties into the dendrimer scaffold by Burn and coworkers [33]. As this moiety is nearly planar, the entire dendrimer structure remains relatively flat only until significant crowding takes place at higher generations. The double bond was constructed usually through Wittig–Horner reaction; alternative approaches included palladium-catalyzed Heck reaction [34] and Ramberg–Backlund reaction [35].

In the initial report [33], up to third-generation *E*-stilbenyl dendrimers with a distyrylbenzene core were synthesized to study the generation effect (Figure 9.4). Due to the *meta* linkage in the molecular design, the absorption and emission spectra of the dendrimers were similar to those of the central distyrylbenzene core. On going from solution to thin film, a red-shifted emission maximum was also observed, but to a much smaller extent than for PA dendrimer. The devices made with **3–5** showed blue EL, with EQE of 0.01% for **3**, 0.07–0.09% for **4**, and 0.03% for **5**, respectively. The device stability increased from the first to the third generation; meanwhile, the third-generation dendrimer had the most similar PL and EL spectrum. The authors thus nicely demonstrated that dendrimer generation could serve as a new tool to control intermolecular interactions in DLED.

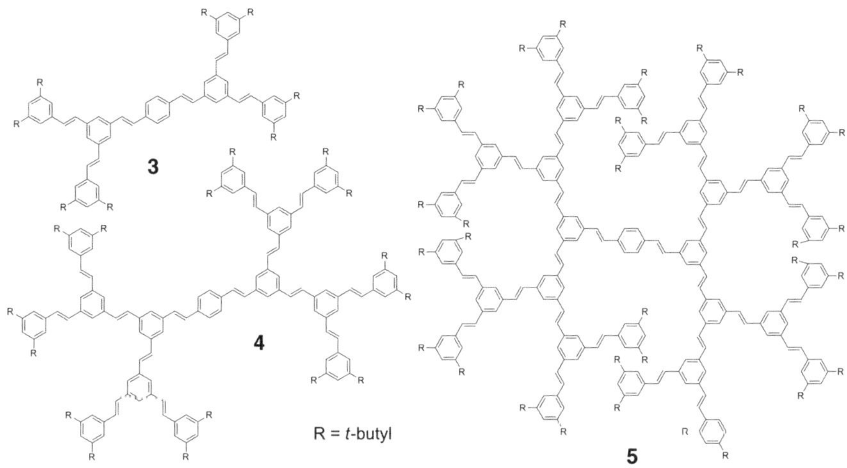

FIGURE 9.4 Chemical structures of *E*-stilbenyl dendrimers **3–5** (from the first to the third generation) [33].

By changing one of the reaction substrates in the final step of the synthesis, *E*-stilbenyl dendrimers with different emission colors could be facilely obtained [34]. For example, introducing a porphyrin core resulted in a red DLED. Likewise, a distyrylanthracene core gave a yellow–green DLED. In other words, the processability and luminescent property were independently provided by the dendron and the core, respectively, demonstrating the advantage of the dendritic architecture.

Generally, dendrimers constructed from *E*-stilbenyl units are hole-dominated materials. Therefore, it is reasonable to expect that introduction of electron-transporting moieties into the dendrimer would balance the charge transport. Following this line of reasoning, Kim et al. have designed and synthesized triazine-functionalized dendrimers [36] such as **6** (Figure 9.5a). However, two-layer devices (ITO/PVK/dendrimer/Al:Li) using **6** as the active material showed only modest EQEs (0.03–0.5%). Another interesting effort was made by Jenekhe and coworkers (Figure 9.5b) [37]. They incorporated diphenylquinoline units, which have been extensively studied as promising electron acceptors [38], into the *E*-stilbenyl dendrimer. Thereby, the dendrimer was converted to an *n*-type material. Unlike previous examples, the chromophores responsible for the emission color are now at the dendrimer surface. But such a design did not seem to work well for active material in DLED. Intimate contact among diphenylquinoline units was extensive in the solid state: the emission color changed from blue to yellow on going from solution to thin film, and the LQY dropped significantly. Consequently, DLED devices using these *n*-type dendrimers alone showed a low EQE (about 0.01%). Nevertheless, these dendrimers can be used as excellent electron-transporting layers. To demonstrate this point, Jenekhe's group fabricated two-layer diodes (ITO/PEDOT/MEH-PPV/dendrimer/Al) by sequential spin coating of MEH-PPV layer and the dendrimer layer. The best performance of thus-obtained devices had a maximum EQE of 5.0%, a power efficiency of 1.3 lm/W, a device efficiency of 1.5 cd/A at 6.5 V, and a brightness level of 240 cd/m^2 (for dendrimer **7**). All the parameters are much better than those from

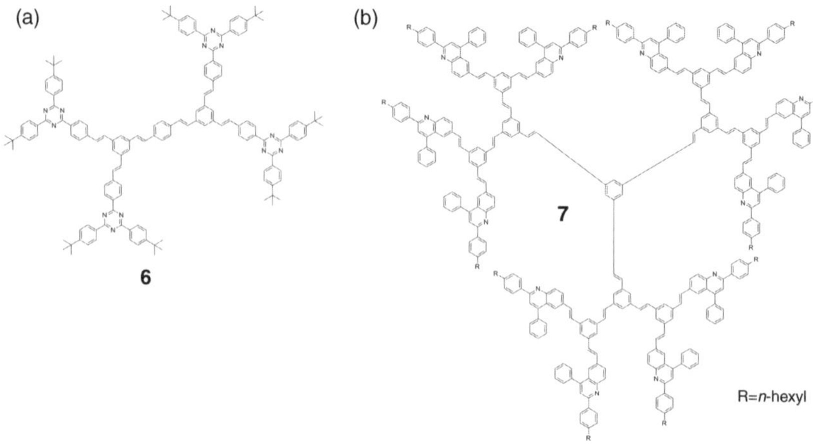

FIGURE 9.5 *E*-stilbenyl dendrimers with electron-transporting units at the periphery. (a) Triazine functionalized *E*-stilbenyl dendrimer [36]. (b) *n*-Type conjugated dendrimer with diphenylquinoline units at the periphery [37].

devices without the electron-transporting dendrimer layer. This work opens a new application avenue for the shape-persistent conjugated dendrimers.

9.2.2.3 Polyphenylene Dendrimers Only a few reports using single-bond-linked chromophores for DLEDs are present in the literature. As a chromophore, polyphenylene (PP) is not a good candidate due to the large torsional angle between benzene rings. But this torsional angle in turn makes PP dendrons good solubilizing groups. In this regard, Cao and coworkers have reported a second-generation dendrimer with an anthracene core as an excellent blue-emitting material (Figure 9.6a) [39]. The dendrimer showed good solubility in various solvents due to the 1,3,5-phenylene-based dendron. In addition, thin films of **8** showed nearly identical PL spectrum with that from solution. This is in great contrast to the PA dendrimer with the same core discussed above. The reason is twofold. First, the branched structure of the 1,3,5-phenylene dendron effectively prevented excimer formation in solid state. Second, the 2-ethylhexyloxy groups at the periphery are much better solubilizing group than the *t*-butyl group. The EL device of **8** exhibited a high EQE of 0.82% at 170 cd/m^2. After inserting a PVK layer between the hole injection layer and emitting layer, the EQE was further improved to 1.05% at 184 cd/m^2.

The syntheses of most of the dendrimers discussed so far relies on metal-catalyzed coupling reactions. A unique and highly efficient synthetic protocol was developed by Müllen and coworkers [40]. Interested readers may refer to a different chapter from this group in this same volume. In a recent work, they introduced triphenylene unit into the dendrimer scaffold to produce pure blue-emitting materials [41]. A key difference between this material and previous examples is that the main chromophores were not confined at the core; rather, they were distributed throughout the entire dendrimer scaffold. Thereby, the ratio of the active component (triphenylene) to

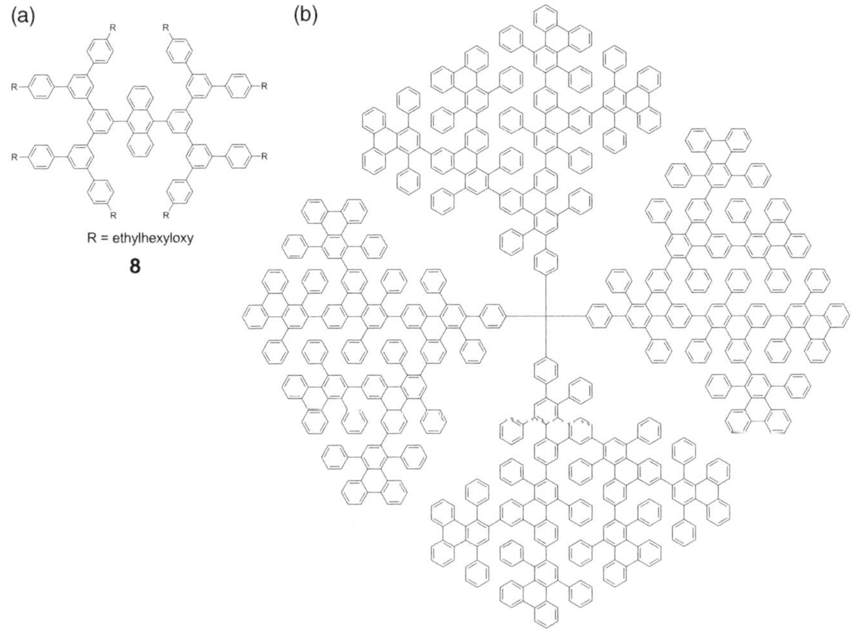

8

9

FIGURE 9.6 Chemical structures of (a) polyphenylene dendrimer with an anthracene core [39]. (b) Polytriphenylene dendrimer [41].

inactive materials (PP backbone) was greatly enhanced. The highly congested backbone results in large torsional angle and thus electronic decoupling among triphenylene units. Therefore, the PL of thin film of the dendrimer showed only a small red-shift, and retained a high LQY. However, DLED device made from this series of dendrimers only showed moderate performance, which was partly explained by the mismatch of HOMO energy level at the interface of the electrode and the active material.

9.2.2.4 Other Dendrimers Yamamoto and coworkers have reported DLED devices based on their phenylazomethine dendrimers [42]. Both symmetric and asymmetric dendrimers were tested, and the influence of incorporation of metal ions was discussed. Other examples include phenothiazine dendrimers [43] and oligofluorene dendrimers [44].

9.2.2.5 Truxene-Based Dendrimers A common challenge in synthesizing shape-persistent conjugated dendrimers is their low solubility, which may cause incomplete transformation in coupling reactions and difficulty in purification. For divergent synthetic strategy, high-generation dendrimers are especially hard to obtain due to the complexity in purification. For convergent synthetic strategy, solubilizing groups have to be incorporated at the periphery of the dendron, but they do so at the expense of blocking the position for further functionalization.

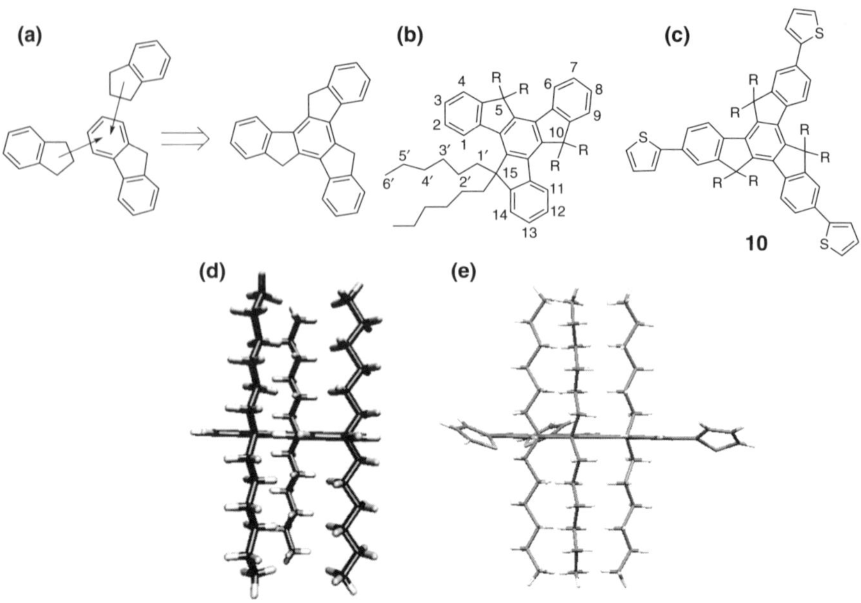

FIGURE 9.7 (a) Illustration of a truxene unit as being annulated by three fluorene moieties. (b) Numbering of the truxene unit. (c) Chemical structure of model compound **10**. (d) Side view of an energy-minimized conformation of alkylated truxene unit. Reproduced by permission of the American Chemical Society [47]. (e) Single crystal structure of the model compound **10**. Reproduced by permission of the American Chemical Society [48].

To circumvent this problem, our group chose 10,15-dihydro-5H-diindeno-[1, 2-*a*:1′,2′-*c*]fluorene, known as truxene, as a building block for π-conjugated dendrimers. Conceptually, the truxene molecule can be seen as three fluorene moieties annulated together by sharing a central benzene ring. Previously, this fused aromatic structure has been mainly used for the synthesis of liquid crystals [45], bowl-shaped polycyclic aromatic hydrocarbons, and C_{60} [46]. To overcome the low solubility of bare truxene, we introduced six alkyl chains to the 5, 10, and 15 positions of the truxene unit. Energy minimization [47] and single crystal structure of model compound **10** (Figure 9.7) [48] showed that these alkyl chains stood upright, perpendicular to the central plane. The presence of the alkyl chains can thus effectively prevent π–π stacking of the truxene plane, which is the main driving force for aggregation [49].

Alkylated truxene exhibits several features as excellent building blocks for DLED materials. First, derivatives of truxene unit usually have LQY approaching unity in solution. Second, the presence of the six alkyl chains greatly improves the solubility of the building block, which in turn makes the dendrimer soluble in many common organic solvents. Also, intermediates in the synthesis can be readily purified by flash chromatography. An important consequence is that no solubilizing group has to be introduced to the periphery of the dendron, so the dendrimer could be further functionalized more easily and completely. Third, the ability of the perpendicular alkyl chains to prevent interchromophore interaction, in combination with the

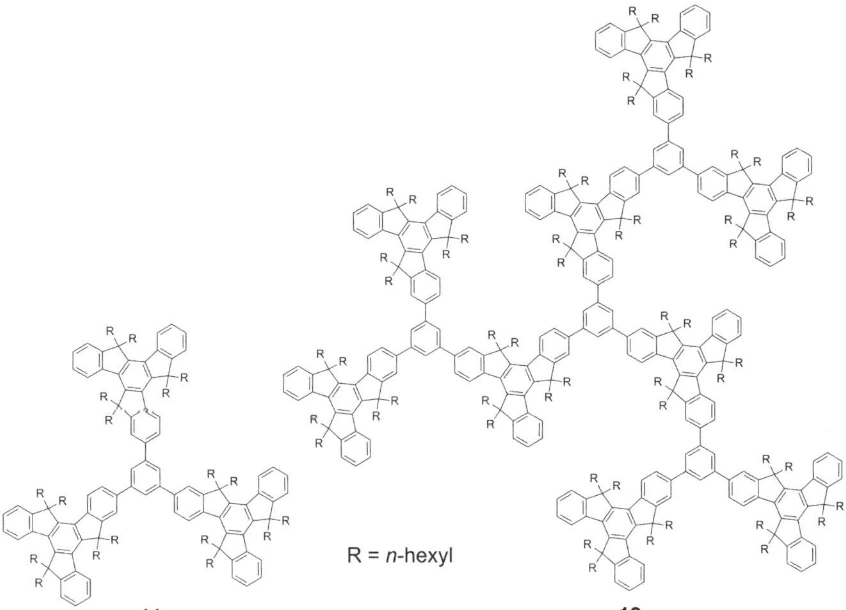

11 **12**

FIGURE 9.8 Chemical structures of the first series of truxene-based dendrimers [50].

dendrimers' architecture, makes truxene-based dendrimer a strong emitter even in the solid state. Finally, derivation of truxene unit is rather simple. The 2, 7, and 12 positions of the truxene can be easily halogenated, allowing various C–C coupling reactions to take place.

In 2003, we reported the synthesis of the first series of dendrimers incorporating truxene units [50], as shown in Figure 9.8. The syntheses of dendrimers **11** (G0) and **12** (G1) were based on repetitive Friedel Crafts acetylation reaction and $SiCl_4$-catalyzed cyclotrimerization reaction. The absorption and emission maxima of the two dendrimers were not much different from the alkylated truxene monomer. Two facts account for this observation. First, as computation study has shown, electronic communication among branches of the truxene is minimum, similar to a 1,3,5-substituted benzene. Therefore, the contribution of a truxene unit to the conjugation length is only slightly larger than a fluorene unit. Second, there are significant torsion angles between the benzene ring and truxene unit due to steric repulsion. Hence, the conjugation length of dendrimer **12** is only slightly longer than a fluorene unit, despite the presence of 40 benzene rings.

In thin films, dendrimers **11** and **12** showed structured absorption and emission bands, only slightly red-shifted compared with those in solution. Hence, the effect of the alkyl chains and the dendritic architecture is evident. With these results in hand, our first DLED devices [47] were fabricated with the configuration of ITO/PEDOT/dendrimer/Ba/Al. The EQE was measured to be 0.16% at 27 V for **11** and 0.16% at 13 V for **12**, respectively. However, the EL emissions were significantly red-shifted in comparison with the PL spectra in solid state, which was explained by the formation of ketonic defects during device operation, similar to polyfluorene-based devices.

Encouraged by these initial results, we continued our search for better truxene-based DLED materials. We then explored an interesting molecular design, that is, dendrimers constructed solely by truxene units. Figure 9.9 shows the chemical structures of all-truxene dendrimers [51]. Compared with the previous series, their absorption peaks red-shifted considerably to 350 nm, indicative of more effective conjugation between truxene moieties. The emission spectra of all-truxene dendrimer showed three peaks at 386, 405, and 430 nm. Again, the generation number does not have a significant influence on their photophysical properties [52].

Smooth thin films of **13** and **14** showed one main peak around 406 nm with high LQYs (60% for **13** and 62% for **14**). Subsequently, devices were fabricated using **13** and **14** with the configuration of ITO/PEDOT:PSS/PVK/dendrimer/TPBI/Ba/Al. The turn-on voltages were 6.0 V and 5.8 V for **13** and **14**, respectively. Both devices emitted pure-blue light with 1931 Commission Internationale de l'Eclairage (CIE) coordinates of (0.16, 0.08). Notably, no ketonic defect or excimer emission was observed in these devices, showing great advantage over devices using polyfluorene. The maximum EQEs were as high as 1.93% at 7.0 V for **13** and 1.99% at 6.3 V for **14**. Such a performance was remarkable compared to previous examples.

Our most successful light-emitting materials developed so far are *E*-stilbenyl-truxene dendrimers (see Figure 9.10 for the chemical structure). Reported in 2006 [53], the synthesis of such dendrimers was based on repetitive Suzuki and Horner–Wadsworth–Emmons reactions. In the absorption spectra in solution (Figure 9.11a), compounds **16** and **17** displayed two absorption peaks at 330 and 378 nm, corresponding to phenyl-functionalized truxene at the periphery and the central *E*-stilbenyl-truxene unit, respectively. Two strong emission bands in the visible region were observed for both **16** and **17**, one at 424 nm and the other at 448 nm. The absorption and emission maxima of **16** and **17** showed certain degree of red-shift to those of model compound **18**.

The absorption behaviors of **16** and **17** in thin films (Figure 9.11b) were quite similar to those in solution. This fact reflects that no obvious interchain and/or intermolecular aggregation existed in the ground state, probably due to the presence of bulky branches and a large number of alkyl groups. The emission spectra of **16** and **17** in the thin films showed slight red shift (about 10–20 nm) in comparison with those in solution. Annealing the films at different temperatures in nitrogen atmosphere did not significantly change its absorption and emission spectra, but did have significant impact on device performances, as will be mentioned later.

Dendrimer **16** was soluble in many common organic solvents, which makes it suitable for many solution processes such as spin-coating, inkjet-printing, and dip-coating.[1] Devices with the configuration of ITO/PEDOT:PSS/PVK/**16**/Ba/Al were fabricated [54]. Because **16** has a HOMO level at 5.8 eV, a thin layer of PVK was introduced on top of PEDOT:PSS as the hole-injection layer. For an annealed device,

[1] Recently, we have optimized the synthesis protocol and were able to obtain **16** on a large scale, setting the basis for its commercialization.

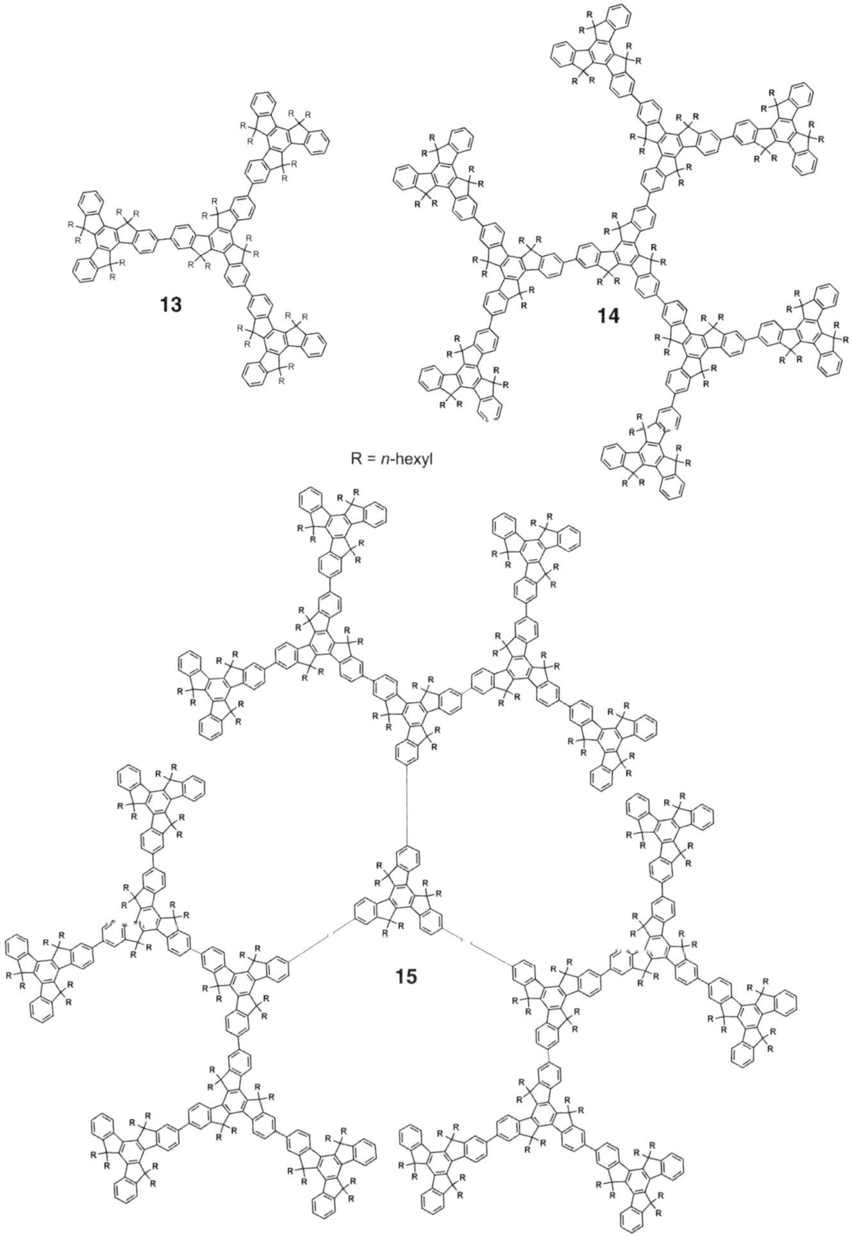

FIGURE 9.9 Chemical structures of dendrimers solely constructed from truxene units [51].

the CIE coordinate of the emitted light (0.155, 0.086) was quite close to saturated blue with the CIE coordinates of (0.14, 0.08) specified in the National Television System Committee standard [55]. More importantly, as the operation voltage increased, the EL spectra remained essentially unchanged. The *J-V-L* characteristics and the

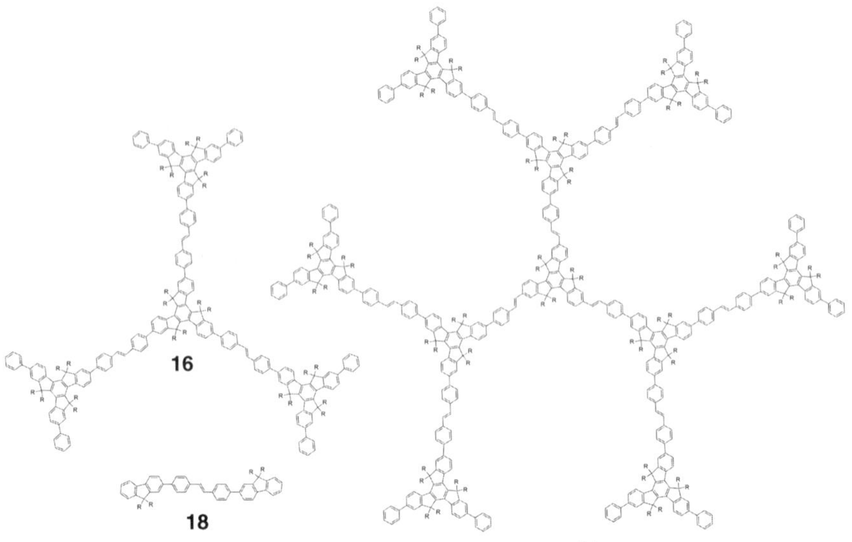

FIGURE 9.10 Chemical structures of *E*-stilbenyl-truxene dendrimers **16** and **17** and model compound **18** (FSF) [53].

dependence of the luminance efficiency on the current density for annealed device of **16** are displayed in Figure 9.12a and 9.12b. The maximum luminance efficiency of the annealed device reached 5.3 cd/A, and the EQE reached 6.6% at a current density of 10.7 mA/cm^2 with the luminance of 550 cd/m^2. To our best knowledge, **16** exhibited the best performance among blue-emitting dendrimer; its performance was also among the best of all pure-blue fluorescent materials. Further experiments confirmed that the enhancement brought about by thermal annealing was due to the improvement of the charge injection and charge transport.

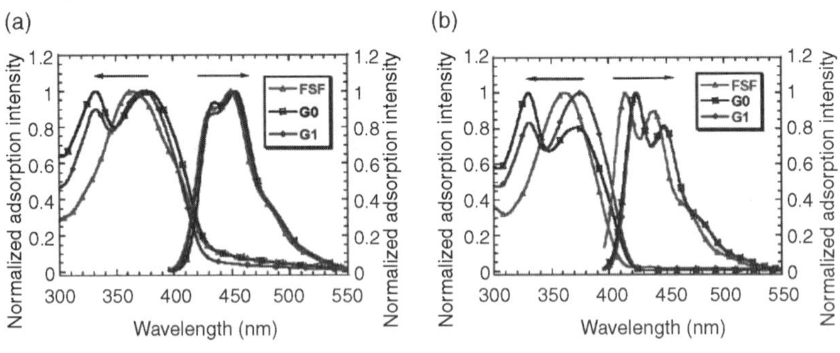

FIGURE 9.11 Absorption and emission spectra of *E*-stilbenyl-truxene dendrimers and model compound in (a) dilute solution; (b) thin film. Reproduced by permission of the American Chemical Society [53].

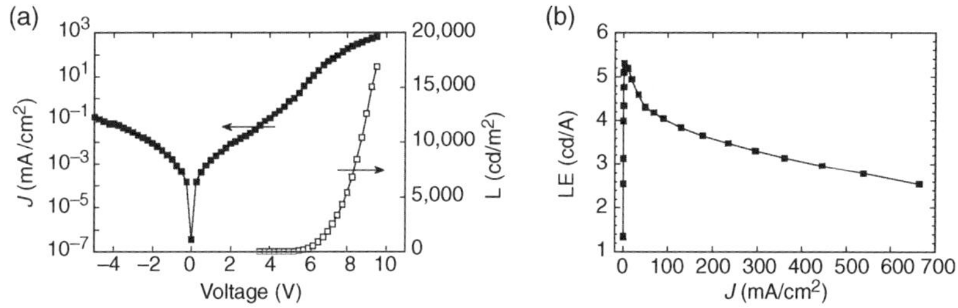

FIGURE 9.12 (a) *J-V-L* characteristics of an annealed device of **16**. (b) Dependence of the luminance efficiency on the current density for an annealed device of **16** [54].

With such high efficiency at high current density and excellent color stability, dendrimer **16** is suitable not only for the application in full color displays but also for white OLEDs in solid-state lighting. To demonstrate these applications, we chose Ir $(mppy)_3$ and $Ir(piq)_2acac$ as green and red phosphorescent emitters. From the energy level diagram (Figure 9.13b), we can see that **16** can transfer the exciton energy to the phosphorescent molecules, making it a good host material. By doping **16** with Ir $(mppy)_3$ or $Ir(piq)_2acac$, and using the same device structure and annealing procedure, highly efficient green and red OLEDs could be successfully obtained. As a step further, by codoping $Ir(mppy)_3$ and $Ir(piq)_2acac$ into **16** as the single host matrix, white OLEDs with three emission peaks of blue ($\lambda_{max} = 430$ nm), green ($\lambda_{max} = 510$ nm), and red ($\lambda_{max} = 620$ nm) color, with CIE coordinates of (0.312, 0.332), were successfully achieved. At the luminance for practical applications (1000 cd/m^2), the forward viewing power efficiency and luminance efficiency were as high as 6.9 lm/W and 16.6 cd/A, respectively. In addition, the CIE coordinates changed only slightly even if the brightness varies by three orders of magnitude, showing very stable white emission. The combination of high efficiency, stable white emission, and a high color rendering index value (a number used to describe the quality of white light), makes **16** based white OLED an ideal candidate for solid-state lighting (Figure 9.14) [54].

In addition to conventional dendrimer structures, our group has also designed and synthesized a series of novel 3-D dendritic structures that possess intriguing properties (Figure 9.15) [56]. They are all excellent blue emitters in both solution and the solid state. A notable feature of these molecules is the stability in the condensed form. Even after exposure to air at 200°C for several hours, the color integrity of the thin film

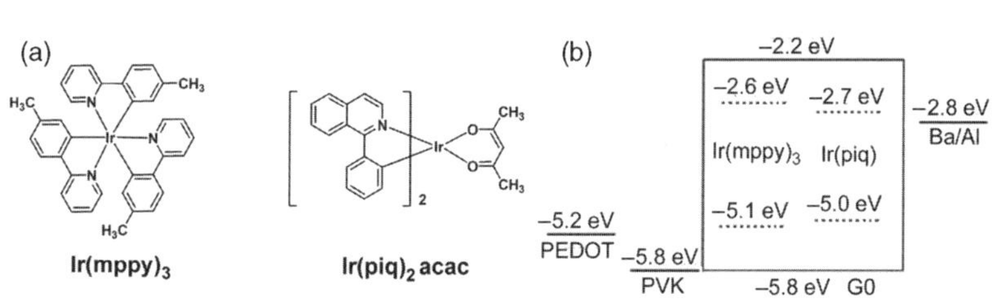

FIGURE 9.13 (a) Chemical structures of phosphorescent dyes for white OLED based on **16**. (b) Energy levels of all materials in the device.

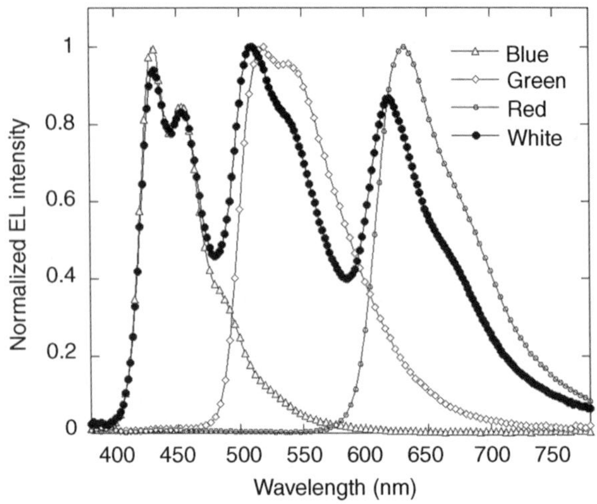

FIGURE 9.14 EL spectra of **16**-based blue, green, red, and white OLEDs [54].

remained unaltered. No long-wavelength emission was observed, unlike other fluorene-based blue-emitting materials [57]. The 3-D architecture proved to be a powerful design to prevent interchromophore interaction in the solid state and to improve the film's resistance to degradation or crystallization [58]. The maximum EQE for devices with a configuration of ITO/PEDOT/PVK/**19**/TPBI/Ba/Al was as high as 2.9% for **19a**, 2.4% for **19b**, 2.0% for **19c**, and 1.4% for **19d**.

9.2.2.6 Shape-Persistent Phosphorescent Dendrimers In OLED devices based on fluorescent materials, at most 25% of all excitons can be used to emit light for small molecules (for conjugated polymers the number might be higher) [59]. The reason is

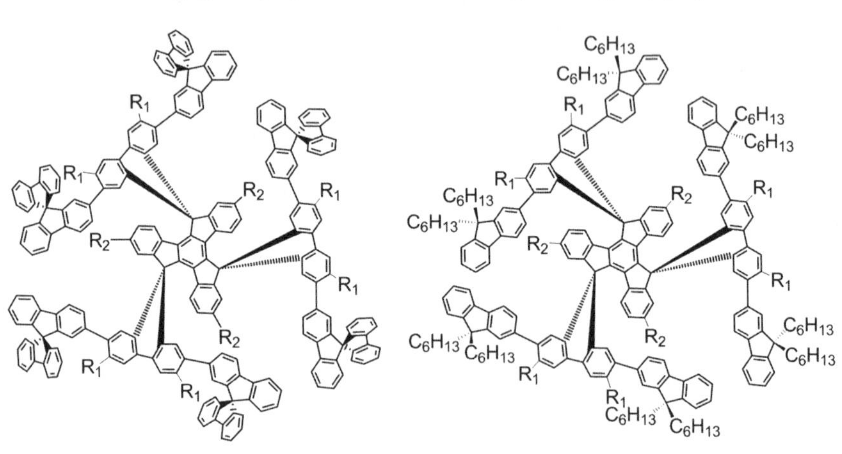

19a: $R_1 = OCH_3$, $R_2 = H$;
19b: $R_1 = H$, $R_2 = OCH_3$.

19c: $R_1 = OCH_3$, $R_2 = H$;
19d: $R_1 = H$, $R_2 = OCH_3$.

FIGURE 9.15 Chemical structures of two series of 3-D dendrimers.

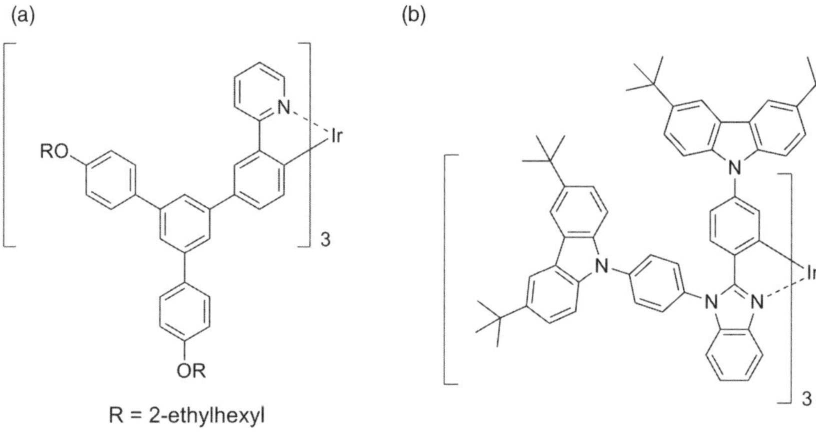

20 21

FIGURE 9.16 Examples of shape-persistent phosphorescent materials. (a) 1,3,5-Phenelyne-based green phosphor **20** [60]. (b) Carbazole-based phosphor **21** [61].

that during charge injection, the spin of the charge cannot be controlled. When electron and hole combine in a molecule, statistically the ratio of singlet to triplet excitons is 1:3. Because fluorescent materials can only utilize the singlet excitons, at least 75% of the excitons are wasted. This is why much interest has been paid to phosphorescent materials that utilize triplet excitons as the emitter [23].

By synthesizing conjugated, dendritic ligands and coordinating them with suitable metal ions such as Ir, shape-persistent phosphorescent dendrimers can be constructed. For example, Figure 9.16a shows the chemical structure of a phosphorescent DLED material, reported by Burn and coworkers [60]. This green emissive dendrimer, when blended in a host and used in a bilayer device, achieved a brightness of 400 cd/m^2 at 4.5 V, with a corresponding EQE of 16% and a power efficiency of 40 lm/W. This is very close to the theoretical limit of efficiency of 20% based on an outcoupling of a fifth of the light generated in the device.

Similar to fluorescent dendrimers discussed earlier, dendrons with electron- or hole-transporting ability could be attached to the periphery of the ligand, to enhance charge transport. For example, Wang and coworkers reported a series of Ir-containing dendrimers with structures shown in Figure 9.16b [61]. Undoped devices using a neat emissive dendrimer film of **21** showed excellent performance with a peak luminous efficiency of 45.7 cd/A (13.4%), high luminance and high current density (180,000 cd/m^2, 360 mA/cm^2 at 12 V). More examples could be found in a recent review by Burn et al. [62]

9.2.3 Outlook

The development of DLEDs has matured to the stage where they are widely recognized as the third class of materials in OLEDs. Device efficiencies have improved dramatically since their first demonstration. As a good example, *E*-stilbenyl-truxene dendrimer has shown device performance comparable to those of some commercialized OLEDs. To realize the commercial potential of DLEDs, several issues should be addressed in future studies. First, the synthesis should be more concise and cost-effective. Second, the lifetime of devices needs to be improved. Solution-processed devices, whether polymer or dendrimer-based, tend to degrade under working conditions. Complicated problems related to interfaces and thin film structures hamper a clear elucidation of the origin of such degradation. Nevertheless, the problem of lifetime must be solved before DLEDs can have significant impact on the market. Third, to better utilize the solution processability of these dendrimers, one would expect to also improve other steps in their fabrication process. For example, silver gels could be readily used to pattern Ag electrodes [63]. More efforts should be made to optimize the device structure and fabrication process, to take full advantage of the dendrimer's properties.

9.3 SHAPE-PERSISTENT CONJUGATED DENDRIMERS AS LIGHT-HARVESTING MATERIALS

9.3.1 Introduction

9.3.1.1 Introduction to the Concept of Light-Harvesting Most life relies directly or indirectly on solar energy. The yearly incidence of solar energy on the earth's surface is many orders of magnitude more than that used by the planet's population as power [64]. Therefore, harnessing the solar energy as a clean and renewable substitute for fossil fuels has been a hot topic for a long time [65]. It is reasonable to seek some biomimetic principles from the natural photosynthetic process, in which the light energy is converted and stored as chemical energy with a high efficiency [66]. The atomic detail of the natural photosynthesis complex has been elucidated during the past 20 years [67]. Hence, scientists are now in a better position to learn the design concept of this efficient solar energy harvesting machine, a schematic representation of which is given in Figure 9.17 [68].

At the center of the photosynthetic unit (PU) is a pair of *chlorophyll* molecules, called reaction center (RC). Immediately surrounding the RC is the LH1 complex, which is composed of a ring-shaped assembly of chlorophyll and carotenoid moieties in a protein matrix. Farther away from the RC are LH2 and LH3, which share with LH1 a similar design [69]. One of the most important functions of these light-harvesting units is to collect solar energy at spectra not covered by the RC, and then transfer the absorbed energy efficiently to the RC, where charge separation and subsequent electrochemical reaction occurs [70]. Also, the mechanism of energy transfer has been studied theoretically in detail [71,72]. Remarkably, it has been shown that excitons that form on any of the chlorophylls within the LH system finally go to RC, much like an antenna.

A dendrimer shares some architectural similarities with the PU. By positioning suitable chromophores at suitable sites, it is possible to design and synthesize

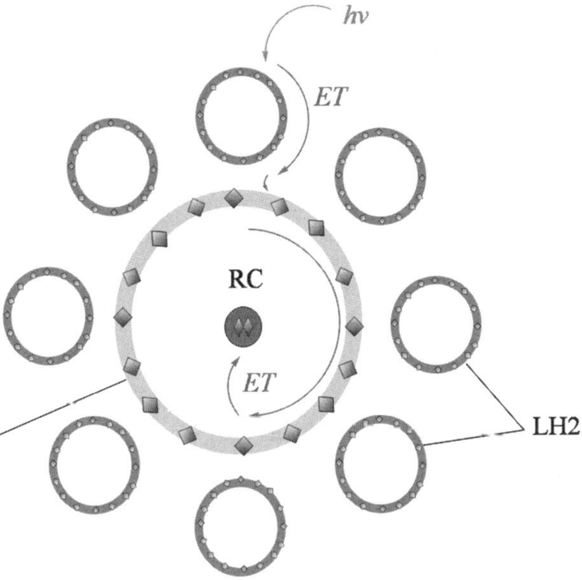

FIGURE 9.17 Schematic representation of a photosynthetic unit [68].

dendrimers that mimic the light-harvesting ability of the PU [73]. The globular shape of a dendrimer provides a large surface that can be functionalized with light-capturing chromophores, which then transferred the absorbed energy toward the core. Electro-active unit could then be placed at the core to finish the final step of the artificial photosynthesis. However, the application of light-harvesting (LH) dendrimers is not limited to energy conversion. As mentioned in the last section, LH dendrimer with highly fluorescent cores could be used in OLED devices [31]. In addition, LH dendrimers could also serve as highly sensitive molecular sensors due to its inherent signal amplification ability [74]. Other potential applications include nonlinear optics [75].

9.3.1.2 Why Use Shape-Persistent Conjugated Dendrimers as LH Materials

Considerable amount of work has been reported on light-harvesting dendrimers using aliphatic or so-called "flexible" linkage, especially on the Fréchet-type dendrons [68,76–78]. However, shape-persistent LH dendrimers have some unique properties in comparison with flexible ones.

First, a rigid scaffold helps to pin down the distance and relative orientation of the transition dipoles of donor and acceptor. This makes theoretical modeling and precise design possible. In nature, the exact allocation of chromophores relies on the well-defined 3-D structure of protein matrix [70]. Here, the rigid dendrimer scaffold is expected to function similarly. Second, in shape-persistent LH dendrimers, both Dexter [79] and Förster [80] energy transfer (ET) mechanism might operate, depending on the exact molecular structure. In flexible dendrimers, disrupted conjugated system precludes the Dexter ET mechanism, which requires orbital coupling. Finally, in most conjugated LH dendrimers, the conjugation length is sensitive to the dendrimer scaffold. This allows easy tuning of the photophysical property of the chromophore, but complicates the interpretation of the spectra, on the other hand.

Below, we review the progress in this field of research with contributions from several groups, including our own. We do not cover transition metal containing LH dendrimers [81].

9.3.2 Examples of Shape-Persistent Conjugated Dendrimer as LH Materials

In 1993, Xu and Moore reported the first shape-persistent conjugated dendrimer, using a PA scaffold [17]. The synthetic protocol was based primarily on Sonogashira reaction between aryl iodide and terminal alkyne. The fourth-generation dendrimer **22** (Figure 9.18), with a molecular weight of 14 kDa, represented the largest pure hydrocarbon at that time. This paper represents the start of the preparation of large dendrimers that are composed of stiff building blocks and have well-defined conformation.

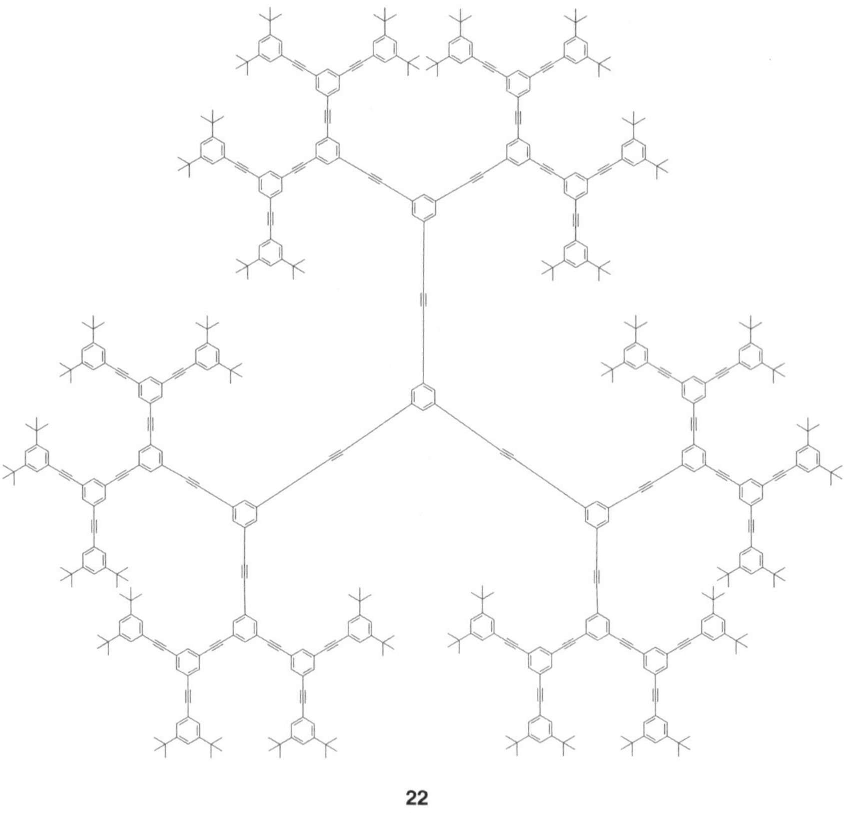

22

FIGURE 9.18 Chemical structure of the fourth-generation PA dendrimer (94 mer) [17].

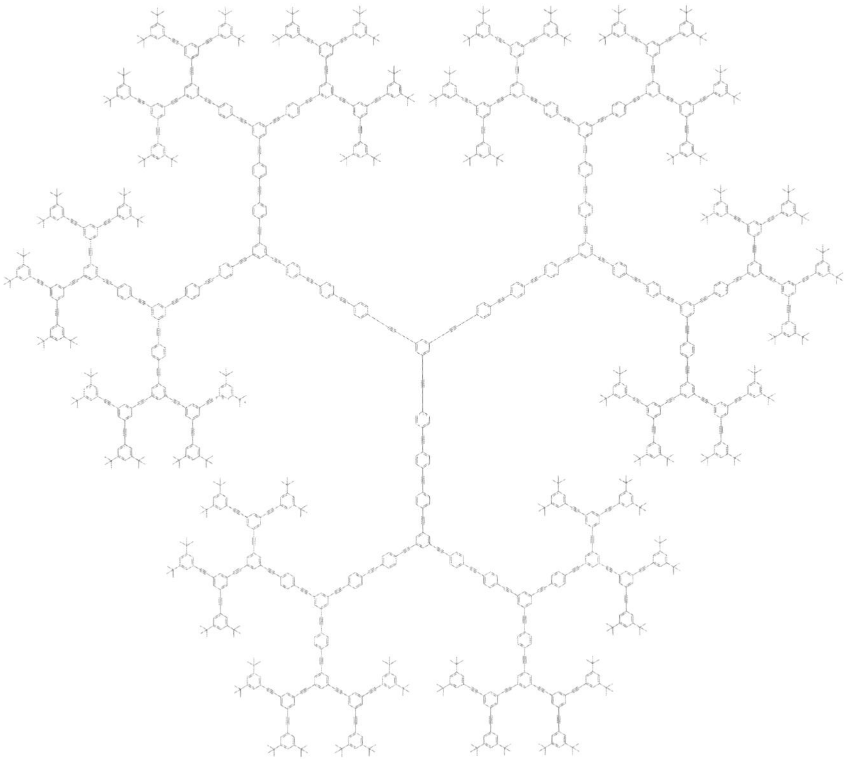

23

FIGURE 9.19 Chemical structure of fourth-generation extended dendrimer reported by Xu and Moore [82].

In a subsequent report, Xu and Moore developed a new series of PA dendrimer with an expanded structure (Figure 9.19) [82]. In these extended dendrimers, the length of the repeating linkage and hence the conjugation length increased linearly from the periphery to the core. The result of this length gradient is multifold. First, in conventional dendrimers with simple geometric progression, the molar mass increases at a rate exceeding the spacial availability [83]. This makes the synthesis of higher generation dendrimer very difficult or practically impossible. By adopting expanded branching structures, such steric congestion is minimized. Second, due to alleviated steric congestion, extended dendrimer takes on a shape completely different from the compact, globular shape of a conventional dendrimer. Third, the branch extension strategy enables large-sized dendrimers to be realized after only a few repetitive cycles. Finally and most importantly, an energy gradient is inherently built into the extended architecture. Its importance in improving ET efficiency will be revealed shortly.

Moore and coworkers first showed that excitons were localized in the small building blocks in these dendrimers, due to the *meta* branching linkage [84]. For

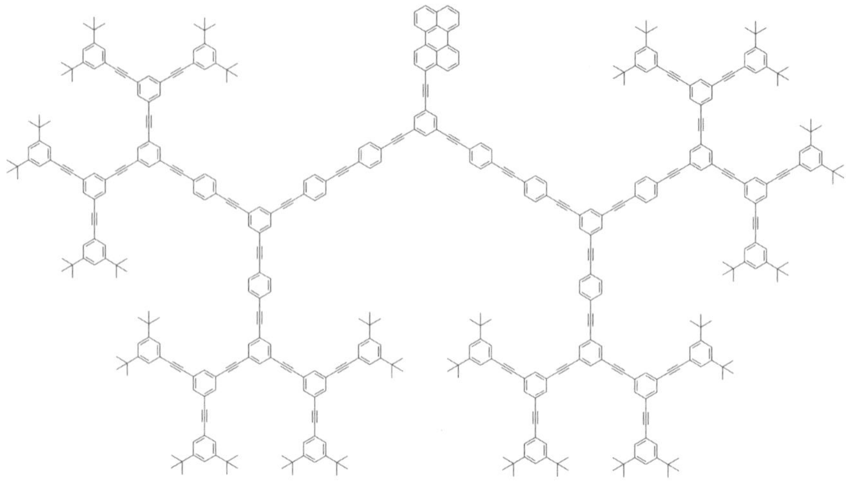

FIGURE 9.20 Chemical structure of an extended dendron **24** with a perylene as an energy "sink" [86].

example, the bandgap of compact dendrimers remains similar to that of the smallest fragment in **22**. However, the excitation energy of the extended structure decreased continuously from the periphery to the core, due to increased conjugation length. This gradient funnels the excitation energy unidirectionally to the central chromophore, much like what happens in the natural LH complex. Note that in the absence of the energy gradient, the exciton migration direction would be random [85].

To study the photophysical process in these dendrimers in detail, Moore's group introduced an energy "sink", for example, a perylene unit, to the PA dendron core (Figure 9.20) [86]. Upon excitation of the PA dendron **24** (310 nm) where perylene has minimum absorption, most of the fluorescence came out of the core, indicating efficient ET. For compact dendrimers, the ET efficiency decreased as the molecular size goes up, due to greater average distance between the donor and the acceptor. In contrast, such a decrease was not observed in extended dendrons. For example, compound **24** shows an ET efficiency of 98%, much higher than that of its compact counterpart with similar molecular size (85%). This difference can be translated to two orders of magnitude increase in the ET rate. Although **24** has better spectral overlap between donor and acceptor, this alone cannot account for the large difference; more likely, the energy gradient in **24** plays a significant role in guiding the excitation energy toward the central perylene. The detail of the ET process is complicated; application of simple Förster mechanism did not give a satisfying result. More interesting photophysical processes such as correlated excimer formation were observed and studied in detail in these dendrimers [87].

The PA dendrimers investigated by Moore's group were geometrically symmetric owing to the *meta* connectivity. By changing the connectivity, Melinger's group developed a series of asymmetrically branched PA dendrimers shown in Figure 9.21 [88]. This change of connectivity has significant impact not only on the overall dendrimer

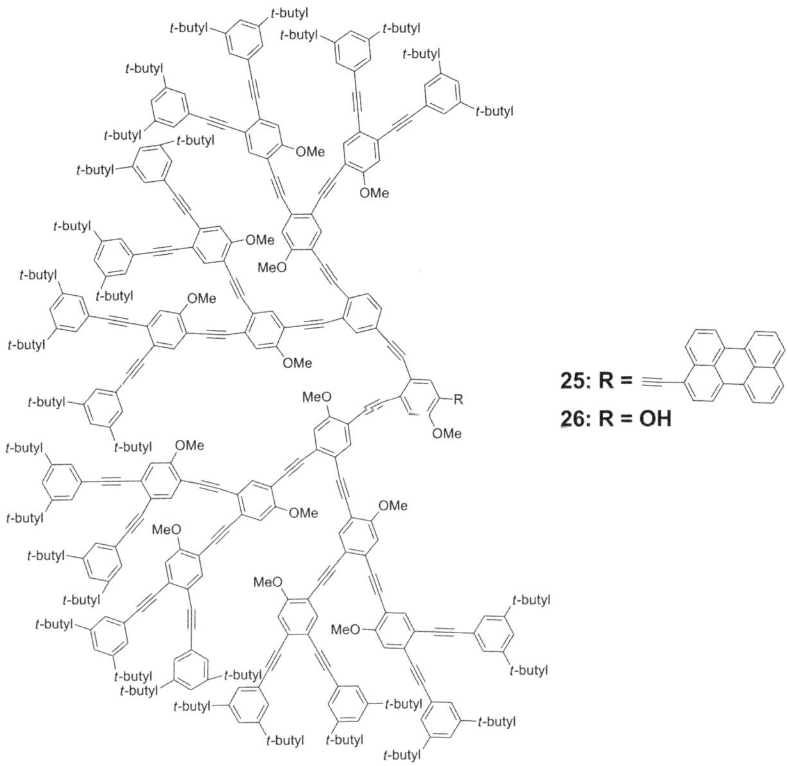

FIGURE 9.21 Chemical structure of asymmetric PA dendrimers with and without a perylene trap [88].

architecture but also on the photophysical properties. The most prominent change is that the assumption of exciton localization is not valid any more; the longest conjugation length increases linearly with generation. For example, the absorption edge of the fourth-generation dendrimer **26** appeared at 453 nm, corresponding to a conjugation length of nine PA units with a "*para* kink." The presence of a large number of chromophores with different conjugation lengths also gave the dendrimer a broad absorption spectrum. Finally, as there are some shortcuts between the peripheral chains and the core, the ET process was also significantly modified.

Efficient ET process was observed in this series of dendrimer with a perylene trap, such as in **27** (with ET efficiency between 85% and 93%), although it was a little slower than in its symmetric counterpart. Ultrafast measurement shows that the ET process from the PA backbone to the perylene traps was completed on a subpicosecond time scale.

The examples presented to this point generally employed phenylene as a building block of the shape-persistent LH dendrimer. Another common choice is thiophene. Oligo- and polythiophenes have been the focus of studies on organic electronic devices (particularly for solar cells and OFETs), due to their excellent optical, redox,

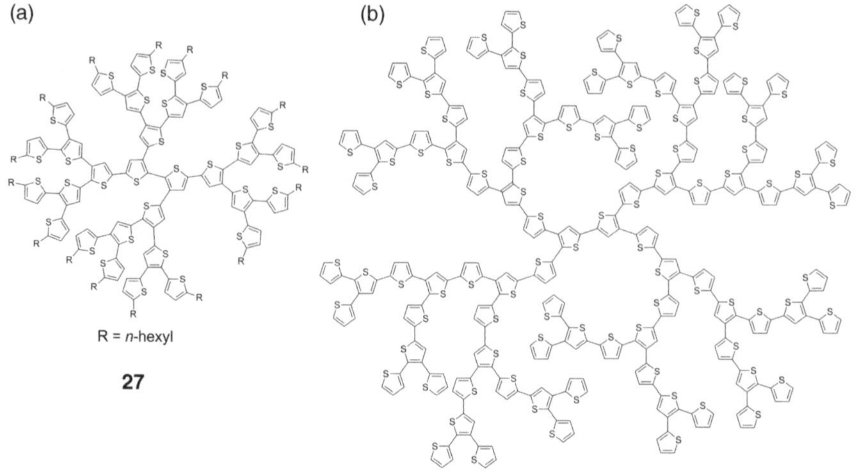

FIGURE 9.22 (a) Third-generation thiophene dendrimer synthesized by Advincula and coworkers [92]. (b) Fourth-generation thiophene dendrimer synthesized by Bäuerle and coworkers. Note their difference in the functional group at the periphery [94].

self-assembly, and transport properties [89]. The easy functionalization of the thiophene unit also makes thiophene-containing molecules especially attractive synthetic targets. Most studies have focused on linear arrangement of thiophenes such as oligo- and polythiophenes. Only recently have branched or crossed thiophene systems been proposed as a third possibility [90,91]. Such unique spatial arrangement and chemical connectivity of thiophene units are expected to bring new properties to this much explored material.

In 2002, Advincula and coworkers reported the synthesis of the first series of thiophene dendrimers (Figure 9.22a) based on Stille coupling reaction [92]. Subsequently, they studied the self-assembly behaviors of the dendrons and dendrimers on solid substrates [93]. In 2007, Bäuerle's group reported another divergent synthetic protocol for thiophene dendrimers, which allowed continuous growth and peripheral functionalization [94]. The new synthetic protocol relied on Suzuki coupling reaction and protection/deprotection of the α-position of the thiophene ring using $SiMe_3$. A fourth-generation dendrimer is shown in Figure 9.22b. Electronically, these thiophene dendrimers resembled the asymmetric PA dendrimer; the conjugation length increased continuously with increasing the generation, because the α–α linkage between thiophene rings allowed effective electron delocalization. Moreover, the dendrimer had a broad absorption spectrum due to the presence of many different conjugation lengths. Unlike their linear counterparts, evident twists between neighboring thiophene rings existed due to steric interaction. As a consequence, these dendrimers were soluble even without peripheral solubilizing group, which is remarkable for a conjugated system with so many aromatic moieties. Regardless of the excitation wavelength, the emission of a dendrimer came from the unit with the longest conjugation length, namely the one with most α–α linked thiophenes. This observation indicates highly efficient ET in such dendrimers. The broad absorption spectrum allows the thiophene dendrimers to be used as active materials in solar cells, as discussed in the next section. This group also demonstrated the application of these thiophene dendrimers as entangles photon sensor materials in a later report [94b].

As can be seen from the PA dendrimer, if the thiophene unit could be separated by an acetylene group, steric crowdedness would be alleviated; and the structure could be made more extended. In this respect, we have initiated a program of synthesizing shape-persistent conjugated LH dendrimers using oligo-(thienylethynylene)s (OTEs) as the building blocks [95]. OTE exhibited several features as ideal branches in LH dendrimer [96,97]: (1) easily tunable lengths through facile chemical synthesis approaches; (2) linear and rigid structures with minimum conformational flexibility; (3) tunable absorption and emission behaviors. As an extension of our work on truxene-based dendrimers, we have developed a series of dendrimers **29** (**G0**), **30** (**G1**), and **31** (**G2**) (Figures 9.23 and 9.24), using OTEs of different lengths as branching units [48]. The use of truxene unit allows the molecular weight to increase rapidly with generation; and the OTE fragments allow the molecular dimension to expand quickly. The second-generation dendrimer **31** has the largest diameter (10 nm) and the highest molecular weight (27 kDa) among all reported second-generation dendrimers. To understand the photophysical properties of these dendrimer, model compounds **32** (**G0m**), **33** (**G1m**), and **34** (**G2m**) that represent the largest conjugation lengths in each generation of dendrimer were also synthesized (Figure 9.25) [98].

The absorption and emission spectra of these compounds are shown in Figure 9.26. All dendrimers showed two absorption peaks, one around 343 nm, assigned to the peripheral units (resembling model compound **10**), along with another one in the longer wavelength region. By comparison with linear model compounds, the latter peak was assigned to the longest OTE-truxene units in each dendrimer. The absorption maxima of these peaks increased progressively toward longer wavelengths due to increased conjugation length in higher generation. Upon excitation at 343 nm, where unit **10** absorbs, nearly quantitative ET to the longest branches was observed. The ET efficiencies were 96% for **29**, 97% for **30**, and 98% for **31**, respectively. The high ET efficiency can be ascribed to the designed energy gradient from the periphery to the core, and the large overlap between the emission of the donor and absorption of the acceptor. Significant "antenna effect" was observed, like that in the natural LH complex; despite similar LQY, the emission intensity (per molecule) of **31** was several times higher than that of model compound **34** when excited at 343 nm.

We went on to address a key issue of shape-persistent conjugated dendrimers that was seldom tested: to what extent is the dendrimer "shape-persistent"? And what is its real conformation? Energy minimization calculations did suggest a relatively flat conformation; but because the rotation barrier around thiophene–acetylene bond and truxene–thiophene bond is small, there may nonetheless be considerable freedom for rotation. To answer these questions, we performed two characterizations. First, we employed tapping mode atomic force microscopy (AFM) to investigate molecule **31** on mica substrates. Figure 9.27 shows the AFM image of **31** on mica surface at 3 nM solution, in which many separated and randomly deposited spots were observed. It is noteworthy that these spots were relatively uniform in width. Further dilution of the

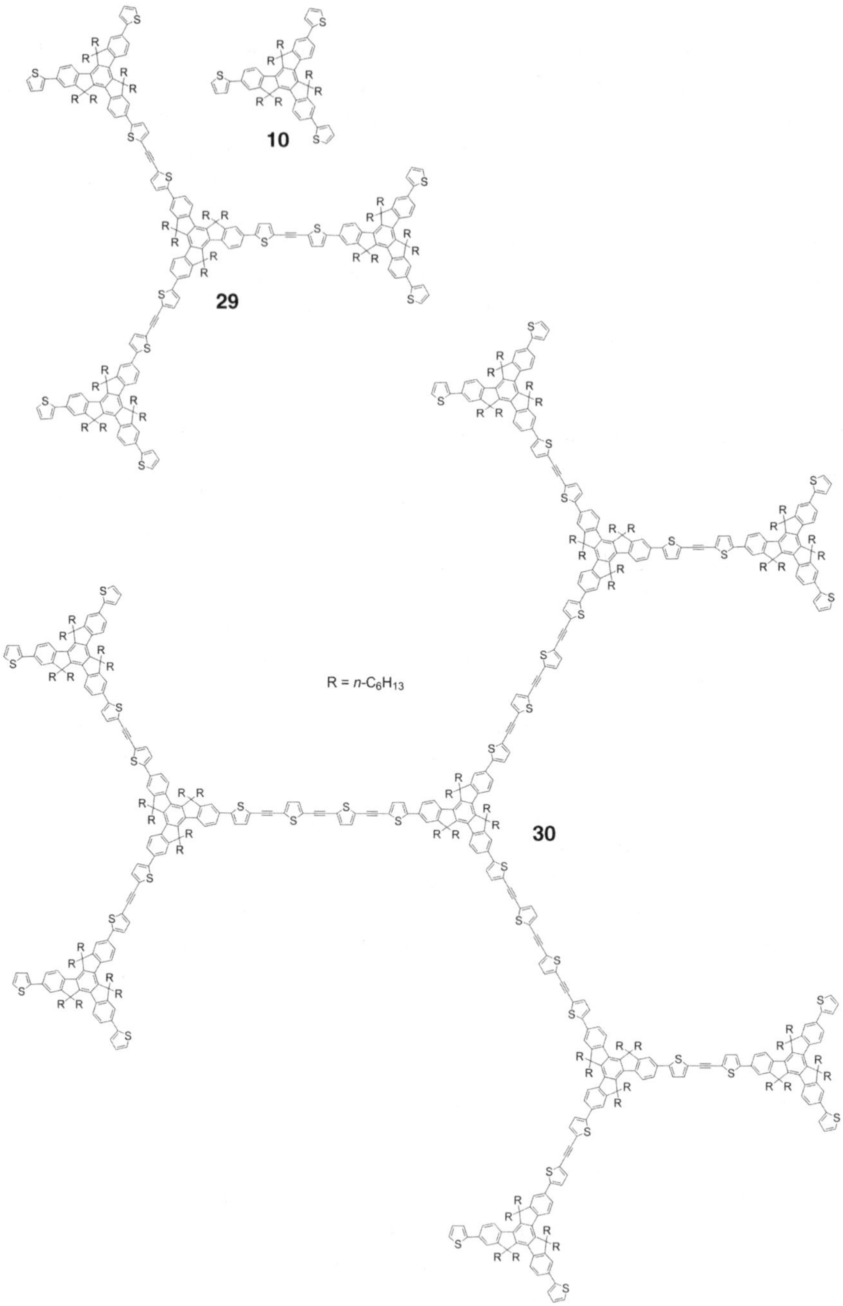

FIGURE 9.23 Chemical structure of the zeroth- and first-generation OTE-truxene dendrimers and reference compound **10** [48].

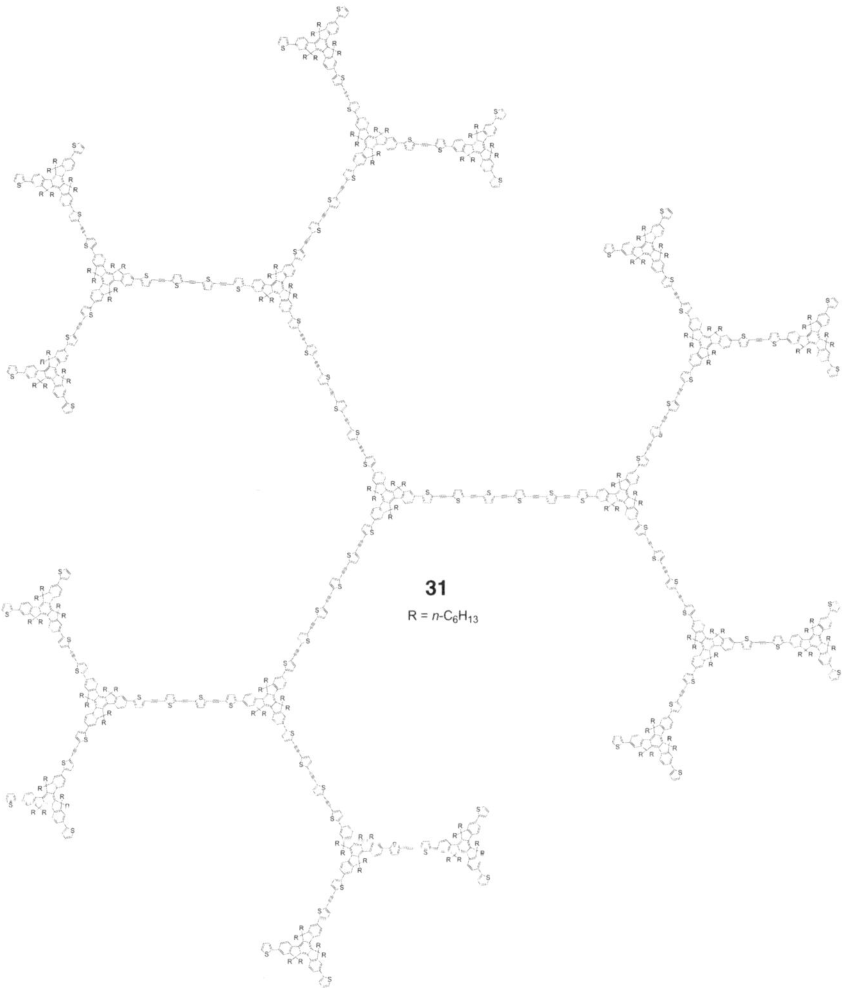

FIGURE 9.24 Second-generation OTE-truxene dendrimer **31** [48].

solution did not change the dimension of these spots. Hence, we postulated that every spot corresponded to a single molecule of **31**. The height of the spots varied in the range from 1.5 to 2.2 nm, which is in fair agreement with the molecular thickness, estimated either from the molecular model or the single crystal structure of **10**. Therefore, it is believe that on mica substrate, molecule **31** assumes a relatively flat conformation, similar to the simulation results. The second piece of evidence comes from dynamic light-scattering (DLS) measurement in solution, where substrate–molecule interaction is absent. We obtained diffusion coefficients of $2.0 \times 10^{-10}\,m^2/s$ for **29**, $1.1 \times 10^{-10}\,m^2/s$ for **30**, and $7.0 \times 10^{-11}\,m^2/s$ for **31**, respectively. Diffusion-ordered NMR spectroscopy gave similar results. By using an oblate spheroid model and Stokes–Einstein equation, a radius of 2.7 nm was obtained for **29**, 5.7 nm for **30**,

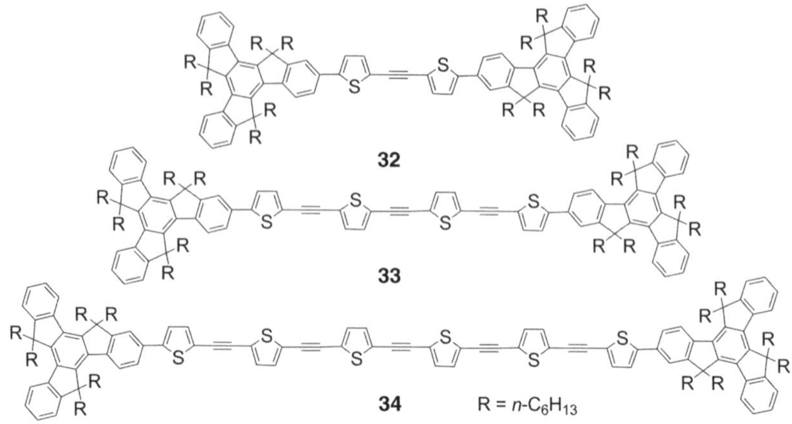

FIGURE 9.25 Chemical structures of model compounds **32 (G0m)**, **33 (G1m)**, and **34 (G2m)** [98].

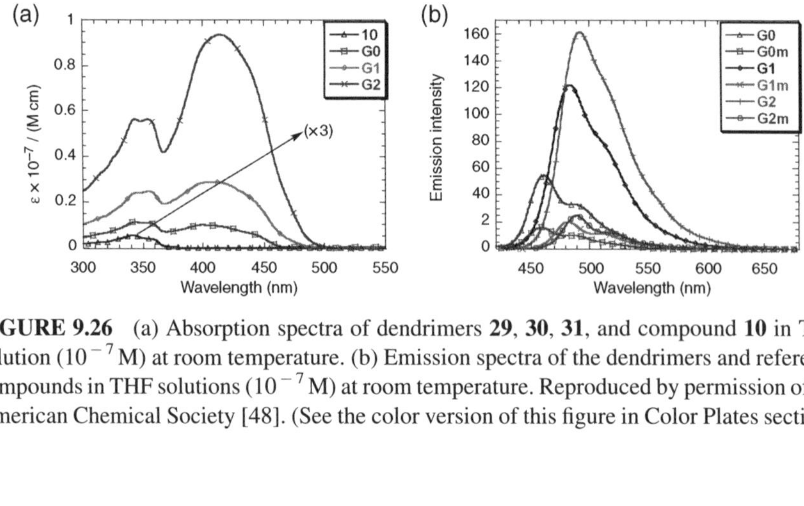

FIGURE 9.26 (a) Absorption spectra of dendrimers **29**, **30**, **31**, and compound **10** in THF solution (10^{-7} M) at room temperature. (b) Emission spectra of the dendrimers and reference compounds in THF solutions (10^{-7} M) at room temperature. Reproduced by permission of the American Chemical Society [48]. (See the color version of this figure in Color Plates section.)

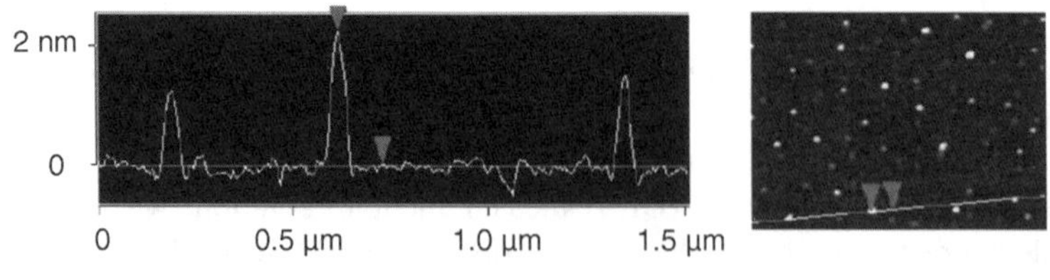

FIGURE 9.27 AFM images and height profiles of the individual dendrimer molecule **31** on mica from a 3 nM solution. Reproduced by permission of the American Chemical Society [48]. (See the color version of this figure in Color Plates section.)

and 9.7 nm for **31**, respectively. These were almost identical to the radii obtained from the molecular modeling. Therefore, it was suggested that even in solution the extended dendrimer molecules retained a relatively flat shape. This contrasts with the globular shape of "flexible" dendrimers and compact conjugated dendrimers, in which steric congestion overcame the stiffness of the building blocks [17]. Again, the length gradient shows its importance; steric congestion was significantly reduced, allowing the molecule to stretch to a flat conformation. This work is a nice validation of the word "shape-persistent."

9.3.3 Outlook

In this section, we reviewed how shape-persistent dendrimers were used to achieve fast, directional ET that mimics some aspects of photosynthesis. The importance is clearly shown of an energy gradient that funnels the peripheral excitons to the central core. However, we are still far away from the grand goal of artificial photosynthesis. To achieve that, instead of releasing the exciton energy as fluorescence, the excited state at the core has to enter a charge separated state. Then, the electrons and holes should favor different direction of movement to bring about a long-lived charge-separated state. We note that several LH dendrimers with flexible chains have shown some primitive but promising sign of this complicated function [99,100]. As a step further, electroactive units should be introduced to take up the electron/hole, converting the light energy into chemical energy and finishing a photochemical cycle.

The complex photophysical process in these dendrimers still needs proper theoretical models; the classical Dexter or Förster mechanism alone seems inadequate to describe the process. For example, the distance between the donor and the acceptor should be represented by a weighted distribution instead of a single value [88]. Clarification of the complex photophysics in these LH dendrimers would also help us to understand the natural photosynthetic process.

Besides mimicking the photosynthetic process, the unique properties of rigid LH dendrimers sparkled many interesting applications, some of which are described in the next section of this chapter. But it should be mentioned that realization of functional devices using these dendrimers is still rare. One reason is the tremendous synthetic efforts needed to produce these large-sized molecules. Closer collaboration among synthetic chemists and material scientists would allow a full exploration of the application potential of these beautiful structures.

9.4 MISCELLANEOUS APPLICATIONS

9.4.1 Introduction

In this section, we will present the application of shape-persistent conjugated dendrimers in other areas of organic electronics not covered so far. It is not aimed to be comprehensive; rather, we wish to illustrate how these large structures could be tailored to meet different requirements in some main areas in organic electronics. Examples include solar cell and nonlinear optics.

9.4.2 Organic Solar Cells

Solar cells are devices that convert light energy into electricity. Because of the increasing cost and decreasing availability of fossil fuels, search for alternative energy source becomes an urgent issue attracting attention all over the world. Among all possible solutions, solar cell seems to be a very promising one. Silicon-based solar cells are well developed, but the cost is still too high compared with the traditional ways of electricity generation. Hence, OSCs have been proposed as an alternative choice for high efficiency, low cost technology for energy conversion [101]. The operation mode of a solar cell is just the opposite of an OLED. After the absorption of light energy and formation of an exciton, the exciton charge separates at the D–A interface [102]. The electrons and holes then follow different pathways to the opposite electrodes. Two types of device configurations have achieved the most success. One is the bulk heterojunction solar cell (BHJSC) [103], and the other is the Grätzel-type dye-sensitized solar cell (DSSC) [104]. In both devices, organic materials play vital roles. In BHJSC, an organic donor material is blended with electron acceptors, usually [6,6]-phenyl-C_{61}-butyric acid methyl ester (PCBM). Ideally, the two materials form a bicontinuous phase upon spin-coating (or after other treatments), in which electrons and holes can flow in respective channels toward the electrode. In a DSSC, the electron acceptor is an inorganic semiconductor such as titanium dioxide. After charge separation, the electrons hop onto the inorganic semiconductor and migrate to the electrode. The organic dyes are then reduced by an electrolyte and ready for next excitation. The most important parameter for a solar cell device is the power conversion efficiency η, which is the ratio between power converted and collected, when a solar cell is connected to an electrical circuit. The highest η for a BHJSC achieved so far is over 6% [105], and over 11% for a DSSC [106].

The working principle of solar cells places several requirements on the active materials. The most important requirement is strong absorbance over a wide range of spectrum. Ideally, a material's absorption spectrum should match the solar spectrum. The second requirement is suitable HOMO and LUMO levels that give the exciton a driving force to charge separate while maintaining a reasonably high open-circuit voltage (V_{oc}) [107]. Next, for BHJSC, the donor material should be able to form good bicontinuous film with the acceptor, and have adequate hole-transporting ability [102]. For DSSC, the organic dye should be able to absorb strongly on the inorganic semiconductors and possess adequate distribution of electron density [108]. Optimizing all these factors is a challenging task.

Because most donor materials in BHJSC studied so far are polythiophenes such as poly(3-hexylthiophene) (P3HT), it is reasonable to expect the above-mentioned thiophene dendrimer to be a good candidate for BHJSC. Indeed, recently, Bäuerle and coworkers have demonstrated η up to 1.72% [109] using their thiophene dendrimers mentioned in the last section. Several general trends were observed. First, thiophene dendrimer without peripheral $SiMe_3$ group have better performances than those with $SiMe_3$ groups. In other word, reducing inactive units increases the device performance. Second, higher generation dendrimers generally have better performance. This observation has two origins. One is that higher generation thiophene dendrimers possess longer α-conjugated pathways and absorb light at longer wavelengths where the solar photon flux is higher. The other reason is that higher generation dendrimers have higher hole mobility. One of the disadvantages of these thiophene dendrimers over P3HT is that their absorption spectra do not cover the red and near-infrared regions, caused by shorter conjugation length and decreased intermolecular interaction. In addition, the hole mobility of the dendrimer is also lower than P3HT. Nevertheless, this work demonstrated that conjugated dendrimer can be used as the third type materials in OSC. Also, such large dendrimer with diameter approaching 10 nm brings up an interesting question. As has been proved in previous work, a nanoscale phase separated morphology is the optimum one for BHJSC [102]. In this respect, the 3-D persistency of the dendrimer sets a lower limit to the domain size, which enables unique control over the degree of the phase separation. Similar work using phenyl cored thiophene dendrimers is reported by Kopidakis et al., and a maximum η of 1.3% is achieved [110]. In a recent work, Bäuerle's group attached the thiophene dendrons to a ruthenium(II) phthalocyanine core using coordination chemistry between pyridine and metal-phthalocyanine. A solution processed BHJSC with η of 1.6% has been achieved [111].

Following our work on OTE-truxene dendrimers, we also investigated their applicability as OSC materials. However, in most cases, their absorption spectra are too narrow. To extend the absorption spectrum to longer wavelength, we have replaced the OTE segments with oligo(thienylenevinylene) (OTV) segments [112], which have lower energy bandgap due to better conjugation through the double bond. Figure 9.28 illustrates the chemical structures of two zeroth-generation extended dendrimer with OTV branches and truxene nodes, along with their model compounds. From the absorption spectra of thin films, it was seen that the OTV-truxene dendrimers covered a much broader spectrum than OTE truxene dendrimers, even though the generation is lower. Typically, poly(thienylenevinylene)-based OSC devices do not show good performances due to the nonluminescent nature of thienylenevinylene unit [113]. In our case, the introduction of truxene unit effectively improved the luminescent ability of the OTV segments. As a result, OSC devices using **37** showed better performance than the poly(3-hexylthienylenevinylene), despite narrower absorption spectrum. The highest η is 0.40% for OSC devices based on **37**. In subsequent works, we have further introduced an electron-accepting unit such as benzothiadiazole into the dendrimer scaffold to further extend the absorption spectrum into the red region, and preliminary results showed that η could be improved to 0.54% (unpublished results) (Figure 9.29) [112].

Shape-persistent conjugated dendrimers also showed application potential in DSSCs. For example, Yamamoto and coworkers used their phenylazomethine dendrimers as a charge separator in DSSC [114]. DSSCs prepared by casting phenylazomethine dendrimers onto dye-sensitized TiO_2 film exhibits higher open-circuit voltage than the bare film, due to the suppression of back electron transfer by the dendrimer.

A general problem for dendrimer-based OSC is their narrower light absorption range and lower absorption coefficient in the red and near-infrared regions, compared

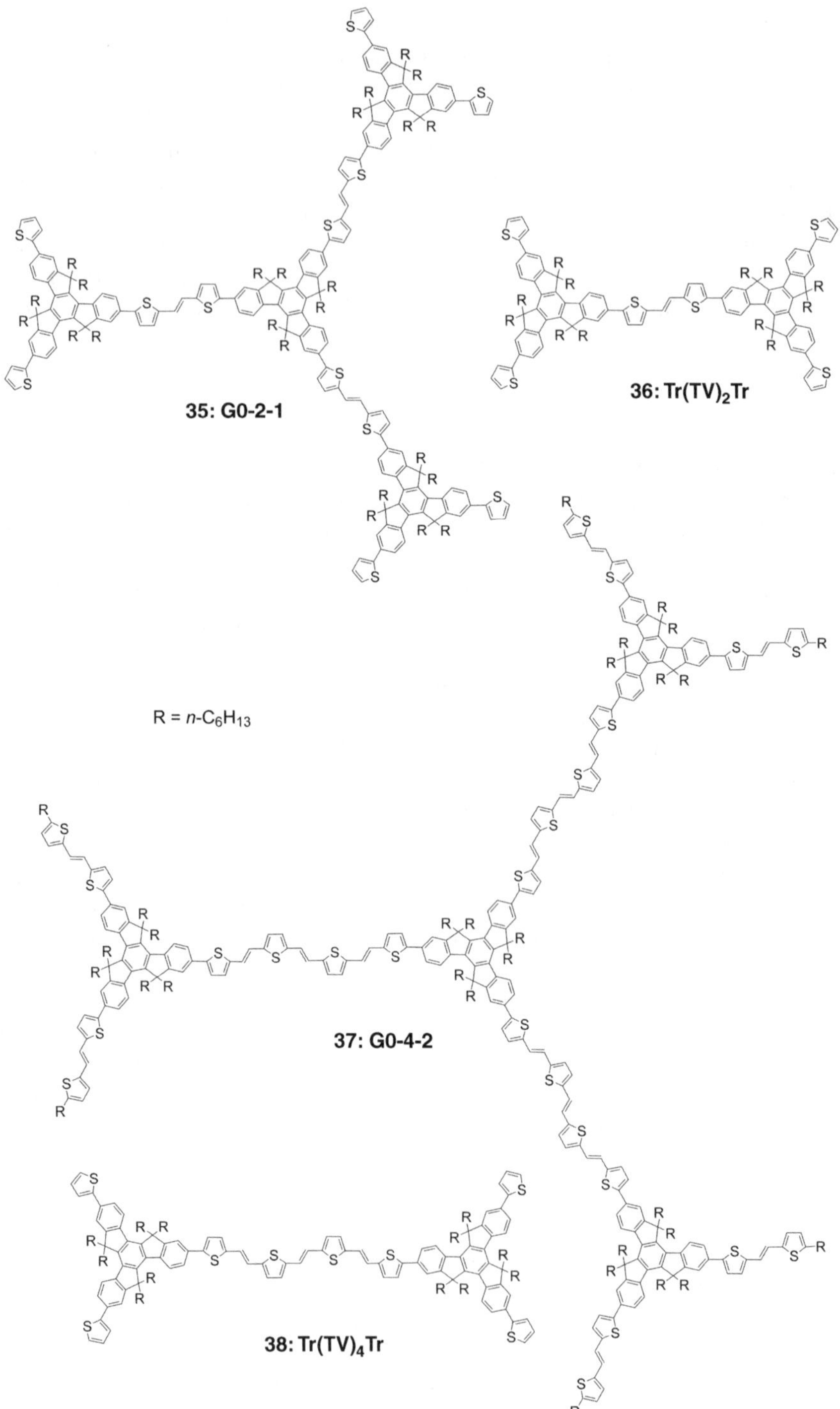

FIGURE 9.28 Chemical structures of OTV-truxene dendrimers for OSC and their model compounds [112].

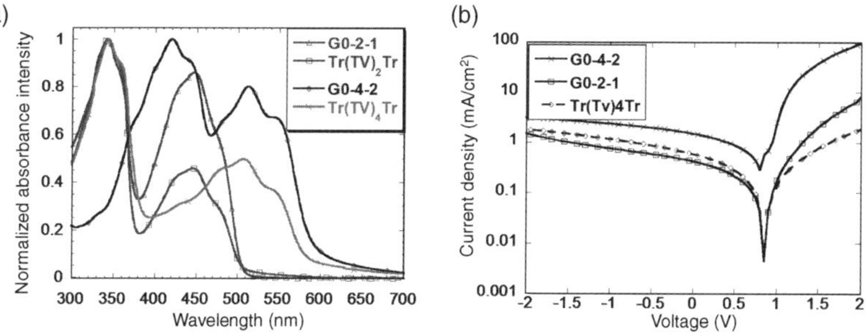

FIGURE 9.29 (a) Absorption spectra of OTV-truxene dendrimers and model compounds. (b) *J-V* curves of a single layer BHJSC with these dendrimers and PCBM as photoactive layers under illumination of AM 1.5, 100 mW/cm^2 [112].

to linear polymer. The reason is that dendrimers usually have shorter conjugation length than corresponding polymers. This drawback can be overcome by introducing small units that are able to absorb light in that region, such as diketopyrrolopyrrole [115]. We expect to see dendrimers to emerge as the third type of materials for OSCs, just as they did as DLEDs.

9.4.3 Two-Photon Absorption Material

Nonlinear optics (NLO) describes the behavior of light when the dielectric polarization P responds nonlinearly to the electric field E of the light [116]. Here we will mainly focus on the phenomenon of two-photon absorption (TPA), a third-order nonlinear optical process [117]. TPA is the process that a molecule simultaneously absorbs two photons and gets excited from one state to a higher energy electronic state. Consequently, the strength of absorption depends on the square of the light intensity. TPA materials have demonstrated many potential applications including optical data storage, multiphoton microscopy, 3-D microfabrication, optical power limiting, and frequency up-conversion lasing [118].

The most important parameter of a TPA material is TPA cross-section σ_2. Its unit is GM, where 1 GM is 10^{-50} cm^4 s/photon. Great effort has been made to obtain materials with large TPA cross-section, usually with a long conjugated system with electron-donating group and electron-accepting group at two ends [119]. Compared to linear polymers, a conjugated dendrimer has the ability to concentrate a large number of chromophores, whose number grows exponentially with generation. Meanwhile, interchromophore coupling in some conjugated dendrimers has also been shown to be beneficial to nonlinear optical enhancement.

Goodson's group has extensively studied various photophysical processes in triphenylamine-based dendrimers [120]. In contrast to 1,3,5-benzene or truxene branching centers, the sp^2-hybridized nitrogen provides much stronger interbranch coupling, resulting in large enhancement of TPA properties. For example, Figure 9.30 shows a series of triphenylamine dendrimers using acetylene linkage [121]. With the generation growing from zeroth to second, the σ_2 increases

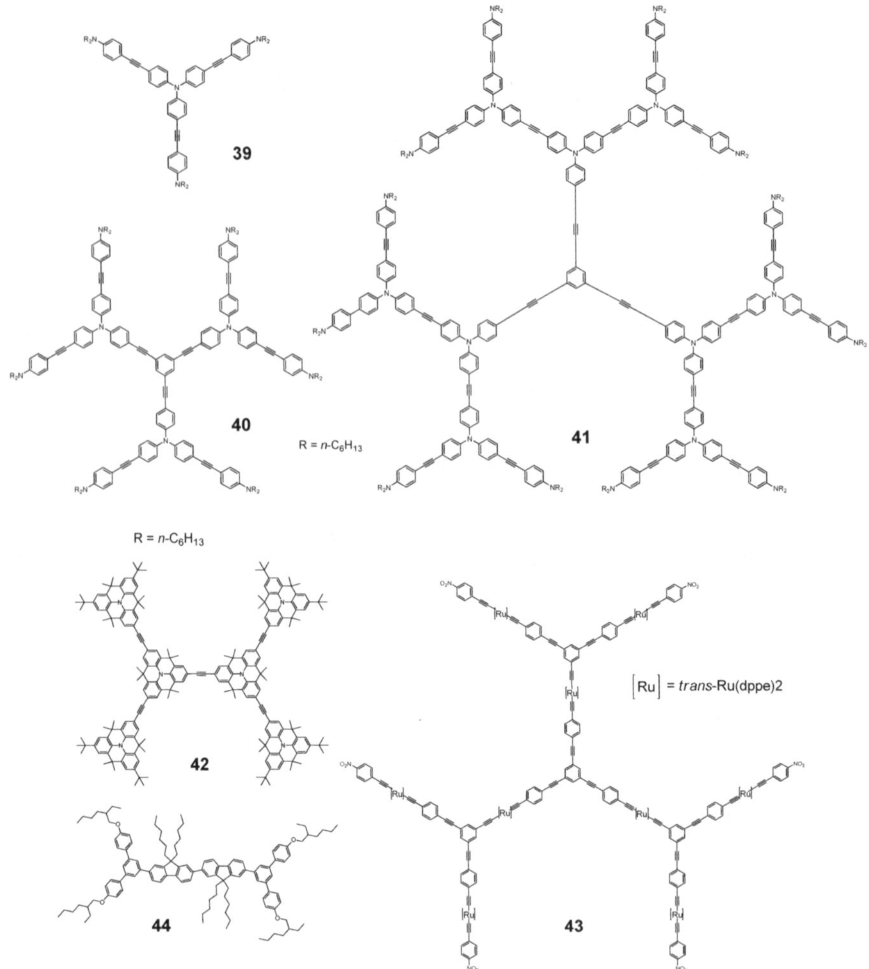

FIGURE 9.30 Chemical structures of several TPA dendrimers [121–124].

steadily. The maximum σ_2 reaches up to 7000 GM for the largest dendrimer. In such dendrimers, fast energy migration and exciton delocalization are observed, which can be related to enhanced NLO effect. To ensure the planarity and therefore electron delocalization around the nitrogen atom, Fang et al. introduced methylene bridges to constrain the flexibility of the triphenylamine group [122]. As a result, large cross-sections (up to 6100 GM) was obtained for structure **42**. Besides pure organic dendrimer, organometallic unit can also be employed. For example, Humphrey's group reported a Ru-containing organometallic dendrimer **43** that showed not only good TPA property but also remarkable three-photon absorption ability [123]. In a recent report, Burn's group used their bisfluorene-biphenyl dendrimer (structure **44**) to realize two-photon pumped lasing, showing great application potentials in up-conversion lasers [124].

9.5 CONCLUSION

In this chapter, we have reviewed recent progress in applying shape-persistent dendrimers in organic electronics. In the areas of DLEDs, conjugated dendrimers are already established as a promising type of material, even for commercialization. In other areas, they remain at an early stage of development. For each application, the dendrimer structure should be carefully designed to meet the specific requirements of the device. Thanks to the increasing amount of synthetic tools and conjugated building blocks available, we expect shape-persistent conjugated dendrimers to play a more significant role in the development of organic electronics in the near future.

ACKNOWLEDGMENT

We are grateful to the National Basic Research Program (2007CB808000 and 2002CB613402) of the Ministry of Science and Technology and National Natural Science Foundation of China. We thank our students and colleagues for their devoted research and valuable discussion that made this contribution possible.

REFERENCES

[1] H. Klauk (Ed.), *Organic Electronics: Materials, Manufacturing, and Application*, Wiley-*VCH, Weinheim*, **2006**.

[2] R. F. Service, *Science* **2005**, *310*, 1762–1763.

[3] S. R. Forrest, *Nature* **2004**, *428*, 911–918.

[4] M. J. Currie, J. K. Mapel, T. D. Heidel, S. Goffri, M. A. Baldo, *Science* **2008**, *321*, 226–228.

[5] S. C. Lo, P. L. Burn, *Chem. Rev.* **2007**, *107*, 1097–1116.

[6] V. C. Sundar, J. Zaumseil, V. Podzorov, E. Menard, R. L. Willett, T. Someya, M. E. Gershenson, J. A. Rogers, *Science* **2004**, *303*, 1644–1646.

[7] E. Menard, M. A. Meitl, Y. G. Sun, J. U. Park, D. J. L. Shir, Y. S. Nam, S. Jeon, J. A. Rogers, *Chem. Rev.* **2007**, *107*, 1117–1160.

[8] S. Allard, M. Forster, B. Souharce, H. Thiem, U. Scherf, *Angew. Chem.* **2008**, *120*, 4138–4167; *Angew. Chem. Int. Ed.* **2008**, *47*, 4070–4098.

[9] N. Tessler, Y. Preezant, N. Rappaport, Y. Roichman, *Adv. Mater.* **2009**, *21*, 2741–2761.

[10] (a) G. R. Newkome, C. N. Moorefield, F. Vögtle, *Dendritic Molecules: Concepts, Syntheses, Perspectives*, VCH, Weinheim, **1996**; (b) J. M. J. Fréchet, D. A. Tomalia (Eds.), *Dendrimers and Other Dendritic Polymers*, J. Wiley & Sons Ltd., Chichester, **2001**.

[11] (a) C. C. Lee, J. A. MacKay, J. M. J. Fréchet, F. C. Szoka, *Nat. Biotechnol.* **2005**, *23*, 1517–1526; (b)B. Helms, E. W. Meijer, *Science* **2006**, *313*, 929–930.

[12] V. Coropceanu, J. Cornil, D. A. da Silva, Y. Olivier, R. Silbey, J. L. Brédas, *Chem. Rev.* **2007**, *107*, 926–952.

[13] C. C. Kwok, M. S. Wong, *Thin Solid Films* **2002**, *417*, 136–142.

[14] V. R. Neuwahl, R. Righini, A. Adronov, P. R. L. Malenfant, J. M. J. Fréchet, *J. Phys. Chem. B* **2001**, *105*, 1307–1312.

[15] J. S. Moore, *Acc. Chem. Res.* **1997**, *30*, 402–413.

[16] (a) A. Mishra, C. Q. Ma, P. Bäuerle, *Chem. Rev.* **2009**, *109*, 1141–1276; (b)A. J. Berresheim, M. Muller, K. Müllen, *Chem. Rev.* **1999**, *99*, 1747–1785.

[17] Z. F. Xu, J. S. Moore, *Angew. Chem.* **1993**, *105*, 261–264; *Angew. Chem. Int. Ed. Engl.* **1993**, 32, 246–248.

[18] T. M. Figueira-Duarte, S. C. Simon, M. Wagner, S. I. Druzhinin, K. A. Zachariasse, K. Müllen, *Angew. Chem.* **2008**, *120*, 10329–10332; *Angew. Chem. Int. Ed.* **2008**, *47*, 10175–10178.

[19] A. Herrmann, T. Weil, V. Sinigersky, U. M. Wiesler, T. Vosch, J. Hofkens, F. C. De Schryver, K. Müllen, *Chem. Eur. J.* **2001**, *7*, 4844–4853.

[20] C. W. Tang, S. A. VanSlyke, *Appl. Phys. Lett.* **1987**, *51*, 913–915.

[21] (a) B. W. D'Andrade, S. R. Forrest, *Adv. Mater.* **2004**, *16*, 1585–1595; (b) S. J. Su, E. Gonmori, H. Sarabe, J. Kido, *Adv. Mater.* **2008**, *20*, 4189–4194.

[22] J. H. Burroughes, D. D. C. Bradley, A. R. Brown, R. N. Marks, K. Mackay, R. H. Friend, P. L. Born, A. B. Holmes, *Nature* **1990**, *347*, 539–541.

[23] (a) C. Adachi, M. A. Baldo, S. R. Forrest, S. Lamansky, M. E. Thompson, R. C. Kwong, *Appl. Phys. Lett.* **2001**, *78*, 1622–1624; (b) M. A. Baldo, S. Lamansky, P. E. Burrows, S. R. Forrest, M. E. Thompson, *Appl. Phys. Lett.* **1999**, *75*, 4–6.

[24] K. Danel, T. H. Huang, J. T. Lin, Y. T. Tao, C. H. Chuen, *Chem. Mater.* **2002**, *14*, 3860–3865.

[25] A. L. Kanibolotsky, R. Berridge, P. J. Skabara, I. F. Perepichka, D. D. C. Bradley, M. Koeberg, *J. Am. Chem. Soc.* **2004**, *126*, 13695–13702.

[26] S. A. Jenekhe, J. A. Osaheni, *Science* **1994**, *265*, 765–768.

[27] Q. B. Pei, Y. Yang, *J. Am. Chem. Soc.* **1996**, *118*, 7416–7417.

[28] J. S. Moore, Z. F. Xu, *Macromolecules* **1991**, *24*, 5893–5894.

[29] R. Beavington, M. J. Frampton, J. M. Lupton, P. L. Burn, I. D. W. Samuel, *Adv. Funct. Mater.* **2003**, *13*, 211–218.

[30] Y. Shirota, *J. Mater. Chem.* **2005**, *15*, 75–93.

[31] P. W. Wang, Y. J. Liu, C. Devadoss, P. Bharathi, J. S. Moore, *Adv. Mater.* **1996**, *8*, 237–241.

[32] S. W. Cha, S. H. Choi, K. Kim, J. I. Jin, *J. Mater. Chem.* **2003**, *13*, 1900–1904.

[33] M. Halim, J. N. G. Pillow, I. D. W. Samuel, P. L. Burn, *Adv. Mater.* **1999**, *11*, 371–374.

[34] J. N. G. Pillow, M. Halim, J. M. Lupton, P. L. Burn, I. D. W. Samuel, *Macromolecules* **1999**, *32*, 5985–5993.

[35] H. F. Chow, M. K. Ng, C. W. Leung, G. X. Wang, *J. Am. Chem. Soc.* **2004**, *126*, 12907–12915.

[36] C. K. Kim, E. S. Song, H. J. Kim, C. Park, Y. C. Kim, J. K. Kim, J. W. Yu, C. Kim, *J. Polym. Sci. Part A Polym. Chem.* **2006**, *44*, 254–263.

[37] T. W. Kwon, M. M. Alam, S. A. Jenekhe, *Chem. Mater.* **2004**, *16*, 4657–4666.

[38] A. P. Kulkarni, S. A. Jenekhe, *J. Phys. Chem. C* **2008**, *112*, 5174–5184.

[39] L. Zhao, C. Li, Y. Zhang, X. H. Zhu, J. B. Peng, Y. Cao, *Macromol. Rapid Commun.* **2006**, *27*, 914–920.

[40] C. G. Clark, R. J. Wenzel, E. V. Andreitchenko, W. Steffen, R. Zenobi, K. Müllen, *J. Am. Chem. Soc.* **2007**, *129*, 3292–3301.

[41] T. S. Qin, G. Zhou, H. Scheiber, R. E. Bauer, M. Baunigarten, C. E. Anson, E. J. W. List, K. Müllen, *Angew. Chem.* **2008**, *120*, 8416–8420; Angew. Chem. Int. Ed. **2008**, 47, 8292–8296.

[42] (a) J. S. Cho, A. Kimoto, M. Higuchi, K. Yamamoto, *Macromol. Chem. Phys.* **2005**, *206*, 635–641; (b)A. Kimoto, J. S. Cho, M. Higuchi, K. Yamamoto, *Macromolecules* **2004**, *37*, 5531–5537.

[43] D. H. Choi, K. I. Han, I. H. Chang, S. H. Choi, X. H. Zhang, K. H. Ahn, Y. K. Lee, J. Jang, *Synth. Met.* **2007**, *157*, 332–335.

[44] P. A. Levermore, R. Xia, W. Lai, X. H. Wang, W. Huang, D. D. C. Bradley, *J. Phys. D Appl. Phys.* **2007**, *40*, 1896–1901.

[45] K. Isoda, T. Yasuda, T. Kato, *Chem. Asian J.* **2009**, *4*, 1619–1625.

[46] (a) L. T. Scott, M. M. Boorum, B. J. McMahon, S. Hagen, J. Mack, J. Blank, H. Wegner, A. de Meijere, *Science* **2002**, *295*, 1500–1503; (b) B. Gómez-Lor, Ó. de Frutos, A. M. Echavarren, *Chem. Commun.* **1999**, 2431–2432; (c) T. S. Perova, J. K. Vij, *Adv. Mater.* **1995**, *7*, 919–922.

[47] X. Y. Cao, X. H. Liu, X. H. Zhou, Y. Zhang, Y. Jiang, Y. Cao, Y. X. Cui, J. Pei, *J. Org. Chem.* **2004**, *69*, 6050–6058.

[48] J. L. Wang, J. Yan, Z. M. Tang, Q. Xiao, Y. Ma, J. Pei, *J. Am. Chem. Soc.* **2008**, *130*, 9952–9962.

[49] (a) Recently, we were able to utilize such a strong π–π interaction to create novel one-dimensional nano/microstructures. See ref: J. Y. Wang, J. Yan, Z. D. Li, J. M. Han, Y. Ma, J. Bian, J. Pei, *Chem. Eur. J.* **2008**, *14*, 7760–7764; (b) J. Y. Wang, J. Yan, L. Ding, Y. Ma, J. Pei, *Adv. Funct. Mater.* **2009**, *19*, 1746–1752.

[50] X. Y. Cao, W. B. Zhang, J. L. Wang, X. H. Zhou, H. Lu, J. Pei, *J. Am. Chem. Soc.* **2003**, *125*, 12430–12431.

[51] Y. Jiang, Y. X. Lu, Y. X. Cui, Q. F. Zhou, Y. Ma, J. Pei, *Org. Lett.* **2007**, *9*, 4539–4542.

[52] Y. Jiang, L. Wang, Y. Zhou, Y. X. Cui, J. Wang, Y. Cao, J. Pei, *Chem. Asian J.* **2009**, *4*, 548–553.

[53] Y. Jiang, J. Y. Wang, Y. Ma, Y. X. Cui, Q. F. Zhou, J. Pei, *Org. Lett.* **2006**, *8*, 4287–4290.

[54] L. Wang, Y. Jiang, J. Luo, Y. Zhou, J. Zhou, J. Wang, J. Pei, Y. Cao, *Adv. Mater.* **2009**, *21*, 4854–4858.

[55] C. Poynton, *Digital Video and HDTV*, Morgan Kaufmann, San Francisco, **2003**.

[56] J. Luo, Y. Zhou, Z. Q. Niu, Q. F. Zhou, Y. Ma, J. Pei, *J. Am. Chem. Soc.* **2007**, *129*, 11314–11315.

[57] E. J. W. List, R. Guentner, P. S. de Freitas, U. Scherf, *Adv. Mater.* **2002**, *14*, 374–378.

[58] Recently, we have also succeeded in obtaining highly fluorescent one-dimensional nanowires using the three-dimensional scaffold. See reference: J. Luo, T. Lei, L. Wang, Y. Ma, Y. Cao, J. Wang, J. Pei, *J. Am. Chem. Soc.* **2009**, *131*, 2076–2077.

[59] (a) M. A. Baldo, D. F. O'Brien, M. E. Thompson, S. R. Forrest, *Phys. Rev. B* **1999**, *60*, 14422–14428; (b) C. Adachi, M. A. Baldo, S. R. Forrest, S. Lamansky, M. E. Thompson, R. C. Kwong, *Appl. Phys. Lett.* **2001**, *78*, 1622–1624.

[60] S. C. Lo, N. A. H. Male, J. P. J. Markham, S. W. Magennis, P. L. Burn, O. V. Salata, I. D. W. Samuel, *Adv. Mater.* **2002**, *14*, 975–979.

[61] (a) J. Q. Ding, B. Wang, Z. Y. Yue, B. Yao, Z. Y. Xie, Y. X. Cheng, L. X. Wang, X. B. Jing, F. S. Wang, *Angew. Chem.* **2009**, *121*, 6792–6794; *Angew. Chem. Int. Ed.* **2009**, 48, 6664–6666; (b) J. Q. Ding, J. H. Lu, Y. X. Cheng, Z. Y. Xie, L. X. Wang, X. B. Jing, F. S. Wang, *Adv. Funct. Mater.* **2008**, *18*, 2754–2762.

[62] P. L. Burn, S. C. Lo, I. D. W. Samuel, *Adv. Mater.* **2007**, *19*, 1675–1688.

[63] S. I. Park, Y. Xiong, R. H. Kim, P. Elvikis, M. Meitl, D. H. Kim, J. Wu, J. Yoon, C. J. Yu, Z. Liu, Y. Huang, K. C. Hwang, P. Ferreira, X. Li, K. Choquette, J. A. Rogers, *Science* **2009**, *325*, 977–981.

[64] J. A. Merrigan, *Sunlight to Electricity: Prospects for Solar Energy Conversion by Photovoltaics*, MIT Press, Cambridge, **1975**.

[65] N. S. Lewis, *Science* **2007**, *315*, 798–801.

[66] A. L. Lehninger, D. L. Nelson, M. M. Cox, *Principles of Biochemistry*, Worth Publishers Inc., New York, **1993**.

[67] W. Kühlbrandt, D. N. Wang, Y. Fujiyoshi, *Nature* **1994**, *367*, 614–621.

[68] A. Adronov, J. M. J. Fréchet, *Chem. Commun.* **2000**, 1701–1710.

[69] G. Mcdermott, S. M. Prince, A. A. Freer, A. M. Hawthornthwaite-Lawless, M. Z. Papiz, R. J. Cogdell, N. W. Isaacs, *Nature* **1995**, *374*, 517–521.

[70] J. Barber, B. Andersson, *Nature* **1994**, *370*, 31–34.

[71] N. J. Turro, *Modern Molecular Photochemistry*, University Science Books, Sausalito, **1991**.

[72] B. W. Van Der Meer, G. CokerIII, S.-Y. Simon Chen, *Resonance Energy Transfer, Theory and Data*, VCH, Weinheim, **1994**.

[73] A. Nantalaksakul, D. R. Reddy, C. J. Bardeen, S. Thayumanavan, *Photosynth. Res.* **2006**, *87*, 133–150.

[74] L. Pu, *Chem. Rev.* **2004**, *104*, 1687–1715.

[75] Z. H. Peng, J. S. Melinger, V. Kleiman, *Photosynth. Res.* **2006**, *87*, 115–131.

[76] A. Adronov, S. L. Gilat, J. M. J. Fréchet, K. Ohta, F. V. R. Neuwahl, G. R. Fleming, *J. Am. Chem. Soc.* **2000**, *122*, 1175–1185.

[77] D. L. Jiang, T. Aida, *J. Am. Chem. Soc.* **1998**, *120*, 10895–10901.

[78] D. L. Jiang, T. Aida, *Nature* **1997**, *388*, 454–456.

[79] D. L. Dexter, *J. Chem. Phys.*, **1953**, *21*, 836–850.

[80] T. Förster, *Ann. Phys.*, **1948**, *2*, 55–75.

[81] V. Balzani, S. Campagna, G. Denti, A. Juris, S. Serroni, M. Venturi, *Acc. Chem. Res.* **1998**, *31*, 26–34.

[82] Z. F. Xu, J. S. Moore, *Angew. Chem.* **1993**, *105*, 1394–1396; *Angew. Chem. Int. Ed. Engl.* **1993**, *32*, 1354–1357.

[83] P. G. de Gennes, H. Hervet, *J. Phys. Lett.* **1983**, *44*, L351–L360.

[84] R. Kopelman, M. Shortreed, Z. Y. Shi, W. H. Tan, Z. F. Xu, J. S. Moore, A. Bar-Haim, J. Klafter, *Phys. Rev. Lett.* **1997**, *78*, 1239–1242.

[85] S. F. Swallen, R. Kopelman, J. S. Moore, C. J. Devadoss, *J. Mol. Struct.* **1999**, *485–486*, 585–597.

[86] C. Devadoss, P. Bharathi, J. S. Moore, *J. Am. Chem. Soc.* **1996**, *118*, 9635–9644.

[87] S. F. Swallen, Z. G. Zhu, J. S. Moore, R. Kopelman, *J. Phys. Chem. B* **2000**, *104*, 3988–3995.

[88] J. S. Melinger, Y. C. Pan, V. D. Kleiman, Z. H. Peng, B. L. Davis, D. McMorrow, M. Lu, *J. Am. Chem. Soc.* **2002**, *124*, 12002–12012.

[89] (a) J. Roncali, *Chem. Rev.* **1997**, *97*, 173–205; (b) P. Bäuerle, in *Electronic Materials: The Oligomer Approach* (Eds.: K. Müllen, G. Wegner), Wiley-VCH, Weinheim, **1998**, pp. 105–197; (c) D. Fichou (Ed.), *Handbook of Oligo- and Polythiophenes*, Wiley-VCH, Weinheim, **1999**.

[90] A. Zen, A. Bilge, F. Galbrecht, R. Alle, K. Meerholz, J. Grenzer, D. Neher, U. Scherf, T. Farrell, *J. Am. Chem. Soc.* **2006**, *128*, 3914–3915.

[91] A. Bilge, A. Zen, M. Forster, H. Li, F. Galbrecht, B. S. Nehls, T. Farrell, D. Neher, U. Scherf, *J. Mater. Chem.* **2006**, *16*, 3177–3182.

[92] C. J. Xia, X. W. Fan, J. Locklin, R. C. Advincula, *Org. Lett.* **2002**, *4*, 2067–2070.

[93] C. J. Xia, X. W. Fan, J. Locklin, R. C. Advincula, A. Gies, W. Nonidez, *J. Am. Chem. Soc.* **2004**, *126*, 8735–8743.

[94] (a) C. Q. Ma, E. Mena-Osteritz, T. Debaerdemaeker, M. M. Wienk, R. A. J. Janssen, P. Bäuerle, *Angew. Chem.* **2007**, *119*, 1709–1713; Angew. Chem. Int. Ed. **2007**, 46, 1679–1683; (b) M. R. Harpham, O. Suzer, C. Q. Ma, P. Bäuerle, T. Goodson, *J. Am. Chem. Soc.* **2009**, *131*, 973–979.

[95] J. L. Wang, J. Luo, L. H. Liu, Q. F. Zhou, Y. Ma, J. Pei, *Org. Lett.* **2006**, *8*, 2281–2284.

[96] J. M. Tour, *Acc. Chem. Res.* **2000**, *33*, 791–804.

[97] J. M. Tour, *Chem. Rev.* **1996**, *96*, 537–553.

[98] J. L. Wang, Z. M. Tang, Q. Xiao, Q. F. Zhou, Y. Ma, J. Pei, *Org. Lett.* **2008**, *10*, 17–20.

[99] M. Kozaki, K. Akita, K. Okada, *Org. Lett.* **2007**, *9*, 1509–1512.

[100] K. R. J. Thomas, A. L. Thompson, A. V. Sivakumar, C. J. Bardeen, S. Thayumanavan, *J. Am. Chem. Soc.* **2005**, *127*, 373–383.

[101] R. F. Service, *Science* **2008**, *319*, 718–720.

[102] B. C. Thompson, J. M. J. Fréchet, *Angew. Chem.* **2008**, *120*, 62–82; *Angew. Chem. Int. Ed.* **2008**, *47*, 58–77.

[103] G. Yu, J. Gao, J. C. Hummelen, F. Wudl, A. J. Heeger, *Science* **1995**, *270*, 1789–1791.

[104] B. Oregan, M. Grätzel, *Nature* **1991**, *353*, 737–740.

[105] J. Y. Kim, K. Lee, N. E. Coates, D. Moses, T. Q. Nguyen, M. Dante, A. J. Heeger, *Science* **2007**, *317*, 222–225.

[106] Y. Chiba, A. Islam, Y. Watanabe, R. Komiya, N. Koide, L. Y. Han, *Jpn. J. Appl. Phys.* **2006**, *45*, L638–L640.

[107] G. Dennler, M. C. Scharber, T. Ameri, P. Denk, K. Forberich, C. Waldauf, C. J. Brabec, *Adv. Mater.* **2008**, *20*, 579–583.

[108] H. Choi, C. Baik, S. O. Kang, J. Ko, M. S. Kang, M. K. Nazeeruddin, M. Grätzel, *Angew. Chem.* **2008**, *120*, 333–336; *Angew. Chem. Int. Ed.* **2008**, *47*, 327–330.

[109] C. Q. Ma, M. Fonrodona, M. C. Schikora, M. M. Wienk, R. A. J. Janssen, P. Bäuerle, *Adv. Funct. Mater.* **2008**, *18*, 3323–3331.

[110] N. Kopidakis, W. J. Mitchell, J. van de Lagemaat, D. S. Ginley, G. Rumbles, S. E. Shaheen, W. L. Rance, *Appl. Phys. Lett.* **2006**, *89*, 103524.

[111] M. K. R. Fischer, I. Lopez-Duarte, M. M. Wienk, M. V. Martinez-Diaz, R. A. J. Janssen, P. Bäuerle, T. Torres, *J. Am. Chem. Soc.* **2009**, *131*, 8669–8676.

[112] J. L. Wang, C. Zhong, Z. M. Tang, H. Wu, Y. Ma, Y. Cao, J. Pei, *Chem. Asian J.* **2010**, *5*, 105–113.

[113] (a) J. Hou, Z. Tan, Y. He, C. Yang, Y. Li, *Macromolecules* **2006**, *39*, 4657–4662; (b) Y. Li, Y. Zou, *Adv. Mater.* **2008**, *20*, 2952–2958; (c) I.-W. Hwang, Q.-H. Xu, C. Soci, B. Chen, A. K.-Y. Jen, D. Moses, A. J. Heeger, *Adv. Funct. Mater.* **2007**, *17*, 563–568.

[114] N. Satoh, T. Nakashima, K. Yamamoto, *J. Am. Chem. Soc.* **2005**, *127*, 13030–13038.

[115] A. B. Tamayo, B. Walker, T. Q. Nguyen, *J. Phys. Chem. C* **2008**, *112*, 11545–11551.

[116] R. W. Boyd, *Nonlinear Optics*, 2nd edition, Academic Press, San Diego, **2003**.

[117] G. S. He, L. S. Tan, Q. Zheng, P. N. Prasad, *Chem. Rev.* **2008**, *108*, 1245–1330.

[118] (a) S. J. Chung, K. S. Kim, T. H. Lin, G. S. He, J. Swiatkiewicz, P. N. Prasad, *J. Phys. Chem. B* **1999**, *103*, 10741–10745; (b) W. H. Zhou, S. M. Kuebler, K. L. Braun, T. Y. Yu, J. K. Cammack, C. K. Ober, J. W. Perry, S. R. Marder, *Science* **2002**, *296*, 1106–1109; (c) O. Mongin, L. Porrès, L. Moreaux, J. Mertz, M. Blanchard-Desce, *Org. Lett.* **2002**, *4*, 719–722; (d) S. J. K. Pond, O. Tsutsumi, M. Rumi, O. Kwon, E. Zojer, J. L. Brédas, S. R. Marder, J. W. Perry, *J. Am. Chem. Soc.* **2004**, *126*, 9291–9306.

[119] M. Albota, D. Beljonne, J. L. Brédas, J. E. Ehrlich, J. Y. Fu, A. A. Heikal, S. E. Hess, T. Kogej, M. D. Levin, S. R. Marder, D. McCord-Maughon, J. W. Perry, H. Rockel, M. Rumi, C. Subramaniam, W. W. Webb, X. L. Wu, C. Xu, *Science* **1998**, *281*, 1653–1656.

[120] T. G. Goodson, *Acc. Chem. Res.* **2005**, *38*, 99–107.

[121] O. Varnavski, X. Z. Yan, O. Mongin, M. Blanchard-Desce, T. Goodson, *J. Phys. Chem. C* **2007**, *111*, 149–162.

[122] Z. Fang, T. L. Teo, L. P. Cai, Y. H. Lai, A. Samoc, M. Samoc, *Org. Lett.* **2009**, *11*, 1–4.

[123] M. Samoc, J. P. Morrall, G. T. Dalton, M. P. Cifuentes, M. G. Humphrey, *Angew. Chem.* **2007**, *119*, 745–747; *Angew. Chem. Int. Ed.* **2007**, 46, 731–733.

[124] G. Tsiminis, J. C. Ribierre, A. Ruseckas, H. S. Barcena, G. J. Richards, G. A. Turnbull, P. L. Burn, I. D. W. Samuel, *Adv. Mater.* **2008**, *20*, 1940–1944.

10

FINE-CONTROLLED METAL ASSEMBLY IN DENDRIMERS

TAKANE IMAOKA AND KIMIHISA YAMAMOTO
Chemical Resources Laboratory, Tokyo Institute of Technology, Yokohama 226-8503, Japan

10.1 INTRODUCTION

The dendrimer architecture [1–3] is suitable for the preparation of finely structured metal–organic hybrid frameworks, which is an advanced model of conventional macromolecular–metal complexes [4–6]. These materials are capable for use in catalysts [7–11], electronics [12,13], and other biological processes [14]. Furthermore, the precision in the metal–organic structure by the dendritic architecture provided a new possibility for drastic improvement in these characteristics or creation of a novel functionality in some cases.

Millions of organic compounds made of carbon (C), nitrogen (N), oxygen (O), and hydrogen (H) are now available. The structural variety of these organic compounds is due to many reactions that provide methods for cutting and pasting atoms at will. Even if the target molecule has a complicated structure, it could be easily prepared at the molecular level in most cases. Although the chemistry of higher order architectures (e.g., self-organizing structure, self-assembling structure, or helical structure) is now developing, the convenience is a significant advantage in employing the methodology of organic chemistry for the preparation of finely defined nanostructures [15].

Several bottom-up strategies to build a mesophase structure from organometallic complexes have actually been proposed. For example, macromolecule–metal hybridization is the most elementary procedure for constructing an assembly of metal complexes [16,17]. It is generally made by the complexation of metal ions with presynthesized macromolecular multiligands. As a result, metal ions or their complexes

Designing Dendrimers, Edited by Sebastiano Campagna, Paola Ceroni, Fausto Puntoriero.
© 2012 John Wiley & Sons, Inc. Published 2012 by John Wiley & Sons, Inc.

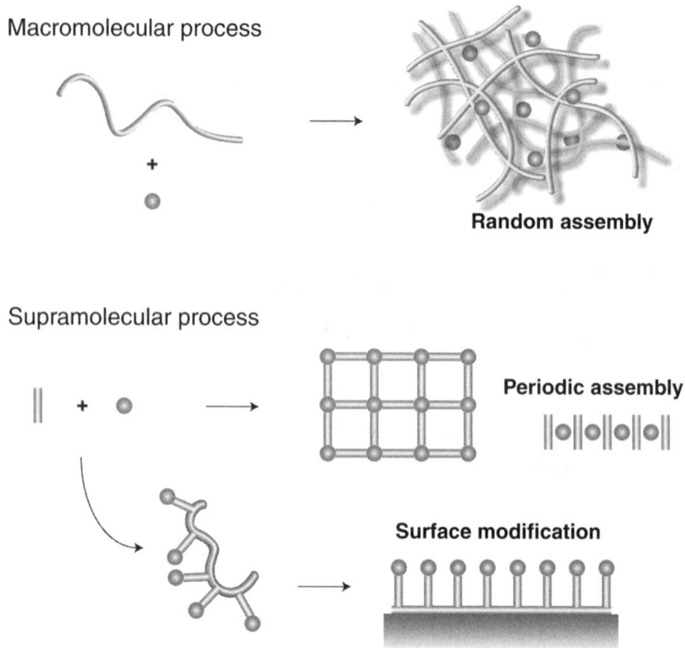

FIGURE 10.1 Fabrication techniques for metal—organic hybrid nanostructures. Conventional macromolecule-metal complexes provide a hybrid structure in which metal complexes are condensed in a polymer matrix. This procedure cannot regulate their mesophase structures because the matrix consists of length-dispersed polymer chains. Recently, several fabrication techniques (self-assembly and surface modification), which can form ordered molecular assemblies, have been proposed.

acquire the enhanced functions of redox communication or excitation energy transduction among the metal complex centers as they retain their character such as catalytic activity or photosensitivity. A steric effect of the mesophase structure also occasionally appears. However, this structure is often complicated because the length or topology of the molecular threads is not unity and not defined. In addition, the molecules are usually twisted around each other like boiled spaghetti strands; therefore, the nanoenvironment around the metal complex center should have a statistical distribution caused by the disorder in the mesophase (Figure 10.1). Decreasing the disorder around the metal center is necessary to enhance and sharpen the functionality of the complex. Further extension of these defined structures to a mesoscopic scale (2–50 nm) has been significantly developed because it is essential for solid-state materials to improve their function and efficiency.

The strategy using the macromolecule–metal architecture has now moved to the next stage. The self-assembly [18–22] or surface modification techniques (self-assembling monolayer (SAM), LB [23,24], or LBL (layer-by-layer) assembly [25–28]) are currently employed in most cases for the fine-tuning of the polymer aggregation structure (Figure 10.1). Recently, a new approach using a dendritic

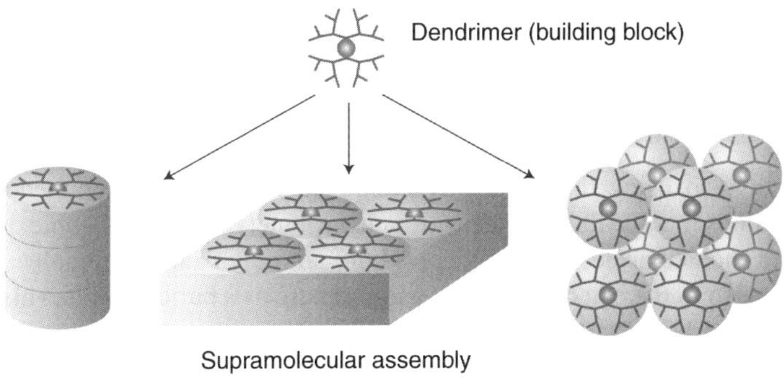

FIGURE 10.2 The dendrimers can be used as the building blocks for the fabrication of mesophase molecular packing structures. The self-assembling in general depends on the terminal units or the entire conformation of the dendrimer.

architecture has been developed as a building strategy toward a well-organized supramolecular assembly (Figure 10.2) [29–37]. This approach is similar to that of proteins in a biological system. Metal-containing proteins are hybridized with lipid bilayers to construct a cell as a functional unit of life.

This review describes recent advances in the production of metal-assembling dendrimers through a fine complexation process. Because the dendrimer has a discrete topology with a defined molecular weight and structure, the metal-assembling dendrimer could be used as a discrete molecular reactor for the fine synthesis of inorganic (metal or metal-oxide) nanomaterials. This strategy [38] first proposed by Crooks et al., could be enhanced to a perfect control of the number of metal atoms composed of one nanoparticle by using the fine-controlled metal assembly in specific dendrimers, which enabled atomic-level control of the templating [39,40]. As a result, we could elucidate the unexplored properties of subnanometer-inorganic clusters, which are much different from those of the larger clusters or bulk metals.

These materials are also applied to electronics as hole- or electron-transporting materials [17]. Many compounds have been developed for photochemical applications, such as solar cells or light-emitting devices. The advantages of using the architecture of a macromolecular metal complex are as follows: (1) Utilization of various excitations including MLCT (metal-to-ligand charge transfer), LMCT (ligand-to-metal charge transfer), and IVCT (intervalence charge transfer) in addition to the inherent excitation pathways of metal and organic ligands, such as the d–d and π–π^* transitions. (2) Energy transduction and electron transfer accompanying the excitation are available in nanoscopic-condensed systems in which a donor–acceptor pair is closely placed within a few nanometers. (3) Not only the chemical composition but also the higher order nano- to microscopic geometry can be designed on the basis of the established architecture of the macromolecules. We also discuss the common photochemical devices, such as OLED (organic light-emitting diode) or solar cells, highlighting the strategy for the material development of the device components.

10.2 SYNTHESIS AND PRINCIPLES OF THE METAL-ASSEMBLING DENDRIMERS

10.2.1 Architecture

An advantage to using the dendrimer architecture for the macromolecule–metal hybridization is that the microdomain around each metal center is tunable during the topological molecular design. They contribute not only to the intramolecular-level control but also to the extended mesophase structure as a building unit. This fact has recently attracted much attention because it should become a powerful tool to build a hierarchical structure on nanometer and micrometer scales. Needless to say, the metallodendrimer would be positioned in a unit cell for macromolecule–metal hybrid composites. The most important feature of this procedure is almost complete independence between the designing of each phase in the molecular unit (functional center), local domain (dendrimer), and extended structure (molecular assembly). To establish the entire process, a universal way to finely decorate various metallic elements in the dendrimer is a key process.

Metallodendrimers are roughly categorized in two different groups by the binding mode of the metal element in the whole architecture. One is metallodendrimers having metal complexes incorporated as the building units [41–44]. The others are metal-assembling dendrimers having metal ions attached to the dendrimer ligands [40,45–62]. The former way provides a completely defined structure of the organometallic hybrids, but has the limitation that there are few available structures and applications. The metal ligation should be strong enough so as not to exchange or decompose. The latter way can be applied to various metallic elements not only in static complexes but also in dynamic coordination systems, such as metal cluster synthesis [63–68]. However, the hybrid structure of metals and dendrimers in a substance cannot be defined to one limited form until all the coordination sites in the dendrimer were filled by metal ions (Scheme 10.1). Because the environment for each generation (layer) of the dendrimer is not the same, the complexes in each generation are chemically different. This means a loss of structural unity in the metal–dendrimer hybrid. In addition, even the number of loaded metal ions in the dendrimer molecules should have a statistical dispersity. When metal clusters were synthesized using normal dendrimer ligands, one should find some dispersity in the cluster size. This review mainly focuses on this point based on a unique multicoordinating dendrimer in which metal complexes significantly recognize the site to coordinate.

10.2.2 Multicoordination Equilibria

If there were macromolecular ligands that can assemble metal ions in a finely controlled stepwise process, the resulting metal assembly should have a much lower dispersity than one that results from a random coordination system. As a typical example, we assume a model system in which the coordination sites are arranged to form a dendritic chain. The number of coordination sites is doubled by stepping the generations (max fourth generation) in the dendrimer model. The coordination sites in

COLOR PLATES

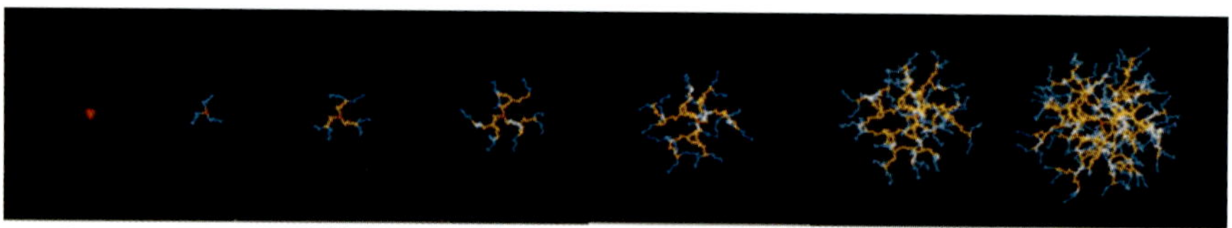

Core | 1nm G=0 | 1.58nm G=1 | 2.2 nm G=2 | 3.10 nm G=3 | 4.0 nm G=4 | 5.3 nm G=5

FIGURE 1.9 Molecular simulations for [core:NH_3]; ($G = 0$–5); {*dendri*-poly(amidoamine)-$(NH_2)_n$} (PAMAM) dendrimers and a generational comparison of hydrodynamic diameters.

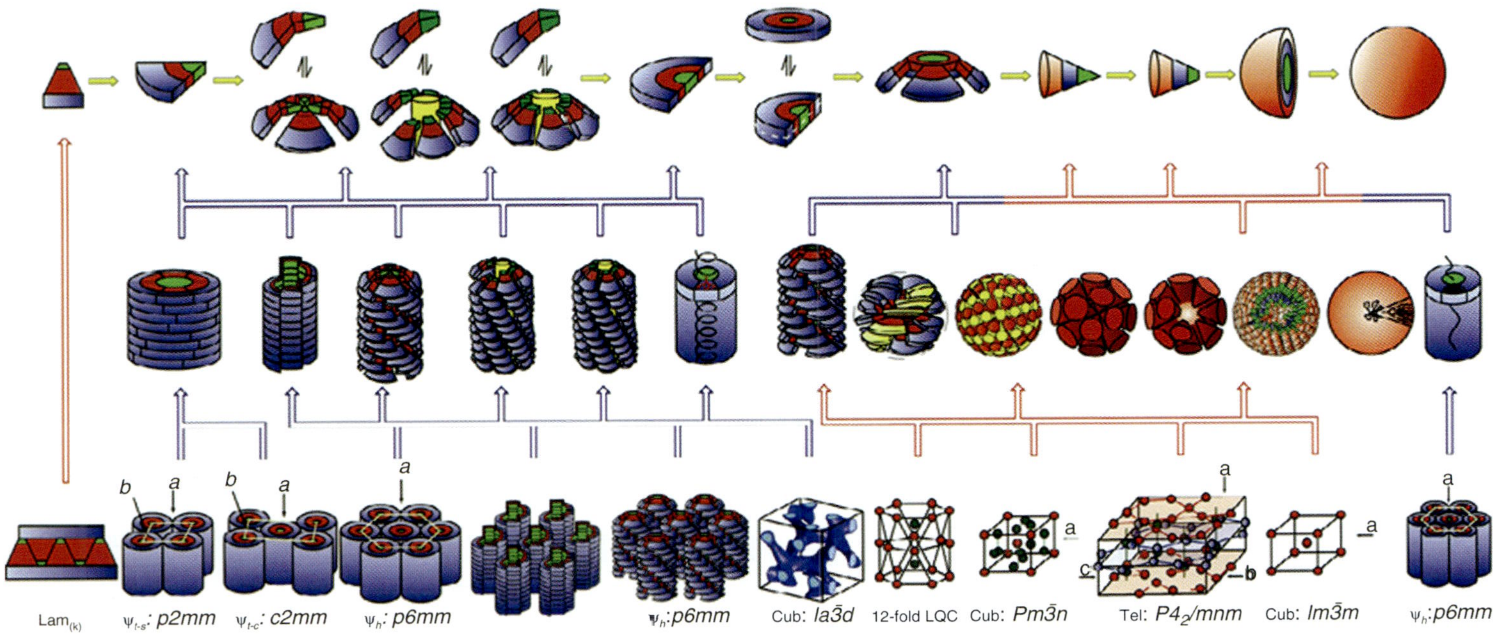

FIGURE 1.12 Periodic and quasiperiodic lattices formed via the self-assembly of Percec-type dendrons [45].

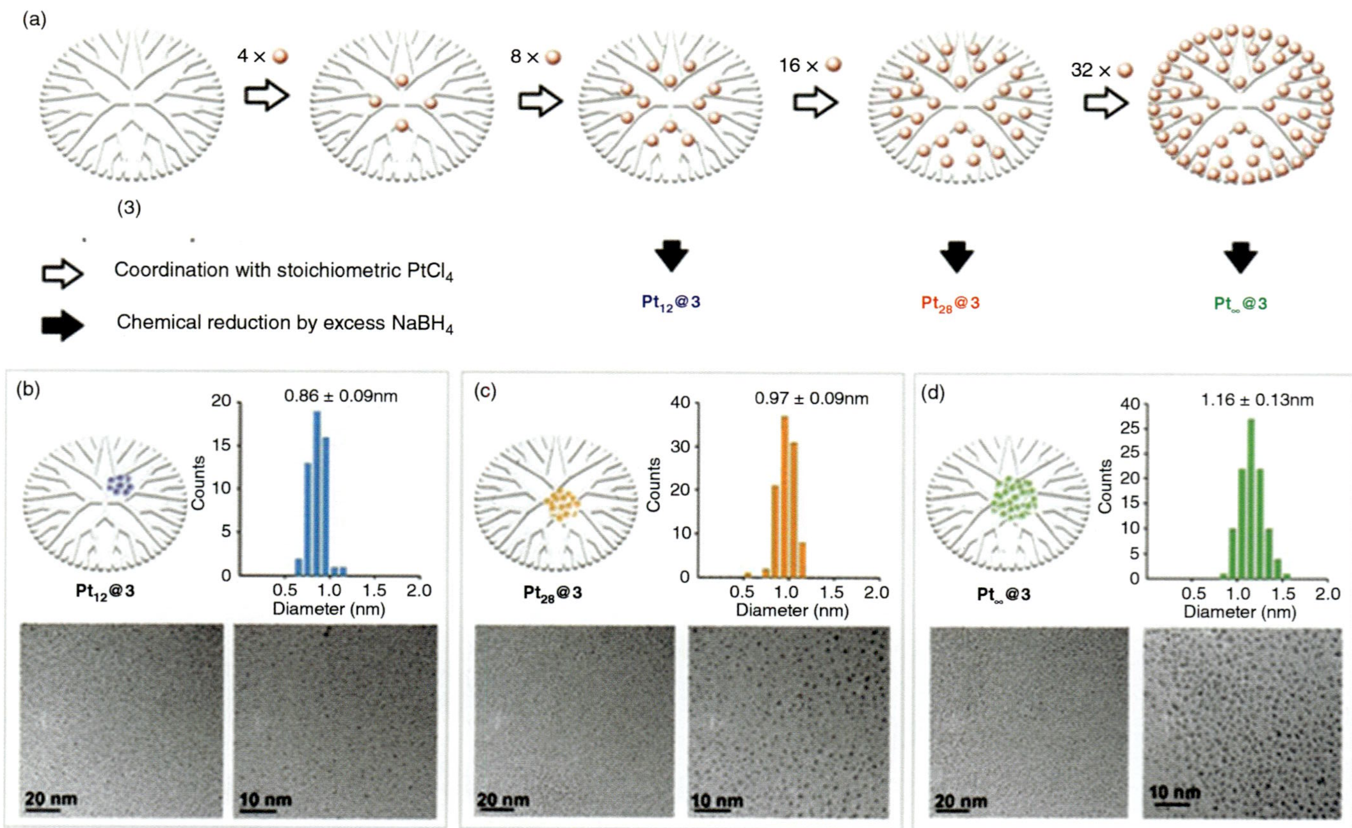

FIGURE 1.21 (a) Sequential complexation of $PtCl_4$ with a Yamamoto-type dendrimer, (3), to produce $Pt_{12}@3$, $Pt_{28}@3$, and $Pt_{60}@3$; $[H\text{-}3]_n$:[S-1] core–shell type nano-compounds, respectively, after reduction with $NaBH_4$. (*See text for full caption.*)

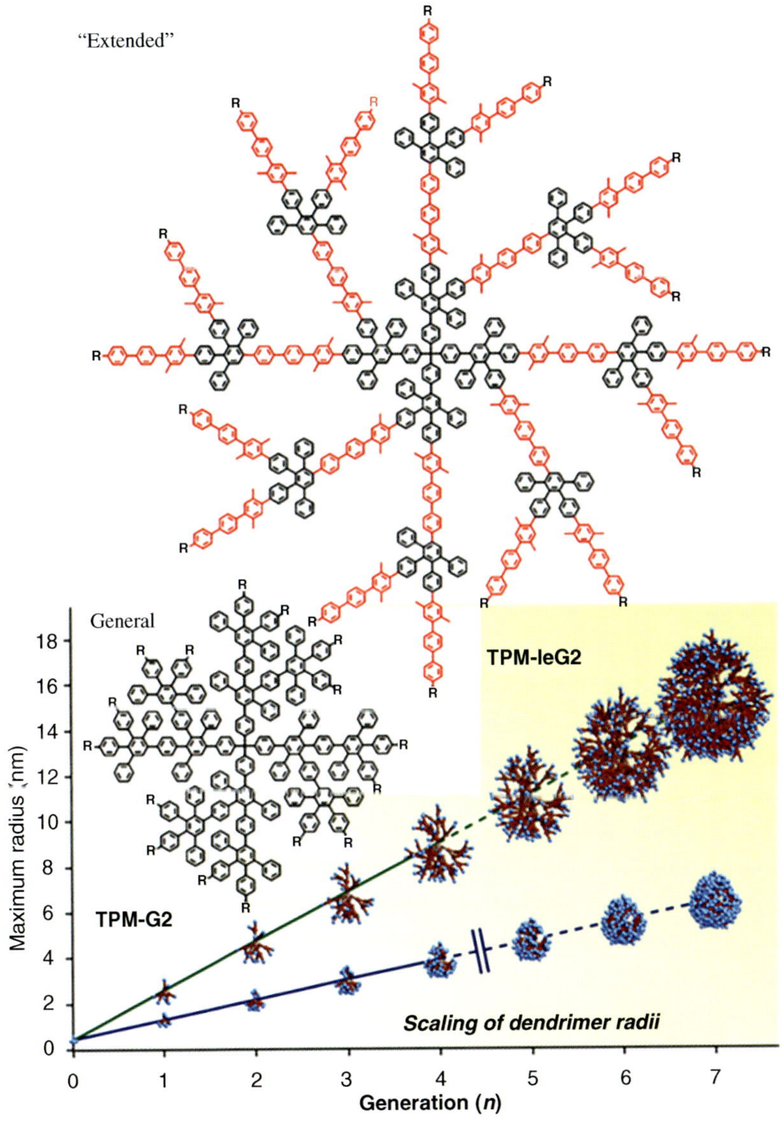

FIGURE 5.4 The generation-induced changes of the maximum radius of *standard* polyphenylenes TPM-G1–5 and those with linear triphenylene extended dendrons TPM-leG1–7.

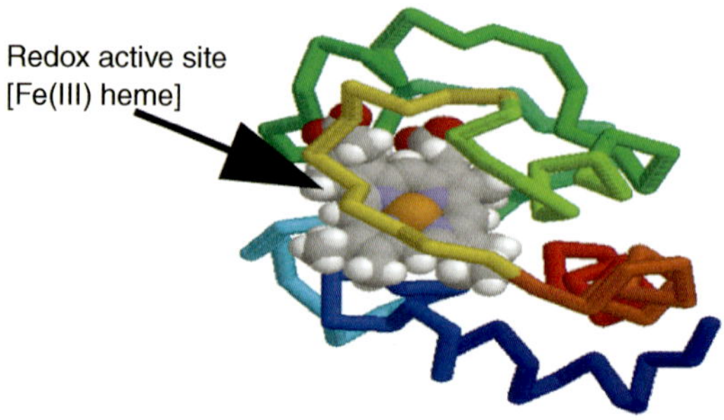

FIGURE 7.2 A schematic representation of horse heart cytochrome *c* showing the location of its redox active site, partially buried inside the protein framework.

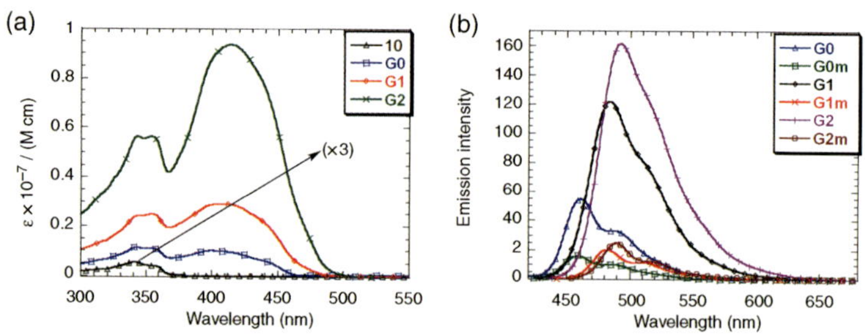

FIGURE 9.26 (a) Absorption spectra of dendrimers **29**, **30**, **31**, and compound **10** in THF solution (10^{-7} M) at room temperature. (*See text for full caption.*)

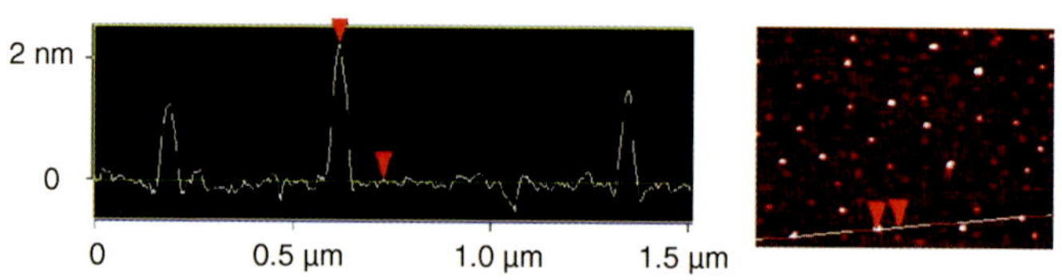

FIGURE 9.27 AFM images and height profiles of the individual dendrimer molecule **31** on mica from a 3 nM solution. Reproduced by permission of the American Chemical Society [48].

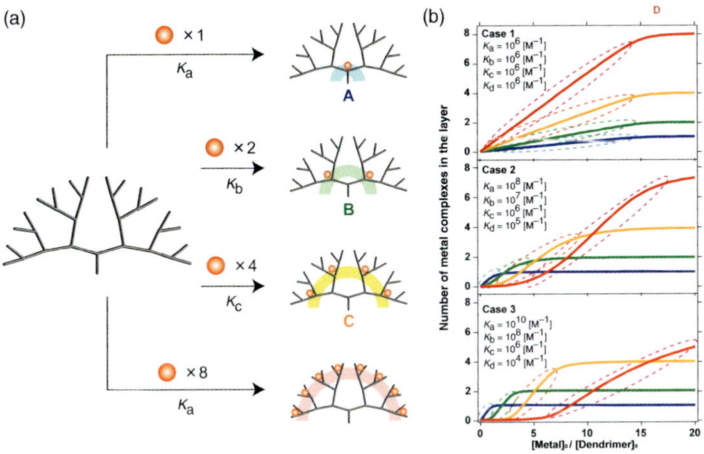

FIGURE 10.3 (a) A model of a multicoordination equilibrium system based on a four-layered dendrimer. (*See text for full caption.*)

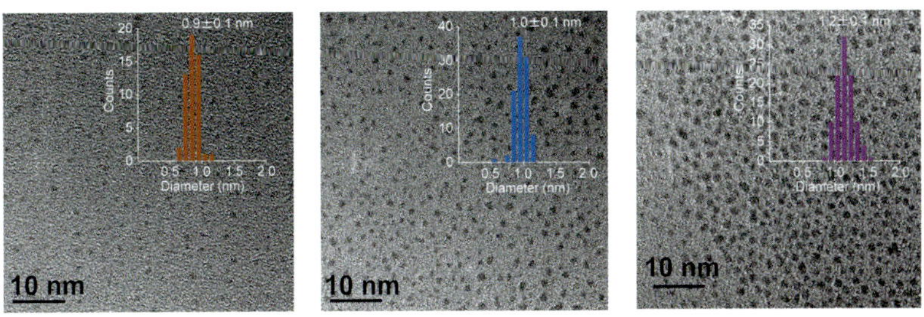

FIGURE 10.8 Transmission electron micrograph (TEM) image of Pt clusters (left: Pt_{12}, Center: Pt_{28}, right: Pt_{60}) synthesized in a phenylazomethine dendrimer template (**TPM-DPA-G4**).

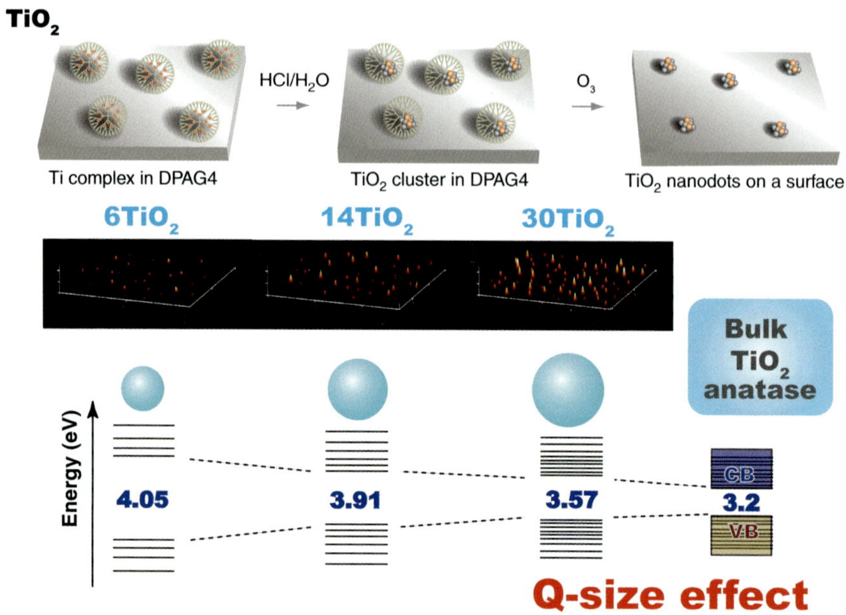

FIGURE 10.10 Schematic representation of the quantum-size TiO_2 nanodot on a surface synthesized using a dendrimer template.

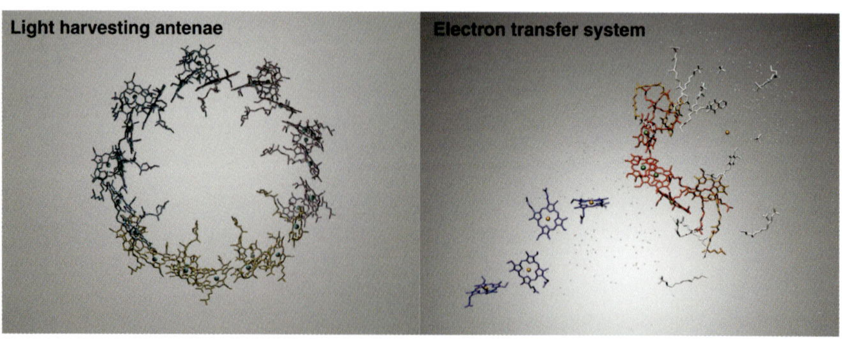

FIGURE 10.14 Molecular 3-D coordination structures of metalloproteins playing an important role in the photosynthetic reaction center. (*See text for full caption.*)

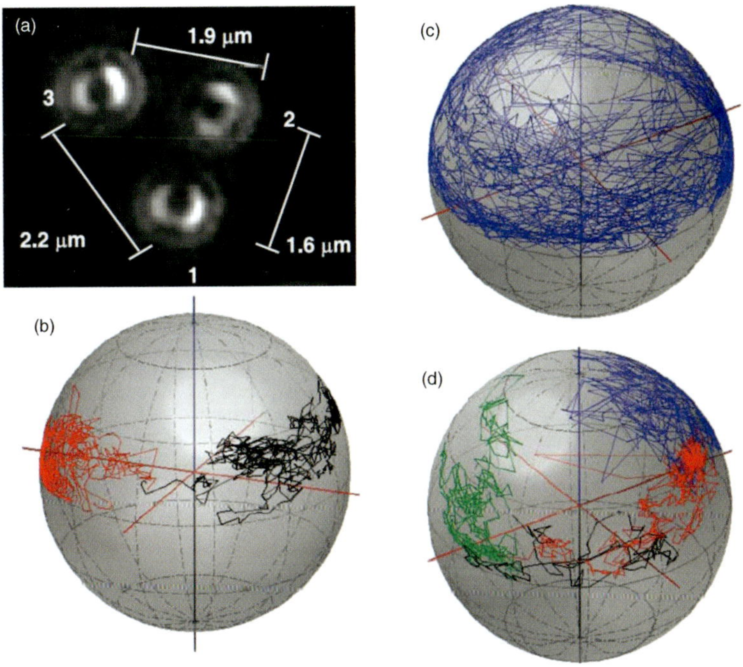

FIGURE 12.3 (a) Three single molecules of compound **2** embedded in a PMA polymer film at 295 K. (b–d) Projection maps for molecule 1–3, respectively. (*See text for full caption.*)

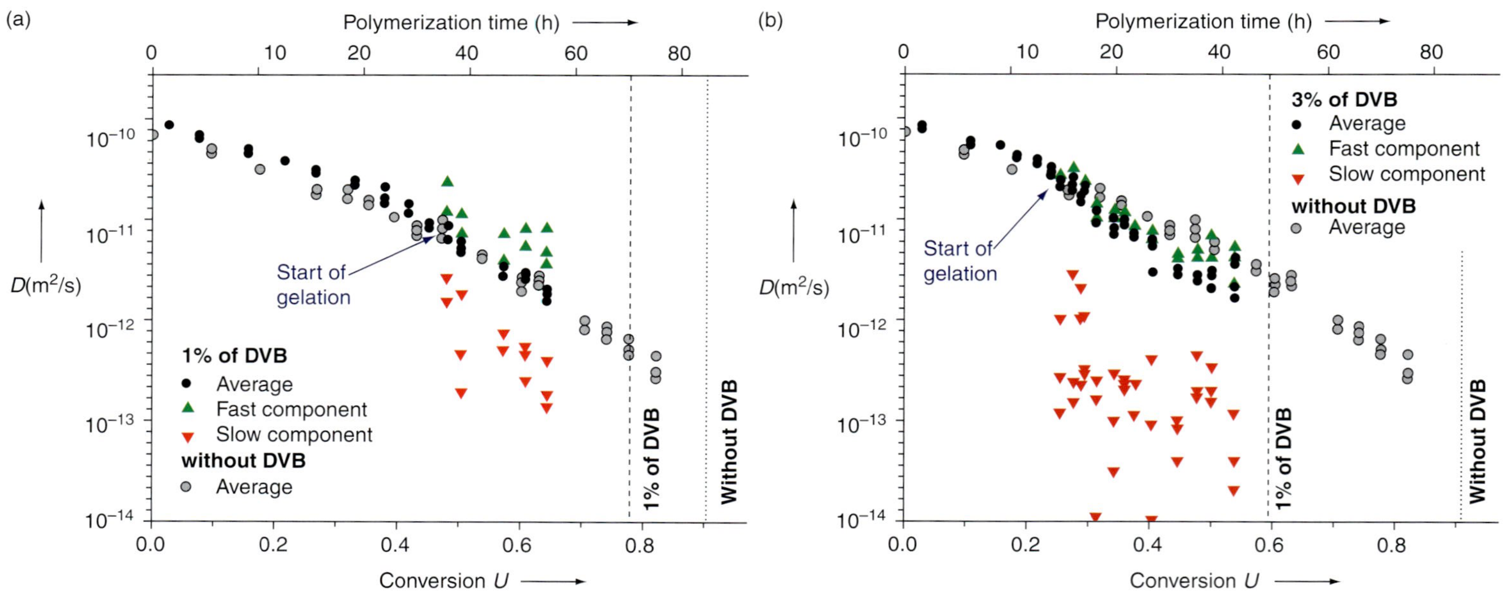

FIGURE 12.4 Dependence of the diffusion constant D obtained by FCS and wide-field microscopy on U and on the polymerization time using compound **4** as probe: (a, b) Polymerization of styrene probed with compound **4** in the presence of 0%, 1%, and 3% w/w of 1,4-divinylbenzene (DVB) cross-linker. The dashed line indicates the point at which all molecules were immobilized as determined by wide-field microscopy. Copyright from Ref. 44.

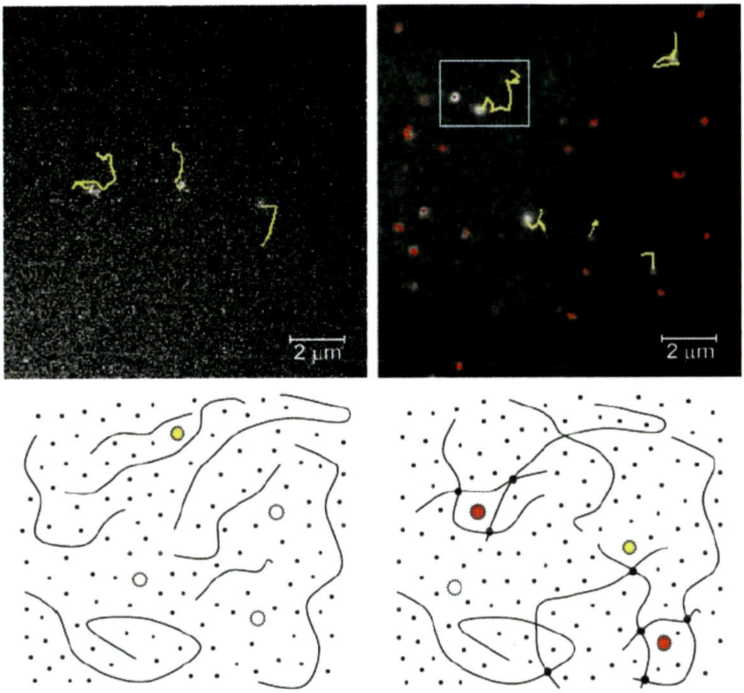

FIGURE 12.5 (Left side) Wide-field image at 0.64*U* and schematic representation of compound **4** without the presence of DVB cross-linker. (*See text for full caption.*)

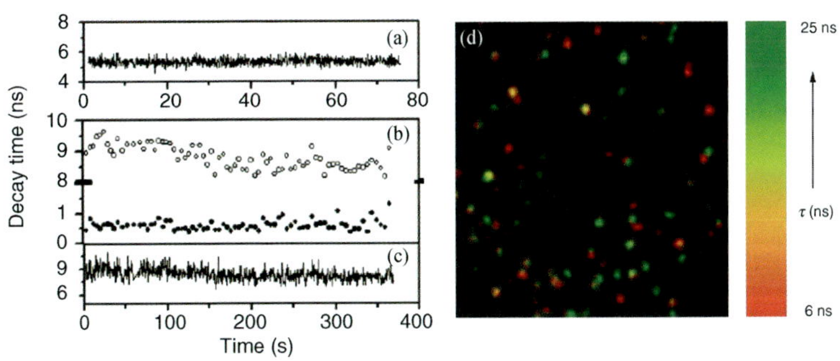

FIGURE 12.8 (a) Single molecule of compound **7** in Zeonex binned at 500 photons per decay and analyzed as a monoexponential. (*See text for full caption.*)

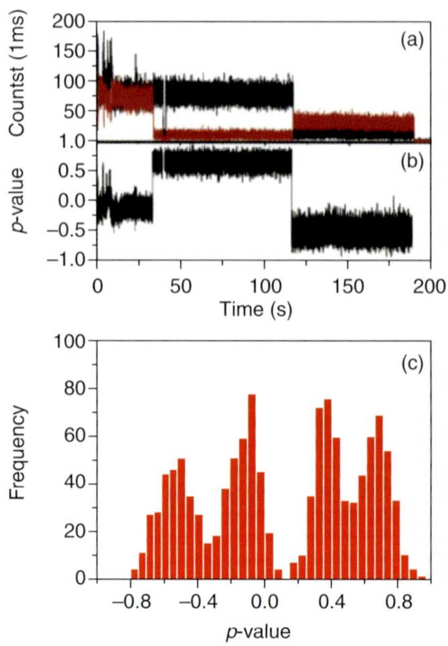

FIGURE 12.12 (a) Example of a fluorescence intensity trajectory of a single molecule of compound **11** where the parallel and perpendicular component where detected separately. (*See text for full caption.*)

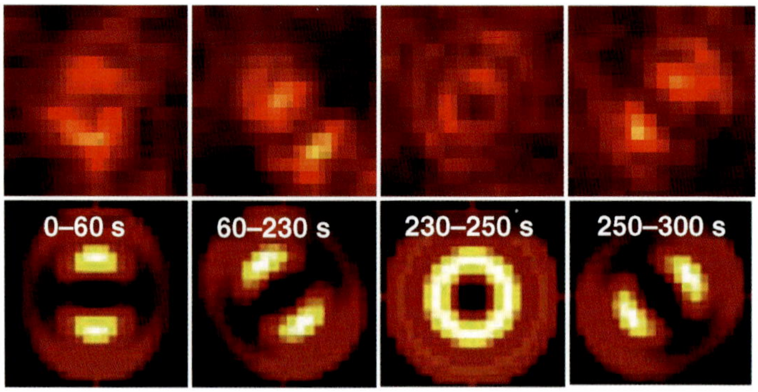

FIGURE 12.13 Series of emission patterns and corresponding simulated patterns for a single molecule of compound **11** in a 25 nm Zeonex film. (*See text for full caption.*)

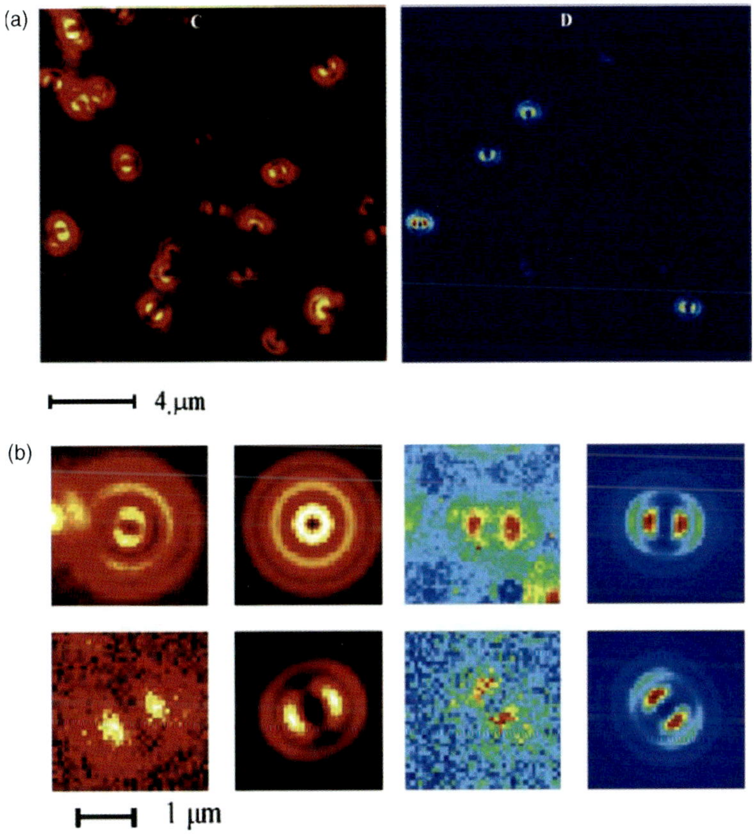

FIGURE 12.19 (a) Defocused wide-field images of the same sample area but different emission wavelengths, showing the emission and orientation of TDI (left) and PMI chromophores (right). (*See text for full caption.*)

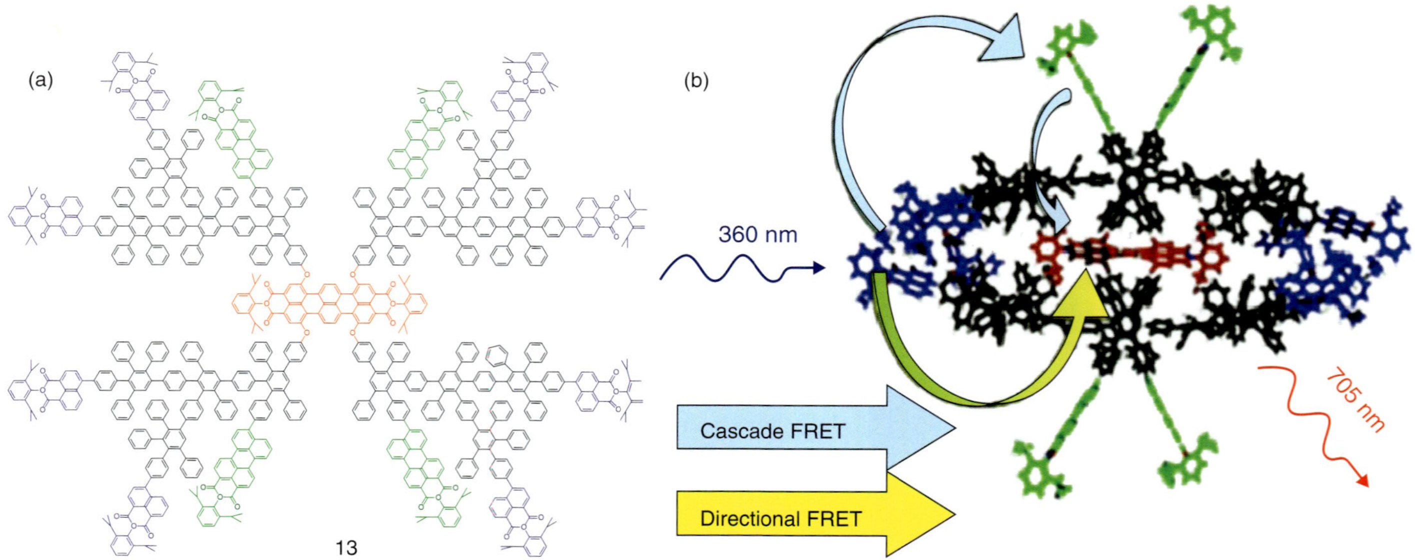

FIGURE 12.20 (a) Chemical structure of compound **13** containing 8 NMI chromophores at the outer shell, 4 PMI chromophores within the scaffold and 1 TDI in the center. (b) 3D structure of compound **13**. The arrows illustrate energy transfer processes after exciting the NMI chromophores.

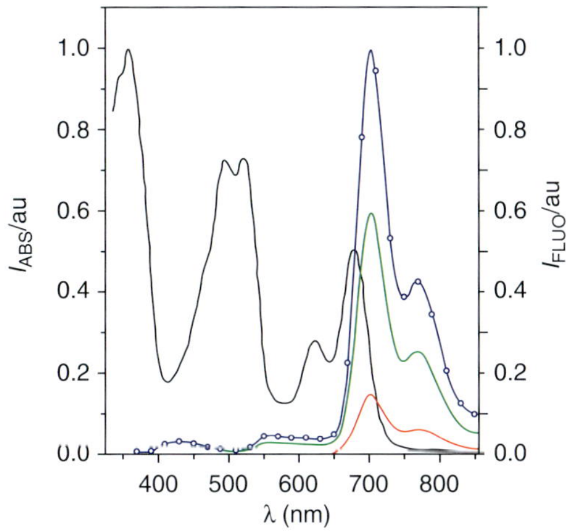

FIGURE 12.21 Absorption (black line) and fluorescence (line + symbol) spectra of compound **13** in toluene. (*See text for full caption.*)

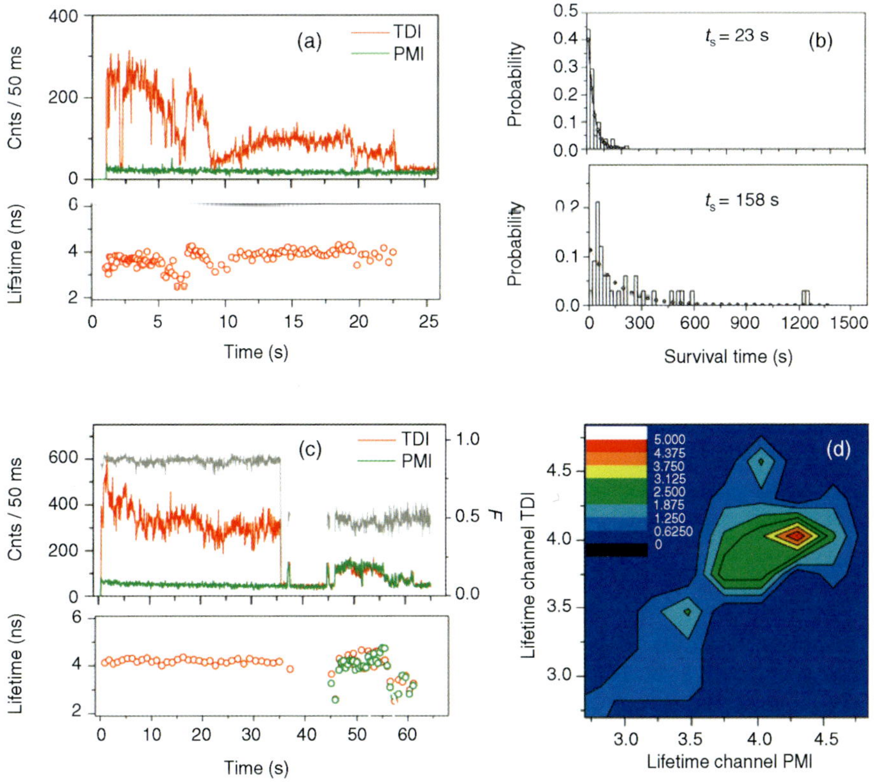

FIGURE 12.23 Single-molecule fluorescence intensity trajectories on 360 nm excitation of two molecules of compound **13** in Zeonex. (*See text for full caption.*)

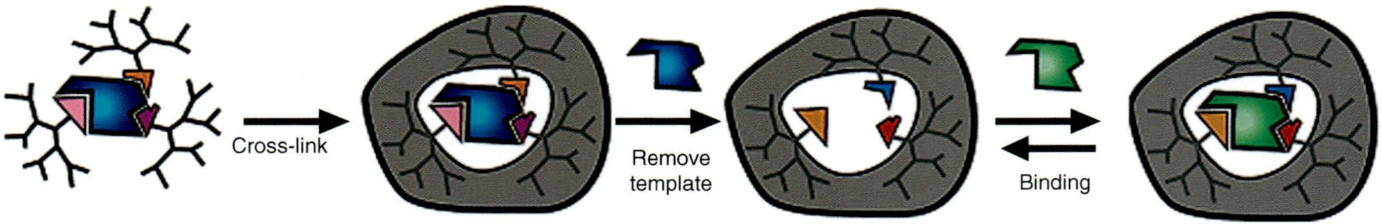

SCHEME 13.14 General concept of monomolecular imprinting inside dendrimers.

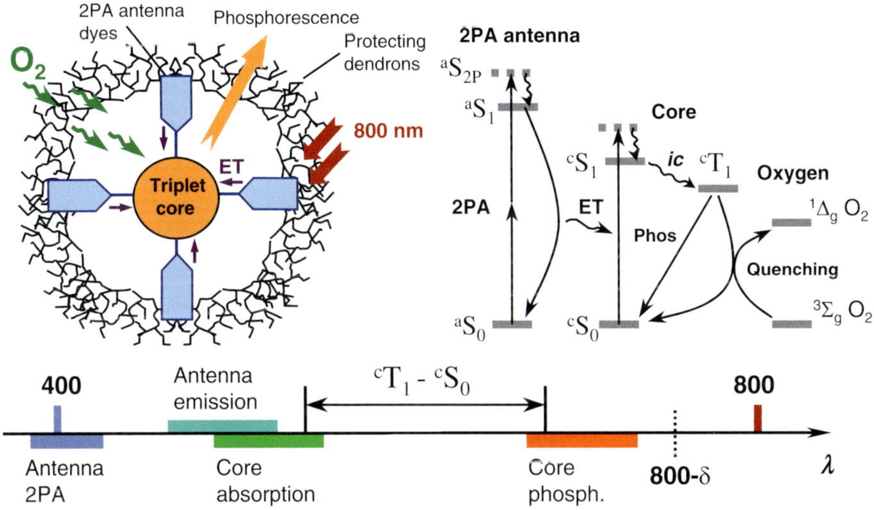

FIGURE 14.15 Scheme of a two-photon-enhanced phosphorescent oxygen sensor, its Jablonski diagram and related wavelength scale.

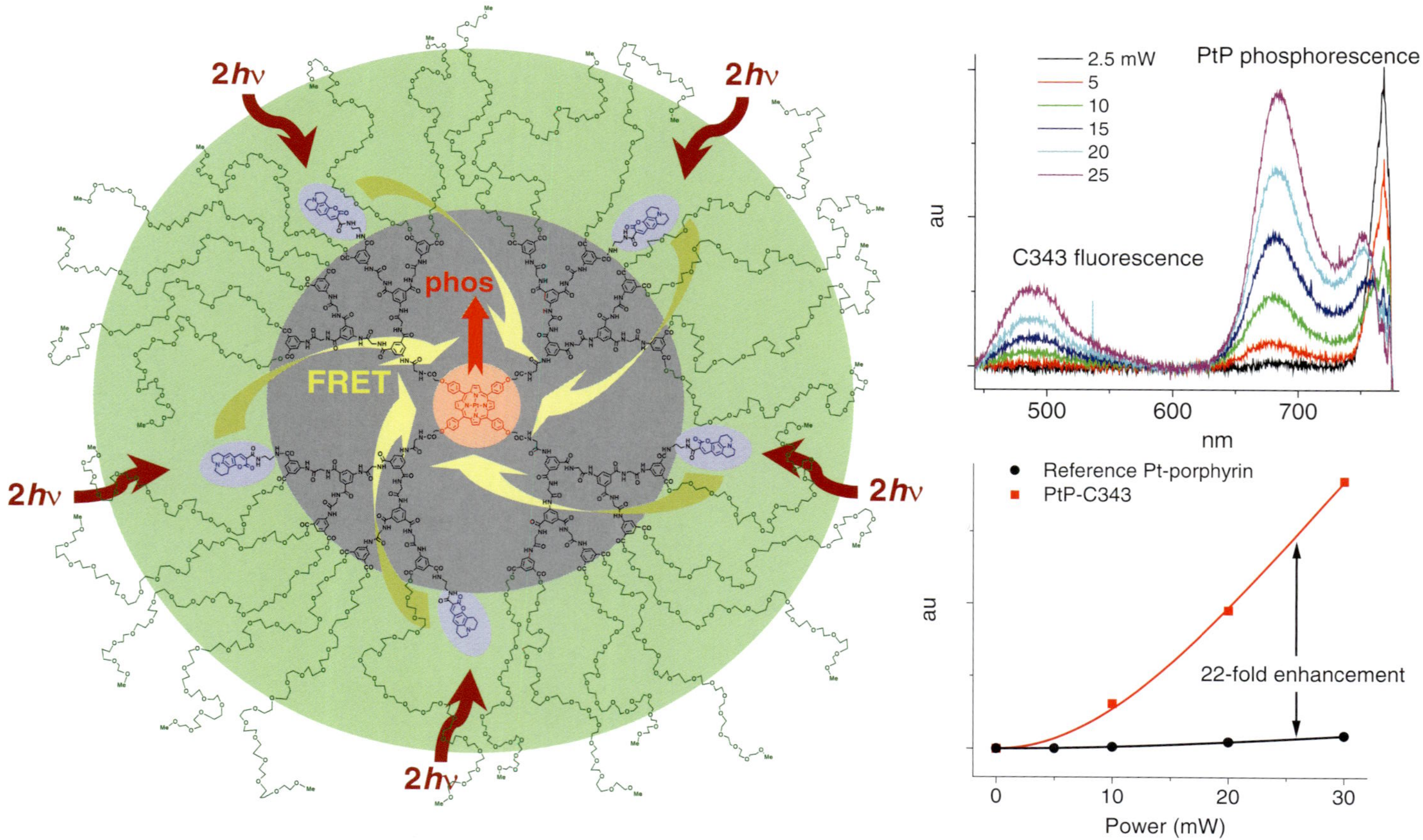

FIGURE 14.20 Two-photon-enhanced oxygen probe PtP–C343 consisting of phosphorescent Pt-*meso*-tetra-*para*-alkoxyphenylporphyrin (PtP, red), several coumarin-343 units (C343, blue), poly(aryl-glycine) dendrimer (black) and peripheral oligoethyleneglycol residues (green). (*See text for full caption.*)

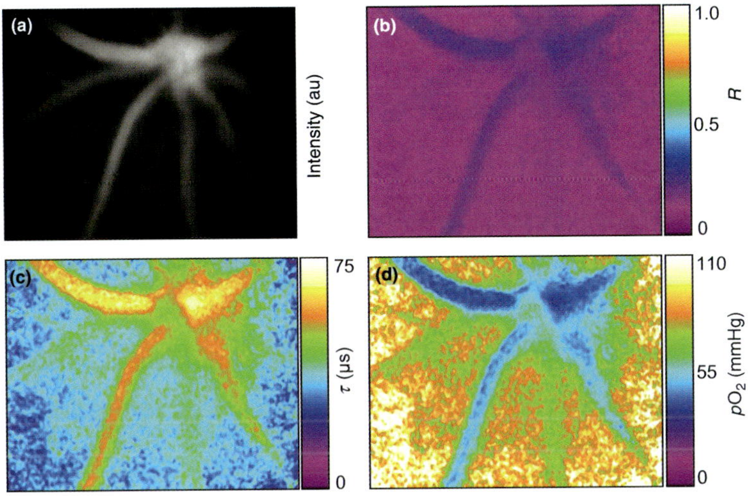

FIGURE 14.21 Maps of the phosphorescence intensity (a), goodness of fit shown by R-factor (b), phosphorescence lifetime (c), and oxygen pressure (d). The maps correspond to a region whose physical dimensions are 0.92 mm × 1.2 mm [121].

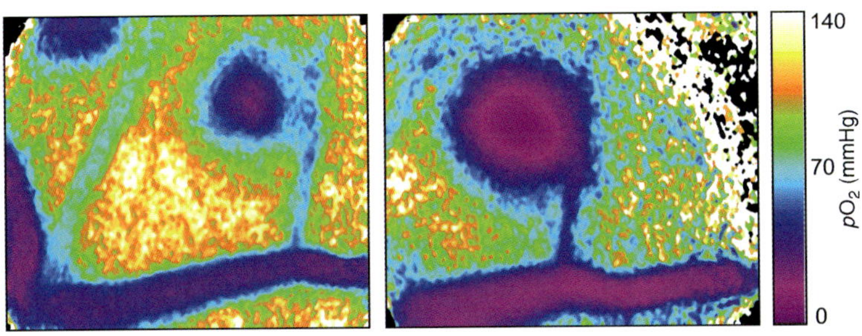

FIGURE 14.22 Oxygen pressure maps of the region containing two laser coagulation spots (left) and a region centered on one of the spots (right).

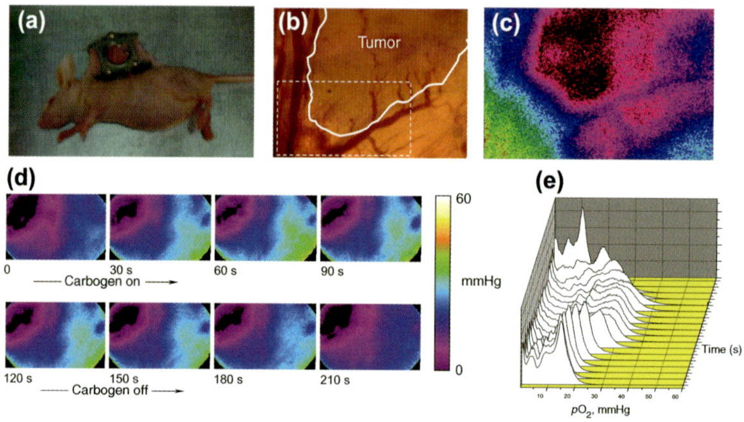

FIGURE 14.23 Imaging tumor hypoxia in a mouse dorsal window tumor model using Oxyphor G3 (gen 2 Pd tetrabenzoporphyrin–AG dendrimer). (*See text for full caption.*)

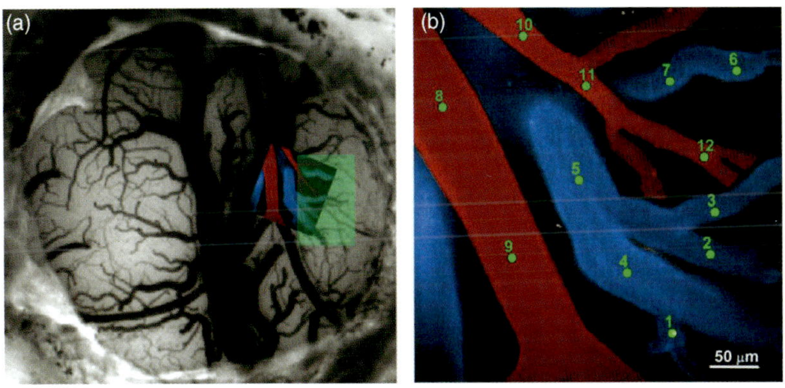

FIGURE 14.25 (a) CCD image of cranial window in a rat with confocal angiogram overlaid and region of functional activation identified in green. (b) Color-coded angiogram of microvessels in rat somatosensory cortex, with identified locations for pO_2 measurement [18].

(a)

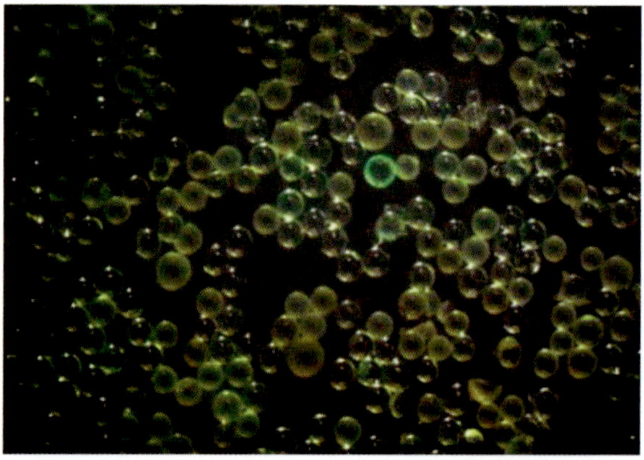

(b)

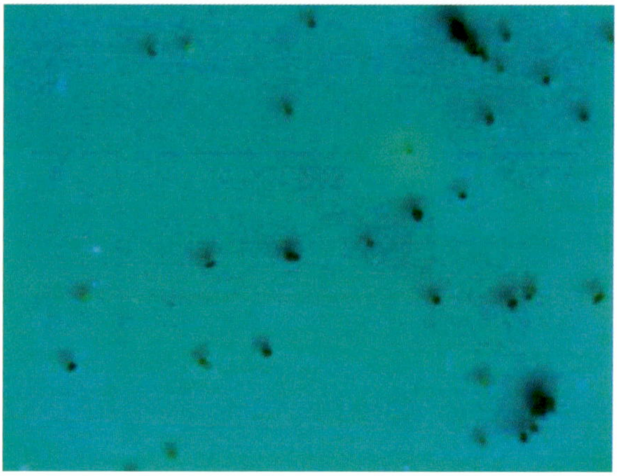

FIGURE 15.2 (a) On-bead assay with 8-butyryloxypyrene-1,3,6-trisulfonate **1** leading to the identification of dendrimer **C11** [14]. (*See text for full caption.*)

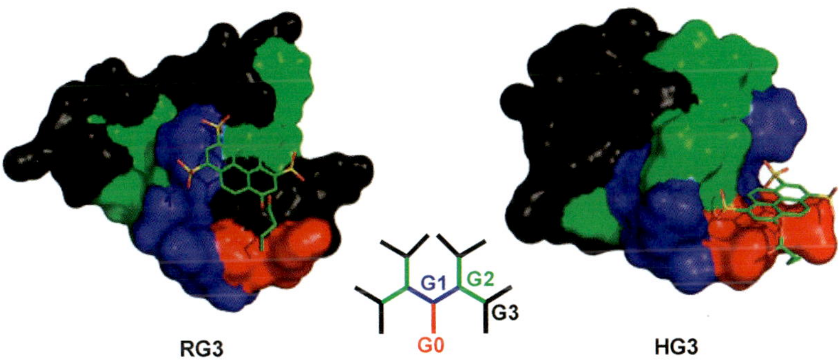

FIGURE 15.11 Docking poses of substrate **1** to peptide dendrimer **RG3** and **HG3**. The dendrimer is shown in surface representation color-coded by generation number (red = G0, blue = G1, green = G2, black = G3). Substrate carbons are represented in green, nitrogens in blue, oxygens in red, and sulfurs in yellow.

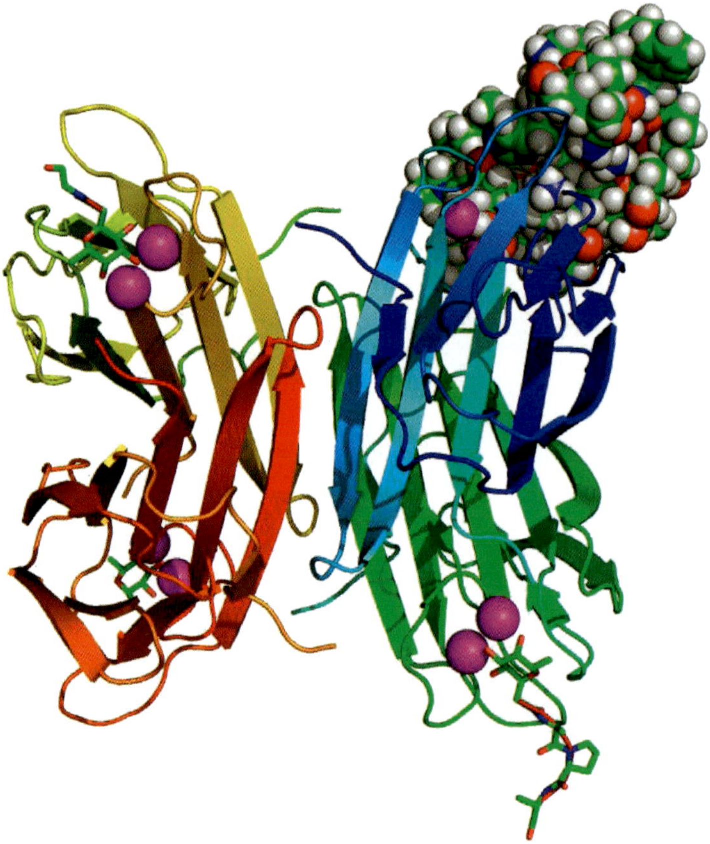

FIGURE 15.12 Structural model of the dendrimer **FD2**–LecB complex obtained by molecular dynamics and energy minimization. The Ca^{2+} ions are shown in magenta and indicate the location of the fucose-binding pocket. Three binding sites are shown with the resolved fucosyl residues of the tripeptide **2G0** cFucLysProLeuNH$_2$ and the fourth site is modeled with dendrimer **FD2**.

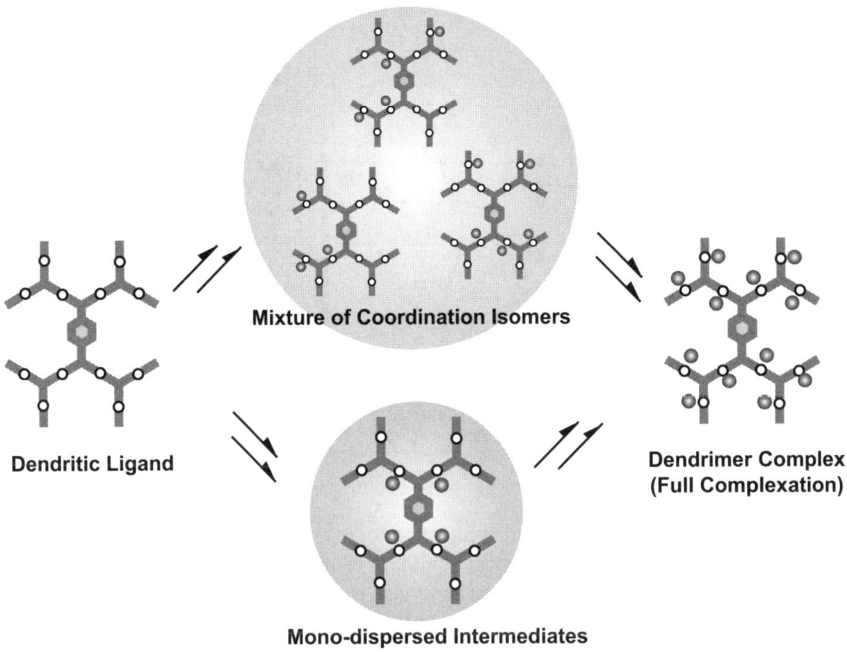

SCHEME 10.1

the same generation are placed in the same coordination environment due to their topological character, while the environments in the different generations are not identical. Here we define four individual coordination reactions and their constants (K_a–K_d), which correspond to each generation (layer) in the dendrimer model.

$$A + M \rightleftharpoons AM \quad (K_a = [AM]/[A][M],\ [A]_0 = [\text{dendrimer}]) \tag{10.1a}$$

$$B + M \rightleftharpoons BM \quad (K_b = [BM]/[B][M],\ [B]_0 = 2[\text{dendrimer}]) \tag{10.1b}$$

$$C + M \rightleftharpoons CM \quad (K_c = [CM]/[C][M],\ [C]_0 = 4[\text{dendrimer}]) \tag{10.1c}$$

$$D + M \rightleftharpoons DM \quad (K_d = [DM]/[D][M],\ [D]_0 = 8[\text{dendrimer}]) \tag{10.1d}$$

As shown in a previous paper [39], the concentration profiles of all the free ligands ([A]–[D]) and metal ion ([M]) after reaching equilibrium can be calculated for the given initial concentrations of the dendrimer ($[\text{dendrimer}]_0$) and metal ions ($[M]_0$). The profiles of the coordinated ligands against the added metal ion are shown in Figure 10.3. If all the coordination constants were identical (Case 1: $K_a = K_b = K_c = K_d = 10^6\ \text{mol}^{-1}\ \text{L}$), the metal ions would be equally distributed to all ligands (A, B, C, and D) at any molar amount of addition. In this case, the number of metals in one dendrimer molecule would have a statistical distribution until the coordination sites are completely filled by the added metal ions. For example, the metal ions are equally distributed to each layer depending on the number of coordination sites in each layer on

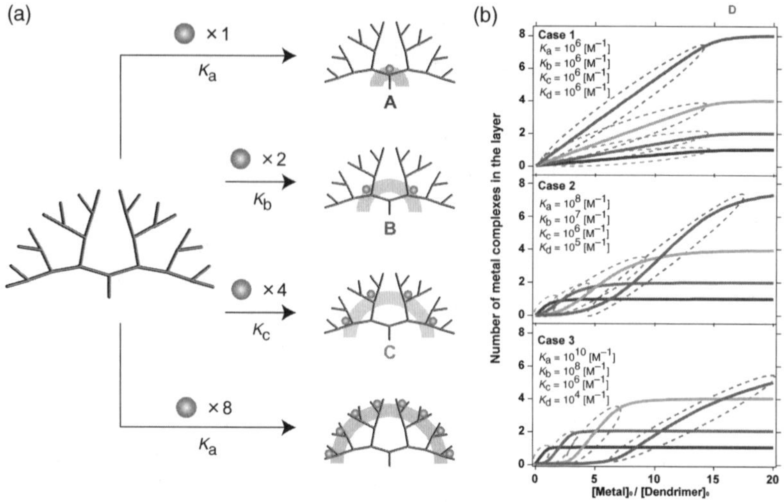

FIGURE 10.3 (a) A model of a multicoordination equilibrium system based on a four-layered dendrimer. Definition of four independent equilibria enables calculations of the entire complexation status for the given coordination constants. (b) Calculated profiles of the number of metal complexations in each layer versus the added equivalent ratio of the metal ion for three different situations. The assumed concentration of the dendrimer is $2 \times 10^5\,\text{mol}^{-1}\,\text{L}$ and the number of coordination sites in each layer is 1, 2, 4, and 8. (See the color version of this figure in Color Plates section.)

the midway (5 M amounts) to the full complexation (Table 10.1). In contrast, the coordination profiles were changed by downward trend in the coordination constants (Case 2: $K_a = 10K_b = 100K_c = 1000K_d = 10^7\,\text{mol}^{-1}\,\text{L}$). At the initial point of the addition, the distribution priority to the core-side layer is higher than that to the outer layer, and then the prior coordination site is gradually moved to the outer layer. If 5 M amounts of metal ion were added to the dendrimer, 89% (4.46/5) of the added metal ion is distributed to the inner three layers. The behavior became striking when a larger change in the constants is assumed (Case 3: $K_a = 10^2K_b = 10^4K_c = 10^6K_d = 10^{10}\,\text{mol}^{-1}\,\text{L}$). In this case, the added metal ions clearly distinguished the sites to which they should coordinate. During the initial stage ($0 < [\text{M}]_0/[\text{dendrimer}]_0 < 1$), most of the metal ions

TABLE 10.1 Number of metals distributed to each layer of the dendrimers assumed in Fig. 10.3 upon the quantitative addition of 5 molar amounts

Layer	Ligands[a]	1	2	3
A	1 (1)	1.00	0.99	0.33
B	2 (2)	1.98	1.76	0.66
C	4 (2)	1.90	1.68	1.33
D	8 (0)	0.07	0.54	2.66

[a] Values in the parentheses are the number of metals distributed to each layer if the ideal stepwise coordination was occurred from the core side.

were coordinated to the core-side layer (A). After the filling of layer A, the coordination site moved to the next layer (B) until that layer was filled ($1 < [M]_0/[dendrimer]_0 < 3$). Again it moved to layer C ($3 < [M]_0/[dendrimer]_0 < 7$) and finally to D ($7 < [M]_0/[dendrimer]_0 < 15$). The coordination process was completely stepwise; therefore, most metal ions (99%) would be trapped within the inner three layers if 5 M amounts of the free metal ions were added to the dendrimer (Table 10.1). The dendrimer model assumed in Case 3 can distribute metal ions to core-side layers with higher priority. In principle, this should result in the monodispersity of the number of metal ions distributed to each dendrimer molecule because the statistical factor becomes negligible at each section of the molar ratio between the dendrimer and the metal ion.

10.2.3 Fine-Controlled Metal Assembling in DPAs

As shown in the previous section, the complexation randomness within the dendritic architecture should be removed if the constants of the metal coordination could be gradated by layers. A design strategy for such an unusual system is to utilize the unique dense-shell property of rigid dendrimers [69,70]. All dendrimers reported as ligands for multimetal ion assemblies have flexible backbones composed of single covalent (σ) bonds. However, they are not suitable for precise metal assembly because their conformational changes and back folding with thermal vibrations would prevent the precise discrimination of the coordination sites [71]. We reported some dendritic ligands that have fully aromatic π conjugating backbones [72,73]. The dendritic phenylazomethines (DPAs) shown in Figure 10.4 behave as unique dendritic ligands, in which various metal ions were coordinated to the imine nitrogen (C=N) from the core-side layer in an orderly fashion.

An example of the stepwise complexation about the simplest DPA (**Ph-DPA G4**) with $SnCl_2$ is explained below [39]. $SnCl_2$ acts as a Lewis acid and forms a 1:1 complex with each imine (C=N) in the **Ph-DPA G4**. This is confirmed by the model compound (**Ph-DPA G1**) bearing only one imine unit. Because the complexation accompanies a significant change in the π–π^* transition energy, it can be monitored by measurements of the UV-vis absorption spectra upon the addition of $SnCl_2$. The spectral change saturating up to the addition of 1 M amount of $SnCl_2$ means a 1:1 complexation between the phenylazomethine unit and $SnCl_2$. It can also be determined by the Job titration method, showing the maximum absorption change at the 1:1 mixing ratio. A similar UV-vis titration was also applied to the **Ph-DPA G4** bearing 15 imines within 1 dendrimer (Figure 10.5). This material showed a significant change in the UV-vis absorption upon the addition of $SnCl_2$. The spectrum of **Ph-DPA G4** gradually changed with an isosbestic point at 371 nm up to the addition of 1 equimolar amount versus the molar concentration of the dendrimer. The isosbestic point then shifted upon the further addition of $SnCl_2$, and appeared at 368 nm between 1 and 3 equimolar amounts. During the addition of $SnCl_2$ from 3 to 7 equimolar amounts, the isosbestic point appeared at 363 nm, and finally it moved to 360 nm with the addition of more than 7 equimolar amounts. Overall, the isosbestic point shifted about 11 nm from 371 to 360 nm.

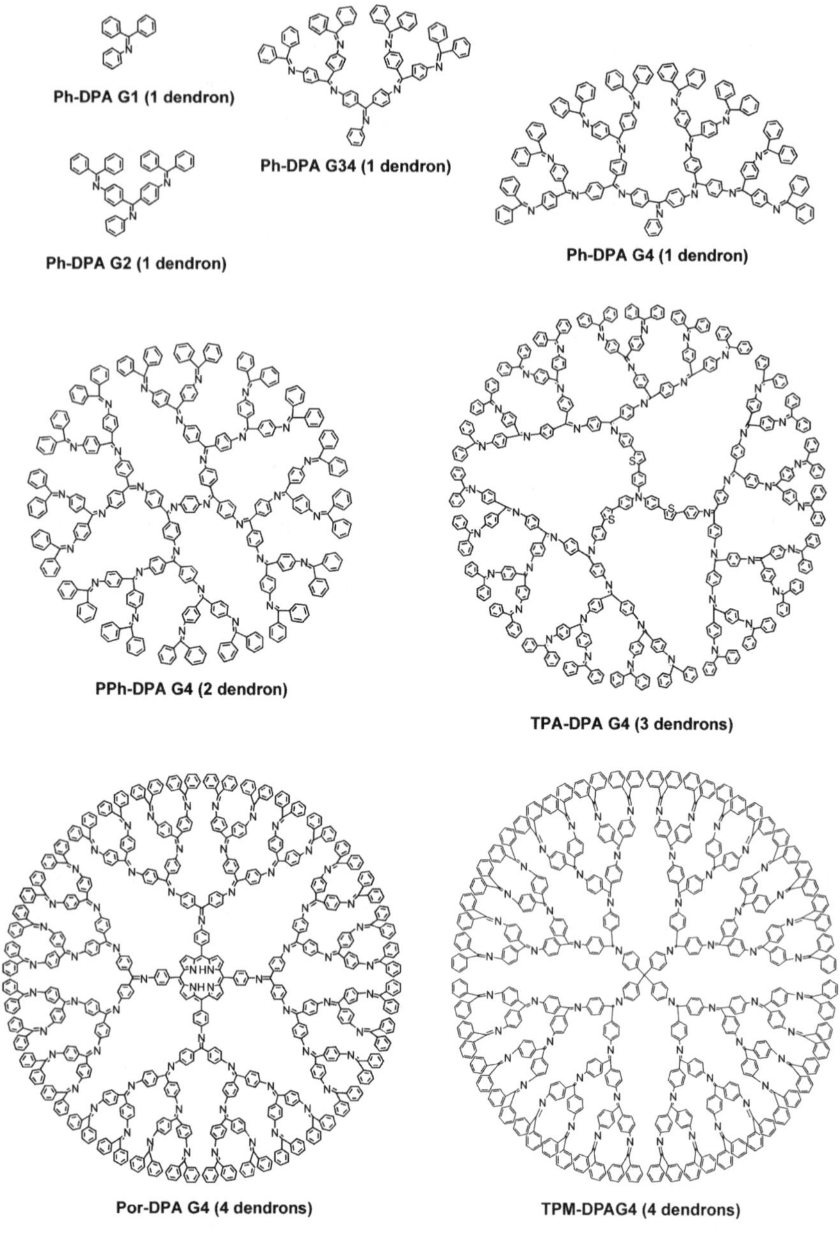

FIGURE 10.4 Structures of **DPA G4** having various core units.

These results observed during the titration strongly suggest a stepwise process of multiple coordination reactions. If the entire coordination simultaneously proceeded, only one isosbestic point should be observed during the titration because the spectral change would appear as an average of all the reactions. In this system, four different

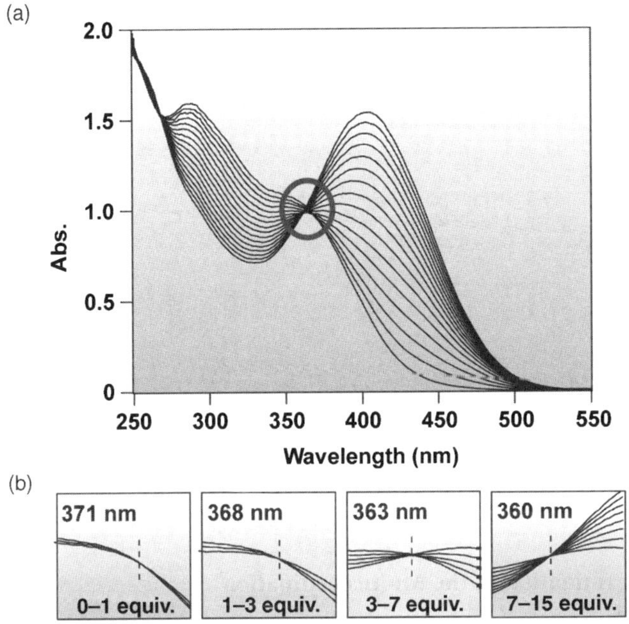

FIGURE 10.5 (a) The UV-vis spectral change upon the addition of $SnCl_2$ to **Ph-DPA G4**. (b) Enlargements of the isosbestic points successively appeared with the displayed equimolar amounts of $SnCl_2$.

isosbestic points appeared in turn starting from each section at 0, 1, 3, and 7 equimolar amount additions, and continued until the next section (finished at 15 M amounts). In addition, the number of equimolar amounts between each section (1, 2, 4, and 8) completely agreed with the number of imines in each layer of the **Ph-DPA G4**. The coordination profiles about each layer should be those shown in Figure 10.3.

A similar stepwise radial complexation was observed for the DPA derivatives (Figure 10.4). When $SnCl_2$ was added to a solution of these dendrimers, four isosbestic points were successively observed in the spectra. In **PPh-DPA G4** bearing two dendrons (Figure 10.4), each shift occurred with the repeated additions of 2, 4, 8, and 16 equimolar amounts of $SnCl_2$ [40,52]. On the other hand, **TPA-DPA G4** bearing three dendrons (Figure 10.4) showed distinct shifts with the repeated additions of 3, 6, 12, and 24 equimolar amounts [74], while **Por-DPA G4** [53,54], **TPM-DPA G4** [75,76], and **Cyc-DPA G4** [77] (Figure 10.4) bearing four dendrons showed the distinct shifts with 4, 8, 16, and 32 equimolar added amounts. In each dendrimer, the equimolar amounts of $SnCl_2$ added to the solution needed to shift the isosbestic point also agreed with the number of imines in the first to fourth shells. These results support the conclusion that the complexation in the DPA derivatives proceeds from the core-side imines to the terminal ones, as shown in Scheme 10.2. The stepwise reaction is also evidenced by several methods (transmission electron microscopy, electrochemical measurement of coordinating iron complexes, and product analysis of metal assisting reduction of imines) [52,54,56].

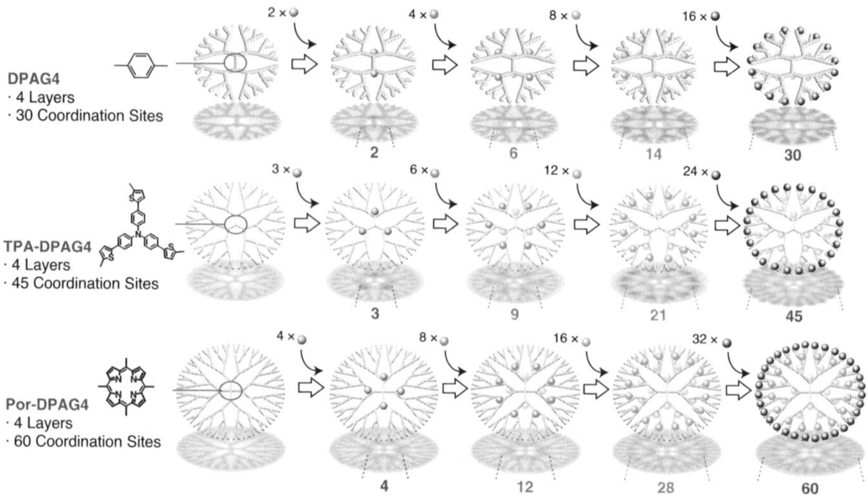

SCHEME 10.2

10.2.4 Determination of the Multicoordination

For the quantitative evaluation of the complexation, we have to know how much of a difference between neighboring coordination constants (K_C) is required for the stepwise shift in the isosbestic point. To estimate the profile of the isosbestic point, we again used the theoretical concentration profile again. A UV-vis absorption coefficient (ε) is generally written as a function of the wavenumber with a Gaussian expression. The function of the wavelength (λ) is given by

$$\varepsilon(\lambda) = \varepsilon^\circ \exp\left\{\frac{-\delta(1/\lambda^\circ - 1/\lambda)^2}{kT}\right\} \tag{10.2}$$

where λ° is the wavelength at the absorption maximum, ε° is the absorption coefficient at $\lambda = \lambda^\circ$, and δ is the thermal vibration parameter of the chromophore. The observed absorbance (A) should be the sum of these functions, therefore, we assigned one function to each phenylazomethine unit in every layer (either the complex and free base). For ease of calculation, the parameters of these absorptions (ε° and δ) were set to the same value for each chromophore. The absorbance change (ΔA) that involved in the small addition of the metal ion ($\Delta[\mathrm{M}]_0$) is then given by

$$\Delta A(\lambda) = \varepsilon^\circ \sum_{\mathrm{N}} \left[\Delta[\mathrm{N}]\exp\left\{\frac{-\delta(1/\lambda^\circ_{\mathrm{N}} - 1/\lambda)^2}{kT}\right\} + \Delta[\mathrm{NM}]\exp\left\{\frac{-\delta(1/\lambda^\circ_{\mathrm{NM}} - 1/\lambda)^2}{kT}\right\}\right] \tag{10.3}$$

as the sum of each component (N = A, B, C, and D, denoting the coordination sites in each layer) using the calculated value of the concentration profiles (Figure 10.3).

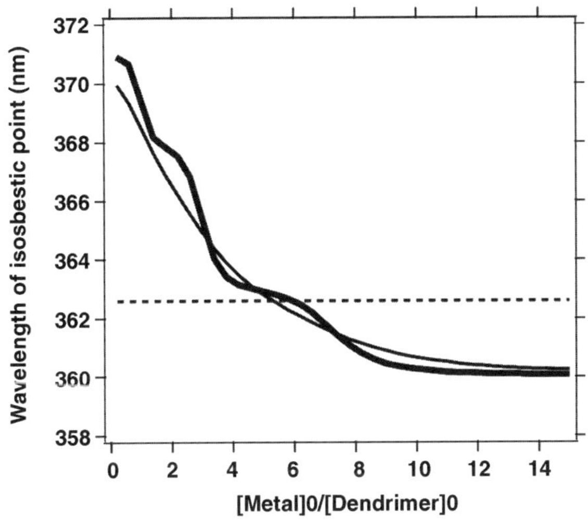

FIGURE 10.6 Calculated changes in the isosbestic point for three different conditions in the coordination constants assuming the same system treated in Figure 10.3. The bold line corresponds to the strongly stepwise system in which the coordination constants in each layer is gradated by a 100-times difference ($K_a = 10^2 K_b = 10^4 K_c = 10^6 K_d = 10^{10}\ \mathrm{mol}^{-1}\ \mathrm{L}$). The thin solid line corresponds to a stepwise system in which the coordination constants in each layer is gradated by a 10-times difference ($K_a = 10K_b = 100K_c = 1000K_d = 10^8\ \mathrm{mol}^{-1}\ \mathrm{L}$). The dotted line corresponds to the equally coordinating system in which all the coordination sites have the same constants ($K_a = K_b = K_c = K_d = 10^6\ \mathrm{mol}^{-1}\ \mathrm{L}$).

Based on the definition, the isosbestic points were calculated as the λ solution of the equation $[\Delta A(\lambda) = 0]$. The simulated profiles of the isosbestic point for the assumed $\lambda°$ values are shown in Figure 10.6. When the coordination constants (K_C) for each layer are the same value (Case 1), the isosbestic point was fixed for every molar amount of the added metal ion. In contrast, the isosbestic point shows a stepwise change during the addition if the K_C values were set to Case 3, in which each K_C for the layers were gradated by a 100 times difference. If the difference in K_C is smaller (Case 2), the change in the isosbestic point becomes continuous.

The results shown in the simulation demonstrate that the coordination constants (K_C) in the DPA experiments should have a considerable difference when the constants of the neighboring layers are separated by more than two orders of magnitude.

10.3 FINE SYNTHESIS OF METAL- OR METAL-OXIDE NANOPARTICLES

10.3.1 Strategy

Metal nanoparticles have attracted much attention because the surface area percent becomes higher with the decreasing particle size. From the standpoint of catalysis [78], this increase in the surface area is very important for improving the catalytic

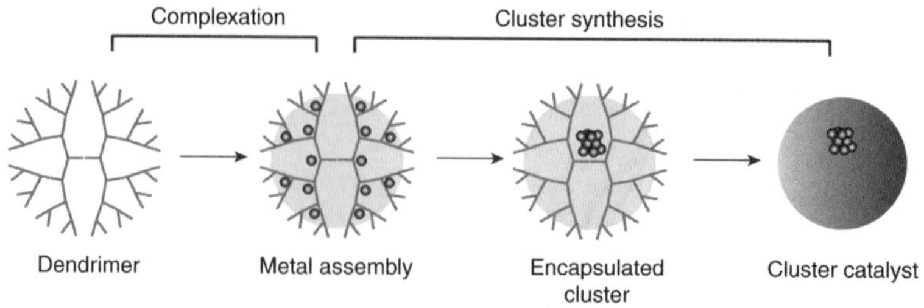

FIGURE 10.7 Schematic representation of metal nanocluster synthesis in a dendrimer template.

activity. If the intrinsic character of the particle would not depend on the size, the size should be minimized as far as it downsized. However, in practice, the character is strongly affected by the size of the metal clusters in the nanometer- or sub-nanometer-size domain. The size-specific character is dominated by electronic effects caused by the d-band charge density, surface coordination, and quantum size effect. This is not only in metal nanoparticles but also in metal-oxide nanoparticles. In some cases, the specific size effect provides drastic enhancement in the catalytic activity as observed especially for gold nanoclusters [79]. Thus, complete understanding of the specific effect should be required for further design of novel inorganic (metal or metal-oxide) nanomaterials. However, there have been no available methods to prepare subnanoparticles of which size and number of metal atoms are finely controlled in the solution phase (Figure 10.7).

The dendrimer complex can be regarded as a precursor of metal nanoparticles. Because metal ions are strongly entrapped in the dendrimer architecture and the diffusion is limited in the interior, the number of metal atoms composing one cluster particle is equal to the number of metal precursors (ions) in the dendrimer template. The accuracy of the fine synthesis is, therefore, depends on the accuracy of the metal-assembling process. However, the metal assembly in a typical metal-assembling dendrimer, such as poly(amidoamine) (PAMAM), is not defined, but includes the statistical distribution in the number of metal ions. As described in the previous section, the number of assembled metal ions in one dendrimer has a significant distribution when the metal ions equally coordinate to every ligation site in the dendrimer. When the fine metal-assembling dendrimer was used, the statistical factor should be drastically minimized. This would improve the precision of the metal nanoparticle synthesis with atomic-level control.

10.3.2 Synthesis and Characterization of Pt Nanoparticles

The ultimate miniaturization into nano- or a subnanoscale makes metal particles much more active due to the drastic increase of their specific surface area and the significant change in their electronic structure. In addition, metal clusters are thus expected to have a higher activity than the corresponding bulk metal, exhibiting unique emission, magnetism, and catalytic properties. However, there remain numerous issues that must be considered for the precise size control and utilization of the discrete metal nanoclusters. First of all, the large-scale synthesis of such a fine nanocluster in the solution phase is unprecedented. In the case of gas-phase synthesis, size control is possible [80,81], yet mass production and stabilization are difficult to achieve. Furthermore, even if we succeeded in the fine synthesis with micellar surfactants used for the protection of these clusters from the aggregation [78], these surfactants, such as alkanethiols, often inhibit an inherent catalytic activity by covering the active metal surface. Modern catalysis technology places the highest priority on fine platinum nanoparticles because they can be used as electrode catalysts in fuel cells [82]. On the other hand, there are no established methods for the synthesis of subnanoscale metal clusters. Thus, the size control of Pt nanoparticles at atomic level and the elucidation of their functionality is a significant challenge. Presently, colloidal particles of platinum with a 2–3 nm diameter are considered to be useful as the electrode catalyst in fuel cells [83–85]. However, as platinum is in limited supply, it has become important to evaluate the performance of catalysts with minimum particle sizes.

Dendrimers with a tetraphenylmethane (TPM) core are particularly suitable as templates for the synthesis of metal clusters. **TPM-DPA G4** (Figure 10.4) undergoes a stepwise radial complexation from the inner imines, allowing perfect control of the number of metallic components encapsulated in the cage-like structure [75]. Indeed, at the point of 4, 12, 28, and 60 equiv. ([metal]/[dendrimer]) during the titration, a discrete metal–organic assembly can be produced without any distribution of the number. Compared with conventional dendrimers, the present dendrimers are rigid and create an inner cavity due to a strong shell effect based on dense-shell packing.

The addition of a reducing agent (sodium borohydride: $NaBH_4$) to a solution of the dendrimer complex resulted in a two-step UV-vis absorption change involving fast kinetics followed by slow kinetics. The dendrimer-encapsulated platinum clusters, **Pt$_n$@TPM-DPA G4** ($n = 12$, 28, or 60), were prepared with a very small size distribution observed by transmission electron microscopy (TEM). The particle sizes of the $n = 12$, 28, and 60 clusters were 0.86 ± 0.09, 0.97 ± 0.09, and 1.16 ± 0.13 nm, respectively (Figure 10.8) [76].

The platinum subnanoparticle with a very narrow size distribution was examined as a catalyst for the oxygen reduction reaction (ORR). Many previous research studies have reported that smaller nanoparticles with a size under 2 nm exhibited a lower catalytic ORR performance. However, the obtained platinum particle synthesized in the

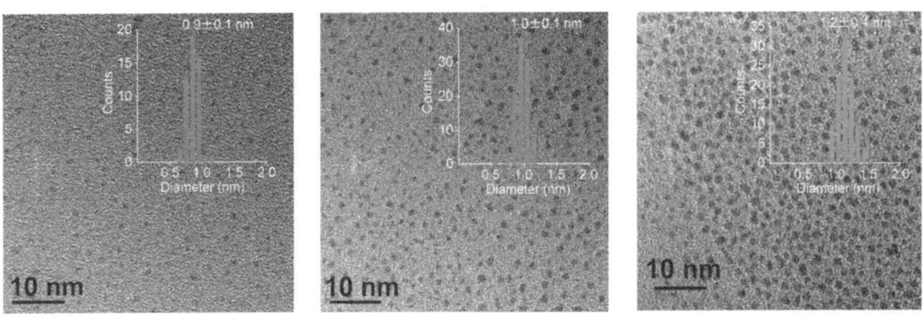

FIGURE 10.8 Transmission electron micrograph (TEM) image of Pt clusters (left: Pt_{12}, Center: Pt_{28}, right: Pt_{60}) synthesized in a phenylazomethine dendrimer template (**TPM-DPA-G4**). (See the color version of this figure in Color Plates section.)

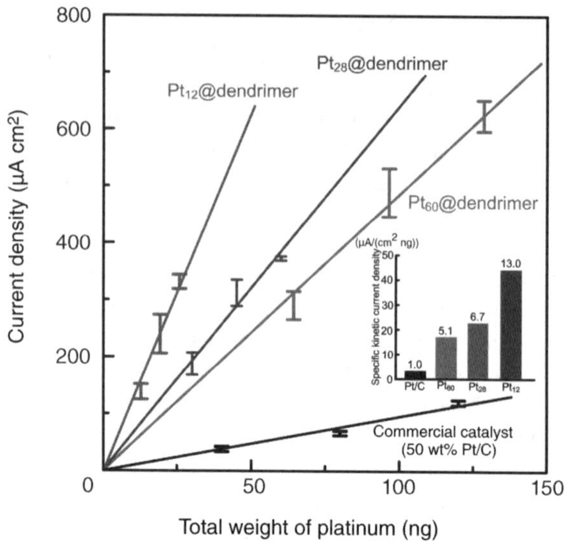

FIGURE 10.9 Correlation between kinetic current density of ORR and the total weight of platinum loaded on a glassy carbon electrode (GCE). The inset figure shows the specific activity normalized by the weight.

dendrimer template with a size smaller than 1 nm exhibited a very high catalytic activity per platinum atom weight [76]. A higher mass activity was observed for the smaller particles. The kinetic limiting current normalized by the weight of platinum was increased, and reached a 13 times higher activity than the commercial Pt/C catalyst (platinum nanopoarticle on carbon support) with a size of about 2–4 nm (Figure 10.9). While the kinetic limiting current was increased with the downsizing, the redox potential at which the ORR starts was shift to the negative potential region, which means that a higher overpotential is required for the ORR.

10.3.3 Synthesis and Characterization of TiO_2 Nanoparticles

Because the dendrimers can act as a fine metal-assembling template for various metal elements, extensive applications are expected. Not only metallic nanoparticles but also metal-oxide nanoparticles can be synthesized in the dendrimer template [86]. An example of the synthesis of TiO_2 subnanoparticles with the predicted quantum size effect on TiO_2, which has not been experimentally determined. In the present case, a precursor complex ($Ti(acac)Cl_3$) was assembled in the dendrimer template with a defined stoichiometry, and the subsequent conversion into TiO_2 was carried out using two different methods. One is the hydrolysis typically conducted with HCl [87], and the other is thermolysis over 400°C [88]. The clusters with a different number of TiO_2 unit cells and showed a different optical bandgap, suggested that the bandgap becomes larger for smaller nanoparticles as predicted by the quantum chemistry theory. In addition, the fine synthesis enabled a precise discussion about the crystal phase of the subnanopar-

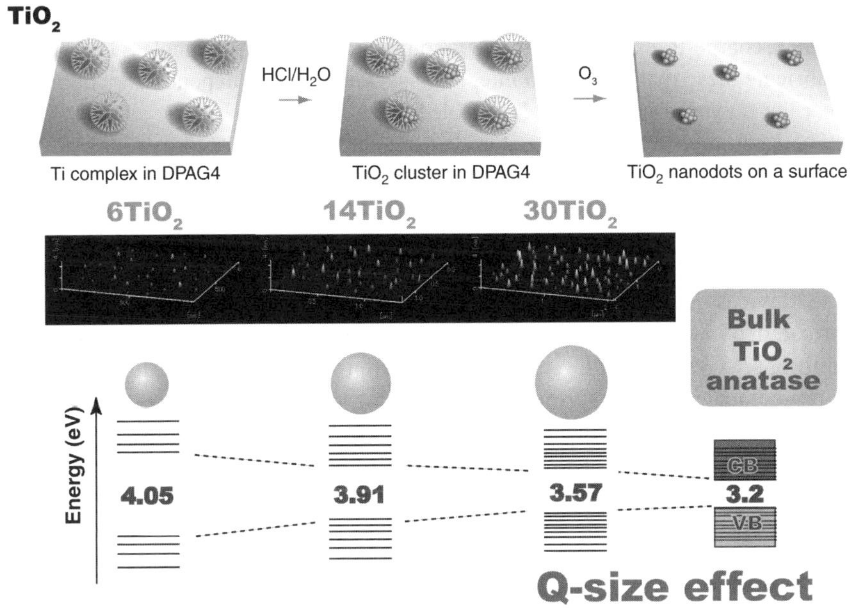

FIGURE 10.10 Schematic representation of the quantum-size TiO_2 nanodot on a surface synthesized using a dendrimer template. (See the color version of this figure in Color Plates section.)

ticles. Results obtained from the optical bandgap measurements of these TiO_2 nanoparticles prepared by different synthetic methods (hydrolysis and thermolysis) suggested that the crystal phase in TiO_2 could remain at the crystal size smaller than 1 nm. A theoretical prediction of the bandgap suggests a continuous variation with the crystal particle size [89]. The experimental bandgap obtained for the TiO_2 subnanoparticles with the dendrimer template could also match the possible theory, suggesting that the crystal phase is almost the same as that of the bulk crystal (Figure 10.10).

The new synthetic method to access the subnanometer particles enabled the quantitative evaluation of the activity, which has not been investigated. The synthetic strategy provides potential application of these subnanometer materials as novel nanocatalysts, which have not been prepared in the solution phase.

10.4 PHOTOCHEMISTRY OF METAL-ASSEMBLING MACROMOLECULES

10.4.1 General Strategies

There are basically two kinds of materials, which can be employed for photochemical applications in OLED [13] and organic thin-layer solar cells [90,91] One is a polymer material including a linear polymer, grafting polymer, hyper branch polymer and dendrimers. The other is a low-molecular-weight material, such as the metal–ligand complex, charge-transfer complex and π-conjugating organic molecules. The strategy to fabricate these materials in an integrated system (device) depends on whether we select the polymer materials or low-molecular-weight material.

The low-molecular-weight materials are generally fabricated by a dry process, such as the chemical vapor deposition (CVD) method. One advantage of using this kind of material (method) is its applicability for a multilayer device structure, which consists of metal, organic, and metal–complex composites. The CVD method requires a high-vacuum condition; therefore, low cost and mass production of these devices is rather difficult. In contrast, the polymer materials could be fabricated by wet processes, which do not require any special equipment and conditions, except for the solvent. The fabrication of a homogeneous film composed of polymer materials on a surface is much easier even if the surface area is very large. Although multilayer modification through the wet process is unfavorable, the recent development of a molecular-level implementation of multiple functions in one layer enables the achievement of various device functions with a much simpler device architecture.

The strategy to design polymer materials for electric or photochemical devices has been limited by the selection of the chemical composition. The electron transfer and luminescent property of the material were added as the unit structure was implemented on the graft part or main chain of the polymer. However, the solid-state structure of these conventional polymers is complicated because the main chains entwine with each other. The statistical character of the polymer has been prevented by fine structuring with geometric or orientational control. Under this situation, a new class of macromolecule "dendrimers" was proposed by Tomalia and Naylor [3] and Newkome et al. [92] in 1985. These molecules can be defined as highly regulated branched polymers possessing a focal point [2,69,93–95]. The "divergent" method was employed for the synthesis of the early dendrimers. It was later improved by Fréchet et al. and called as the "convergent method" [1,96] that enables the preparation of a completely monodispersed dendrimer of over 10 kDa molecular weight without any disorder in the terminal number.

One of the characteristics of the dendrimers is the "encapsulating effect" [2,97,98] Based on their topology, the density of the molecular units should be higher than that of the inner shells. Utilization of the nanospace inside the dendrimer shell is wide, for example, as a microenvironment for shape-selective catalysis [99–102], a template for metal nanoparticle synthesis [38,63,66,68], a molecular storage area in supramolecular host–guest systems [103–106], and so on.

Standard dendrimers consist of three different moieties called the core, dendron, and terminus (Figure 10.11). Each part is important for characterizing the properties of the dendrimers. In general, the core is located at the center of the dendrimer where the environment is much different from that of the exterior due to steric factors from the peripheral parts. This position is suitable for applications that utilize the unique microenvironment, for example, a shape-selective catalyst [99–102], a protected photoexcitation center [107–111], or chemical storage [112,113]. In contrast, the terminal possesses a character opposite to the core. It can be used as a reactive

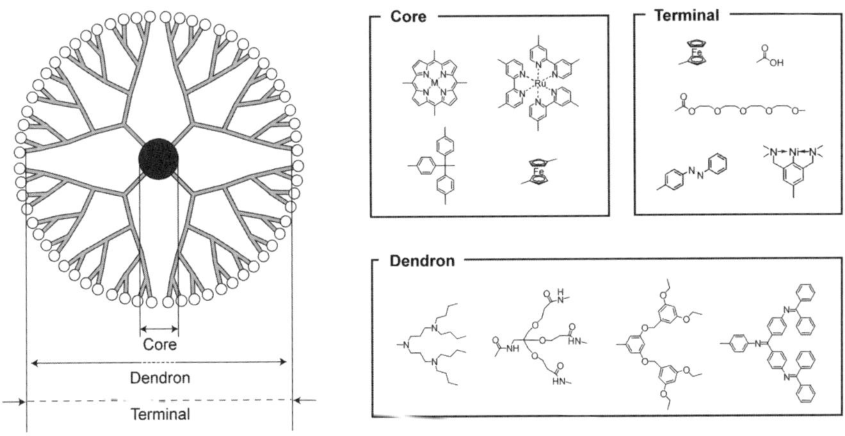

FIGURE 10.11 Typical examples of dendrimers containing core, dendron, and terminal moieties. Each part plays an important role in characterizing the fundamental properties of the dendrimers.

catalytic center to the external environment [9–11,114,115]. This position often causes a condensation effect based on their topological character. This results in an enhancement of the intrinsic functionality of the modification center. For example, a drastic enhancement in the molecular sensing property was reported by Astruc and coworkers [116–119]. In addition, the terminal can be used as a linker to the other dendrimer molecules [120,121]. An example of a redox-driven self-assembly has been shown by Abruña et al. [122]. Apart from the core and terminal, the dendron moiety does not have an outstanding character as the modification centers. However, it is the most important part of the dendrimer because the structure of the dendron influences the fundamental properties and the entire conformation. This position also plays a role in characterizing the environments of the core and terminals. The dendron serves as a mediator of the electron or energy transfer through the dendritic architecture [5,42,123–125]. The modification of the dendron moiety therefore should result in a drastic change in the absolute properties and functions of the dendrimers.

Therefore, the most important factor that characterizes the fundamental physical properties of the dendrimers is the rigidity of the monomers constituting the dendron structure. Ballauff and Likos showed that the entire conformation of the dendrimer depends on its rigidity based on a theoretical analysis [69,70]. If the dendrimer was made from flexible backbones, the terminal monomers may be folded back onto the core side. This phenomenon, called “Back Folding,” disturbs the LBL structure of the dendrimers (Figure 10.12). It can be experimentally confirmed as the generation number dependence of the intrinsic viscosity of the dendrimers in solution [126,127]. The intrinsic viscosity of the flexible dendrimer increases with the generation number and reaches a maximum at generation 2 or 3 followed by a decrease with the generation number. This means that the back folding of the terminal chains in the dendrimer results in a smaller particle size of the molecule [128–130]. A

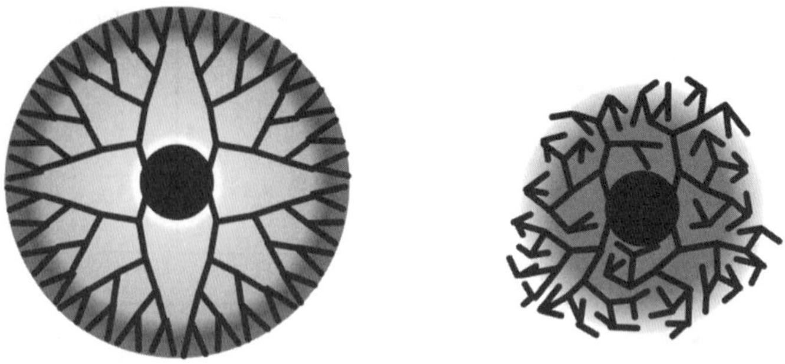

FIGURE 10.12 Schematic representation of dendrimers composed of rigid and flexible backbones. The rigid dendrimer can retain its topological architecture that has a dense shell and sparse interior. In contrast, the termini of the flexible dendrimer intrude in its inner space (back folding) and are be more compact.

π-conjugating backbone effectively reduces the back folding due to the rigidity. As a result, stiff dendrimers can retain their topological character in the actual conformation [131]. In addition, they should have a relatively larger space than the flexible dendrimers. This would allow these dendrimers to encapsulate large guest molecules [103,132] and metal clusters [38].

10.4.2 Light-Emitting Materials

Many conjugated molecules and metal complexes have been developed as light-emitting materials in photochemical devices such as OLEDs. The most important requirement for the material is a bright emission even in the solid state. Although many luminescent compounds show a strong fluorescence in a dilute solution, the fluorescence intensity drastically decreases in their solid state due to the excimer formation or intermolecular energy migration that leads to fluorescence quenching. Thus, low-molecular-weight luminescent materials should be dispersed in photochemically inert host materials, which allows spacing between the chromophores for their efficient emission in photochemical devices (Figure 10.13). Macromolecular light-emitting materials, which includes a fluorescent (phosphorescent) core as the main chain or graft groups, have been extensively studied by many researchers. Because the polymer chain helps with the homogeneous dispersion of the chromophores into the solid state by suppressing their local crystallization, the luminescence properties improved in many cases. We can modulate the property by modifying the composition of the luminescent part and other inert moieties [133–139]. As noted in the previous section, the design significantly depends on a statistical factor based on the undefined geometry. As solid-state emitters, dendrimers with a luminescent molecular group are attracting much attention because their molecular architecture allows the design of the exact geometry of its key components, that is, the "core and terminus."

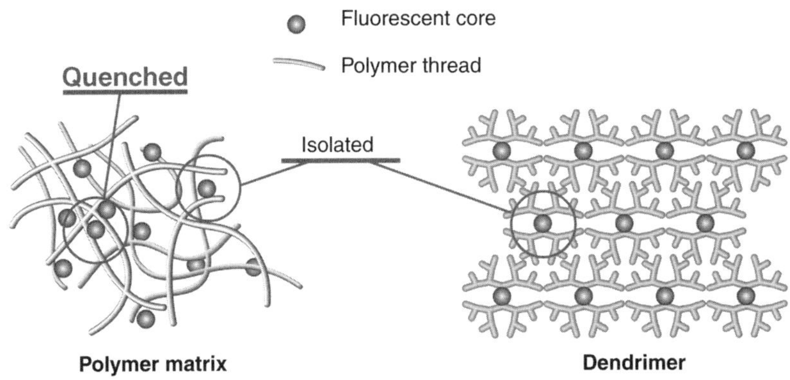

FIGURE 10.13 Fluorescent components are statistically doped in a matrix of a conventional linear polymer. Therefore, a partial quenching is inevitable depending on the concentration. A fluorescent component placed at the core of a rigid dendrimer can retain the intermolecular distance.

Dendrimers can possess only one core unit. However, the luminescent core based on a conjugated structure or metal complexes exhibits an excellent luminescent property in the solid state as well as in the solution phase [140]. Many reports showed that the dendritic shell effectively suppresses the concentration quenching of the fluorescence in the solution phase. This architecture also suppresses the quenching process by the other quenching species such as the oxygen molecule or electron acceptors. Phosphorescent molecules, which are generally quenched by triplet–triplet annihilation, also retain their luminescence property in a dendritic nanostructure [141]. As a result, the solid-state emission through a photoexcitation and electric excitation (in light-emitting diode devices) is drastically improved by the dendrimer formation [142–149].

10.4.3 Materials for Energy Transduction

Management of an excitation energy transfer (EET) or fluorescent resonance energy transfer (FRET) is very important when designing photochemical devices employing a molecular chromophor [136,137,150]. For luminescence devices, it is necessary to restrict any unfavorable energy dissipation for the maximum use of the excitation energy as mentioned in the last section. The luminescent color can be modulated through the interchromophor energy transfer that realizes multicolor emissions observed as a white light. In principle, a long-range FRET is attenuated by the distance between the excitation energy donor (D) and acceptor (A) as expressed by Eq. (10.4) [151].

$$k_t = \frac{\left(\frac{R_0}{r}\right)^6}{\tau_0} \qquad (10.4)$$

$$R_0 = (8.8 \times 10^{23} k^2 2Jn^{-4} Q_0)^{\frac{1}{6}} [\text{Å}]$$

where r is distance between the donor and acceptor, κ^2 is the orientation factor, J is the overlap integral of the fluorescence intensity of the donor and the molar extinction coefficient of the acceptor normalized by the frequency expressed in wavenumbers, n is the index of refraction of the solvent, and Q_0 is excitation quantum efficiency of the donor.

Precise design of the luminescent property requires the fine geometry of these chromophors in the solid state. Even if the luminescent film is composed of a mixture of several chromophors, the dendrimer architecture can produce a well-defined intermolecular distance allowing the precise control of through space (FRET) or through bond energy transfer [152–154]. We can also take advantage of the highly regulated nanostructure to optimize the photochemical processes in photovoltaic devices. The charge-separating reaction upon the photoexcitation of the chromophor is a well-established process as the initial reaction during photosynthesis. The photoexcitation is not direct, but transmitted through a light-harvesting antenna that is composed of a multiple P850 cytochrome placed near the center (Figure 10.14). Stacking of an organic and metal complex chromophor in close proximity can be achieved by the dendrimer synthesis. The maximum energy transduction in a multiporphyrin system is achieved by the oligomeric structure based on the dendrimer topology. At the topology level, the dendrimer architecture is suitable for collecting the photoexcitation energy because the core unit is located at the focal point [155–164]. At the geometry level, the summation of the distances between the core and every terminus is minimized if the symmetric multidendronized core, which leads to a core-centered model, was used. Several experimental results showed that a close-packing dendrimer architecture enables complete repression of the energy loss [108].

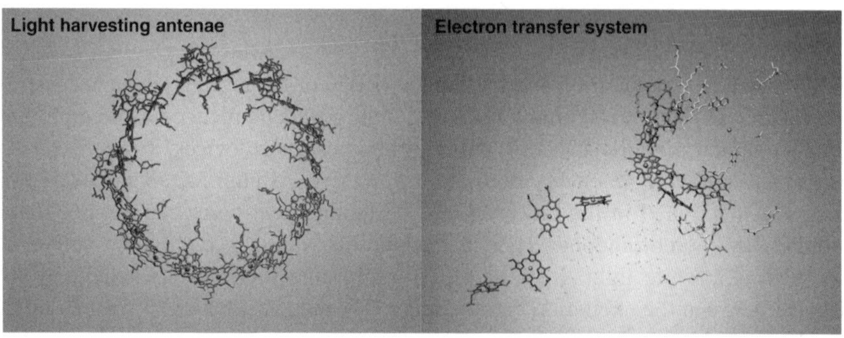

FIGURE 10.14 Molecular 3-D coordination structures of metalloproteins playing an important role in the photosynthetic reaction center. Each picture is based on a crystal structure, visualizing only its functional components without any protein. (Right) Light-harvesting complex from *Rhodopseudomonas acidophila* (pdb code: 1kzu) [191]. The photosynthetic reaction center from *Rhodopseudomonas viridis* (pdb code: 2prc) [192]. (See the color version of this figure in Color Plates section.)

10.4.4 Materials for Electron or Hole Transporting

Photochemical devices including OLEDs and solar cells require materials that can transport electrons and holes. Recently, PEDOT-PSS is one of the most utilized polymeric materials for hole transport [165]. Although there is no established compound providing easy film fabrication and high performance, aryl amine derivatives that contain the triphenylamine (TPA) or carbazole unit are some of the potent candidates superior to PEDOT-PSS [166,167]. Dendrimers that contain these aryl amine components are quite remarkable materials. One of the advantages of using this material is its excellent film-forming property based on its low viscosity and low crystallinity. Due to their spherical morphology, the dendrimers that have hole-transporting units as the termini allows high carrier mobility, regardless of the core structure and the functionality [35,168–170]. Thus, a flexible design is available based on the luminescence and charge transport. Examples of the application in OLED devices are presented in the following section.

Bulk charge transport through a semiconducting film is an essential process in organic electronics. However, a molecular-level charge transfer and its management is also an important key for any improvement. Especially, for photovoltaic devices, the initial electron transfer (charge separation) upon photoexcitation determines most of the photoenergy conversion efficiency. If back-electron transfer against the charge separation occurs, the total conversion efficiency will be significantly reduced. Although several strategies to suppress the back-electron transfer have been reported, nanostructure construction based on the dendrimer is a powerful tool to optimize the distance between the electron donor and acceptor. The photosynthetic reaction center contains a well-established redox sequence in order to facilitate long-range charge separation for an efficient photoenergy conversion. The dendrimer structure enables the fine construction of the function based on a topological and geometrical molecular design.

10.5 APPLICATIONS FOR PHOTOCHEMICAL DEVICES

The function of devices is established as an integration of material properties such as photoexcitation, luminescence, energy transfer, and electron transfer. In addition to the molecular-level design, all chemical processes including the formation of multilayer, heterojunction, fine-tuning of the electron- and hole injection should be precisely designed for the device construction. Not only the electric property but also the fabrication process of ultrathin films must be taken into account. In the following sections, examples of using the dendrimer structure in actual devices are introduced.

10.5.1 Fabrication of an Ultrathin Organic Film for Molecular Devices

In general, molecules with the dendritic architecture show a very good solubility in most organic solvents. In the case of a linear polymer, the solubility drastically decreases if the backbone structure is composed of a rigid π-conjugating architecture. In sharp contrast, the dendrimer is still dissolved even if the rigid structure is the main

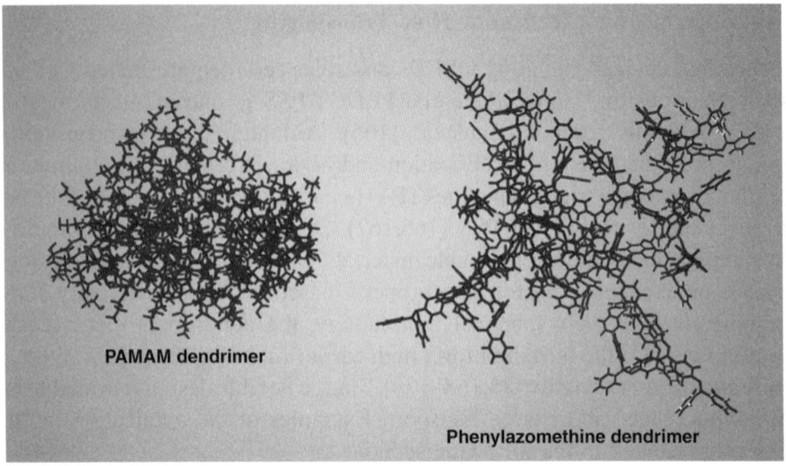

FIGURE 10.15 Molecular 3-D structures of different dendrimers. The structure of the PAMAM dendrimer (left) with optimization of the molecular dynamics calculations is from the literature. The structure of the phenylazomethine dendrimer (right) with optimization of the semiempirical orbital calculations is from the literature.

component. This observation is due to the spherical morphology that allows the least contact between individual molecules. As an example of the phenylazomethine dendrimer, the 3-D structure obtained through a molecular modeling calculation is shown in Figure 10.15. The peripheral terminals of the DPA are hardly bent to the core side, thus the molecule retains its star shape. The nanospace in the dendrimer molecule is greater than that of the flexible PAMAM dendrimers [3,171–173]. This characteristic also appeared in the experimental values obtained by a hydrodynamic measurement in the solution phase (THF). The hydrodynamic radii (R_h) of the DPAs with PPh and Por cores are 2.0 and 2.2 nm, respectively. These values are consistent with the estimated terminal-to-terminal length (~5 nm) based on the molecular modeling of **DPAG4-Por**. The intrinsic viscosities of the DPAs ($[\eta]$) increase from G1 (generation 1) to G3, corresponding to the density decrease ($[\eta] \propto [M]/V_h$). They reach a constant value for dendrimers larger than G3. This behavior is much different from that of the flexible dendrimers such as the polybenzylether (Fréchet type) [126,127] dendrimers or PAMAMs [3]. The $[\eta]$ values of these flexible dendrimers decrease as the generation number increases after reaching the maximum value at G3. These results reflect the fact that the backbones of the DPAs are not folded back to the core even in a THF solution (Figure 10.15).

The π-conjugation in the DPAs contributes not only to the rigidity but also to the stability of the chemical bonding. Durability of the DPAs against heat decomposition is nearly equal to that of engineering plastics. The temperatures for a 10% weight decrease under a N_2 atmosphere ($T_{d10\%}$) exceeds 500°C for each DPA (Table 10.2) [72,174].

The encapsulating effect of the DPA shell has been clearly observed. The DPAs having a redox active core (porphyrin, triphenylamine) showed the attenuation effect

TABLE 10.2 10% weight decreasing ($T_{d10\%}$) and glass transition temperatures (T_g) of phenylazomethine dendrimers

Dendrimer	$T_{d10\%}$ (°C)	T_g (°C)
PPh-DPA G1	~ 400	50
PPh-DPA G2	514	148
PPh-DPA G3	511	174
PPh-DPA G4	521	188
Por-DPA G1	531	[a]
Por-DPA G2	528	[a]
Por-DPA G3	523	195
Por-DPA G4	531	201

[a] No clear phase transition was observed.

of electron transfer between the core and an electrode surface [53,54,74]. In the cyclic voltammograms, the oxidation and reduction peak separation became greater with the increasing generation number, suggesting a slower electron transfer rate constant (k^0). The UV-vis absorption measurements and spin-lattice relaxation times (T_1) from the NMR measurements revealed that the solvent molecules are gradually excluded from the interior of these larger DPAs [54].

The physical properties noted above significantly contribute to the formation of a homogeneous film and further processing. Indeed, a spin-cast film on a flat substrate has a very low roughness, and the dendrimer molecules were assembled in a nearly hexagonal closest packed structure (Figure 10.16). In the next section, examples of the application in an OLED device will be introduced.

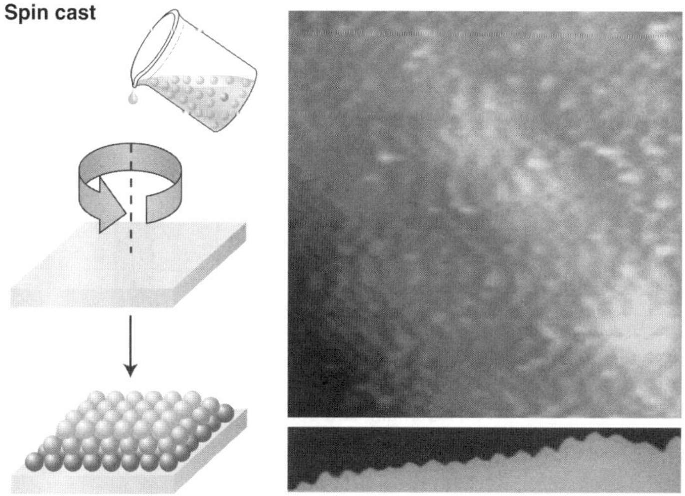

FIGURE 10.16 An atomic force microscope (AFM) image of the spin-cast film of the phenylazomethine dendrimer on a mica surface. The bottom part on the right-hand side shows a cross-sectional AFM image corresponding to the dashed line in the top view (upper part).

10.5.2 Light-Emitting Diodes (OLEDs)

Normal OLED devices require hole (positive charge) and electron injection from the anode and cathode, respectively. These charges are passed through the hole- or electron-transporting layer, leading to the 1:1 coupling in the emitting layer. If coupling occurred, a singlet exciton, leads to a bright luminescence from the fluorescent material (Figure 10.17). Occasionally, phosphorescence would be observed as a result of an intersystem crossing [175]. From a macroscopic viewpoint, the first requirement is control of injection barrier at the interface between the electrode and material. Thickness control of the carrier transport layers is effective to some extent not to induce a significant voltage drop. In addition, the methodology for tuning of the barrier using an intermediate layer is well established. However, this strategy requires very complicated device architecture based on the multilayered processing.

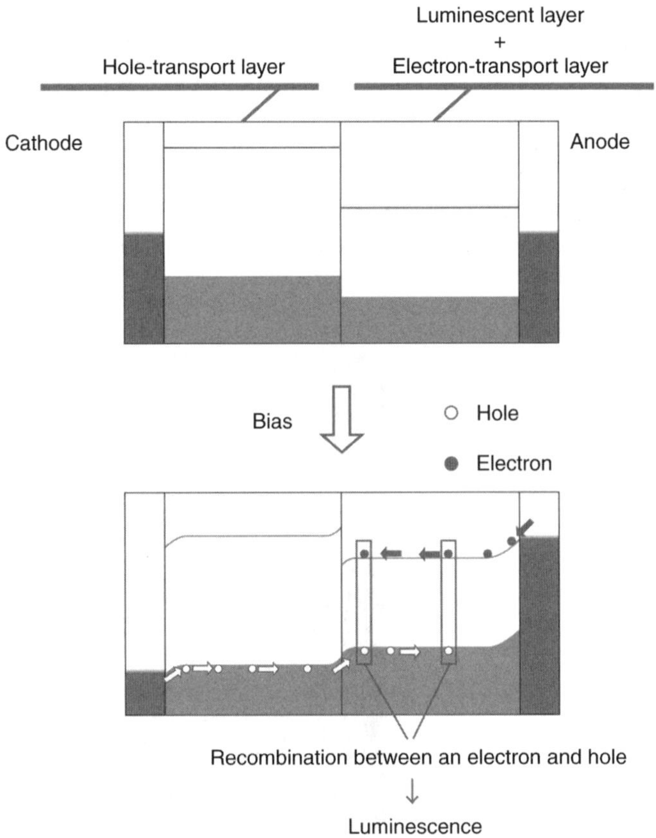

FIGURE 10.17 Potential diagram of an electroluminescence (EL) device before and after biasing.

Therefore, the recent trend in OLED chemistry has moved to the development of materials for much simplified single-layered or quasisingle-layered devices []. An advantage of macromolecular materials is the flexible design of multifunctional films, which can be fabricated by a simple cast process.

Indeed, dendrimers composed of carbazole units exhibit a high hole-transporting performance in OLED devices. A carbazole dendrimer (CzD) with a cyclic phenylazomethine (CPA) core acts as an efficient hole-transport material [178]. Simple solvent casting (spin-cast method) affords a homogeneous transparent film with a 50–100 nm thickness. Further fabrication on the cast film is available using CVD or a wet process (after insolubilization through photo-cross-linking). The CPA unit is a metal collecting site similar to the DPAs. The dendrimer can collect the metal ion in the dendrimer core. For the two-layer OLEDs consisting of **CzDG3-CPA** with Alq, the turn-on voltage was simply reduced from 4.3 to 3.7 V, while the maximum luminescence was enhanced from 786 to 932 cd/m^2 only by complexation with Eu(OTf)$_3$. The current performance of the **CzDG3–CPA** complex (181 mA/cm^2) was over three times that of the free-base form (54 mA/cm^2) at the same applied voltage (10 V). Insertion of a thin solid CsF layer between the anode and the electron-transport layer (Alq) results in a significant improvement in the device performance based on the enhancement of the electron injection. The advanced cell using the CsF layer also showed the complexation effect on the electroluminescence (EL) performance. The maximum luminescence was increased from 3717 to 6718 cd/m^2 by applying a lower voltage after the complexation with Eu(OTf)$_3$ even under nonoptimized conditions (Figure 10.18). At least, this hole-transporting material with metal complexes at the core exhibits a higher performance in an OLED device. This is an example of material enhancement by complexation with a metallic compound as the dopant, which improves the carrier number and mobility.

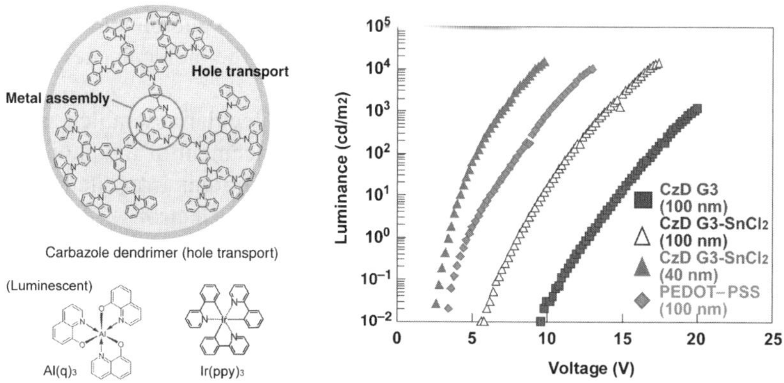

FIGURE 10.18 (Left) A carbazole dendrimer (G3) with metal collecting site. (Right) The OLED property using **CzD-G3** as a hole-transport layer. The OLED performance was enhanced by complexation with $SnCl_2$.

A similar effect was also reported for DPAs having a TPA core [74]. This material showed a high hole-transport property in its film based on a low crystallinity. The OLED cell assembled with the dendrimer and tris(8-hydroxyquinoline)aluminum (Alq) showed ca. 1200 cd/m^2 of luminescence with a 10 V applied voltage. Interestingly, the luminescent performance increased over 20 times only by the complexation with 1 equiv. of $SnCl_2$. This means that the effective doping of metal ions by finely controlled hybridization using the DPA system enhanced the hole-transport property. The role of the precise assembly in hole-transport materials is to repress aggregation or crystallization. This synergetic effect was also available in other topological systems (DPA-modified polymer, asymmetric DPA-CzD) based on the DPA architecture.

An advantage of using these dendrimer-based metal–organic semiconducting materials is their easy fabrication as a thin solid film only by a spin-coating method. An amorphous film of ca. 50–100 nm thickness could be prepared without any partial crystallization. This could be extended to the fabrication of films with a larger area that cannot be produced by a vapor deposition method using monomeric molecules.

10.5.3 Solar Cells

Although the common components for OLED devices can be used, specially developed materials are usually employed for use in solar cell devices. Materials based on a macromolecular complex are also useful for photovoltaics and solar energy conversion. There are several types of photovoltaic devices employing organic materials; that is, an organic thin-layer solar cell [165] and a dye-sensitized solar cell (DSSC) [179,180] that are currently in the advanced stage of development for practical use. In both cases, efficient long-range carrier (hole and electron) transport toward each electrode (anode and cathode) is important [181], as well as the initial charge separation upon photoexcitation [182]. The charge separation efficiency and the stability contribute to the internal conversion efficiency that corresponds to a number of opposite charge pairs per one photon absorbed by the dye. This efficiency can be significantly improved by the molecular-level nanostructure as noted in the previous section.

In addition to the molecular-level design, the molecular assembled structure on a submicrometer scale is also essential for enhancement of the photovoltaic efficiency. First of all, the maximum absorption of photons irradiated from the outside is necessary to maximize the output (photoelectrons). In order to increase the absorption, a large amount of dyes as photon antennas and the reaction center of the photo-induced electron transfer should be implemented in the photovoltaic cell. Recently, a multichromophore system that allows absorption over a wide range of wavelengths was developed. Of course, an increase in the film thickness is demanded for assembling a large number of chromophore molecules onto a substrate. This strategy inevitably leads to an increase in the cell resistance, thus electron- and hole-transporting materials are required to have a very high carrier mobility.

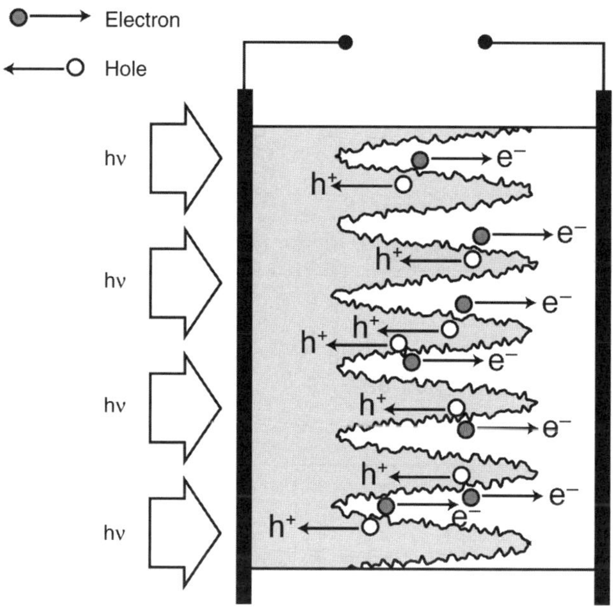

FIGURE 10.19 Illustration of a bulk hetero-junction structure that maximizes the interface area between a hole- and an electron-transporting material.

A bulk-heterojunction architecture [91,183,184] (Figure 10.19) can afford the maximum interface area in which dyes can absorb photons and produce a pair of carriers (electron and hole). Materials that exhibit high carrier mobility are one of the keys to taking advantage of this interface architecture. In most cases of the organic thin-layer solar cells, oligo- or polythiophene derivatives are used as the hole-transporting materials [185]. Electron-transporting materials, such as the fullerene derivatives, are frequently used with the organic hole-transporting materials. Wong et al. reported that the complexation of thiophene derivatives with a metal is an effective strategy to enhance the conversion efficiency [186]. In general, chemical doping by metallic or nonmetallic components often fails because the heterogeneous distribution of the dopant prevents the carrier mobility in an organic semiconducting thin layer. The result strongly suggests that the homogeneous hybridization at the molecular level allows the maximum use of the photoenergy.

Similarly, chemical doping can also enhance the total performance of the DSSC. Yamamoto et al. demonstrated that the molecular-level hybridization between a macromolecular host and metal complexes improved the total cell performance. Triphenylamine with dendritic phenylazomethine (TPA-DPA) has a rigid sphere-like structure of which the hydrodynamic radius is 2.43 nm [187]. The complexation of TPA-DPA with $SnCl_2$ proceeded in a stepwise fashion from the core to the terminal imine following the basicity gradient among the imine groups in each dendron shell. The DSSCs prepared by casting these freebase dendrimers onto a dye-sensitized TiO_2 film exhibited a higher open-circuit voltage than the bare film

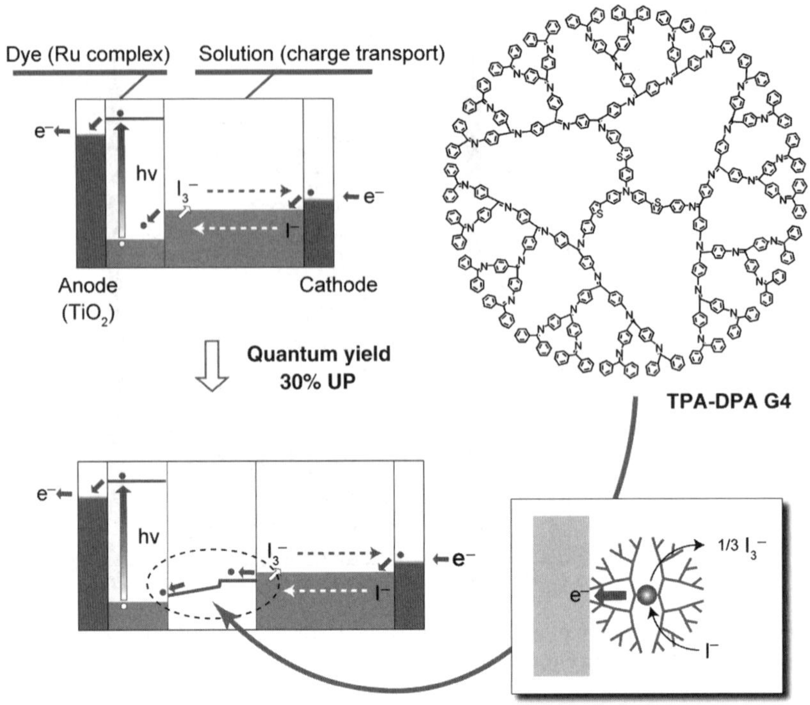

FIGURE 10.20 Potential diagram of a dye-sensitized solar cell employing the phenylazomethine dendrimer with a trithienylphenyl amine core. The dendrimer component acts as an electron rectifier suppressing the back-electron transfer reaction.

through suppression of the back-electron transfer. The insulating effect affording a better performance of the DSSC was further enhanced by the generation-number increase. Moreover, the complexation with $SnCl_2$ reduced the resistance of the TPA-DPA and improved the fill factor (*ff*). Accordingly, the energy conversion efficiency of the DSSC prepared using a fifth-generation TPA-DPA is 21% higher than that for the bare film and, when complexed with $SnCl_2$, provides a 34% improvement (Figure 10.20).

The micellar structure, such as dendrimers or nanoparticles coated by a SAM, would provide a segregating channel (for the electron and hole), which might become part of the heterojunctions. Imahori et al. reported organic solar cells prepared using the quaternary self-organization of porphyrin (donor) and fullerene (acceptor) units by clusterization with gold nanoparticles on nanostructured SnO_2 electrodes [188,189] The power conversion efficiency of the composite electrode reached as high as 1.5%, which is 45 times higher than that of the reference system consisting of both the single components of the porphyrin and fullerene. Li et al. [190] demonstrated a charge-separating reaction based on a molecular hybrid composed of a porphyrin-terminated dendrimer and fullerene derivative. Due to the extended electron- and hole-channel around the dendrimer architecture, an efficient and stable charge separation could be achieved on the basis of the molecular-level design. Not only for a photovoltaic application but the dendrimer architecture also contributed various photochemical processes. For example, the photochemical hydrogen evolution reaction was achieved using a molecular sensitizer based on the dendrimer architecture.

10.6 CONCLUSION

The chemistry of a well-defined organic nanostructure has been developed over the past decade. The advent of the dendrimer architecture provided a new strategy to construct a fine nanostructure of discrete macromolecules, and possibility of broad applications. In particular, a new class of π-conjugating dendrimers, which can coordinate to multiple metal ions showed a very unique property in its metal-assembling process. Although the dendrimer has many (30 or 60) ligation sites, the metal ions are preferentially assembled at the ligation sites closer to the core. As a result, metal ions are orderly assembled in the dendrimer scaffold through a step-by-step fashion. This phenomenon allows suppression of the statistical distribution in the number of assembled metal ions.

The metal assembly in a discrete molecular capsule can be converted to a size-regulated metal cluster with a size smaller than 1 nm. Due to the well-defined number of metal clusters in the subnanometer size region, its property is much different from that of bulk or general metal nanoparticles. For example, the platinum subnanocluster was determined to be a more active catalyst for the ORR than the larger cluster. The size-specific variation in the onset potential where the ORR started indicates that the poor catalytic performance as a fuel cell catalyst mainly originated from the electron transfer process to the oxygen molecule. Recent studies have demonstrated that the electron transfer process can be improved by the synthesis of a bimetallic composition. A similar synthetic protocol can be applied to the synthesis of metal-oxide subnanoparticles, such as TiO_2. Estimation of the optical bandgap elucidated that the quantum size effect also appeared in TiO_2 nanoparticles smaller than 1 nm. This is the first investigation of subnanoclusters properties that have not yet been determined.

We can now utilize this architecture in photochemical and electric devices, such as light-emitting diodes or solar cells. Photochemical events (excitation, luminescence, quenching or charge separation) should be implemented in the photoelectrochemical devices as molecular-level functions. Although these essential functions are obtained with organic π-extended molecules or metal complexes, it is not enough to use most of the energy because the annihilation process (charge recombination) occurs during the electron-transporting process through the films. A mechanism suppressing this process can be realized by building a higher order structure of which the scale is greater than single nanometers. In other words, the hierarchical structure from the molecular level (functional unit) to the molecular assembly should be constructed along the hierarchical electron transfer process from the photoreaction center (dye and donor–acceptor system) to the electrodes.

Dendritic macromolecules including their metal complexes are some of the promising candidates for the next-generation organic semiconducting materials. They are not simply hole- and electron-transporting materials, but also functional molecular modules assisting the photochemical reactions and electron (hole) rectification. Their well-defined macromolecular architecture on a single-nanometer scale is expected to control the local electron transfer, and the molecular assembling structure up to a micrometer scale plays an important role in fabricating long-range electron-transporting channels. In addition, the topological architecture of the dendrimer may provide an intermolecular electrostatic gradient across the LBL architecture. A new strategy taking this paradigm into account might accelerate the development of organic semiconducting materials for the next generation.

REFERENCES

[1] S. M. Grayson, J. M. Fréchet, *Chem. Rev.* **2001**, *101*, 3819–3868.

[2] S. Hecht, J. M. Fréchet, *Angew. Chem. Int. Ed.* **2001**, *40*, 74–91.

[3] D. A. Tomalia, A. M. Naylor, *Angew. Chem. Int. Ed.* **1990**, *29*, 138–175.

[4] V. Balzani, G. Bergamini, P. Ceroni, F. Voegtle, *Coord. Chem. Rev.* **2007**, *251*, 525–535.

[5] V. Balzani, A. Juris, M. Venturi, S. Campagna, S. Serroni, *Chem. Rev.* **1996**, *96*, 759–834.

[6] G. R. Newkome, E. He, C. N. Moorefield, *Chem. Rev.* **1999**, *99*, 1689–1746.

[7] B. Helms, J. M. J. Fréchet, *Adv. Synth. Catal.* **2006**, *348*, 1125–1148.

[8] D. Méry, D. Astruc, *Coord. Chem. Rev.* **2006**, *250*, 1965–1979.

[9] R. van Heerbeek, P. C. Kamer, P. W. van Leeuwen, J. N. Reek, *Chem. Rev.* **2002**, *102*, 3717–3756.

[10] D. Astruc, F. Chardac, *Chem. Rev.* **2001**, *101*, 2991–3024.

[11] L. J. Twyman, A. S. King, I. K. Martin, *Chem. Soc. Rev.* **2002**, *31*, 69–82.

[12] A. C. Grimsdale, K. Müllen, *Angew. Chem. Int. Ed.* **2005**, *44*, 5592–5629.

[13] S. R. Forrest, *Nature* **2004**, *428*, 911–918.

[14] T. Darbre, J.-L. Reymond, *Acc. Chem. Res.* **2006**, *39*, 925–934.

[15] G. M. Whitesides, B. Grzybowski, *Science* **2002**, *295*, 2418–2421.

[16] F. Ciardelli, E. Tsuchida, D. Wohrle, *Macromolecule-metal Complexes*, Springer, New York, **1996**.

[17] G. R. Whittell, I. Manners, *Adv. Mater.* **2007**, *19*, 3439–3468.

[18] S. Kitagawa, R. Kitaura, S. Noro, *Angew. Chem. Int. Ed.* **2004**, *43*, 2334–2375.

[19] P. R. Andres, U. S. Schubert, *Adv. Mater.* **2004**, *16*, 1043–1068.

[20] B. J. Holliday, C. A. Mirkin, *Angew. Chem. Int. Ed.* **2001**, *40*, 2022–2043.

[21] I. Beletskaya, V. S. Tyurin, A. Y. Tsivadze, R. Guilard, C. Stern, *Chem. Rev.* **2009**, *109*, 1659–1713.

[22] M. Fujita, M. Tominaga, A. Hori, B. Therrien, *Acc. Chem. Res.* **2005**, *38*, 369–378.

[23] N. Fukuda, M. Mitsuishi, A. Aoki, T. Miyashita, *J. Phys. Chem. B* **2002**, *106*, 7048–7052.

[24] H. Tanaka, M. Mitsuishi, T. Miyashita, *Langmuir* **2003**, *19*, 3103–3105.

[25] E. J. Calvo, A. Wolosiuk, *J. Am. Chem. Soc.* **2002**, *124*, 8490–8497.

[26] G. Decher, J. D. Hong, J. Schmitt, *Thin Solid Films* **1992**, *210*, 831–835.

[27] M. Huang, Y. Shao, X. Sun, H. Chen, B. Liu, S. Dong, *Langmuir* **2005**, *21*, 323–329.

[28] G. M. Rahman, D. M. Guldi, R. Cagnoli, A. Mucci, L. Schenetti, L. Vaccari, M. Prato, *J. Am. Chem. Soc.* **2005**, *127*, 10051–10057.

[29] B. K. Cho, A. Jain, S. M. Gruner, U. Wiesner, *Science* **2004**, *305*, 1598–1601.

[30] M. C. Coen, K. Lorenz, J. Kressler, H. Frey, R. Mulhaupt, *Macromolecules* **1996**, *29*, 8069–8076.

[31] M. Enomoto, A. Kishimura, T. Aida, *J. Am. Chem. Soc.* **2001**, *123*, 5608–5609.

[32] J. M. Fréchet, *Proc. Natl. Acad. Sci. USA* **2002**, *99*, 4782–4787.

[33] S. D. Hudson, H. Jung, V. Percec, W. Cho, G. Johansson, G. Ungar, V. S. K. Balagurusamy, *Science* **1997**, *278*, 449–452.

[34] V. Percec, C. Ahn, G. Ungar, D. J. P. Yeardley, M. Moller, S. S. Sheiko, *Nature* **1998**, *391*, 61–64.

[35] V. Percec, M. Glodde, T. K. Bera, Y. Miura, I. Shiyanovskaya, K. D. Singer, V. S. K. Balagurusamy, P. A. Heiney, I. Schnell, A. Rapp, H. W. Spiess, S. D. Hudson, H. Duan, *Nature* **2002**, *419*, 384–387.

[36] D. C. Tully, K. Wilder, J. M. J. Fréchet, A. R. Trimble, C. F. Quate, *Adv. Mater.* **1999**, *11*, 314–318.

[37] S. C. Zimmerman, F. Zeng, D. E. C. Reichert, S. V. Kolotuchin, *Science* **1996**, *271*, 1095.

[38] R. M. Crooks, M. Zhao, L. Sun, V. Chechik, L. K. Yeung, *Acc. Chem. Res.* **2001**, *34*, 181–190.

[39] K. Yamamoto, M. Higuchi, A. Kimoto, T. Imaoka, K. Masachika, *Bull. Chem. Soc. Jpn.* **2005**, *78*, 349–355.

[40] K. Yamamoto, M. Higuchi, S. Shiki, M. Tsuruta, H. Chiba, *Nature* **2002**, *415*, 509–511.

[41] P. Ceroni, F. Paolucci, C. Paradisi, A. Juris, S. Roffia, S. Serroni, S. Campagna, A. J. Bard, *J. Am. Chem. Soc.* **1998**, *120*, 5480–5487.

[42] M.-S. Choi, T. Yamazaki, I. Yamazaki, T. Aida, *Angew. Chem. Int. Ed.* **2004**, *43*, 150–158.

[43] M. Marcaccio, F. Paolucci, C. Paradisi, S. Roffia, C. Fontanesi, L. J. Yellorlees, S. Serroni, S. Campagna, G. Denti, V. Balzani, *J. Am. Chem. Soc.* **1999**, *121*, 10081–10091.

[44] V. Balzani, P. Ceroni, A. Juris, M. Venturi, S. Campagna, F. Puntoriero, S. Serroni, *Coord. Chem. Rev.* **2001**, *219*, 545–572.

[45] J. D. Epperson, L. J. Ming, G. R. Baker, G. R. Newkome, *J. Am. Chem. Soc.* **2001**, *123*, 8583–8592.

[46] C. Giansante, P. Ceroni, V. Balzani, F. Vögtle, *Angew. Chem. Int. Ed.* **2008**, *47*, 5422–5425.

[47] R. J. M. Klein-Gebbink, A. W. Bosman, M. C. Feiters, E. W. Meijer, R. J. M. Nolte, *Chem. Eur. J.* **1999**, *5*, 65–69.

[48] M. F. Ottaviani, S. H. Bossmann, N. J. Turro, D. A. Tomalia, *J. Am. Chem. Soc.* **1994**, *116*, 661–671.

[49] M. F. Ottaviani, F. Montalti, N. J. Turro, D. A. Tomalia, *J. Phys. Chem. B* **1997**, *101*, 158–166.

[50] M. Tominaga, J. Hosogi, K. Konishi, T. Aida, *Chem. Commun.* **2000**, 719–720.

[51] F. Vögtle, S. Gestermann, C. Kauffmann, P. Ceroni, V. Vicinelli, V. Balzani, *J. Am. Chem. Soc.* **2000**, *122*, 10398–10404.

[52] M. Higuchi, M. Tsuruta, H. Chiba, S. Shiki, K. Yamamoto, *J. Am. Chem. Soc.* **2003**, *125*, 9988–9997.

[53] T. Imaoka, H. Horiguchi, K. Yamamoto, *J. Am. Chem. Soc.* **2003**, *125*, 340–341.

[54] T. Imaoka, R. Tanaka, S. Arimoto, M. Sakai, M. Fujii, K. Yamamoto, *J. Am. Chem. Soc.* **2005**, *127*, 13896–13905.

[55] T. Imaoka, R. Tanaka, K. Yamamoto, *Chem. Eur. J.* **2006**, *12*, 7328–7336.

[56] R. Nakajima, M. Tsuruta, M. Higuchi, K. Yamamoto, *J. Am. Chem. Soc.* **2004**, *126*, 1630–1631.

[57] K. Yamamoto, T. Imaoka, *Bull. Chem. Soc. Jpn.* **2006**, *79*, 511–526.

[58] V. Balzani, P. Ceroni, S. Gestermann, M. Gorka, C. Kauffmannb, F. Vögtle, *Dalton Trans.* **2000**, 3765–3771.

[59] V. Balzani, F. Vögtle, *C. R. Chimie* **2003**, *6*, 867–872.

[60] I. Grabchev, J. Chovelon, X. Qianc, *New J. Chem.* **2003**, *27*, 337–340.

[61] V. Vicinelli, P. Ceroni, M. Maestri, V. Balzani, M. Gorka, F. Vögtle, *J. Am. Chem. Soc.* **2002**, *124*, 6461–6468.

[62] C. Saudan, V. Balzani, M. Gorka, S. K. Lee, M. Maestri, V. Vicinelli, F. Vögtle, *J. Am. Chem. Soc.* **2003**, *125*, 4424–4425.

[63] L. Balogh, D. A. Tomalia, *J. Am. Chem. Soc.* **1998**, *120*, 7355–7356.

[64] V. Chechik, R. M. Crooks, *J. Am. Chem. Soc.* **2000**, *122*, 1243–1244.

[65] H. Lang, R. A. May, B. L. Iversen, B. D. Chandler, *J. Am. Chem. Soc.* **2003**, *125*, 14832–14836.

[66] M. Zhao, R. M. Crooks, *Adv. Mater.* **1999**, *11*, 217–220.

[67] M. Zhao, R. M. Crooks, *Angew. Chem. Int. Ed.* **1999**, *38*, 364–366.

[68] M. Zhao, L. Sun, R. M. Crooks, *J. Am. Chem. Soc.* **1998**, *120*, 4877–4878.

[69] M. Ballauff, C. N. Likos, *Angew. Chem. Int. Ed.* **2004**, *43*, 2998–3020.

[70] S. Rosenfeldt, N. Dingenouts, D. Pötschke, M. Ballauff, A. J. Berresheim, K. Müllen, P. Lindner, *Angew. Chem. Int. Ed.* **2004**, *43*, 109–112.

[71] T. L. Chasse, R. Sachdeva, Q. Li, Z. Li, R. J. Petrie, C. B. Gorman, *J. Am. Chem. Soc.* **2003**, *125*, 8250–8254.

[72] M. Higuchi, S. Shiki, K. Ariga, K. Yamamoto, *J. Am. Chem. Soc.* **2001**, *123*, 4414–4420.

[73] M. Higuchi, K. Yamamoto, *Bull. Chem. Soc. Jpn.* **2004**, *77*, 853–874.

[74] N. Satoh, J. S. Cho, M. Higuchi, K. Yamamoto, *J. Am. Chem. Soc.* **2003**, *125*, 8104–8105.

[75] O. Enoki, H. Katoh, K. Yamamoto, *Org. Lett.* **2006**, *8*, 569–571.

[76] K. Yamamoto, T. Imaoka, W. Chun, O. Enoki, H. Katoh, M. Takenaga, A. Sonoi, *Nat. Chem.* **2009**, *1*, 397–402.

[77] O. Enoki, T. Imaoka, K. Yamamoto, *Org. Lett.* **2003**, *5*, 2547–2549.

[78] D. Astruc, F. Lu, J. R. Aranzaes, *Angew. Chem. Int. Ed.* **2005**, *44*, 7852–7872.

[79] T. Ishida, M. Haruta, *Angew. Chem. Int. Ed.* **2007**, *46*, 7154–7156.

[80] R. Alayan, L. Arnaud, M. Broyer, E. Cottancin, J. Lerme, J. L. Vialle, M. Pellarin, *Phys. Rev. B* **2006**, *73*, 125444.

[81] H. Yasumatsu, T. Hayakawa, S. Koizumi, T. Kondow, *J. Chem. Phys.* **2005**, *123*, 124709.

[82] B. C. Steele, A. Heinzel, *Nature* **2001**, *414*, 345–352.

[83] Y. Verde, G. Alonso-Nuñez, M. Miki-Yoshida, M. JoséYacamán, V. H. Ramos, A. Keer, *Catal. Today* **2005**, *107–108*, 826–830.

[84] F. Raimondi, G. G. Scherer, R. Kötz, A. Wokaun, *Angew. Chem. Int. Ed.* **2005**, *44*, 2190–2209.

[85] P. J. Ferreira, G. J. la O, Y. Shao-Horn, D. Morgan, R. Makharia, S. Kocha, H. A. Gasteiger, *J. Electrochem. Soc.* **2005**, *152*, A2256–A2271.

[86] N. Satoh, T. Nakashima, K. Kamikura, K. Yamamoto, *Nat. Nanotechnol.* **2008**, *3*, 106–111.

[87] M. C. Yan, F. Chen, J. L. Zhang, M. Anpo, *J. Phys. Chem. B* **2005**, *109*, 8673–8678.

[88] J. Tang, F. Redl, Y. Zhu, T. Siegrist, L. E. Brus, M. L. Steigerwald, *Nano. Lett.* **2005**, *5*, 543–548.

[89] C. N. R. Rao, A. Müller, A. K. Cheetham, *The Chemistry of Nanomaterials: Synthesis, Properties and Applications*, Chapter 11, Wiley-VCH Verlag GmbH, Weinheim, **2004**.

[90] P. Peumans, S. Uchida, S. R. Forrest, *Nature* **2003**, *425*, 158–162.

[91] G. Yu, J. Gao, J. C. Hummelen, F. Wudl, A. J. Heeger, *Science* **1995**, *270*, 1789–1791.

[92] G. R. Newkome, Z. Yao, G. R. Baker, V. K. Gupta, *J. Org. Chem.* **1985**, *50*, 2003–2004.

[93] A. W. Bosman, H. M. Janssen, E. W. Meijer, *Chem. Rev.* **1999**, *99*, 1665–1688.

[94] F. Vögtle, S. Gestermann, R. Hesse, H. Schwierz, B. Windisch, *Prog. Polym. Sci.* **2000**, *25*, 987–1041.

[95] F. Zeng, S. C. Zimmerman, *Chem. Rev.* **1997**, *97*, 1681–1712.

[96] C. J. Hawker, J. M. J. Fréchet, *J. Am. Chem. Soc.* **1990**, *112*, 7638–7647.

[97] C. M. Cardona, S. Mendoza, A. E. Kaifer, *Chem. Soc. Rev.* **2000**, *29*, 37–42.

[98] C. B. Gorman, J. C. Smith, *Acc. Chem. Res.* **2001**, *34*, 60–71.

[99] P. Bhyrappa, J. K. Young, J. S. Moore, K. S. Suslick, *J. Am. Chem. Soc.* **1996**, *118*, 5708–5711.

[100] M. Kimura, T. Shiba, M. Yamazaki, K. Hanabusa, H. Shirai, N. Kobayashi, *J. Am. Chem. Soc.* **2001**, *123*, 5636–5642.

[101] M. Uyemura, T. Aida, *J. Am. Chem. Soc.* **2002**, *124*, 11392–11403.

[102] M. Uyemura, T. Aida, *Chem. Eur. J.* **2003**, *9*, 3492–3500.

[103] V. Balzani, P. Ceroni, C. Giansante, V. Vicinelli, F. G. Klärner, C. Verhaelen, F. Vögtle, U. Hahn, *Angew. Chem. Int. Ed.* **2005**, *44*, 4574–4578.

[104] W. Ong, M. Gomez-Kaifer, A. E. Kaifer, *Chem. Commun.* **2004**, 1677–1683.

[105] Y. Tomoyose, D.-L. Jiang, R. Jin, T. Aida, T. Yamashita, K. Horie, E. Yashima, Y. Okamoto, *Macromolecules* **1996**, *29*, 5236–5238.

[106] S. C. Zimmerman, Y. Wang, P. Bharathi, J. S. Moore, *J. Am. Chem. Soc.* **1998**, *120*, 2172–2173.

[107] W. D. Jang, N. Nishiyama, G. D. Zhang, A. Harada, D. L. Jiang, S. Kawauchi, Y. Morimoto, M. Kikuchi, H. Koyama, T. Aida, K. Kataoka, *Angew. Chem. Int. Ed.* **2005**, *44*, 419–423.

[108] D.-L. Jiang, T. Aida, *J. Am. Chem. Soc.* **1998**, *120*, 10895–10901.

[109] T. Sato, D.-L. Jiang, T. Aida, *J. Am. Chem. Soc.* **1999**, *121*, 10658–10659.

[110] F. Vögtle, M. Plevoets, M. Nieger, G. C. Azzellini, A. Credi, L. De Cola, V. De Marchis, M. Venturi, V. Balzani, *J. Am. Chem. Soc.* **1999**, *121*, 6290–6298.

[111] X. Zhou, D. S. Tyson, F. N. Castellano, *Angew. Chem. Int. Ed.* **2000**, *39*, 4301–4305.

[112] M. Enomoto, T. Aida, *J. Am. Chem. Soc.* **2002**, *124*, 6099–6108.

[113] D.-L. Jiang, T. Aida, *Chem. Commun.* **1996**, 1523–1524.

[114] D. de Groot, B. F. de Waal, J. N. Reek, A. P. Schenning, P. C. Kamer, E. W. Meijer, P. W. van Leeuwen, *J. Am. Chem. Soc.* **2001**, *123*, 8453–8458.

[115] R. van de Coevering, M. Kuil, R. J. Gebbink, G. van Koten, *Chem. Commun.* **2002**, 1636–1637.

[116] M. C. Daniel, J. Ruiz, J. C. Blasie, N. Daro, D. Astruc, *Chem. Eur. J.* **2003**, *9*, 4371–4379.

[117] M. C. Daniel, J. Ruiz, S. Nlate, J. C. Blasie, D. Astruc, *J. Am. Chem. Soc.* **2003**, *125*, 2617–2628.

[118] C. Valério, E. Alonso, J. Ruiz, J. C. Blasie, D. Astruc, *Angew. Chem. Int. Ed.* **1999**, *38*, 1747–1751.

[119] C. Valério, J. Fillaut, J. Ruiz, J. Guittard, J. C. Blasie, D. Astruc, *J. Am. Chem. Soc.* **1997**, *119*, 2588–2589.

[120] M. T. Reetz, D. Giebel, *Angew. Chem. Int. Ed.* **2000**, *39*, 2498–2501.

[121] N. Tomioka, D. Takasu, T. Takahasi, T. Aida, *Angew. Chem. Int. Ed.* **1998**, *37*, 1531–1534.

[122] K. Takada, D. J. Díaz, H. D. Abruña, I. Cuadrado, B. González, C. M. Casado, B. Alonso, M. Morán, J. Losada, *Chem. Eur. J.* **2001**, *7*, 1109–1117.

[123] M.-S. Choi, T. Aida, T. Yamazaki, I. Yamazaki, *Angew. Chem. Int. Ed.* **2001**, *40*, 3194–3198.

[124] M.-S. Choi, T. Aida, H. Luo, Y. Araki, O. Ito, *Angew. Chem. Int. Ed.* **2003**, *42*, 4060–4063.

[125] M.-S. Choi, T. Aida, T. Yamazaki, I. Yamazaki, *Chem. Eur. J.* **2002**, *8*, 2668–2678.

[126] M. S. Matos, J. Hofkens, W. Verheijen, F. C. D. Schryver, S. Hecht, K. W. Pollak, J. M. J. Fréchet, B. Forier, W. Dehaen, *Macromolecules* **2000**, *33*, 2967–2973.

[127] T. H. Mourey, S. R. Turner, M. Rubinstein, J. M. J. Fréchet, C. J. Hawker, K. L. Wooley, *Macromolecules* **1992**, *25*, 2401–2406.

[128] D. Boris, M. Rubinstein, *Macromolecules* **1996**, *29*, 7251–7260.

[129] R. L. Lescanec, M. Muthukumar, *Macromolecules* **1990**, *23*, 2280–2288.

[130] W. Ortiz, A. E. Roitberg, J. L. Krause, *J. Phys. Chem. B* **2004**, *108*, 8218–8225.

[131] M. Wind, K. Saalwähter, U. Wiesler, K. Müllen, H. W. Spiess, *Macromolecules* **2002**, *35*, 10071–10086.

[132] M. Kimura, Y. Saito, K. Ohta, K. Hanabusa, H. Shirai, N. Kobayashi, *J. Am. Chem. Soc.* **2002**, *124*, 5274–5275.

[133] J. P. Cross, M. Lauz, P. D. Badger, S. Petoud, *J. Am. Chem. Soc.* **2004**, *126*, 16278–16279.

[134] R. Shunmugam, G. N. Tew, *J. Am. Chem. Soc.* **2005**, *127*, 13567–13572.

[135] S. Kimata, D.-L. Jiang, T. Aida, *J. Polym. Sci. A: Polym. Chem.* **2003**, *41*, 3524–3530.

[136] J. L. Wang, J. Luo, L. H. Liu, Q. F. Zhou, Y. Ma, J. Pei, *Org. Lett.* **2006**, *8*, 2281–2284.

[137] T. H. Xu, R. Lu, X. P. Qiu, X. L. Liu, P. C. Xue, C. H. Tan, C. Y. Bao, Y. Y. Zhao, *Eur J. Org. Chem.* **2006**, 4014–4020.

[138] F. S. Precup-Blaga, J. C. Garcia-Martinez, A. P. Schenning, E. W. Meijer, *J. Am. Chem. Soc.* **2003**, *125*, 12953–12960.

[139] J. Qu, N. G. Pschirer, D. Liu, A. Stefan, F. C. De Schryver, K. Müllen, *Chem. Eur. J.* **2004**, *10*, 528–537.

[140] S. Lo, T. D. Anthopoulos, E. B. Namdas, P. L. Burn, I. D. W. Samuel, *Adv. Mater.* **2005**, *17*, 1945–1948.

[141] R. Kunieda, M. Fujitsuka, O. Ito, M. Ito, Y. Murata, K. Komatsu, *J. Phys. Chem. B* **2002**, *106*, 7193–7199.

[142] P. L. Burn, S. Lo, I. D. W. Samuel, *Adv. Mater.* **2007**, *19*, 1675–1688.

[143] X. Y. Cao, X. H. Liu, X. H. Zhou, Y. Zhang, Y. Jiang, Y. Cao, Y. X. Cui, J. Pei, *J. Org. Chem.* **2004**, *69*, 6050–6058.

[144] A. W. Freeman, S. C. Koene, P. R. L. Maalenfant, M. E. Thompson, J. M. J. Fréhet, *J. Am. Chem. Soc.* **2001**, *122*, 12385–12386.

[145] J. Zheng, J. T. Petty, R. M. Dickson, *J. Am. Chem. Soc.* **2003**, *125*, 7780–7781.

[146] A. B. Padmaperuma, G. Schmett, D. Fogarty, N. Washton, S. Nanayakkara, L. Sapochak, K. Ashworth, L. Madrigal, B. Reeves, C. W. Spangler, *Mater. Res. Soc. Symp. Proc.* **2000**, *621*, Q3.9.1-6.

[147] J. F. Huang, H. Luo, C. Liang, I. W. Sun, G. A. Baker, S. Dai, *J. Am. Chem. Soc.* **2005**, *127*, 12784–12785.

[148] C. W. Wu, C. M. Tsai, H. C. Lin, *Macromolecules* **2006**, *39*, 4298–4305.

[149] T. Qin, J. Ding, L. Wang, M. Baumgarten, G. Zhou, K. Müllen, *J. Am. Chem. Soc.* **2009**, *131*, 14329–14336.

[150] W. R. Dichtel, J. M. Serin, C. Edder, J. M. Fréchet, M. Matuszewski, L. S. Tan, T. Y. Ohulchanskyy, P. N. Prasad, *J. Am. Chem. Soc.* **2004**, *126*, 5380–5381.

[151] C. N. Fleming, K. A. Maxwell, J. M. DeSimone, T. J. Meyer, J. M. Papanikolas, *J. Am. Chem. Soc.* **2001**, *123*, 10336–10347.

[152] T. G. GoodsonIII, *Acc. Chem. Res.* **2005**, *38*, 99–107.

[153] Y. Wang, M. I. Ranasinghe, T. G. Goodson III, *J. Am. Chem. Soc.* **2003**, *125*, 9562–9563.

[154] R. Métivier, F. Kulzer, T. Weil, K. Müllen, T. Basché, *J. Am. Chem. Soc.* **2004**, *126*, 14364–14365.

[155] H. Imahori, *J. Phys. Chem. B* **2004**, *108*, 6130–6143.

[156] F. Loiseau, S. Campagna, A. Hameurlaine, W. Dehaen, *J. Am. Chem. Soc.* **2005**, *127*, 11352–11363.

[157] M. Maus, R. De, M. Lor, T. Weil, S. Mitra, U. M. Wiesler, A. Herrmann, J. Hofkens, T. Vosch, K. Mülen, F. C. De Schryver, *J. Am. Chem. Soc.* **2001**, *123*, 7668–7676.

[158] N. D. McClenaghan, R. Passalacqua, F. Loiseau, S. Campagna, B. Verheyde, A. Hameurlaine, W. Dehaen, *J. Am. Chem. Soc.* **2003**, *125*, 5356–5365.

[159] J. S. Melinger, Y. Pan, V. D. Kleiman, Z. Peng, B. L. Davis, D. McMorrow, M. Lu, *J. Am. Chem. Soc.* **2002**, *124*, 12002–12012.

[160] Y. J. Mo, D. L. Jiang, M. Uyemura, T. Aida, T. Kitagawa, *J. Am. Chem. Soc.* **2005**, *127*, 10020–10027.

[161] N. Nishiyama, H. R. Stapert, G. D. Zhang, D. Takasu, D. L. Jiang, T. Nagano, T. Aida, K. Kataoka, *Bioconjug. Chem.* **2003**, *14*, 58–66.

[162] M. I. Ranasinghe, Y. Wang, T. G. Goodson III, *J. Am. Chem. Soc.* **2003**, *125*, 5258–5259.

[163] O. P. Varnavski, J. C. Ostrowski, L. Sukhomlinova, R. J. Twieg, G. C. Bazan, T. G. GoodsonIII, *J. Am. Chem. Soc.* **2002**, *124*, 1736–1743.

[164] X. Yan, T. G. Goodson III, T. Imaoka, K. Yamamoto, *J. Phys. Chem. B* **2005**, *109*, 9321–9329.

[165] B. C. Thompson, J. M. Fréchet, *Angew. Chem. Int. Ed.* **2007**, *47*, 58–77.

[166] Y. Shirota, *J. Mater. Chem.* **2000**, *10*, 1–25.

[167] Y. Shirota, H. Kageyama, *Chem. Rev.* **2007**, *107*, 953–1010.

[168] H. John, R. Bauer, P. Espindola, P. Sonar, J. Heinze, K. Müllen, *Angew. Chem. Int. Ed.* **2005**, *44*, 2447–2451.

[169] M. J. Frampton, H. L. Anderson, *Angew. Chem. Int. Ed.* **2007**, *46*, 1028–1064.

[170] J. Wu, M. Baumgarten, M. G. Debije, J. M. Warman, K. Müllen, *Angew. Chem. Int. Ed.* **2004**, *43*, 5331–5335.

[171] M. Han, P. Chen, X. Yang, *Polymer* **2005**, *46*, 3481–3488.

[172] C. L. Jackson, H. D. Chanzy, F. P. Booy, B. J. Drake, D. A. Tomalia, B. J. Bauer, E. J. Amis, *Macromolecules* **1998**, *31*, 6259–6265.

[173] P. K. Maiti, T. Çagin, G. Wang, W. A. Goddard III, *Macromolecules* **2004**, *37*, 6236–6254.

[174] R. Tanaka, T. Imaoka, K. Yamamoto, *J. Photopolym. Sci. Technol.* **2004**, *17*, 323–326.

[175] Y. Sun, N. C. Giebink, H. Kanno, B. Ma, M. E. Thompson, S. R. Forrest, *Nature* **2006**, *440*, 908–912.

[176] T. H. Huang, J. T. Lin, L. Y. Chen, Y. T. Lin, C. C. Wu, *Adv. Mater.* **2006**, *18*, 602.

[177] M. Y. Lai, C. H. Chen, W. S. Huang, J. T. Lin, T. H. Ke, L. Y. Chen, M. H. Tsai, C. C. Wu, *Angew. Chem. Int. Ed.* **2008**, *47*, 581–585.

[178] A. Kimoto, J. Cho, M. Higuchi, K. Yamamoto, *Macromolecules* **2004**, *37*, 5531–5537.

[179] M. Grätzel, *Nature* **2001**, *414*, 338–344.

[180] A. Hagfeldt, M. Grätzel, *Acc. Chem. Res.* **2000**, *33*, 269–277.

[181] S. Westenhoff, I. A. Howard, J. M. Hodgkiss, K. R. Kirov, H. A. Bronstein, C. K. Williams, N. C. Greenham, R. H. Friend, *J. Am. Chem. Soc.* **2008**, *130*, 13653–13658.

[182] M. Lor, J. Thielemans, L. Viaene, M. Cotlet, J. Hofkens, T. Weil, C. Hampel, K. Müllen, J. W. Verhoeven, M. Van Der Auweraer, F. C. De Schryver, *J. Am. Chem. Soc.* **2002**, *124*, 9918–9925.

[183] J. K. Lee, W. L. Ma, C. J. Brabec, J. Yuen, J. S. Moon, J. Y. Kim, K. Lee, G. C. Bazan, A. J. Heeger, *J. Am. Chem. Soc.* **2008**, *130*, 3619–3623.

[184] C. Yang, J. Y. Kim, S. Cho, J. K. Lee, A. J. Heeger, F. Wudl, *J. Am. Chem. Soc.* **2008**.

[185] A. R. Murphy, J. M. J. Fréchet, *Chem. Rev.* **2007**, *107*, 1066–1096.

[186] W. Y. Wong, X. Z. Wang, Z. He, A. B. Djurisić C. T. Yip, K. Y. Cheung, H. Wang, C. S. Mak, W. K. Chan, *Nat. Mater.* **2007**, *6*, 521–527.

[187] N. Satoh, T. Nakashima, K. Yamamoto, *J. Am. Chem. Soc.* **2005**, *127*, 13030–13038.

[188] T. Hasobe, H. Imahori, P. V. Kamat, T. K. Ahn, S. K. Kim, D. Kim, A. Fujimoto, T. Hirakawa, S. Fukuzumi, *J. Am. Chem. Soc.* **2005**, *127*, 1216–1228.

[189] H. Imahori, Fujimoto, S. Kang, H. Hotta, K. Yoshida, T. Umeyama, Y. Matano, S. Isoda, *Adv. Mater.* **2005**, *17*, 1727–1730.

[190] W. S. Li, K. S. Kim, D. L. Jiang, H. Tanaka, T. Kawai, J. H. Kwon, D. Kim, T. Aida, *J. Am. Chem. Soc.* **2006**, *128*, 10527–10532.

[191] S. M. Prince, M. Z. Papiz, A. A. Freer, G. McDermott, A. M. Hawthornthwaite-Lawless, R. J. Cogdell, N. W. Isaacs, *J. Mol. Biol.* **1997**, *268*, 412–423.

[192] C. R. Lancaster, H. Michel, *Structure* **1997**, *5*, 1339–1359.

11

ENLIGHTENING STRUCTURE AND PROPERTIES OF DENDRIMERS BY FLUORESCENCE DEPOLARIZATION

GIACOMO BERGAMINI, ENRICO MARCHI, PAOLA CERONI, AND VINCENZO BALZANI

Dipartimento di Chimica "G. Ciamician", Università di Bologna, via Selmi 2, I-40126 Bologna, Italy

11.1 INTRODUCTION

Dendrimers containing luminescent components [1,2] are particularly interesting since (i) luminescence signals offer a handle to better understand the dendritic structures and superstructures, (ii) cooperation among the photoactive components can allow the dendrimer to perform useful functions such as light harvesting, (iii) changes in the photophysical properties can be exploited for sensing purposes with signal amplification, and (iv) photochemical reactions can change the structure and other properties of dendrimers. Moreover, dendrimers can form internal dynamic niches in which small molecules or ions can be hosted [3,4], because of their tree-like multibranched structure.

Luminescent dendrimers are also suitable for fluorescence anisotropy measurements [5] that can give information on specific properties, such as (a) rotational reorientation times and hydrodynamic properties [6], (b) individual molecule rotational isomerism [7], (c) local motions of subunits [8], (d) mechanisms and rate of delocalization of the excitation energy [9], and (e) hosting ability [10].

Designing Dendrimers, Edited by Sebastiano Campagna, Paola Ceroni, Fausto Puntoriero.
© 2012 John Wiley & Sons, Inc. Published 2012 by John Wiley & Sons, Inc.

The fluorescence anisotropy is defined as

$$r = (I_{\|} - I_{\perp})/(I_{\|} + 2I_{\perp}) \tag{11.1}$$

where $I_{\|}$ and $I_{\perp}$ are the emission intensities registered when the emission and excitation polarizers are oriented parallelly or perpendicularly, respectively. For randomly oriented molecules containing a single fluorophore, directly after excitation ($t = 0$), the anisotropy r_0 will be 0.4 if the absorption and emission transition dipole moments have the same orientation and decreases if they are not collinear and when the molecule undergoes a change in orientation during the excited state lifetime [11].

The fluorescence anisotropy decay can be fitted by a monoexponential model for a spherical rotor:

$$r(t) = r_0\, e^{-t/\theta} \tag{11.2}$$

where θ is the rotational relaxation time.

An estimate of the θ value can also be obtained for a spherical rotor with a monoexponential decay of luminescence by the value of steady-state anisotropy (r_{ss}), the maximum anisotropy value (r_0), and the lifetime τ, according to the following equation [11]:

$$r_{ss} = \frac{r_0}{1 + \tau/\theta} \tag{11.3}$$

The rotational relaxation time θ can be related to the hydrodynamic volume V_h by the Stokes–Einstein–Debye equation. In particular, for compounds having a van der Waals volume much bigger than the volume of the solvent molecules, "sticking" boundary conditions are applicable, that is, the form of the rotor has no influence since it moves together with solvent molecules, and θ can be expressed as:

$$\theta = V_h\, \eta/k_B T \tag{11.4}$$

where η is the viscosity of the solvent, k_B the Boltzmann constant, and T the absolute temperature.

An ideal fluorophore for fluorescence anisotropy measurements should possess a high emission quantum yield and a lifetime with the same order of magnitude as the rotational correlation time.

If a supramolecular system contains *only one* fluorescent unit, fluorescence depolarization takes place, as for simple molecular species, if the absorption and emission transition moments are intrinsically noncollinear, and as a result of change in orientation of the fluorophore during the excited state lifetime. Such a dynamic change in orientation may be caused by (i) rotation of the supramolecular system as a whole, and (ii) local motions of the fluorophore, in the case of a flexible supramolecular architecture. For a supramolecular system that contains *more than one* identical fluorescent unit with different orientations, fluorescence depolarization can also occur by (iii) energy migration from the originally excited unit to a differently oriented one. Of course, the rates of global rotation, local movements, and energy migration must compete with the rate of deactivation of the fluorophore excited state. At constant temperature, the speed of global rotation depends on the size of the species and the viscosity of the solvent, the rate of local movements depends on the specific position of the fluorophore in the supramolecular structure and on the nature of the bonds that link the fluorophore to the supramolecular scaffold, and the rate of energy migration depends on the overlap between the absorption and emission spectra of the fluorophores and their distance. Energy migration among distinct supramolecular structures should be considered only for very short intermolecular distances, such as in the powder state, because of the short lifetime of the fluorescent excited state (nanosecond timescale).

In the specific case of dendrimers, fluorescent units can be incorporated in predetermined sites of their architecture or hosted inside the dendritic cavities. In the following, we will discuss selected examples to illustrate the potentialities of the fluorescence anisotropy technique to enlighten structure and properties of luminescent dendrimers. These examples can be divided on the basis of the position of the luminescent unit(s) in the dendritic structure (Scheme 11.1): (a) a single luminescent unit in the core; (b) multiple luminescent units at the periphery; (c) luminescent units in the branching points; (d) luminescent units encapsulated in the dendrimer dynamic cavities.

Fluorescence anisotropy is not limited to artificial macromolecules; it is gaining increasing attention also to study biological macromolecules inside cells [12], but this is outside the scope of the present work.

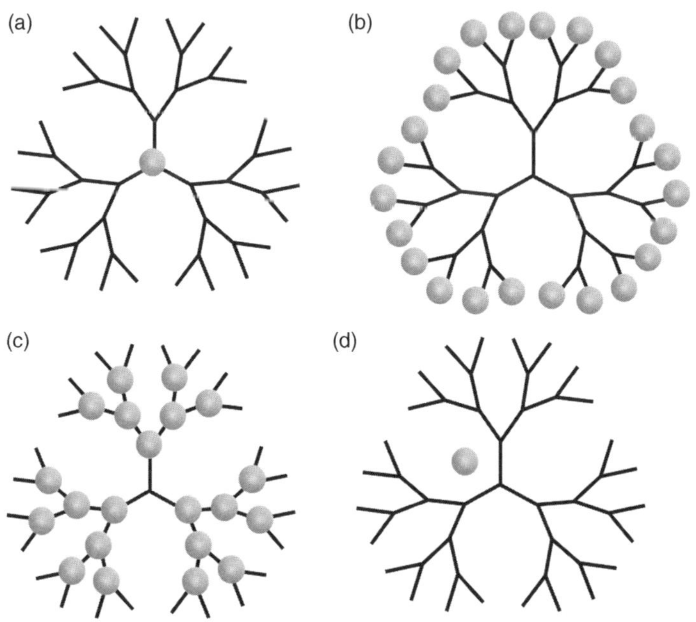

SCHEME 11.1 Schematic representation of dendrimers containing luminescent units (gray circle) located in the core (a), periphery (b), branching points (c), and hosted inside the dendrimer (d).

11.2 DENDRIMERS WITH A SINGLE LUMINESCENT UNIT AS A CORE

In the present case, fluorescence anisotropy measurements shine light on the hydrodynamic properties of the dendrimers as a function of generation, and on the local motion of the core chromophore, that is, the rigidity of its connections to the dendrons. In particular, the rotational behavior depends on the dimension and form of the rotor, interactions between probe and solvent, viscosity of the solvent, and temperature. Interaction of the dendrimer with solvent can indeed lead to changes in the dendrimer structure, depending on the nature of the core and the dendrons.

11.2.1 Dendrimers with a p-Terphenyl Core

These dendrimers are constituted by a *p*-terphenyl core appended with two *n*-octyl units and two oligosulfonimide branches of increasing size from G0T up to G2T (Scheme 11.2). Their absorption and emission bands are slightly red-shifted compared to those of *p*-terphenyl T [13]. In dichloromethane the fluorescence quantum yield of the dendrimers is similar and very high (0.7–0.8) and the fluorescence decay is short ($\tau = 0.8$ ns).

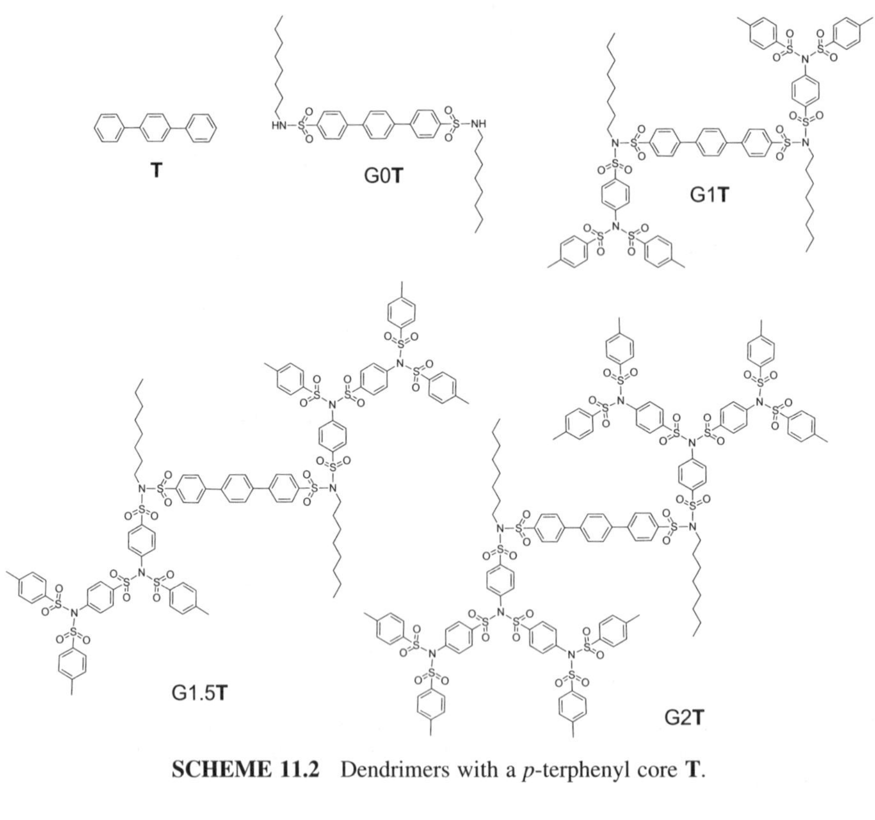

SCHEME 11.2 Dendrimers with a *p*-terphenyl core **T**.

11.2.1.1 Steady-State Properties The steady-state anisotropy r_{ss} of **T** and dendrimers G0**T**, G1**T**, G1.5**T**, and G2**T** in dichloromethane is constant throughout the emission band, is practically zero for **T** and increases with increasing dimension of the dendrimer (Figure 11.1a). The terphenyl-cored dendrimers show a noticeably high steady-state anisotropy taking into account the experimental conditions used (a low viscosity solvent at room temperature). Indeed, the maximum anisotropy value (r_0), determined by the relative orientation of the absorption and emission dipoles, is 0.33

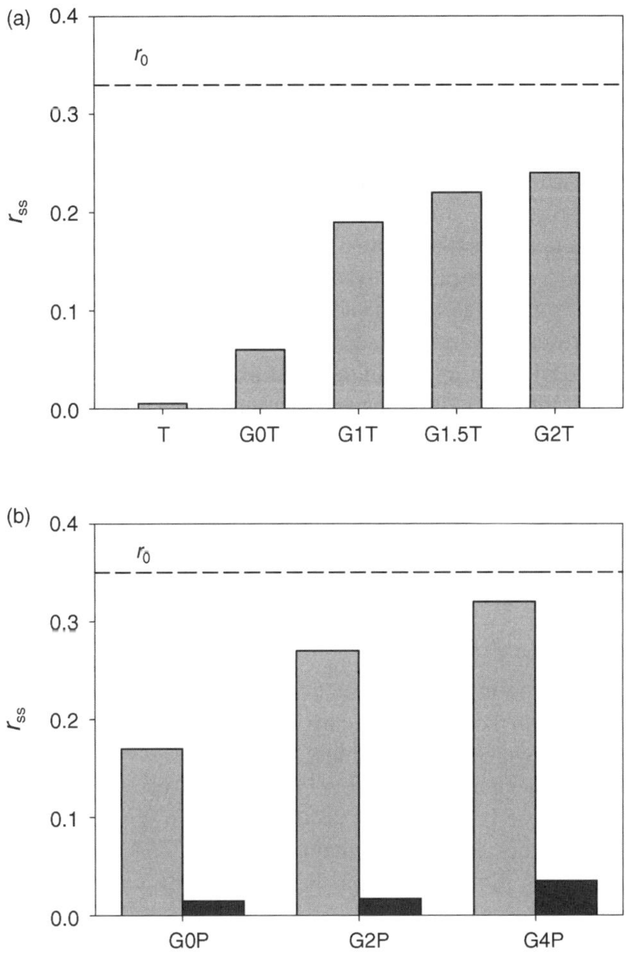

FIGURE 11.1 Steady-state fluorescence anisotropy r_{ss} in dichloromethane solutions at 298 K (light gray bars) of compounds (a) G0**T**–G2**T**, and **T** [13], and (b) G0**P**, G2**P**, and G4**P** [16]. The dashed line corresponds to the corresponding r_0 values in rigid matrix. For the pentaphenyl cored dendrimers the r_{ss} values observed for powder on a glass slide at 298 K are also reported (dark gray bars).

TABLE 11.1 Steady-State and Time-Resolved Anisotropy of Fluorescence, and Hydrodynamic Volumes Calculated by Eq. 11.4 in Dichloromethane Solution at 293 K

	r_{ss}	θ (ns)	V_h (nm^3)
T	<0.01	<0.5	
G1**T**	0.19	1.1	10.1
G1.5**T**	0.22	1.7	15.6
G2**T**	0.24	2.8	25.6
G0**P**	0.17 (0.015)[a]	–[b]	
G2**P**	0.27 (0.017)[a]	–[b]	
G4**P**	0.32 (0.035)[a]	–[b]	

[a] In the powder on a glass slide.
[b] Rotational relaxation time θ cannot be measured accurately because of the very short lifetime of the corresponding emitting excited state (about 0.7 ns).

for terphenyl measured in polystyrene matrix [14] and it corresponds to a lack of rotational depolarization.

In these compounds, depolarization channels related to energy migration are not available in solution since the terphenyl core is the unique fluorescent component and the lowest excited state of the system and interdendrimer energy transfer is prevented by the short fluorescence lifetime and dendrimer concentration. Furthermore, fluorescence depolarization via local motions is unlikely because the fluorescent unit is the core of the dendritic structure and even rotation of the terphenyl moiety around its molecular axis would not substantially change the relative orientation of the longitudinal absorption and emission transition moments [15]. Therefore, fluorescence anisotropy has to be related to global rotation of the dendrimer that is affected by solvent viscosity.

11.2.1.2 Time-Dependent Properties The fluorescence anisotropy decay of the dendrimers can be fitted to a monoexponential model (Eq. 11.2) for all compounds. As expected, rotational relaxation times θ increase with dendrimer generation (Table 11.1), demonstrating a slower rotation for larger molecules. These values are in good agreement with those calculated by Eq. 11.3. The rotational relaxation time θ can be related to the hydrodynamic volume V_h by Eq. 11.4. The calculated V_h values increase with dendrimer generation (Table 11.1) without reaching a plateau value; this demonstrates that there is no collapse of the dendrimer structure upon increasing generation. These values are in fair agreement with the ones obtained by molecular modeling, considering the nonspherical shape of the examined compounds [13].

11.2.2 Dendrimers with a p-Pentaphenyl Core

Dendrimers G0**P**, G2**P**, and G4**P** are constituted by a *p*-pentaphenyl core appended with two *n*-octyl chains and two oligosulfonimide branches of increasing generation (Scheme 11.3). These dendrimers show an absorption band in dichloromethane solution with maximum at 325 nm and an intense emission band at 410 nm

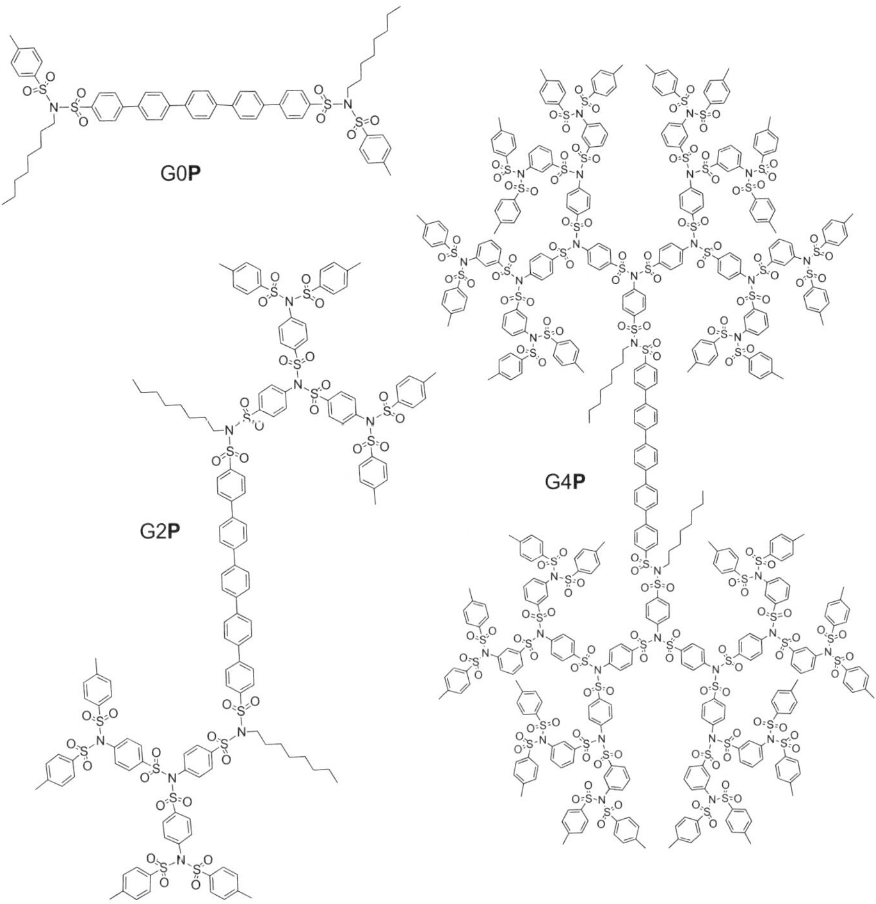

SCHEME 11.3 Dendrimers with a *p*-pentaphenyl core **P**.

($\Phi_{em} = 0.7$–0.8, $\lambda_{ex} = 325$ nm) that can be assigned to the *p*-pentaphenylene core [16]. The fluorescence decay is strictly monoexponential and the lifetime is practically the same in all cases (0.7 ns). In the solid state (powder on a glass slide), a strong violet emission is observed with a band maximum red-shifted of about 10 nm compared with the solution spectrum, whereas the lifetime remains the same.

11.2.2.1 Steady-State Properties In dichloromethane solution at 293 K, the anisotropy value increases along the series G0**P**, G2**P**, and G4**P** (Figure 11.1b, light gray bars, Table 11.1), that is, with increasing volume of the three compounds, showing that depolarization takes place by dendrimer rotation. The fluorescence anisotropy value (0.32) exhibited by G4**P** is very high and close to that observed for all the dendrimers in diluted rigid matrix at 77 K (0.35), which can be taken as the r_0 value determined by the relative orientation of the absorption and emission transition dipole moments. This result shows that for this very bulky dendrimer rotation is almost prevented, within the excited state lifetime, even in a low viscosity solvent like dichloromethane.

In the solid state at room temperature, where dendrimer rotation is of course completely precluded as it is in rigid matrix, the three compounds show anisotropy values (Figure 11.1b, dark gray bars, Table 11.1) much smaller than those found in solution. Such an apparently surprising result suggests that a new depolarization channel is effective in the solid state, namely intermolecular energy migration among nonparallelly oriented chromophores. The increase in the anisotropy value in going from G0**P** to G4**P** is likely related to a decrease in the interdendrimer energy migration rate with increasing bulkiness of the insulating branches.

11.2.3 Dendrimers with a DPP Core

Dendrimers containing a 1,4-dioxo-3,6-diphenylpyrrolo[3,4-*c*]pyrrole (DPP) core and Fréchet-type dendrons appended at the nitrogen (Gn**DPPN**) or at the carbon (Gn**DPPC**) skeleton of the core (see e.g., G3**DPPN** and G3**DPPC** in Scheme 11.4) are studied in various solvents to study the effect of the solvent on the dendrimer conformation and volume [17]. The **DPP** chromophore has a high quantum yield of fluorescence (0.8–0.9) for all the dendrimers and a single excited state lifetime ($\tau = 6$–7 ns depending on the solvent and slightly increasing with dendrimer generation).

11.2.3.1 Time-Dependent Properties The fluorescence anisotropy shows a monoexponential decay with rotational relaxation time θ increasing with dendrimer generation and dependent on the solvent nature. By applying Eq. 11.4, the hydrodynamic volumes have been calculated (Table 11.2). For both families of dendrimers, hydrodynamic volumes in acetonitrile are smaller than in toluene, suggesting that the former is not a good solvent. Moreover, hydrodynamic volumes are larger for G1**DPPC** than G1**DPPN**, but they are quite close for the third generation, indicating that in the Gn**DPPC** family back-folding takes place for larger dendrimers.

TABLE 11.2 Steady-State and Time-Resolved Anisotropy of Fluorescence, and Hydrodynamic Volumes Calculated by Eq. 11.4

	θ (ns)			V_h (nm^3)		
	Toluene	Acetone	Acetonitrile	Toluene	Acetone	Acetonitrile
G1**DPPN**[a]	0.23	0.14	0.13	1.6	1.9	1.5
G2**DPPN**[a]	0.59	0.30	0.30	4.1	4.0	3.5
G3**DPPN**[a]	1.25	0.67	0.64	8.7	9.0	7.6
G4**DPPN**[a]	2.45	1.34	1.34	17.0	18.0	15.8
G1**DPPC**[b]	0.62	0.30	–	4.3	4.0	–
G3**DPPC**[b]	1.47	0.96	–	10.2	10.8	–
G4**DPPC**[b]	2.32	–	–	16.1	–	–

[a] $\lambda_{ex} = 420$ nm; $\lambda_{em} = 520$ nm.
[b] $\lambda_{ex} = 488$ nm; $\lambda_{em} = 540$ nm.

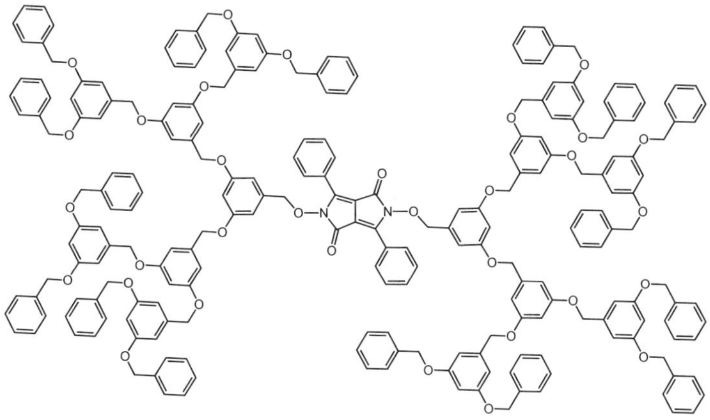

G3**DPPN**

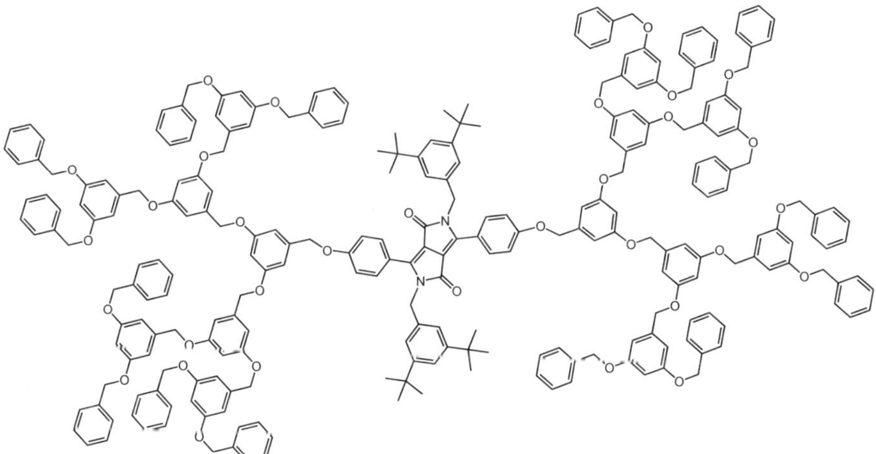

G3**DPPC**

SCHEME 11.4 Dendrimers with a 1,4-dioxo-3,6-diphenylpyrrolo[3,4-*c*]pyrrole (**DPP**) core functionalized at the N atom (G3**DPPN**) or C atom (G3**DPPC**) of the core skeleton with third-generation Frechét dendrons.

11.3 DENDRIMERS WITH MULTIPLE LUMINESCENT UNITS AT THE PERIPHERY

In the present case, fluorescence depolarization can take place by global rotation of the dendrimer, local motions of the fluorophores, and energy migration. The last mechanism is strongly dependent on the distance and nature of the luminescent units. Indeed, energy hopping is favorable for small distances and strong overlap of absorption and emission spectra of the fluorophores.

Furthermore, in dendritic structure containing several fluorophores multiple excitation can lead to singlet–singlet annihilation and thus to fluorescence depolarization. This channel of depolarization is not discussed in the present paper; for an example, see Ref. [18].

11.3.1 Dendrimers Containing Multiple Dansyl Units at the Periphery

Poly(propylene amine) dendrimers G2**D**, G3**D**, G4**D**, and G5**D** (Scheme 11.5) contain 8, 16, 32, and 64 peripheral dansyl units **D**, respectively. The photophysical properties are very close to that of the dansyl model compound [19].

Concerning the mechanisms of fluorescence depolarization, energy migration is not highly efficient in these compounds because dansyl shows a very large Stokes shift between the absorption ($\lambda_{max} = 340$ nm) and emission ($\lambda_{max} = 515$ nm) bands. Indeed, according to Förster equation [20], the overlap integral between absorption and emission spectra J is 3×10^{-18} cm^3/M and the distance value where the rate of energy transfer and of intrinsic deactivation are equal (R_0) is about 1.1 nm, compared to an average estimated distance of about 1.3 nm for G4**D** (a radius of about 2 nm has been reported) [21]. Therefore, energy migration is not very efficient compared to intrinsic deactivation and is not supposed to significantly contribute to fluorescence depolarization.

11.3.1.1 Steady-State Properties The fluorescence anisotropy value in dichloromethane (Figure 11.2, dark gray bars) is practically zero for **D** and increases with increasing dendrimer generation [22]. In a dichloromethane/propylene glycol 1:30 (v/v) mixture, r_{ss} (Figure 11.2, light gray bars) is higher than in pure dichloromethane because of the increased viscosity. Indeed, also in the case of model compound **D** and the smallest investigated G2**D** dendrimer, nonzero values of r_{ss} have been measured, different from the results in dichloromethane solution. The steady-state anisotropy measured for the dendrimers Gn**D** in dichloromethane/propylene glycol 1:30 (v/v) is (i) much lower (about 0.2) than the maximum anisotropy value $r_0 = 0.31$ for dansyl fluorophore [23], suggesting that motion is still present under these experimental conditions, (ii) increasing with dendrimer molecular mass, (iii) but much less sensitive to this parameter, compared to dichloromethane solution (the same r_{ss} values have been obtained for G3**D**–G5**D**, Figure 11.2). These results suggest that fluorescence depolarization takes place by two processes, a slower one whose rate decreases with increasing dendrimer generation, and a faster one that is not largely dependent on dendrimer generation and that is the dominant one in more viscous solvents. We assign the slow process to the global rotation of the dendrimer, and the fast one to local motions of the peripheral dansyl units appended to the dendritic scaffold. The close similarity of steady-state anisotropy observed for G3**D**, G4**D**, and G5**D** in the more viscous mixture of solvents is because local motions are not strongly influenced by the dendrimer generation and they are mainly responsible for fluorescence depolarization.

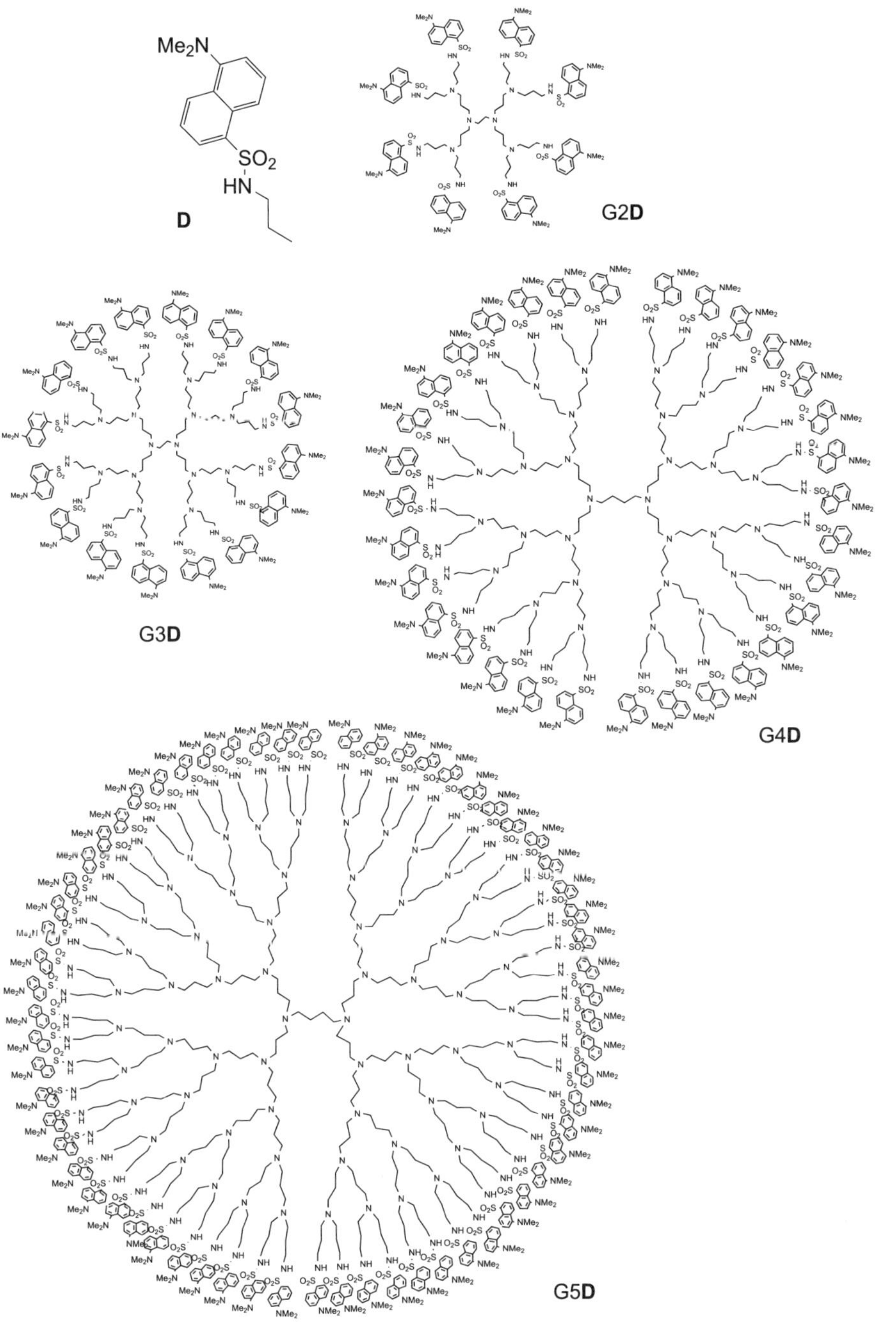

SCHEME 11.5 Poly(propylene amine) dendrimers from the second up to the fifth generation with dansyl units (**D**) at the periphery.

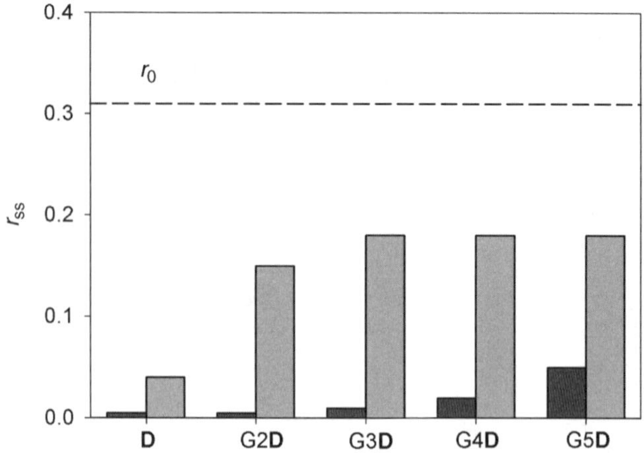

FIGURE 11.2 Steady-state fluorescence anisotropy r_{ss} of compound **D**, and dendrimers G2**D**–G5**D** in dichloromethane (dark gray bars) and dichloromethane/propylene glycol 1:30 (v/v) (light gray bars) solution at 298 K [22].

11.3.1.2 Time-Dependent Properties In dichloromethane solution, the anisotropy decay of **D** is too fast to be measured with our equipment since it undergoes full depolarization during the excited state lifetime. For the dendrimers, the rotational relaxation time θ (Table 11.3) can be obtained by fitting the anisotropy data with a monoexponential decay and upon setting the time-zero anisotropy r_0 equal to that of the dansyl unit, 0.31. The obtained θ values are not strongly different going from G2**D** to G5**D** (from 0.6 to 2.4 ns, Table 11.3), despite the large variation in molecular mass and volume. Furthermore, these values are not in agreement with those predicted for a spherical rotor from Eq. 11.3. The hydrodynamic volume V_h calculated from Eq. 11.4

TABLE 11.3 Steady-State and Time-Resolved Anisotropy of Fluorescence in Dichloromethane (DCM) and Dichloromethane/Propylene Glycol (DCM/PGly) 1:30 (v/v)

	Solvent	r_{ss}	θ (ns)	V_h (nm^3)
D	DCM	<0.01	<0.5	
	DCM/PGly 1:30	0.04	–[a]	
G2**D**	DCM	<0.01	0.6	5.8
	DCM/PGly 1:30	0.15	–[a]	–
G3**D**	DCM	0.01	1.3	12.6
	DCM/PGly 1:30	0.18	–[a]	–
G4**D**	DCM	0.02	1.8	17.4
	DCM/PGly 1:30	0.18	–[a]	–
G5**D**	DCM	0.05	2.4	23.3
	DCM/PGly 1:30	0.18	–[a]	–

[a] Double-exponential decay of fluorescence precluded an accurate determination of the corresponding θ values.

increases with dendrimer generation (Table 11.3), but the corresponding values are quite close and too small for the larger dendrimers. For example, in the case of G4**D** a radius of 1.6 nm is estimated by the corresponding hydrodynamic volume, while molecular model suggest a more expanded structure (radius about 2 nm) [21]. These results indicate that this family of dendrimers may not be considered as simple spherical rotors. Indeed, as evidenced by steady-state measurements, the local motion is an additional process that contributes to depolarize fluorescence and shorten rotational relaxation times.

Time-dependent data are not presented in dichloromethane/propylene glycol mixtures since in this solvent mixture a biexponential decay of fluorescence intensity is observed, which prevents fitting of the corresponding anisotropy decays.

11.3.2 Dendrimer with Stilbenyl Units at the Periphery

Similarly to the Gn**D** family, the poly(propylene amine) dendrimer G2**S** contains identical fluorescent units (in this case, stilbenyl units **S**) appended in the periphery (Scheme 11.6) [24]. Therefore, also for G2**S** depolarization can in principle occur by (i) global rotation of the dendrimer, (ii) local motions of the peripheral

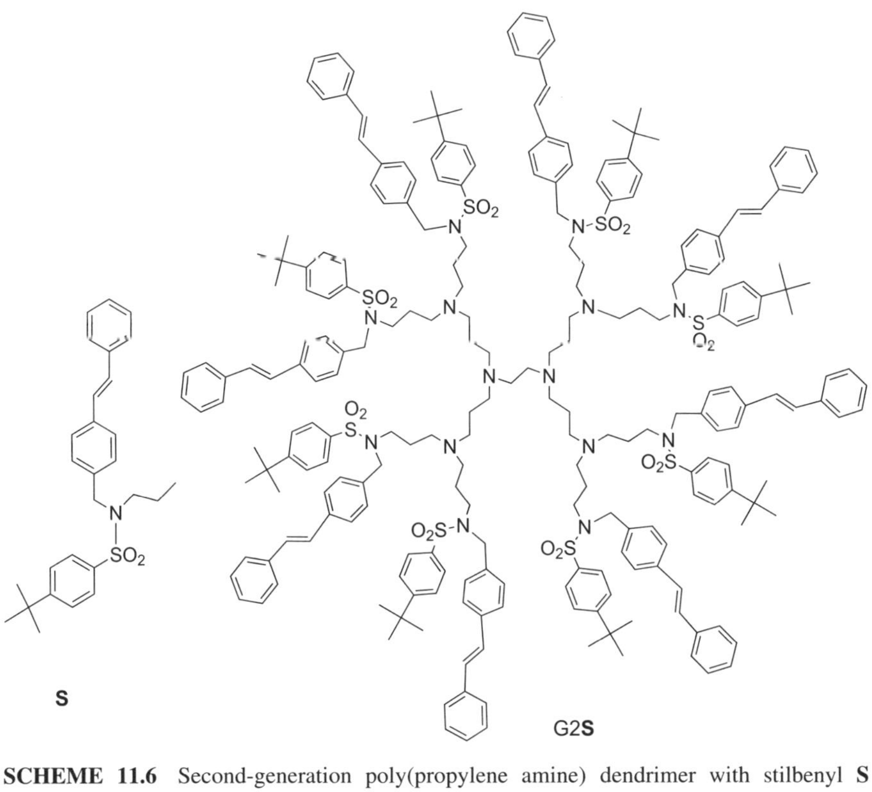

SCHEME 11.6 Second-generation poly(propylene amine) dendrimer with stilbenyl **S** peripheral units.

S fluorophores, and (iii) energy migration among the **S** units. Contrary to what happens for the Gn**D** dendrimers, in the case of G2**S** energy migration is expected to be fast because of the strong overlap between the absorption ($\lambda_{max} = 310$ nm) and emission ($\lambda_{max} = 353$ nm) bands of the stilbenyl moiety, as demonstrated by the much higher values of J (3.4×10^{-15} cm^3/M) and R_0 (1.7 nm) reported for stilbenyl rather than dansyl fluorophores.

Both the stilbenyl model compound **S** and the second-generation dendrimer G2**S** are fluorescent with a very short excited state lifetime (<200 ps), so that fluorescence anisotropy decay could not be measured with our equipment.

11.3.2.1 Steady-State Properties As shown in Figure 11.3a, the fluorescence anisotropy of the reference compound **S** increases upon increasing the fraction of propylene glycol in the solvent mixtures (from pure acetonitrile to 1:30 (v/v) acetonitrile/propylene glycol). This is an expected result since in such a molecular

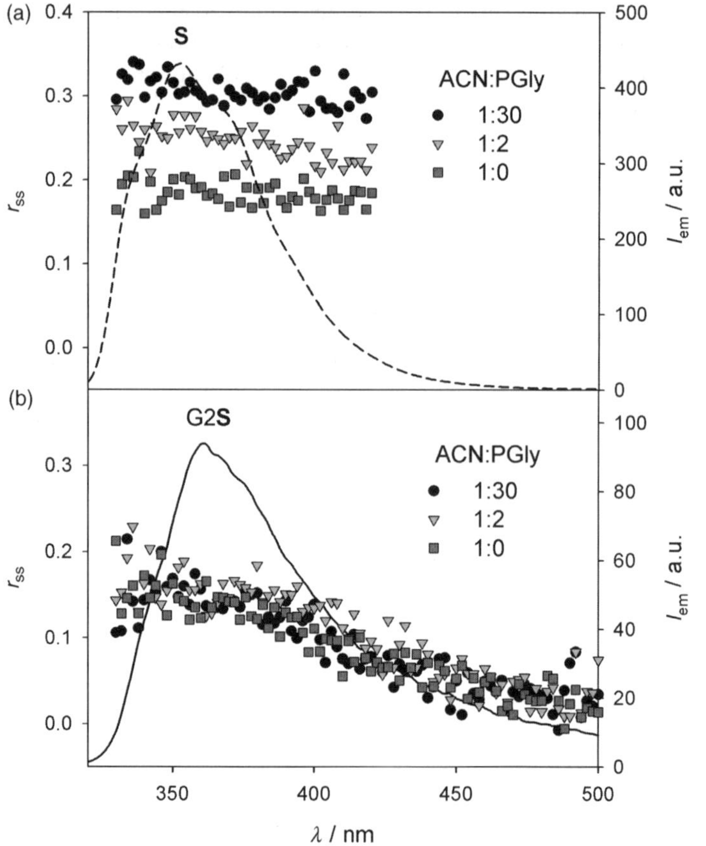

FIGURE 11.3 Steady-state anisotropy r_{ss} in acetonitrile (squares), acetonitrile/propylene glycol 1:2 (v/v) (triangles), and 1:30 (v/v) (circles) solutions at 298 K of compounds **S** (a) and G2**S** (b). Emission spectra in acetonitrile solution are also reported [22].

system the fluorescence anisotropy depends on the competition between excited state lifetime and rotational relaxation time, with the latter increasing upon increasing solvent viscosity (Eq. 11.4). On the other hand, for dendrimer G2**S** an increase in solvent viscosity does not cause any appreciable increase in fluorescence anisotropy (Figure 11.3b), maintaining a r_{ss} value around 0.16 at 330 nm. The fact that the fluorescence anisotropy of the dendrimer is not affected by rotational dynamics means that another mechanism is responsible for depolarization, that is, local motions of the **S** fluorophores appended at the dendrimer periphery and/or energy migration among the eight **S** units. In order to elucidate which mechanism is actually involved, we studied the behavior of the corresponding dansylated second-generation dendrimer, G2**D**, in the same mixture of solvents. This dendrimer is supposed to have roughly the same dimensions as G2**S**, and the same local flexibility for random motions of the fluorophores. We have found that, contrary to what happens to G2**S**, in going from acetonitrile to acetonitrile/propylene glycol 1:30 (v/v), the fluorescence anisotropy of G2**D** is strongly affected and it increases approximately by a factor of 10. These results show that the lack of increase in the fluorescence anisotropy of G2**S** with increasing solvent viscosity can be attributed to a faster depolarization mechanism independent of solvent viscosity, that is, energy migration among the **S** units, consistent with the previously reported formation of excimers between nearby **S** units [24].

Another interesting feature of G2**S** fluorescence anisotropy spectrum (Figure 11.3b) is that it is not constant all over the investigated spectral region; it decreases from 0.16 at 330 nm to about 0.02 at 500 nm. This result is consistent with the fact that in the spectral region from 330 to 550 nm two different species, stilbenyl monomers and excimers, are emitting light with different emission lifetimes ($\tau_1 < 200$ ps for the emission band at 350 nm and $\tau_2 = 3.5$ ns corresponding to the tail at lower energy). The lower r_{ss} value observed for G2**S** (Figure 11.3b) compared to **S** (Figure 11.3a) in acetonitrile solution is indicative of the fact that the rate of energy migration among stilbenyl units in G2**S** is higher than that of rotation of **S** in acetonitrile solution.

11.3.3 Dendrimers with Two-Photon Absorbing Chromophores at the Periphery

Phosphorus dendrimers from generation 1 (G1**TP** in Scheme 11.7) to 4, which contains up to 96 two-photon absorbing (**TP**) chromophores at the periphery, show strong one-photon absorption in the near UV with a nearly linear increase of the molar absorption coefficient with the number of chromophores, suggesting no significant ground state interaction [25]. As to the emission properties, a modest red-shift and broadening of the band upon increasing generation are observed together with a significant decrease of the emission quantum yield ($\Phi_{em} = 0.75$ for G1**TP** and 0.48 for G4**TP** in toluene solution), testifying an interchromophoric interaction in the excited state. These dendrimers have found interesting applications as luminescent sensors of TNT with signal amplification and near IR two-photon excitation [26].

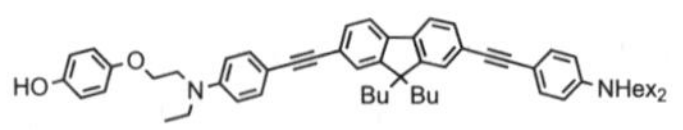

TP

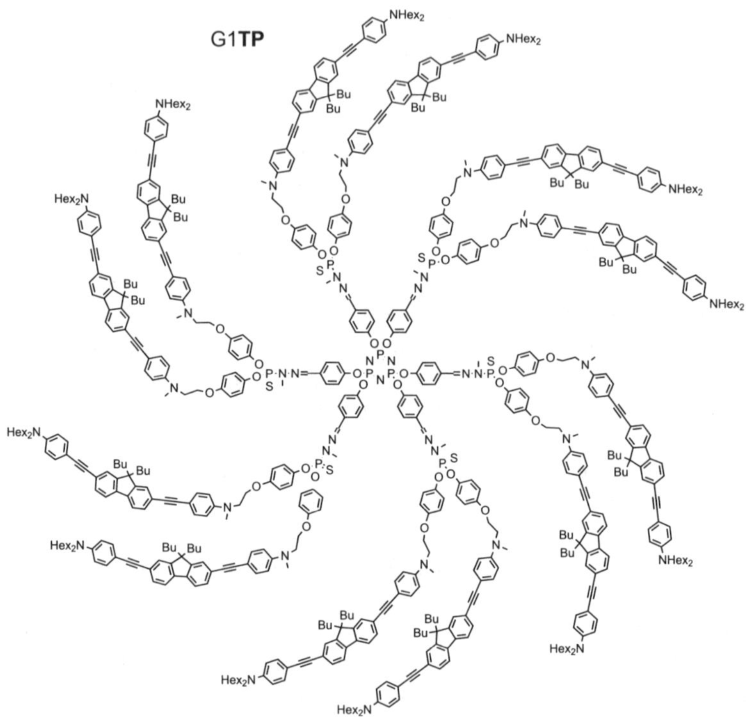

SCHEME 11.7 First-generation phosphorus dendrimer functionalized with two-photon absorbing peripheral chromophores (**TP**).

11.3.3.1 Steady-State Properties The r_{ss} values of dendrimers in toluene solution at room temperature decrease with dendrimer generation ($r_{ss} = 0.032$ for G1**TP** and 0.012 for G4**TP**), but they are, in all cases, much lower than the model **TP** chromophore ($r_{ss} = 0.178$). This result has been interpreted on the basis of a rapid energy migration between **TP** chromophores in the dendritic structures.

11.3.3.2 Time-Dependent Properties The fluorescence anisotropy decays of dendrimers are fitted by a two-exponential model. In the case of G3**TP**, the resulting values of θ are of the order of hundreds of femtoseconds and one hundred of picoseconds with slight variations upon changing the solvent. The first component is shorter upon increasing dendrimer generation and is attributed to incoherent hopping exciton transport, due to the increase in chromophore density upon increasing dendrimer generation. The slower decay is weakly dependent on dendrimer generation and can be associated with less densely packed chromophores. The fluorescence anisotropy decay is thus utilized to monitor the exciton migration over the dendrimer surface: exciton migration covers about 20 units in G4**TP**, providing a more detailed description of the mechanism of sensory amplification.

11.4 DENDRIMERS WITH LUMINESCENT UNITS IN THE BRANCHING POINTS

As in the previous case, fluorescence depolarization can give us information on global rotation of the dendrimer, local motion of subunits, and energy migration. For the last mechanism, comparison between structurally different dendrimers can elucidate differences in the rates of intra- and interdendron energy migration, as exemplified in the following.

11.4.1 Dendrimers Containing Zn(II)-Porphyrins in the Branching Points

Morphology-dependent antenna properties have been reported for a series of dendrimers $P(L)_n$ ($n = 1$, 4, Scheme 11.8) constituted by a free-base porphyrin core appended with up to four, much larger dendrons L, each containing seven Zn(II)-porphyrin units [27]. The presence of poly(benzylether) dendritic wedges at the periphery makes such dendrimers soluble in common organic solvents. In the star-shaped $P(L)_4$ dendrimer, energy transfer from the excited singlet states of dendrons L to the focal P core takes place with rate constant $1.0 \times 10^9\ s^{-1}$ and 71% efficiency, whereas in the conically shaped $P(L)_1$ dendrimer the energy transfer rate constant was 10 times smaller and the efficiency 19%. This result shows that morphology has indeed a noticeable effect on the energy-transfer rate.

11.4.1.1 Steady-State Properties Excitation of $P(L)_4$ at 544 nm with polarized light in a viscous medium of polyethylene glycol (MW = 200) resulted in a highly depolarized fluorescence from the Zn-porphyrin units ($r_{ss} = 0.02$, to be compared with 0.14 of a monomeric reference compound)[1], indicating an efficient energy migration among the Zn-porphyrin units before energy transfer to the free base core. In the case of the conically shaped $P(L)_1$ compound, the fluorescence anisotropy was much higher (0.07). Because of the highly viscous medium, molecular motions are significantly slowed down and fluorescence depolarization has been assigned to energy migration within the dendritic array. As previously postulated [28], these results suggest a cooperation of the four dendrons of $P(L)_4$ in facilitating the energy migration among the Zn-porphyrin units. Clearly, the $P(L)_4$ system, which incorporates 28 light absorbing Zn-porphyrin units into a dendritic scaffold having

[1] The polarization values P reported in Ref. [27] have been converted to anisotropy value r by the formula: $r = 2P/(3 - P)$ [11], for the sake of clarity.

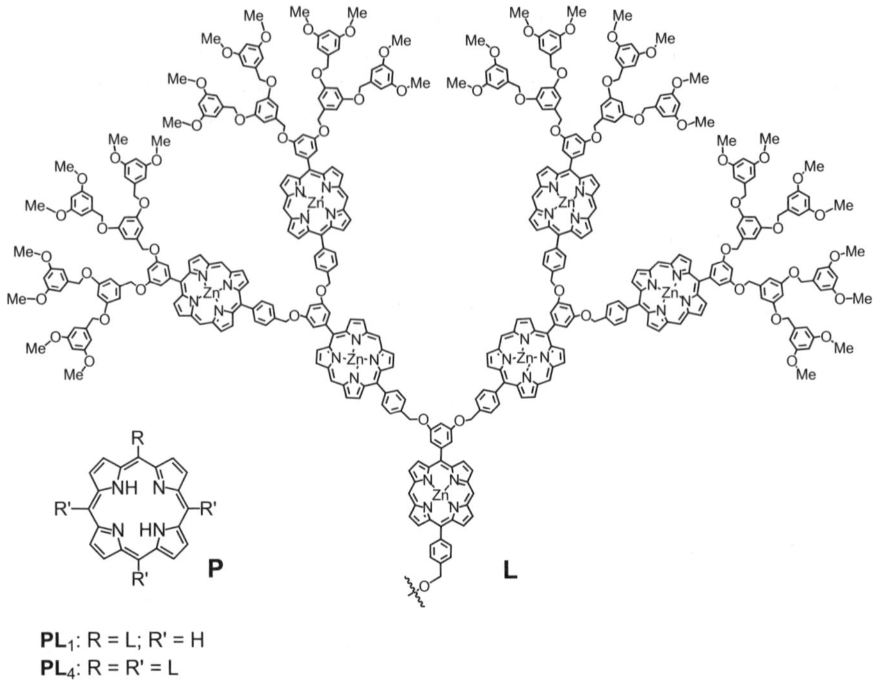

SCHEME 11.8 Dendrimers containing a free-base porphyrin core and 1, or 4 dendrons L constituted by 7 Zn(II)-porphyrins and dimethoxybenzene units.

an energy accepting core, mimics several aspects of the natural light-harvesting LH1 complex.

11.5 DENDRIMERS WITH LUMINESCENT UNITS ENCAPSULATED IN THE DENDRIMER DYNAMIC CAVITIES

Fluorescence depolarization of a guest hosted inside a dendrimer give us information about the association process, the guest mobility, and/or energy migration among guest molecules if more than one guest is encapsulated per dendrimer.

11.5.1 Eosin Molecules Enclosed in a Dendrimer

It is well-known [29,30] that poly(propylene amine) dendrimers in dichloromethane solution can extract eosin from aqueous solution. In particular, the fourth-generation poly(propylene amine) dendrimer G4**B** (Scheme 11.9) decorated at the periphery with 32 dimethoxybenzene groups can host one or more eosin molecules, depending on the relative concentrations of dendrimer and eosin in the dichloromethane and aqueous solutions used for the extraction. In all cases, the average number of eosin molecules

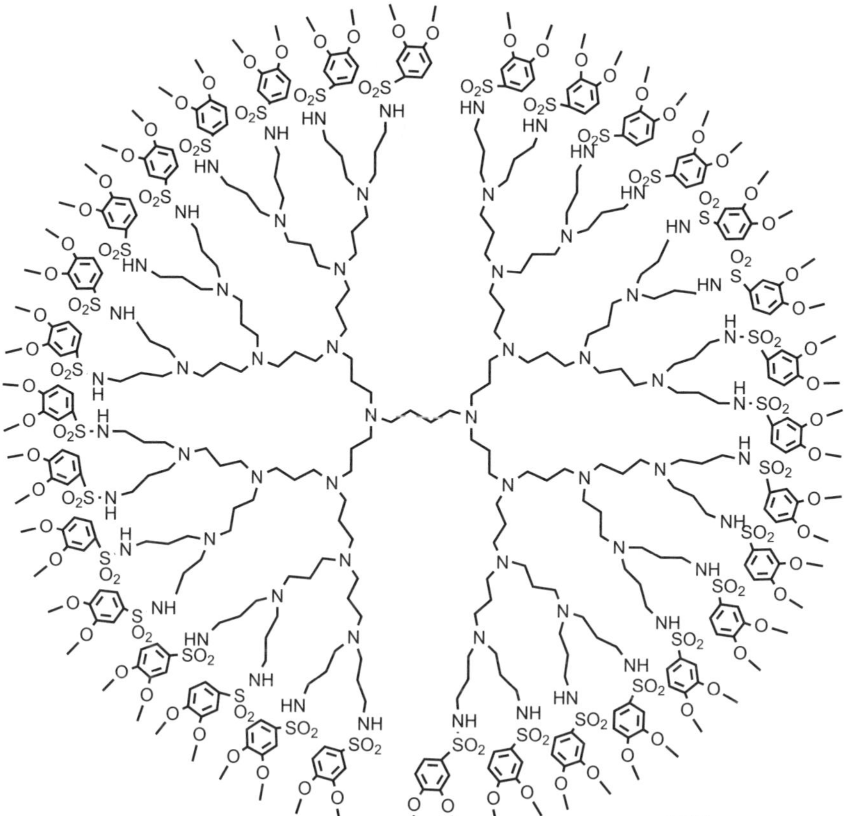

G4**B**

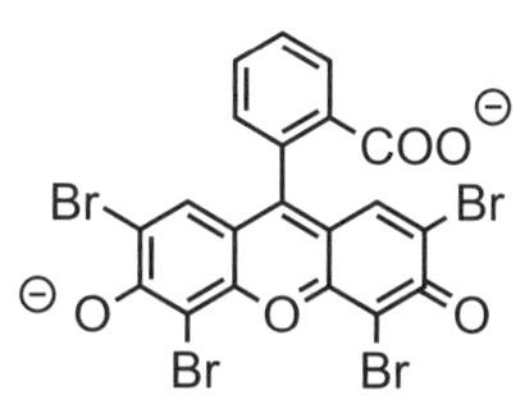

E

(Encapsulated)

SCHEME 11.9 Fourth-generation poly(propylene amine) dendrimer functionalized at the periphery with dimethoxybenzene units. This dendrimer can encapsulate eosin molecules in its dynamic cavities.

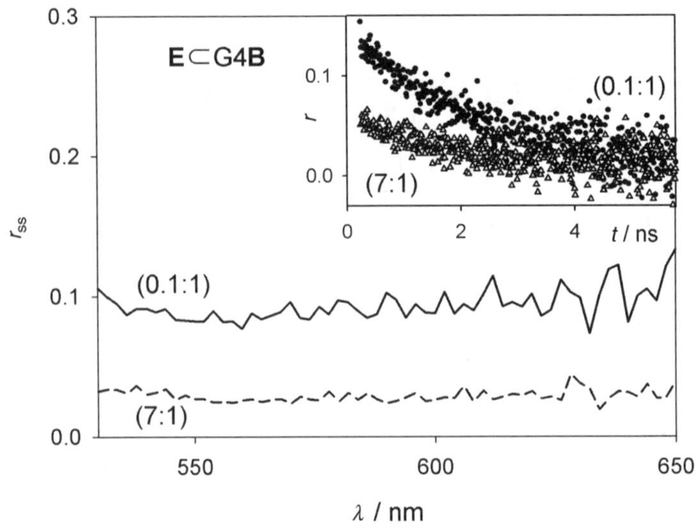

FIGURE 11.4 Steady-state fluorescence anisotropy r_{ss} and anisotropy decay as functions of time (inset) in dichloromethane solution of **E** ⊂ G4**B** with ratio 0.1:1 and 7:1 at 298 K [22].

per dendrimer can be determined by measuring the absorbance of the dichloromethane solution at 515 nm, where the encapsulated eosin shows an intense absorption band ($\varepsilon = 70{,}000\ \mathrm{M}^{-1}\,\mathrm{cm}^{-1}$).

11.5.1.1 Steady-State Properties Eosin (as tetrabutyl ammonium salt) in dichloromethane does not show any fluorescence anisotropy ($r_{ss} < 0.01$). This means that rotational depolarization occurs faster than the excited state lifetime. The steady-state anisotropy values for dichloromethane solutions of eosin hosted in G4**B** with 0.1:1 and 7:1 eosin/G4**B** concentration ratios are shown in Figure 11.4. For solutions with 0.1:1 eosin/G4**B** ratio, where no more than one eosin molecule is hosted inside a G4**B** dendrimer, a steady-state fluorescence anisotropy of 0.09 can be observed. However, solutions with 7:1 eosin/G4**B** ratio, in which seven eosin molecules, as an average, are encapsulated in each dendrimer, exhibit a much smaller (0.03) steady-state anisotropy. Since there is no reason that rotational depolarization is faster when more eosin molecules are contained into the dendrimer cavities, the strongly reduced anisotropy observed for the solution with 7:1 eosin/G4**B** concentration ratio is assigned to the occurrence of a fast energy migration among the eosin molecule hosted in the same dendrimer. This result is consistent with the small Stokes shift between absorption ($\lambda_{max} = 515$ nm) and emission ($\lambda_{max} = 546$ nm) [29] and the consequent high values of J ($5.5 \times 10^{-14}\ \mathrm{cm}^3/\mathrm{M}$) and R_0 (4.2 nm).

11.5.1.2 Time-Dependent Properties In dichloromethane solution, the anisotropy decay of **E** (as tetrabutyl ammonium salt) is too fast to be measured with our equipment, since it undergoes full depolarization during the excited state lifetime. The anisotropy decays for eosin encapsulated in G4**B** with 0.1:1 or 7:1 ratio are shown in the inset of Figure 11.4. The rotational relaxation time is lower when more eosin molecules are encapsulated inside a dendrimer, but estimation of reliable θ values is precluded by the biexponential nature of the eosin fluorescence decay, as previously observed [29]. These qualitative time-resolved results and the steady-state measurements suggest that (i) when, at most, one eosin molecule is encapsulated inside G4**B** its motion is slowed down, so that r_{ss} and θ values are higher than in pure dichloromethane solution, and (ii) upon encapsulation of more than one eosin molecule per dendrimer an additional mechanism, that is, energy migration, contributes to depolarize fluorescence.

11.5.2 Dendrimer with a Cyclam Core Associated to a Molecular Clip by a Zn(II) Ion

Dendrimer **D** (Scheme 11.10), consisting of a 1,4,8,11-tetraazacyclotetradecane (cyclam) core with appended 12 dimethoxybenzene and 16 naphthyl units [31], is able to coordinate Zn^{2+} metal ions by its cyclam core [32]. This is demonstrated by dramatic changes in the fluorescence spectra of the dendrimer. On the other hand, a molecular clip **C** (Scheme 11.10), constituted by two anthracene sidewalls and a benzene bridging units containing two sulfate groups in the *para* position [33], is able to bind Zn^{2+} ions giving rise to the complex **C**(Zn^{2+}), as demonstrated by small changes in the absorption and emission spectra [34]. By addition of Zn^{2+} ion to an equimolar solution of **D** and **C** in dichloromethane/acetonitrile 1:1 (v/v), a self-assembled system constituted by a dendrimer, a Zn^{2+} ion, and a clip is formed: **C**(Zn^{2+})**D**, as demonstrated by efficient energy transfer from naphthalene fluorophores of **D** to anthracene groups of **C** [35]. The self-assembly process can also be proved by fluorescence anisotropy measurements on the clip fluorescence. The measured r_{ss} value (Figure 11.5) (a) is very low ($r_{ss} \approx 0.01$) in the case of the clip **C** in solution at room temperature, (b) it does not show any appreciable change upon addition of 1 equiv. of Zn^{2+} ions, and (c) it is doubled ($r_{ss} = 0.02$) upon addition of 1 equiv. of Zn^{2+} and 1 equiv. of dendrimer **D**. For the systems reported in Figure 11.5, fluorescence depolarization can take place by one or more of the following mechanisms: global rotation, local motion, energy migration between the two anthracene sidewalls of the clip. The last process is not supposed to change rate constant since just one molecular clip **C** is present in the self-assembled complex **C**(Zn^{2+})**D**. Local motions of the anthracene sidewalls are not expected to change in the three situations. On the other hand, global rotation is significantly slowed down because of an increase of the rotor hydrodynamic volume in going from **C** to **C**(Zn^{2+}) to **C**(Zn^{2+})**D**. Therefore, association between dendrimer **D** and clip **C** is easily revealed by changes in steady-state fluorescence anisotropy in solution. This example is reminiscent of the biological applications of fluorescence anisotropy, where association between biological macromolecules is revealed by this technique [12].

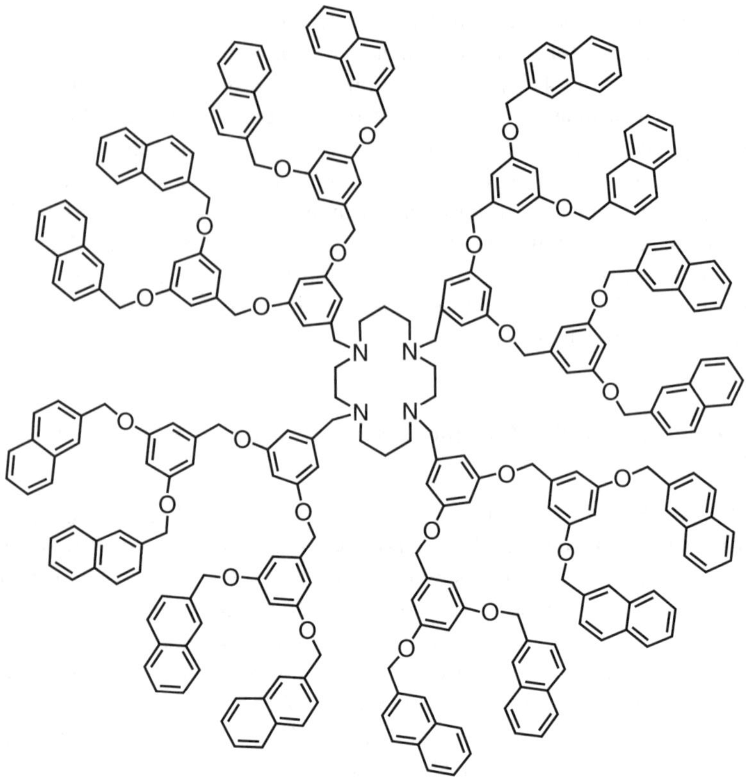

D

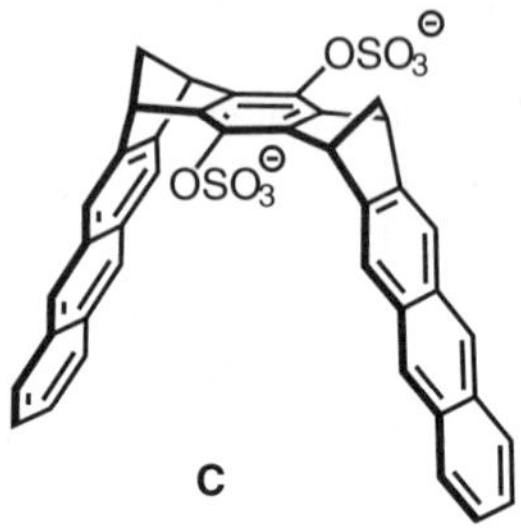

C

SCHEME 11.10 Dendrimer with a 1,4,8,11-tetraazacyclotetradecane (cyclam) core **D**, and a molecular clip **C** with two anthracene sidewalls.

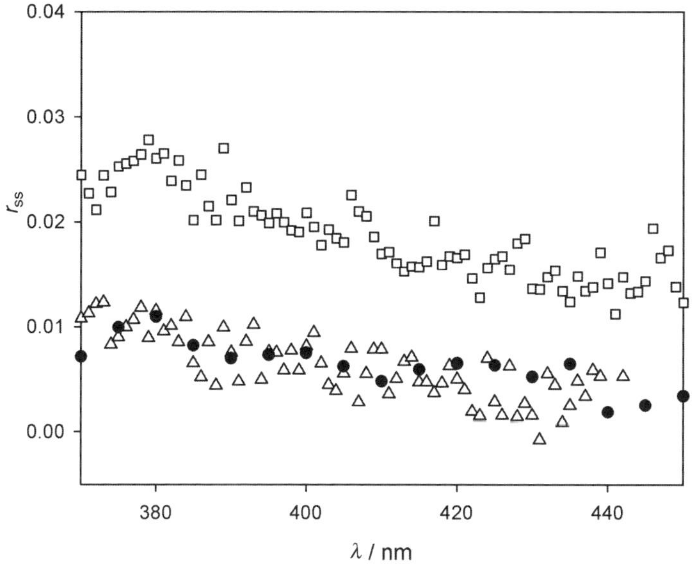

FIGURE 11.5 Steady-state anisotropy of a 3.5×10^{-6} M solution of clip $\mathbf{C}^{2-}$ before (solid circles) and after successive additions of 1 equiv. of $Zn(CF_3SO_3)_2$ (open triangles) and 1 equiv. of **D** (open squares) in acetonitrile/dichloromethane 1:1 (v/v) solution at 298 K. $\lambda_{ex} = 355$ nm [35].

11.6 CONCLUSIONS

The dendritic systems presented in this paper, based on various luminophores located in selected positions of the dendritic architecture, provide representative examples of different mechanisms of fluorescent depolarization.

In the case of a single fluorophore located at the dendritic core, fluorescence depolarization may be due to global rotation of the dendrimer in solution or to interdendrimer energy migration in the powder. By examining rotational relaxation times in solution, information on the hydrodynamic volume of the dendrimers and influence of the solvent on dendron back folding are obtained.

In the case of multiple fluorescent units located at the periphery or in the branching points of the dendrimer, mechanisms of fluorescence depolarization are: global rotation of the macromolecule, local motion of the luminescent subunits, and energy migration between identical chromophores. The last path should be considered or not depending on the distance and nature of the fluorescent units. It is worth noting that energy migration cannot be evidenced by conventional fluorescence intensity decay since it does not lead to a quenching of the emitting excited state. Therefore, fluorescence depolarization offers a handle to study the rate and the mechanism (coherent or incoherent hopping) of energy migration and the average number of chromophoric units involved in this process. This feature is particularly important in view of studying signal amplification in dendritic luminescent sensors.

In the last case, the investigated luminophore is not covalently linked to the dendritic structure, but it is encapsulated in its cavities. Therefore, the study of fluorescence anisotropy of encapsulated chromophores may evidence the possibility of restricted rotation of the chromophore inside the dendrimer cavities, the energy migration processes when more than one luminophore is encapsulated in the same dendrimer, and, in some specific cases, the association process.

Fluorescence anisotropy reveals to be once more a very useful tool to investigate dendrimer structures and energy migration processes.

ACKNOWLEDGMENTS

This work has been supported by Fondazione Carisbo ("Dispositivi nanometrici basati su dendrimeri e nanoparticelle").

REFERENCES

[1] (a) F. Vögtle, G. Richardt, N. Werner, *Dendrimer Chemistry*, Wiley-VCH, Chichester, **2009**; (b) G. R. Newkome, F. Vögtle, *Dendrimers and Dendrons*, Wiley-VCH, Weinheim, **2001**; (c) J. M. J. Fréchet, D. A. Tomalia (Eds.), *Dendrimers and Other Dendritic Polymers*, John Wiley & Sons, Chichester, UK, **2001**.

[2] For some recent reviews, see: (a) X. H. Peng, Q. M. Pan, G. L. Rempel, *Chem. Soc. Rev.* **2008**, *37*, 1619; (b) D. Astruc, C. Ornelas, J. Ruiz, *Acc. Chem. Res.* **2008**, *41*, 841; (c) J.-P. Majoral (Ed.), Special Issue on Dendrimers in *New J. Chem.* 2007, 31; (d) P. L. Burn, S. C. Lo, I. D. W. Samuel, *Adv. Mater.* **2007**, *19*, 1675; (e) B. Donnio, S. Buathong, I. Bury, D. Guillon, *Chem. Soc. Rev.* **2007**, *36*, 1495; (f) M. Gingras, J.-M. Raimundo, Y. M. Chabre, *Angew. Chem. Int. Ed.* **2007**, *46*, 1010.

[3] For some reviews, see: (a) M. Gingras, J.-M. Raimundo, Y. M. Chabre, *Angew. Chem. Int. Ed.* **2007**, *46*, 1010; (b) T. Darbre, J.-L. Reymond, *Acc. Chem. Res.* **2006**, *39*, 925; (c) C. A. Schalley, B. Baytekin, H. T. Baytekin, M. Engeser, T. Felder, A. Rang, *J. Phys. Org. Chem.* **2006**, *19*, 479; (d) D. Astruc, F. Chardac, *Chem. Rev.* **2001**, *101*, 2991; (e) M. W. P. L. Baars, E. Meijer, W. Top. *Curr. Chem.* **2000**, *210*, 131.

[4] For some recent papers, see: (a) M. Chai, A. K. Holley, M. Kruskamp, *Chem. Commun.* **2007**, 168; (b) R. van Heerbeek, P. C. J. Kamer, P. N. M. W. van Leeuwen, J. N. H. Reek, *Org. Biomol. Chem.* **2006**, *4*, 211; (c) R. M. Versteegen, D. J. M. van Beek, R. P. Sijbesma, D. Vlasspoulos, G. Fytas, E. W. Mejer, *J. Am. Chem. Soc.* **2005**, *127*, 13862; (d) J. P. Cross, M. Lauz, P. D. Badger, S. Petoud, *J. Am. Chem. Soc.* **2004**, *126*, 16278; (e) R. Nakajima, M. Tsuruta, M. Higuchi, K. Yamamoto, *J. Am. Chem. Soc.* **2004**, *126*, 1630; (f) M. A. C. Broeren, J. L. J. van Dongen, M. Pittelkow, J. B. Chrisensen, M. H. P. van Genderen, E. W. Meijer, *Angew. Chem. Int. Ed.* **2004**, *43*, 3557; (g) A. Dirksen, U. Hahn, F. Schwanke, M. Nieger, J. N. H. Reek, F. Vögtle, L. De Cola, *Chem. Eur. J.* **2004**, *10*, 2036.

[5] T. G. Goddson III, *Acc. Chem. Res.* **2005**, *38*, 99.

[6] (a) R. Vestberg, A. Nyström, M. Lindgren, E. Malmström, A. Hult, *Chem. Mater.* **2004**, *16*, 2794; (b) M. S. Matos, J. Hofkens, W. Verheijen, F. C. De Schryver, S. Hecht, K. W. Pollak, J. M. J. Fréchet, B. Forier, W. Dehaen, *Macromolecules* **2000** *33*, 2967.

[7] T. Vosch, J. Hofkens, M. Cotlet, F. Köhn, H. Fujiwara, R. Gronheid, K. Van Der Biest, T. Weil, A. Herrmann, K. Müllen, S. Mukamel, M. Van der Auweraer, F. C. De Schryver, *Angew. Chem. Int. Ed.* **2001**, *40*, 4643.

[8] (a) B. Huang, M. A. Prantil, T. L. Gustafson, J. R. Parquette, *J. Am. Chem. Soc.* **2003**, *125*, 14518; (b) C. M. Cardona, J. Alvarez, A. E. Kaifer, T. D. McCarley, S. Pandey, G. A. Baker, N. J. Bonzagni, F. V. Bright, *J. Am. Chem. Soc.* **2000**, *122*, 6139.

[9] (a) M. Guo, O. Varnanski, A. Narayanan, O. Mongin, J.-P. Majoral, M. Blanchard-Deche, T. Goodson III, *J. Phys. Chem. A* **2009**, *113*, 4763; (b) S. Cho, W.-S. Li, M.-C. Yoon, T. K. Ahn, D.-L. Jang, J. Kim, T. Aida, D. Kim, *Chem. Eur. J.* **2006**, *12*, 7576; (c) T.-S. Ahn, A. Nantalaksakul, R. R. Dasari, R. O. Al-Kaysi, A. M. Müller, S. Thayumanavan, C. J. Bardeen, *J. Phys. Chem. B* **2006**, *110*, 24331; (d) J. Larsen, J. Andersson, T. Polívka, J. Sly, M. J. Crossley, V. Sundström, E. Åkesson, *Chem. Phys. Lett.* **2005**, *403*, 205; (e) M. I. Ranasinghe, M. W. Hager, C. B. Gorman, T. Goddson III, *J. Phys. Chem. B* **2004**, *108*, 8543; (f) W.-S. Li, D.-L. Jiang, T. Aida, *Angew. Chem. Int. Ed.* **2004**, *43*, 2943; (g) B. Huang, M. A. Prantil, T. L. Gustafson, J. R. Parquette, *J. Am. Chem. Soc.* **2003**, *125*, 14518; (h) O. P. Varnavski, J. C. Ostrowski, L. Sukhomlinova, R. J. Twieg, G. C. Bazan, T. Goodson III, *J. Am. Chem. Soc.* **2002**, *124*, 1736; (i) M. Lor, R. De, S. Jordens, G. De Belder, G. Schweitzer, M. Cotlet, J. Hofkens, T. Weil, A. Herrmann, K. Müllen, M. Van Der Auweraer, F. C. De Schryver, *J. Phys. Chem. A* **2002**, *106*, 2083; (l) M.-S. Choi, T. Aida, T. Yamazaki, I. Yamazaki, *Chem. Eur. J.* **2002**, *8*, 2668; (m) M. Maus, S. Mitra, M. Lor, J. Hofkens, T. Weil, A. Herrmann, K. Müllen, F. C. De Schryver, *J. Phys. Chem. A* **2001**, *105*, 3961; (n) K. L. Yeow, K. P. Ghiggino, J. N. H. Reek, M. J. Crossley, A. W. Bosman, A. P. H. J. Schenning, E. W. Meijer, *J. Phys. Chem. B* **2000**, *104*, 2596; (o) D.-L. Jiang, T. Aida, *J. Am. Chem. Soc.* **1998**, *120*, 10895.

[10] P. M. R. Paulo, R. Gronheid, F. C. De Schryver, S. M. B. Costa, *Macromolecules* **2003**, *36*, 9135.

[11] J. R. Lakowicz, *Principles of Fluorescence Spectroscopy*, 3rd edition, Springer, New York, **2006**.

[12] J. A. Levitt, D. R. Matthews, S. M. Ameer-Beg, K. Suhling, *Curr. Opin. Biotechnol.* **2009**, *20*, 28.

[13] G. Bergamini, P. Ceroni, V. Balzani, M. Del Mar Villavieja, R. Kandre, I. Zhun, O. Lukin, *ChemPhysChem.* **2006**, *7*, 1980.

[14] B. Schartel, V. Wachtendorf, M. Grell, D. D. C. Bradley, M. Hennecke, *Phys. Rev. B* **1999**, *60*, 277.

[15] H. H. Jaffé, M. Orchin, *Theory and Application of Ultraviolet Spectroscopy*, Chapter 12, John Wiley and Sons, Inc. London, **1964**.

[16] G. Bergamini, P. Ceroni, V. Balzani, R. Kandre, O. Lukin, *ChemPhysChem.* **2009**, *10*, 265.

[17] W. Verheijen, J. Hofkens, B. Metten, J. Vercammen, R. Shukla, M. Smet, W. Dehaen, Y. Engelborghs, F. De Schryver, *Macromol. Chem. Phys.* **2005**, *206*, 25.

[18] G. De Belder, G. Schweitzer, S. Jordens, M. Lor, S. Mitra, J. Hofkens, S. De Feyter, M. Van der Auweraer, A. Herrmann, T. Weil, K. Müllen, F. C. De Schryver, *ChemPhysChem.* **2001**, *2*, 49.

[19] F. Vögtle, S. Gestermann, C. Kauffmann, P. Ceroni, V. Vicinelli, L. De Cola, V. Balzani, *J. Am. Chem. Soc.* **1999**, *121*, 12161.

[20] T. Förster, *Discuss. Faraday Trans.* **1959**, *27*, 7.

[21] G. Teobaldi, F. Zerbetto, *J. Am. Chem. Soc.* **2003**, *125*, 7388.

[22] V. Vicinelli, G. Bergamini, P. Ceroni, V. Balzani, F. Vögtle, O. Lukin, *J. Phys. Chem. B* **2007**, *111*, 6620.

[23] L. Serrano, A. Horovitz, B. Avron, M. Bycroft, A. R. Fersht, *Biochemistry* **1990**, *29*, 9343.

[24] V. Vicinelli, P. Ceroni, M. Maestri, M. Lazzari, V. Balzani, S.-K. Lee, J. van Heyst, F. Vögtle, *Org. Biomol. Chem.* **2004**, *2*, 2207.

[25] O. Mongin, T. R. Krishna, M. H. V. Wets, A.-M. Caminade, J.-P. Majoral, M. Blanchard-Desce, *Chem. Commun.* **2006**, 915.

[26] M. Guo, O. Varnavski, A. Narayanan, O. Mongin, J.-P. Majoral, M. Blanchard-Desce, T. Goodson III, *J. Phys. Chem. A* **2009**, *113*, 4763.

[27] (a) M.-S. Choi, T. Aida, T. Yamazaki, I. Yamazaki, *Chem. Eur. J.* **2002**, *8*, 2668; (b) M.-S. Choi, T. Aida, T. Yamazaki, I. Yamazaki, *Angew. Chem. Int. Ed.* **2001**, *40*, 3194.

[28] D.-L. Jiang, T. Aida, *J. Am. Chem. Soc.* **1998**, *120*, 10895.

[29] (a) U. Hahn, M. Gorka, F. Vögtle, V. Vicinelli, P. Ceroni, M. Maestri, V. Balzani, *Angew. Chem. Int. Ed.* **2002**, *41*, 3595; (b) V. Balzani, P. Ceroni, S. Gestermann, M. Gorka, C. Kauffmann, F. Vögtle, *Tetrahedron* **2002**, *58*, 629; (c) V. Balzani, P. Ceroni, S. Gestermann, M. Gorka, C. Kauffmann, M. Maestri, F. Vögtle, *ChemPhysChem.* **2000**, *4*, 224.

[30] (a) F. Köhn, J. Hofkens, U.-M. Wiesler, M. Cotlet, M. van der Auweraer, K. Müllen, F. C. De Schryver, *Chem. Eur. J.* **2001**, *7*, 4126; (b) M. W. P. L. Baars, R. Kleppinger, M. H. J. Koch, S.-L. Yeu, E. W. Meijer, *Angew. Chem. Int. Ed.* **2000**, *39*, 1285; (c) Z. Sideratou, D. Tsiourvas, C. M. Paleos, *Langmuir* **2000**, *16*, 1766; (d) G. Pistolis, A. Malliaris, D. Tsiourvas, C. M. Paleos, *Chem. Eur. J.* **1999**, *5*, 1440; (e) J. F. G. A. Jansen, E. M. M. de Brabander-van den Berg, E. W. Meijer, *Science* **1994**, *266*, 1226.

[31] C. Saudan, V. Balzani, P. Ceroni, M. Gorka, M. Maestri, V. Vicinelli, F. Vögtle, *Tetrahedron* **2003**, *59*, 3845.

[32] C. Saudan, V. Balzani, M. Gorka, S.-K. Lee, J. van Heyst, M. Maestri, P. Ceroni, V. Vicinelli, F. Vögtle, *Chem. Eur. J.* **2004**, *10*, 899.

[33] B. Branchi, P. Ceroni, V. Balzani, M. C. Cartagena, F.-G. Klärner, T. Schrader, F. Vögtle, *New J. Chem.* **2009**, *33*, 397.

[34] B. Branchi, P. Ceroni, V. Balzani, G. Bergamini, F.-G. Klärner, F. Vögtle, *Chem. Eur. J.* **2009**, *15*, 7876.

[35] B. Branchi, G. Bergamini, L. Fiandro, P. Ceroni, A. Alvino, G. Doddi, F. Vögtle, F.-G. Klärner, *Dalton Trans.* **2011**, *40*, 1356.

12

SINGLE-MOLECULE SPECTROSCOPY OF DENDRIMER SYSTEMS

Tom Vosch

Nano-Science Center, Department of Chemistry, University of Copenhagen, Universitetsparken 5, 2100 Copenhagen, Denmark

12.1 SINGLE-MOLECULE DETECTION AND SPECTROSCOPY

The most commonly used way to optically detect a single molecule is through fluorescence, and hence this method is known as single-molecule fluorescence spectroscopy (SMFS) [1–3]. SMFS is generally performed by using confocal or wide-field fluorescence microscopy, which both have their specific advantages and disadvantages. In addition, scanning near-field optical microscopy (SNOM or NSOM) can also be used, which can provide a subdiffraction optical resolution (between 10 and 100 nm) and topological information at the same time [4–7]. Drawbacks of SNOM are the less user-friendly operation and the limited depth information (maximum up to 300 nm). Recent developments have made it possible to also achieve subdiffraction resolution images with confocal and wide-field microscopy [8–14]. The only other common detectable optical signal from single molecules is surface enhanced Raman scattering but this technique requires the presence of a metallic (commonly silver) plasmonic particle [15–17]. We will limit ourselves here to confocal and wide-field fluorescence microscopy. In order to detect fluorescence from a single molecule, the fluorescence intensity of the molecule should be well above the background. This background can originate from inelastic scattering and from impurity fluorescence in the solvents, cover

Designing Dendrimers, Edited by Sebastiano Campagna, Paola Ceroni, Fausto Puntoriero.
© 2012 John Wiley & Sons, Inc. Published 2012 by John Wiley & Sons, Inc.

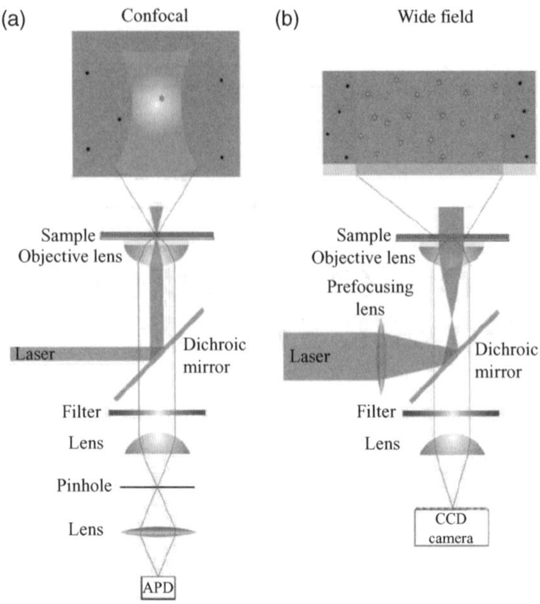

FIGURE 12.1 (a) Confocal fluorescence microscopy setup. The excitation light is focused via an objective lens into a diffraction limited spot. The fluorescence light from the spot is collected by the same objective and passed through optical filters to remove excitation and scattering light. A pinhole is inserted at an image plane to reject scattering/fluorescence light that is not incident from the focal point. Finally, photons are detected by an APD. (b) Wide-field fluorescence microscopy setup. A collimated laser beam is focused into the back-focal plane of the objective to realize a uniform field of illumination. The emission is collected through the same objective and, after passing through filters to remove excitation and scattering light, imaged onto the CCD detector. Reproduced from Ref. [26] with permission of the Royal Society of Chemistry.

glasses, optics and of course from the dark counts generated by the detector. Only a limited number of fluorescent molecules possesses the (photo)physical properties like molar absorption coefficient (ε), quantum yield of fluorescence (Φ_{fl}) and photostability necessary for single-molecule experiments. In general, the lower limit of the $\varepsilon \times \Phi_{fl}$ product, at the applied excitation wavelength, is set at 20,000 $M^{-1}\,cm^{-1}$. However, as a result of autofluorescence, optical density, and so on, these spectroscopic requirements are also a function of the specific sample and wavelength used. It is fair to say that nowadays the two most widely used fluorescence microscopy modes to detect and study single molecules are confocal microscopy and wide-field microscopy (Figure 12.1). In confocal microscopy, the objective lens focuses the excitation light to a diffraction limited spot. The fluorescence generated by a single molecule residing in the confocal volume is collected by the same objective lens. Before reaching the detector, the fluorescence light passes appropriate filters and a pinhole, rejecting background fluorescence and out of focus light, respectively. The emitted fluorescence can be measured by a point detector, such as an avalanche photodiode (APD), which very efficiently detects photons while having a low dark current, resulting in a good signal-to-noise ratio (SNR). The good SNR is also resulting from the extremely small excitation volume (femtoliters) since scattering is minimized in this way. With the confocal technique one molecule is studied at a time, and when imaging of extended areas is required, this is possible by scanning the sample with respect to the laser focus point or vice versa. In addition to measuring fluorescence intensities of single molecules, using pulsed excitation light and proper electronics allows to measure fluorescence decay times of single molecules (at the subnanosecond time scale) at the same time. Emission spectra can be measured with a CCD (charge coupled device) camera combined with a polychromator. The confocal approach is mainly useful to get highly temporally resolved (from microseconds on) dynamic data on individual molecules for extended periods of time, if the photostability of the system under study allows for this. Nowadays, measuring many parameters at once (lifetime, intensity, spectra, polarization, etc.) in a multiparametric approach is standard [18]. The confocal approach is also used in fluorescence correlation spectroscopy (FCS) [19,20]. In FCS, the excitation beam is focused in solution and one waits for fluorescent molecules to pass by. The basic principle behind FCS is that the detected fluorescence will fluctuate as molecules diffuse through the confocal volume element. A correlation analysis of the fluctuating intensity signal yields the diffusion constant of the system under study averaged over a large population of molecules [21,22].

In wide-field microscopy a CCD camera is used to record movies of the events taking place in the several hundred square micrometers large excitation area. The resolvable spots are again diffraction limited, but now represent a larger excitation volume due to a decreased resolution, mainly in the z-direction, reducing the SNR. The time resolution for large domain imaging is limited by the frame transfer rate of the used CCD camera (millisecond or submillisecond range) and is lower than for the aforementioned point detectors. A way to increase the SNR ratio in wide-field microscopy is by using total internal reflection (TIRF) illumination, which implies excitation by an evanescent field [23]. The reduced thickness of the excitation volume results in an increased SNR. Wide-field microscopes are most commonly used to study diffusion and other dynamic processes taking place over an extended area. An additional advantage is that many molecules can be studied in parallel; hence statistically relevant numbers of molecules can be sampled much faster than in a confocal approach.

Many modifications exist on the two general schemes elaborated above. For confocal microscopy pulsed interleaved excitation [24], that is, the sequential excitation of the sample with different wavelengths, for example, can be readily implemented. In FCS one can work with two foci at a fixed distance in order to get artifact-free diffusion data [25]. For wide field, TIRF (or quasi-TIRF) has already been mentioned. Other modifications such as a stroboscopic illumination [11] or deliberately working out of focus (vide infra) are becoming routine modes of microscopy. To conclude one can say that the system under interrogation and the response one wants to obtain dictate how the two fundamental approaches, wide field or confocal, are adapted to accommodate a specific experiment.

12.2 DENDRIMERS CONTAINING SINGLE CHROMOPHORES AS NANOSCALE REPORTERS

12.2.1 Introduction

By monitoring their fluorescence, single chromophores can serve as sensitive reporters of the local environment [27–32]. In the same way, dendrimers containing single chromophores in the core or at the rim can also serve as sensitive reporters of motion as well as polarity changes of their immediate surroundings. The covalent incorporation of chromophores into the center leads to macromolecular architectures of well-defined sizes (see Figure 12.2 for an example) [33]. By adjusting the polyphenylene shell, nanoemitters of well-defined size can be obtained. Figure 12.2 depicts the transition from a molecular perylene diimide (PDI) emitter **1** to a third-generation dendrimer with a PDI in the center **4**. An increase of the number of

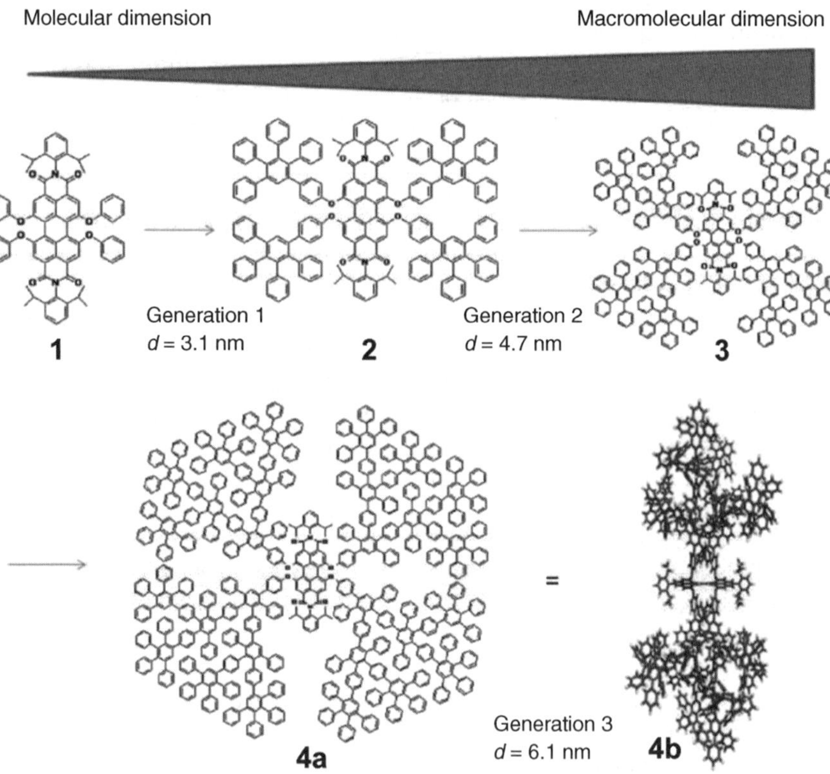

FIGURE 12.2 2D structures of the molecular PDI emitter. The transition from the emitter **1** to the nanoscopic dendritic emitters **2** with four branches of first generation, **3** with second generation, and **4** with third-generation polyphenylene dendrimer branches is shown. **4a** shows the 2D structure of a third-generation dendrimer with PDI in the center and **4b** represents a visualization of the 3D structure of the same molecule.

polyphenylene rings results in a step-wise augmentation of the sizes of the nanoemitters from 3.1 nm for the first, 4.7 nm for the second and 6.1 nm for the third-generation dendrimers (Figure 12.2, PDI chromophores **2**, **3**, and **4**, respectively).

Since the diffusion coefficient reflects the capacity of a single nanoemitter to diffuse freely in a given environment, the ability to synthesize emitters of defined sizes allows the detection of changes in the surrounding nanoenvironment such as viscosity and free volume.

12.3 DEFOCUSED WIDE-FIELD IMAGING OF REPORTER MOLECULES FOR PROBING LOCAL MOBILITY IN POLYMER FILMS

There are several ways of determining the 3D orientation of single molecules [34–37]. One of them is by using defocused wide-field imaging, which has the advantage that it allows for collecting data and determining the exact 3D orientation for multiple molecules at the same time [36,38–40]. Compared to the previously described wide-field microscopy technique, here the sample is positioned 1 μm from where it is focused toward the objective. As a result of the defocusing the images display the angular distribution of the emitted fluorescence of the single molecule, which is mapped as a spatial distribution of intensity, reflecting the 3D molecular orientation [38]. The obtained defocused images are analyzed in order to determine molecular orientation in each image frame by pattern matching. In this approach the obtained frames are fitted to a database of calculated orientation patterns by least-square fitting. The accuracy of the determined orientation depends on the signal to noise ratio [38]. Figure 12.3 shows three single PDI containing dendrimer molecules of compound **2** embedded in a polymethylacrylate (PMA) polymer (glass transition temperature $T_g = 281$ K) film at 295 K that are relatively close to each other. Figure 12.3b–d shows the projection maps for molecules 1–3 shown in Figure 12.3a, respectively. The defocused images were obtained with 1 s integration and 0.4 s interval time.

Two different types of behavior can be observed in the maps. Molecule 2 (Figure 12.3c) exhibits fast rotations with no preference for any orientation. On the other hand, molecules 1 and 3 are temporarily locked in one orientation and occasionally jumps to a completely different orientation occur. For example, molecule 1 is first oriented in the region indicated by the black line but undergoes a jump in orientation, indicated by the red line, after 970 s. Molecule 3 shows an even more complex behavior: though it is locked in three different orientations for 1760 s, it starts rotating in a rather wide range of angles afterwards. From that point on, completely different rotational dynamics are observed. This clearly reflects temporal heterogeneity of polymer relaxations. Most interestingly, molecule 3 passes through the initial region on the projection map, observed between 0 and 190 s and indicated in black, when jumping from the green region into the red region after 890 s. These results indicate temporal heterogeneity and may point to a memory effect in polymer

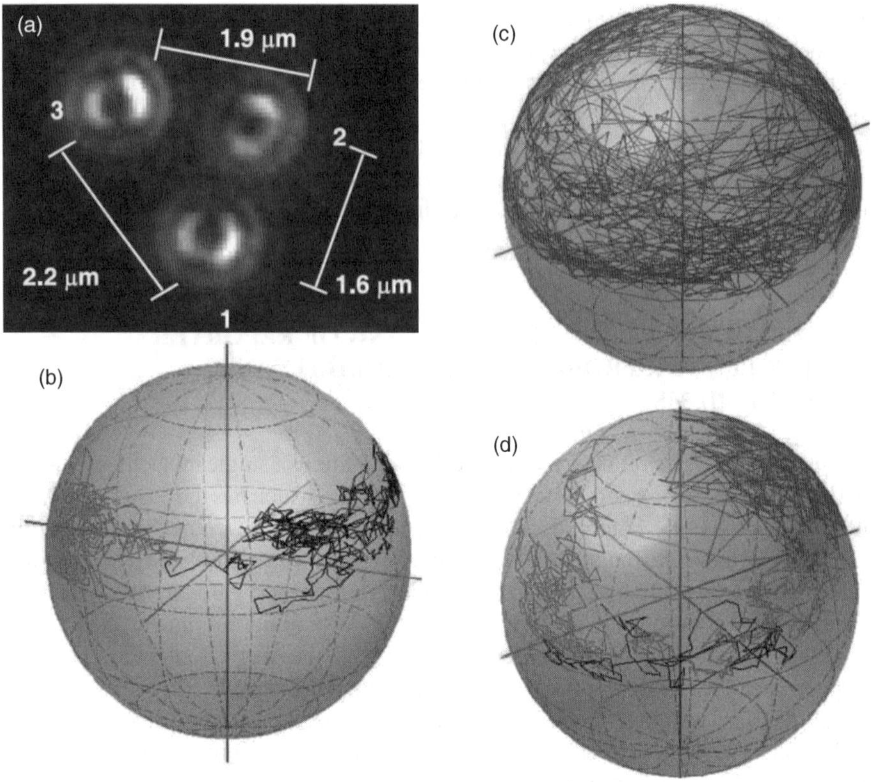

FIGURE 12.3 (a) Three single molecules of compound **2** embedded in a PMA polymer film at 295 K. (b–d) Projection maps for molecule 1–3, respectively. The red lines indicate the *x*–*y* image plane and blue line is the optical (*z*) axis: (b) 0–960 s (black) and 960–2380 s (red). (c) 0 until end (blue). (d) 0–190 s (black), 190–895 s (green), 895–1763 s (red), and 1763–2940 (blue). Copyright from Ref. [36]. (See the color version of this figure in Color Plates section.)

relaxation at the dimension of the probe molecule, meaning approximately 3 nm. By using single-molecule fluorescence detection techniques and the appropriate reporter molecules, more information about polymer dynamics, relaxation and the drastic viscosity changes that occur around the glass-transition temperature can be studied [26,30,36,41–43].

12.4 REPORTER MOLECULES FOR PROBING DIFFUSION IN POLYMERIZATION REACTIONS

Recently, single-molecule spectroscopy was used to study the bulk radical polymerization of polystyrene by detecting changes of the diffusion constants of PDI containing dendrimers acting as probes [44]; a combination of FCS and wide-field single-molecule tracking was used to collect data. In connection with the above-mentioned basic notion that the mobility of molecules changes during the polymerization process, the diffusion times of the macromolecular emitter compound **4** was studied. The dynamic range that can be covered by both methods overlaps, allowing following the full conversion (denoted here as U, ranging from 0 to 1, 1 being full conversion) in the polymerization process based on single-molecule fluorescence measurements (see Figure 12.4).

Compound **4** was used for studying the polymerization of styrene in the absence and presence of a 1,4-divinylbenzene cross-linker [44]. In the absence of a cross-linker, compound **4** diffuses freely in the surrounding medium (see Figure 12.5). The FCS autocorrelation curves can be fitted with one diffusion time up to high conversion (0.83). The resulting values of the diffusion constant D, decreasing with increasing U, are shown as gray circles in Figure 12.4a and b (same data set). The acceptable fits with one D value indicate that translational diffusion of the dyes is rather homogeneous.

A second set of experiments focused on the diffusion in polymer networks. For that purpose, the above-mentioned experimental conditions were repeated with addition of 1% and 3% DVB cross-linker, respectively. At low U when using 1% of DVB, autocorrelation curves and diffusion constants were similar to the experiment without DVB. However, when gelation started, the FCS curves could only be fitted with two diffusion constants as indicated by the triangles (fast: green, slow: red; black circles: weighted average) in Figure 12.4. The fast component can be related to still freely moving reporter molecules while the slow component is related to the trapped reporter molecules. When using 3% of DVB, gelation started earlier than at 1% and the D values that were lower at similar U. Also here the FCS curves could be best fitted with two diffusion constants.

As the speed of diffusion decreases as a function of the conversion rate, a point will be reached when FCS measurements are no longer adequate and the single molecules can be tracked by the wide-field detection (dotted line in Figure 12.4). The yellow tracks in the left-side image of Figure 12.5 indicate the location the individual molecules of compound **4** for the case when no cross-linker is present. They can be seen still moving unhindered at 0.64U. In the presence of a cross-linker at similar conversion (0.65U), the motion of the compound **4** reporter molecules describes the onset of heterogeneity that arises during the formation of a network. In the wide-field measurements this heterogeneity can be directly visualized as a fraction of the molecules is clearly trapped by its nanoenvironment while the remainder of the molecules seems to diffuse unhindered at the same time. The latter is depicted in the right-side image of Figure 12.5 showing reporter molecules barely move (red tracks) and some that are still mobile (yellow tracks).

These investigations can be extended to other polymerization systems such as interpenetrating networks and nanocomposites during their formation process. This particular kind of study will provide a deeper understanding of the factors that control heterogeneities in a polymerization process. With this knowledge it may be possible to attain a better control over polymerization and the properties of the resulting polymers and polymer networks.

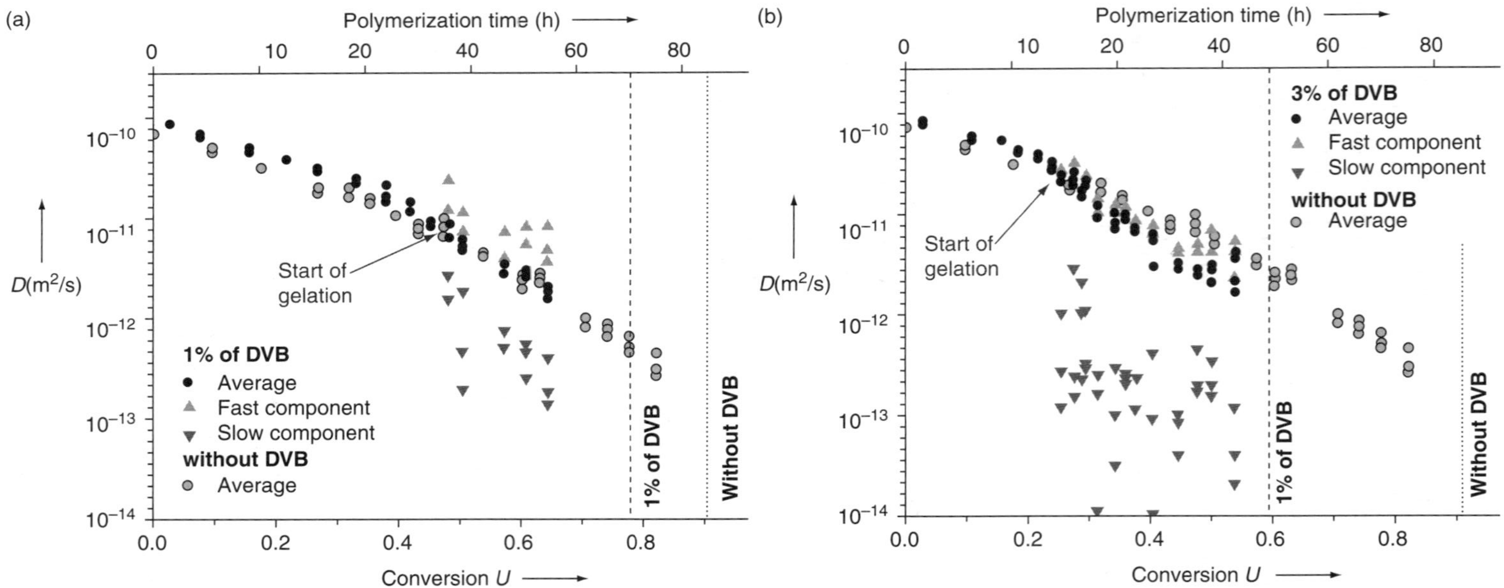

FIGURE 12.4 Dependence of the diffusion constant D obtained by FCS and wide-field microscopy on U and on the polymerization time using compound **4** as probe: (a, b) Polymerization of styrene probed with compound **4** in the presence of 0%, 1%, and 3% w/w of 1,4-divinylbenzene (DVB) cross-linker. The dashed line indicates the point at which all molecules were immobilized as determined by wide-field microscopy. Copyright from Ref. [44]. (See the color version of this figure in Color Plates section.)

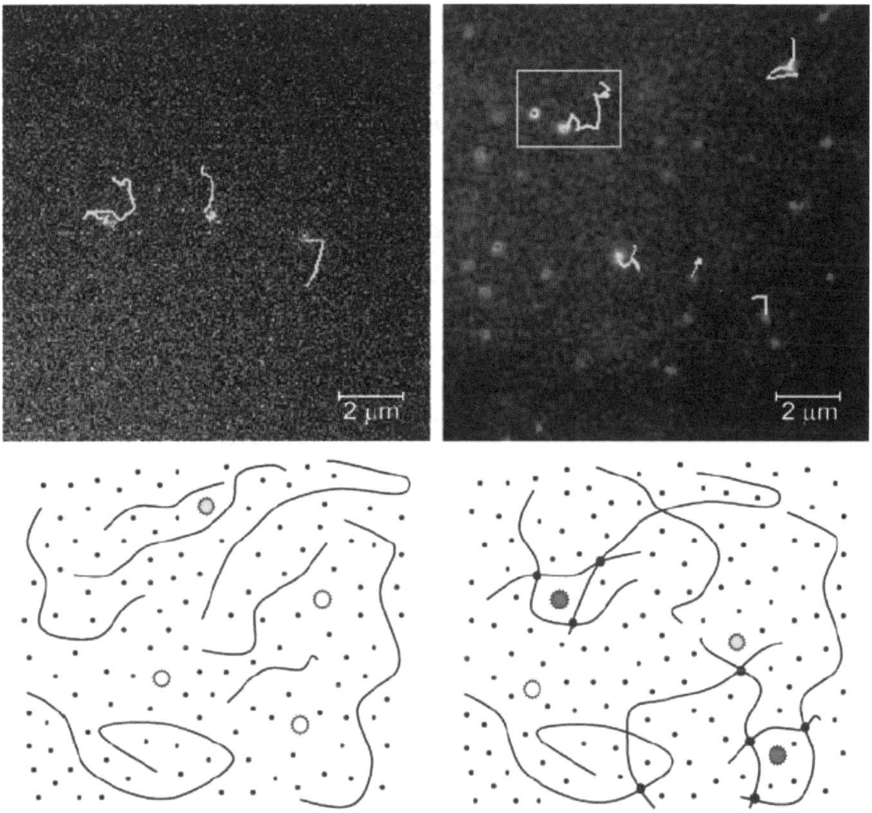

FIGURE 12.5 (Left side) Wide-field image at 0.64U and schematic representation of compound **4** without the presence of DVB cross-linker. (Right side) Wide-field image at 0.65U and schematic representation of compound **4** with 1% of DVB cross-linker present. For the schematic representation, the dyes are shown as circles for which the color indicates their current velocity (white: too fast for detection by wide-field microscopy, yellow: slow enough for wide-field microscopy, red: very slow/immobilized). Copyright from Ref. [44]. (See the color version of this figure in Color Plates section.)

12.5 REDOX/ELECTRON TRANSFER PROBES

Studying electron transfer with single-molecule fluorescence microscopy is challenging since in contrast to fluorescence resonance energy transfer where fluorescence is emitted mainly from the acceptor chromophore, a highly efficient electron transfer usually leads to a strong quenching of the fluorescence of the emitting chromophore. Thus, reports on single-molecule photo-induced electron transfer are rather limited in number [45–47]. A particular situation of electron transfer refers to the case when the locally excited state (LES) and the charge-separated state (CSS) are close in energy. In this case, upon excitation, the LES deactivates mainly via forward

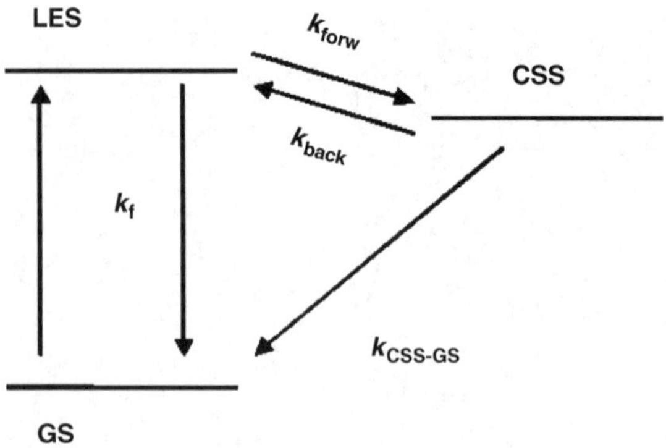

FIGURE 12.6 (a) Scheme of forward and backward electron transfer resulting in delayed fluorescence: LES: local excited state, GS: ground state, CSS: charge-separated state, k_{forw}: rate constant forward electron transfer, k_{back}: rate constant backward electron transfer, $k_{CSS\text{-}GS}$: rate constant from charge-separated state to ground state. Copyright from Ref. [51].

electron transfer to the CSS and, if the radiationless deactivation of the CSS to the ground state (GS) is inefficient (see Figure 12.6), the CSS decays via the LES by backward electron transfer. As a net result, the emitted fluorescence is delayed but can retain a high quantum yield [48–50].

In order to explore photo-induced electron transfer via delayed fluorescence, polyphenylene dendrimers can be used, which allow for the introduction of electron donating and accepting groups in the center as well as in the periphery [48,49,52–55]. In view of the exponential distance dependence of the electron transfer process, the structural rigidity provided by the polyphenylenes represents a key feature to control distance between the donor and acceptor moieties. Redox-active triphenylamine (TPA) groups, which are widely used as electron-donor materials [56,57], can be introduced into the bay region of PDI (see Figure 12.7) [52]. The distance between the PDI electron acceptor and the TPA donors was varied at the nanoscale with increasing dendrimer generations.

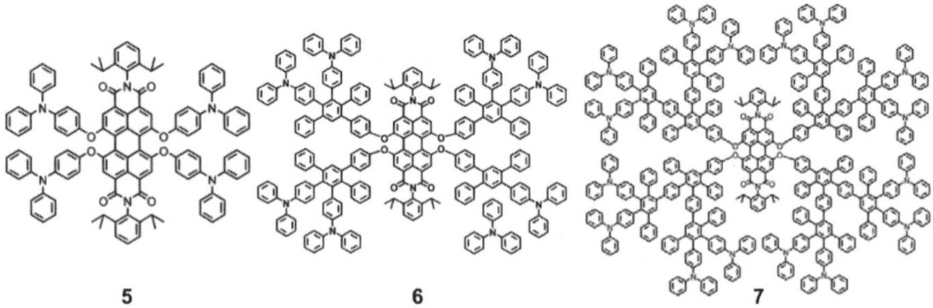

5 6 7

FIGURE 12.7 First-, second-, and third-generation polyphenylene dendrimers (**5**, **6**, **7**) based on a PDI core with 4, 8, 16 TPA groups at the periphery, respectively.

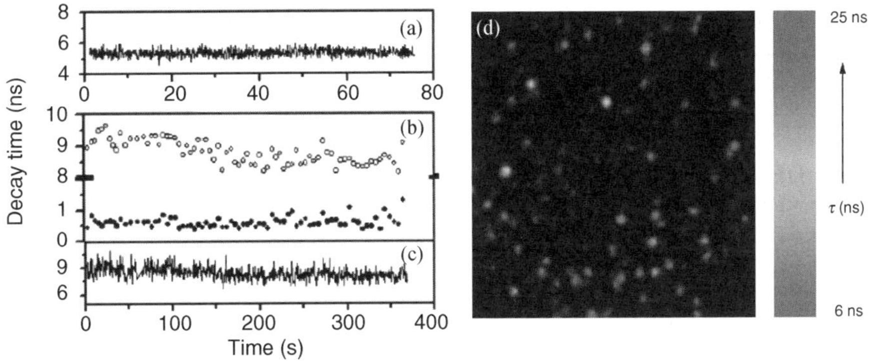

FIGURE 12.8 (a) Single molecule of compound **7** in Zeonex binned at 500 photons per decay and analyzed as a monoexponential. (b) Decay time of a single molecule of compound **7** binned at 10,000 photons per decay and analyzed with biexponential fit. (c) The long time component of the decay in figure (b) binned at 500 photons per decay. (d) Lifetime image of single molecules of compound **6** in polystyrene. Copyright from Ref. [51]. (See the color version of this figure in Color Plates section.)

Figure 12.8a shows the fluorescence decay time trajectory of a single molecule of compound 7 in Zeonex, in which the molecule does not undergo electron transfer, using bins of 500 photons for which decay times were estimated by maximum likelihood estimation (MLE) fitting with a monoexponential model [58,59]. The trace shows no decay time fluctuations during the measurement. In contrast, when a single molecule of compound **7** is measured in polystyrene, fluctuations are seen in both forward and backward electron transfer-related decay times (Figure 12.8b) [53,55]. Here, single-molecule decay times were estimated from bins of 10,000 photons by MLE fitting with a biexponential model accounting for forward and backward electron transfer to determine accurately both short and long decay times. Fluctuations are even clearer to see in the backward electron transfer-related decay times when analyzed in bins of 500 photons per decay (Figure 12.8c). They are much broader than the expected standard deviation from MLE and than the decay time fluctuation of single molecules of compound **7** in Zeonex. These observed fluctuations were shown to be induced, on the one hand, by conformational changes in the dendrimer structure and, on the other hand, by polymer chain reorientations in the vicinity of the observed single molecules. Conformational changes have been linked to libration [60]. For example, changes in the torsional angle of adjacent phenyl rings located in the dendritic branches near the donor moiety transferring the charge. This torsional motion has as a net result a change in the through bond D–A coupling and thus induces decay time fluctuations in milliseconds time scale. Polymer chain reorientation resulted in changes in the local polarity and hence to changes in the solvation of the CSS. As a result, switching between different donor moieties within a single molecule was enabled and was shown contributing to fluctuations in decay time in the time scale of seconds.

By using single molecules of compound 6 as example, fluorescence lifetime images (FLIM, see Figure 12.8d) [61] can be constructed, allowing to visualize directly that molecules display delayed fluorescence [49] Molecules with significant longer decay time than 6 ns can be considered as having backward electron transfer (colored yellow/green).

12.6 DENDRIMERS AS SCAFFOLDS FOR STUDYING ENERGY TRANSFER PROCESSES IN MULTICHROMOPHORIC SYSTEMS

12.6.1 Introduction

In this section, we will see how dendrimers can act as a synthetic multichromophoric model system that can be used to study energy transfer processes and excited-state interactions. The study of these excited-state dynamics and multichromophoric interactions plays an important role in the understanding of biological systems (like the Light Harvesting 2 complex, DsRed, allophycocyanines, etc.) [62–67] and material science applications (like for example MEH-PPV for organic light emitting diodes LEDs) [68–70]. Again the single-molecule detection and spectroscopy aspect will be highlighted.

12.6.2 Förster Type Energy Transfer Processes

Before going into detail on the experimental data it is maybe first important to briefly review the processes that will be mentioned here. Also the role that the dendrimer architecture plays will be highlighted. Förster type energy transfer processes occur when chromophores are in close proximity to each other (1–10 nm), and this transition dipole–dipole interaction decreases with the distance to the sixth power [71]. The rate constant for energy transfer k_{ET} can be expressed by Eq. 12.1 where k_D is the decay rate constant of the excited state of the donor, R is the interchromophoric distance, and R_0 is the Förster radius. The Förster radius is the distance at which the rate constants for energy transfer and decay of the excited state of the donor are equal. This R_0 value can be calculated from experimental data by Eq. 12.2

$$k_{ET} = k_D \left[\frac{R_0}{R}\right]^6 \tag{12.1}$$

$$R_0^6 = \frac{9(\ln 10)}{128\pi^5 N_A} \frac{\kappa^2 \Phi_D}{n^4} J \tag{12.2}$$

In Eq 12.2, besides the presence of a number of constants, κ^2 is the orientation factor, Φ_D is the quantum yield of fluorescence of the donor in absence of the energy transfer, J is the overlap integral between the donor emission spectrum (area under the spectrum normalized to 1) and the absorption spectrum of the acceptor, n^4 is the refractive index of the medium to the fourth power. More details on the Förster type energy transfer equations and possible pitfalls can be found in a recent article [72].

The two important parameters with respect to studying accurately energy transfer processes are the distance R and the orientation factor κ^2. In order to determine

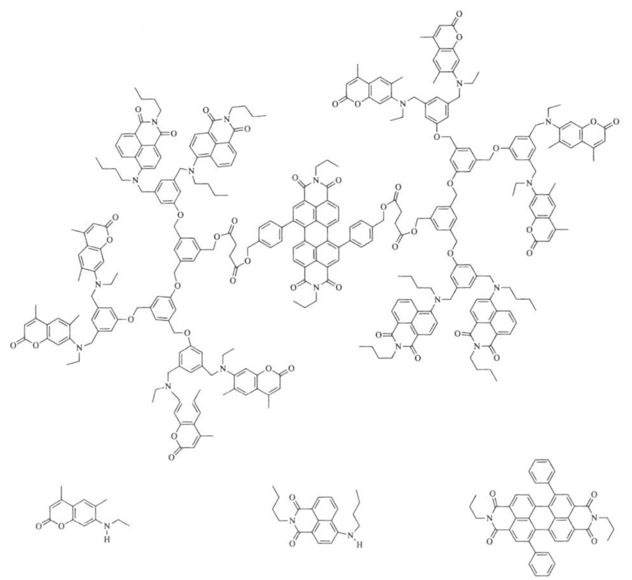

Coumarin 2	**Fluorol 7GA**	**Perylene core**
Donor 1 (D1)	Donor 2 (D2)	Acceptor (Ac)
Abs Max = 350 nm	Abs Max = 415 nm	Abs Max = 555 nm
Em Max = 440 nm	Em Max = 520 nm	Em Max = 610 nm

FIGURE 12.9 Flexible polyarylether dendrimer studied by Serin et al. Reproduced from Ref. [73] with permission of the Royal Society of Chemistry.

exactly the rate constant for energy transfer, it is important to use rigid dendritic structures since the rate constant for energy transfer depends on the sixth power of the distance. The orientation factor κ^2, which varies between 0 and 4, represents the orientation of the two transition dipole moments with respect to each other. Although energy transfer will also occur in flexible systems like the polyarylether dendrimer studied by Serin et al. [73] (see Figure 12.9), one will obtain here average values and one can expect a large distribution of rates when such system is studied at the single-molecule level.

By using benzene units as building blocks, rigid and shape persistent polyphenylene dendrimers were synthesized by the research group of Prof. Müllen [74–76]. These dendrimers have a high degree of structural organization, possible geometry architectures, and offer the possibility to introduce chromophores at the rim, center, or in the scaffold. Figure 12.10 shows a scheme of the design concepts to achieve dendritic multichromophoric systems with varying numbers of PMI chromophores, predefined geometries, different chromophore orientations and varying chromophore distances [39,51,77–91]. The tetraphenylmethane core ensures a higher structural rigidity as well as the formation of a globular architecture, whereas the biphenyl core results in a higher degree of mobility along the biphenyl unit as well as a more "pancake-like" architecture. In addition to the number of chromophores, the orientation of the chromophores can be varied by

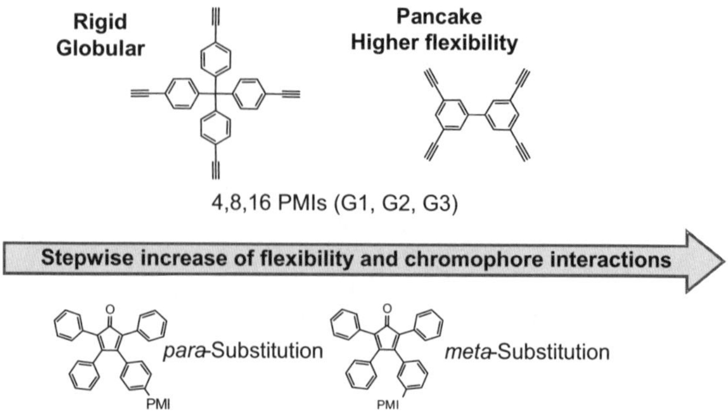

FIGURE 12.10 Overview over some of the building blocks to achieve dendritic multichromophoric systems with varying numbers of PMI chromophores, predefined geometries, different chromophore orientations and varying chromophore distances. In the top row the core elements are depicted for creating the first-generation (G1) globular and pancake-like dendritic structures containing four PMI chromophores. Second (G2) and third (G3) generations of this core elements allow the attachment of 8 and 16 PMI chromophores, respectively. The bottom row displays the building blocks used for attaching the PMI chromophores to the core elements.

attaching the PMI chromophores in the *meta* or *para* position of the outer phenyl groups of the dendritic scaffold.

A further advantage of polyphenylene dendrimers is the absence of functional groups within the dendritic scaffold. This makes them highly chemically and thermally stable. Also, the opportunity to modify the substitution pattern of the chromophores as well as the scaffold of polyphenylene dendrimers allows for the precise adjustment of chromophore distances and orientations. Polyphenylene dendrimers are optically inert in the visible part of the spectrum, due to the twisting between the phenyl rings in the dendritic building blocks. Polyphenylene dendrimers absorb light up to 350 nm, depending on the generation considered and do not emit fluorescence above 450 nm [92]. As a result, they are an ideal scaffold since they do not interfere with the energy transfer processes themselves, provided that the chromophores involved are chosen to absorb and emit in the visible part of the spectrum. In this way, complex and highly ordered multichromophore architectures were designed in which the energy is collected and can be transferred in a directed fashion [77,93].

12.7 EXCITATION ENERGY TRANSFER BETWEEN STRUCTURALLY IDENTICAL CHROMOPHORES

We will demonstrate in this section that excitation energy transfer processes can be studied at the single-molecule level by making use of a first-generation polyphenylene dendrimer system containing a sp^3 carbon atom as a core and 1–4

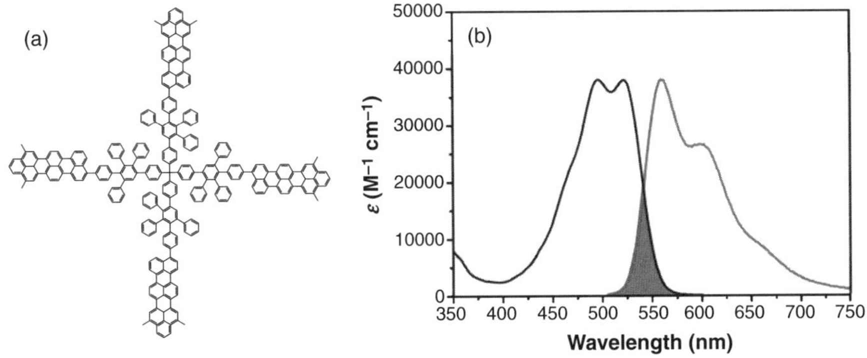

FIGURE 12.11 (a) Chemical structure of a first-generation polyphenylene dendrimer with a carbon atom as core decorated with 4 PMI chromophores. The structures containing 1, 2, 3, and 4 PMI chromophores will be named compound **8**, **9**, **10**, and **11**, respectively. (b) Ground-state absorption and S_1 emission spectrum of the PMI chromophore (compound **8**). The overlap between the two spectra (dark area) indicates that energy hopping is possible when multiple PMI chromophores are present (note that this is a schematic figure, and that for determining the J value the area under the fluorescence curve needs to be normalized to 1).

PMI chromophores at the rim (compounds **8**–**11**, respectively, see Figure 12.11 for chemical structure).

12.7.1 Energy Hopping

When two fluorescent chromophores are in close proximity (1–10 nm range) and there is a weak interaction between their transition dipole moments, the Forster model is applicable [71]. The physical basis for this so-called Förster resonant energy transfer (FRET) involves the transfer of excitation energy from a donor to an acceptor of two weakly coupled transition dipole moments. The process is best known for dedicated donor and acceptor molecules; however, the original theory was derived for a system in which the acceptor is chemically identical to the donor molecule. This phenomenon is now commonly known as energy hopping or homo-FRET. For nonparallel oriented chromophores, energy hopping can be experimentally determined by time-resolved fluorescence depolarization experiments [94].

On the single-molecule level, hopping can be indirectly demonstrated by the change in the orientation of emission polarization or emission dipole moment. By measuring intensity trajectories of the parallel and perpendicular emitted fluorescence of individual molecules of compound **11**, one can visualize the energy hopping process at the single-molecule level. The parallel and perpendicular components of the stepwise changes in fluorescence intensity can be seen in Figure 12.12a. These stepwise changes here can be explained by consecutive bleaching of the four PMI chromophores. Figure 12.12b shows four distinct levels in the polarization trajectory, p-value, of a single molecule of compound **11** embedded in Zeonex [51]. The

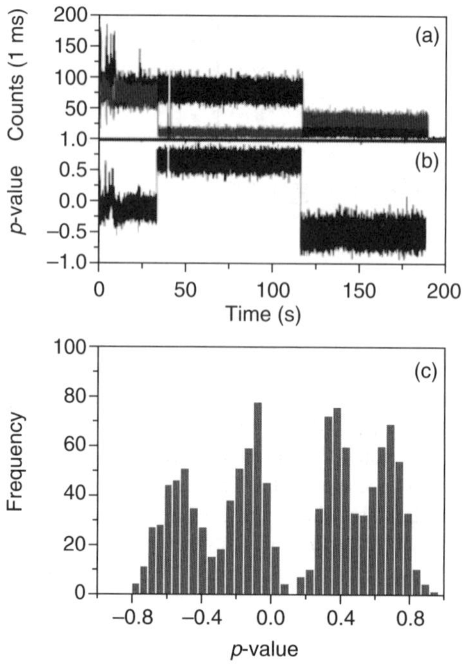

FIGURE 12.12 (a) Example of a fluorescence intensity trajectory of a single molecule of compound **11** where the parallel and perpendicular component were detected separately. (b) Polarization value of the trace in panel a. (c) Histogram of the four p-value levels of the trace in panel b. Copyright from Ref. [51]. (See the color version of this figure in Color Plates section.)

polarization trajectory and corresponding histogram (Figure 12.12c) displaying four distinct p-values of this spatially well-defined system indicate that emission occurs always from the chromophore that is lowest in energy at each point in time [80,95]. This energetically lowest PMI chromophore acts as a fluorescent trap while the other chromophores are communicating via energy hopping to it, leading to discrete p-values in the polarization trajectory. Based on the overlap between the ensemble absorption and emission spectra (see Figure 12.11b), a Förster radius of 4.78 nm can be calculated [51,84,87]. The interchromophoric distances in compound **11** are well within this radius, indicating that indeed energy hopping is very efficient. An alternative way of demonstrating energy hopping can be done by rotating the excitation polarization during the experiment (at a rate of a few Hz) and looking whether the parallel and perpendicular components of the emission are in phase or not [84,85,89]. If they are not in phase that means the chromophores emit independently and there is no hopping to a specific emissive trap. If they are in phase and the modulation of the parallel and perpendicular component of the emission does not go down to zero, this indicates that there are multiple absorbers and one chromophore acts as an emissive trap. If the modulation goes down to zero, this indicates that there are either multiple parallel oriented chromophores that absorb and a single fluorescent trap or that there is only one chromophore present. For example, when performing this

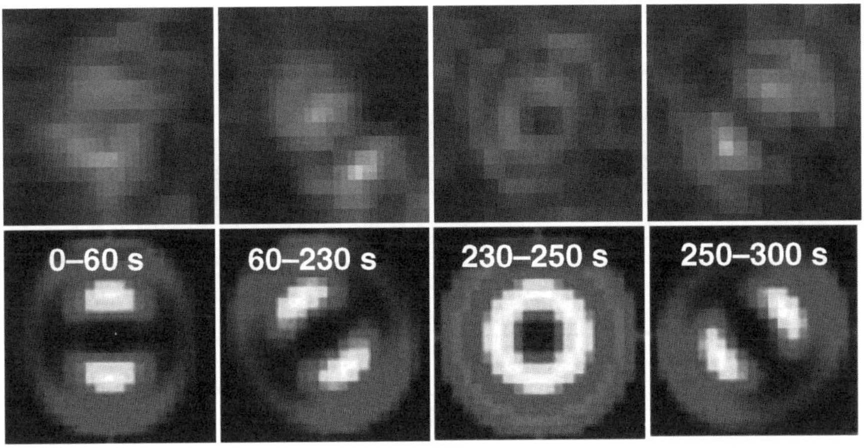

FIGURE 12.13 Series of emission patterns and corresponding simulated patterns for a single molecule of compound **11** in a 25 nm Zeonex film. Images were recorded at 1 μm defocusing and 10 s exposure time. Copyright from Ref. [39]. (See the color version of this figure in Color Plates section.)

experiment on single molecules of compound **9**, 93% displayed in phase modulation with an offset and 6% in phase modulation without an offset, demonstrating that hopping to a fluorescent trap is clearly present [84].

That each chromophore in turn can eventually act as an energy sink can be further demonstrated by measuring the orientation of the fluorescent trap by using a wide-field defocusing technique that allows direct probing of the emission dipole orientation [38]. Individual molecules of compound **11** were embedded in a Zeonex film. The patterns shown in Figure 12.13 were observed during the indicated consecutive period, while each image was obtained by integrating the signal during an exposure time of 10 s. The emission pattern of a single molecule of compound **11** changes as a function of time. The patterns provide direct evidence that the different chromophores of compound **11** emit in time as different dipole orientations are observed successively. By comparison of the recorded experimental patterns with simulated patterns [38], we attribute, following the sequence shown in Figure 12.13, the first pattern is due to a chromophore with an emission dipole oriented in plane along the x-axis. The second pattern is due to another in-plane emission dipole with an angle of $-45°$ with respect to the x-axis. The third and fourth patterns correspond to an out-of-plane emission dipole and to an in-plane emission dipole with an angle of $+45°$ with respect to the x-axis, respectively [39].

12.8 SINGLET–SINGLET ANNIHILATION

When a multichromophoric system undergoes optical excitation under high photon flux, two excited states (S_1) can be present in a single molecule at the same time. If the S_1–S_0 transition of one chromophore is in resonance with a S_1–S_n transition of the

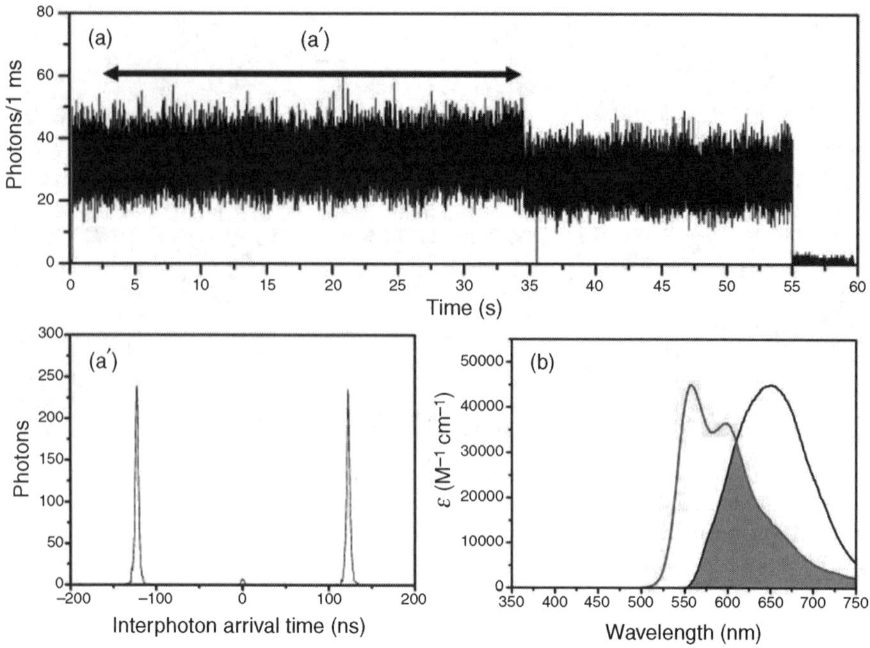

FIGURE 12.14 (a) Fluorescence intensity trajectory of a single molecule of compound **9** in Zeonex, together with the coincidence graph of the first (a') intensity level. (b) S_1–S_n absorption and S_1 emission spectra of compound **8** in toluene. The overlap between the two spectra (dark area) indicates that singlet–singlet annihilation is possible when multiple PMI chromophores are present (note that this is a schematic figure, and that for determining the *J* value the area under the fluorescence curve needs to be normalized to 1).

other chromophore, energy transfer between the excited singlet states can occur. As a result only one excited state remains in the multichromophoric system and this process is often referred to as singlet–singlet annihilation. As depicted in Figure 12.14b, there is a large overlap between the S_1–S_n absorption spectrum and S_1 emission spectrum for the PMI chromophores in compound **11** (the singlet absorption spectrum was measured by femtosecond transient absorption measurements) [84]. The presence of singlet–singlet annihilation was proven before at the ensemble level by means of femtosecond fluorescence upconversion and time-resolved polychromatic femtosecond transient absorption measurements [96]. The observation of singlet–singlet annihilation as a fast, excitation-intensity-dependent decay component in time-resolved fluorescence measurements is impossible in single-molecule spectroscopy due to the limited time resolution (300 ps) presently provided by avalanche photodiodes used for detection. Unambiguous evidence for the annihilation phenomenon can be obtained because for a single chromophore the probability of emitting two consecutive photons drops to zero for time intervals shorter than the excited-state decay time [97–99]. This property of the photon arrival time statistics, termed photon antibunching, has been measured at room temperature under pulsed laser illumination for individual molecules by measuring and histogramming the interphoton arrival times [84,85]. Hence, for a single emitter, the zero peak, which represents pairs of fluorescence photons generated during the same laser pulse, is necessarily vacant as long as the laser pulse width is much smaller than the fluorescence decay time of the molecule. Thus, for multichromophoric systems under pulsed excitation conditions, the absence of the peak at zero in the interphoton arrival time distribution would indicate the existence of an efficient singlet–singlet annihilation process. Figure 12.14a shows a fluorescence intensity trajectory of a single molecule of compound **9** in Zeonex. Two intensity levels can be observed indicating the presence of 2 consecutive bleaching chromophores. An interphoton arrival histogram of the intensity level where the two chromophores are emitting is given in Figure 12.14a′. The absence of the central peak shows that compound **9** acts as a single emitter.

The quality of a multichromophoric structure as deterministic single photon sources can be determined by calculating the so-called Mandel parameter and has been reported extensively, not only for PMI but also for PDI chromophore containing dendrimer systems [82–85,90,100,101].

12.9 SINGLET–TRIPLET AND SINGLET–RADICAL ION ANNIHILATION

Direct excitation of one of the chromophores can lead also to intersystem crossing (ISC) to the triplet state. Because of the relatively long lifetime of the triplet state, a second chromophore can be excited while a first chromophore is still in the triplet state. As a result, two excited states are present, a S_1 and a T_1 state. If the triplet state exhibits transitions into higher excited triplet states, T_n, that are in resonance with the S_1 to S_0 transition, energy transfer from the excited singlet state to the energetically lower lying triplet state, can occur [84]. This process is often called singlet–triplet annihilation. Such a process is not easy to see in bulk experiments but manifests itself directly in single-molecule fluorescence trajectories as collective on/off steps [84].

Figure 12.15a shows the spectral overlap of this singlet–triplet annihilation process and in Figure 12.16c examples are presented for collective on/off jumps related to singlet–triplet annihilation in compound **11**. Note that careful analysis of the on/off jumps corresponding to different intensity levels in multichromophoric systems can result in quantitative information on intersystem crossing and even intersystem crossing of higher excited states [82]. The latter will be demonstrated by single-molecule data of compound **11**. Figure 12.16a shows the fluorescence intensity and lifetime trajectory of a single molecule of compound **11** that exhibits subsequent photobleaching of the four chromophores. The stable fluorescence lifetime of 4.4 ns throughout the trajectory indicates that strong coupling interactions are absent in compound **11**. Intensity fluctuations due to triplet formation in the different parts of the trajectory can be analyzed by a fluorescence intensity correlation [102–105]. As can be seen in Figure 12.16b, on-times increase from 2.2 ms for the highest intensity

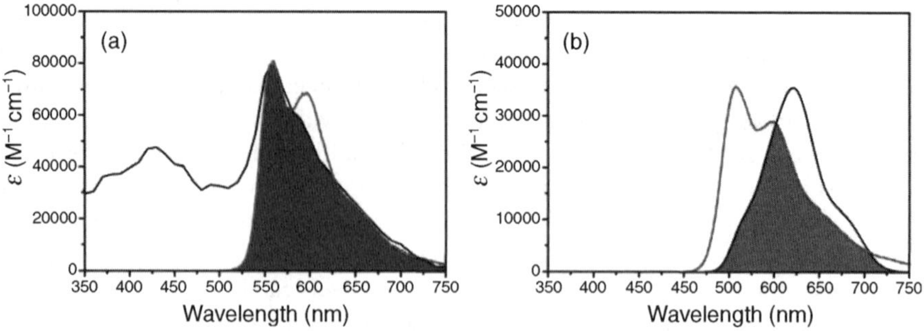

FIGURE 12.15 (a) Triplet absorption (T_1–T_n) and S_1 emission spectrum of compound **8** in toluene. The overlap between the two spectra (dark area) indicates that singlet–triplet annihilation is possible when multiple PMI chromophores are present (note that this is a schematic figure, and that for determining the *J* value the area under the fluorescence curve needs to be normalized to 1). (b) Radical anion absorption spectrum and S_1 emission spectrum of compound **8** in toluene. The overlap between the two spectra (dark area) indicates that singlet–radical anion annihilation is possible when multiple PMI chromophores are present (note that this is a schematic figure, and that for determining the *J* value the area under the fluorescence curve needs to be normalized to 1).

level (I) to 120 ms for the remaining chromophore (level IV), that is, an apparent approximately 55-fold increase in on-times, which is substantially larger than the factor of four arrived at based on the change in intensity. Evidently, triplet formation is much more frequent in the first intensity levels when no photobleaching of the chromophores has occurred and becomes less frequent with stepwise photobleaching of the PMI chromophores. Concomitantly, off-times also increase from 208 μs for level I to 610 μs for level IV.

To explain the observed phenomena, a classical three level scheme involving only the S_0, S_1, and T_1 states is not sufficient. A nonlinear process involving higher excited states S_n might play an important role in opening up new triplet-formation pathways. Two such processes have been described in the literature: intersystem crossing is sometimes very efficient from S_n, due to fortuitous matching of energies in higher excited states [106–108], and singlet fission, a process known to occur in multichromophoric systems such as organic crystals, conjugated polymers, and even light-harvesting complexes [109–113]. Singlet fission refers to the conversion of one singlet S_n into two triplets: $S_n \rightarrow T_1 + T_1$. Although singlet fission can provide very high triplet-formation rates, the relatively large distances between the chromophores in the dendrimers investigated here renders this mechanism less likely. Here intersystem crossing from S_n states seems the most likely explanation for the increased number of on/off jumps in the fluorescence intensity trajectory, since high excitation powers are used in single-molecule experiments, increasing the probability to generate S_n states via singlet–singlet annihilation. As mentioned above, the triplet lifetime τ_{off}, decreases with the number of chromophores (Figure 12.16b). Photoinduced reverse intersystem crossing from $T_n \rightarrow S_n$ has been demonstrated to be a potential mechanism for the shortening of triplet lifetimes at the single-molecule

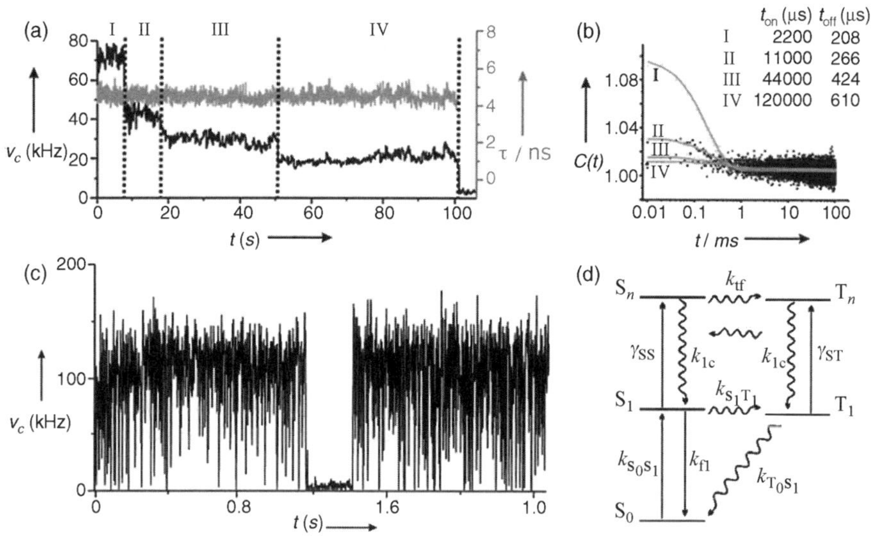

FIGURE 12.16 (a) Fluorescence intensity (count rate v_c (kHz) vs. time t (s)) and decay time (fluorescence decay time, τ (ns) vs. time, t (s)) trajectory of a single molecule of compound **11**. (b) Second-order intensity correlation $C(t)$ of the different intensity levels clearly demonstrating different off-times (triplet related) for different intensity levels. (c) Expanded view of the first intensity level of a single molecule of compound **11**, demonstrating short off-times (triplet related) and one long off event (related to formation of a radical anion). (d) Kinetic scheme of ground- and excited-state pathways of multichromophoric systems like compound **11**. Copyright from Ref. [82].

level [114–116]. Although this mechanism is already possible in systems containing a single PMI chromophore (like compound **8**) due to the $T_1 \rightarrow T_n$ absorbance at 488 nm (see also Figure 12.15a) [84], the decrease of τ_{off} in compound **11** is promoted by efficient singlet–triplet annihilation ($S_1 + T_1 \rightarrow S_0 + T_n$) followed by reverse intersystem crossing ($T_n \rightarrow S_n$). For the assignment of the additional triplet-formation mechanism it is necessary to estimate its probability and rate constants based on a five-level model as shown in Figure 12.16d. The model includes the singlet levels S_0, S_1, S_n, and the triplet levels T_1 and T_n, and the following pathways are present: (1) excitation $S_0 \rightarrow S_1$ ($k_{S_0S_1}$), (2) fluorescence $S_1 \rightarrow S_0$ (k_{fl}), (3) singlet–singlet annihilation $S_1 + S_1 \rightarrow S_0 + S_n$ (γ_{SS}), (4) intersystem crossing $S_1 \rightarrow T_1$ ($k_{S_1T_1}$), (5) triplet formation from higher singlet states $S_n \rightarrow T_n$ (k_{tf}), (6) singlet–triplet annihilation $S_1 + T_1 \rightarrow S_0 + T_n$ (γ_{ST}), (7) reverse intersystem crossing $T_1 \rightarrow S_0$ ($k_{T_1S_0}$), (8) higher excited-state reverse intersystem crossing $T_n \rightarrow S_n$ ($k_{T_nS_n}$), and (9) internal conversion from higher excited states (k_{IC}).

Besides the above-mentioned on–off blinking process, there is an additional rare event leading to longer fluorescence intermittence than the triplet off-times. The cause of this is probably the formation of a radical anion on one of the PMIs [82,84]. This radical anion can quench the fluorescence of the other PMIs by singlet–radical anion energy transfer. Figure 12.15b shows the spectral overlap of this process and an example of a long collective off event from 1.2 until 1.4 s is presented in Figure 12.16c. This rare long off has also been observed for PDI and other chromophores, where similarly a long off state can be observed. Depending on the environment, these longer off events follow a power-law or exponential distribution [28,117–120].

12.10 EXCITATION ENERGY TRANSFER BETWEEN STRUCTURALLY DIFFERENT CHROMOPHORES

The above-mentioned Förster resonant energy transfer processes are also applicable between structurally different chromophores. Here we will mainly focus on energy transfer from chromophores with a high S_0–S_1 gap (donor) to chromophores with a lower S_0–S_1 gap (acceptor). The term dyad/triad that will be used refers to a bi- or multichromophoric system that bears two/three different types of chromophores, respectively.

Polyphenylene dendrimers can again provide a rigid framework for the introduction of chromophores into the center, the scaffold or the periphery, by using similar design concepts as presented in the previous paragraph. Different rylene dyes such as naphthalenemonoimides (NMI), PMI, and TDI chromophores can be attached in this way and act as light-harvesting arrays for directional transfer of excitation energy [87,121–125]. Nature has brought this principle to perfection, as can be seen from the crystal structure of the light-harvesting antenna complex (LH_2) of the purple bacterium *Rhodopseudomonas acidophila* [126]. In biological systems a combination of Förster type energy transfer and strongly coupled chromophores arrays is used to transfer energy over large distances [127,128]. On top, the chromophore systems are protected from photobleaching by the presence of other molecules like carotenoids.

Figure 12.17a shows the chemical structure of compound **12**, a dendritic dyad with a TDI chromophore in the center and PMI chromophores at the periphery, which has a high photostability suitable for single-molecule experiments. An efficient, vectorial energy transfer via an energy gradient from the periphery toward the center of the

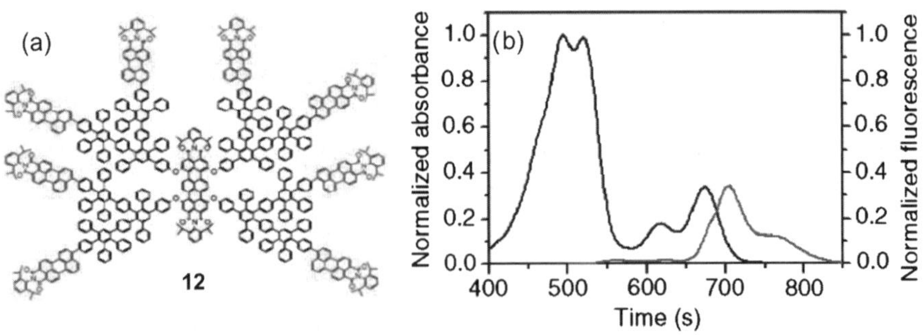

FIGURE 12.17 (a) Chemical structure of compound **12** containing one TDI chromophore as a core and 8 PMI chromophores at the rim. (b) Steady-state absorption and emission spectra (excitation wavelength at 500 nm) of compound **12** in toluene.

dendrimer was obtained for multichromophores based on this design concept. Figure 12.17b shows the absorption and emission spectra of compound **12**. Absorption features around 500 and 675 nm can be attributed to the PMI and TDI parts, respectively. When exciting the PMI part at 500 nm, the emission is almost uniquely coming from the TDI, indicating a very efficient energy transfer efficiency (see Figure 12.17b). At the single-molecule level, the use of APDs with their limited time resolution makes the detection of fast kinetic components of directional FRET from the PMI to the TDI, as suggested by the ensemble solution time-resolved data, impossible [129]. However, due to the excellent photostability and brightness of compound **12**, a multichannel plate detector (~23 ps response time) could be used, which is more suited to detect a fast rise term if one uses appropriate reconvolution procedures [91]. For each single molecule, all the detected photons from the acceptor were used to build a decay histogram that was fitted with a biexponential model function with positively and negatively contributing time constants by MLE fitting. The residual graphs in Figure 12.18a show that a monoexponential fit does not fit the data well, whereas a biexponential fit with a negative pre-exponential factor for the short time constant gives a good result. In all cases, the ratio of the pre-exponential factors accounting for the contributions of the recovered time constants was around -1.0 ± 0.2. This demonstrates that the excited state of the TDI part in compound **12** is built up only from FRET from the PMI chromophores. A typical analyzed single-molecule decay from an individual molecule of compound **12** is given in Figure 12.18a. The histogram of the rise components detected from 100 different single molecules of compound **12** shows a broad distribution with a mean value of 48 ps (Figure 12.18b). This finding points to a broad range of FRET time constant values at the single-molecule level that implies that there is a range of interchromophoric distances and orientations between the donor and acceptor chromophores in the dendrimers when immobilized in Zeonex [91].

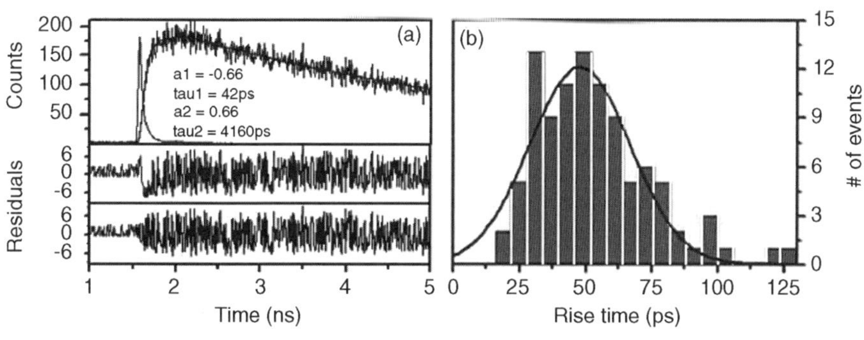

FIGURE 12.18 (a) Top panel shows fluorescence decay histogram recorded with an MCP–PMT detector of the acceptor emission from a single molecule of compound **12** in Zeonex upon selective 488 nm excitation of the donors. The instrumental response function of the system and the biexponential fit curve are also shown. The middle and bottom panels show residual graphs for mono- (middle panel) and biexponential (bottom panel) fits. (b) Histogram of the rise times of 100 single molecules of compound **12**. Copyright from Ref. [91].

The close proximity of the PMI to the TDI chromophores as well as the overlap between the emission and absorption spectra facilitates energy transfer from the PMI to the TDI chromophores as was demonstrated by the steady state and single-molecule data. In fact, the efficiency of this process is such that only TDI emission can be observed in ensemble-level experiments, but no donor PMI emission, irrespective of the excitation wavelength [130]. However, at high excitation intensities simultaneous donor and acceptor emission can sometimes be observed [91,123]. An explanation of this simultaneous emission could be exciton blockade. Exciton blockade occurs if two donor chromophores are excited simultaneously then only one of these can transfer its energy to the central TDI chromophore. The second excited PMI chromophore, however, cannot transfer its energy to the excited TDI, and emits normally. The net result is that both the donor and acceptor emit. This explanation however did not fully explain the observed data at the single-molecule level [131]. It was found that simultaneous donor and acceptor emission was more likely to occur after some of the donor molecules in compound **12** had photobleached. Finally, it was also observed that the donor and acceptor emission were uncorrelated, that is, the emission of a donor or acceptor photon occurs in a stochastic fashion, with no apparent relation between them. These findings showed that exciton blockade by itself was not a satisfactory explanation for the simultaneous donor–acceptor emission. To fully explain these observations, defocused wide-field imaging was performed, exciting only the PMI chromophores, but separating the emission of the PMI and TDI chromophores over two identical CCD cameras [124,131]. The defocused imaging displayed that three different types of emission could be recovered: only acceptor, only donor (after photobleaching of the acceptor), and simultaneous donor and acceptor emission. After the acceptor had bleached, the donor emission would usually be present as a single emission pattern only, indicating emission from a single donor chromophore. However, there would be a progressive change to other emission patterns until the last of the donor chromophores bleached. This can be explained by the fact that some chromophores will have a lower energy than the others, and will act as a preferential energy acceptor (see above). After the photobleaching of this lowest energy chromophore another will take its place, until only a single donor chromophore remains. Figure 12.19 shows images from two examples of single molecules of compound **12** showing both emission from the PMI and TDI part, where the left column shows the acceptor emission, and the right column shows the donor emission. Details from these frames, showing emission patterns corresponding to simultaneous donor and acceptor emission from the same molecule are shown in Figure 12.19b, along with their fitted orientational patterns. In all the 15 measured molecules of compound **12** that displayed this dual emissive behavior, the orientation of the donor chromophore was perpendicular to that of the acceptor chromophore, which would seriously hinder energy transfer between them.

Based on the confocal [91,131] and defocused imaging [124,131] the following picture can be constructed: in intact dendrimers (no photobleached chromophores), energy hopping among the donors ensures that the excitations end up at the acceptor, even if unfavorably oriented donors are present in the dendritic molecule. Multiple excitations within the dendrimer are quenched by very efficient singlet PMI–singlet

(a)

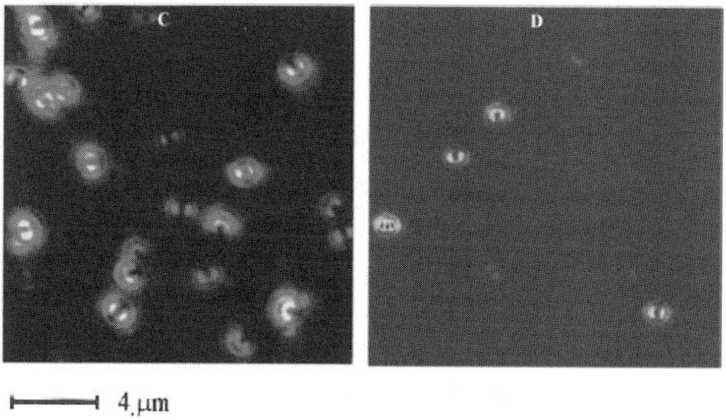

(b)

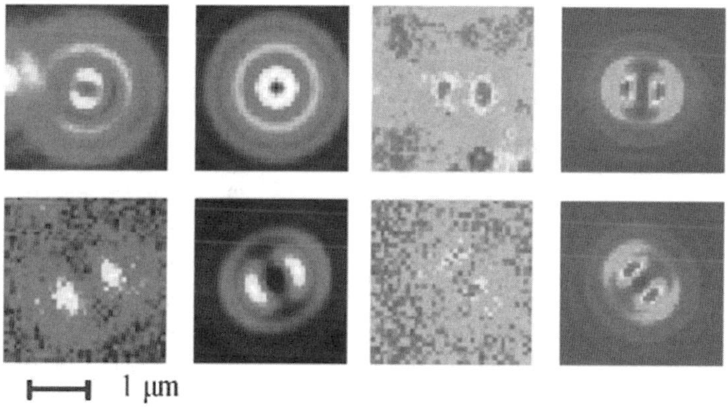

FIGURE 12.19 (a) Defocused wide-field images of the same sample area but different emission wavelengths, showing the emission and orientation of TDI (left) and PMI chromophores (right). (b) Simultaneous emission from the dendrimer chromophores: each row represents a different molecule. On the left the acceptor emission and fitted orientational pattern are shown, while on the right side the donor emission and fitted pattern are shown. The patterns indicate that the donor and acceptor chromophores are oriented perpendicular to one another. Copyright from Ref. [131]. (See the color version of this figure in Color Plates section.)

PMI annihilation or by singlet PMI–singlet TDI annihilation, which leads to the existence of only one excited chromophore, the acceptor. After photobleaching of a number of donors, unfavorably oriented donors can become "isolated." When these donors are excited, they will not undergo FRET nor will they annihilate with the excited acceptor molecule, eventually populated via other donors, but relax to the ground state by emission of fluorescence. Thus, the simultaneous donor–acceptor emission in the dendritic systems arises mainly from unfavorably oriented donor molecules and only to a minor extent from an exciton blockade resulting from multiple excitations.

A system devised for cascade energy transfer is the third-generation globular polyphenylene dendrimer compound **13** (see Figure 12.20) bearing a TDI

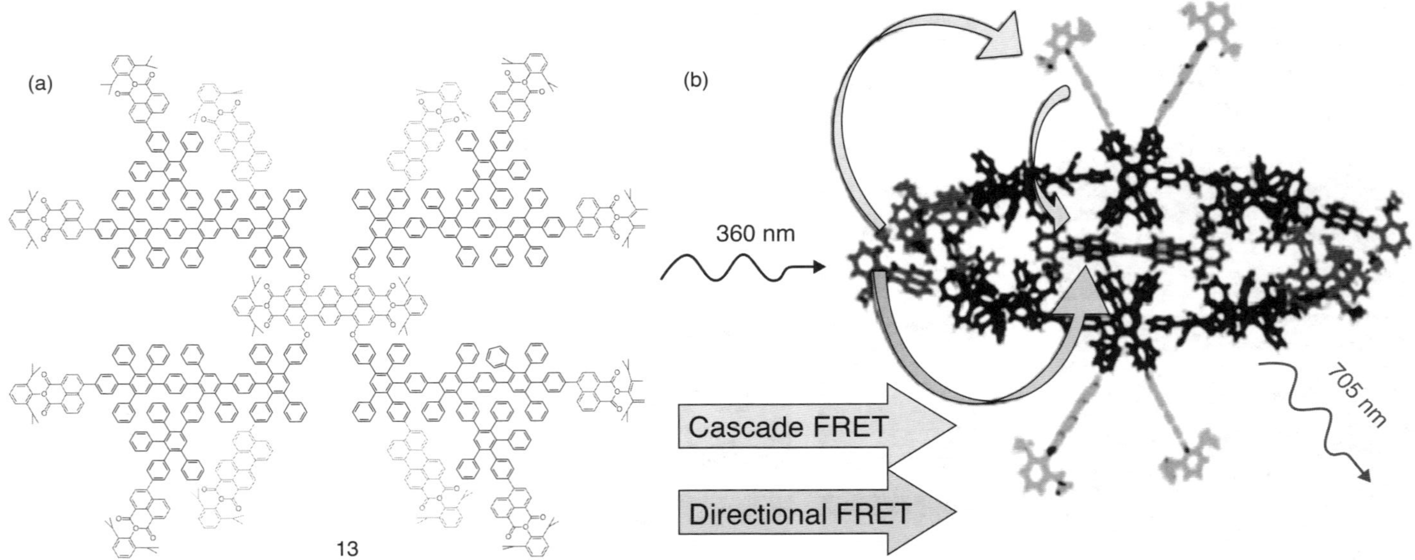

FIGURE 12.20 (a) Chemical structure of compound **13** containing 8 NMI chromophores at the outer shell, 4 PMI chromophores within the scaffold and 1 TDI in the center. (b) 3D structure of compound **13**. The arrows illustrate energy transfer processes after exciting the NMI chromophores. Copyright from Ref. [123] (See the color version of this figure in Color Plates section.)

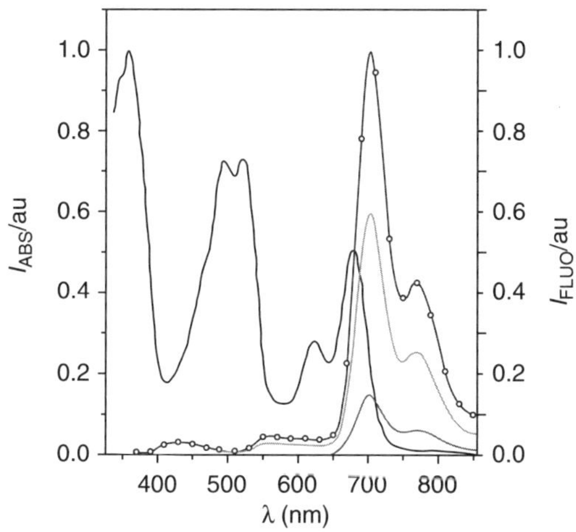

FIGURE 12.21 Absorption (black line) and fluorescence (line + symbol) spectra of compound **13** in toluene. Fluorescence spectra were recorded upon excitation at 360 nm (blue), 490 nm (green), or 640 nm (red). Copyright from Ref. [123]. (See the color version of this figure in Color Plates section.)

chromophore as a core, 4 PMI chromophores in the scaffold and 8 NMI chromophores at the periphery. This nanosized-emitter reveals a high degree of complexity since all spatial locations, distances between the chromophores and their absorption and emission spectra were adjusted carefully [132]. In this way, an efficient dendritic triad was obtained absorbing light over the whole visible spectrum and showing stepwise energy transfer over long distances from the periphery via the scaffold toward the center of the dendrimer. The absorption spectrum and emission spectra exciting in the NMI (360 nm), PMI (490 nm) and TDI (640 nm) region are given in Figure 12.21. The low amount of fluorescence at 450 nm (NMI) and 550 nm (PMI) indicate that very efficient energy transfer to the TDI is present.

According to the ensemble FRET data [123], highly efficient FRET is expected to occur when probing single molecules of compound **13** with 490 nm laser light. When probing 100 molecules of compound **13**, 83% shows exclusive emission in the TDI channel, before complete photobleaching occurs (see Figure 12.22a, upper and lower panels). For such molecules, the fractional intensity *F* (TDI channel divided by the sum of the TDI and PMI channel) shows a constant value around unity (Figure 12.22a), therefore proving that FRET occurs efficiently from the PMI units to the TDI core. Following the trajectory of the fluorescence decay time and of the fluorescence spectrum of these molecules, both parameters stay constant during the measurement, suggesting emission is related to a single excited species, here the TDI chromophore [91,130]. The fluorescence intensity in the TDI channel decreases in a stepwise fashion because of successive photobleaching of the PMI donors (Figure 12.22a, upper panel) [91,130] Fluorescence spectra detected from such molecules peak around 680 nm (Figure 12.22b) and resemble, both in shape and in peak position, the ensemble solution spectrum of the bare TDI chromophore [91,129].

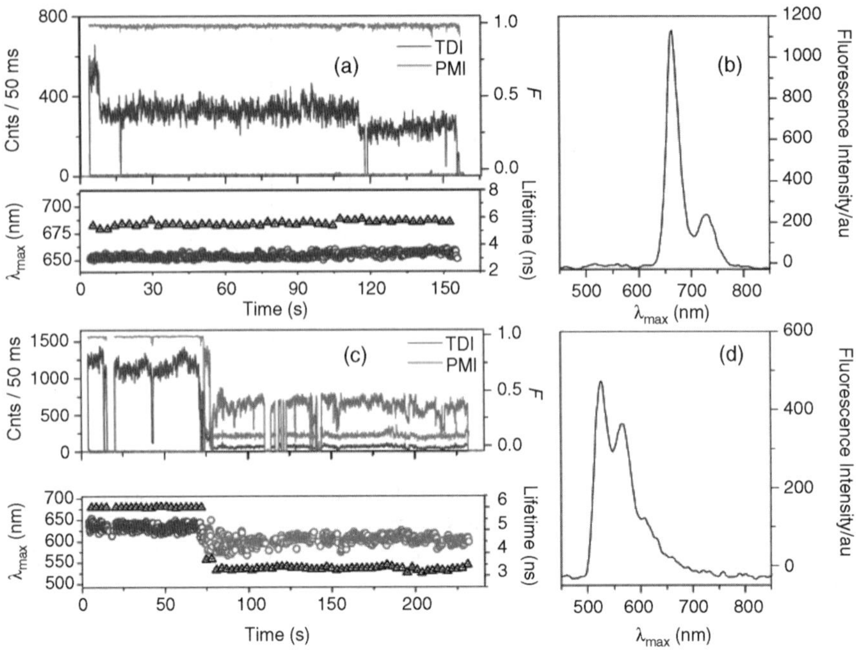

FIGURE 12.22 Single-molecule fluorescence intensity trajectories on 490 nm excitation of two molecules of compound **13** in Zeonex. Detected photons in the TDI channel are colored red, and those from the PMI channel are green. (a) Upper panel: trajectories of the fluorescence intensity (red and green lines) and fractional intensity (gray line) from a single molecule of compound **13**. Lower panel: trajectories of the fluorescence decay time (red circles) and emission maximum (black triangles) from the single molecule of compound **13** displayed in the upper panel. (b) Fluorescence spectrum (3 s integration time) detected from the single molecule displayed in panel a. (c) Upper panel: trajectories of the fluorescence intensity (red and green lines) and fractional intensity (gray line) from a single molecule of compound **13**. Lower panel: trajectories of the fluorescence decay time (red and green circles) and emission maximum (black triangles) from the single molecule of compound **13** displayed in the upper panel. (d) Fluorescence spectrum (3 s integration time) detected from the single molecule displayed in panel c after the bleaching of TDI. Copyright from Ref. [123].

Fluorescence decay times detected in the TDI channel show values ranging from 3 ns (fluorescence decay time of the bare TDI chromophore) up to 9 ns, similar to those observed previously for single dyad molecules immobilized in Zeonex, that is, dendrimers based on a TDI core and 4 PMI chromophores at the periphery [91]. For the dyad, a broad distribution of single-molecule fluorescence decay times detected from the TDI core was connected with the presence of a distribution of ground-state TDI conformers as a result of the torsion induced by the dendritic arms attached to the core chromophore; therefore, similar reasoning can be invoked for compound **13**. The second subpopulation of triad molecules (17%) shows emission in the PMI channel only after the bleaching of the TDI core (Figure 12.22c, upper and lower panels). Bleaching of TDI is followed by the appearance of a single-molecule fluorescence spectrum peaking around 540 nm (Figure 12.22d). Such emission and fluorescence decay time values of about 4 ns are characteristic for the PMI chromophore [91,129]. The bleaching of TDI followed by emission from PMI is seen as a change in the value of *F* from 1 to nearly 0 (Figure 12.22c, upper panel) and therefore as a change in FRET efficiency from 100% to zero because of the disappearance (photobleaching) of the acceptor chromophore.

In 5% of the probed single molecules of compound **13**, the bleaching of the TDI core is not followed by the emission from PMIs but from an intermediate species with a fluorescence spectrum peaking around 590 nm and as a result of this *F* values around 0.5 are obtained for this intermediate species. This intermediate species is probably an oxidized TDI chromophore with a reduced extent of conjugation and therefore with spectrally blue-shifted absorption and emission properties compared to an undamaged TDI chromophore. After bleaching of the intermediate, fluorescence is mainly detected in the PMI channel.

In order to probe the FRET dynamics from the NMI chromophores, single molecules of compound **13** were excited with 360 nm laser light and the fluorescence emitted by the PMI and TDI chromophores was simultaneously monitored. Within a population of 100 single molecules, compound **13** showed either exclusive TDI emission (80%, Figure 12.23a) or TDI emission followed by activity in both channels (20%, Figure 12.23c). None of the interrogated compound **13** molecules showed emission solely in the PMI channel. Also it was observed that, for similar excitation power, the survival time of single molecules of compound **13** excited at 360 nm is by far shorter than that of the same molecules excited at 490 nm. Figure 12.23b shows the probability densities of the survival times for single molecules of compound **13** undergoing excitation at 360 nm (upper panel) and 490 nm (lower panel). Exponential fits to the histograms from Figure 12.23b give time constants of 23 s (360 nm excitation) and 158 s (490 nm excitation). This difference in survival times can be explained by the difference in fluorescence quantum yield of 0.98 and 0.1 for PMI and NMI, respectively. For NMI, assuming that nonradiative deactivation is mainly related to triplet formation [133], for a quantum yield of fluorescence of about 0.1 and a fluorescence lifetime of 0.33 ns, the rate constant and the quantum yield for ISC are estimated to be around $3 \times 10^{10}\ s^{-1}$ and 0.9, respectively. When NMI is part of the triad and 360 nm excitation is used, NMI fluorescence is quenched from 0.33 to 0.04 ns by energy transfer to the TDI core, either directly or in a cascade fashion. Therefore, for a single triad molecule, optical excitation of an NMI chromophore will open competition mainly between triplet formation and direct energy transfer. Having such a high quantum yield of triplet formation, the probability of photobleaching of an NMI chromophore on 360 nm excitation becomes considerably higher than the probability of photobleaching of a PMI chromophore on 490 nm excitation. Consequently, the survival time of a single triad molecule becomes substantially shorter on 360 nm excitation than on 490 nm excitation (Figure 12.23b).

Shown in Figure 12.23c is an example of a single triad molecule where, on 360 nm laser excitation, bleaching of the TDI is followed by activity in both TDI and PMI channels. For this molecule, bleaching of TDI leads to a drop in the fractional intensity from 1 to 0.5 (Figure 12.23c), pointing to the intermediate species that was also

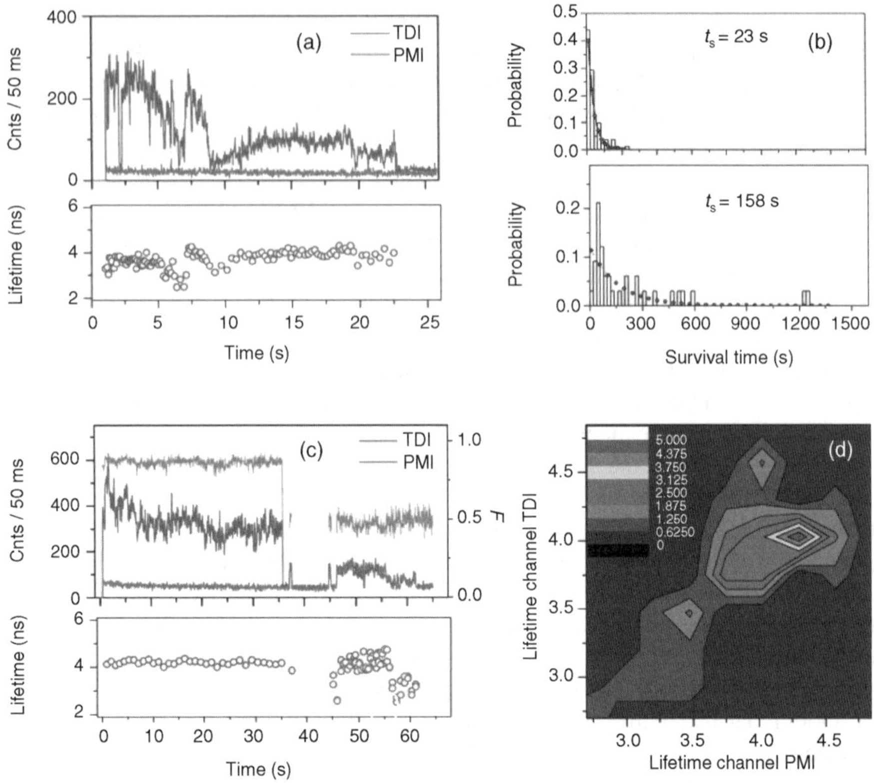

FIGURE 12.23 Single-molecule fluorescence intensity trajectories on 360 nm excitation of two molecules of compound **13** in Zeonex. Detected photons in the TDI channel are colored red, and those from the PMI channel are green. (a) Trajectories of the fluorescence intensity (upper panel, red and green lines) and fluorescence decay time (lower panel, red circles) from a single molecule of compound **13**. (b) Normalized probability densities of the survival time of single triad molecules on 360 nm pulsed excitation (upper panel) and 490 nm pulsed excitation (lower panel). Also shown are exponential fits (line + symbol) to the respective histograms. (c) Trajectories of the fluorescence intensity (upper panel, red and green lines), fractional intensity (upper panel, gray line), and fluorescence decay time (lower panel, red and green circles) from another single molecule of compound **13**. (d) Contour plot showing the correlation between the single-molecule fluorescence decay times detected in the TDI and PMI channels in the region from 45 to 58 s shown in panel c, upper part. Copyright from Ref. [123]. (See the color version of this figure in Color Plates section.)

observed in the single-molecule experiments at 490 nm laser excitation [123]. On 360 nm excitation, 20% of the interrogated single triad molecules showed such an intermediate species. The fluorescence emitted from the TDI side is the result of excitation energy transfer from the NMI chromophores, either by direct transfer or via stepwise transfer through a PMI. NMI chromophores have a high probability for triplet formation and therefore a high chance to rapidly photobleach. The higher ISC efficiency to the triplet state of NMI might lead to the creation of singlet oxygen that can react with the core TDI, leading to the formation of the intermediate species. Consequently, the chance for formation of the intermediate species will be higher if a single triad molecule undergoes excitation at 360 nm than at 490 nm.

REFERENCES

[1] P. Tinnefeld, M. Sauer, *Angew. Chem. Int. Ed.* **2005**, *44*, 2642–2671.

[2] F. Kulzer, M. Orrit, *Ann. Rev. Phys. Chem.* **2004**, *55*, 585–611.

[3] W. E. Moerner, D. P. Fromm, *Rev. Sci. Instrum.* **2003**, *74*, 3597–3619.

[4] D. W. Pohl, W. Denk, M. Lanz, *Appl. Phys. Lett.* **1984**, *44*, 651–653.

[5] E. Betzig, J. K. Trautman, *Science* **1992**, *257*, 189–195.

[6] J. K. Trautman, J. J. Macklin, L. E. Brus, E. Betzig, *Nature* **1994**, *369*, 40–42.

[7] U. Durig, D. W. Pohl, F. Rohner, *J. Appl. Phys.* **1986**, *59*, 3318–3327.

[8] E. Betzig, G. H. Patterson, R. Sougrat, O. W. Lindwasser, S. Olenych, J. S. Bonifacino, M. W. Davidson, J. Lippincott-Schwartz, H. F. Hess, *Science* **2006**, *313*, 1642–1645.

[9] S. Bretschneider, C. Eggeling, S. W. Hell, *Phys. Rev. Lett.* **2007**, 98.

[10] S. W. Hell, *Sci.* **2007**, *316*, 1153–1158.

[11] C. Flors, J., Hotta, H. Uji-I, P. Dedecker, R. Ando, H. Mizuno, A. Miyawaki, J. Hofkens, *J. Am. Chem. Soc.* **2007**, *129*, 13970–13977.

[12] M. Heilemann, S. van de Linde, M. Schiittpelz, R. Kasper, B. Seefeldt, A. Mukherjee, P. Tinnefeld, M. Sauer, *Angew. Chem. Int. Ed.* **2008**, *47*, 6172–6176.

[13] M. G. L. Gustafsson, *Proc. Natl. Acad. Sci. USA* **2005**, *102*, 13081–13086.

[14] R. Heintzmann, T. M. Jovin, C. Cremer, *J. Opt. Soc. Am, A-Opt. Image Sci. Vis.* **2002**, *19*, 1599–1609.

[15] S. M. Nie, S. R. Emery, *Science* **1997**, *275*, 1102–1106.

[16] K. Kneipp, Y. Wang, H. Kneipp, L. T. Perelman, I. Itzkan, R. Dasari, M. S. Feld, *Phys. Rev. Lett.* **1997**, *78*, 1667 1670.

[17] K. Kneipp, H. Kneipp, I. Itzkan, R. R. Dasari, M. S. Feld, *Chem. Rev.* **1999**, *99*, 2957–2975.

[18] V. Kudryavtsev, S. Felekyan, A. K. Wozniak, M. Konig, C. Sandhagen, R. Kuhnemuth, C. A. M. Seidel, F. Oesterhelt, *Anal. Bioanal. Chem.* **2007**, *387*, 71–82.

[19] E. L. Elson, D. Magde, *Biopolymers* **1974**, *13*, 1–27.

[20] M. Douglas, L. E. Elliot, W. W. Watt, *Biopolymers* **1974**, *13*, 29–61.

[21] R. Rigler, U. Mets, J. Widengren, P. Kask, *Eur. Biophys. J. Biophys. Lett.* **1993**, *22*, 169–175.

[22] P. Schwille, J. Korlach, W. W. Webb, *Cytometry* **1999**, *36*, 176–182.

[23] T. Funatsu, Y. Harada, M. Tokunaga, K. Saito, T. Yanagida, *Nature* **1995**, *374*, 555–559.

[24] B. K. Muller, E. Zaychikov, C. Brauchle, D. C. Lamb, *Biophys. J.* **2005**, *89*, 3508–3522.

[25] T. Dertinger, V. Pacheco, I. von der Hocht, R. Hartmann, I. Gregor, J. Enderlein, *Chemphyschem* **2007**, *8*, 433–443.

[26] D. Wöll, E. Braeken, A. Deres, F. C. De Schryver, H. Uji-I, J. Hofkens, *Chem. Soc. Rev.* **2009**, *38*, 313–328.

[27] W. E. Moerner, M. Orrit, *Science* **1999**, *283*, 1670–1676.

[28] J. N. Clifford, T. D. M. Bell, P. Tinnefeld, M. Heilemann, S. M. Melnikov, J. Hotta, M. Sliwa, P. Dedecker, M. Sauer, J. Hofkens, E. K. L. Yeow, *J. Phys. Chem. B* **2007**, *111*, 6987–6991.

[29] F. Stracke, C. Blum, S. Becker, K. Mullen, A. J. Meixner, *Chemphyschem* **2005**, *6*, 1242–1246.

[30] E. Braeken, P. Marsal, A. Vandendriessche, M. Smet, W. Dehaen, R. A. L. Vallee, D. Beljonne, M. Van der Auweraer, *Chem. Phys. Lett.* **2009**, *472*, 48–54.

[31] M. W. Holman, R. C. Liu, D. M. Adams, *J. Am. Chem. Soc.* **2003**, *125*, 12649–12654.

[32] J. Hofkens, W. Verheijen, R. Shukla, W. Dehaen, F. C. De Schryver, *Macromolecules* **1998**, *31*, 4493–4497.

[33] A. Herrmann, T. Weil, V. Sinigersky, U. M. Wiesler, T. Vosch, J. Hofkens, F. C. De Schryver, K. Mullen, *Chem.-Eur. J.* **2001**, *7*, 4844–4853.

[34] B. Sick, B. Hecht, L. Novotny, *Phys. Rev. Lett.* **2000**, *85*, 4482–4485.

[35] A. P. Bartko, K. W. Xu, R. M. Dickson, *Phys. Rev. Lett.* **2002**, *89*, 4.

[36] H. Uji-I, S. M. Melnikov, A. Deres, G. Bergamini, F. De Schryver, A. Herrmann, K. Mullen, J. Enderlein, J. Hofkens, *Polymer* **2006**, *47*, 2511–2518.

[37] J. A. Veerman, M. F. Garcia-Parajo, L. Kuipers, N. F. Van Hulst, in *5th International Conference on Near Field Optics and Related Techniques (NFO-5)*, Blackwell Science Ltd, Shirahama, Japan, **1998**, pp. 477–482.

[38] M. Bohmer, J. Enderlein, *Journal of the Optical Society of America B-Optical Physics* **2003**, *20*, 554–559.

[39] W. Schroeyers, R. Vallee, D. Patra, J. Hofkens, S. Habuchi, T. Vosch, M. Cotlet, K. Mullen, J. Enderlein, F. C. De Schryver, *J. Am. Chem. Soc.* **2004**, *126*, 14310–14311.

[40] J. Hohlbein, C. G. Hubner, *Appl. Phys. Lett.* **2005**, *86*, 3.

[41] N. Tomczak, R. A. L. Vallee, E. van Dijk, M. Garcia-Parajo, L. Kuipers, N. F. van Hulst, G. J. Vancso, in *3rd International Conference on Scanning Probe Microscopy of Polmers (SPMP 2003)*, Pergamon-Elsevier Science Ltd, Rolduc, Netherlands, **2003**, pp. 1001–1011.

[42] C.-Y., Wei, C.-Y., Lu, Y., Kim, D., Vanden Bout *J. Fluorescence* **2007**, *17*, 797–804.

[43] A. Schob, F. Cichos, J. Schuster, C. von Borczyskowski, in *3rd International Conference on Scanning Probe Microscopy of Polmers (SPMP 2003)*, Pergamon-Elsevier Science Ltd, Rolduc, Netherlands, **2003**, pp. 1019–1026.

[44] D. Woell, H. Uji-I, T. Schnitzler, J. I. Hotta, P. Dedecker, A. Herrmann, F. C. De Schryver, K. Muellen, J. Hofkens, *Angew. Chem. Int. Ed.* **2008**, *47*, 783–787.

[45] H. P. Lu, X. S. Xie, *J. Phys. Chem. B* **1997**, *101*, 2753–2757.

[46] L. Zang, R. C. Liu, M. W. Holman, K. T. Nguyen, D. M. Adams, *J. Am. Chem. Soc.* **2002**, *124*, 10640–10641.

[47] H. Yang, G. B. Luo, P. Karnchanaphanurach, T. M. Louie, I. Rech, S. Cova, L. Y. Xun, X. S. Xie, *Science* **2003**, *302*, 262–266.

[48] T. D. M. Bell, A. Stefan, S. Masuo, T. Vosch, M. Lor, M. Cotlet, J. Hofkens, S. Bernhardt, K. Mullen, M. van der Auweraer, J. W. Verhoeven, F. C. De Schryver, *Chemphyschem* **2005**, *6*, 942–948.

[49] M. Cotlet, S. Masuo, M. Lor, E. Fron, M. Van der Auweraer, K. Mullen, J. Hofkens, F. De Schryver, *Angew. Chem. Int. Ed.* **2004**, *43*, 6116–6120.

[50] M. Cotlet, S. Masuo, G. Luo, J. Hofkens, M. van der Auweraer, J. Verhoeven, K. Müllen, X. S. Xie, F. de Schryver, *Proc. Natl. Acad. Sci. USA* **2004**, *101*, 14343–14348.

[51] F. C. De Schryver, T. Vosch, M. Cotlet, M. Van der Auweraer, K. Mullen, J. Hofkens, *Acc. Chem. Res.* **2005**, *38*, 514–522.

[52] J. Qu, J. Zhang, A. C. Grimsdale, K. Müllen, F. Jaiser, X. Yang, D. Neher, *Macromolecules* **2004**, *37*, 8297–8306.

[53] M. Cotlet, S. Masuo, G. B. Luo, J. Hofkens, M. Van der Auweraer, J. Verhoeven, K. Mullen, X. L. S. Xie, F. De Schryver, *Proc. Natl. Acad. Sci. USA* **2004**, *101*, 14343–14348.

[54] M. Lor, J. Thielemans, L. Viaene, M. Cotlet, J. Hofkens, T. Weil, C. Hampel, K. Mullen, J. W. Verhoeven, M. Van der Auweraer, F. C. De Schryver, *J. Am. Chem. Soc.* **2002**, *124*, 9918–9925.

[55] R. Gronheid, A. Stefan, M. Cotlet, J. Hofkens, J. Q. Qu, K. Mullen, M. Van der Auweraer, J. W. Verhoeven, F. C. De Schryver, *Angew. Chem. Int. Ed.* **2003**, *42*, 4209–4214.

[56] J. E. Moser, *Nat. Mater.* **2005**, *4*, 723–724.

[57] Y. Shirota, Y. Kuwabara, H. Inada, T. Wakimoto, H. Nakada, Y. Yonemoto, S. Kawami, K. Imai, *Appl. Phys. Lett.* **1994**, *65*, 807–809.

[58] P. Hall, B. Selinger, *J. Phys. Chem.* **1981**, *85*, 2941–2946.

[59] M. Maus, M. Cotlet, J. Hofkens, T. Gensch, F. C. De Schryver, J. Schaffer, C. A. M. Seidel, *Anal. Chem.* **2001**, *73*, 2078–2086.

[60] W. B. Davis, M. A. Ratner, M. R. Wasielewski, *J. Am. Chem. Soc.* **2001**, *123*, 7877–7886.

[61] T. Philip, B. Volker, H. Dirk-Peter, H. Kyung-Tae, S. Markus, *Single Mol.* **2000**, *1*, 215–223.

[62] P. Horton, A. V. Ruban, R. G. Walters, *Annu. Rev. Plant Physiol. Plant Mol. Biol.* **1996**, *47*, 655–684.

[63] R. Y. Tsien, *Annu. Rev. Biochem.* **1998**, *67*, 509–544.

[64] R. MacColl, *J. Struct. Biol.* **1998**, *124*, 311–334.

[65] S. Jansson, *Encycl. Biol. Chem.* **2004**, *2*, 567–570.

[66] M. Cotlet, J. Hofkens, S. Habuchi, G. Dirix, M. Van Guyse, J. Michiels, J. Vanderleyden, F. C. De Schryver, *Proc. Natl. Acad. Sci. USA* **2001**, *98*, 14398–14403.

[67] D. Loos, M. Cotlet, F. De Schryver, S. Habuchi, J. Hofkens, *Biophys. J.* **2004**, *87*, 2598–2608.

[68] D. H. Hu, J. Yu, P. F. Barbara, *J. Am. Chem. Soc.* **1999**, *121*, 6936–6937.

[69] P. F. Barbara, A. J. Gesquiere, S. J. Park, Y. J. Lee, *Acc. Chem. Res.* **2005**, *38*, 602–610.

[70] A. J. Gesquiere, Y. J. Lee, J. Yu, P. F. Barbara, *J. Phys. Chem. B* **2005**, *109*, 12366–12371.

[71] T. Förster, *Ann. Phys.* **1948**, *2*, 55.

[72] S. E. Braslavsky, E. Fron, H. B. Rodriguez, E. S. Roman, G. D. Scholes, G. Schweitzer, B. Valeur, J. Wirz, *Photochem. Photobiol. Sci.* **2008**, *7*, 1444–1448.

[73] J. M. Serin, D. W. Brousmiche, J. M. J. Frechet, *Chem. Commun.* **2002**, 2605–2607.

[74] R. E. Bauer, C. G. Clark, Jr., M. Klapper, L. Zhi, K. Müllen, *Polym. Preprints (American Chemical Society, Division of Polymer Chemistry)* **2007**, *48*, 49–50.

[75] H. Zhang, P. C. M. Grim, P. Foubert, T. Vosch, P. Vanoppen, U. M. Wiesler, A. J. Berresheim, K. Müllen, F. C. De Schryver, *Langmuir* **2000**, *16*, 9009–9014.

[76] H. Zhang, P. C. M. Grim, T. Vosch, U. M. Wiesler, A. J. Berresheim, K. Müllen, F. C. De Schryver, *Langmuir* **2000**, *16*, 9294–9298.

[77] A. C. Grimsdale, T. Vosch, M. Lor, M. Cotlet, S. Habuchi, J. Hofkens, F. C. De Schryver, K. Mullen, *J. Luminescence* **2005**, *111*, 239–253.

[78] T. Weil, U. M. Wiesler, A. Herrmann, R. Bauer, J. Hofkens, F. C. De Schryver, K. Mullen, *J. Am. Chem. Soc.* **2001**, *123*, 8101–8108.

[79] T. Gensch, J. Hofkens, A. Heirmann, K. Tsuda, W. Verheijen, T. Vosch, T. Christ, T. Basche, K. Mullen, F. C. De Schryver, *Angew. Chem. Int. Ed.* **1999**, *38*, 3752–3756.

[80] J. Hofkens, M. Maus, T. Gensch, T. Vosch, M. Cotlet, F. Kohn, A. Herrmann, K. Mullen, F. De Schryver, *J. Am. Chem. Soc.* **2000**, *122*, 9278–9288.

[81] J. Hofkens, L. Latterini, G. De Belder, T. Gensch, M. Maus, T. Vosch, Y. Karni, G. Schweitzer, F. C. De Schryver, A. Hermann, K. Mullen, *Chem. Phys. Lett.* **1999**, *304*, 1–9.

[82] P. Tinnefeld, J. Hofkens, D. P. Herten, S. Masuo, T. Vosch, M. Cotlet, S. Habuchi, K. Mullen, F. C. De Schryver, M. Sauer, *Chemphyschem* **2004**, *5*, 1786–1790.

[83] S. Masuo, T. Vosch, M. Cotlet, P. Tinnefeld, S. Habuchi, T. D. M. Bell, I. Oesterling, D. Beljonne, B. Champagne, K. Mullen, M. Sauer, J. Hofkens, F. C. De Schryver, *J. Phys. Chem. B* **2004**, *108*, 16686–16696.

[84] T. Vosch, M. Cotlet, J. Hofkens, K. Van der Biest, M. Lor, K. Weston, P. Tinnefeld, M. Sauer, L. Latterini, K. Mullen, F. C. De Schryver, *J. Phys. Chem. A* **2003**, *107*, 6920–6931.

[85] P. Tinnefeld, K. D. Weston, T. Vosch, M. Cotlet, T. Weil, J. Hofkens, K. Mullen, F. C. De Schryver, M. Sauer, *J. Am. Chem. Soc.* **2002**, *124*, 14310–14311.

[86] T. Vosch, J. Hofkens, M. Cotlet, F. Kohn, H. Fujiwara, R. Gronheid, K. Van Der Biest, T. Weil, A. Herrmann, K. Mullen, S. Mukamel, M. Van der Auweraer, F. C. De Schryver, *Angew. Chem. Int. Ed.* **2001**, *40*, 4643.

[87] M. Maus, R. De, M. Lor, T. Weil, S. Mitra, U. M. Wiesler, A. Herrmann, J. Hofkens, T. Vosch, K. Mullen, F. C. De Schryver, *J. Am. Chem. Soc.* **2001**, *123*, 7668–7676.

[88] J. Hofkens, W. Schroeyers, D. Loos, M. Cotlet, F. Kohn, T. Vosch, M. Maus, A. Herrmann, K. Mullen, T. Gensch, F. C. De Schryver, *Spectrochim. Acta Part A-Mol. Biomol. Spectrosc.* **2001**, *57*, 2093–2107.

[89] T. Gensch, J. Hofkens, J. Kohn, T. Vosch, A. Herrmann, K. Mullen, F. C. De Schryver, *Single Mol.* **2001**, *2*, 35–44.

[90] T. D. M. Bell, S. Habuchi, S. Masuo, I. Osterling, K. Mullen, P. Tinnefeld, M. Sauer, M. van der Auweraer, J. Hofkens, F. C. De Schryver, in *Conference of the Physical Chemistry Division of the Royal-Australian-Chemistry-Institute* Hobart, Australia, **2004**, pp. 1169–1173.

[91] M. Cotlet, R. Gronheid, S. Habuchi, A. Stefan, A. Barbafina, K. Mullen, J. Hofkens, F. C. De Schryver, *J. Am. Chem. Soc.* **2003**, *125*, 13609–13617.

[92] D. J. Liu, S. De Feyter, M. Cotlet, A. Stefan, U. M. Wiesler, A. Herrmann, D. Grebel-Koehler, J. Q. Qu, K. Mullen, F. C. De Schryver, *Macromolecules* **2003**, *36*, 5918–5925.

[93] H. Wolf-Klein, C. Kohl, K. Müllen, H. Paulsen, *Angew. Chem. Int. Ed.* **2002**, *41*, 3378–3380.

[94] M. Maus, S. Mitra, M. Lor, J. Hofkens, T. Weil, A. Herrmann, K. Mullen, F. C. De Schryver, *J. Phys. Chem. A* **2001**, *105*, 3961–3966.

[95] J. Hofkens, T. Vosch, S. De Feyter, F. C. De Schryver, *Macromol. Symp.* **2002**, *178*, 1–10.

[96] G. De Belder, G. Schweitzer, S. Jordens, M. Lor, S. Mitra, J. Hofkens, S. De Feyter, M. Van der Auweraer, A. Herrmann, T. Weil, K. Mullen, F. C. De Schryver, *Chemphyschem* **2001**, *2*, 49–55.

[97] R. H. Brown, R. Q. Twiss, *Nature* **1956**, *177*, 27–29.

[98] T. Basche, W. E. Moerner, M. Orrit, H. Talon, *Phys. Rev. Lett.* **1992**, *69*, 1516–1519.

[99] B. Lounis, W. E. Moerner, *Nature* **2000**, *407*, 491–493.

[100] M. Sliwa, C. Flors, I. Oesterling, J. Hotta, K. Mullen, F. C. De Schryver, J. Hofkens, *J. Phys. Condens. Matter* **2007**, 19.

[101] C. Flors, I. Oesterling, T. Schnitzler, E. Fron, G. Schweitzer, M. Sliwa, A. Herrmann, M. van der Auweraer, F. C. de Schryver, K. Mullen, J. Hofkens, *J. Phys. Chem. C* **2007**, *111*, 4861–4870.

[102] J. Bernard, L. Fleury, H. Talon, M. Orrit, *J. Chem. Phys.* **1993**, *98*, 850–859.

[103] K. D. Weston, P. J. Carson, H. Metiu, S. K. Buratto, *J. Chem. Phys.* **1998**, *109*, 7474–7485.

[104] P. Tinnefeld, D. P. Herten, M. Sauer, *J. Phys. Chem. A* **2001**, *105*, 7989–8003.

[105] W. T. Yip, D. H. Hu, J. Yu, D. A. Vanden Bout, P. F. Barbara, *J. Phys. Chem. A* **1998**, *102*, 7564–7575.

[106] I. Ruckmann, A. Zeug, R. Herter, B. Roder, *Photochem. Photobiol.* **1997**, *66*, 576–584.

[107] M. R. V. Sahyun, D. K. Sharma, *Chem. Phys. Lett.* **1992**, *189*, 571–576.

[108] G. G. Aloisi, F. Elisei, L. Latterini, U. Mazzucato, M. A. J. Rodgers, *J. Am. Chem. Soc.* **1996**, *118*, 10879–10887.

[109] C. Zenz, G. Cerullo, G. Lanzani, W. Graupner, F. Meghdadi, G. Leising, S. De Silvestri, *Phys. Rev. B* **1999**, *59*, 14336–14341.

[110] B. Kraabel, D. Hulin, C. Aslangul, C. Lapersonne-Meyer, M. Schott, *Chem. Phys.* **1998**, *227*, 83–98.

[111] R. Osterbacka, M. Wohlgenannt, D. Chinn, Z. V. Vardeny, *Phys. Rev. B* **1999**, *60*, 11253–11256.

[112] H. Rademaker, A. J. Hoff, R. Vangrondelle, L. N. M. Duysens, *Biochim. Biophys. Acta* **1980**, *592*, 240–257.

[113] C. C. Gradinaru, J. T. M. Kennis, E. Papagiannakis, I. H. M. van Stokkum, R. J. Cogdell, G. R. Fleming, R. A. Niederman, R. van Grondelle, *Proc. Natl. Acad. Sci. USA* **2001**, *98*, 2364–2369.

[114] P. Tinnefeld, V. Buschmann, K. D. Weston, M. Sauer, *J. Phys. Chem. A* **2003**, *107*, 323–327.

[115] D. S. English, E. J. Harbron, P. F. Barbara, *J. Phys. Chem. A* **2000**, *104*, 9057–9061.

[116] C. I. Richards, J. C. Hsiang, D. Senapati, S. Patel, J. H. Yu, T. Vosch, R. M. Dickson, *J. Am. Chem. Soc.* **2009**, *131*, 4619–4621.

[117] M. Haase, C. G. Hubner, E. Reuther, A. Herrmann, K. Mullen, T. Basche, *J. Phys. Chem. B* **2004**, *108*, 10445–10450.

[118] E. K. L. Yeow, S. M. Melnikov, T. D. M. Bell, F. C. De Schryver, J. Hofkens, *J. Phys. Chem. A* **2006**, *110*, 1726–1734.

[119] J. P. Hoogenboom, J. Hernando, E. van Dijk, N. F. van Hulst, M. F. Garcia-Parajo, *Chemphyschem* **2007**, *8*, 823–833.

[120] F. Cichos, C. von Borczyskowski, M. Orrit, *Curr. Opin. Colloid Interface Sci.* **2007**, *12*, 272–284.

[121] R. Metivier, T. Christ, F. Kulzer, T. Weil, K. Mullen, T. Basche, in *325th Wilhelm and Else Heraeus Seminar on Organic Molecular Solids*, Elsevier Science Bv: Bad Honnef, GERMANY, **2004**, pp. 217–224.

[122] R. Metivier, F. Kulzer, T. Weil, K. Mullen, T. Basche, *J. Am. Chem. Soc.* **2004**, *126*, 14364–14365.

[123] M. Cotlet, T. Vosch, S. Habuchi, T. Weil, K. Mullen, J. Hofkens, F. De Schryver, *J. Am. Chem. Soc.* **2005**, *127*, 9760–9768.

[124] P. Dedecker, B. Muls, A. Deres, H. Uji-I, J. Hotta, M. Sliwa, J. P. Soumillion, K. Mullen, J. Enderlein, J. Hofkens, *Adv. Mater.* **2009**, *21*, 1079–1090.

[125] C. Hippius, I. H. M. van Stokkum, M. Gsanger, M. M. Groeneveld, R. M. Williams, F. Wurthner, *J. Phys. Chem. C* **2008**, *112*, 2476–2486.

[126] G. McDermott, S. M. Prince, A. A. Freer, A. M. Hawthornthwaitelawless, M. Z. Papiz, R. J. Cogdell, N. W. Isaacs, *Nature* **1995**, *374*, 517–521.

[127] G. D. Scholes, *Annu. Rev. Phys. Chem.* **2003**, *54*, 57–87.

[128] R. Jimenez, S. N. Dikshit, S. E. Bradforth, G. R. Fleming, *J. Phys. Chem.* **1996**, *100*, 6825–6834.

[129] G. Schweitzer, R. Gronheid, S. Jordens, M. Lor, G. De Belder, T. Weil, E. Reuther, M. Mullen, F. C. De Schryver, *J. Phys. Chem. A* **2003**, *107*, 3199–3207.

[130] R. Gronheid, J. Hofkens, F. Kohn, T. Weil, E. Reuther, K. Mullen, F. C. De Schryver, *J. Am. Chem. Soc.* **2002**, *124*, 2418–2419.

[131] S. M. Melnikov, E. K. L. Yeow, H. Uji-I, M. Cotlet, K. Mullen, F. C. De Schryver, J. Enderlein, J. Hofkens, *J. Phys. Chem. B* **2007**, *111*, 708–719.

[132] T. Weil, E. Reuther, K. Mullen, *Angew. Chem. Int. Ed.* **2002**, *41*, 1900–1904.

[133] V. Wintgens, P. Valat, J. Kossanyi, A. Demeter, L. Biczok, T. Berges, *New J. Chem.* **1996**, *20*, 1149–1158.

13

DEGRADABLE DENDRIMERS

MARC GINGRAS

CNRS, Aix-Marseille II University, UPR 3118 Interdisciplinary Center on Nanosciences of Marseille, 163 Av. de Luminy, Case 913, 13288 Marseille Cedex 09, France

The present chapter is a complement to a previous review on cleavable dendrimers [1] and stands from an academic–industrial research activity on dendrimer–drug conjugates for drug delivery [2]. The purpose of this review is to emphasize a new stage in the chemistry of dendrimers: their degradation, which came after many efforts to assemble them in an efficient manner. A well-documented survey of the literature will be presented for the first time in this fastly expanding field by using some selected examples and applications. The degradability of dendrimers is a modern topic, in our era of green chemistry and environmental concerns. The current progress in this field suggests highly important uses to come from degradable dendrimers, especially in life sciences, in nanoscience and in materials chemistry.

13.1 INTRODUCTION AND HISTORICAL PERSPECTIVES

Dendrimers are considered as a modern and an elegant class of hyperbranched macromolecules in polymer sciences [3]. They resemble tree-like molecular architectures built from repetitive monomers or chemical units with branching points, radially connected to a template core. Among their main characteristics are the possibilities for incorporating a large number of functions (outward and/or inward) or some chemical units and exploiting a dense chemical environment and some internal voids, which could serve to encapsulate some molecular entities (see below) (Figure13.1).

Designing Dendrimers, Edited by Sebastiano Campagna, Paola Ceroni, Fausto Puntoriero.
© 2012 John Wiley & Sons, Inc. Published 2012 by John Wiley & Sons, Inc.

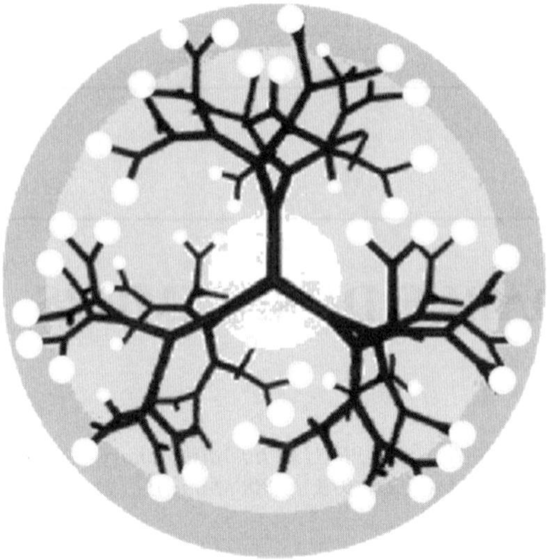

FIGURE 13.1 A dendrimer with its branching units, its core, and its internal voids. Modified image taken from: http://www.nisenet.org/image/dendrimer (unknown author)

A hydrodynamic volume that could reach the nanometer scale is dictated by the solvation conditions [4]: the core, the monomer units, the branching points, the dendrons, the generation numbers, and the peripheral or the internal functions. Their synthesis and a control of the "rigidity" of the chemical backbone units (with a restricted degree of freedom) render possible a modulation of the shapes and their persistence. As a consequence, they became some interesting nanosize objects and ideal building blocks in the elaboration of higher chemical structures [5–7]. They are currently found as nano objects, as nano templates in numerous applications, including in nanomedicine. The links between dendrimers and nanochemistry in the construction and in the manipulation of higher and more functional nano objects and devices became an evidence over the recent years [8].

In contrast to most polymers, dendrimers are often considered as well-defined, discrete macromolecules with a high degree of molecular uniformity and monodispersity. In theory, those macromolecules can possess a defined molecular structure and a precise chemical formula especially at lower generations, like many small molecules. Those features make them attractive for an easier characterization and further evaluations [9] of their tailor-made properties, after building them from numerous established synthetic strategies. A strong advantage over polydispersed polymers is a better control of dispersity and less dependence on the polymerization conditions. The latter are sometimes not easily reproducible, but needed for polymer therapeutics [10]. This fact is especially important in nanomedicine and in drug delivery (see below). The loading of active ingredients could also be more important and precise, and the conjugation offers a better control over a slow drug release, a better tissue targeting, and even a passive absorption in specific interfaces [11]. Moreover, they are often solvent soluble macromolecules with a lower viscosity than for linear polymers, with a somewhat controlled hydrophilic, lipophilic, or amphiphilic character, without too much defects when made from a convergent synthesis. Dense outward (and/or inward) functions generate specific properties and utilities such as a polyvalent ligand, a molecular receptor, a molecular confinement, a molecular translocator (vector), and so on. Multivalency, adhesive processes, amplification, recognition, additive or cooperative effects have all been demonstrated. The expression "dendritic effects" was coined to illustrate those specific cooperative behaviors coming from a multiple functionalization [12].

Historically [13], the discovery of dendrimers occurred at the end of the 1970s via a repetitive sequence of addition reactions toward a molecular growth and branching [14], but the reports of the first discrete dendrimers with a somewhat monodisperse high molecular weight occurred in the 1980s [15–17]. They remained for many years a curiosity with a real synthetic challenge. Their purity and the establishment of their structural integrity were questioned many times among the scientific community. As a consequence, those exigencies for a structural proof and purity indirectly helped to improve our modern analytical methods (such as mass spectrometry and other biophysical techniques) [18] and to the structural characterization of mid-size macromolecules, which were at the frontier of polymer sciences, organic chemistry, or chemical biology (a field sometimes called macromolecular organic chemistry). For about two decades, many efforts in dendrimer chemistry have been primarily focused on the synthesis, the evaluation of the properties, and the quest for useful applications in various fields: organic, supramolecular chemistry, materials and polymers science, life sciences, and nanosciences [19].

After many years of dendrimeric construction (even stated as "Lego© chemistry") [20,21], a few scientists realized their potential for encapsulating guest molecules or functions [22], such as in the classic example of a "dendritic box" [23]. On the other hand, it was already noticed that solvent molecules occupied the internal dendritic voids of those hyperbranched macromolecules. As a consequence, everything was set to compare dendrimers to some "molecular sponges" and nanocarriers of smaller guest molecules. In biomimicry, it is a situation similar to a substrate being bound to the catalytic site of an enzyme, with a very specific internalization and chemical environment (change of solvation, proximity of functional groups, molecular ordering, etc.) [24,25]. It was the beginning of a new era demonstrating that dendrimers are not only aesthetic molecules but also they could possess intriguing properties, as vectors or as translocators. Among the pioneering applications of encapsulation, were some assays for the binding of nucleic acids (DNA) and a delivery inside cells [26–28]. Dendrimers could be compared to a nanocarrier of polynucleic acids helping for crossing a cell membrane [29]. From the increasing popularity of gene therapy, dendrimers have been quickly noticed as a powerful and interesting class of DNA, si-RNA, and RNA vectors.

After such an exponential progress in the mid-1990s, the field of dendrimer chemistry also took a leading research direction toward some biological applications, as mentioned in earlier reviews [30–37]. Recent general reviews indicated a similar trend [38–44], and new books [3]. Those large, complex, and multifunctional entities could tackle the field of important biological objects, sometimes at a similar molecular scale, and also because of their capacity for encapsulation. At a first glance, it seemed simpler to load molecules inside commercial or tailor-made dendrimers. After all, many efforts were put forward to simplify the synthesis of dendrimers in the quest for future applications, with a need for a larger scale production. The (bio)degradation or the covalent cleavage of dendrimers was not seen as a priority until the beginning of the twenty-first century.

From a historical perspective, the field of degradable dendrimers followed a period after some major studies were achieved in the polymer field, including degradable polymer therapeutics. Such a degradable model for polymers was originally proposed by Ringsdorf in 1975 [45]. Other important reviews on that topics are available [46–48]. The cleavage of dendrimers started during their early days when the scientific community requested more structural evidences of the dendritic structures. The involvement of modern mass spectrometry techniques (such as MALDI-TOF, ESI, etc.) often present a major parent dendritic molecular ion. However, in a few cases, it could fragment to a smaller ion. It remained a pure analytical observation of a degradation. It was also found during the synthesis of PAMAM dendrimers that a retro-Michael addition could lead to methyl acrylate and an amine [49]. Similar observations of dendrimer defects can be observed during the synthesis of PPI dendrimers, as indicated by some MALDI-TOF [50] or ESI [51] studies. An interesting analytical study using ^{1}H, ^{13}C NMR, and GC, with a G0.5 PAMAM (EDA-core) dendrimer in methanol at various temperatures from -15 to 50°C concluded for an equilibrium between a Michael, and a retro-Michael addition in solution starting even at 4°C (Scheme 13.1) [52]. Higher temperatures shifted this equilibrium toward the retro-Michael products and an increased degradation. It is a demonstration that care must be taken against a degradation upon PAMAM storage for a long period of time in order to avoid some changes in properties, such as for vectorization. All in all, the degradability of dendrimers was clearly noticed many years ago but it was not until the mid-1990s that some defined applications suggested that their degradation could offer new perspectives in dendrimer chemistry.

In 1993, an early report mentioned the degradation of a hyperbranched polyester dendrimer in the quest for a better characterization under a strongly basic conditions (NaOH, THF, EtOH, H_2O, reflux) [53]. After protecting the chain ends by stable methoxy functions, the quantification of the recovered units permitted to evaluate the degree of branching. However, the uncontrolled and irregular nature of the polymerization provided a polydispersity of 2.1, which is not representative of a monodisperse, discrete, macromolecule (Scheme 13.2).

The first enzymatic degradations of dendrimers were disclosed in 1996, after a depolymerization of a chiral polyester dendrimer made from (R)-3-hydroxy-butanoic

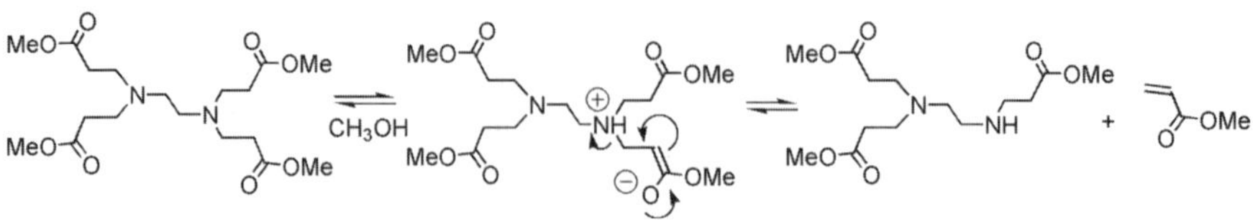

SCHEME 13.1 Degradation of PAMAM dendrimers at temperatures above 4°C by a retro-Michael reaction [52].

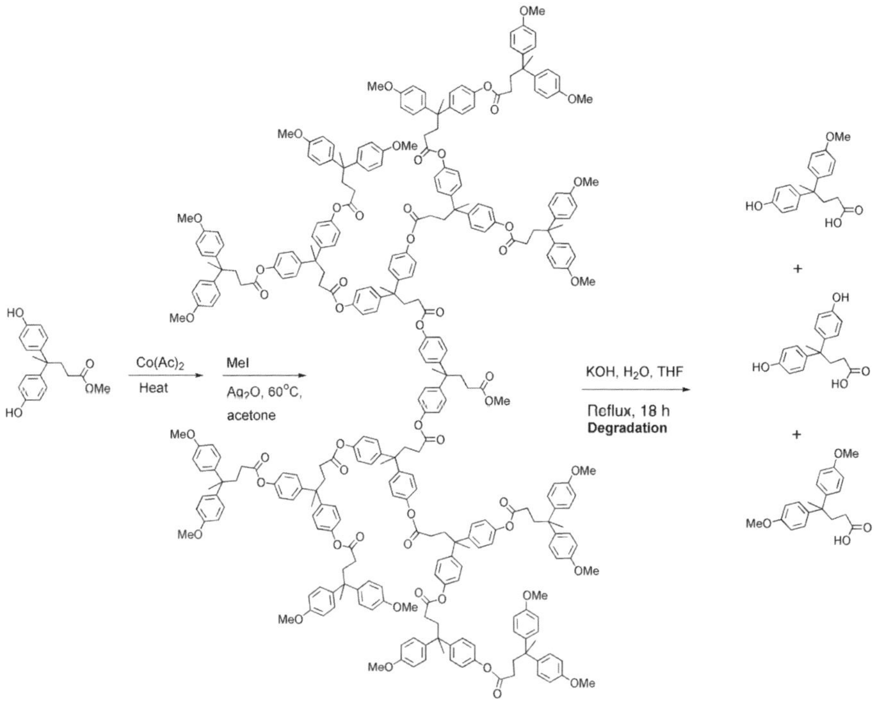

SCHEME 13.2 Analysis of the degree of branching by degradation of poly(aryl ester) dendrimers.

acid and trimesic acid [54]. Seebach and coworkers investigated the mode of cleavage of several dendrimers using various types of enzymes (Scheme 13.3). More details will be given elsewhere in this chapter. In the same year, followed an *in vitro* gene delivery by some degraded poly(amidoamine) (PAMAM) dendrimers, called "fractured dendrimers," obtained after a heat-treatment in a solvent [55]. As described above, we may suppose that many retro-Michael additions and other side-reactions took place. Some degraded PAMAM dendrimers were more effective than the native dendrimers. Starting from the experience on a fully degradable hyperbranched poly

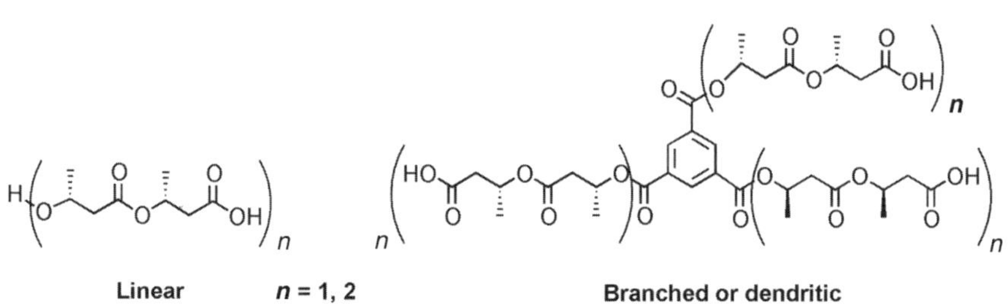

SCHEME 13.3 Enzymatic degradation of (R)-3-hydroxy-butanoic acid derivatives and dendrimers.

(organosiloxane) polymer in 1995 [56], a degradable polysiloxane dendrimer was embedded into a methacrylate resin and an acidic hydrolysis allowed the degradation of the dendrimer for making some nanocavities, in an analogy to molecular imprinting [57]. Those four events made the early experimental history of a real intention to use the degradable properties of a dendrimer or a hyperbranched polymer.

Early in the 1920s, only a few reports existed on degradable dendrimers. Some new interests slowly grew up in 2000 from the first photorelease of some molecular species by a photo-induced covalent fragmentation where dendrimers could be seen as a "covalent reservoir" of those species [58]. McGrath reinforced the concept of "cleavable dendrimers" [59].

However, several circumstantial events almost happened in the same period, and really propulsed the field of cleavable dendrimers toward an exponential production of publications and patents, after breaking the news in many journals in 2003. Three independent research groups lead by de Groot, Shabat, and McGrath investigated some controlled chemical and biochemical means to cleave dendrimers by some well-thought strategies involving cascade reactions and some initiators or triggers (see below). A few expressions were put forward in many headlines of journals: exploding dendrimers [60], self-immolative dendrimers [61], and self-destructive dendrimers [62].

In 2007, the first compilation of literature data dedicated to degradable or cleavable dendrimers was published by us [1]. Nowadays, a literature search (ISI Web of Knowledge©, all databases) estimated to be more than 135 publications the results of a search with the terms "degradable or biodegradable dendrimers," where degradation and covalent cleavage of chemical bonds were declared as important factors. We can roughly estimate to more than 1070 manuscripts (excluding patents) using the search terms "dendrimers and drug delivery," including 210 reviews with 25% (52) of the production in 2008 and only a few ones published before 1999 (4). The number of manuscripts exponentially increased from 1993, when a transfection of DNA with PAMAM dendrimers was demonstrated [26]. Solely in 2008, 208 publications (20%) were produced, compared to 25 in the year 2000 (2.3%). The vast majority of the studies concerned the following domains: pharmacy, biochemistry, chemistry, materials science, and oncology. However, the number of publications describing the release of drugs by covalent bonds dissociation via a dendrimer–drug conjugate is much less important, compared to a noncovalent drug delivery via an encapsulation (or an inclusion complex). As a comparative study [48], even if dendrimers are excellent vectors and vehicles for drug transportation, the field of polymer therapeutics [10,63–67] largely bypassed the number of publications in dendrimer-based delivery (roughly 7000 versus 1070). Relative to dendrimers, the field of degradable polymers already acquired a certain scientific maturity many years ago [68].

The development of degradable dendrimers could only progress after mastering the synthetic strategies and the methodologies for making them, where many chemical reactions for assembling dendrimers were specifically chosen for their simplicity, for a neglected amount of by-products, for an easy purification and for some high-yielding processes over multiple bonds formation, without too much dependence on steric effects (hopefully). The convergent, the divergent methods, and other strategies for building dendrimers needed to be explored in depth. Having now in hand some dendritic materials, degradability has been a next leading concept, especially when green chemistry has become so popular. It already represents a modern topic having lead to some major advancements in a varieties of applications: tissue engineering, nanomedicine, drug delivery, photolithography, release of fragrances, and so on.

In a nut shell, the main objective of this chapter is to highlight an important new stage in the development of dendrimers by providing a critical summary of the state-of-the-art on cleavable or degradable dendrimer-conjugates using *covalent bonds dissociation* and to provide a brief survey of some applications. Supramolecular encapsulation of guest molecules or any supramolecular releases of a smaller guests from a dendritic inclusion complex will not be at the center of the discussions in this chapter, but will be mentioned when necessary. Due to their recent importance and a fast development, this review will mainly focus on discrete dendritic entities, with a well-defined monodisperse molecular structure, in such a way that polymers, dendronized or hyperbranched polymers [69–73] or related dendronized supermolecules [74–76] with a somewhat higher polydispersity will voluntarily be put aside in our survey of the literature. However, some analogies and comparisons to those domains will be made when needed, for instance, when comparing molecular masses, sizes, and architectures upon their properties or their biological activities.

13.2 TOWARD A "CONTROLLED" DENDRIMER-CONJUGATE DISASSEMBLY

13.2.1 A Few Modes of Dendrimer-Conjugate Disassembly

Many creative thoughts involving dendritic degradation could be imagined. As shown in Figures 13.2–13.8, we selected the most common dendrimer-conjugate disassemblies, which could be conceptually divided into seven main modes or limiting cases [77]. This conceptual division is not exhaustive, other specific or borderline cases might be operating.

13.2.1.1 Encapsulation of a Conjugate in a Dendrimer and Disassembly of the Inclusion Complex Figure 13.2 represents some intradendritic guests, which are entrapped by using a dendritic supramolecular encapsulation (noncovalent). After a

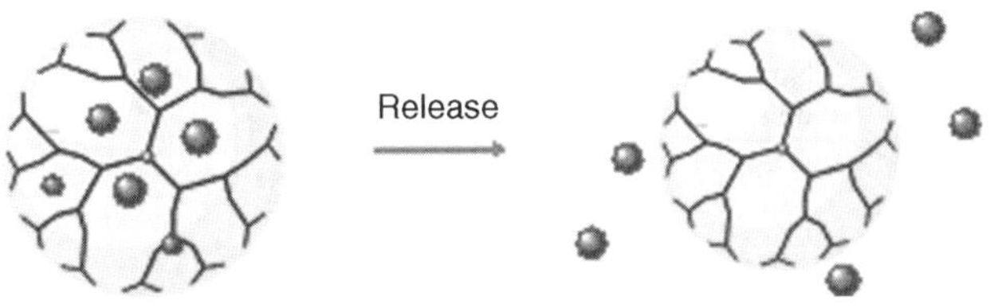

FIGURE 13.2 Encapsulation of a conjugate in a dendrimer and disassembly of the inclusion complex.

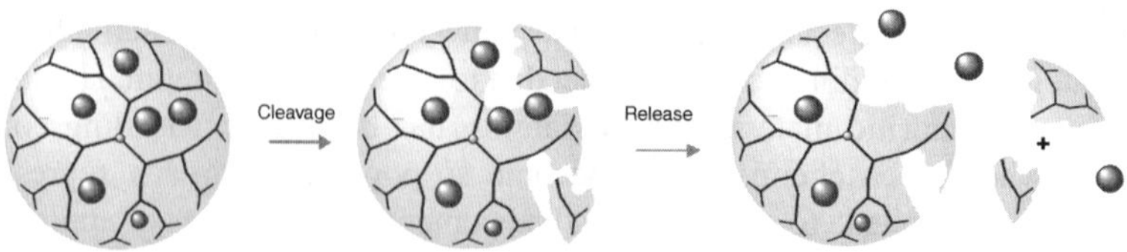

FIGURE 13.3 Encapsulation of a conjugate in a dendrimer and disassembly of the inclusion complex after cleavage of the dendritic scaffold.

disassembly of this complex, the guest will be released. The substrates (guests) may be neutral, charged (ionic), polar or nonpolar [78]. They could be located in the dendritic internal voids, in the periphery or in a peripheral shell, such as in a concept of a core–shell architecture [79].

13.2.1.2 Encapsulation of a Conjugate in a Dendrimer and Disassembly of the Inclusion Complex After Cleavage of the Dendritic Scaffold The approach in Figure 13.3 relies on the loading of a dendrimer with a substrate to make an inclusion complex and then initiate the release of the guests from a covalent dissociation of the dendrimer backbone, or the scaffold [80–82].

13.2.1.3 Supramolecular Assembly of the Conjugates on a Dendritic Surface and Disassembly of the Complex This mode of release in Figure 13.4 mainly relies on some electrostatic interactions at the surface of a dendrimer between its peripheral functions and a complementary function from the substrate. It may involve some separated charged species or some strong H-bondings such as in a salt bridge (as in a weak base–weak acid interaction) [83].

13.2.1.4 Supramolecular Assembly of Dendrimers and Entrapment of the Conjugates Followed by a Disassembly of the Architecture In Figure 13.5, a supramolecular assembly of dendrimers is formed with some substrates entrapped in an interdendritic manner [84–86]. A noncovalent dislocation could deliver the entrapped substrates [87].

13.2.1.5 Covalent Disassembly of a Dendrimer-Conjugate A covalent disassembly of a dendritic architecture releases smaller chemical "fragments" or molecular species, which could be neutral or charged (ionic). Several chemical

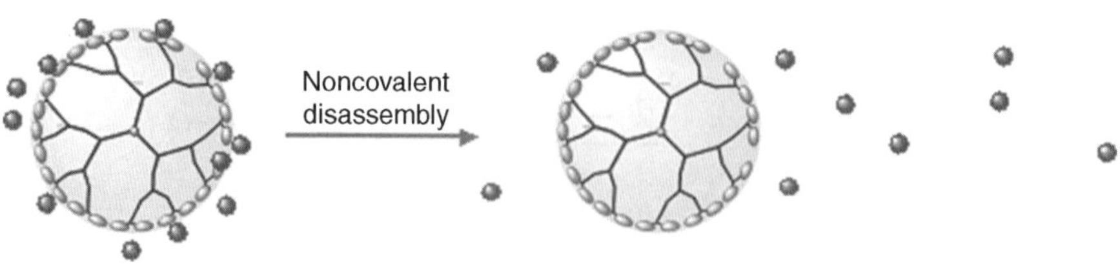

FIGURE 13.4 Supramolecular assembly of the conjugates on a dendritic surface and disassembly of the complex.

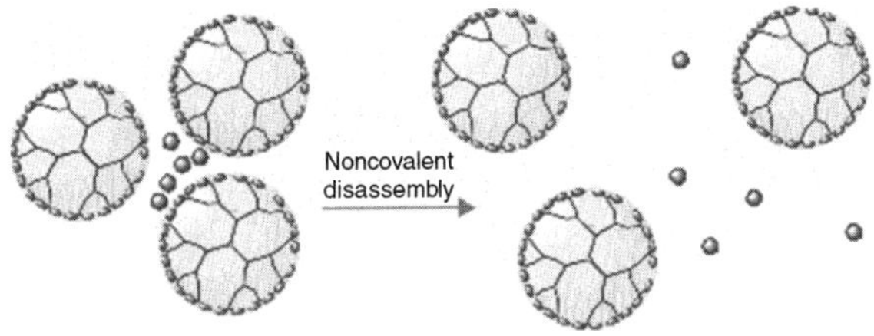

FIGURE 13.5 Supramolecular assembly of dendrimers and entrapment of the conjugates followed by a disassembly of the architecture.

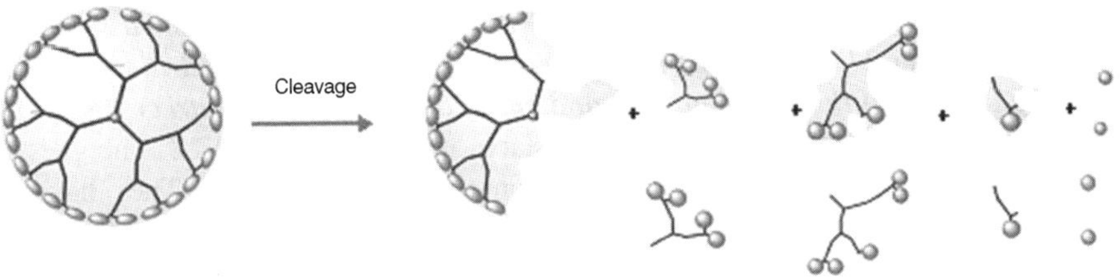

FIGURE 13.6 Covalent disassembly of a dendrimer-conjugate.

reactions must occur to break some bonds. It could also be called a chemical degradation, via some elimination reactions, and the dendrimer fragmentation could be complete or not. The degradation would usually be stepwise, as shown in Figure 13.6.

13.2.1.6 Encapsulation of the Conjugates in a Core–Shell Dendrimer and a Covalent Disassembly of the Shell will Release the Conjugates A core–shell architecture concept is presented in Figure 13.7. A covalent mode of degradation will promote the cleavage of some bonds and a disassembly of a dendritic shell. It could then initiate a supramolecular disassembly of an inclusion complex. The dendritic shell could be made from dense, tight dendritic peripheral labile functions [88–90] (such as in a dendritic box [91]) or from a degradable covalent matrix [92].

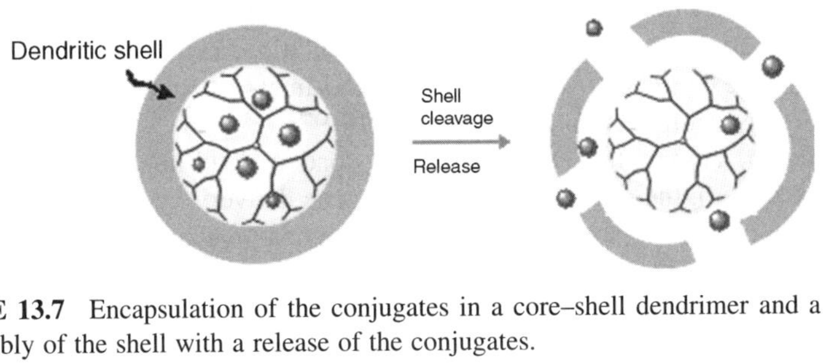

FIGURE 13.7 Encapsulation of the conjugates in a core–shell dendrimer and a covalent disassembly of the shell with a release of the conjugates.

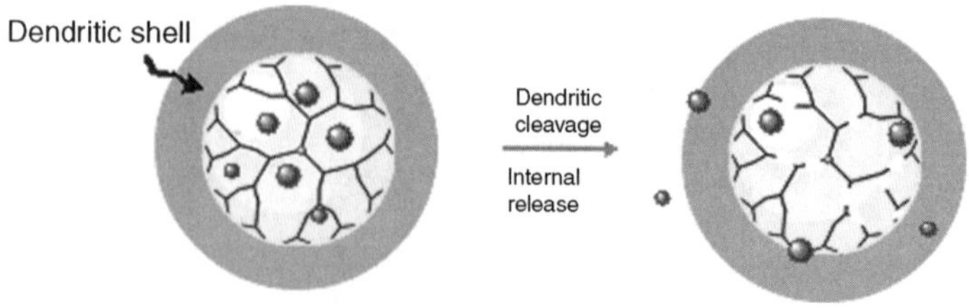

FIGURE 13.8 Covalently-bound or encapsulated conjugate in a core–shell dendrimer and disassembly of the complex or cleavage of the internal dendritic scaffold with a release of the conjugates in and out of a stable shell.

13.2.1.7 Covalently Bound or Encapsulated Conjugate in a Core–Shell Dendrimer and Disassembly of the Complex or Cleavage of the Internal Dendritic Scaffold with the Release of the Conjugates in and out of a Stable Shell (Matrix) A core-dendrimer is embedded into a chemically stable matrix in Figure 13.8. The guests can be present in this core by encapsulation (inclusion complex). They have a possibility for a slow diffusion inside and outside of the matrix. A second variant is a possibility for a covalently bound conjugate to a core-dendrimer embedded into the matrix. The cleavage of the core-dendrimer conjugate and/or the degradation of the dendritic scaffold promotes a slow release of the substrate inside and outside of the matrix cavity, after a selective degradation of the core-dendrimer [93].

In summary, the encapsulation mode *(a)* Figure 13.2 is probably the most common; especially in drug delivery; then follows mode *(d)* Figure 13.6, which is often initiated from a chemical, a biochemical, a photochemical, or a biological degradation. The other modes *(b)* Figure 13.3, *(c)* Figure 13.4, *(e)* Figure 13.7, and *(f)* Figure 13.8 are more recent and less reports could be found, although mode *(e)* (Figure 13.7) is becoming very popular for a transfection involving a change to an acidic pH, such as in a lysosome, after an endocytosis of the dendrimer-conjugate (pH 5.4 instead of 7.4 at a physiological pH). This communication will mainly focus on the cleavage of covalent bonds toward a degradation, excluding dendritic encapsulations, because many of them were already reviewed, especially in the field of drug-delivery and nano oncology.

13.2.2 Nonconvalent Versus Covalent Modes of Dendrimer-Conjugate Disassembly

In this section, we will simplify the discussion on two main modes of dissociation: by a noncovalent approach (a supramolecular encapsulation, an inclusion complex) [94] or by a cleavage of some covalent bonds (a dendrimer–substrate conjugate). A few reviews already compared both approaches [95–97].

Historically, among the early studies to load a dendrimer with a substrate (covalently or not), the use of dendrimers for gene delivery by DNA encapsulation has been one of the first applications. The commercial availability of PAMAM and PEI dendrimers plus the efficiency in the transfection often encouraged such experiments. The main advantage of encapsulating a guest molecule, compared to a dendrimer-conjugate, is that no chemical reaction is needed to ligate the substrate to the dendrimer. It simplifies the process of loading and translocating molecules. Both molecular partners need to be mixed together in a solution at a defined molar ratio. It also avoids some complicated chemoselective reactions between proper functional groups of the dendrimers and the conjugate. Additionally, those functional groups might not be present or inappropriately reactive on one or both partners. The functional and chemical complexity of some substrates (for instance, a drug) might discourage a covalent approach. As a consequence, only some specific classes of conjugates could be used in a covalent mode, along with some specific functionalized dendrimers.

On the other hand, the main drawbacks of a supramolecular encapsulation remain the supramolecular complementarity and the specificity in binding between a dendrimer and a molecular guest. In other words, there could be an incompatibility, a disfavorable thermodynamic association; it then confers a specific association constant for a dendrimers/guest assembly under specific conditions. Finding the right partners is needed and it presents a limiting factor. To facilitate and to put in practice such encapsulations, a surface modification [98] of the dendrimer by a PEGylation (functionalization by some poly(ethylene glycol) linkers) [99–103] or by other means, toward a delicate balance for hydrophobic/hydrophilic, polar/nonpolar, ionic/neutral interactions might be needed. As for other classic examples of common surface modifications, some acetylations or esterifications [104] were often used to increase the loading capacity of the resulting dendrimer. It could also render more biocompatible, less immunogenic, and less toxic PAMAM or other dendrimers, especially those decorated by some free amino functions. To this end, we should also cite a recent work of Haag and coworkers toward some "universal nanocarriers" by new dendritic core–shell architectures [105] and a review amongst many others [106]. Additionally, a chemical surface functionalization could help in the loading capacity of the resulting dendrimer but will also affect the transport ability, the water-solubility, the cellular uptake, the biodistribution (by passive or active targeting) and the degradability by the living system. The dendritic protection of the encapsulated drug (substrate) against some biological degradations could favorably modify the pharmacokinetics, the pharmacodynamics, and has a strong influence on the drug persistence in the blood stream. Those effects have been well known for many decades in polymer therapeutics and in peptide chemistry.

One of the most important disadvantages of encapsulation include the variable composition of the inclusion complex and the stability of the resulting supramolecular assembly carried out in various biological environments. Most often any perturbations of the dendritic shape, the ionic strength, pH, the conformation or the hydrodynamic volume of the dendrimers will modify the kinetics of release and a disassembly of the inclusion complex. Those factors might be hardly predictable, neither well controlled in a complex living organism.

On the other hand, the covalent approach will have its substrate tightly bound to the dendritic architecture and some chemical reactions will have to occur before its release. It confers a greater stability, and a predictability (at least *in vitro*) in the degradation of the dendrimers, combined to a selection of labile bonds, branchings, hydrophobicity, hydrophilicity, and other factors governing the degradation of dendrimers. The substrate loading is often more predictable and superior with covalent reactions. Some purifications could also be more easily envisaged. Depending on the applications, the hydrophilicity/hydrophobicity, wetting properties, and the water-solubility will often have to be controlled, depending on the substrate. A surface functionalization by a PEGylation or the use of other hydrophilic linkers, combined to the proper dendritic scaffold, might ensure a final water-soluble dendrimer-conjugate.

Overall, both approaches are complementary. The supramolecular approach looks simpler at a first glance but the instability and the less predictable thermodynamic supramolecular assembly might be discouraging compared to the covalent approach. On the other hand, the covalent approach suffers from further synthetic efforts but might provide a better control of the resulting effects. A chemical surface modification renders the supramolecular approach at a similar level in terms of synthetic efforts and purifications.

13.2.3 Some Parameters for Degradability (In Vitro)

This section will mainly focus on the hydrolytic degradability of dendrimers by chemical, enzymatic means, or by using catalytic antibodies, but not via biodegradation, which is directly related to the metabolism operating in living organisms. There are a few reports on the stability and the degradation of dendrimers in life sciences, including some studies of their metabolism and their biodegradation using radioactive markers, and some novel imaging techniques using near-infrared nanoprobes conjugated to dendritic aliphatic polyesters [107]. Additionally, most often this section will not focus on a common deprotection or protection scheme used in dendrimer synthesis, as some examples of degradation.

13.2.3.1 Stimulus or Triggers Up to the present, the degradation of dendrimers has been initiated by common stimulus for breaking chemical bonds: (a) pH variation, (b) enzymes, (c) catalytic antibodies, (d) light (single or double photon [108] absorption), and various reactions with specific reagents. They comprise some metal-mediated, redox, and thermal reactions. It is anticipated that any selective and modern activation methods for cleaving a covalent bond could be useful, especially if they are chemoselective and mild. Only recently, new dendrimers preparations employed ultrasounds and microwaves [109,110]. Degradation using those "nonconventional activation techniques" to voluntarily cleave or degrade dendrimers are presently rare in the literature.

Hydrolysis by means of a pH variation or by the enzymatic uses of a depolymerase, an esterase, a lipase, and a protease were among the first reports in the literature for cleaving some monodisperse, discrete dendrimers [54]. Well before its time in dendrimer chemistry, this report also described the first enzymatic degradations in the literature and it clearly indicated that ester functions were adequate for such enzymatic cleavages (as it was known for some degradable polyesters). Some dendritic moieties could find their way to an enzymatic catalytic site in spite of their peculiar degree of branching, their steric hindrance and their globular topology (Scheme 13.3).

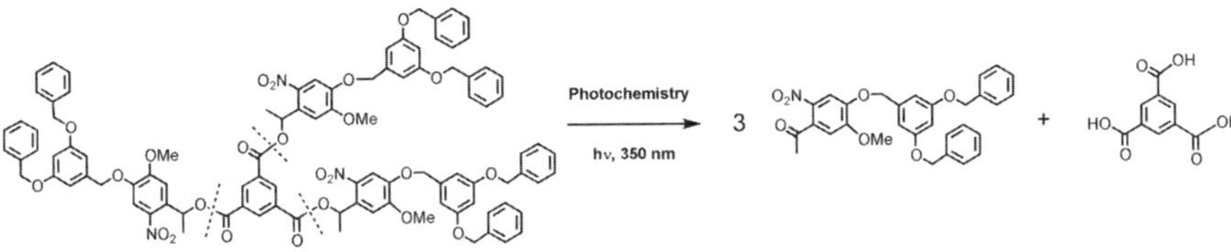

SCHEME 13.4 The first photolabile dendrimers [58].

o-Nitrobenzyl ether groups, already well-known as protective groups in synthesis, were utilized as site-specific UV-photolabile linkages for the photolytic degradation of dendrimers. The photocleavable dendrimers were first reported by McGrath and coworkers in 2000 (Scheme 13.4) [58]. More details and references will be given in the next sections. This field of study in dendrimer chemistry should find more importance in the future for any 2-D and 3-D applications in submicro- and nanotechnology (for instance, in resists chemistry and nanophotolithography or any other light-driven smart nanodevices), and also in nanomedicine and cancer therapy. Other ways to trigger some chemical degradations will be reported in details in the next sections.

13.2.3.2 Labile Linkers and Functional Groups We can list a few common and useful labile functional groups helping in the cleavage of dendrimers: esters, imino groups (imines, hydrazones, etc.), acetals and ketals, carbamates [111], carbonates [111], disulfides, amides, siloxanes, and so on. In general, it is known in tissue engineering that polyester are fastly degraded and hydrolyzed compared to polyamide or polyether [112]. Due to the large number of labile functions and to the numerous publications on dendrimers, we will only report a few examples of linkages. The author already apologizes if some works are not fully represented in this section.

13.2.3.3 Esters, Carbonates, and Carbamates Generally, esters are by far the most used and potent labile linkers under physiological conditions (pH 7.4). Most often they offer a sustained release of an active ingredient over several hours without a severe accumulation of the dendrimer or its degradation products in living organisms, because of a sufficient biodegradation rate. Some biocompatible and degradable polyester dendrimers (sometimes containing ketal units) [113] can be prepared [114,115], or were made from commercial dendritic polyols such as PAMAM-OH, Boltorn® (Perstorp) or polyglycerol dendrimers. Many of them can also be synthesized from some natural metabolites or less toxic units [116–121]. In addition to a pH variation, some enzymes (esterases, lipases, and few proteases) [122] were often successful for degrading some architectures, including dendritic or star-shape polyesters previously mentioned [54]. The work of Kambouris and Hawker used strongly basic hydrolytic conditions (NaOH, EtOH/H_2O, reflux conditions) to fully degrade a hyperbranched poly(aryl ester) dendrimer [53]. Other polyarylesters might be degradable [123].

An interesting and systematic study from Hills and coworkers on some parameters responsible for an enzymatic degradation of some hyperbranched polyesters polymers was recently published [122]. Some comparisons could be made in branched dendritic polyester chemistry [124,125]. In short, a Boltorn®-type core polyol reacted with several types of fatty acids (with various chain lengths) under variable stoichiometric ratios. The analysis of their degradation was followed after using a series of six lipases. The parameters considered were end groups, temperatures, lipases, degree of esterification, and the dendritic core degradation alone. The latter was not significantly degraded probably because of some steric hindrance and the absence of hydrophobic binding interactions. There was an influence on the type of alkane chains and their length. The degradation was dependent on the number of fatty acid chains bound to the dendritic core versus the remaining hydroxyl functions, it may be related to a good balance for hydrophilicity/hydrophobicity.

Another study from some poly(ε-caprolactone)s with a dendritic PAMAM-OH core was pursued [126]. The conclusion reinforced the results from Hill's on the delicate balance of a hydrophilic/lipophilic composition of the dendrimer, upon its degradation. A certain number of hydroxyl functions on the core helped in the degradation rate.

A PEGylated bow-tie polyester dendrimer also containing some carbamate linkages was tested against a slow hydrolytic degradation under physiological pH at 7.4 or at pH 5.0 (at 37°C) [127]. The PEG chains linked via a carbamate function to a Boltorn®-type core were probably cleaved at pH 5.0. However, at higher pH, it seems that some ester and carbamate functions could have been cleaved. The important point is a selective excision of some constitutive elements (PEG chains) from the dendrimer according to the pH and the choice of the labile functions.

Several drugs conjugated to some polyester dendrimers were tested for a slow drug release. The syntheses of some star-shape and polyester dendrimers incorporating the drug naproxen were achieved. However, the authors mentioned some enzymatic assays but they were not included in the communication [128]. Similarly, some poly(ester) star-shape dendrimers functionalized by L-Dopa at their periphery were synthesized with some hydrolyzable diester linkages [129]. L-Dopa is a drug prescribed to treat the Parkinson disease. However, the authors only mentioned a sequential degradation of this prodrug. An anticancer drug such as paclitaxel conjugated to a G4 PAMAM-OH dendrimer via a succinic acid diester linker was released with the help of an esterase [130]. Finally, an antidepressant drug, venlafaxine, was conjugated to a G2.5 PAMAM-CO_2H via some esters linkages for releasing the drug at pH 7.4 [93]. Additionally, this conjugate was incorporated to an hydrogel, made from a polyacrylamide cross-linked with PEG, for a drug diffusion through this matrix.

As shown in a Section 13.3.11 on some applications in carbohydrate chemistry, some examples of hydrolytic cleavages using a Boltorn®-type core as a soluble support for carbohydrate synthesis and combinatorial chemistry were reported. Another example on the cleavage of some carbohydrate units from a solid phase was also reported with some photocleavable *o*-nitrobenzylic groups (see Section 13.3.11).

Among other researchers, Shabat and coworkers made uses of carbonate linkers incorporated in some cleaving units for a cascade disassembly [131,132]. Similarly, carbamates were also often described in the literature [133]. The latter two functions are complementary and as useful as the ester linkage in the degradation of dendrimers [134].

Amides and Mixed Functions Some hyperbranched poly(ester-amide)s made from gallic acid and some natural amino acids were investigated toward a hydrolytic (pH variation) or an enzymatic degradation by proteinase K. They could be degraded by both means *in vitro* at 37.5°C [135]. A recent study with amide, carbamate, and ester linkers was reported. A poly(L-lysine) dendrimer was used as a template and it was decorated with PEG chains, and 20(S)-camptothecin (CPT) as a common anticancer drug. PEG chains were linked using a carbamate function and CPT by using esters [136]. An *in vitro* hydrolysis in two buffers at a physiological pH 7.4 and pH 5.0 indicated a slow release of CPT over time; the drug seemed to be selectively cleaved over carbamate and amide functions. In human plasma, a similar release was observed at pH 7.4.

In a similar type of multifunctional peptide dendrimer containing only amide functions, a methotrexate drug (a common anticancer drug) was bound to a poly(L-lysine) dendrimer core, decorated by some PEG chains [137]. The latter were linked via some amide functions. A thorough and extensive *in vivo* study was achieved and the dendrimers were described as biodegradable.

Some degradations were reviewed for many (peptide) dendrimers containing some amide linkers [138], however, this function is sometimes too stable for a sustained drug delivery over time [139]. Nevertheless, it has found its way to deliver drugs from a PAMAM or a polyglycerol [140] scaffold.

Amir and Shabat used phenylacetamide function and a penicillin-G-amidase to trigger a cascade reaction to the complete degradation of a dendrimer in his concept of a "self-immolative dendrimer" [141]. More details and references will be given in Section 13.2.4.

Another work used the conjugation of a PAMAM or PEI core functionalized by a shell of short poly(L-glutamic acid) as a dendritic biodegradable carrier, a folic acid (as a target unit) and a near-infrared dye (as an imaging unit) [142]. A conjugation of methotrexate and PAMAM dendrimers was achieved at a low drug payload using amide linkages [143].

Acetals and Ketals (pH Responsive Dendrimers) Because acetals can usually be cleaved under very mild acidic conditions, they have found their way as useful labile linker for degradable polymers [144,145] and dendrimers. A study of the cleavage of the acetal linkages for drug delivery was published [146]. For an instance, the release of an anticancer drug such as doxorubicin can be triggered by a change of pH [147]. A similar approach was used, but with a dendritic Boltorn® core linked to a PEG chain for making micelles. Some acetals were installed at the periphery for making a shell-type structure. Their hydrolytic cleavage under acidic conditions makes those dendrimers pH sensitive for the release of encapsulated doxorubicin [148]. A study of the degradation at pH 5.0, 7.4, and 9.0 (37°C, water:methanol 1:1) of such a dendronized structure (fourth generation) indicated from SEC analysis that the molecular weight of the dendrimer steadily decreased with time under neutral, and basic conditions but no change in the polymer MW was observed under acidic conditions. The conclusion was that the degradation mechanism involved random ester cleavages [149].

A pH responsive polyglycerol dendrimer was acetylated and used as a supramolecular carrier of a Congo red dye. The latter inclusion complex was stable at basic or neutral pH for months but the molecular cargo was released at pH <3 when the acetylated shell was cleaved [90]. Ketals have been less often reported in the field of cleavable dendrimers. They were interesting intermediates and protective groups in the synthesis of "bow-tie dendrimers" [150].

Imino Functions (pH Responsive Dendrimers) Hydrazones were also useful as a pH-sensitive linker for the release of methotrexate as an anticancer drug or folic acid from some poly(aryl ether) dendrimers [151]. Another work using Boltorn®-type polyester dendrimers also indicated that doxorubicin could be released using hydrazone linkers [115,152].

Imines are usually easily cleaved under acidic and almost neutral pH (especially alkylated imines). Haag and coworkers used water-soluble pH responsive dendritic core–shell nanocarriers of polar dyes based on the covalent attachment of PEG chains to a PEI core dendrimer via some imino functions. This model helped to release the encapsulated guest by cleaving the PEG chains [89]. In a similar manner, they used imines with long alkyl chains with a Congo red dye as a guest. The sensitivity to pH was higher than for acetals [90].

Disulfides (Redox Responsive Dendrimers) The disulfide linker has been reviewed in the field of drug delivery, albeit without the involvement of dendrimers [153]. It has already found several uses in medicinal chemistry and related biological fields.

Very recently, disulfides appeared as a useful linker in dendrimer chemistry that can be cleaved under mild reductive conditions with thiols (glutathione, cysteine, dithiothreitol, etc.). Most examples involved PAMAM-cystamine dendrimers. For instance, some low molecular weight dendrons were used to bind DNA and release their cargo by an *in vitro* reduction with dithiothreitol (DTT), as a mimic to glutathione (GSH) found in the cytoplasm of a cell [154]. In this way, it could provide a new generation of vectors for gene delivery with DNA. A PAMAM dendrimer conjugated to *N*-acetylcysteine (NAC) via a disulfide linkage was reported [155]. NAC is a neuroantiinflammatory and antioxidant therapeutic. The release of NAC after reacting with cysteine and GSH was successfully investigated at various pH. The release was faster at physiological pH 7.4 compared to lysosomal pH 5. However, bovine serum albumin containing a cysteine residue did not cleave the conjugate. In a similar study, *N*-acetyl-L-cysteine (NAC) was conjugated to PAMAM dendrimers as cationic G4-NH_4 or anionic G3.5-CO_2H PAMAM dendrimers and tested against intracellular GSH. NAC was released within 1 h at GSH 10 mM. The antioxidant properties were also followed as an indirect measure of NAC release in microglial cells. A PAMAM dendrimer was decorated with PEG and cysteines that could form disulfides at their periphery [156]. Inclusion complexes were tried with rose bengal.

Cystamine core PAMAM dendrimers are commercially available. Their cleavage of the disulfide function produces two thiolated dendrons. Curcumin, a spice in South Asian cooking, was found to have anticancer, antiinflammatory, and antioxidant properties. It was conjugated to a cystamine core PAMAM dendrimer [157].

Reduction and cleavage of the disulfide lead to two thiolated dendrons having conjugated curcumin. Additionally, the hydrolysis of the PAMAM–curcumin conjugate could in theory release the spice by a cleavage of some ester functions. Cystamine core PAMAM was also cleaved and the thiolated dendrons were linked to biotin and MRI (magnetic resonance imaging) and fluorescent agents [158]. During the preparation of some antibody-conjugated PAMAM dendrimers, a disulfide function was used [159].

Beside the use of PAMAM, melamine-base dendrimers incorporated a disulfide linkage to a fluorescent dansyl group were reported [160]. The rate of exchange of thiol-disulfide with dithiotreitol was dependent on the structures. Disulfides near the core experienced a slower exchange because of steric hindrance and peripheral disulfide functions exchanged more quickly.

Siloxanes and Other Silicon Derivatives Siloxane functions and some derivatives are known to be biocompatible and cleavable under specific pH [161]. Among some works, a hyperbranched poly(siloxysilane) degradation was already reported in 1995 [56] and other degradations of hyperbranched poly(silyl ester)s [162], a dendritic carboranylsilanes [163], and a hybrid xerogel [164] were published.

Other Labile Functions Recently, 1,3,5-triazaadamantanes core dendrimers were shown to be degradable under basic or acidic conditions [80,165]. Thiazolidine linkages [166] and urea functions were also used [106].

13.2.3.4 Degree of Branching and Topology As for poly(ester) dendrimers, especially from the work of Seebach et al. [54], it seems that a cleavage near a branching point or a dendritic core is not always favorable for some enzymatic cleavages. It may be explained because of a steric effect limiting the complexation of the substrate to the enzyme and an access of the substrate (or a part of it) to the catalytic center. As a consequence, the hydrolytic degradation started in the linear portion of the star-shape substrate, the arms [167,168]. Furthermore, one might assume that a star molecule will be enzymatically hydrolyzed at the periphery, because of less steric hindrance and a better access to the catalytic site. In another case related to the experiment of Hills and coworkers [122] the poly(ester) Boltorn®-type core dendrimer was not significantly degraded in the presence of lipases. It might be due to some difficulties to cleave sterically hindered ester functions, but they were also in the proximity of a high degree of branching (the dendritic core).

Other poly(ester-amide) dendrimers were synthesized where the core contained some amide functions. The synthesis and the degradation of some polylactide (PLA) attached to a PAMAM core was investigated. It was found that unlike linear PLA of similar molecular weight, they had a higher degradation rate which was explained by shorter PLA chains, polar end groups, and a higher hydrophilic environment (including PAMAM hydrophilic effects) [169]. Some fragrances (citronellol and L-menthol) could be released from the dendritic periphery from the cleavage of ester groups by the uses of a lipase (pH 7.2, 37°C) or a cutinase (pH 7.8, 37°C) [170]. The ester groups were selectively cleaved in some dendrimers. The important conclusion is the decreasing ability of the enzymes to interact with the substrates as the branching degree increases. It also seemed related to the rigidity of dendrimers and to the steric hindrance. Higher generation dendrimers behave poorly as a function of the generation number.

As for poly(amides) or peptide dendrimers, a general trend is that branching causes of a slower degradation. There is a case where a linear poly(L-glutamic acid) was degraded faster than a cored-PAMAM poly(L-glutamic acid) dendrimer in the presence of the protease cathepsin B [142]. Another example described a greater difficulty for proteases to cleave dendrimers. Tetrabranched peptides were shown to be more stable and even protease and peptidase resistant to degradation. Monomeric versus tetrabranched dendrimeric mimotope peptides were tested for a degradation in plasma, in serum or with trypsin and chymotrypsin [171]. Another report on antimicrobial dendrimeric peptides demonstrated the greater proteolytic stability of a dendrimeric structure compared to the linear tetra- or octapeptides in the presence of trypsin or chymotrypsin [172].

From a very limited set of data, the following trends are a general guidelines for the enzymatic degradation of some dendritic macromolecules of similar molecular weights, linkages, and repetitive monomeric composition:

(a) A cleavage in the proximity of a branching point is slower compared to a cleavage of a linear substrate because of steric effects.
(b) An open dendritic structure of low generation number or a dendron might be cleaved faster than a higher generation dendrimer because of an easier access to a catalytic site of an enzyme and a larger conformational flexibility.
(c) The rate of cleavage of an open structure of a low generation dendrimer might be slower than for a linear polymer, depending on a conformational flexibility and the hydrophilicity.
(d) Higher dendritic generations with a dense periphery in a close shell will be cleaved very slowly because of serious steric effects and a high degree of branching. A conformational flexibility is also limited.
(e) A star-shape dendrimer, which has an intermediate topology and conformational flexibility between a linear polymer and a low generation dendrimer would behave in a borderline case, but cleavage will first occur faster on a linear structure, distant from the branching core.
(f) As for proteases, a proteolytic cleavage is slower for peptide dendritic structures, compared to peptide linear structures.

Further data and model dendrimers would be needed to assert those trends solely based on a few experimental data. Other effects might overcome those factors, especially the hydrophilic/lipophilic or hydrophobic interactions, causing some important conformational issues in dendrimers and polymer chemistry.

13.2.3.5 Surface Functionalization In the recent years, the modern concepts of "multifunctional dendrimers" or "multifunctional dendritic nanodevices" were often put forward. The reader is referred to some excellent reviews [106,173,174]. That implies a profound surface functionalization of the dendrimers with some functional units toward a "refinement" of the dendritic structures for a better structure/activity relationship, cellular uptake, biocompatibility, targeting, imaging, and bioactivity. Among many examples, the work of Schlüter and coworkers used a new type of poly(amidoamine) dendrimers decorated by amines, ammoniums, t-Boc, carbobenzyloxy groups, natural amino acids, ethylene diamine, or dansyl fluorescent labels [175]. They concluded that the surface functionalization grossly influences the dendrimer toxicity and that the internal structure does not seem to play a major role despite the fact that the interior of low generation dendrimers is accessible. Another interesting nanodevice for cancer therapy was from Baker, Jr. and coworkers using PAMAM as a template for introducing units such as methotrexate (MTX—anticancer drug), fluorescein isothiocyanate (FITC—fluorescent unit), and folic acid [176].

In spite of many new complex dendritic structures, further studies will be necessary to better understand their *in vitro* hydrolysis and their *in vivo* degradation. Some general trends are proposed. For instance, a surface functionalization by a PEGylation of PAMAM [177,178] poly(L-lysine) [137], or triazine dendrimers [179] could have a drastic consequence on water-solubility, steric stabilization for preventing aggregation, cellular uptake (internalization) [180], targeting tissues, targeting organs, or dendrimer-drug persistence in the blood. In short, it could provide a positive effect while modifying the pharmacodynamics and the pharmacokinetics [181]. However, a good balance between hydrophilicity and hydrophobicity of the surface is often needed for many enzymes. All of those factors will directly affect the rate of release of a drug *in vivo* and *in vitro*.

In general, a dendrimer surface modification by a PEGylation or an esterification with long aliphatic chains usually provides a protective shell to the dendrimer and prevent the hydrolytic enzymes to work efficiently. In the recent years, the expression "core–shell dendrimers" was used in the literature to describe such architectures.

Another example of such degradable dendrimers was recently reported. They had a Boltorn®-type polyester core and were covered by a mixed shell of PEG chains and amino acids. The latter provided the anchoring points to PEG chains and for new functions at the surface (such as doxorubicin) [182]. The hydrolysis of that dendrimer was tested in a buffer at pH 7.4 and in human plasma at 37°C. Even if a degradation occurred, the results and conditions were not optimized (a reconstituted plasma at about pH 9). Another example is given by using α-chymotrypsin for cleaving a peptide dendrimer [93]. A recent work from PEGylated poly(L-lysine) dendrimers was previously mentioned, albeit under hydrolytic conditions, without the addition of enzymes [136].

13.2.3.6 Hydrophilicity and Charges It is reported that hydrophilic biomaterials are quickly degraded compared to biomaterials containing hydrophobic units. For instance, a glycolic acid–based dendrimer is known to degrade faster than a caproic acid–based dendrimer [112]. Hydration and charge effects also play a role in increasing the rate of hydrolysis by promoting some local polar environments of the dendrimer [149].

13.2.3.7 Generation Number and Size Hydrolysis of dendrimers is often faster for lower generation dendrimers. Dendrons are even degraded faster than dendrimers because of their open structure. It is found that the degradation rate depends upon the steric hindrance, and the bulkiness of the structural elements. To some extent, a dense and compact higher generation dendrimer apparently protects labile functions from hydrolysis. For instance, the synthesis of some cationic dendrimers having ammonium groups at their periphery and some ester functions as labile groups were studied. Their hydrolysis occurred at 37°C in D_2O. The important point is that the dendrons were all cleaved after a month, but some dendrimers remained. A faster degradation occurred with lower generation dendrimers from G0 to G2 [183]. Dendrons with open structures and a higher conformational mobility were thus hydrolyzed faster than dendrimers. A similar general trend was observed in the release of some fragrances by Hayes and coworkers [170]. As for some innovative means to effect the degradation of dendrimers, we can cite the work of Cordova and Janda in the surface modification and the degradation of modified, polyester Boltorn®-type dendrons [184]. Catalytic antibody 38C2 catalyzed a retro-aldol reaction for delivering some aldehyde functions by an elimination reaction. To conclude, some parameters for the *in vitro* degradation of dendrimers by a chemical or an enzymatic mean were recently defined for a few classes of dendrimers, especially for polyester dendrimers. There is a need to better define those parameters for various classes of dendrimers and linkers. Some rare systematic studies already opened the way to this field. It could provide an important and selective way to modulate the properties of dendrimers, and a better mode of using their ability as covalent nanocarriers. Additionally, if dendrimers are building blocks for assembling higher architectures, there would be a need to sculpture them in a proper way at a supramolecular level and the general rules to do it must be clear and well identified.

13.2.4 Cleavable Units for a Cascade Disassembly

More specific and complex cleavable units, also called "chemical adaptor systems" [185] incorporating some "molecular triggers" have been designed for some specific cascade or domino reactions in the cleavage of dendrimers. Those cleavable units have previously been described in the literature, before their uses in dendrimer chemistry. For instance, Monneret and coworkers published some anticancer prodrugs with similar self-immolative spacers (cleavable units) as those below, for the slow release of doxorubicin (DOX) and 5-fluorouracil (5-FU), albeit without a dendritic scaffold [186]. Another work was related to CPT [187]. The concept of cleavable units, even in a cascade reaction, was already reported in some studies on an antibody-directed enzyme prodrug therapy (ADEPT) [188] and in improved chemical strategies for a better targeting in cancer therapy [189].

An important historical event for the promotion of molecular triggers involved three independent research groups reporting some "chemical adaptors units" designed to release a molecular species (often a drug) via some cascade reactions from many contiguous bond-breaking processes. By definition, a chemical adaptor or molecular trigger is a molecular component inclusive of the dendrimer, that is especially designed to initiate a directional cleavage of the latter at a starting point in the macromolecule, after obeying to a stimuli (light, pH, metals, redox, etc.). In other words, the resulting dendrimer is seen as a smart material or a molecular device ready to fragment and to release a species when a specific signal is given. The specificity of the signal, the efficiency in the sequential bond-breaking processes and the avoidance of side reactions, are the key factors in those processes. It is important for some medical applications such as in drug delivery, where the molecular fragments are well defined and should be biocompatible, nontoxic, and easily degradable. Another important point is their synthesis, which should avoid complicated strategies with such sophisticated molecular systems.

Those concepts definitively launched the field of cleavable dendrimers and the disassembly of dendritic backbones [190]. By analogy to polymers, it could be compared to a depolymerization. Those chemical adaptors were incorporated in a specific functional sequence and they can trigger a cascade dendritic cleavage in a linear or in a geometric way, by using an external stimuli. This domino-type cleavage could be promoted by a pH variation [90,115,135,139,148,191,192], photochemistry [58,108,193], transition metals [194,195], enzymes [54,134,141,142,196] (hydrolytic enzymes of esters, amides, and carbamates, etc.), catalytic antibodies [184,191,197], a redox reaction [198], or a thermal process. Scheme 13.5 reports a brief literature survey of some chemical adaptors, it was adapted from a previous article [1].

The cleavage of dendrimers could proceed from partial removal of a few functions, removal of specific sequences leading to a functional macromolecule, removal of dendrons (or a part of them), or some advanced cleavages of the backbone toward a full (bio)degradation into simple chemical species. In most cases, the release of a molecule for a specific application was sought (such as an anticancer drug delivery). Alternatively, the remaining dendritic backbone could be of some utilities for further applications.

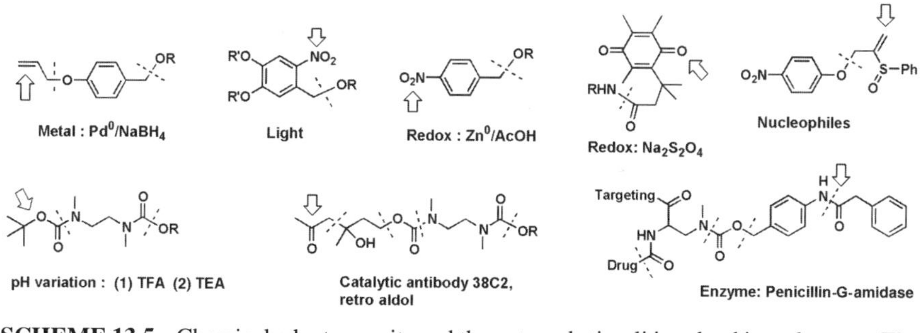

SCHEME 13.5 Chemical adaptors units and the external stimuli involved in a cleavage. The arrow represents the initiating part of the cascade reactions.

In order to illustrate the concept of chemical adaptators, an example among many others is a receiver-amplifier as a self-immolative dendritic device based on some multiple triggering sites that are converging toward a final release of several identical fluorescent products, called the reporters (6-aminoquinoline). They were produced after a chemical decomposition of the dendritic backbone via some cascade reactions. An amplifying signal of the degradation was thus observed after the multiple molecular events, from the large number of fluorescent reporters released. The triggering of this series of events was achieved after the cleavage of an amide bond by an enzyme, penicillin-G-amidase [199].

13.3 A BRIEF SURVEY OF APPLICATIONS FROM DEGRADABLE DENDRIMERS

Until now many applications involving cleavable dendrimers were sought, where the degradability or the covalent cleavages are immediately responsible for the desired properties. As for the survey of applications, mainly fundamental or academic applications will be reported. As shown in Table 13.1, some recent applications of cleavable dendrimers were found as nanocarriers for drug [200] and gene delivery, pH responsive macromolecules [201], smart materials, (bio)degradable materials, scaffolds for the release of fragrance and flavors, a matrix in tissue repairs, and supramolecular nanocontainers. They were also used in diagnostic and imaging, molecular imprinting, photoresist chemistry, or a scaffold for solid-phase carbohydrate chemistry. Among all of these reports, the main applications were related to the fields of life sciences, such as in drug delivery and (bio)nanotechnology.

13.3.1 Dendrimer–Drug Conjugates to Fight Cancer

In contrast to classic low molecular weight pharmaceutical drugs, macromolecular prodrugs are now being better considered and evaluated in medicinal chemistry [48,202]. Besides polymer therapeutics, dendrimer–drug conjugates have recently found a niche in the plethora of drug delivery systems, for instance to fight cancer. The advantages of cell and tissue targeting, polytherapy, drug persistence in the blood, enhanced permeation and retention (EPR) effect (see below), drug solubilization, and drug nanocarriers brought by dendrimer-conjugates are now shaking the ground of traditional medicine in the area of controlled drug release and nanomedicine [203] (nano oncology) [204,205]. Several routes for a drug administration are also sought, including intravenous, skin penetration, and oral administration [11]. The purpose of this section is to illustrate the main applications, the properties, and the advantages of degradable dendrimers as chemotherapeutics [206]. At first, we will provide a brief summary of the many important parameters and concepts in the pharmacodynamics and pharmacokinetics of dendrimer–drug conjugates [207]. Some general guidelines and selected examples for an efficient dendrimer-based drug delivery will be presented.

TABLE 13.1 Nonexhaustive List of Applications Related to Cleavable Dendrimers and Cleaving Methods

Approx. Dates of Important Activities	Applications	Cleaving Methods	Research Activities (Rated by No. of Publications)
1996	First gene transfer with degraded dendrimers and enzymatic cleavage of dendrimers	Hydrolytic, pH dependent	*
2004	Glycodendrimers–solid-phase synthesis	Hydrolytic, photolytic, metal-catalyzed	*
2002–2006	Degradable–biocompatible materials	Photolabile, enzymatic, hydrolytic	****
2003–	Tissue repairs—ocular sealant for sutureless eye-surgery	Biodegradation, hydrolytic	*
2002–2009	Anticancer—chemotherapeutics-drug release polytherapy, drug solubilization, prodrugs, drug nanocarriers, and so on	Enzymatic, catalytic antibodies, hydrolytic, metal-mediated, photolabile, chemical trigger	*****
2003	Fundamental studies (chemical adaptor units—chemical triggers)	Redox, hydrolytic, photolabile, enzymatic, metal-catalyzed, and so on	***
2000	Chemically amplified photoresists	Photolytic	*
1995–1996	First molecular imprints with dendrimers	Hydrolytic	*
2002	Fragrance release	Hydrolytic	*
2006	Gas release	Hydrolytic	*
2001	Pesticide release	Hydrolytic	*
2003–2006	Nonsteroidal antiinflammatory drugs	Hydrolytic, enzymatic	**
1999	Steroids	Hydrolytic	*

13.3.1.1 Importance of Dendritic Architectures and Constituting Units Due to their simplified synthesis, hyperbranched or linear polymers have been used as macromolecular scaffolds for many drug release applications. However, polydispersity and reproducibility in the preparation may lead to irreproducible pharmacokinetic behaviors due to the variation of the molecular weight distribution profile and the crystallinity. Nowadays, dendrimers are a promising class of drug scaffolds because they are well-defined, monodisperse, easily soluble in many mediums, and well characterized. Moreover, many classes of dendrimers are commercially available as a scaffold (PAMAM, PEI, PPI, Boltorn,® and polyglycerol dendrimers). Additionally, their architecture became a platform incorporating multiple drugs, targeting and imaging units, and solubilizing groups. Size, generation number, and shape of dendrimers will have a tremendous effect on the toxicity, their *in vivo* degradability, and their rate of clearance in blood and tissues. Overall, the main objective is to use low toxicity dendrimers with a high therapeutic index with common drugs already on the market.

13.3.1.2 Biocompatibility and Toxicity of Dendrimers Biocompatibility was defined as "the ability of a material to perform with an appropriate host response in a specific application" [208]. A biocompatibility and a low toxicity must be insured by the peripheral functionalization (end groups) of the dendrimers, a sufficient rate of degradation, and by a renal elimination, which depend on the dendrimer generation, size, shape, and branching [209]. A delicate balance between clearance, degradability, biodistribution, drug persistence, and drug accumulation is of high importance in the pharmacodynamics and the pharmacokinetics. As for degradability and release of a drug, a cleavable dendritic backbone often comes from the hydrolysis of some linkages like amides, esters, and carbamates. The last two are preferred because of their higher hydrolytic rates. For a low toxicity of the degradation products, some constitutive units of the dendrimers are often natural metabolites (succinic acid, glycerol, etc.). End groups and charges also play a major role in the toxicity. Capping of dendrimers with PEG chains, glycerol, acetyl groups is often beneficial. PEG chains often offer nonimmunogenic, biocompatible, water-soluble, and less toxic dendrimers. As a general rule, neutral (acetyl, acetamide, glycidol, glycerol, PEG, benzyloxycarbonyl, etc.) or anionic surface groups (carboxylates, sulfonates, phosphates, etc.) in dendrimers are preferred from cationic ones (free amines, some amino acids, guanidine, ammonium groups, etc.) for less toxicity, for avoiding membrane cell disruption and cell lyses [208]. Neutral or charged surfaces are often responsible for a strong interaction with the membrane, as a first step toward drug intake by a permeable cell membrane or by other mechanisms of cellular internalization. A few recent systematic studies on the variation of the dendritic end groups versus toxicity were achieved by the groups of Schlüter [175] and Simanek [210].

13.3.1.3 Cellular Uptake of Dendrimer-Conjugates In contrast to low molecular weight drugs commonly used on the market, most dendrimers and macromolecules are believed to be internalized in cells via the endocytic pathways without involving a membrane permeability mechanism, although some exceptions are known. As studied by AFM and fluorescence spectroscopy, on some lipid bilayers and on Rat2 cells, some cationic seventh-generation PAMAM dendrimers induced membrane damages and the formation of holes [211,212]. Until now, a few systematic studies were achieved but it is known that the surface functionalization, the conjugate architecture, the molecular size, and the branching would be determinant for a cellular uptake mechanism by an endosomal process. Confocal fluorescence microscopy has been of some help in the understanding of cellular uptake and cell trafficking of the prodrug. Duncan and coworkers studied the effect of linear or dendrimeric structures of PAMAM and PEI on the binding, the endocytic capture, and the intracellular trafficking with a Oregon green-conjugate [213–215]. Among other works, studies on surface-modified PAMAM dendrimers with long chain esters (lauroyl chains) or propanolol were studied with FITC [104]. Other effects of surface functionality with charged species were undertaken on the intracellular trafficking with A549 lung epithelial cells [98]. El-Sayed et al. studied G0–G4 PAMAM dendrimers with a fluorescent label across some epithelial and endothelial barriers [216]. They found that the increase in molecular weight and size resulted in an exponential increase of the extravasation time across a microvascular endothelium.

13.3.1.4 Improving the Therapeutic Index of a Drug: Targeting, Biodistribution, and Drug Persistence The targeting delivery of a drug prevents some side effects of a chemotherapeutic drug toward healthy tissues and organs. It increases the therapeutic index of an anticancer drug. As for targeting tumors, it was demonstrated many years ago by Maeda and Matsumura [217,218] the concept of enhanced permeability and retention of a drug in tumors (called EPR effect) as a passive targeting mode based on the physical size of a macromolecular prodrug (5–100 nm) versus the large inter-endothelial junction in cancerous tissue (estimated to be 500 nm) and its prolonged residence time. In his original article, Maeda reported a fivcfold increase in tumor uptake by a polymeric chemotherapeutic. An EPR phenomena in tumors is physically favorable compared to healthy tissues having smaller interendothelial junction (5–8 nm), if the macromolecular size is sufficiently large for long blood circulation time and a decrease rate of renal filtration and clearance (MW about 3040 kDa). As for the rate of clearance, it is mainly determined by the size, the shape, and the conformation of the dendrimers versus the size and geometry of the glomerulus pore of the kidneys. In other words, there is a physical accumulation of the drug in a tumor because of a less efficient lymphatic drainage and more permeable endovascular tissues of a tumor, permitting the uptake of a larger prodrug, in contrast to healthy tissues. The combined factors affecting clearance and biodistribution in organs have been recently reviewed [219,220]. In summary, appropriate molecular weights, sizes, molecular conformation, flexibility, and branching can increase blood circulation time and the therapeutic index of a drug. Fréchet and coworkers proposed the following general guidelines based on a limited number of *in vivo* studies:

(a) Increasing the molecular weight of macromolecules diminishes the renal filtration while increasing the persistence of a prodrug in the tumor and in the blood stream.

(b) An increase of branching diminishes the renal filtration and increases the persistence of the dendritic prodrug.
(c) Rigid, globular macromolecules decrease the renal filtration and increase the persistence of the prodrug in the blood stream compared to flexible and elongated (chain) polymers.
(d) More rigid, tubular macromolecules could have a better pore penetration and a higher accumulation in tumors.

As for an active targeting, several strategies have been put forward from the conjugation of some monoclonal antibodies or from some specific dendritic surface functionalization with some targeting units, favoring cancer cell adhesion, recognition of the prodrugs and cell internalization. Among some common targeting units are folic acid, biotin [221], neurotensin, and a cyclic peptide (cRGD) and some glycosides [208]. These units (also called epitopes) are based on a specific recognition of some receptors often overexpressed in some cancer cell lines (for instance, folic acid is overexpressed in brain, human breast, and lung cancers).

13.3.1.5 Anticancer Dendrimer–Drug Conjugates An excellent review summarized a collection of covalently bound anticancer drugs to a dendritic scaffold [222]. Some general reviews related to dendrimers in drug delivery are available, including an important section on anticancer dendrimer conjugates [36,40,44,97,223–228].

Until now, many common anticancer drugs conjugated to dendrimers were reported in the literature: doxorubicin (DOX), methotrexate (MTX), camptothecin (CPT), etoposide, 5-fluorouracil (5-FU), and paclitaxel (taxol). The following sections will delineate a few representative examples of some anticancer dendritic prodrugs using the covalent disassembly modes presented in Section 13.2.1. The examples will also include some cleavable units for a cascade disassembly (molecular triggers) by using a specific stimuli (for instance, a pH variation). A highlight article [229] on self-immolative dendrimers [61] as some anticancer drug delivery platforms based on an enzymatic triggering was reported by Shabat. Some cascade cleavages were presented for effecting the delivery of a single, double, or triple type of drugs for a possible bi- or tritherapy.

Doxorubicin

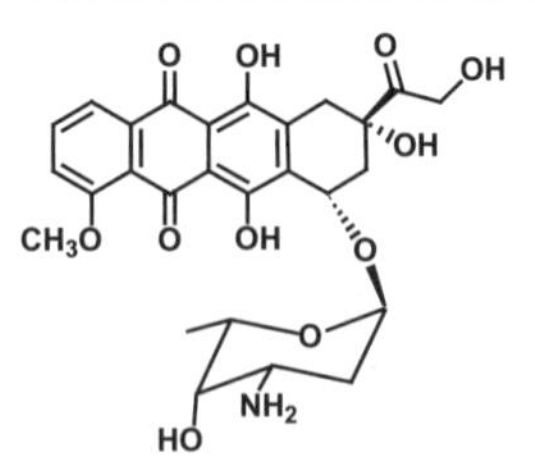

Doxorubicin

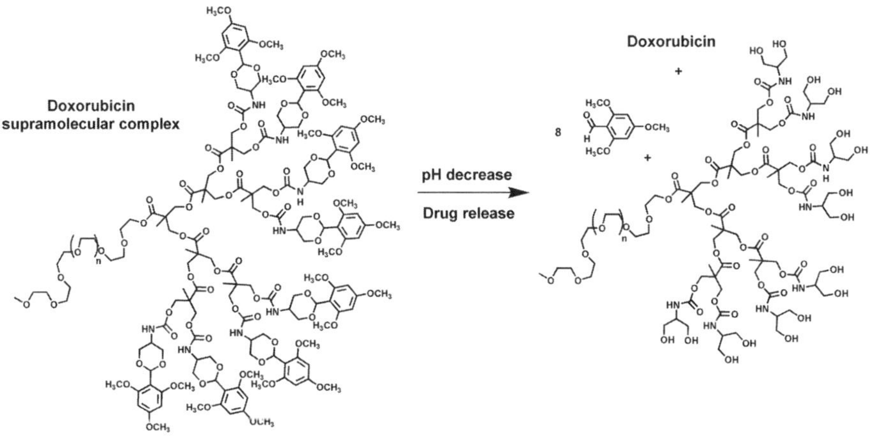

SCHEME 13.6 A doxorubicin polyester dendrimer complex having a cleavable acetal shell for liberating the anticancer drug.

A few systems combined supramolecular encapsulation of a drug and a release after some covalent bond cleavages of the dendritic shell [79]. Micelles made from dendrimers involved supramolecular encapsulation and delivery of doxorubicin (Scheme 13.6). The system comprised a PEG polymer coupled to a dendritic wedge having cleavable esters, carbamates, and a shell of ketal/acetal functions. The drug could be delivered after hydrolysis of the ketal/acetal functions under mild acidic conditions, similar to the pH in endosomes, with the hope for a better delivery in cancer cells. Another work made use of polyglycerol or PEI dendrimers functionalized by ketals or imines that could also be hydrolyzed after reactions of the dendritic shell.

A doxorubicin dendritic prodrug was synthesized as a conjugate from a three-arm PEG star and polyester dendrons (Scheme 13.7) [152]. A free phenolic group at the core allowed the use of radiolabeling (radioiodination). An acid labile hydrazone linkage was used for the covalent attachment of doxorubicin. This system had a stable polymeric backbone, a low polydispersity, some water solubility, and a negligible toxicity. Some biological evaluations were achieved *in vitro* and *in vivo* (with mice). Cell viability, biodistribution, and the use of confocal microscopy were reported [115,152].

A degradable molecular system with a cleavable unit based on a *para* benzyloxy phenyl ether and some carbamate linkers using a double trigger with logic gate "OR" was proposed. Depending on the choice of two enzymes as a trigger for the molecular events, either penicillin-G-amidase or antibody 38C2, it could promote the release of DOX. The biological tests were achieved on the inhibition growth of the human erythroleukemia cell line (HEL) [230].

Methotrexate (MTX)

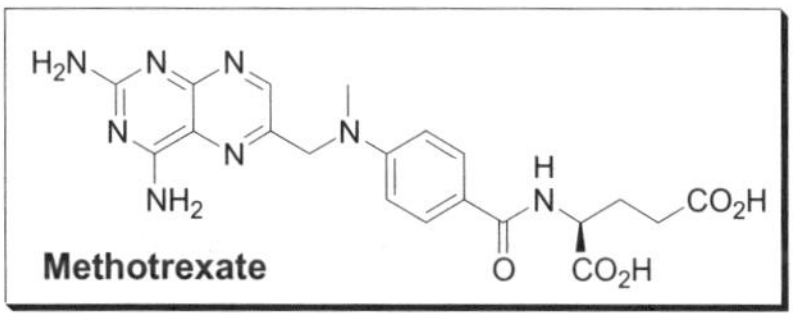

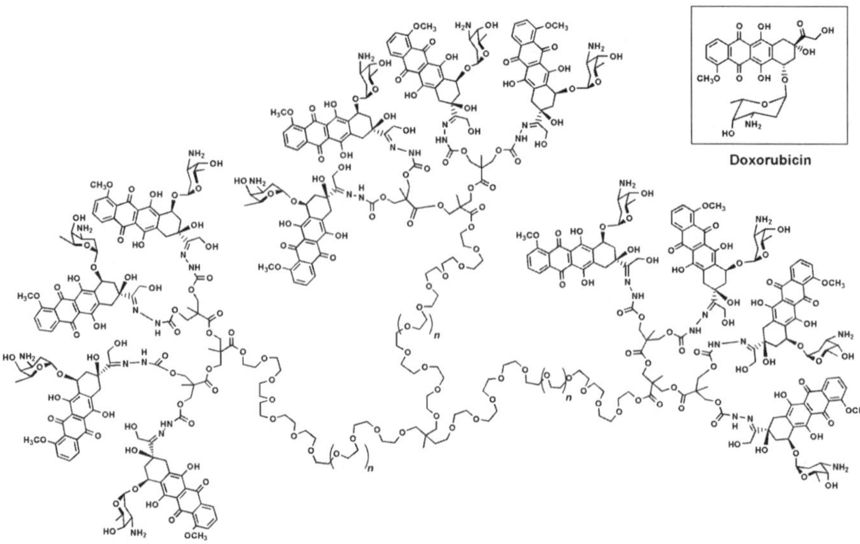

SCHEME 13.7 Cleavable polyester-PEG dendrimer as a macromolecular doxorubicin prodrug.

Methotrexate is an antifolate drug used in the treatment of many kinds of cancers and autoimmune diseases. It affects the metabolism of folic acid by inhibiting the dihydrofolate reductase. Therefore, its mode of action is the inhibition of folic acid, which is needed in the synthesis of nucleoside thymidine, required for DNA, RNA synthesis, and consequently, it affects more rapidly dividing malignant cells.

Several poly(aryl ether) dendrons and G1, G2 dendrimers were synthesized and functionalized by hydrazide functions at their periphery. The latter groups served to conjugate folic acid (FA) or methotrexate (MTX) by the use of a pH-sensitive hydrazone linker. This pioneer system was built from a G2 dendrimer for investigating the release of methotrexate [151].

A more elaborate methotrexate conjugate made from PAMAM–MTX dendrimers were reported as a statistical mixture when MTX and PAMAM were covalently bound by using amide functions. The results of PAMAM conjugation diminished the cytotoxicity and improved the therapeutic index of the drug toward sensitive or resistant cancer cell lines [143].

Camptothecin (CPT)

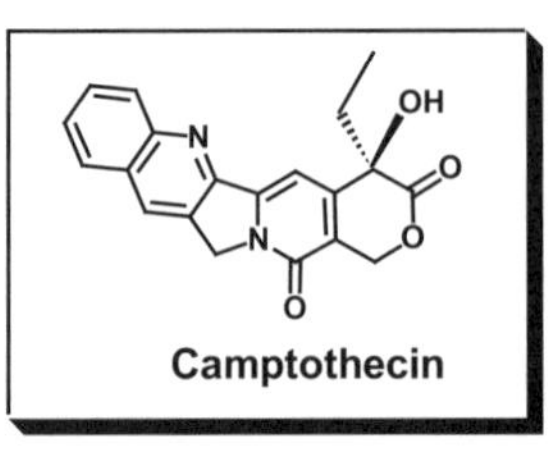

Camptothecin (CPT) is a cytotoxic, poorly water-soluble polycyclic quinoline alkaloid that inhibits the DNA enzyme topoisomerase I (topo I). CPT binds to topo I and causes DNA damage, which results in apoptosis. The search for new drug delivery systems for a selective chemotherapy lead to several articles by Shabat and coworkers. It first started from the synthesis of a CPT-conjugate prodrug with a self-immolative linker activated by an antibody catalysis [187] or an enzyme [197]. Penicillin-G-amidase or catalytic antibody 38C2 triggered the release of CPT by hydrolysis of a carbamate function. Cell growth inhibition assays indicated up to a 2250-fold decrease in toxicity for the prodrug relative to the drug.

As a comparison to a dendritic prodrug, here is a description of a HPMA (*N*-(2-hydroxypropyl)-methacrylamide) copolymer with some grafted molecular platforms comprising a drug (CPT), a trigger site, and a targeting ligand (Scheme 13.8) [196e].

A self-immolative second-generation dendritic prodrug incorporating CPT, PEG chains, and some cleavable units were synthesized with carbamate functional groups. The latter PEG chains were installed via click chemistry. They helped in decreasing the hydrophobic properties and avoided aggregates formation. The cleavage of the dendrimer led to the release of CPT after an enzymatic triggering by penicillin-G-amidase [231].

Paclitaxel (Taxol)

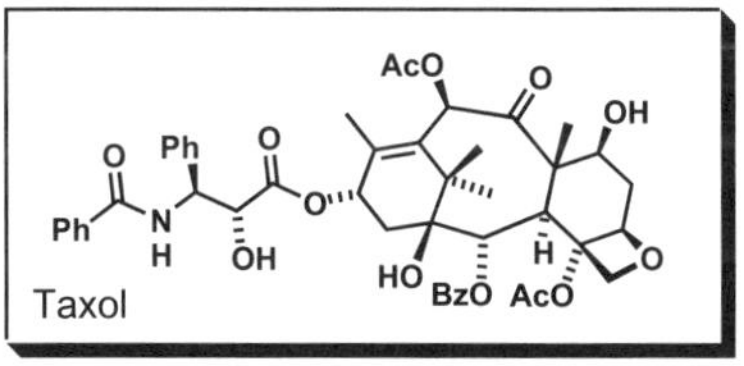

Preliminary studies on the cascade disintegration of elongated cleavable units were achieved by de Groot et al. for releasing paclitaxel or DOX [134]. In this article, they used an enzymatic triggering from human plasmin to cleave some linear prodrugs via a double or a triple benzyloxycarbonyl spacer incorporating carbonate or carbamate functions, and a combination with a bisamine cyclization spacer. It was shown that the catalytic rates of the degradation were faster for a double or a triple spacer compared to a single one. Dendrimers were not used in this study but prepared the groundwork for a subsequent publication on dendrimer–drug conjugates. Similarly, another work described the test of many prodrugs for their toxicity against many human tumor cell lines. They could release paclitaxel after a bioreductive activation from nitro- or azido-benzene derivatives and a carbonate linker [232].

de Groot et al. introduced some "cascade-release dendrimers" liberating paclitaxel (taxol) with an amazing long range self-elimination chemical adaptor unit (a double 1,8-elimination). A reduction of a nitro to an amino function served as a redox trigger to release two taxol molecules. This concept was extended to a higher generation dendrimer releasing simultaneously four molecules of paclitaxel (Scheme 13.9) [198b].

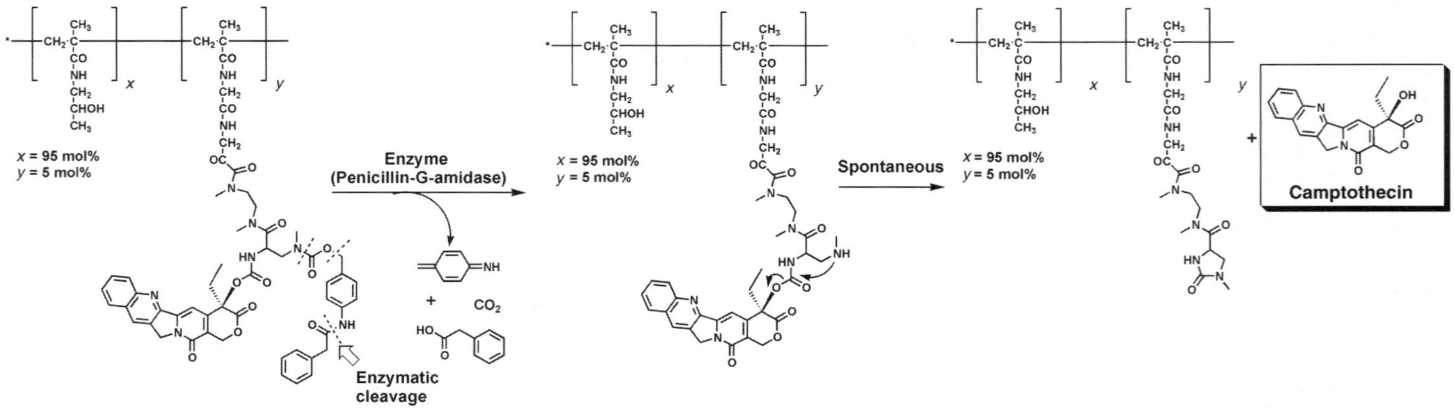

SCHEME 13.8 Branched macromolecules delivering camptothecin anticancer drug after cleavage by penicillin-G-amidase.

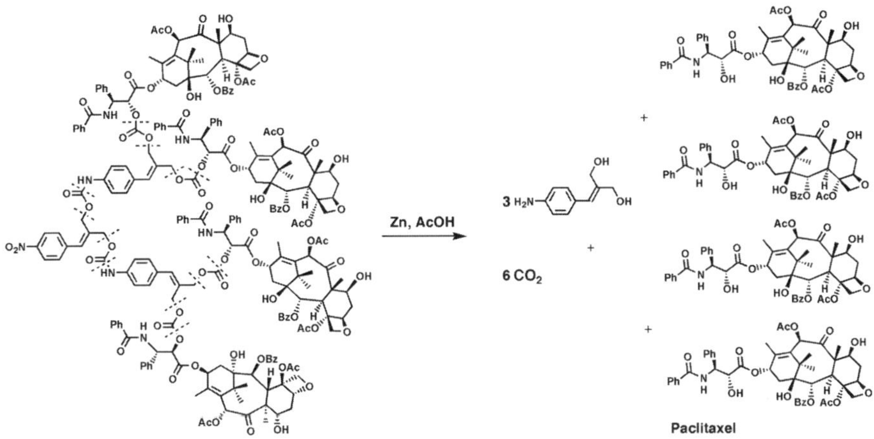

SCHEME 13.9 Cascade-release dendrimer liberating four paclitaxel (taxol) molecules upon reduction.

A dendritic system based on the conjugation of G4 PAMAM-OH dendrimer with paclitaxel was reported [130]. The linkage unit between paclitaxel and the dendrimer used succinic acid for a better degradation via an ester function. A comparison of the cytotoxicity of A2780 human ovarian carcinoma cells from the dendrimer, a linear PEG–paclitaxel conjugate and the free drug was investigated. A 10-fold increase of the dendrimer cytotoxicity was found compared to the free drug.

5-Fluorouracil

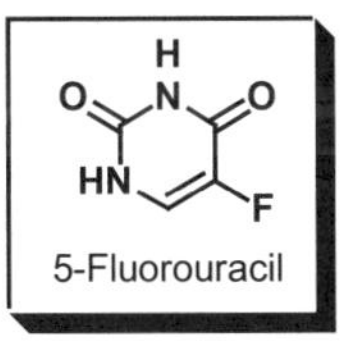

5-Fluorouracil is a classic anticancer drug widely used for the treatment of colorectal and pancreatic cancer for about 40 years. It mainly acts as a thymidylate synthase inhibitor by interrupting this enzyme and blocks the synthesis of the pyrimidine and thymidine, thus blocking DNA replication. The effect relies heavily upon quickly dividing cells, which used their nucleotide synthesis. G0.5 to G5.5 PAMAM-type dendrimers having a cyclam core were synthesized as a basic molecular scaffold for making a 5-FU dendrimer-conjugate [192]. The hydrolysis of an activated amide function in a phosphate buffer solution at pH 7.4 was studied but no study *in vitro* or *in vivo* was achieved yet. A previous prodrug system, not using dendrimers, based on a self-immolative glucoronide spacer (a cleavable unit) released 5-FU after treatment with a *E. coli* β-glucoronidase [186].

Etoposide

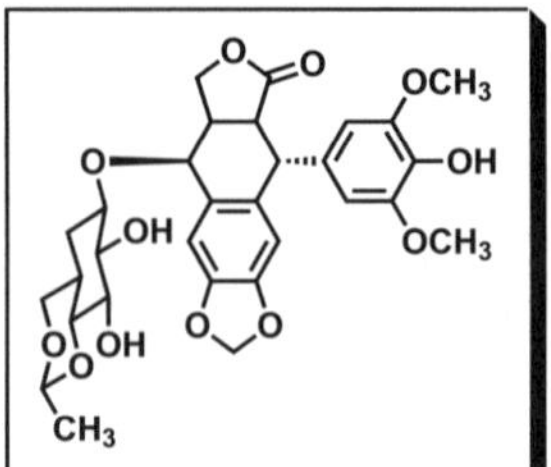

Etoposide

This anticancer drug is an inhibitor of topoisomerase II and is currently used in the therapy of lungs cancer, testicules cancer, leukemia, and glioblastoma. A first *in vivo* study on a linear prodrug incorporating a cleavable unit released etoposide via some cascade reactions [233]. The multiple events were triggered by a catalytic antibody 38C2 (a retro aldol). Another study prior to use dendrimers was achieved in the presence of a carboxylesterase toward the hydrolysis of a prodrug containing a carbonate function to be hydrolyzed [234].

13.3.1.6 Cancer Polytherapy with Dendrimers The concept of tritherapy against cancer cells was reported in an elegant work of Shabat and coworkers using camptothecin, doxorubicin, and etoposide attached to single chemical adaptor unit [191]. Triggering the process by catalytic antibody 38C2, it was possible to release three different drugs (CPT, DOX, and etoposide) in a cascade reaction. The heterodendritic trimeric prodrug was more potent than individual monomeric drugs when incubated with the antibody and tested against three cancer cell lines responsible for leukemia (Scheme 13.10). This work was an extension of a similar bioactivation method using self-immolative heterodendritic prodrugs in bitherapy (DOX and CPT) [197b].

13.3.1.7 Multifunctional Anticancer Dendritic Platforms as Nanodevices The most advanced and complex syntheses of dendritic nanodevices as a platform for a selective drug delivery include several functional units bound to a commercial dendritic scaffold (PAMAM, DAB, PEI, PPI, polyglycerol, Boltorn®) as shown in Figure 13.9.

Each units could have a function for imaging (with a fluorescent ligand such as FITC or a NIR ligand), for targeting cancer cells and tumors (folic acid, etc.), for solubilizing the device (with some nontoxic, nonimmunogenic groups), for transporting or for encapsulating a drug (PEG, alkyl chain, acetyl groups, etc.), or for destroying cancer cells with an anticancer drug conjugate such as MTX, DOX, CPT, and so on (Figure 13.10). For an excellent review in this field, Paleos et al. gathered a nice compilation of such dendritic devices [106]. Tomalia et al. also reported some guidelines for a multipurpose nanodevice for cancer therapy and imaging [204].

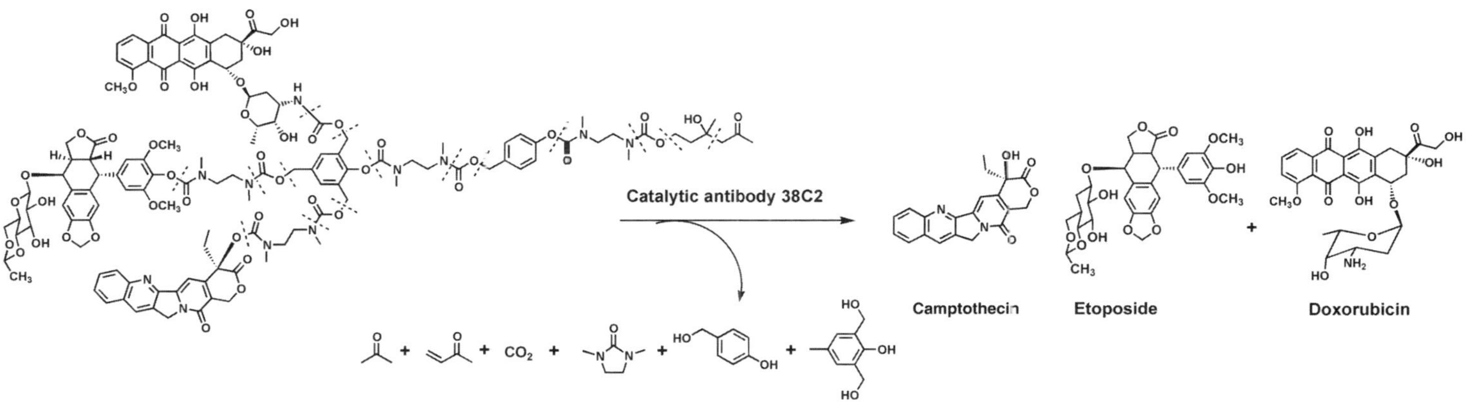

SCHEME 13.10 A heterodendritic dendrimer prodrug demonstrating tritherapy by a simultaneous delivery of three anticancer drugs catalyzed by antibody 38C2 as a trigger (retro-aldol reaction).

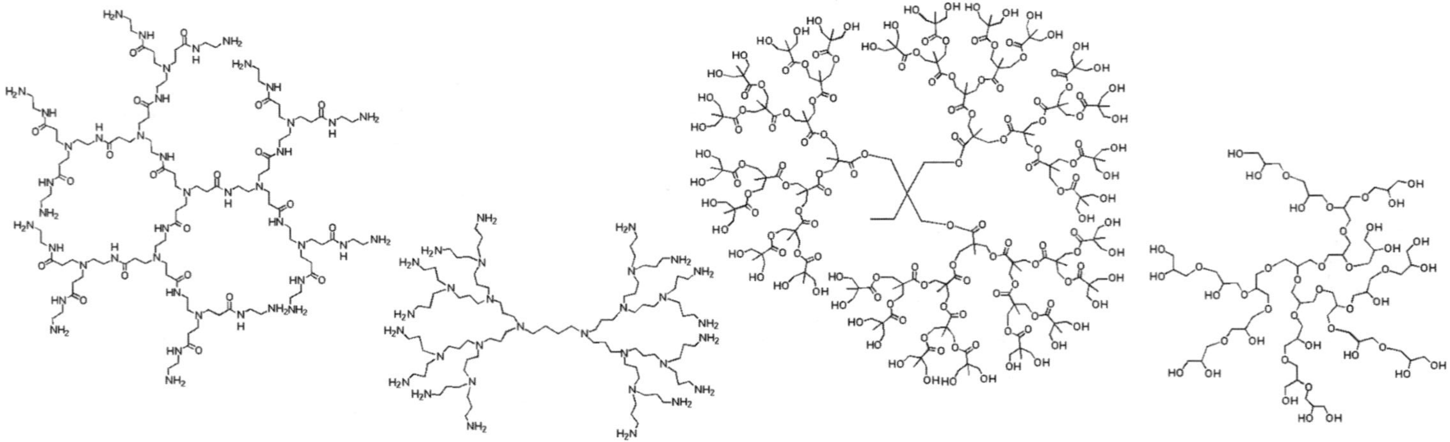

PAMAM DAB Boltorn Polyglycerol (PG)

FIGURE 13.9 Some commercial dendritic cores for building the nanodevices.

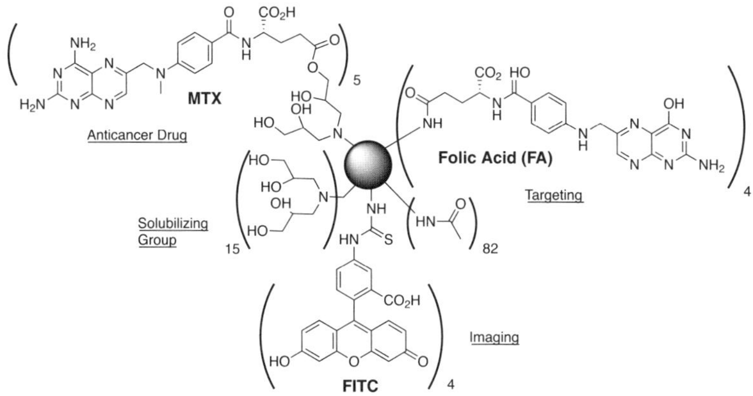

FIGURE 13.10 Example of an assembly of the functional units on a dendritic platform with a G5 PAMAM core.

As for the targeting units, folic acid was often reported. It interacts with overexpressed folate receptors on cancer cells. Neurotensin (a peptide) was also used in a similar platform for fighting colon, pancreatic, prostate, and lung carcinomas [235]. A cyclic peptide (cRGD) also made its way as a new targeting unit from the work of Baker, Jr. and coworkers [236].

The nanodevice model containing folic acid and MTX linked to a PAMAM scaffold is probably one of the most studied covalent combination in such a dendritic platform from Baker et al. Another one made uses of a partially acetylated G5 PAMAM template, combined to an imaging agent (fluorescein isothiocyanate), folic acid, and paclitaxel (taxol, an anticancer drug) [196a]. Ester linkages were used for the delivery of taxol. A similar platform used a combination of the same units as above, excepted MTX as an anticancer drug and a partial acetylation of G5 PAMAM [176]. The ester linkage of MTX could be achieved via a glycidol linker. The latter also helped to eliminate the peripheral NH_2 groups for avoiding unwanted nonspecific targetings. An amide link was also used to directly conjugate MTX to PAMAM.

A recent nanodevice model of diagnostic and imaging was based on a biodegradable drug carrier from the coupling of PAMAM or PEI dendritic cores with some arms of poly(L-glutamic acid), folic acid residues for targeting cancer cells, and a near-infrared indocyanine dye [196c]. The resulting conjugate polymers could be degraded by the endosomal enzyme cathepsin B and could selectively bind to tumor cells expressing folate receptors. Radioiodination and fluorescence confocal microscopy were also demonstrated for imaging cancer cells and the location and transport of a dendritic platform incorporating doxorubicin (cell mapping of the platform) [148].

13.3.1.8 In Vivo *Studies of Some Anticancer Dendritic Platforms as Nanodevices*

DOX-Boltorn® Systems Among the first advanced *in vivo* studies on the biological evaluation of dendrimer were some reports on the design and the synthesis of Boltorn®-type polyester dendrimers for DOX delivery (Figure 13.11) [152].

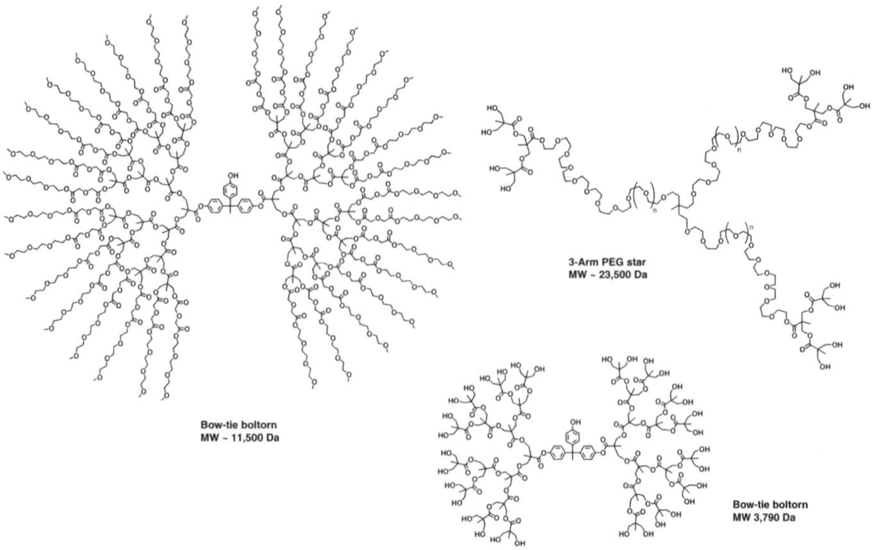

FIGURE 13.11 Example of some degradable polyester dendrimers (Boltorn®-PEG type).

The importance of this dendritic model relies on the access of several monodisperse architectures of various molecular weight, size, and shapes with a low toxicity, a good biocompatibility, and a nonimmunogenicity. They incorporated hydrophilic PEG chains and hydrolyzable ester linkages for insuring a metabolic elimination of dendritic fragments.

At first, the dendritic polyesters degradation *in vitro*, their toxicity, and the *in vivo* biodistribution in mice were studied [127]. The *in vitro* cytotoxicity of the polyesters was negligible and the blood clearance was evaluated for three types of dendrimers. The biodistribution in mice using radiolabeling (radioiodination) of all dendrimers indicated some accumulation of those macromolecules in liver and intestines for the dendrimers of generation 1, but not in vital organs such as heart, lungs, spleen, or stomach. All other generations did not show an accumulation. The clearance rate in the blood circulation was faster for smaller polyesters, but higher at macromolecular weight >30–40 kDa and branching helped for longer plasma circulation times. In this case, an accumulation of the dendrimer in tumors is important according to a model of renal filtration through some glomerular pores and the EPR effect. It represents one of the first systematic studies between the molecular weight/architectures of dendrimers and the pharmacokinetics for well-defined macromolecules. A summary of this work was compiled in a recent review [220].

Among many applications, it was described that a single dose of an asymmetric polyester bow-tie dendrimers loaded with 8–10 wt.% DOX and conjugated to PEG chains could cure mice bearing C-26 colon carcinomas [237]. It is believed that one of the important roles of PEG with a narrow polydispersity is to avoid aggregation, as a steric stabilizer. The drug was linked by some pH-sensitive hydrazone functions. The DOX–dendrimer conjugate was eliminated from the serum with a half-life of 16 ± 1 h

(<10 min for free DOX) and its tumor uptake was ninefold higher than for the free drug after 48 h. The maximum tolerated dose of the prodrug was 20 mg/kg compared to 6 mg/kg for the free DOX. Overall, the therapeutic index of the DOX prodrug was significantly better than for the free drug.

CPT-Poly(L-Lysine) Systems A more recent study involved a PEGylated core–shell dendritic CPT conjugate with an excellent therapeutic index. A poly(L-lysine) core dendrimer was functionalized by aspartic acid as a linker to which PEG chains and CPT were covalently bound [136]. The conjugate had a molecular weight of 40 kDa and a blood circulation half-life of 30.9 ± 8.8 h and a tumor uptake of $4.2\% \pm 2.3$ of the injected dose per gram tissue. The free drug content was less than 1% in the blood after 30 min and a tumor accumulation of $0.29 \pm 0.04\%$ was evaluated. A superior efficacy was demonstrated in murine (C26) and human colon carcinoma (HT-29) tumor models compared to no treatment. The *in vitro* hydrolysis was already reported.

MTX-Poly(L-Lysine) Systems A similar dendrimeric system from G3 to G5 incorporating a poly(L-lysine) core surrounded by covalently bound PEG chains and the MTX drug was synthesized in order to study the pharmacokinetics and the tumor drug retention in rats and mice [137]. The dendrimers were radiolabeled using tritium (^3H). Conjugation of MTX generally increased the blood clearance compared to a dendritic system with PEG alone. In general, those dendrimer conjugates increase the lifetime of MTX in the blood circulation. The increase of dendrimer generation number decreased the blood clearance, the degradation, and the renal elimination. Some studies were achieved with a dendritic molecular weight >20 kDa. Biodistribution indicated that G3 and G4 dendrimers were mostly retained in the kidneys whereas G5 dendrimers were concentrated in the spleen and in the liver. As for the EPR effect on tumors, retention of the prodrug was shown in solid Walker256 and HT1080 tumors in both rats (1–2% dose/g in tumor) and mice (11% dose/g in tumor).

Dendrimer Platinate A G3.5 PAMAM-CO_2H scaffold was used for the conjugation of cisplatin, a common anticancer drug generating biologically active diaqua-Pt species. It was among the first *in vivo* studies in the literature. The biological activity of the dendrimer conjugate was compared to cisplatin [238]. In short, the advantages of the dendrimer conjugate are as follows: a higher load of Pt (25% versus 3–8% in other linear macromolecules), a higher maximum tolerated dose of Pt-dendrimer (15 mg/kg) compared to cisplatin (1 mg/kg), a selective accumulation of Pt in solid tumors, a lower toxicity, and a higher biological activity in some cancers. For instance, melanoma B16F10 tumor models in mice were affected by Pt-dendrimer in contrast to cisplatin.

13.3.2 Nonsteroidal Antiinflammatory Dendrimer–Drug Conjugates

Nonsteroidal antiinflammatory drugs (NSAIDs) are part of the most important drugs on the market for fighting symptoms of inflammation in rheumatoid arthritis, osteoarthritis, postoperative pains, headaches, and so on. Rheumatoid arthritis is often a severe, chronic, systemic inflammatory disorder that affects many tissues and organs, but especially the joints and cartilages, causing some mobility problems. For a sustain relief and a better drug delivery, the development of dendritic prodrugs was sought after a covalent coupling of common drugs to a dendritic scaffold.

Several examples are known with ibuprofen [196c], naproxen [128,139,239], 5-amino salicylic acid (5-ASA) [240], and salicylic acid [241]. We have already discussed about *N*-acetyl cystamine (NAS) in a previous section on disulfide linkers. In the case of salicylic, the dendrimer was made of many salicylic acid units in its backbone, and also at its periphery. It was able to sustain a slow delivery of the drug for several days. As for ibuprofen, a G4 PAMAM scaffold was used and the dendrimer-conjugate concentrated into the cytoplasm of the A549 cells. It provided a higher activity, a higher local drug level at the inflammatory site.

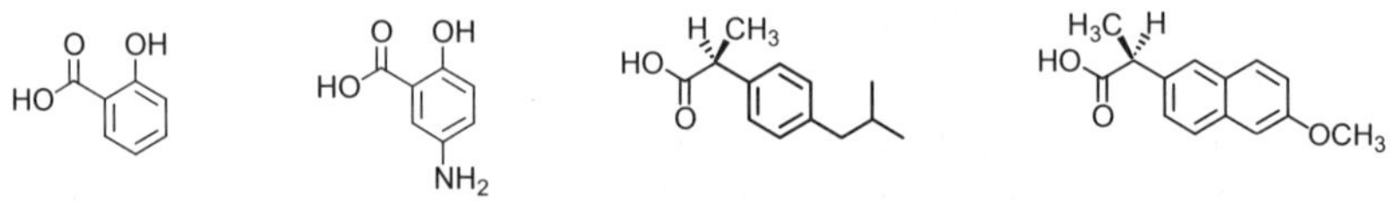

Salicylic acid 5-Amino salicylic acid (S)-ibuprofen (S)-(+)-naproxen

The antiinflammatory naproxen drug was covalently bound on a PAMAM core for making a polyester or a polyamide dendritic prodrug that could be cleaved by esterases in plasma or by a pH variation (Figure 13.12). Naproxen was directly linked via some amide or ester bonds, using a spacer as a solubilizing group such as PEG or L-lactic acid. Again, this work confirmed that the amide linkage was too stable for a slow release of the drugs but the ester functions were slowly cleaved after many hours. A test of a transepithelial permeability gave an enhancement of the drug permeation. Another study reported the synthesis of a series of naproxen drugs conjugated to some small polyols, using 3,5-dihydroxybenzoic acid for creating a degree of branching [128]. There was a mention of some enzymatic studies but they were not yet disclosed.

5-ASA modified PAMAM dendrimers were studied for their colonic delivery behavior. The localized amount of drug, due to the enzyme azoreductase in the colon, favored some colon-specific properties with a local release. Overall, the conjugation of the NSAIDs with some dendrimers helped in their transport, their localization, and

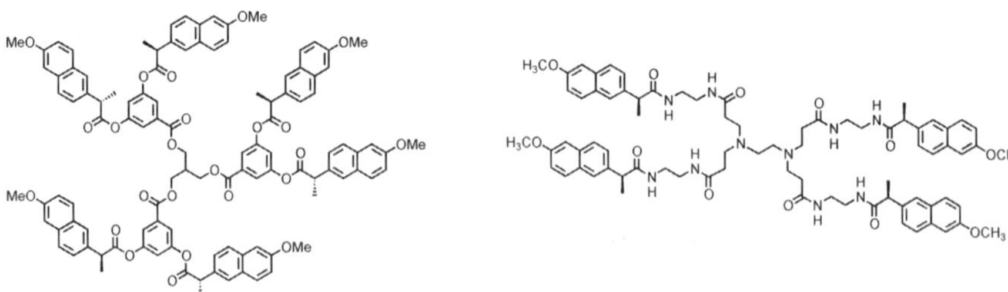

FIGURE 13.12 Nonsteroidal-dendrimer conjugates containing (*S*)-(+)-naproxen.

offer a better control of drug release over time that could not be achieved by a dendrimer encapsulation. Their sustained release was consistently better compared to the free drugs or to the dendrimer inclusion complexes.

13.3.3 Steroidal–Dendrimer Conjugates

A few publications appeared in this field, one described the synthesis of star-like poly (aryl ether) dendrimers with water-soluble PEG units, as potential drug carriers of cholesterol. However, there was no demonstration of a drug-release activity [111]. Another one reported the conjugation of methylprednisolone to PAMAM dendrimers via glutaric acid as a linker. Methylprednisolone is a steroid with some antiinflammatory properties, also useful to fight rheumatoid arthritis or some inflammatory states related to an immunological response (it is also an immunosuppressor). Fluorescence emission and flow cytometry indicated that the conjugate was rapidly taken into some A549 human lung epithelial carcinoma cells, but the pharmacological behaviors were similar to the free drug.

13.3.4 Antimicrobial–Dendrimer Conjugates

Many amine-terminated dendrimers incorporating ammonium groups at their periphery were found to have an antibacterial activities, without a delivery of an active ingredient. However, penicillin V was covalently linked to G2.5 and G3 PAMAM dendrimers via some PEG spacers by using amide or ester bonds. The bioactivity of the conjugate was similar to the free penicillin V, even if the antimicrobial release did not show any peculiarities [242].

13.3.5 Dendrimer-Based Degradable Biomaterials for Medical Applications

Even if the main topic of this chapter concerns discrete, monodisperse dendrimers, this section will provide a quick and brief summary of some degradable dendritic polymers and biomaterials when needed. Degradable functional polymers have been investigated for the past 50 years as medical devices and biomaterials [243–245]. However, new trends in dendrimer chemistry indicated some promising uses in those directions in tissue engineering [112,246]. For instance, a review described some dendritic macromers for hydrogel formation toward some materials for ophthalmic, orthopedic, and biotech applications [247]. Another review reported some applications in tissue engineering: bone and cartilage tissues, corneal tissues, bioadhesive sealants, and collagen mimics based on degradable dendrimers and hydrogels [248]. Finally, a review on ocular dendritic drug delivery systems was published [249].

New biodegradable and biocompatible materials are important for some biomedical applications. Recently, Grinstaff and coworkers promoted the synthesis of some "biodendrimers" incorporating some biocompatible and less toxic natural metabolites or nonimmunogenic PEG chains, succinic acid, lactic acid, glycerol, myristic acid, and some natural L-amino acids (L-phenyl alanine, L-lysine, L-cysteine, β-alanine, etc.) as shown in Figure 13.13. Degradable dendritic poly(glycidols) were

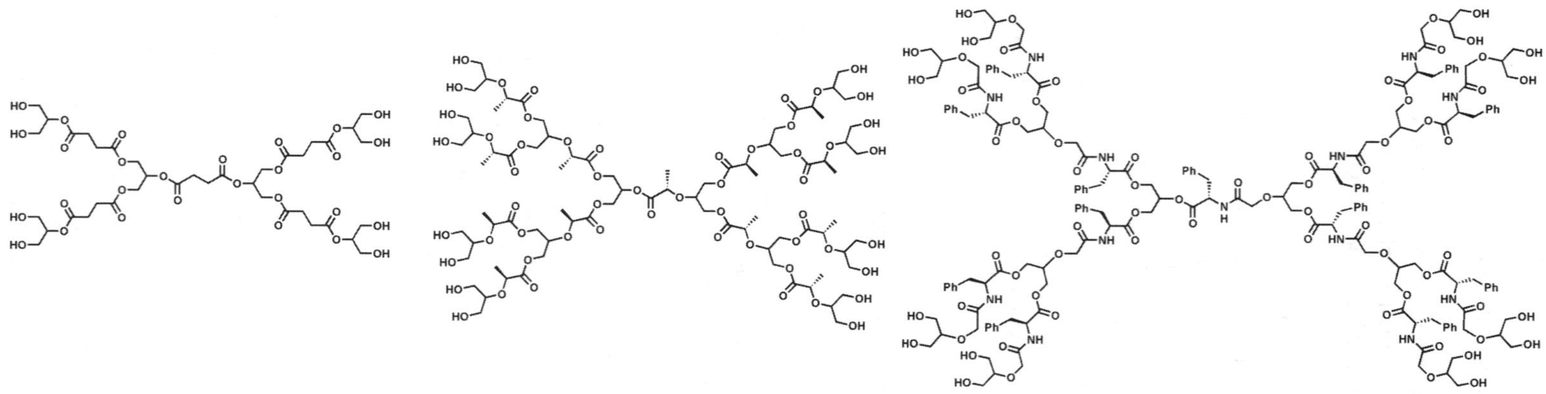

Glycerol-succinic acid biodendrimers   Glycerol-(+)-lactic acid biodendrimers   Glycerol-(L)-phenyl alanine biodendrimers

FIGURE 13.13 Representative structures of some biodendrimers.

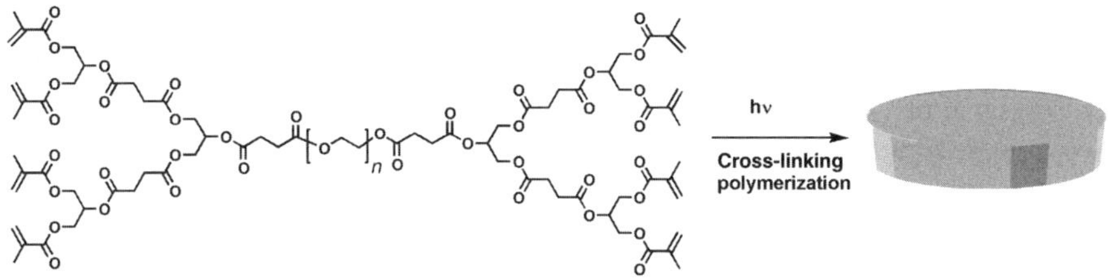

SCHEME 13.11 Biodendrimer-based hydrogel scaffolds for cartilage tissue repair by photo-cross-linkable methacrylated poly(glycerol succinic acid) dendrimers.

also described [250]. In general, the biodendrimer/biopolymer must provide a temporary matrix for cells, before the regeneration of tissues; it directs the arrangement of some cells type to specific location within a matrix; it modulates the biological activity in tissue repairs, and it must possess physical and mechanical properties similar to the host-damaged tissue. Swelling of the biopolymer must be minimized for geometrical reasons. Hydration is also important for the growth of the tissues. A polyester matrix is known to quickly degrade and hydrolyze, compared to polyamide or polyether. Depending on the degree of cross-linking, and because dendrimers are good gelators, a hydrogel network usually satisfies the requisite properties as a resorbable 3-D porous structure.

As for a cartilage tissue repair [251], a high water content, biocompatibility, and the rate of degradation of the matrix material must be tuned to the synthesis of the extracellular matrix. The role of photo-cross-linkable biodendrimer-based hydrogel scaffolds was promoted as a mean to use methacrylated dendrimers for increasing cross-linking and avoiding solvent swelling, which could be detrimental in the adjustment of precise geometric entities. Multiple branching of a PEG_{3400} core to polyester wedges offered a dendrimer with considerable degradability before the cross-linked network breaks down, thus providing a better mechanical strength of the materials (Scheme 13.11).

A similar concept for osteochondrial repair was used with some photo-cross-linkable carbamate dendrimers incorporating PEG, succinic acid, glycerol, or β-alanine with methacrylate units at the periphery [252].

As for some applications in ophthalmology, a review can be found on hydrogels for corneal wound repair [253]. Grinstaff and coworkers used methacrylated biodendrimer wedges and PEG, which assisted in a corneal tissue repair or as an ocular sealant for sutureless eye surgery [254–257]. It could also help to secure cataract incisions [258] when using another synthetic strategy based on cross-linked hydrogels made from some thiazolidine linkages [168] coming from an aldehyde and polylysine/polycysteine peptide-based dendrons.

13.3.6 Dendrimer for the Release of Fragrances

The Firmenich company filed a patent in 2002 and published a first communication in 2003 on dendrimers able to slowly release a fragrance by the cleavage of chemical bonds at the periphery of a dendrimer [259]. This strategy used a neigboring group participation (as a 2-carbamoylbenzoate) for releasing of a tertiary alcohol, after the hydrolysis at pH 7.6 of some poly(propylene imine) dendrimers. This concept of a slow release of a fragrance and an odor is particularly important in this field. Their investigation was highly detailed in a subsequent publication [260,261]. They synthesized various dendritic shapes (linear and globular dendrimers) in order to analyze a shape selective release. The following conclusions were drawn: (a) the size and the topology of the dendrimers had only a limited impact on the kinetic of release of the fragrance from this intramolecular cleavage; (b) suitable catalyzing groups and solubility or dispersibility of the dendrimers in the reaction medium were important.

Branched cleavable poly(amide)s were used in the release of citronellol and L-menthol (Scheme 13.12). The combination of amide functions incorporated in a dendritic backbone, and the covalent attachment of the organoleptic compounds by ester functions, enable the use of several enzymes. In that peculiar case, a choice was made for a lipase (*Candida cylindracea*) and a cutinase (*Fusarium solani pisii*). The investigations revealed a partial enzymatic cleavage of the ester functionalities, which became more difficult as the branching, the rigidity, or the bulkiness of the polyamide–fragrance conjugates increase [260].

The release of some volatile aldehydes and ketones were produced by a reversible hydrazone formation in a dynamic manner [262]. Even if dendrimers were not used, it lead the way to future applications of some dynamic combinatorial fragrance studies when using multivalent dendrimers. Other publications mentioning dendrimers in the release of fragrances are related to inclusion complexes or encapsulation [263].

13.3.7 Dendrimer–Pesticide Conjugates

There are a very limited number of applications of dendrimers involving pesticides, conjugated in a covalent manner. One of the reasons is that pesticides do not often possess the proper chemical functions or they might not be reactive enough to promote a coupling to a dendrimer. Additionally, many of them are not much water-soluble for making a formulation. A high loading of them onto a dendrimeric scaffold might even lead to a worst solubility. A scarce and interesting work of Majoral and coworkers indicated that phosphorus-based dendrimers could be linked to fipronil (Aventis CropScience) and served as a slow release of pesticides over time in the presence of water [264].

13.3.8 Thermolabile Dendrimers and Photoresists

As mentioned in the introduction, some PAMAM dendrimers could be thermally degraded in a methanol solution by a retro-Michael addition at temperatures above 4°C. After heating hydroxyl-terminated G3 and G4 PAMAM at 150°C on some metal oxides supports, FTIR indicated a decrease of the amide band intensity and some carboxylate species [265]. The thermal decomposition of

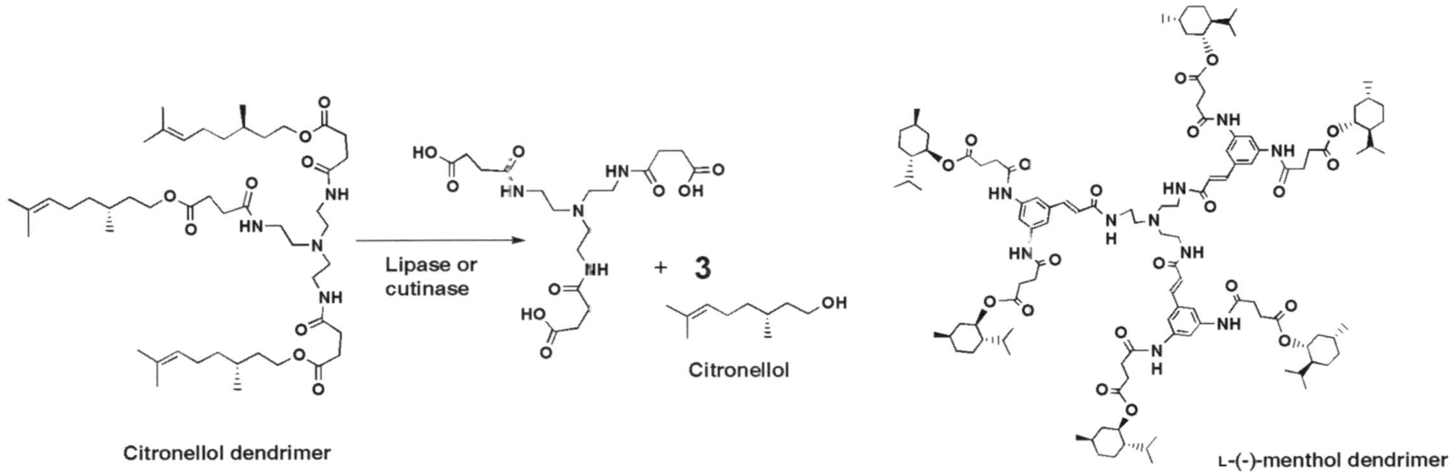

SCHEME 13.12 Fragrance release dendrimers liberating citronellol or L-menthol after lipase or cutinase ester hydrolysis.

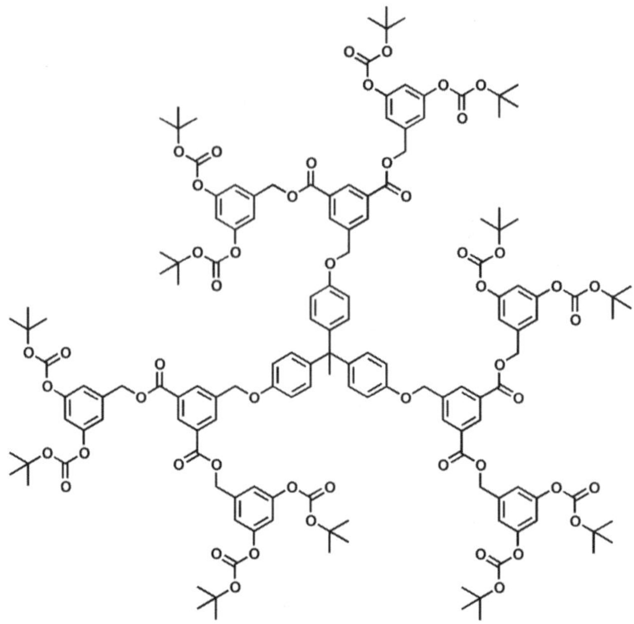

FIGURE 13.14 Dendrimers with thermally labile t-Boc groups are of use in chemically amplified resist materials for sub-100 nm lithography.

G4.5 PAMAM (carboxylate surface) dendrimer containing platinum-complexes was studied [266]. Similarly, G4 PAMAM-OH encapsulating platinum particles were described [267].

Resist materials are used in lithography to fabricate electronic circuitry and DRAM (dynamic random access memories). Using dendrimers with thermally labile end groups (t-Boc), the first example of a chemically amplified lithography resist materials based on dendrimers was created. With conventional e-beam irradiation, sub-100 nm lithography produced lines in the 50–100 nm range. It is believed that the dendritic shape compared to long polymeric chains is responsible for the increase in resolution (Figure 13.14) [268]. A review on novel organic resists for nanoscale imaging with chemically amplified cycloaliphatic resists was reported [269].

A new class of thermally responsive dendrimers were recently described. They are based on a thermally reversible furan-maleimide Diels–Alder reaction with some poly(aryl ether) dendrons [270,271]. The reversible reaction occurred above 90°C, but the reassembly of the dendrons was possible at 65°C for several days. Other dendritic thermal decompositions and their mechanisms have been studied but the applications were not described [272].

13.3.9 Redox Dendrimers

Redox-driven shaving dendrimers use a redox stimuli (with $Na_2S_2O_4$) for cleaving peripheral substituted quinone end groups. Reduction of the latter induced an

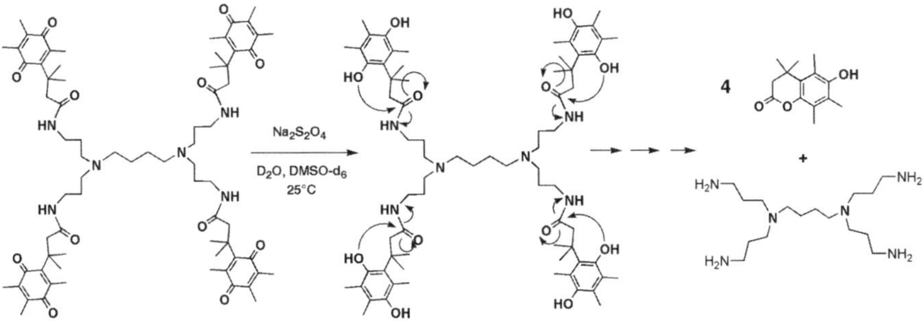

SCHEME 13.13 Redox-shaving dendrimers based on the reduction of quinone derivatives as some chemical adaptor units.

intramolecular nucleophilic cyclization which leads to the cleavage products in less than 20 min. This strategy could potentially be applied for the release of an encapsulated guest (Scheme 13.13) [198a].

We can also include in this section dendrimers mentioning the reductive cleavage of a disulfide linkage, as previously stated in the section on labile functions, and any other dendrimers degraded by the use of molecular triggers activated from an oxidation or a reduction.

13.3.10 Molecular Imprinting Using Core Dendrimers

In 1995 Moller and coworkers utilized a degradable hyperbranched poly(organosiloxane) polymer as a template for making molecular cavities in a matrix, albeit without mentioning molecular imprinting [56]. A poly(organosiloxane) dendrimer was embedded into a methacrylate resin and an acidic hydrolysis cleaved the core dendrimer for making nanocavities [57]. This example is among the first ones on molecular imprinting with dendrimers.

Since 1999, a series of reports were published by Zimmerman and coworkers on the synthesis of functionalized cored dendrimers designed to produce a peripheral shell and a matrix from some intramolecular ring-closing metathesis (RCM) reactions [273,274]. Some reviews presenting this concept were published [275,276]. It involved a "monomolecular imprinting" combined to a metathesis polymerization of homoallyloxy terminated dendrimers with some interdendron cross-linkings and the formation of a dendritic matrix incorporating the dendritic core (a porphyrin, a trimesitoic acid, or other cores) [277–280]. It could eventually be cleaved after hydrolyzing the ester functionalities under basic aqueous conditions. The result is a molecular imprint (MIP) of a dendritic core (a porphyrin or trimesoic acid) as a hollow nanosphere. The novelties in the field of molecular imprints relied on a controlled, reversible metathesis polymerization, self-adjusting with the template dendritic core (as a scaffold) because of an RCM equilibrium. It could produce less defects and more regular imprints compared to a classic MIP formation. An impressive quantitative removal of the dendritic core template was mentioned. The resulting imprint could be compared to a receptor and the removed dendritic core, as a substrate (Scheme 13.14).

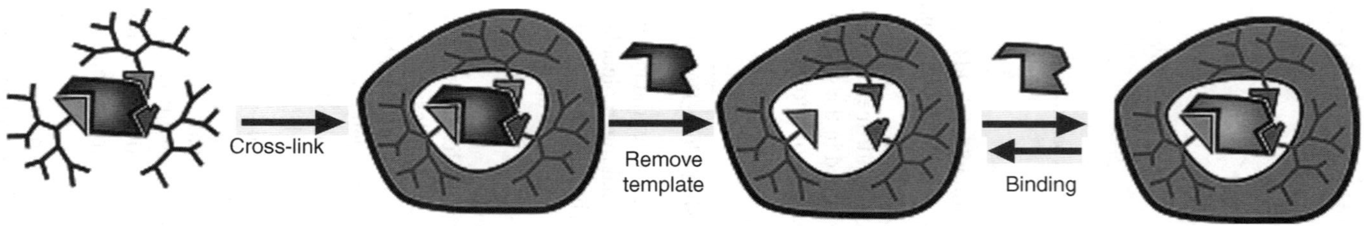

SCHEME 13.14 General concept of monomolecular imprinting inside dendrimers. (See the color version of this figure in Color Plates section.)

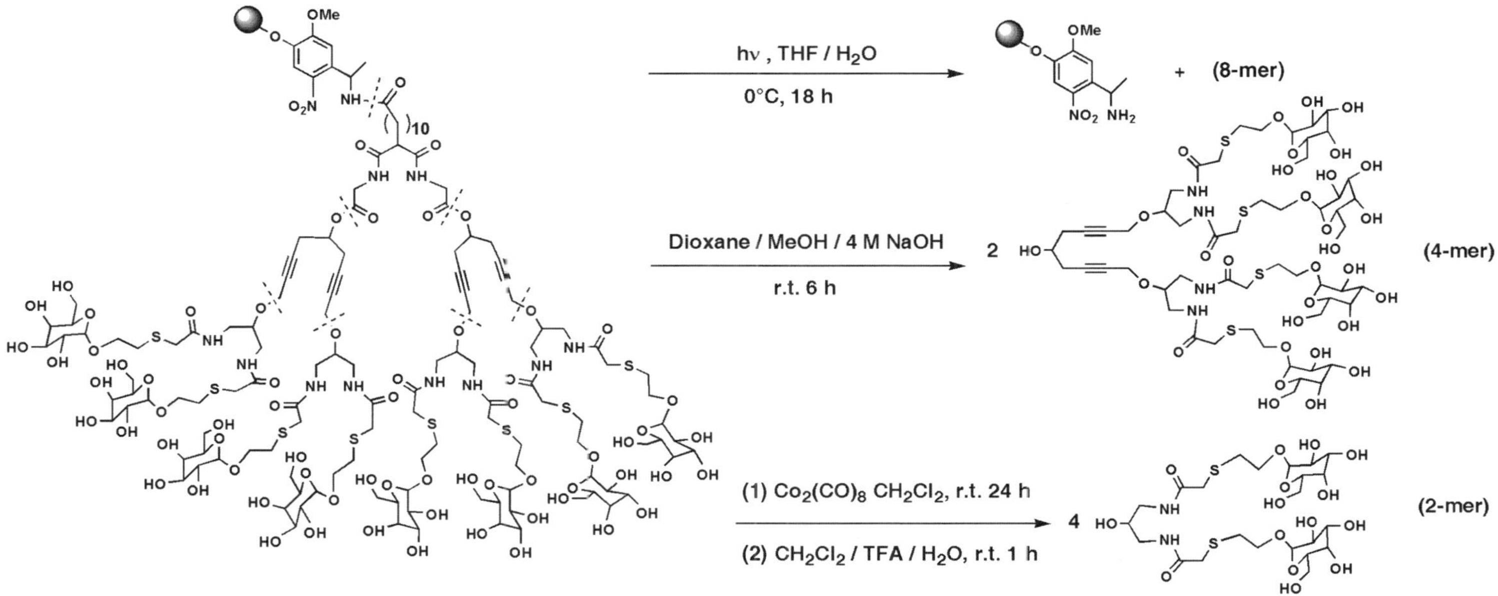

SCHEME 13.15 Solid-phase synthesis of dendritic carbohydrate clusters with three types of orthogonally cleavable fragments.

Using the same principle, some cored dendrimers made from internal cross-links were reported [281]. These nanomaterials could act as selective synthetic hosts of guests similar to the extracted dendritic core, or as integrated chemosensors of amines [282,283].

13.3.11 Carbohydrate-Based Dendrimers

Combinatorial solid-phase syntheses of dendritic glycoclusters were described as a promising approach for studying the cluster effects in glycobiology and in carbohydrate–protein interactions. As shown in Scheme 13.15, a dendritic wedge containing peripheral galactosides was presented [284]. Using photochemistry, it was possible to release the 8-mer glycodendrimer from the resin by a covalent bond dissociation. A basic hydrolysis of the esters produced the 4-mer compounds. The propargyloxy group functions could be cleaved by using a cobalt carbonyl complex for generating the 2-mer products. In short, some orthogonal methods for the selective cleavage of a resin-bound glycodendrimer were reported.

Another versatile and soluble combinatorial dendritic system based on a polyester Boltorn® dendritic support, used an *o*-nitrobenzylic photocleavable spacers and some linked carbohydrate units. It was possible to cleave the dendrimer support and spacer by using laser in MALDI-TOF MS or by a basic hydrolytic cleavage of the polyester support. As for the last method, the carbohydrates residues can be recovered [285].

13.3.12 Dendrimers and Gas Release

An interesting and recent application was the generation of gases using degradable dendrimers as a covalent reservoir of that gas. As for some biological applications and studies, the release of nitric oxide was reported. It was achieved from a PAMAM dendrimer containing some *S*-nitrosothiol functions at their periphery [286,287]. The mechanism of a diazeniumdiolate decomposition indicated a release under physiological conditions (pH 7.4, 37°C) with a duration >16 h. As for another application, although there was no conjugate synthesized and hence no real cleavage of dendrimers, singlet oxygen was generated via a two-photon excited FRET method using the proper chromophores and a porphyrin core dendrimer in an oxygen atmosphere [288].

13.4 CONCLUSION

In the past few years, a new stage was set in dendrimer chemistry from the development of a modern class of dendrimers, which are cleavable or degradable with defined, monodisperse and well-characterized structures. Because of those features, they are now highly regarded compared to dendritic polymers, which often have a higher polydispersity and the associated problems of synthetic reproducibility. The next step in this development has focused on some "chemical adaptor units" (or cleavable units) for dendrimer disassembly by cascade reactions triggered by several external stimuli such as light, redox reactions, enzymes, pH variation, and so on. Such results could be compared to "smart materials" of use in various applications. The most important ones are in the fields of anticancer prodrugs and drug delivery. Recent biological evaluations are extremely encouraging toward some new modes of biodistribution, cell targeting, controlled release of drugs, and polytherapy while taking advantage of the EPR effect and the dendritic scaffold. Dendrimers could be seen as some covalent nanocarriers of some (bio)molecules where the properties of the molecular scaffold can be tailor-made for achieving a specific function. They have also brought new insights about tissue repairs, diagnostic and imaging, molecular imprints, and photoresists. From the recent exponential number of publications per year, it is anticipated that cleavable dendrimers will bring a new dimension in dendrimer chemistry, especially in the fields of nano oncology, nanomedicines, (bio)nanotechnology, supramolecular, and materials chemistry. This review is among the first ones dedicated to cleavable dendrimers from covalent bonds disassembly and many more results are yet to come.

ACKNOWLEDGMENT

The author acknowledges CNRS, the French Ministry of Research and Education, and Aix-Marseille II University. Dr. Yoann M. Chabre is greatly acknowledged for helping in the literature search. We acknowledge Angewandte Chemie and John Wiley & Sons for providing the permission to use previous figures and schemes as some basic structural elements which were modified in this chapter.

REFERENCES

[1] M. Gingras, J.-M. Raimundo, Y. M. Chabre, *Angew. Chem. Int. Ed.* **2007**, *46*, 1010–1017.

[2] M. Gingras, Y. Chabre, J.-M. Raimundo, M. Boix, M. Do, C. Sarazin, L. Delaye, U.S. Pat./World PCT Dendrimer–Drug Conjugates—Serial No. 11/458,901, University of Nice-Sophia Antipolis, Nice, France/Allergan Research Center, Riverside, CA, USA.

[3] Among some books and reviews on dendrimers: (a) F. Vögtle, G. Richardt, N. Werner, A. J. Rackstraw (Eds.), *Dendrimer Chemistry: Concepts, Syntheses, Properties, Applications*, Wiley-VCH Verlag Gmbh, Weinheim, **2009**, 354 pp.; (b) R. Meijboom (Ed.), *Carbosilane Dendrimers and Their Use in Polymerisation*, VDM Verlag Dr. Muller Aktiengesellschaft & Co KG, **2009**, 152 pp.; (c) P. Dvornic, M. J. Owen (Eds.), *Silicon-Containing Dendritic Polymers*, Springer-Verlag, New York, **2008**, 428 pp.; (d) I. Majoros, J. R. Baker (Eds.), *Dendrimer-Based Nanomedicine*, World Scientific Publishing Co., **2008**, 440 pp.; (e) G. R. Newkome (Ed.), *Handbook of Dendrimers: Synthesis, Nanoscience and Applications*, Wiley-VCH Verlag Gmbh, **2007**, 1000 pp.; (f) F. Vögtle, G. Richardt, N. Werner (Eds.), *Dendritische Moleküle*, Teubner Verlag B. G. Gmbh, Wiesbaden, Germany, **2007**, 393 pp.; (g) B. Klajnert, M. Bryszewska (Eds.), *Dendrimers in Medicine*, Nova Science Publishers, **2007**, 141 pp.; (g) L.H. Gade (Ed.),

Dendrimer Catalysis, Top. Organometal. Chem., Springer-Verlag, Berlin, **2006**, 189 pp.; (h) U. Boas, J. B. Christensen, P. M. H. Heegard, L. Peng (Eds.), *Dendrimers in Medicine and Biotechnology: New Molecular Tools*, Royal Society of Chemistry, **2006**, 182 pp.; (i) D. A. Tomalia, J. M. Fréchet (Eds.), *Dendrimers and Dendritic Polymers, Prog. Polym. Sci.*, Elsevier, Amsterdam, **2005**, 288 pp.; (j) A. M. Caminade, J. P. Majoral, *Acc. Chem. Res.* **2004**, *37*, 341–348; (k) C. A. Schalley, F. Vögtle (Eds.), *Dendrimers V: Functional and Hyperbranched Building Blocks, Photophysical Properties, Applications in Materials and Life Sciences, Top. Curr. Chem.*, Springer-Verlag, Berlin, **2003**, 273 pp.; (l) G. R. Newkome (Ed.), *Advances in Dendritic Macromolecules: 2002*, Jai Press Inc., **2002**, 272 pp.; (m) J. M. J. Fréchet, D. A. Tomalia (Eds.), *Dendrimers and Other Dendritic Polymers*, John Wiley & Sons, **2004**, 688 pp.; (n) G. R. Newkome, C. N. Moorefield, F. Vögtle (Eds.), *Dendrimers and Dendrons: Concepts, Syntheses, Applications*, 2nd edition, Wiley-VCH, **2001**, 635 pp.; (o) C. A. Schalley, F. Vögtle (Eds.), *Dendrimers IV: Metal Coordination, Self Assembly, Catalysis, Top. Curr. Chem.*, Springer-Verlag, Berlin, **2001**, 243 pp.

[4] A. Topp, B. J. Bauer, D. A. Tomalia, E. J. Amis, *Macromolecules* **1999**, *32*, 7232–7237.

[5] P. R. Dvornic, D. A. Tomalia, *Macromol. Symp.* **1994**, *88*, 123–148.

[6] D. A. Tomalia, *Prog. Polym. Sci.* **2005**, *30*, 294–324.

[7] K. L. Wooley, C. J. Hawker, *Top. Curr. Chem.* **2005**, *245*, 287–305.

[8] D. A. Tomalia, *J. Nanoparticle Res.* **2009**, *11*, 1251–1310.

[9] A. M. Caminade, R. Laurent, J. P. Majoral, *Adv. Drug Deliv. Rev.* **2005**, *57*, 2130–2146.

[10] R. Duncan, H. Ringsdorf, R. Satchi-Fainaro, *J. Drug Target.* **2006**, *14*, 337–341.

[11] Y. Cheng, Z. Xu, M. Ma, T. Xu, *J. Pharm. Sci.* **2008**, *97*, 123–143.

[12] C. Liang, J. M. J. Fréchet, *Prog. Polym. Sci.* **2005**, *30*, 385–402.

[13] D. A. Tomalia, J. M. J. Fréchet, *J. Polym. Sci. Part A Polym. Chem.* **2002**, *40*, 2719–2728.

[14] E. Buhleier, W. Wehner, F. Vögtle, *Synthesis* **1978**, 155–159.

[15] D. A. Tomalia, H. Baker, J. Dewald, M. Hall, G. Kallos, S. Martin, J. Roeck, J. Ryder, P. Smith, *Polym. J.* **1985**, *17*, 117–132.

[16] G. R. Newcome, Z.-Q. Yao, G. R. Baker, V. K. Gupta, *J. Org. Chem.* **1985**, *50*, 2003–2004.

[17] R. G. Denkewalter, J. Kolk, W. J. Lukasavage, U.S. Pat. 4,289,872, 1981, *Chem. Abstr.* **1984**, *100*, 103907.

[18] J. C. Blais, C. O. Turrin, A. M. Caminade, J. P. Majoral, *Anal. Chem.* **2000**, *72*, 5097–5105.

[19] D. A. Tomalia, S. Uppuluri, D. R. Swanson, J. Li, *Pure Appl. Chem.* **2000**, *72*, 2343–2358.

[20] V. Maraval, J. Pyzowski, A. M. Caminade, J. P. Majoral, *J. Org. Chem.* **2003**, *68*, 6043–6046.

[21] N. Launay, A. M. Caminade, J. P. Majoral, *J. Am. Chem. Soc.* **1995**, *117*, 3282–3283.

[22] S. Hecht, J. M. J. Fréchet, *Angew. Chem. Int. Ed.* **2001**, *40*, 74–91.

[23] J. Jansen, E. M. M. Debrabandervandenberg, E. W. Meijer, *Science* **1994**, *266*, 1226–1229.

[24] J. P. Majoral, C. Larre, R. Laurent, A. M. Caminade, *Coord. Chem. Rev.* **1999**, *192*, 3–18.

[25] C. Larre, B. Donnadieu, A. M. Caminade, J. P. Majoral, *J. Am. Chem. Soc.* **1998**, *120*, 4029–4030.

[26] J. Haensler, F. C. SzokaJr., *Bioconj. Chem.* **1993**, *4*, 372–379.

[27] M. X. Tang, C. T. Redemann, F. C. SzokaJr., *Bioconj. Chem.* **1996**, *7*, 703–714.

[28] J. F. Kukowska-Latallo, A. U. Bielinska, J. Johnson, R. Spindler, D. A. Tomalia, J. R. Baker, Jr., *Proc. Natl. Acad. Sci. U.S.A.*, **1996**, *93*, 4897–4902.

[29] A. M. Caminade, C. O. Turrin, J. P. Majoral, *Chem. Eur. J.* **2008**, *14*, 7422–7432.

[30] R. F. Service, *Science* **1995**, *267*, 458–459.

[31] J. P. Majoral, A. M. Caminade, R. Laurent, P. Sutra, *Heteroatom. Chem.* **2002**, *13*, 474–485.

[32] M. J. Cloninger, *Curr. Opin. Chem. Biol.* **2002**, *6*, 742–748.

[33] R. Esfand, D. A. Tomalia, *Drug Discov. Today* **2001**, *6*, 427–436.

[34] Y. Kim, S. C. Zimmerman, *Curr. Opin. Chem. Biol.* **1998**, *2*, 733–742.

[35] B. Klajnert, M. Bryszewska, *Acta Biochim. Polonica* **2001**, *48*, 199–208.

[36] U. Boas, P. M. H. Heegaard, *Chem. Soc. Rev.* **2004**, *33*, 43–63.

[37] A. U. Bielinska, A. Yen, H. L. Wu, K. M. Zahos, R. Sun, N. D. Weiner, J. R. BakerJr., B. J. Roessler, *Biomaterials* **2000**, *21*, 877–887.

[38] S. F. M. van Dongen, H.-P. De Hoog, R. J. R. W. Peters, M. Nallani, R. J. M. Nolte, J. C. M. van Hest, *Chem. Rev.* **2009**, *109*, 6212–6274.

[39] B. Helms, E. W. Meijer, *Science* **2006**, *313*, 929–930.

[40] J. M. J. Fréchet, *Prog. Polym. Sci.* **2005**, *30*, 844–857.

[41] C. C. Lee, J. A. MacKay, J. M. J. Fréchet, F. C. Szoka, *Nat. Biotechnol.* **2005**, *23*, 1517–1526.

[42] S. Svenson, D. A. Tomalia, *Adv. Drug Deliv. Rev.* **2005**, *57*, 2106–2129.

[43] P. D. Senter, D. A. Scheinberg, *Mol. Pharmaceutics* **2009**, *6*, 82–85.

[44] H. Yang, W. J. Kao, *J. Biomater. Sci. Polym. Ed.* **2006**, *17*, 3–19.

[45] H. Ringsdorf, *J. Polym. Sci. Polym. Symp.* **1975**, *51*, 135–153.

[46] R. Duncan, *Nat. Rev. Drug Discov.* **2003**, *2*, 347–360.

[47] L. Gros, H. Ringsdorf, H. Schupp, *Angew. Chem. Int. Ed.* **1981**, *20*, 305–325.

[48] M. Micha-Screttas, H. Ringsdorf, *Curr. Top. Med. Chem.* **2008**, *8*, 1161–1164.

[49] P. B. Smith, S. J. Martin, M. J. Hall, D. A. Tomalia, A characterization of the structure and synthetic reactions of polyamidoamine "Starburst" polymers, in *Applied Polymer Analysis and Characterization* (Ed.: J. Mitchell Jr.), Hanser, Munchen/New York, **1987**, pp. 357–385.

[50] M. Liu, J. M. J. Fréchet, *Polym. Bull.* **1999**, *43*, 379–386.

[51] J. C. Hummelen, J. L. J. vanDongen, E. W. Meijer, *Chem. Eur. J.* **1997**, *3*, 1489.

[52] J. Peterson, V. Allikmaa, T. Pehk, M. Lopp, *Proc. Estonian Acad. Sci. Chem.* **2001**, *50*, 167–172.

[53] P. Kambouris, C. J. Hawker, *J. Chem. Soc. Perkin Trans 1*, **1993**, 199S, 2717–2721.

[54] D. Seebach, G. F. Herrmann, U. D. Lengweiler, B. M. Bachmann, W. Amrein, *Angew. Chem. Int. Ed.* **1996**, *35*, 2795–2796.

[55] M. X. Tang, C. T. Redemann, F. C. SzokaJr., *Bioconj. Chem.* **1996**, *7*, 703–714.

[56] A. M. Muzafarov, M. Golly, M. Moller, *Macromolecules* **1995**, *28*, 8444–8446.

[57] A. M. Muzafarov, E. A. Rebrov, O. B. Gorbacevich, M. Golly, H. Gankema, M. Moller, *Macromol. Symp.* **1996**, *102*, 35–46.

[58] M. Smet, L.-X. Liao, W. Dehaen, D. V. McGrath, *Org. Lett.* **2000**, *2*, 511–513.

[59] D. V. McGrath, *Mol. Pharmaceutics* **2005**, *2*, 253–263.

[60] S. Borman, *Chem. Eng. News* **2003**, *81*(39), 4–5.

[61] R. J. Amir, N. Pessah, M. Shamis, D. Shabat, *Ang. Chem. Int. Ed.* **2003**, *42*, 4494–4499.

[62] E. W. Meijer, M. H. P. van Genderen, *Nature* **2003**, *426*, 128–129.

[63] R. Duncan, H. Ringsdorf, R. Satchi-Fainaro, in *Polymer Therapeutics I: Polymers as Drugs, Conjugates and Gene Delivery Systems*, Vol. 192, Springer, **2006**, pp. 1–8.

[64] R. Duncan, *Nat. Rev. Cancer* **2006**, *6*, 688–701.

[65] R. Haag, F. Kratz, *Angew. Chem. Int. Ed.* **2006**, *45*, 1198–1215.

[66] J. Khandare, T. Minko, *Prog. Polym. Sci.* **2006**, *31*, 359–397.

[67] R. Duncan, *J. Drug Target.* **2006**, *14*, 333–335.

[68] Among many reviews: M. Okada, *Prog. Polym. Sci.* **2002**, *27*, 87–133.

[69] C. Gao, D. Yan, *Prog. Polym. Sci.* **2004**, *29*, 183–275.

[70] A. D. Schlüter, in *Functional Molecular Nanostructures, Top. Curr. Chem.* **2005**, *245*, 151–191.

[71] A. F. Zhang, L. J. Shu, Z. S. Bo, A. D. Schlüter, *Macromol. Chem. Phys.* **2003**, *204*, 328–339.

[72] A. D. Schlüter, J. P. Rabe, *Angew. Chem. Int. Ed.* **2000**, *39*, 865–883.

[73] C. R. Yates, W. Hayes, *Eur. Polym. J.* **2004**, *40*, 1257–1281.

[74] V. Percec, P. W. Chu, G. Ungar, J. P. Zhou, *J. Am. Chem. Soc.* **1995**, *117*, 11441–11454.

[75] V. Percec, C. H. Ahn, G. Ungar, D. J. P. Yeardley, M. Moller, S. S. Sheiko, *Nature* **1998**, *391*, 161–164.

[76] V. Percec, W. D. Cho, G. Ungar, D. J. P. Yeardley, *Angew. Chem. Int. Ed.* **2000**, *39*, 1598–1602.

[77] A. K. Patri, J. F. Kukowska-Latallo, J. R. BakerJr., *Adv. Drug Deliv. Rev.* **2005**, *57*, 2203–2214.

[78] (a) A. D'Emanuele, D. Attwood, *Adv. Drug Deliv. Rev.* **2005**, *57*, 2147–2162; (b) D. K. Smith, *Chem. Commun.* **2006**, 34–44.

[79] R. Haag, *Angew. Chem. Int. Ed.* **2004**, *43*, 278–282.

[80] Even if this concept was not demonstrated for the encapsulated guests, the conceptual idea is present in this communication: A. M. Balija, R. E. Kohman, S. C. Zimmerman, *Angew. Chem. Int. Ed.* **2008**, *47*, 8072–8074.

[81] D. J. Welsh, S. P. Jones, D. K. Smith, *Angew. Chem. Int. Ed.* **2009**, *48*, 4047–4051.

[82] G. P. Vlasov, *Russ. J. Bioorg. Chem.* **2006**, *32*, 205–218.

[83] R. Göller, J.-P. Vors, A.-M. Caminade, J.-P. Majoral, *Tetrahedron Lett.* **2001**, *42*, 3587–3590.

[84] J. M. J. Fréchet, *Proc. Natl. Acad. Sci. U.S.A.*, **2002**, *99*, 4782–4787.

[85] T. Emrick, J. M. J. Fréchet, *Curr. Opin. Colloid Interface Sci.* **1999**, *4*, 15–23.

[86] V. Percec, C.-H. Ahn, G. Ungar, D. J. P. Yeardley, M. Möller, S. S. Sheiko, *Nature* **1998**, *391*, 161–164.

[87] M. R. Radowski, A. Shukla, H. von Berlepsch, C. Bottcher, G. Pickaert, H. Rehage, R. Haag, *Angew. Chem. Int. Ed.* **2007**, *46*, 1265–1269.

[88] S. J. Xu, M. Kramer, R. Haag, *J. Drug Target.* **2006**, *14*, 367–374.

[89] S. Xu, Y. Luo, R. Haag, *Macromol. Biosci.* **2007**, *7*, 968–974.

[90] M. Krämer, J.-F. Stumbé, H. Türk, S. Krause, A. Komp, L. Delineau, S. Prokhorova, H. Kautz, R. Haag, *Angew. Chem. Int. Ed.* **2002**, *41*, 4252–4256.

[91] J. F. G. A. Jansen, E. W. Meijer, *J. Am. Chem. Soc.* **1995**, *117*, 4417–4418.

[92] H. Hui, F. Xiao-dong, C. Zhong-lin, *Polymer* **2005**, *46*, 9514–9522.

[93] H. Yang, S. T. Lopina, *J. Biomed. Mater. Res. A* **2005**, *72*, 107–114.

[94] H. L. Crampton, E. S. Simanek, *Polym. Int.* **2007**, *56*, 489–496.

[95] A. K. Patri, J. F. Kukowska-Latallo, J. R. BakerJr., *Adv. Drug Deliv. Rev.* **2005**, *57*, 2203–2214.

[96] Y. Cheng, T. Xu, *Eur. J. Med. Chem.* **2008**, *43*, 2291–2397.

[97] Y. Cheng, J. Wang, T. Rau, X. He, T. Xu, *Front. Biosci.* **2008**, *13*, 1447–1471.

[98] O. P. Perumal, R. Inapagollla, S. Kannan, R. M. Kannan, *Biomaterials* **2008**, *29*, 3469–3476.

[99] A. Abuchowski, J. R. McCoy, N. C. Palczuk, T. van Es, F. F. Davis, *J. Biol. Chem.* **1977**, *252*, 3582–3586.

[100] F. M. Veronese, G. Pasut, *Drug Discov. Today* **2005**, *10*, 1451–1458.

[101] D. M. Sweet, R. B. Kolhatkar, A. Ray, P. Swaan, H. Ghandehari, *J. Control. Rel.* **2009**, *138*, 78–85.

[102] Y. Kim, A. M. Klutz, K. A. Jacobson, *Bioconj. Chem.* **2008**, *19*, 1660–1672.

[103] M. Berna, D. Dalzoppo, G. Pasut, M. Manunta, L. Izzo, A. T. Jones, R. Duncan, F. M. Veronese, *Biomacromolecules* **2006**, *7*, 146–153.

[104] A. Saovapakhiran, A. D'Emanuele, D. Attwood, J. Penny, *Bioconj. Chem.* **2009**, *20*, 693–701.

[105] M. R. Radowski, A. Shukla, H. von Berlepsch, C. Bottcher, G. Pickaert, H. Rehage, R. Haag, *Angew. Chem. Int. Ed.* **2007**, *46*, 1265–1269.

[106] Among some reviews: C. M. Paleos, D. Tsiourvas, Z. Sideratou, L. Tziveleka, *Curr. Top. Med. Chem.* **2008**, *8*, 1204–1224.

[107] A. Almutairi, W. J. Akers, M. Y. Berezin, S. Achilefu, J. M. J. Fréchet, *Mol. Pharmaceutics* **2008**, *5*, 1103–1110.

[108] J. R. R. Maijigapu, A. N. Kurchan, R. Kottani, T. P. Gustafson, A. G. Kutateladze, *J. Am. Chem. Soc.* **2005**, *127*, 12458–12459.

[109] F. Wiesbrock, C. Patteux, T. K. Olszewski, A. Blanrue, G. A. Heropoulos, B. R. Steele, M. Micha-Scretas, T. Calogeropoulou, *Eur. J. Org. Chem.* **2008**, 4344–4349.

[110] D. T. Rijkers, G. W. van Esse, R. Merckx, A. J. Brouwer, H. J. Jacobs, R. J. Pieters, R. M. Liskamp, *Chem. Comm.* **2005**, 4581–4583.

[111] M. Liu, K. Kono, J. M. J. Fréchet, *J. Polym. Sci. Part A Polym. Chem.* **1999**, *37*, 3492–3503.

[112] M. W. Grinstaff, *Chem. Eur. J.* **2002**, *8*, 2838–2846.

[113] H. Ihre, A. Hult, J. M. J. Fréchet, I. Gitsov, *Macromolecules* **1998**, *31*, 4061–4068.

[114] H. Ihre, A. Hult, E. Soderlind, *J. Am. Chem. Soc.* **1996**, *118*, 6388–6395.

[115] O. L. P. De Jesus, H. R. Ihre, L. Gagne, J. M. J. Fréchet, F. C. Szoka, *Bioconj. Chem.* **2002**, *13*, 453–461.

[116] M. A. Carnahan, M. W. Grinstaff, *Macromolecules* **2006**, *39*, 609–616.

[117] N. R. Luman, K. A. Smeds M. W. Grinstaff, *Chem. Eur. J.* **2003**, *9*, 5618–5626.

[118] M. A. Carnahan, M. W. Grinstaff, *Macromolecules* **2001**, *34*, 7648–7655.

[119] M. A. Carnahan, M. W. Grinstaff, *J. Am. Chem. Soc.* **2001**, *123*, 2905–2906.

[120] N. R. Luman, M. W. Grinstaff, *Org. Lett.* **2005**, *7*, 4863–4866.

[121] N. R. Luman, T. Kim, M. W. Grinstaff, *Pure Appl. Chem.* **2004**, *76*, 1375–1385.

[122] R. R. Mallepally, I. Smirnova, W. Arlt, M. Seiler, S. K. Klee-Laquai, G. Hills, *J. Appl. Polym. Sc.* **2009**, *112*, 1873–1881.

[123] Among them: S. K. Potluri, A. R. Ramulu, M. Pardhasaradhi, *Tetrahedron* **2004**, *60*, 3739–3744.

[124] W. K. Lee, J. A. Gardella, *Langmuir* **2000**, *16*, 3401–3406.

[125] A. Breitenbach, Y. X. Li, T. Kissel, *J. Control. Rel.* **2000**, *64*, 167–178.

[126] Z.-M. Miao, S.-X. Cheng, X.-Z. Zhang, Q.-R. Wang, R.-X. Zhuo, *J. Biomed. Mater. Res. Part B Appl. Biomater.*, **2007**, *81B*(1), 40–49.

[127] E. R. Gillies, E. Dy, J. M. J. Fréchet, F. C. Szoka, *Mol. Pharmaceutics* **2005**, *2*, 129–138.

[128] S. K. Potluri, A. R. Ramulu, M. Pardhasaradhi, *Tetrahedron* **2004**, *60*, 10915–10920.

[129] S. Tang, L. J. Martinez, A. Sharma, M. Chai, *Org. Lett.* **2006**, *8*, 4421–4424.

[130] J. J. Khandare, S. Jayant, A. Singh, P. Chandna, Y. Wang, N. Vorsa, T. Minko, *Bioconj. Chem.* **2006**, *17*, 1464–1472.

[131] R. Erez, S. Ebner, B. Attali, D. Shabat, *Bioorg. Chem. Med. Lett.* **2008**, *18*, 816–820.

[132] W. Wrasidlo, U. Schröder, K. Bernt, N. Hübener, D. Shabat, G. Gaedicke, H. Lode, *Bioorg. Med. Chem. Lett.* **2002**, *12*, 557–560.

[133] M. Avital-Shmilovici, D. Shabat, *Bioorg. Med. Chem. Lett.* **2009**, *19*, 3959–3962.

[134] F. M. H. de Groot; W. J. Loos, R. Koekkoek, L. W. A. van Berkom; G. F. Busscher; A. E. Seelen, C. Albrecht, P. de Bruijn, H. W. Scheeren, *J. Org. Chem.* **2001**, *66*, 8815–8830.

[135] X. Li, Y. Su, Q. Chen, Y. Lin, Y. Tong, Y. Li, *Biomacromolecules* **2005**, *6*, 3181–3188.

[136] M. E. Fox, S. Guillaudeu, J. M. J. Fréchet, K. Jerger, N. Macaraeg, F. C. Szoka, *Mol. Pharmaceutics* **2009**, *6*, 1562–1572.

[137] L. M. Kaminskas, B. D. Kelly, V. M. McLeod, B. J. Boyd, G. Y. Krippner, E. D. Williams, C. J. H. Porter, *Mol. Pharmaceutics* **2009**, *6*, 1190–1204.

[138] H. Yang, S. T. Lopina, *J. Biomed. Mater. Res. Part A* **2006**, *76A*, 398–407 and references therein.

[139] M. Najlah, S. Freeman, D. Attwood, A. D'Emanuele, *Int. J. Pharm.* **2006**, *308*, 175–182.

[140] M. Calderon, R. Graeser, F. Kratz, R. Haag, *Bioorg. Med. Chem. Lett.* **2009**, *19*, 3725–3728.

[141] R. J. Amir, D. Shabat, *Chem. Commun.* **2004**, 1614–1615.

[142] W. Tansey, S. Ke, X.-Y. Cao, M. J. Pasuelo, S. Wallace, C. Li, *J. Control. Rel.* **2004**, *94*, 39–51.

[143] S. Gurdag, J. Khandare, S. Stapels, L. H. Matherly, R. M. Kannan, *Bioconj. Chem.* **2006**, *17*, 275–283.

[144] Dextran: E. M. Bachelder, T. T. Beaudette, K. E. Broaders, J. Dashe, J. M. J. Fréchet, *J. Am. Chem. Soc.* **2008**, *130*, 10494–10495.

[145] Chitosan: H. Sashiwa, H. Yajima, S.-i. Aiba, *Biomacromolecules* **2003**, *4*, 1244–1249.

[146] E.R. Gillies, A.P. Goodwin, J.M.J. Fréchet, *Bioconj. Chem.* **2004**, *15*, 1254–1263.

[147] R. Tomlinson, J. Heller, S. Brocchini, R. Duncan, *Bioconj. Chem.* **2003**, *14*, 1096–1106.

[148] E. R. Gillies, J. M. J. Fréchet, *Bioconj. Chem.* **2005**, *16*, 361–368.

[149] C. C. Lee, S. M. Grayson, J. M. J. Fréchet, *J. Polym. Sci. Part A Polym. Chem.* **2004**, *42*, 3563–3578.

[150] E. R. Gillies, J. M. J. Fréchet, *J. Am. Chem. Soc.* **2002**, *124*, 14137–14146.

[151] K. Kono, M. Liu, J. M. J. Fréchet, *Bioconj. Chem.* **1999**, *10*, 1115–1121.

[152] H. R. Ihre, O. L. Padilla De Jesus, F. C. SzokaJr., J. M. J. Fréchet, *Bioconj. Chem.* **2002**, *13*, 443–452.

[153] G. Saito, J. A. Swanson, K.-D. Lee, *Adv. Drug Deliv. Rev.* **2003**, *55*, 199–215 and references therein.

[154] M. A. Kostiainen, H. Rosilo, *Chem. Eur. J.* **2009**, *15*, 5656–5660.

[155] Y. E. Kurtoglu, R. S. Navath, B. Wang, S. Kannan, E. Romero, R. M. Kannan, *Biomaterials* **2009**, *30*, 2112–2121.

[156] C. Kojima, Y. Haba, T. Fukui, K. Kono, T. Takagishi, *Macromolecules* **2003**, *36*, 2183–2186.

[157] W. Shi, S. Dolai, S. Rizk, A. Hussain, H. Tariq, S. Averick, W. L'Amoreaux, A. El Idrissi, P. Banerjee, K. Raja, *Org. Lett.* **2007**, *9*, 5461–5464.

[158] H. Xu, A. S. R. Celeste, K. Yoshinori, H. Yukihiro, A. J. Gunn, M. Bernardo, H. Kobayashi, P. L. Choyke, M. W. Brechbiel, *Bioconj. Chem.* **2007**, *18*, 1474–1482.

[159] T. P. Thomas, A. K. Patri, A. Myc, M. T. Myaing, J. Y. Ye, T. B. Norris, J. R. Baker Jr., *Biomacromolecules* **2004**, *5*, 2269–2274.

[160] W. Zhang, S. E. Tichy, L. M. Pérez, G. C. Maria, P. A. Lindahl, E. E. Simanek, *J. Am. Chem. Soc.* **2003**, *125*, 5086–5094.

[161] For a review: R. Bischoff, S. E. Cray, *Prog. Polym. Sci.* **1999**, *24*, 185–219.

[162] M. Wang, D. Gan, K. L. Wooley, *Macromolecules* **2001**, *34*, 3215–3223.

[163] A. Gonzalez-Campo, C. Vinas, F. Teixidor, R. Nunez, R. Sillanpaa, R. Kivekas, *Macromolecules* **2007**, *40*, 5644–5652.

[164] B. Boury, R. J. P. Corriu, R. Nunez, *Chem. Mater.* **1998**, *10*, 1795–1804.

[165] R. E. Kohman, S. C. Zimmerman, *Chem. Commun.* **2009**, 794–796.

[166] M. Wathier, P. J. Jung, M. A. Carnahan, M. W. Grinstaff, *J. Am. Chem. Soc.* **2004**, *126*, 12744–12745.

[167] R. W. Lenz, R. H. Marchessault, *Biomacromolecules* **2005**, *6*, 1 and references therein.

[168] M. R. Timmins, R. W. Lenz, F. R. Clinton, *Polymer* **1997**, *38*, 551–562.

[169] Q. Cai, Y. Zhao, J. Bei, F. Xi, S. Wang, *Biomacromolecules* **2003**, *4*, 828–834.

[170] F. Aulenta, M. G. B. Drew, A. Foster, W. Hayes, S. Rannard, D. W. Thornthwaite, T. G. A. Youngs, *Molecules* **2005**, *10*, 81–97.

[171] L. Bracci, C. Falciani, B. Lelli, L. Lozzi, Y. Runci, A. Pini, M. Graziella De Montis, A. Tagliamonte, P. Neri, *J. Biol. Chem.* **2003**, *278*, 46590–46595.

[172] J. P. Tam, Y.-A. Lu, J.-L. Yang, *Eur. J. Biochem.* **2002**, *269*, 923–932.

[173] H. L. Crampton, E. E. Simanek, *Polym. Int.* **2007**, *56*, 489–496.

[174] C. M. Paleos, D. Tsiourvas, Z. Sideratou, *Mol. Pharmaceutics* **2007**, *4*, 169–188.

[175] S. Fuchs, T. Kapp, H. Otto, T. Schoneberg, P. Franke, R. Gust, A. D. Schlüter, *Chem. Eur. J.* **2004**, *10*, 1167–1192.

[176] I. J. Majoros, T. P. Thomas, C. B. Mehta, J. R. Baker Jr., *J. Med. Chem.* **2005**, *48*, 5892–5899.

[177] D. M. Sweet, R. B. Kolhatkar, A. Ray, P. Swaan, H. Ghandehari, *J. Control. Rel.* **2009**, *138*, 78–85.

[178] Y. Kim, A. M. Klutz, K. A. Jacobson, *Bioconj. Chem.* **2008**, *19*, 1660–1672.

[179] A. P. Umali, H. L. Crampton, E. E. Simanek, *J. Org. Chem.* **2007**, *72*, 9866–9874.

[180] A. Saovapakhiran, A. D'Emanuele, D. Attwood, J. Penny, *Bioconj. Chem.* **2009**, *20*, 693–701.

[181] L. M. Kaminskas, B. J. Boyd, P. Karellas, G. Y. Krippner, R. Lessene, B. Kelly, C. J. H. Porter, *Mol. Pharmaceutics* **2008**, *5*, 449–463.

[182] S. J. Guillaudeu, M. E. Fox, Y. M. Haidar, E. E. Dy, F. C. Szoka, J. M. J. Fréchet, *Bioconj. Chem.* **2008**, *19*, 461–469.

[183] J.-S. Lee, J. Huh, C.-H. Ahn, M. Lee, T. G. Park, *Macromol. Rapid Commun.* **2006**, *27*, 1608–1614.

[184] A. Cordova, K. D. Janda, *J. Am. Chem. Soc.* **2001**, *123*, 8248–8259.

[185] For a summary: D. Shabat, R. J. Amir, A. Gopin, N. Pessah, M. Shamis, *Chem. Eur. J.* **2004**, *10*, 2626–2634.

[186] R. Madec-Lougerstay, J.-C. Florent, C. Monneret, *J. Chem. Soc. Perkin Trans. 1* **1999**, 1369–1375.

[187] D. Shabat, C. Rader, B. List, R. A. Lerner, C. F. Barbas, *Proc. Natl. Acad. Sci. U.S.A.*, **1999**, *96*, 6925–6930.

[188] I. Niculescu-Duvaz, C. J. Springer, *Adv. Drug Deliv. Rev.* **1997**, *26*, 151–172.

[189] W. Maison, J. V. Frangioni, *Angew. Chem. Int. Ed.* **2003**, *42*, 4726–4728.

[190] E. W. Meijer, M. H. P. van Genderen, *Nature* **2003**, *426*, 128–129.

[191] K. Haba, M. Popkov, M. Shamis, R. A. Lerner, C. F. BarbasIII, D. Shabat, *Angew. Chem. Int. Ed.* **2005**, *44*, 716–720.

[192] R. X. Zhuo, B. Du, Z. R. Lu, *J. Control. Rel.* **1999**, *57*, 249–257.

[193] (a) R. M. Kevwitch, D. V. McGrath, *New. J. Chem.* **2007**, *31*, 1332–1336; (b) I. Grabchev, J. M. Chovelon, *Polym. Degrad. Stab.* **2007**, *92*, 1911–1915; (c) M. A. Kostiainen, D. K. Smith, O. Ikkala, *Angew. Chem. Int. Ed.* **2007**, *46*, 7600–7604; (d) N. Nishiyama, A. Iriyama, W.-D. Jang, K. Miyata, K. Itaka, Y. Inoue, H. Takahashi, Y. Yanagi, Y. Tamaki, H. Koyama, K. Kataoka, *Nat. Mater.* **2005**, *4*, 934–941; (e) I. Grabchev, R. Betcheva, V. Bojinov, D. Staneva, *Eur. Polym. J.* **2004**, *40*, 1249–1254; (f) M. L. Szalai, D. V. McGrath, *Tetrahedron* **2004**, *60*, 7261–7266; (g) I. Grabchev, V. Bojinov, J.-M. Chovelon, *Polymer* **2003**, *44*, 4421–4428; (h) R. M. Kevwitch, D. V. McGrath, *Synthesis* **2002**, 1171–1175. See also the examples in the carbohydrate-based dendrimers Section 13.3.11 using a photolytic cleavage.

[194] S. Li, M. L. Szalai, R. M. Kevwitch, D. V. McGrath, *J. Am. Chem. Soc.* **2003**, *125*, 10516–10517.

[195] M. L. Szalai, R. M. Kevwitch, D. V. McGrath, *J. Am. Chem. Soc.* **2003**, *125*, 15688–15689.

[196] (a) I. J. Majoros, A. Myc, T. Thomas, C. B. Mehta, J. R. Baker Jr., *Biomacromolecules* **2006**, *7*, 572–579; (b) H. Yang, S. T. Lopina, *J. Biomed. Mater. Res. Part A* **2006**, *76A*, 398–407; (c) P. Kolhe, J. Khandare, O. Pillai, S. Kannan, M. Lieh-Lai, R. M. Kannan, *Biomaterials* **2006**, *27*, 660–669; (d) J. Khandare, P. Kolhe, O. Pillai, S. Kannan, M. Lieh-Lai, R. M. Kannan, *Bioconj. Chem.* **2005**, *16*, 330–337; (e) A. Gopin, C. Rader, D. Shabat, *Bioorg. Med. Chem.* **2004**, *12*, 1853–1858; (f) M. Ternon, M. Bradley, *Chem.*

Commun. **2003**, 2402–2403; (g) D. A. Sarracino, C. Richert, *Bioorg. Med. Chem. Lett.* **2001**, *11*, 1733–1736.

[197] (a) M. Shamis, H. N. Lode, D. Shabat, *J. Am. Chem. Soc.* **2004**, *126*, 1726–1731; (b) M. Shamis, H. N. Lode, D. Shabat, *J. Am. Chem. Soc.* **2004**, *126*, 1726–1731; (c) N. Pessah, M. Reznik, M. Shamis, F. Yantiri, H. Xin, K. Bowdish, N. Shomron, G. Ast, D. Shabat, *Bioorg. Med. Chem.* **2004**, *12*, 1859–1866.

[198] (a) W. Ong, R. L. McCarley, *Chem. Commun.* **2005**, 4699–4701; (b) F. M. H. de Groot, C. Albrecht, R. Koekkoek, P. H. Beusker, H. W. Scheeren, *Angew. Chem. Int. Ed.* **2003**, *42*, 4490–4494; (c) M. L. Szalai, D. V. McGrath, *Polymer Preprint.* **2004**, *45*, 110–111.

[199] R. J. Amir, E. Danieli, D. Shabat, *Chem. Eur. J.* **2007**, *13*, 812–821.

[200] M. Cloninger, *Drug Discov. Today* **2004**, *9*, 111–112.

[201] K. Ulbrich, V. Subr, *Adv. Drug Deliv. Rev.* **2004**, *56*, 1023–1050.

[202] K. E. Uhrich, S. M. Cannizzaro, R. S. Langer, K. M. Shakesheff, *Chem. Rev.* **1999**, *99*, 3181–3198.

[203] O. M. Koo, I. Rubinstein, H. Onyuksel, *Nanomed. Nanotech. Biol. Med.* **2005**, *1*, 193–212.

[204] D. A. Tomalia, L. A. Reyna, S. Svenson, *Biochem. Soc. Trans.* **2007**, *35*, 61–67.

[205] N. G. Portney, M. Ozkan, *Anal. Bioanal. Chem.* **2006**, *384*, 620–630.

[206] M. Liu, J. M. J. Fréchet, *Pharm. Sci. Technol. Today* **1999**, *2*, 393–401.

[207] J. B. Wolinsky, M. W. Grinstaff, *Adv. Drug Deliv. Rev.* **2008**, *60*, 1037–1055.

[208] D. F. Williams, *J. Biomed. Eng.* **1989**, *11*, 185–191.

[209] R. Duncan, L. Izzo, *Adv. Drug Deliv. Rev.* **2005**, *57*, 2215–2237.

[210] H. T. Chen, M. F. Neeman, R. R. Parrish, E. E. Simanek, *J. Am. Chem. Soc.* **2004**, *126*, 10044–10048.

[211] A. Mecke, S. Uppuluri, T. M. Sassanella, D. K. Lee, A. Ramamoorthy, J. R. Baker Jr., B. G. Orr, M. M. Banaszak Holl, *Chem. Phys. Lipids* **2004**, *132*, 3–14.

[212] S. P. Hong, A. U. Bielinska, A. Mecke, B. Keszler, J. L. Beals, X. Y. Shi, L. Balogh, B. G. Orr, J. R. BakerJr., M. M. B. Holl, *Bioconj. Chem.* **2004**, *15*, 774–782.

[213] F. P. Seib, A. T. Jones, R. Duncan, *J. Control. Rel.* **2007**, *117*, 291–300.

[214] S. C. W. Richardson, K. L. Wallom, E. L. Ferguson, S. P. E. Deacon, M. W. Davies, A. J. Powell, R. C. Piper, R. Duncan, *J. Control. Rel.* **2008**, *127*, 1–11.

[215] A. T. Jones, M. Gumbleton, R. Duncan, *Adv. Drug Deliv. Rev.* **2003**, *55*, 1353–1357.

[216] M. El-Sayed, C. A. Rhodes, M. Ginski, H. Ghandehari, *Int. J. Pharm.* **2003**, *265*, 151–157 and references therein.

[217] H. Maeda, Y. Matsumura, *Cancer Res.* **1986**, *46*, 6387–6392.

[218] R. Duncan L. Seymour, *J. Drug Target.* **2007**, *15*, 456.

[219] F. Alexis, E. Pridgen, L. K. Molnar, O. C. Farokhzad, *Mol. Pharmaceutics* **2008**, *5*, 505–515.

[220] M. E. Fox, F. C. Szoka, J. M. J. Fréchet, *Acc. Chem. Res.* **2009**, *42*, 1141–1151.

[221] W. Yang, Y. Cheng, T. Xu, X. Wang, L.-P. Wen, *Eur. J. Med. Chem.* **2009**, *44*, 862–868.

[222] Y. Cheng, Y. Gao, T. Rao, Y. Li, T. Xu, *Comb. Chem. High Throughput Screen.* **2007**, *10*, 336–349.

[223] F. Aulenta, W. Hayes, S. Rannard, *Eur. Polym. J.* **2003**, *39*, 1741–1771.

[224] E. R. Gillies, J. M. J. Fréchet, *Drug Discov. Today* **2005**, *10*, 35–43.

[225] I. J. Majoros, C. R. Williams, J. R. BakerJr., *Curr. Top. Med. Chem.* **2008**, *8*, 1165–1179.

[226] C. Villalonga-Barber, M. Micha-Screttas, B. R. Georgopoulos, C. Demetzos, *Curr. Top. Med. Chem.* **2008**, *8*, 1294–1309.

[227] S. H. Medina, M. E. H. El-Sayed, *Chem. Rev.* **2009**, *109*, 3141–3187.

[228] R. K. Tekade, P. V. Kumar, N. K. Jain, *Chem. Rev.* **2009**, *109*, 49–87.

[229] D. Shabat, *J. Polym. Sci. Part A Polym. Chem.* **2006**, *44*, 1569–1578.

[230] R. J. Amir, M. Popkov, R. A. Lerner, C. E. Barbas, D. Shabat, *Angew. Chem. Int. Ed.* **2005**, *44*, 4378–4381.

[231] A. Gopin, S. Ebner, B. Attali, D. Shabat, *Bioconj. Chem.* **2006**, *17*, 1432–1440.

[232] E. W. P. Damen, T. J. Nevalainen, T. J. M. van den Bergh, F. M. H. de Groot, H. W. Scheeren, *Bioorg. Med. Chem.* **2002**, *10*, 71–77.

[233] D. Shabat, H. N. Lode, U. Pertl, R. A. Reisfeld, C. Rader, R. A. Lerner, C. F. Barbas, *Proc. Natl. Acad. Sci. U.S.A.*, **2001**, *98*, 7528–7533.

[234] W. Wrasidlo, U. Schröder, K. Bernt, N. Hübener, D. Shabat, G. Gaedicke, H. Lode, *Bioorg. Med. Chem. Lett.* **2002**, *12*, 557–560.

[235] C. Falciani, M. Fabbrini, A. Pini, L. Lozzi, B. Lelli, S. Pileri, J. Brunetti, S. Bindi, S. Scali, L. Bracci, *Mol. Cancer Ther.* **2007**, *6*, 2441–2448.

[236] R. Shukla, T. P. Thomas, J. Peters, A. Kotlyar, A. Myc, J. R. Baker Jr., *Chem. Commun.* **2005**, 5739–5741.

[237] C. C. Lee, E. R. Gillies, M. E. Fox, S. J. Guillaudeu, J. M. J. Fréchet, E. E. Dy, F. C. Szoka, *Proc. Natl. Acad. Sci. U.S.A.*, **2006**, *103*, 16649–16654.

[238] N. Malik, E. G. Evagorou, R. Duncan, *Anti-Cancer Drugs* **1999**, *10*, 767–776.

[239] M. Najlah, S. Freeman, D. Attwood, A. D'Emanuele, *Int. J. Pharm.* **2007**, *336*, 183–190.

[240] R. Wiwattannapatapee, L. Lomlim, K. Saramunee, *J. Control. Rel.* **2003**, *88*, 1–9.

[241] S. Z. Tang, S. M. June, B. A. Howell, M. H. Chai, *Tetrahedron Lett.* **2006**, *47*, 7671–7675.

[242] H. Yang, S. T. Lopina, *J. Biomater. Sci. Polym. Ed.* **2003**, *14*, 1043–1056.

[243] J. Jagur-Grodzinski, *React. Funct. Polym.* **1999**, *39*, 99–138.

[244] B. D. Ratner, A. S. Hoffman, F. J. Schoen, J. E. Lemons, *Biomaterials Science: An Introduction to Materials in Medicine*, Academic Press, San Diego, **2000**.

[245] R. P. Lanza, R. Langer, W. L. Chick, in *Chick, Principles of Tissue Engineering* (Ed.: R. G. Landes), Academic Press, San Diego, CA, **1997**.

[246] M. Goldberg, R. Langer, X. Jia, *J. Biomater. Sci. Polym. Ed.* **2007**, *18*, 241–268.

[247] M. W. Grinstaff, *J. Polym. Sci. Part A Polym. Chem.* **2008**, *46*, 383–400.

[248] N. Joshi, M. Grinstaff, *Curr. Top. Med. Chem.* **2008**, *8*, 1225–1236.

[249] J. M. Loutsch, D. Ong, J. M. Hill, Dendrimers: an innovative and enhanced ocular drug delivery system, in *Ophthalmic Drug Delivery System* (Ed.: A. K. Mitra), Chapter 15, Marcel Dekker Inc., New York, **2003**, pp. 467–492.

[250] H. Keul, M. Moller, *J. Polym. Sci. Part A Polym. Chem.* **2009**, *47*, 3209–3231.

[251] S. H. M. Söntjiens, D. L. Nettles, M. A. Carnahan, L. A. Setton, M. W. Grinstaff, *Biomacromolecules* **2006**, *7*, 310–316.

[252] L. Degoricija, E. N. Bansal, S. H. M. Sontjens, N. S. Joshi, M. Takahashi, B. Snyder, M. W. Grinstaff, *Biomacromolecules* **2008**, *9*, 2863–2872.

[253] For a review: M. W. Grinstaff, *Biomaterials* **2007**, *28*, 5205–5214.

[254] M. A. Carnahan, M. W. Grinstaff, *J. Am. Chem. Soc.* **2001**, *123*, 2905–2906.

[255] A. J. Velazquez, M. A. Carnahan, J. Kristinsson, S. Stinnett, M. W. Grinstaff, *Arch. Ophthalmol.* **2004**, *122*, 867–870.

[256] L. Degoricija, C. S. Johnson, M. Wathier, T. Kim, M. W. Grinstaff, *Investig. Ophthalmol. Visual Sci.* **2007**, *48*, 2037–2042.

[257] M. A. Carnahan, C. Middleton, J. Kim, T. Kim, M. W. Grinstaff, *J. Am. Chem. Soc.* **2002**, *124*, 5291–5293.

[258] P. C. Khang, M. A. Carnahan, M. Wathier, M. W. Grinstaff, *J. Cataract. Refrac. Surg.* **2005**, *31*, 1208–1212.

[259] E. Frérot, K. Herbal, A. Herrmann, *Eur. J. Org. Chem.* **2003**, 967–971.

[260] F. Aulenta, M. G. B. Drew, A. Foster, W. Hayes, S. Rannard, D. W. Thornthwaite, T. G. A. Youngs, *Molecules* **2005**, *10*, 81–97.

[261] A. Trachsel, J.-Y. de Saint Laumer, O. P. Haefliger, A. Hermann, *Chem. Eur. J.* **2009**, *15*, 2846–2860.

[262] B. Levrand, Y. Ruff, J.-M. Lehn, A. Herrmann, *Chem. Commun.* **2006**, 2965–2967.

[263] C. Ternat, L. Ouali, H. Sommer, W. Fieber, M. I. Velazco, C. J. G. Plummer, G. Kreutzer, H.-A. Klok, J.-A. E. Manson, A. Herrmann, *Macromolecules* **2008**, *41*, 7079–7089.

[264] R. Göller, J.-P. Vors, A.-M. Caminade, J.-P. Majoral, *Tetrahedron Lett.* **2001**, *42*, 3587–3590.

[265] D. S. Deutsch, A. Siani, P. T. Fanson, H. Hirata, S. Matsumoto, C. T. Williams, M. D. Amiridis, *J. Phys. Chem. C* **2007**, *111*, 4246–4255.

[266] B. A. Howell, D. Fan, L. Rakesh, *J. Thermal Anal. Calorimetry* **2006**, *85*, 17–20.

[267] O. Ozturk, T. J. Black, K. Perrine, K. Pizzolato, C. T. Williams, F. W. Parsons, J. S. Ratliff, J. Gao, C. J. Murphy, H. Xie, H. J. Ploehn, D. A. Chen, *Langmuir* **2005**, *21*, 3998–4006.

[268] D. C. Thully, A. R. Trimble, J. M. J. Fréchet, *Adv. Mater.* **2000**, *12*, 1118–1122.

[269] D. Pasini, Q. J. Niu, R. P. Meagley, D. C. Tully, A. R. Trimble, J. M. J. Fréchet, *J. Photopolymer Sci. Tech.* **1999**, *12*, 405–416.

[270] M. L. Szalai, D. V. McGrath, D. R. Wheeler, T. Zifer, J. R. McElhanon, *Macromolecules* **2007**, *40*, 818–823.

[271] J. R. McElhanon, D. R. Wheeler, *Org. Lett.* **2001**, *3*, 2681–2683.

[272] Z. Huang, W. Shi, *Polym. Degrad. Stab.* **2007**, *92*, 1193–1198.

[273] M. S. Wendland, S. C. Zimmerman, *J. Am. Chem. Soc.* **1999**, *121*, 1389–1390.

[274] S. L. Elmer, S. C. Zimmerman, *J. Org. Chem.* **2004**, *69*, 7363–7366.

[275] S. C. Zimmerman, N. G. Lemcoff, *Chem. Commun.* **2004**, 5–14.

[276] Y. Ge, A. P. F. Turner, *Trends Biotechnol.* **2008**, *26*, 218–224.

[277] S. C. Zimmerman, M. S. Wendland, N. A. Rakow, I. Zhaov, K. S. Suslick, *Nature* **2002**, *418*, 399–403.

[278] S. C. Zimmerman, I. Zharov, M. S. Wendland, N. A. Rakow, K. S. Suslick, *J. Am. Chem. Soc.* **2003**, *125*, 13504–13518.

[279] E. Mertz, S. C. Zimmerman, *J. Am. Chem. Soc.* **2003**, *125*, 3424–3425.

[280] S. C. Zimmerman, J. R. Quinn, E. Burakowska, R. Haag, *Angew. Chem. Int. Ed.* **2007**, *46*, 8164–8167.

[281] L. G. Schultz, Y. Zhao, S. C. Zimmerman, *Angew. Chem. Int. Ed.* **2001**, *40*, 1962–1966.

[282] J. B. Beil, S. C. Zimmerman, *Chem. Commun.* **2004**, 488–489.

[283] E. Mertz, S. L. Elmer, A. M. Balija, S. C. Zimmerman, *Tetrahedron* **2004**, *60*, 11191–11204.

[284] T. Amaya, H. Tanaka, T. Takahashi, *Synlett* **2004**, 503–507.

[285] A. B. Kantchev, J. R. Parquette, *Tetrahedron Lett.* **1999**, *40*, 8049–8053.

[286] N. A. Stasko, T. H. Fischer, M. H. Schoenfisch, *Biomacromolecules* **2008**, *9*, 834–841.

[287] N. A. Stasko, M. H. Schoenfisch, *J. Am. Chem. Soc.* **2006**, *128*, 8265–8271.

[288] W. R. Dichtel, J. M. Serin, C. Edder, J. M. J. Fréchet, M. Matuszewski, L.-S. Tan, T. Y. Ohulchanskyy, P. N. Prasad, *J. Am. Chem. Soc.* **2004**, *126*, 5380–5381.

14

PORPHYRIN DENDRIMERS AS BIOLOGICAL OXYGEN SENSORS

SERGEI A. VINOGRADOV AND DAVID F. WILSON

Department of Biochemistry and Biophysics, University of Pennsylvania, PA 19104, Philadelphia, USA

14.1 INTRODUCTION

The growing interest in dendrimers over the past years has been due to their unique molecular design, which offers numerous possibilities in catalysis, construction of nanoscale electro-optical devices, drug delivery, and other applications [1]. Although very attractive from the point of view of basic research, the mass monodispersity and monodispersity of physical properties of dendrimers comes at a high cost of multistage synthesis, which often diminishes the practicality of dendritic materials. Indeed, in only a few applications have the unique structural characteristics of dendrimers been able to justify the tedious synthetic and purification procedures, which often lead to a preference for imperfect, but much less expensive hyperbranched polymers.

In such applications, monodispersity must prevail over cost considerations, and this requirement is often met in biological sensing and medicine. For example, in medical imaging it is highly desirable that physical properties of imaging agents are distributed within a narrow range; otherwise, heterogeneity of image tracers can result in reduced resolution of the imaging modality. It is also important that exogenous probes or contrast agents have well-defined, reproducible and predictable biodistributions, which is most readily achieved through the use of monodisperse materials. In this context, it is not surprising that magnetic resonance imaging using dendritic contrast agents was one of the first proposed uses of dendrimers [2].

Designing Dendrimers, Edited by Sebastiano Campagna, Paola Ceroni, Fausto Puntoriero.
© 2012 John Wiley & Sons, Inc. Published 2012 by John Wiley & Sons, Inc.

Two unique features of dendritic architecture are especially appealing for design of imaging tracers: the ability to encapsulate and "protect" the central core motif, and the ability to enhance a selected molecular property through multiplying functionalities at the dendrimer periphery (or in the interior). Over the years, dendritic encapsulation has received tremendous attention in different areas of chemistry [3,4]. For design of imaging agents it offers a convenient way to isolate the sensor moiety from components of biological systems, improve its solubility and minimize potential toxicity. On the other hand, the ability to place multiple functional units at the periphery of a dendrimer makes it possible to modulate the core nanoenvironment, and by doing so fine-tune the core properties to meet the requirements of the imaging application. In addition, peripheral modification with groups of different charge and hydrophobicity provides a method for controlling solubility, *in vivo* distribution and excretability of imaging agents.

In this chapter we focus on the construction of molecular probes for imaging oxygen in biological systems, which has been the central topic of our research in the past several years. Phosphorescent imaging probes are based on porphyrin dendrimers—a class of functional macromolecules reminiscent of heme-containing proteins, in which hemes are buried deep inside polypeptide macrostructures and protected from direct interactions with solvent and solutes. Mimicking the protein matrix, hydrophobic dendritic wedges in porphyrin dendrimers modulate properties of the encapsulated cores and control the accessibility of the cores to small molecules. Not surprisingly, porphyrin dendrimers have continuously attracted attention in light harvesting, electron transfer (ET), host–guest chemistry, photodynamic therapy and many other areas of research. Their synthesis and use in these applications have been recently covered in excellent comprehensive reviews [5,6].

In porphyrin-based oxygen probes, dendrimers serve to provide well-defined nanoenvironments for optically active cores, that is, Pt or Pd porphyrins, and to create an interface between phosphorescent cores of the probes and biological systems. Most importantly, folded dendritic matrix enables control over the kinetic accessibility of metalloporphyrins for small quencher molecules (oxygen)—a property much needed for tuning the sensitivity and the dynamic range of the oxygen measurement method.

Below, after a brief description of the principles of the oxygen imaging method based on quenching of phosphorescence, we formulate the requirements for *in vivo* oxygen probes and describe molecular oxygen sensors-based Pd and Pt porphyrin (PtP) dendrimers. We then address probes designed specifically for two-photon (2P) oxygen microscopy—a mode of oxygen imaging *in vivo* in three dimensions with near diffraction-limited resolution. Examples of applications of porphyrin dendrimers in oxygen sensing and imaging are featured in the end of the chapter.

14.2 PRINCIPLES OF THE PHOSPHORESCENCE QUENCHING METHOD AND REQUIREMENTS TO *IN VIVO* OXYGEN PROBES

Methods for measurement and imaging of oxygen play a central role in quantification of cellular metabolism and understanding tissue physiology [7,8]. A number of approaches to oxygen measurements in biological systems have been proposed in the past [9–13], but all of them need continuous development in order to satisfy diverse needs of biomedical research. Oxygen-dependent quenching of phosphorescence [14], or simply phosphorescence quenching, is an optical technique, suitable for oxygen measurements in the physiological pO_2 (partial pressure of oxygen) range. Originally proposed in 1987 for *in vitro* oxygen sensing [14], phosphorescence quenching soon thereafter was implemented for local *in vivo* single-point oxygen measurements [15] and planar 2-D imaging [16]. Currently, high-resolution microscopy [17,18] and 3-D tomography [19–21] of oxygen are being developed. The characteristic features of phosphorescence quenching include excellent specificity, submillisecond temporal response, high sensitivity, especially at low oxygen concentrations and relative simplicity of implementation. This method is being increasingly applied in biological research, driving the development and continuous improvement of phosphorescent probes.

Phosphorescence quenching relies on the ability of molecular oxygen, which is a triplet molecule in the ground state ($O_2X^3\Sigma_g^-$), to react with molecules in their excited states, quenching their luminescence [22]. Collisional quenching is much less probable on the time scale of singlet excited states (nanoseconds) than of triplet states (microseconds to milliseconds), making phosphorescence significantly more sensitive to oxygen than fluorescence. Assuming a large excess of oxygen relative to the concentration of triplet emitters—a condition typically met in biological environments—the dependence of the phosphorescence intensity and lifetime on oxygen concentration follows the Stern–Volmer relationship:

$$I_0/I = \tau_0/\tau = 1 + K_{SV}[O_2] \tag{14.1}$$

where I and τ are the phosphorescence intensity and the lifetime at oxygen concentration $[O_2]$ and in the absence of oxygen (I_0, τ_0); and K_{SV} is the Stern–Volmer quenching constant. In practice, using lifetime τ as the analytical signal for $[O_2]$ is more accurate, because the lifetime is independent of the probe distribution and of any other chromophores present in the biological system.

It is customary to express oxygen content in the units of pressure (mmHg) rather than concentration (M),[1] since in the majority of biological experiments partial pressure of oxygen (pO_2) is the actually controlled experimental parameter. We assume that the Henry's law holds in the physiological range of oxygen concentration: $[O_2] = \alpha \times pO_2$, where α is the oxygen solubility coefficient (M/mmHg) for the bulk phase. Considering $K_{SV} = k_2\tau_0$, where k_2 is the bimolecular rate constant for the quenching reaction, Eq. 14.1 can be rewritten as

$$1/\tau = 1/\tau_0 + k_q \times pO_2 \tag{14.2}$$

where $k_q = \alpha k_2$ and has the units of $mmHg^{-1}\,s^{-1}$.

[1] At 298 K and the air pressure of 760 mmHg (oxygen fraction in the air is 21% or 159.6 mmHg), air-equilibrated aqueous solutions are 252 μM in O_2 [23]. Here for simplicity we neglect the pressure of water vapor above the solution, although at higher temperatures it will rise and, consequently, the partial pressure of oxygen (pO_2) will decrease.

Eq. 14.2 contains two parameters specific to the molecule of the probe: constant k_q and lifetime τ_0. Their interplay defines the sensitivity and the dynamic range of the method. For analytical purposes, it is desirable that the measurement parameter, which in our case is the phosphorescence lifetime τ, spans the largest possible interval of values throughout the range of analyte concentrations, assuring highest possible measurement resolution. To quantify the dynamic range of lifetimes we use parameter $R = (\tau_0 - \tau_{air})/\tau_0$, where τ_0 ($pO_2 = 0$ mmHg) and τ_{air} ($pO_2 = 159.6$ mmHg) are the maximal and the minimal values of the phosphorescence lifetimes in physiological experiments. Another important parameter is the signal-to-noise ratio (SNR), which is obviously higher for probes with larger emission quantum yields (assuming the same emission wavelengths).

Pt and Pd porphyrins are typically used as phosphorescent chromophores for oxygen measurements in biological systems (see below). Triplet lifetimes of Pd porphyrins in deoxygenated solutions at ambient temperatures are in the range of hundreds of microseconds, and their phosphorescence quantum yields in the absence of oxygen (ϕ_0) are typically 0.05–0.1 [24]. For Pt porphyrins the corresponding values are tens of microseconds and 0.10–0.25 [24,25], respectively. Constants k_q for "unprotected" metalloporphyrins in aqueous solutions are $\sim$3000 mmHg^{-1} s^{-1}. To illustrate how quenching constants k_q and lifetimes τ_0 affect oxygen measurements, consider two probes, Pd tetraphenylporphyrin (PdP) and PtP, representing some arbitrary Pd and Pt porphyrins: τ_0(PdP) = 500 μs, τ_0(PtP) = 50 μs; ϕ_0(PdP) = 0.05, ϕ_0(PtP) = 0.10, where subscript "0" indicates $pO_2 = 0$ mmHg. Graphs in Figure 14.1 show how the phosphorescence lifetimes (a) and quantum yields (b) of these probes change throughout the physiological pO_2 range.

When the quenching constant is high, for example, $k_q = 3000$ mmHg^{-1} s^{-1} (solid lines in the graphs), the lifetime of PdP decreases from $\tau_0 = 500$ μs to $\tau_{air} = 2.1$ μs, resulting in a large dynamic range ($R \approx 0.996$). However, between about 20 mmHg and air saturation, the lifetime changes by no more than 3% of the total value, between

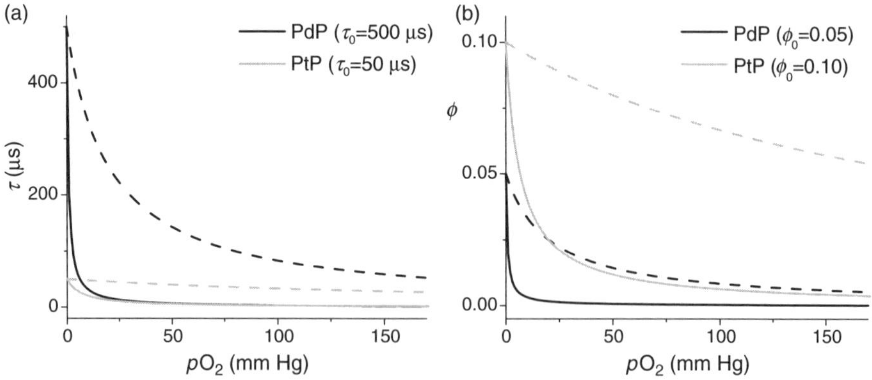

FIGURE 14.1 Dependencies of phosphorescence lifetimes (a) and quantum yields (b) for arbitrary Pd (black) and Pt (gray) porphyrins with two different quenching constants: $k_q =$ 3000 mmHg^{-1} s^{-1} (solid line) and $k_q = 100$ mmHg^{-1} s^{-1} (dashed line).

16 and 2.1 μs (Figure 14.1a). Moreover, already at 10 mmHg the probe's quantum yield becomes extremely low ($\phi < 0.003$) due to the excessive quenching. Consequently, a probe-like PdP would be useful only in a very low range of oxygen concentrations, probably not higher than at $pO_2 \sim 5$ mmHg.

Due to its higher quantum yield and shorter τ_0, probe PtP can be used up to about 50 mmHg ($\tau = 5.9$ μs), but above that limit its lifetime also changes very weakly, by no more than ~3 μs. The lifetime dynamic range of PtP ($R \approx 0.96$) is not very different from that of PdP; but if the quenching constant were to decrease, which is inevitable upon binding of the probe to proteins *in vivo* (see below), this probe would become weakly sensitive to changes in oxygenation. This last scenario is illustrated by the plot corresponding to $k_q = 100\ \text{mmHg}^{-1}\,\text{s}^{-1}$ (Figure 14.1a, dashed gray line), which reveals a rather small dynamic range $R \approx 0.44$. Notably, for probes with short lifetimes, for example, $\tau_0 \sim 1$ μs (typical of $Ru^{2+}(bpy)_3$ and similar complexes [26]), the dynamic range becomes unacceptably small ($R \sim 0.02$), which corresponds to mapping the entire physiological pO_2 range (~0–160 mmHg) onto ~20 ns-interval of lifetimes, on average ~125 ps/mmHg.

It is clear that for higher sensitivity probes with greater τ_0s are preferred, but only if their phosphorescence changes gradually instead of being highly quenched already at very low oxygen concentrations. Such adjustment of sensitivity can be achieved by varying constants k_q as shown in Figure 14.2. For example, at $k_q = 100\ \text{mmHg}^{-1}\,\text{s}^{-1}$ (Figure 14.1a), probe PdP still shows adequate dynamic range ($R = 0.89$), while its quantum yield even at the air saturation is large enough to permit oxygen measurements (Figure 14.1b). Notably, in some cases, when the signal strength is particularly critical, such as in scanning microscopy applications (see below), probes with higher

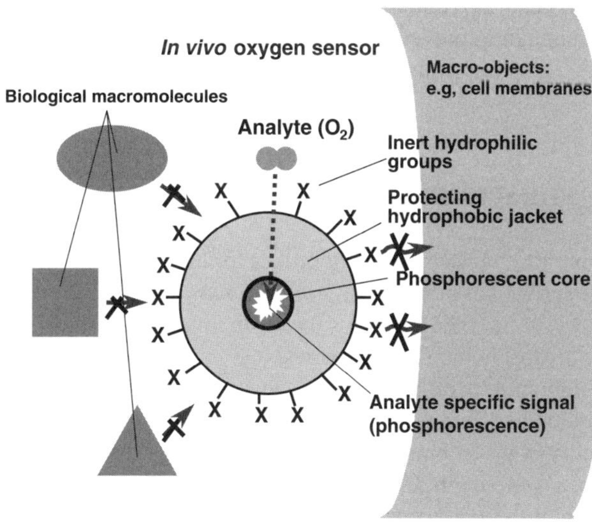

FIGURE 14.2 Cartoon illustrating design of "protected" phosphorescent sensors for *in vivo* oxygen imaging applications.

quantum yields can be preferred even at the expense of sensitivity, especially if only qualitative information about pO_2 is required. Overall, it is clear that control over the values of the quenching constants and the ability to keep them unaltered in measurement environments is the key to accurate oxygen measurements.

Phosphorescence quenching by oxygen typically occurs much faster than the diffusion of the reactants and formation of encounter complexes; and in the majority of cases, diffusion can be considered the rate-limiting step for oxygen quenching reaction [22]. By altering oxygen diffusion coefficients in the local vicinity of the phosphorescent chromophores, constants k_q can be regulated. Such tuning can be achieved by means of dendritic encapsulation.

Additional requirements for phosphorescent probes for biological imaging of oxygen include lack of toxicity and preferably excretability from the blood upon completion of imaging. Globular uncharged molecules with molecular weights up to about 15 kDa are usually excretable by the kidney [27]. If a probe satisfies this criterion and remains confined to the intravascular space (does not diffuse out of the blood vessels), it is likely to be removed from the blood by the kidney-mediated dialysis, and the possibility of long-term toxicity effects can be avoided. Of course, for animal studies, excretability is not as stringent requirement, whereas confinement to a particular tissue compartment (intravascular, interstitial, or intracellular) can be very important. In that case, probes of larger sizes may become advantageous.

The above requirements lead to a molecular design (Figure 14.2), which comprises a bright phosphorescent chromophore with sufficiently long triplet lifetime τ_0 and a protective jacket, whose purpose is to constrain oxygen diffusion in the local environment of the chromophore. The periphery of the probe must be hydrophilic and inert in order to prevent interactions with components of the biological system (e.g., biomacromolecules, cellular membranes). The overall size of the molecule should be large enough to prevent probe leakage through the vascular walls, but yet small enough to allow removal through the kidney, if excretability is desired.

14.3 EARLY PHOSPHORESCENT PROBES

The originally proposed phosphorescent probes for biological oxygen imaging were based on simple derivatives of Pd-porphyrins, Pd-*meso*-tetracarboxyphenylporphyrin (PdTCPP) (Figure 14.3a) being perhaps the most common example [7]. Porphyrins, being relatively hydrophobic chromophores, had to be prebound to serum albumin (usually bovine albumin) in order to enhance their solubility and prevent aggregation. Substantial research has been generated using albumin bound probes, demonstrating broad applicability of the technology, but simultaneously revealing limitations due to the use of the exogenous protein, which became a part of the injected material. Nevertheless, it is important to recognize that practically all modes of phosphorescence quenching method were first demonstrated using simple porphyrin-based albumin bound probes, such as PdTCPP and Green 2W [28,29] (Figure 14.3).

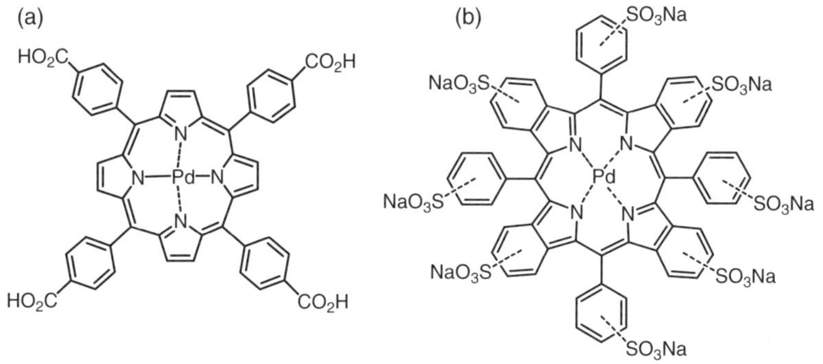

FIGURE 14.3 Simple water-soluble derivatives of phosphorescent Pd porphyrins used originally as complexes with serum albumin for oxygen measurements by phosphorescence quenching: Pd-*meso*-tetracarboxyphenylporphyrin (PdTCPP) (a) and polysulfonato derivative of Pd tetraphenyltetrabenzoporphyrin (Green 2W).

In order to avoid the need for prebinding to albumin, phosphorescent chromophores have to be made water soluble in the physiological pH range. Unfortunately, porphyrins modified with simple ionic groups (SO_3^-, CO_2^-, NH_3^+) or neutral hydrophilic residues, such as small oligoethyleneglycols [30], are still prone to aggregation. In addition, when injected systemically, porphyrins, as well as other small-molecule luminescent probes (e.g., $Ru(bpy)^{3+}$-like complexes) [31], become unevenly distributed throughout the tissue, where they bind to many biological targets. Local environments of probe molecules bound to different proteins are highly heterogeneous, resulting in unpredictable distributions of oxygen quenching constants k_q (Eq. 14.2). As a result, the response of the phosphorescence to oxygen cannot be interpreted quantitatively.

The problem of keeping the diffusional accessibility of the phosphorescent chromophore to oxygen in biological environments constant throughout the studied object and the calibration unchanging, is by far the biggest challenge in the design of phosphorescent oxygen probes. Moreover, this problem is inherent in all methods relying on kinetics of oxygen diffusion. To overcome this difficulty, design of a molecular oxygen sensor entails construction of a well-defined microenvironment around the phosphorescent chromophore in order to isolate it from interactions with other molecules, except for oxygen. Dendritic encapsulation [3,4] arguably provides one of the most straightforward ways to construct monodisperse, well-defined molecular jackets around luminescent chromophores [32,33].

14.4 PHOSPHORESCENT CORES OF DENDRITIC OXYGEN PROBES

The key photophysical properties required for biological oxygen sensing are strong absorption, preferably in the near-infrared region of the spectrum (from

~630 to ~900 nm) to minimize the interference by the natural tissue chromophores (e.g., heme containing proteins, carotenoids), and strong long-lived phosphorescence in solutions at ambient temperatures. Very few chromophores possess such properties. Among them, $Ru^{2+}(bpy)_3$ and related complexes emit from their MLCT states, whose lifetimes are usually in the range of several microseconds [26]. Metalloporphyrins (e.g., Figure 14.3) emit from $(\pi\text{–}\pi^*)^3$ states and exhibit significantly longer triplet lifetimes [24], thus possessing much higher intrinsic sensitivity to oxygen.

Regular *meso*-tetraarylated Pt and Pd porphyrins with various peripheral substituents can be readily synthesized via, for example, the Lindsey [34] or Senge [35] methods. Thermodynamic stabilities of these complexes are extremely high [36], entirely precluding release of free metal ions into biological environments and associated toxicity. The major drawback of using regular Pt and Pd porphyrins for *in vivo* applications is that their absorption bands, the so-called Q-bands, are positioned in the visible range ($\lambda_Q \sim 520$–530 nm; $\varepsilon \sim 20{,}000\ M^{-1}\ cm^{-1}$), thus overlapping with absorption of naturally occurring chromophores. Still, absorption in the visible region may be useful in planar wide-field phosphorescence imaging, where less diffuse nature of excitation serves to improve spatial resolution. In addition, absorption near 500 nm overlaps with emission of many 2P chromophores, which is useful in construction of Förster-type resonance energy transfer (FRET)-enhanced probes for two-photon oxygen microscopy (see below) [17].

Many porphyrinoids possess near-infrared absorption bands [37]; however, very few have characteristics appropriate for oxygen sensing [28,38]. The most useful today are derivatives of so-called laterally π-extended (or π-expanded) porphyrins. Lateral π-extension of Pt and Pd porphyrins by annealing their pyrrole residues with external aromatic rings results in chromophores with dramatically red-shifted absorption bands and strong room-temperature phosphorescence [28,39,40]. Structures and absorption and emission spectra of PdP, Pd tetraphenyltetrabenzoporphyrin (PdTBP), and Pd tetraphenyltetranaphthoporphyrin (PdTNP), are shown as examples in Figure 14.4.[2] The spectra of Pt analogs are very similar in shape, but blue-shifted by ~10–15 nm compared to those of Pd counterparts.

The absorption bands of regular porphyrins, TBPs and TNPs, when taken together, cover practically the entire UV-vis-NIR range, presenting multiple opportunities for excitation. The absorption Q-bands of TBPs and TNPs are shifted to the red, that is, into the region between ~630 and ~950 nm, where the absorption of endogenous chromophores is significantly diminished [41]. The oscillator strengths of these bands are remarkably high ($\varepsilon_Q \sim 100{,}000\ M^{-1}\ cm^{-1}$ for PdTBP; $\varepsilon_Q > 300{,}000\ M^{-1}\ cm^{-1}$ for PdTNP), making TBPs and TNPs especially well suitable for optical tomographic applications [19]. An important feature of all porphyrin-based probes is their very large separation between the absorption and phosphorescence, achievable via excitation at the Soret bands (e.g., 9329 cm^{-1} for

[2] *meso*-Tetrarylated porphyrins, tetrabenzoporphyrins, and tetranaphthoporphyrins are usually abbreviated as Ar_4P, Ar_4TBP, and Ar_4TNP, respectively, in order to distinguish them from *meso*-unsubstituted analogues. Here we refer only to *meso*-tetraarylated macrocycles, and therefore we omit the prefix "Ar_4."

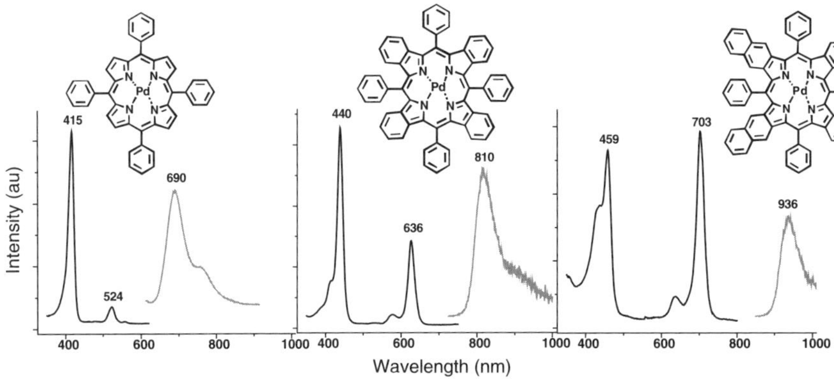

FIGURE 14.4 Absorption (black) and phosphorescence (gray) spectra of Pd-tetraphenylporphyrin (PdP), Pd-tetraphenyltetrabenzoporphyrin (PdTBP), and Pd-tetraphenyltetranaphthoporphyrin (PdTNP).

PdP). Efficient $S_2 \rightarrow S_1$ internal conversion [42], combined with extremely high extinction coefficients of S_0–S_2 transitions ($\sim 3 \times 10^5\,M^{-1}\,cm^{-1}$) makes this pathway superior to the direct S_0–S_1 excitation in those cases when near-UV radiation can be sustained by the biological object.

Until very recently, synthesis of π-extended porphyrins presented a challenging problem. All synthetic methods were based on high temperature condensations between phthalimide (or naphthalimide) and arylacetic acids, or similar donors of benzo- and phenyl groups [43–45]. The harsh conditions of condensation (melting at 300–400°C) allowed for only a few inert substituents, such as alkyl groups or halogens [46], to be introduced into the porphyrin macrocycle. In addition, low yields and complex, inseparable mixtures of products made this approach impractical.

The newly emerged approaches to π-extended porphyrins rely on the Barton–Zard methodology [47] for synthesis of porphyrinogenic pyrroles, which give rise to precursor porphyrins requiring final aromatization. Two methods have been developed into practical synthetic schemes: one making use of retro-Diels–Alder reaction [48,49], another relying on simple and efficient oxidative aromatization strategy [50,51]. The latter method is being used today to synthesize tetrabenzo- and tetranapthoporphyrins [52,53] for construction of phosphorescent oxygen probes, as well as in several other applications [54–57].

Although *meso*-tetraarylated π-extended porphyrins are most common, primarily because of the solubility considerations, *meso*-unsubstituted TBPs and TNPs, which in fact have higher phosphorescence quantum yields and longer lifetimes, can also be synthesized via the oxidative aromatization method [58]. Furthermore, recent introduction of 4,7-dihydroisoindole [59] paved a practical route to 5,15-diaryl-TBPs [60]. The latter porphyrins are especially attractive for construction of oxygen probes, as they combine higher emissivity [61] with the possibility of attaching dendrons to the anchor points in the *meso*-aryl rings, which is lacking in *meso*-unsubstituted porphyrins. Notably, examples of porphyrin dendrimers based on 5,15-diarylporphyrins are well known [62,63].

14.5 POLYGLUTAMIC PD PORPHYRIN DENDRIMERS

Polyglutamic Pd porphyrin dendrimers [64–66] made up the first generation of dendritic oxygen probes. To practitioners in field these probes are known as Oxyphors R2 and G2, which over the years have been used in many biological studies [67]. The idea to use dendritic encapsulation to protect the triplet states of Pd porphyrins from oxygen quenching was inspired by the pioneering work of Aida et al. [68], who showed that dendritic wedges, attached to the *meso*-aryl groups on Zn tetraarylporphyrin, were capable of "caging" the core and shielding it from quenchers of fluorescence. As we began working on the synthesis of polyglutamic Pd porphyrin dendrimers, reports from the Diederich group [62,63] revealed that Newkome-type poly(ester-amide) dendrons, which are similar to polyglutamates in composition, indeed created hydrophobic local environments around encapsulated Fe porphyrins in aqueous solutions. Approximately in parallel with our first publication [64], Balzani and coworkers reported attenuation of oxygen quenching of luminescence in a series of dendritic Ru bypyridines [69], confirming that dendritic encapsulation indeed could be a viable way to build phosphorescent oxygen probes with controllable quenching parameters.

The choice of L-glutamic acid as a building block was largely motivated by its accessibility and biological compatibility. Dendritic polyglutamates resemble in composition heme-proteins, and yet are unlikely to be enzymatically degraded or targeted by the immune system *in vivo*. On the other hand, even if partial decomposition of polyglutamates were to occur, glutamic acid monomers are unlikely to show toxicity. In addition, ease of modification of peripheral carboxyls with various functional groups was attractive from the point of view of fine-tuning the solubility of the construct, depending on the intended application.

Although convergent synthesis of polyglutamic dendrons had been already published by Twyman et al. [70], in our original work we followed the divergent path. On the one hand, the presence of strongly colored porphyrin cores facilitated chromatographic purifications of intermediate dendrimers; on the other hand, it was unclear if higher generation dendrons (>3) would be accessible by the convergent method. Later, however, at least for preparation of lower generation dendrimers [71], we switched entirely to the convergent route, especially after developing a high-yield synthesis of gen 2[3] triglutamic dendron using the CDMT coupling chemistry [72].

The synthesized polyglutamic Pd porphyrin dendrimers [65] are depicted in Figure 14.5. The compounds are described by the general formula PdP-$(Glu^nOH)_4$, where $n = 1$–4 (dendrimer generation) and Glu^nOH is the glutamic layer. Further

[3] Here and throughout the text we use abbreviations "gen 1," "gen 2," etc., to designate dendritic generations.

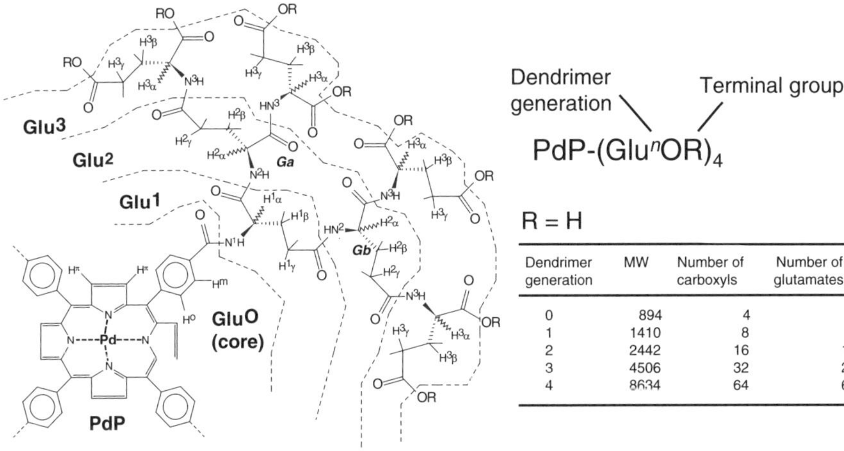

R = H

Dendrimer generation	MW	Number of carboxyls	Number of glutamates
0	894	4	0
1	1410	8	4
2	2442	16	12
3	4506	32	28
4	8634	64	60

FIGURE 14.5 Polyglutamic Pd porphyrin dendrimers, based on PdTCPP (abbreviated as PdP). Molecular data are shown for carboxylic acid terminated dendrimers ($n = 0$–4) [65].

throughout the text we will use similar abbreviations to designate porphyrin dendrimers of different types and generations, terminated by different groups.

Pd porphyrin dendrimers shown in Figure 14.5 exhibit strong phosphorescence in deoxygenated aqueous solutions ($\lambda_{max} \sim 690$ nm, $\tau_0 \sim 700$–800 μs). Considering large separation between the carboxyl groups on the core porphyrin (~18 Å across the porphyrin macrocycle) and relatively small size of the glutamic monomer, the dendrimers were assumed to adopt quite open architectures up to rather high generations ($n > 6$), still allowing for access of oxygen to the core and substantial quenching of the phosphorescence. Indeed, in DMF there was practically no difference in the values of the quenching constants k_q (Figure 14.6a) between the unprotected core (PdTCPP) and all the dendrimers in the series. Oxygen quenching in this case was still presumably limited by the rate of oxygen diffusion through the solvent, and deviations caused by the dendritic shells were negligible (±500 mmHg^{-1} s^{-1}) as compared to the absolute values ($k_q \sim 8000$–10,000 mmHg^{-1} s^{-1}).

In contrast, in water k_q values dropped significantly with increase in the dendrimer size (Figure 14.4b), approximately 1000–1500 mmHg^{-1} s^{-1} per generation. Overall, from gen 0 (PdTCPP) to gen 4 dendrimer, the absolute k_q values decreased 4–5 times.

Large variation between shielding capabilities of dendrimers in water and DMF suggested that depending on the solvent properties, the polyglutamic branches adopted either sparse or compact conformations, altering the barriers of oxygen diffusion to the porphyrin core to a different extent (Figure 14.6). It was therefore reasonable to conclude that the interaction of the dendrimer with the solvent is one of the key factors determining the degree of protection that the dendrimer offers to the core porphyrin. In much the same way as proteins fold in water, hiding their hydrophobic residues inside the macrostructure, more hydrophobic dendrimers fold

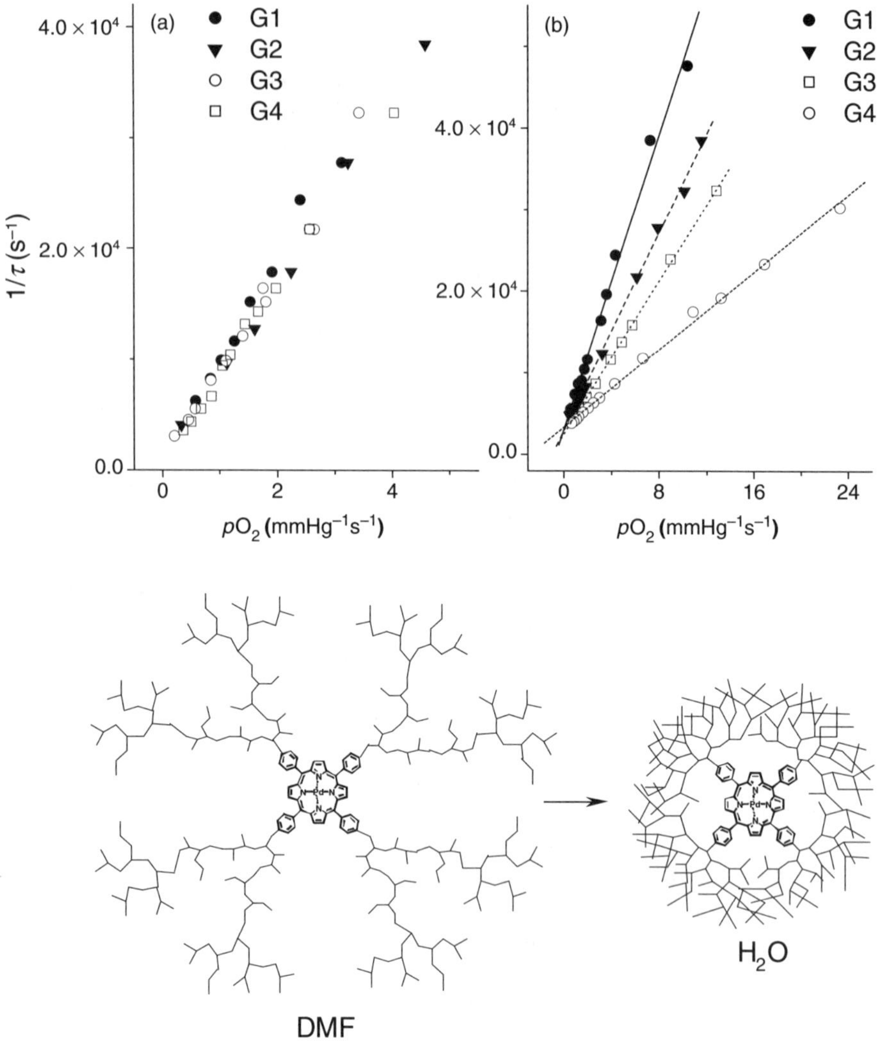

FIGURE 14.6 Oxygen quenching plots of Pd porphyrin-dendrimers PdP-(Glu^{n}OR)$_4$ (n = 1–4) in DMF (a, R = All) and water (b, R = H) solutions at 22°C [65]. G1–G4 refer to dendritic generations (gen 1 – gen 4). Below: hydrophobic folding of dendrons in water restricts oxygen access to encapsulated cores.

more tightly around the core porphyrin, resulting in a decrease of the oxygen quenching constants.

From the practical point of view, the oxygen quenching constant of even gen 4 polyglutamic Pd porphyrin dendrimer ($k_q \sim 1000\,\text{mmHg}^{-1}\,\text{s}^{-1}$ in water) was still significantly higher than required for the biological application. Indeed, at such high quenching constants, only low pO_2 values, about 0–50 mmHg, can be sampled with adequate accuracy. Fortunately, carboxylate terminated polyglutamic dendrimers,

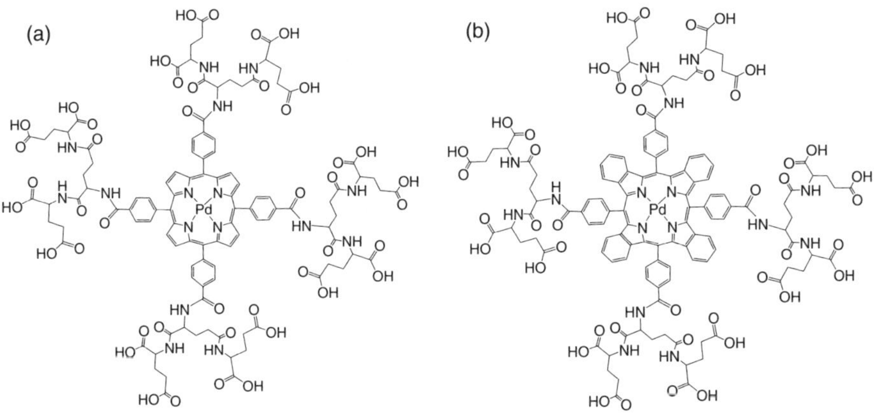

FIGURE 14.7 Gen 2 polyglutamic Pd porphyrin dendrimers, known as Oxyphor R2 (a) [65] and Oxyphor G2 (b) [66]. Molecular weights as polysodium salts, isolated from solutions at pH 7.3: 2794 Da (a), 2994 Da (b).

especially of lower generations, were found to interact strongly with albumin, and that, combined with their excellent aqueous solubility, was the key to their success as oxygen probes. At concentrations of albumin ~1–2% (by mass), which are lower than typically met in blood plasma, and concentrations of porphyrin dendrimers ~10^{-6}–10^{-5} M, which are typically used in oxygen measurements, polyglutamic Pd porphyrin dendrimers form complexes with albumin. These complexes have quenching constants of 200–300 $mmHg^{-1}\,s^{-1}$ and are suitable for measurements in the physiological pO_2 range. Thus, the necessity for prebinding the probes to foreign albumin was eliminated, and gen 2 polyglutamic Pd porphyrin dendrimer, Oxyphor R2 (Figure 14.7) [65], became a popular probe for intravascular oxygen measurements. Later, a similar probe based on PdTBP, termed Oxyphor G2 [66], was synthesized and proved to be a useful near-infrared analog of Oxyphor R2 [67]. Both Oxyphors are stable in aqueous solutions, can be stored for prolonged periods of time and can be injected directly into the blood, where they rapidly bind to endogenous albumin and become distributed throughout the vascular system, permitting physiological oxygen measurements.

The main limitation of polyglutamic oxygen probes is that they can be used only in albumin-rich environments, such as blood plasma. Even in the blood, incomplete binding, typically occurring at higher probe concentrations (above ~10^{-5} M), can lead to heterogeneous responses and compromise oxygen measurements. This situation is especially frequently met in scanning microscopy applications, where low emission intensities encourage use of exceedingly high concentrations of the probe [73], often leading to poorly interpretable decay data.

Another potential difficulty associated with polyglutamic probes is related to the structural inhomogeneity of their adducts with albumin. These are complex macromolecular structures, existing in solutions in multiple conformations. Each conformation is characterized by its own triplet lifetime τ_0 and, more importantly, by its own accessibility of the phosphorescent core to oxygen, bringing about multitude of constants k_q. As a result, phosphorescence decays of polyglutamic probes in the blood are by default nonsingle exponential. The only rigorous way to describe these decays is to carry out phosphorescence lifetime distribution analysis, such as that based on the maximum entropy method (MEM) [74]. However, for the purpose of oxygen measurements it is appropriate to use "apparent" lifetimes, which are the weighted averages derived from underlying lifetime distributions. Averaging can be done, for example, by fitting a nonsingle-exponential decay using two-, three- or higher exponential models and deriving the intensity-weighted average lifetimes. Alternatively, one can use single-exponential fitting, which would generate poorer residuals, but would be more robust due to the fewer number of fitting parameters. Single-exponential fitting in this regard can be considered an alternative form of averaging.

The drawbacks associated with use of dendritic polyglutamic porphyrins clearly show that probes with albumin-independent phosphorescence lifetimes and oxygen quenching constants would greatly simplify data analysis and broaden applicability of the method (e.g., make possible measurements in albumin-free environments). Although polyglutamic dendrimers in the end were incapable of providing sufficient shielding to porphyrins, experiments performed with these molecules implied that the composition of the dendritic matrix is at least as important for shielding as the dendrimer size. This result served as a starting point for more systematic studies of the interplay between the dendrimer composition, size, and encapsulating efficiency, as measured by oxygen diffusion and quenching of phosphorescence.

14.6 INFLUENCE OF SIZE AND COMPOSITION OF DENDRITIC MATRIX ON OXYGEN SHIELDING EFFICIENCY

In order to determine which dendrimers provide optimal attenuation of oxygen quenching of porphyrin phosphorescence a study was performed [75], which involved three types of porphyrin dendrimers: Fréchet-type poly(aryl ethers) [76], Newkome-type poly(ether amides) [77,78], and polyglutamates [65] (Figure 14.8).

Fréchet-type dendrimers (Fr) are composed of less polar monomers, and thus they were assumed to be the most hydrophobic in the selected group. In addition, their folding in aqueous environments was expected to be facilitated by π–π-stacking interactions. Because of the same branching number (BN = 2) and close molecular weights of the monomers, poly(aryl ethers) (MW = 105), and polyglutamates (MW = 112) exhibit essentially the same mass increase per dendritic generation. Newkome poly(ether amides) (MW = 285, BN = 3), on the other hand, are similar to polyglutamates in chemical composition, but have much larger mass increase per generation. All three dendrimers are relatively flexible, and therefore substantial conformational changes could be expected in solvents of different polarities. Pd-*meta*-octahydroxyphenylporphyrin and Pd-*meta*-octacarboxyphenylporphyrin, compatible with either Williamson (Fr) or peptide (Nw, Glu) chemistries, were used as phosphorescent cores. The studies were performed in DMF, THF, and water. To

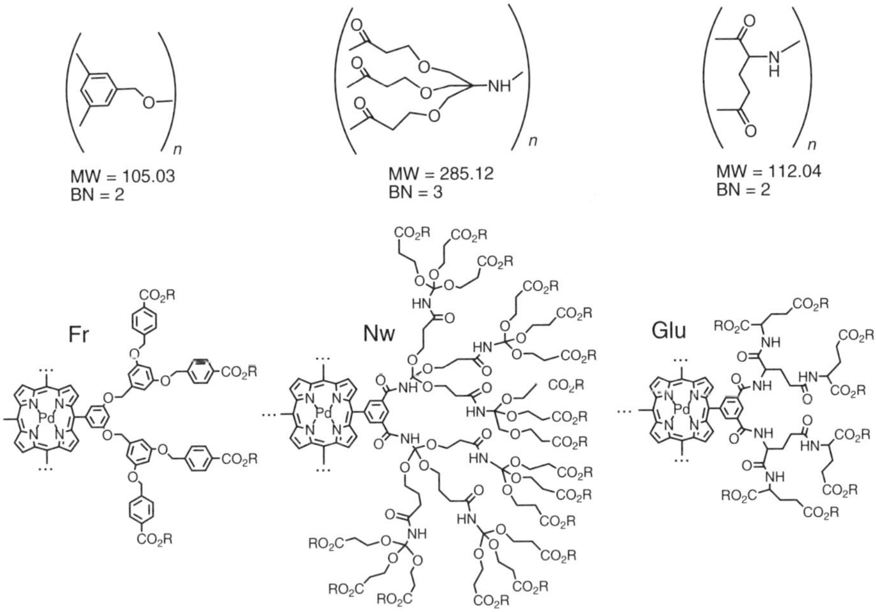

FIGURE 14.8 Monomeric building blocks and examples of corresponding phosphorescent Pd porphyrin dendrimers for oxygen quenching studies. $n = 0$–2, BN: branching number [75]. R = H, Et, PEG350 (polyethyleneglycol monomethyl ether, average MW 350).

ensure appropriate aqueous solubility, termini of the dendrimers were modified with oligoethyleneglycol residues, which rendered uncharged molecules soluble both in organic solvents and in water.

In organic solvents, decrease in the quenching rates was found to be consistent with increase in the dendrimer size, and these changes were rather insignificant. As expected, bulkier poly(ether amides) exhibit a stronger effect, but still the decrease in the quenching was only about 75% for gen 2 dendrimer, whose weight is more than 12 kDa, that is more than 12 times that of the core porphyrin. Notably, in the case of polyglutamic derivatives of PdTCPP (Figures 14.5 and 14.6), dendrons were attached to the *para*-positions on the *meso*-aryl rings of the porphyrin, extending out of the macrocycle instead of covering it from above and below, and thus the shield effect was much weaker for the same size dendrons.

A much larger difference in quenching rates was observed in water. Relative efficiencies of quenching, expressed as ratios of quenching constants of dendrimers versus unprotected cores, are plotted in Figure 14.9 as a function of the dendrimer molecular weight. Poly(aryl ether) dendrimers were found to provide by far greater protection from oxygen than polyglutamates and poly(ester amides). For example, gen 2 Fréchet-type dendrons are about 10 times more effective in shielding than similar in size gen 2 polyglutamates, and even more effective than the significantly larger gen 2 poly(ether amides) (Figure 14.6). Within the Fréchet-series, the quenching constant decreased by ~30 times between gen 0 and gen 3 compounds. Thus, it

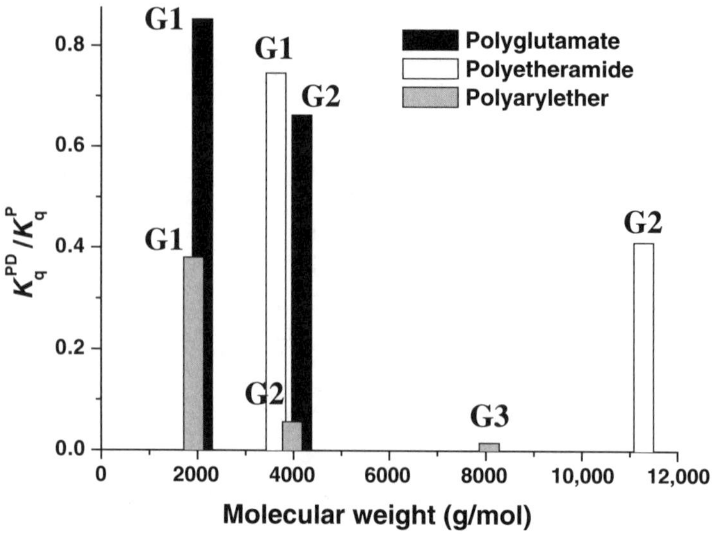

FIGURE 14.9 Relative efficiencies of oxygen quenching for PEGylated dendrimers (Figure 14.8) in aqueous solutions. G1–G3 refer to dendritic generations (gen 1 – gen 3). K_q^{PD} - quenching constant for Pd porphyrin-dendrimer; K_q^P - quenching constant for unprotected (gen 0) porphyrin [75].

became apparent that in polar solvents (water), composition of the dendritic matrix affects the encapsulation efficiency much more than does the dendrimer size.

The overall effect of a hydrophobic dendritic cage surrounding the porphyrin is a decrease in the oxygen quenching constant k_q (Eq. 14.2). It is important to realize that the latter is a product of the quencher solubility and its diffusion coefficient, which are both modulated by the porphyrin nanoenvironment. If the solubility of oxygen in the bulk solvent (e.g., water) is lower than that in the matrix of the dendrimer, the dendrimer may in principle serve as a "concentrator" of oxygen. Nevertheless, the decrease in the diffusion coefficient effectively offsets the increase in its local concentration. Hydrophobic dendritic branches fold, their mobility becomes restricted and that prevents oxygen molecules from freely diffusing to the triplet core (Figure 14.10). As a result, the diffusion rate of oxygen in the body of the dendrimer (k_q^{dend}) becomes significantly lower than that in water ($k_q^{H_2O}$). Importantly, this effect should not be confused with increase in the density. The density of the folded dendrimer is likely lower than that of water; however, the constrained dynamics of the branches has much larger impact on the diffusion than the density.

The practical outcome of these comparative studies was that dendrimers constructed from aromatic building blocks potentially have much higher shielding efficiency in aqueous environments than do other types of dendrimers. In addition, we determined that in order to prevent interaction of dendrimers with biological macromolecules, their periphery should be modified with PEG residues (see below). These findings guided our subsequent development of fully protected phosphorescent oxygen sensors.

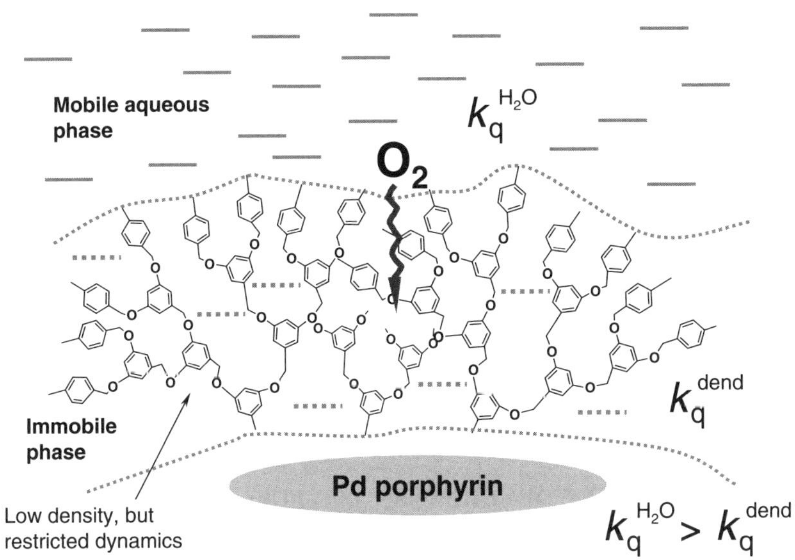

FIGURE 14.10 Cartoon illustrating the effect of a hydrophobically folded dendrimer on oxygen diffusion and phosphorescence quenching rate.

14.7 EFFECT OF THE DENDRIMER OUTER LAYER

As mentioned earlier, it is critically important that quenching properties of phosphorescent probes do not change in the presence of blood plasma proteins and other biological macromolecules. Quenching constants of all carboxylate-terminated porphyrin dendrimers (Figure 14.8), however, were found to be highly sensitive to albumin. For example, k_q of gen 2 Fréchet-type porphyrin dendrimer in albumin solution (2% by mass) is five times lower (30 mmHg^{-1} s^{-1}) than that in the albumin-free aqueous buffer (151 mmHg^{-1} s^{-1}).[4] On the contrary, quenching of porphyrin dendrimers modified with oligoethyleneglycol residues (PEG350, average MW = 350) appeared to be unaffected by serum proteins. Thus, k_q of the PEGylated analog of gen 2 Fréchet-type porphyrin dendrimer was found to have the same value (130 mmHg^{-1} s^{-1}) in the presence and in the absence of albumin [75].

It is important to distinguish between the effect of hydrophobic dendrons and the effect of external PEG residues. While the latter also contribute to the protection from quenching, especially at early dendritic generations, the net effect of PEGs is very small compared to the shielding by dendritic branches. Moreover, the PEG effect rapidly levels off with extension of linear chains. This notion was supported, for example, by oxygen quenching measurements in a series of star-shaped amphiplilic diblock copolymers with Pd porphyrin cores [79]. It was shown by this study that

[4] The value of the quenching constant for PdP-$(Fr_2OH)_8$ in the aqueous buffer was probably affected by aggregation, as revealed by the broadened absorption spectra.

external PEG chains as large as 300,000 Da have practically no effect on oxygen quenching constants of these molecules.

In addition to being able to prevent interactions of dendrimers with proteins, peripheral PEG groups also strongly affect biodistribution and rate of excretion of dendritic macromolecules [80,81]. In pilot experiments we determined that Oxyphor R2 peripherally modified with PEG350 residues was excreted from the blood of mice within minutes after injection. In contrast, Oxyphor R2, which binds to albumin, remained in the blood for several hours. It is expected that increasing the length of the PEG chains will increase *in vivo* circulation lifetimes of the dendritic probes and at the same time prevent their diffusion out of the blood vessels into the interstitial space. Therefore, peripheral PEGylation should make it possible to build phosphorescent probes of virtually any size without affecting their oxygen sensitivity.

14.8 ARYL-GLYCINE DENDRIMERS AND FULLY PROTECTED PHOSPHORESCENT PROBES

From the application point of view, Fréchet-type poly(aryl ether) dendrimers seemed ideal candidates for construction of phosphorescent oxygen probes. However, synthesis of Fréchet-type dendrons is expensive, mainly because their purification involves gradient chromatography. Therefore, we looked for a practical alternative to Fréchet-type dendrimers, that is, dendrimers that could be synthesized in bulk quantities from inexpensive aromatic building blocks and purified without chromatography.

Among known dendrimers with aromatic backbones, dendritic arylamides [82] seemed very attractive because of their high chemical stability, low cost and effective protocols available for their assembly [83,84]. In particular, arylamide dendrimers based on 5-aminoisophthalic acid (5-AIPA) were among the first reported dendritic macromolecules [85,86], although, a number of problems were associated with the use of this building block as well. These problems come mainly from the low nucleophilicity of its amino group and impeded internal rotations of the resulting arylamide skeleton, causing poor solubility and difficulties in characterization of the resulting dendrimers. A way to make the focal functionality more reactive and to simultaneously increase the flexibility of the dendritic backbone is to extend the amino-end in the 5-AIPA molecule by a flexible fragment, terminated with an aliphatic amine. This possibility was explored by constructing aryl-glycine (AG) dendrons and dendrimers [87].

The abbreviations below are used to designate aryl-glycine dendrons and dendrimers.

Dendrons: $X\text{-}AG^nR$, where AG denotes the dendritic aryl-glycine skeleton, *n* is the dendrimer generation number, *X* is the focal functionality and R is the terminal group.

Dendrimers: $C\text{-}(AG^nR)_m$, where C denotes the dendrimer core, AG denotes the dendritic aryl-glycine skeleton, *n* is the generation number, R is the terminal group and *m* is the number of dendritic wedges attached to the core.

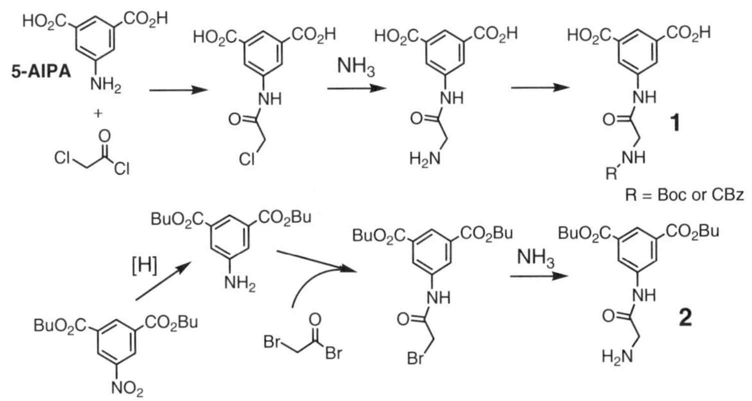

FIGURE 14.11 Synthesis of aryl-glycine (AG) building blocks [87].

The developed synthesis of AG dendrons relies on the classic Fischer's haloacyl halide method [88] for the synthesis of building blocks **1** and **2** in Figure 14.11.

This scheme is inexpensive and easily scalable to multigram quantities. Neither of the intermediate products needs to be isolated and purified. They can be introduced into the following transformations directly from the preceding syntheses, following the precipitation upon acidification or evaporation of the solvent. Building blocks **1** and **2** are obtained in high yields and isolated by precipitation/crystallization.

Synthesis of AG dendrons in principle also can be implemented using the Fischer's method [87], however, a more robust scheme is based on a coupling-deprotection sequence employing the peptide-coupling reactions (Figure 14.12) [89].

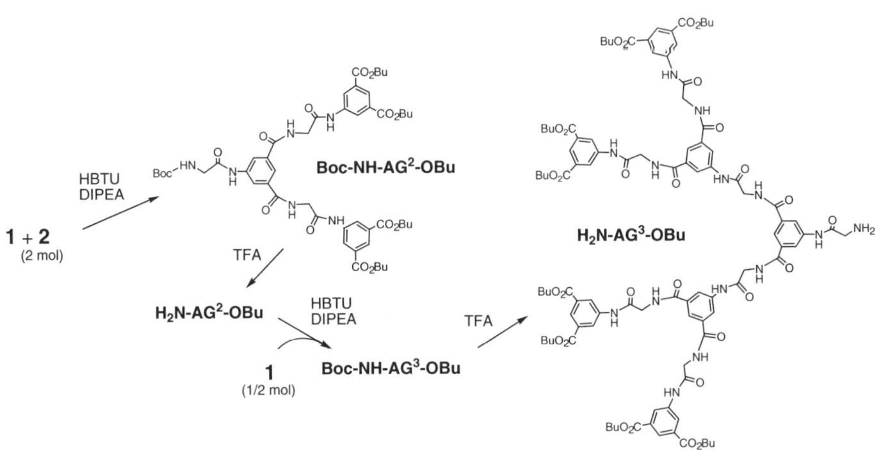

FIGURE 14.12 Synthesis of poly(aryl-glycine) (AG) dendrons. TFA, trifluoroacetic acid; HBTU, *o*-benzotriazole-*N*,*N*,*N'*,*N'*-tetramethyl-uronium hexafluorophosphate; DIPEA, *N*, *N*-diisopropylethylamine [89].

The coupling chemistry used in the original implementation was based on CDMT/NMM coupling (CDMT, 2-chloro-4,6-dimethoxy-1,3,5-triazine; NMM, *N*-methylmorpholine) [90]. It allows synthesis of pure gen 2 dendron in high yield (84%), but HBTU/DIPEA chemistry was found generally more effective, especially at higher generations. Using HBTU/DIPEA, gen 3 AG dendron could be isolated without chromatographic purifications in 97% yield. Overall, the developed approach is inexpensive, has a broad scope and allows modification of the dendrimer periphery, interior, focal functionality and/or internal topology, without changing the basic chemistry. For example, if primary amines R'-NH_2 are used instead of ammonia in the synthesis of building blocks in Figure 14.11, groups R' will automatically become included in the dendrimer interior.

AG dendrons were found to be well suited for encapsulation of metalloporphyrins, giving rise to a new generation of fully protected phosphorescent oxygen probes [89]. The optimized synthetic assembly of the dendritic probes, consisting of the attachment of dendrons to the core porphyrins, hydrolysis of the peripheral ester groups and modification of the peripheral carboxyls with polyethyleneglycol residues, is shown in Figure 14.13. The structure of a selected probe molecule, PdTBP-$(AG^2OPEG)_8$ is also shown in the figure.

Each step of the sequence was carefully optimized in order to ensure maximal yields and monodispersity of the probes. Porphyrins modified with AG^1 and AG^2 dendrons could be isolated in high purity, but modification of porphyrins with AG^3 dendrons proved extremely challenging, always yielding mixtures of dendrimers with six, seven, and eight dendritic branches. The peripheral ester groups were hydrolyzed using a special two-step procedure [89], which in parallel destroys unreacted dendrons, making it possible to entirely avoid chromatographic purifications. The resulting peripheral carboxyl groups on the dendrimers are esterified with PEG residues of the desired length, and simple reprecipitation from THF upon addition of diethyl ether gives pure PEGylated dendrimers.

Photophysical properties (Table 14.1) of AG dendrimers practically match those of the unprotected cores (Figure 14.2). However, both absorption and emission spectra revealed some degree of aggregation in gen 0 and gen 1. In aqueous solutions, the phosphorescence lifetimes (τ_0) and the quantum yields (ϕ) of the probes somewhat increase with an increase in the dendrimer generation.

Phosphorescence quantum yields depend on the rates of singlet internal conversion, intersystem crossing, radiative and nonradiative relaxation of the triplet state, and intra- and intermolecular quenching processes. Since in Pt and Pd porphyrins $S_1 \rightarrow T_1$ intersystem crossing is extremely fast (10^{11}–$10^{12}\ s^{-1}$) [25], the yield of the triplet state is unlikely to be influenced by the substituents. Likewise, the radiative rate of the triplet emission is intrinsic to the chromophore [91] and should not dependent dramatically on the dendrimer generation. In contrast, nonradiative processes are subject to changes induced by the substituents and the environment, which may influence porphyrin's vibrational flexibility and its accessibility to various quenching species. Due to aggregation of gen 0 and gen 1 dendrimers, triplet–triplet self-quenching may be a result of their lower than expected phosphorescence yields. In gen 2 and especially gen 3 dendrimers, although the latter were not ideal, cores are much

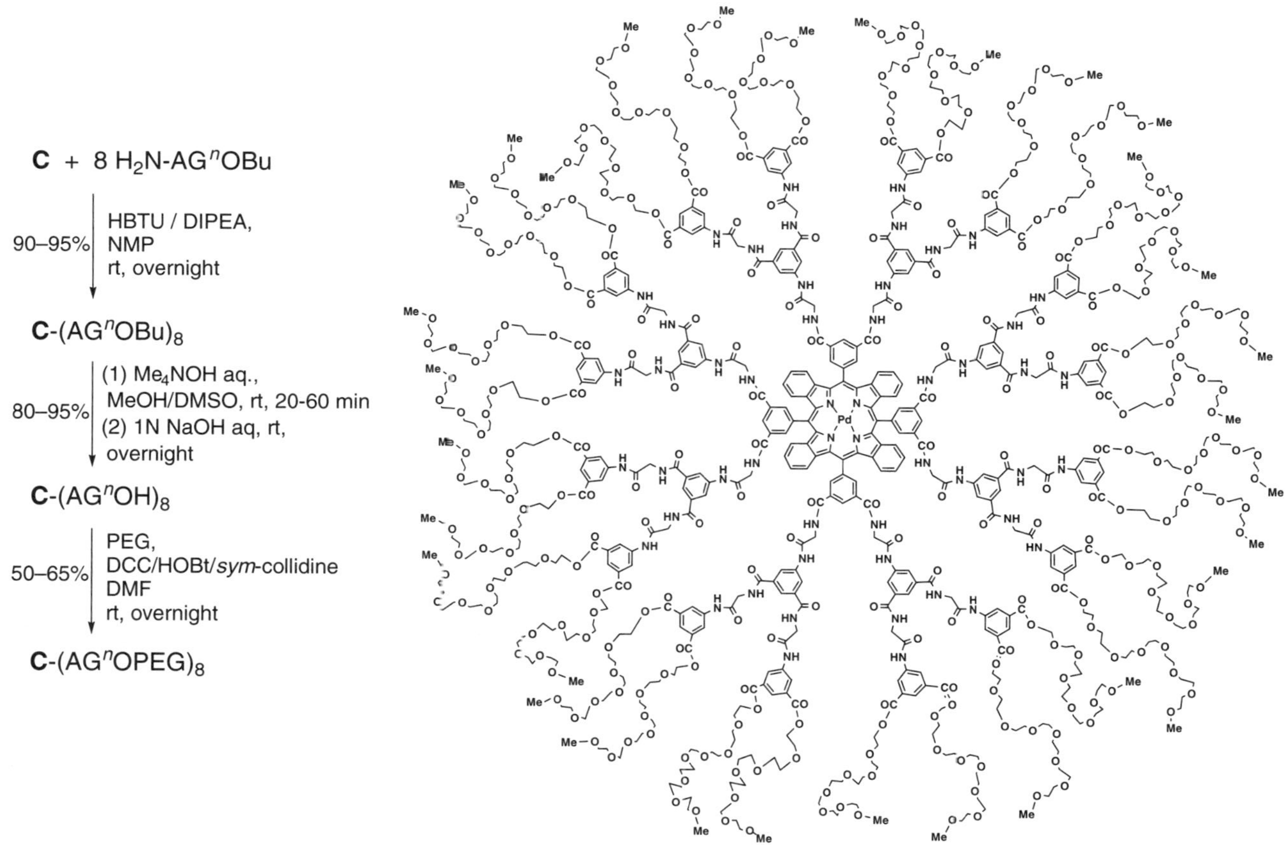

FIGURE 14.13 Synthesis and modification of dendritic aryl-glycine porphyrin dendrimers. **C** (core) designates porphyrin-octacarboxylic acids [89].

TABLE 14.1 Photophysical Properties of Selected Phosphorescent Probes Based on Pt and Pd Porphyrin–AG Dendrimers

Gen	Compound	Absorption, λ (nm)		Phosphorescence[a] λ_{max} (nm)	Phosphorescence[a] ϕ^a	Phosphorescence[a] τ_0 (μs)	k_q ($mmHg^{-1}s^{-1}$)[b]
0	PtP-OPEG[b]	389	520	682	0.05	37	1208 ± 15
1	PtP-$(AG^1OPEG)_8$	390	520	681	0.09	48	521 ± 3
2	PtP-$(AG^2OPEG)_8$	398	516	671	0.14	53	120 ± 3
2	PdP-$(AG^2OPEG)_8$	403	532	708	0.07	720	97 ± 4
3	PtP-$(AG^3OPEG)_8$	409	515	664	0.18	66	66.6 ± 0.2
2	PtTBP-$(AG^2OPEG)_8$	432	621	791	0.24	54	171 ± 5
2	PdTBP-$(AG^2OPEG)_8$	446	635	816	0.03	270	122 ± 5

[a] Quantum yields and lifetimes are measured in deoxygenated aqueous solutions.
[b] Measured using ~2 μM solution of the probe in 50 mM phosphate buffer, pH = 7.4, 23°C.

better encapsulated and consequently are much less likely to experience intermolecular porphyrin-to-porphyrin contacts. In addition, better encapsulation can lead to a less quenching by the solvent and/or less efficient nonradiative $T_1 \rightarrow S_0$ relaxation. The rate of the latter is determined by the magnitude of the spin-orbit coupling, which in π-extended porphyrins is strongly affected by the macrocycle vibrations [61]. Increase in the size of the dendrimer and tighter confinement of the porphyrin inside the folded dendrimer may contribute to the decreased flexibility of the macrocycle and lower the rate of its nonradiative decay.

The effect of dendritic encapsulation on phosphorescence quenching by oxygen in the series of Pt porphyrin–AG dendrimers is exemplified in Figure 14.14a.

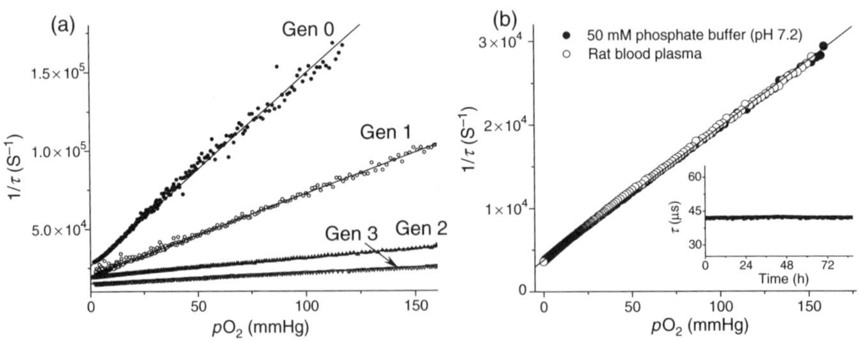

FIGURE 14.14 (a) Stern–Volmer oxygen quenching plots of Pt porphyrin dendrimers in aqueous solutions. The solid lines show the fits of the data to the Stern–Volmer equation. (b) Stern–Volmer plots of gen 2 dendrimer in phosphate buffer solution and in rat blood at 36.9°C. Inset: measurements at air saturation taken at 1 min intervals during 80 h [89].

The slopes of Stern–Volmer plots decrease significantly with an increase in the dendrimer generation. Overall, between gen 0 and gen 3 the quenching rate drops by nearly 20 times, that is 5 times more than in the series of glutamic dendrimers (Figure 14.4). The magnitude of the attenuation effect is similar to that of poly(aryl ether) dendrimers. As a result quenching constants and lifetimes of gen 2 Pd porphyrin dendrimers are well suited for biological oxygen measurements. These probes exhibit good dynamic range and excellent sensitivity. For example, in the case of PdP-based gen 2 dendrimer the relative sensitivity coefficient R (see above) is as high as 0.9.

An important property of PEG-coated AG porphyrin dendrimers is that their oxygen quenching constants k_q and lifetimes τ_0 are completely insensitive to the presence of biological macromolecules. For example, PdTBP-based gen 2 dendrimer was titrated in the rat blood plasma at physiological temperature (36.9°C). To avoid coagulation, the blood was treated with small amount of heparin, and the red blood cells removed by centrifugation in order to facilitate equilibration of the solution with air. The resulting Stern–Volmer plot was superimposed on the plot obtained at identical conditions (36.9°C) in aqueous phosphate buffer at pH 7.2 containing no proteins (Figure 14.14b). This result clearly shows that the probe constants are unaffected by the blood plasma components. Readings by the dendritic probes at constant temperatures and oxygen concentrations are extremely stable (Figure 14.14b, inset). The solutions show no traces of decomposition (change in absorption) when measurements are performed as frequently as every 1 min over 2–3 days periods.

The fully protected dendritic porphyrin-based oxygen probes described in this section open up many new possibilities for noninvasive studies of tissue oxygenation and metabolism using phosphorescence quenching. These probes also can be utilized in various *in vitro* assays and cell culture studies, as they do not require repetitive calibrations and thus provide a superior alternative to commonly used oxygen electrodes and optodes. Further validation of the developed technology will include detailed investigation of probe's biodistributions, phototoxicity, and pharmacokinetics in order to evaluate their potential in clinical imaging.

14.9 TWO-PHOTON-ENHANCED DENDRITIC OXYGEN PROBES

The ability to image oxygen in three dimensions with microscopic resolution would be extremely attractive for a number of areas of biology and medicine, including neuroscience, ophthalmology, diabetes, and cancer. Oxygen levels can vary between individual tissue microcompartments compartments, and the ability to quantify changes in partial oxygen pressure between and within these compartments at the microscopic level would provide invaluable information for physiological research. In spite of this need, no existing technology today provides means for high-resolution oxygen microscopy with 3-D capability. The natural synergy between phosphorescence quenching with 2P microscopy [92] has the potential to fulfill this niche.

2P excitation offers several advantages over linear methods, such as improved depth resolution for 3-D imaging and reduced risk of photodamage [93]. Due to the highly localized nature of the excitation volume in two-photon microscopy, photochemical damage induced by bleaching of the probe and various quenching reactions is confined to the near focus volume—a feature especially important for phosphorescence lifetime imaging. The main difficulty in combining phosphorescence lifetime imaging with two-photon microscopy is that Pt and Pd porphyrins, as well as all other chromophores used for oxygen sensing, possess extremely low two-photon absorption (2PA) cross sections σ_2,[5] necessitating exceedingly high excitation powers, long acquisition periods, and/or unacceptably high probe concentrations [73,94].

The most straightforward way to overcome this limitation would be to use phosphorescent chromophores with higher 2PA cross sections. Much attention has been given in the past years to nonlinear absorption in tetrapyrroles. 2PA cross section of the tetrapyrrole macrocycle can indeed be increased up to thousands of GM units by appropriate substitution, conjugation, and oligomerization [95,96]. Unfortunately, such modifications also typically affect the emissivity of the triplet states and/or hamper triplet production.

An alternative approach to enhancement of 2PA-induced phosphorescence in metalloporphyrins without directly altering their electronic properties would be to harvest the excitation energy by an electronically separate antenna and to pass it onto the phosphorescent core via intramolecular FRET [97]. In such an antenna-core system constructed around a dendrimer, the latter would regulate the rate of oxygen diffusion to the core, just as in the regular dendritic probes (see above), while the dendrimer termini would control the probe biodistribution. A cartoon illustrating this idea is shown in Figure 14.15.

As an example (Figure 14.15) consider a Ti:Sapphire laser producing high intensity radiation at around 800 nm, which excites 2P transitions at twice the frequency, that is, near 400 nm. After the 2P states of the antenna (${}^{a}S_{2P}$) are populated and have internally converted into the lowest excited singlet state ${}^{a}S_1$, the excess energy is transferred to the core. The Förster energy transfer mechanism assumes that the fluorescence (${}^{a}S_1 \rightarrow {}^{a}S_0$) of the donor (2P antenna) overlaps with an absorption band ${}^{c}S_n \leftarrow {}^{c}S_0$ ($n = 1, 2, \ldots$) of the acceptor (core). Therefore, the core must possess linear absorption band(s) somewhere in the region extending to the red from 400 nm. Exact positions of these bands are defined by the Stokes shift of the fluorescence of the antenna relative to its absorption at 400 nm. The FRET from the antenna to the core results either in the population of its singlet excited state ${}^{c}S_1$, which is depopulated via the intersystem crossing (*isc*) to yield the triplet state ${}^{c}T_1$, resulting in either oxygen quenching or phosphorescence.

The original models of two-photon-enhanced phosphorescent probes were based on adducts of Pt tetra- or octacarboxyphenylporphyrins and coumarins [97]. Although coumarins are rather moderate 2P absorbers, their 2PA cross sections (tens of GM units) are still higher than of Pt porphyrins, and the emission of some coumarins, for

[5] The quantity σ_2, commonly referred to as the "two-photon absorption cross-section," is the proportionality coefficient between the photon absorption rate per molecule and the time averaged square of the photon flux. σ_2 is measured in GM (Göppert–Mayer) units: 1 GM = 10^{-50} cm^4 s/photon.

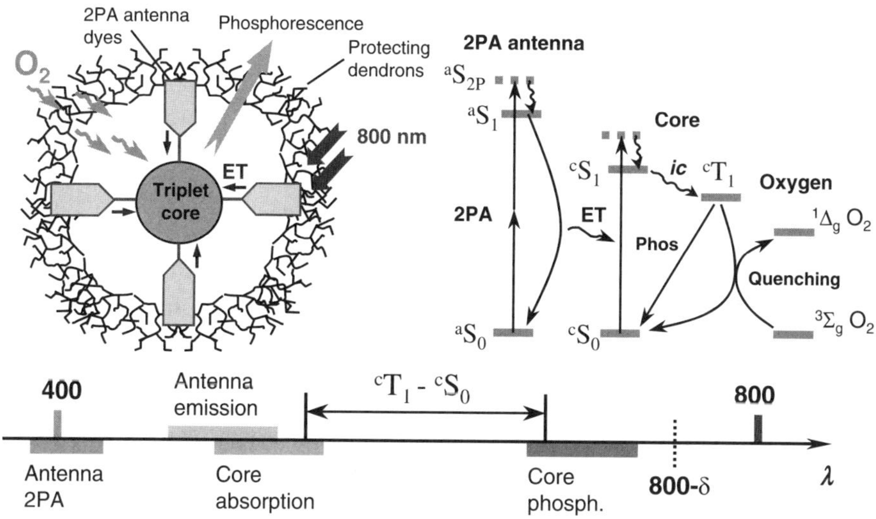

FIGURE 14.15 Scheme of a two-photon-enhanced phosphorescent oxygen sensor, its Jablonski diagram and related wavelength scale. (See the color version of this figure in Color Plates section.)

example of coumarin-343 (C343), is compatible with Q-band absorption of Pt porphyrins ($\lambda_{max} = 510$ nm, $\varepsilon = 25{,}800\ M^{-1}\ cm^{-1}$), suggesting high efficiency FRET. C343 fluoresces with quantum yield $\phi_{fl} = 0.75$ in DMF and its σ_2 value is about 25 GM. Four C343 moieties can be attached to PtTCPP via ethylenediamine linkers using standard peptide chemistry, giving a simple FRET-enhanced two-photon-absorbing phosphorescent probe (Figure 14.16).

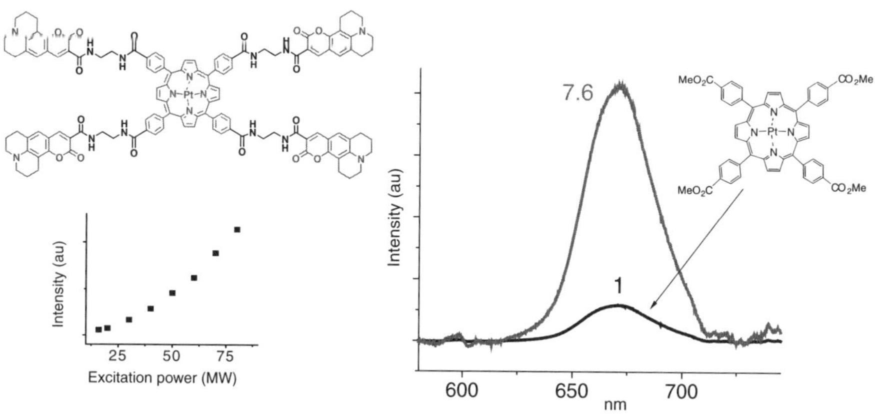

FIGURE 14.16 Pt porphyrin (PtP)–coumarin-343 (C343) adduct and its 2PA-induced phosphorescence spectra. The action cross-section spectrum for the phosphorescence of the adduct is increased 7.6 times compared to the parent Pt porphyrin. The inset shows quadratic power dependence of the phosphorescence upon excitation by 100 fs pulses from a regenerative amplifier (780 nm, 1 kHz, unfocused beam).

In this case, FRET occurred onto the Q-band of the porphyrin with 0.78 efficiency. The phosphorescence quantum yield was found to be 0.65, close to that of the parent Pt porphyrin (0.75), while the phosphorescence upon 2P-excitation was enhanced ~7.6 times. Interestingly, increase in the 2P-induced phosphorescence in this model compound was about four times lower than what was expected from the ratio of 2PA cross sections of the antenna and the core Pt porphyrin. In this regard, the data on dendritic 2P-enhanced triplet constructs for PDT [98,99] complement our findings that only a very small enhancement compared to expected was gained via 2PA-induced FRET onto porphyrins.

Next, a prototype dendritic oxygen sensor was synthesized by attaching four AG dendrons to a tetra-coumarin-343 adduct (Figure 14.17) [97]. As expected, AG dendrons bring the quenching constant into the desired range (332 $mmHg^{-1}\,s^{-1}$), and by increasing the size of the dendrons quenching could be further attenuated.

The FRET efficiency in this molecule was still quite high ($\phi_{FRET} = 0.78$), but the phosphorescence quantum yield ($\phi_{phos} = 0.02$) was significantly lower than

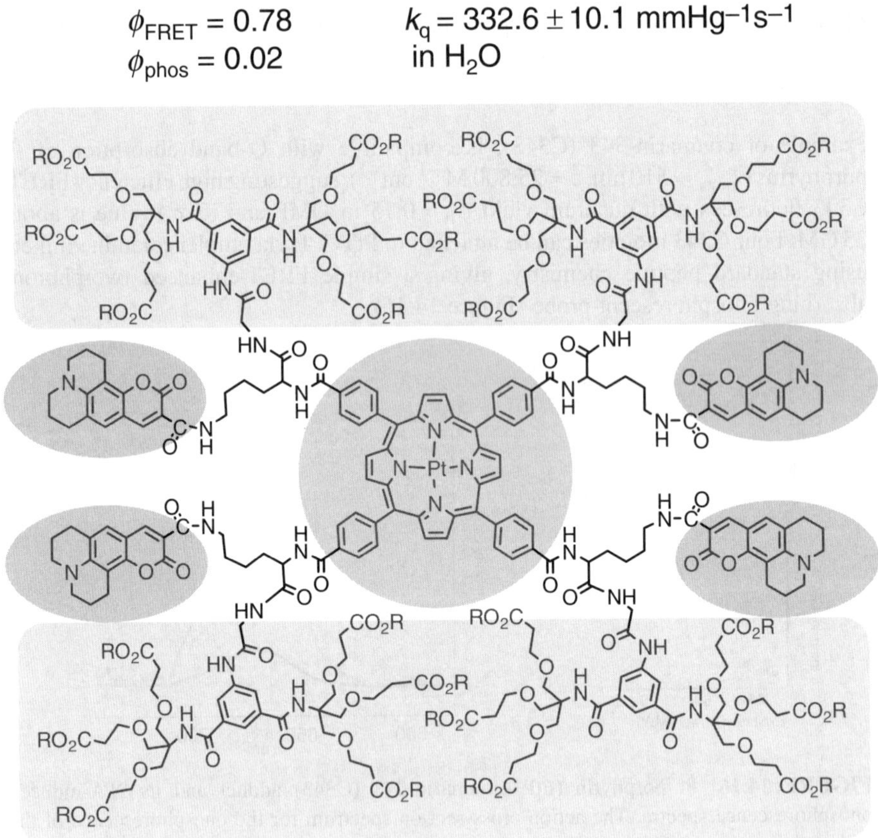

FIGURE 14.17 Prototype of a water-soluble dendritic oxygen sensor with 2PA antenna (R = PEG350).

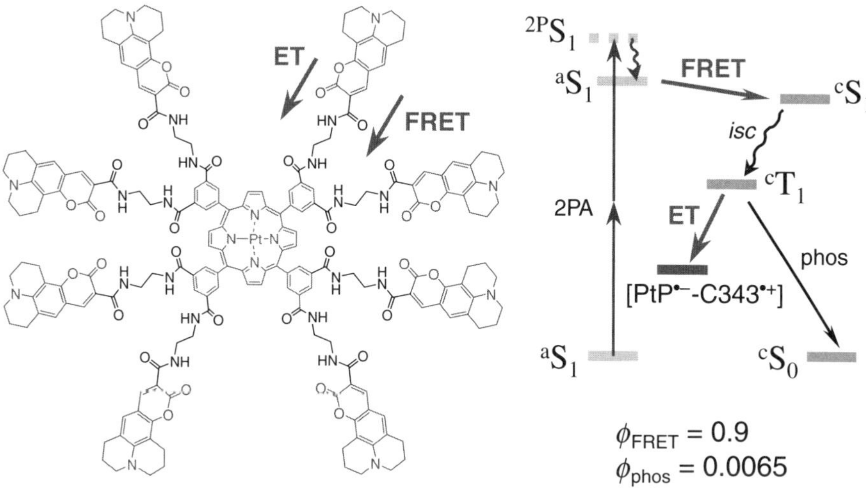

FIGURE 14.18 Pt porphyrin octa-coumarin-343 construct and associated energy pathways [97].

that in the simpler compound in Figure 14.16, suggesting that other pathways of energy dissipation exist and may be important for the design of 2P-enhanced phosphorescent sensors. The measurements performed on the compound with the number of antenna fragments doubled (Figure 14.18) fully agreed with this assumption. While the quantum yield of FRET in this Pt porphyrin octa-coumarin-343 construct was still high ($\phi_{FRET} = 0.9$), the yield of phosphorescence was extremely low ($\phi_{phos} = 0.0065$), most likely due to the competing electron transfer involving the long-lived triplet state of the Pt porphyrin and C343-antenna.

In this molecule, C343 fragments are attached to the *meta*-positions in the *meso*-aryl rings on the PtP core via flexible linkers, while the reduction potential of the octacarboxy core is lower than that of the tetracarboxy core due to the electron-withdrawing effect of carboxyl groups. The combined effect is that the ET process in this case is significantly more effective than in the case of the tetra-C343 adduct. In order to avoid this undesirable ET, the redox potentials of the core and the antenna need to be adjusted; and the antenna fragments need to be placed far enough from the core, but still within the Förster radius r_0, so that the probability of the ET is minimized, while the efficiency of FRET remains high. It is, therefore, reasonable to assign an additional role to the dendrimer in two-photon oxygen probes. Namely, the dendrimer should provide a supporting distance-tuning matrix, in which the active elements of the device (antenna and the core) would be positioned at optimal distances from one another to ensure the desired energy distribution.

The interplay between 2PA, FRET, and ET processes and phosphorescence was studied using model bichromophoric systems based on PtTBPs and rhodamine B (RhB) derivatives [100]. RhB is a significantly stronger 2P absorber than C343.

Its 2PA cross section at 840 nm was originally reported to be 200 GM [101], although later studies report significantly lower values [102]. RhB derivatives strongly fluoresce in aqueous solutions ($\phi_{fl} = 0.3$–0.7), and their fluorescence overlaps with the absorption Q-band of PtTBP ($\varepsilon = 90{,}000\ M^{-1}\ cm^{-1}$). As a result, Förster r_0 for the pair PtTBP–RhB is 54 Å, which gives a plenty of flexibility for positioning the chromophores at a distance from each other but within r_0.

The studies performed on PtTBP–RhB adducts with short linkers (10–15 Å) between the edges of the chromophores revealed that PtTBP phosphorescence in these compounds is severely quenched ($\phi_{phos} < 0.01$), presumably due to electron transfer from the PtTBP triplet state onto RhB. However, when the chromophores were separated by rigid nonconducting linkers, unwanted electron transfer processes could be almost entirely prevented. Using a helix made of 10 proline residues, PtTBP and RhBs could be placed at distances large enough to prevent electron transfer (>40 Å), but still within the $r_0 = 54$ Å. One of the synthesized adducts is shown in Figure 14.19 together with its photophysical data. It is apparent from the numbers that by increasing the distance between the chromophores ET involving the PtTBP triplet was greatly diminished, while the FRET efficiency remained very high.

Systematic photoelectrochemical and photophysical studies involving different phosphorescent metalloporphyrins and different 2PA dyes capable of FRET onto the porphyrins, including dyes specifically designed for 2P applications, have been

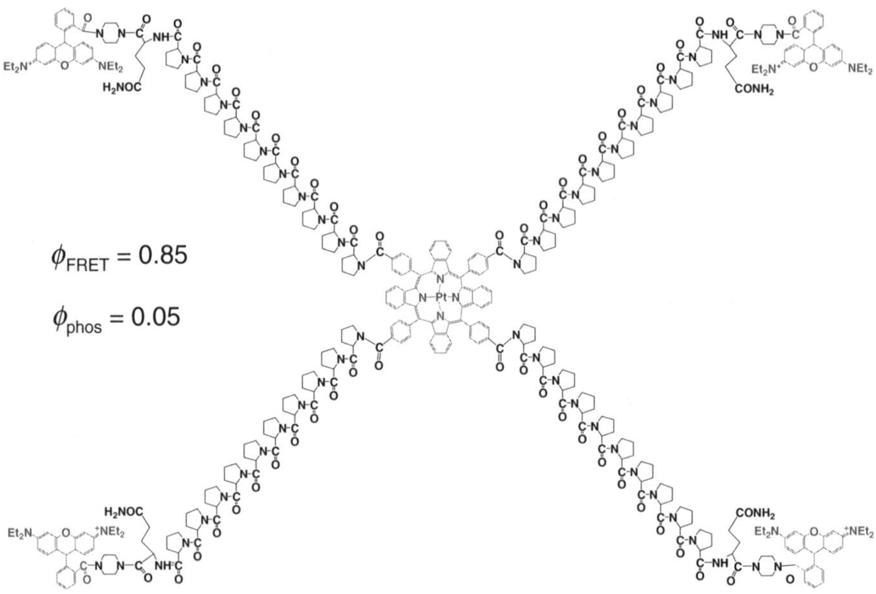

FIGURE 14.19 Pt tetrabenzoporphyrin with rhodamine B antenna fragments mounted on the termini of rigid decaproline spacers [100].

performed in order to find an ultimate pair with minimal driving force for the photo-induced ET [103]. All the special 2P dyes, however, were found to induce strong reductive quenching of porphyrin triplet states via ET, whereas C343 was found to be the least effective quencher. Simultaneously, we found that PtP modified with 4-alkoxyphenyl groups was the least efficient electron acceptor in the photo-induced ET reaction involving C343 as a reducing agent.

To evaluate the effect of dendritic branches as "insulators" from ET, Pt-*meso*-tetra(4-alkoxyphenyl)porphyrin-based AG dendrimers of three successive generations (gen 1 – gen 3) were synthesized and decorated at the periphery with several C343-antenna fragments. Gen 3 branches were determined to be optimal, as they allowed adequate separation between C343 units and the core to prevent most of the ET, while providing necessary protection from oxygen and making it possible to use more of solubilizing methoxypolyethyleneglycol (PEG) groups for modification of the dendrimer periphery. We also found that using PEG350 units did not bring the desired high aqueous solubility, but that longer PEG fragments gave probes with good solubility and no evidence of aggregation in aqueous solutions [104].

Combining distance tuning and redox tuning resulted in a dendritic two-photon-enhanced oxygen probe, PtP–C343, which integrates in a single molecular construct all the characteristics required from a probe for biological oxygen imaging (Figure 14.20) [17].

Molecular modeling revealed that in PtP–C343 even in its fully unfolded state, C343 units are separated from the PtP core by distances shorter than the Förster radius ($r_0 = 46$ Å). Nevertheless, the experimental FRET efficiency did not exceed 75%, which is likely due to the partial quenching of the C343 singlet state by the residual ET (see Figure 14.18). The folded AG dendrimer, surrounding the PtP core, is anticipated to provide an insulating layer between the chromophores. As a result, the phosphorescence quantum yield of Pt-C343 ($\phi_p = 0.10$) is only slightly below that of the PtP dendrimer without C343-antenna ($\phi_p = 0.13$). Upon 2P excitation ($\lambda_{ex} = 840$ nm, 110 fs, 76 MHz repetition rate, average power 0.5 mW, focused beam), the C343-antenna in PtP–C343 enhanced phosphorescence intensity nearly 25-fold compared to the reference Pt porphyrin. The quenching constant of the probe is independent of pH and concentration. A convenient extra feature of PtP–C343 is the residual fluorescence of C343-antenna. C343 fluorescence is well spectrally resolved from PtP phosphorescence and is completely insensitive to oxygen because of its short lifetime ($\tau_{fl} = 0.21$ ns). This signal can provide reference for ratiometric oxygen sensing, but more importantly it can be used in scanning two-photon microscopy applications for fast mapping of the object morphology.

This first example of a two-photon-enhanced dendritic oxygen probe demonstrates how dendrimer can be used for both encapsulation of a functional moiety and distance tuning in functional polychromophoric systems. PtP–C343 has been tested in imaging of oxygen inside cells [17] and currently it is being used extensively *in vivo*.

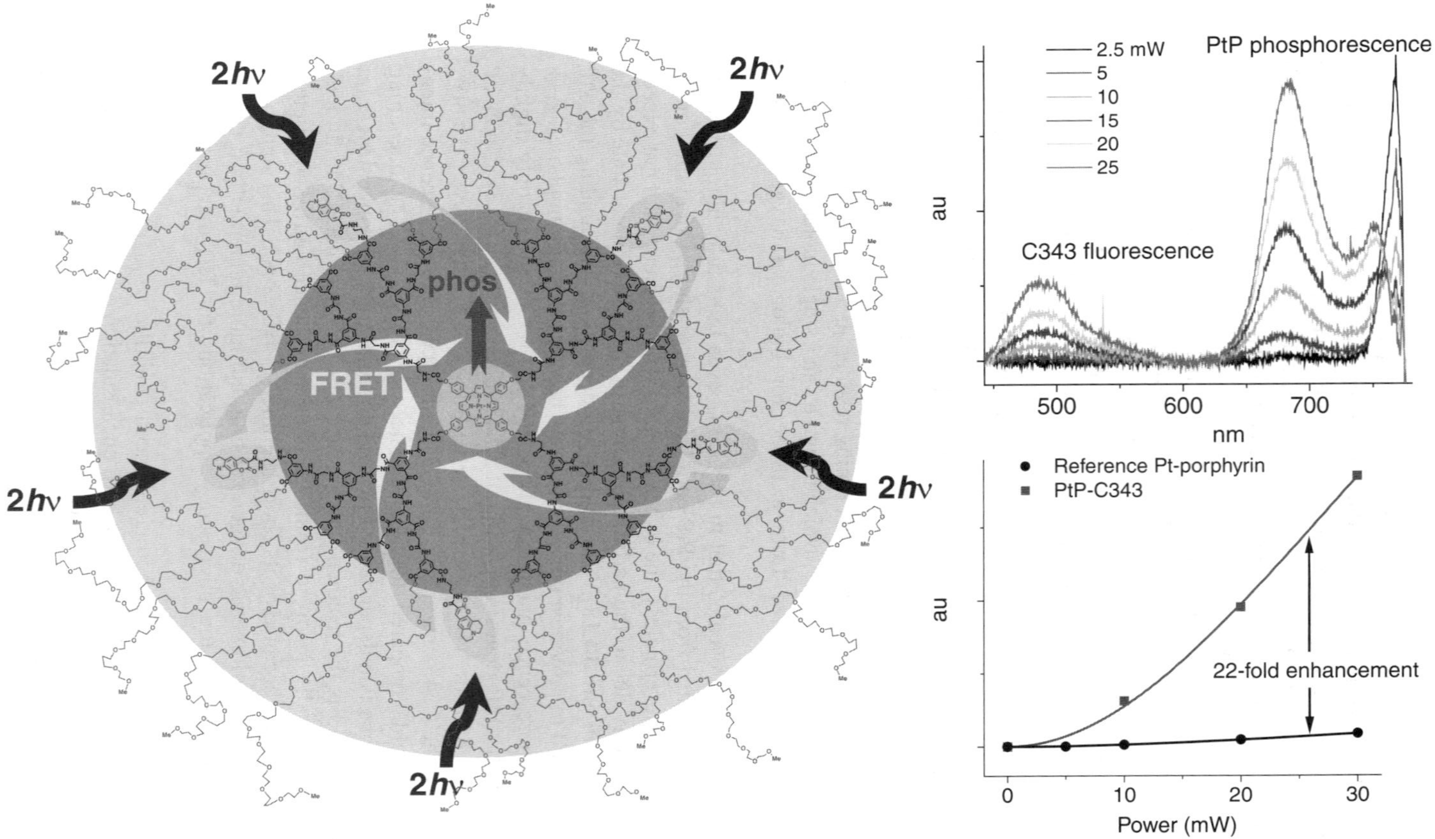

FIGURE 14.20 Two-photon-enhanced oxygen probe PtP–C343 consisting of phosphorescent Pt-*meso*-tetra-*para*-alkoxyphenylporphyrin (PtP, red), several coumarin-343 units (C343, blue), poly(aryl-glycine) dendrimer (black) and peripheral oligoethyleneglycol residues (green). Arrows in the cartoon depict excitation of the C343-antenna via 2PA (brown), FRET (yellow) and phosphorescence of PtP core (red). Graphs show emission spectra of the probe under 2P excitation (840 nm, 110 fs) and the power dependence plot (focused beam, optics transmission ~2%) [17]. (See the color version of this figure in Color Plates section.)

14.10 APPLICATIONS OF DENDRITIC OXYGEN PROBES

14.10.1 Single-Point Oxygen Measurements

The examples of single-point tissue oxygen measurements using dendritic oxygen probes range from sampling relatively large tissue volumes, when near-infrared excitation is delivered to the surface of the object and diffuses deep into the tissue, to highly localized measurements performed in confocal-like fashion. Oxyphor R2 has been used to assess oxygenation in the skeletal muscle and particularly in the diaphragm with the purpose to studying the role oxygen in diaphragm physiology and biochemistry [105]. This probe also has been used in a wide range of studies addressing the effects of hypoxia and ischemia on the brain metabolism [106]. Microscopic highly localized measurements of vascular pO_2 have been performed with an aid of Oxyphor R2 in the brain [73,107], in the rat mesentery [108] and in some other tissues [109] in order to quantify microvascular and tissue oxygen gradients. In addition, Oxyphor R2 has been used to measure oxygen consumption by isolated mouse β-cell islets [110].

Immediately after its appearance, Oxyphor G2 was enthusiastically adopted in hypoxia/ischemia studies, where it has significant advantage over Oxyphor R2 due to the ability of near-infrared excitation to sample deeper brain regions [111]. To study cortical and subcortical microvascular oxygenation in rat kidney, a dual-wavelength phosphorimetric approach has been developed [112], which relies on the difference between the tissue penetration depth of 450 and 635 nm light, both of which can be used to excite phosphorescence of Oxyphor G2 [113]. Subsequently, Oxyphor G2 has been used extensively in renal oxygenation studies [114]. Near-infrared absorbing probes based on polyglutamic and AG dendrimers have been used to measure oxygen distributions in murine tumors [115] and in skeletal muscle [116], including extravascular oxygen distributions. In addition, Pd tetrabenzoporphyrin–AG dendrimer (Oxyphor G3) has been recently used to measure oxygenation in experiments with isolated perfused rat heart [117].

Below some examples of imaging using dendritic phosphorescent probes, including wide-field and confocal imaging, are shown. For examples of two-photon and tomographic imaging the reader referred to the original publications [17,19]. All examples shown are limited by the studies performed by the author's groups independently or in direct collaborations with other laboratories.

14.10.2 Microscopic Imaging of Oxygen in the Mouse Eye

Such diseases of the eye as diabetic retinopathy, retinopathy of prematurity, sickle cell disease, age-related macular degeneration, have regional hypoxia as either a primary causative or early contributory factor. Regions of hypoxia are likely to appear before there is significant irreversible tissue injury. Therefore, information about retinal oxygenation may be extremely valuable both clinically and for understanding the basic physiology of the disease.

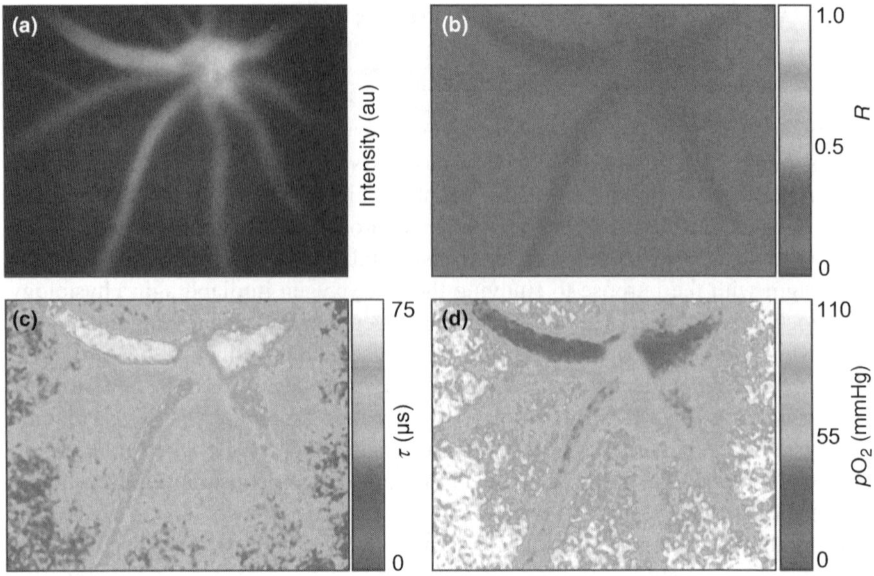

FIGURE 14.21 Maps of the phosphorescence intensity (a), goodness of fit shown by *R*-factor (b), phosphorescence lifetime (c), and oxygen pressure (d). The maps correspond to a region whose physical dimensions are 0.92 mm × 1.2 mm [121]. Reproduced with permission from the Optical Society of America. (See the color version of this figure in Color Plates section.)

Oxygen dependent quenching of phosphorescence has been applied from its early days to image oxygen in the eye [118]. Some more recent examples include application of dendritic polyglutamic probes, such as Oxyphor R2 [119]. An important advance in the technology was made by Shonat et al. [120], who implemented frequency-domain approach to image phosphorescence in wide field in the retinal tissue. This approach was later refined in our laboratory to image oxygen with microscopic resolution in the mouse eye using gen 2 polyglutamic Pd tetrabenzoporphyrin dendrimer (Oxyphor G2, Figure 14.7) [121].

The probe was excited at its Soret band (450 nm) in order to induce phosphorescence only in the superficial layers for higher resolution. The phosphorescence intensity and lifetime images from the retinal vascular bed (Figure 14.21) show that the vessel structure can be easily identified and that the arteriols and venules can be distinguished solely based on their pO_2 levels (Figure 14.21d). The imaging resolution in this case was ~5–10 μm.

One of the goals of this study was to test the method's ability to detect eye injury due to the pathology involving microvessel failure, where blockage of a capillary would result only in local hypoxic regions no more than 10–100 μm in diameter. Localized injuries, about 75 μm in diameter, were created by way of laser photocoagulation, assisted by indocyanine green dye (ICG). The injury was restricted to the microvessels within the focal area. The resulting pO_2 maps are shown in Figure 14.22.

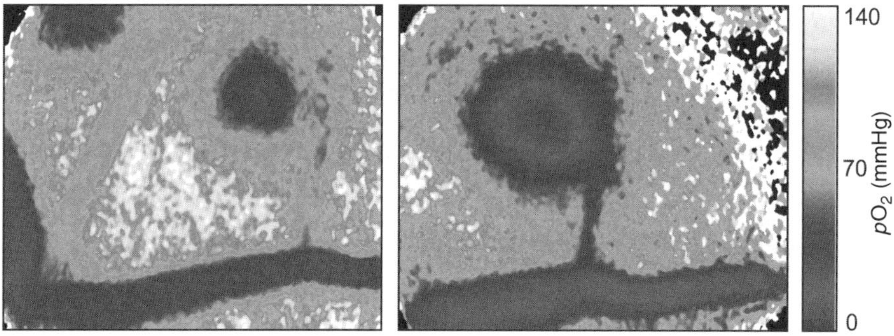

FIGURE 14.22 Oxygen pressure maps of the region containing two laser coagulation spots (left) and a region centered on one of the spots (right). Reproduced with permission from the Optical Society of America. (See the color version of this figure in Color Plates section.)

The regions of the injury can be easily seen as circular areas. The absence of significant leakage of the probe from the vessels was confirmed by measurements made over a 1 h period following the injection, during which the size and the oxygen profile of the region of hypoxia remained constant. As expected, the oxygen pressure in the injury core is extremely low.

These results demonstrate that dendritic phosphors, such as Oxyphor G2, are very useful for obtaining maps of oxygen distribution in the eye, opening up possibilities for studies of experimental rodent models of eye diseases.

14.10.3 Imaging Tumor Hypoxia

There is a large amount of evidence that hypoxia plays a central role in proliferation of cancer. Uncontrollable growth of blood vessels in tumors is associated with poor tissue oxygenation, the degree of which reflects physiological status of tumors, correlates with their metastatic potential and affects outcomes of various therapies. Hypoxia is widely spread in both primary tumors and in their metastases [122]. During tumor growth, an increasingly compromised blood supply generally results in a tumor oxygen tension of 0–20 mmHg, compared to 30–70 mmHg in normal tissues [123]. In addition to chronic diffusion-limited hypoxia, tumors also frequently exhibit transient hypoxia—a transient state caused by fluctuations in individual vessel perfusion.

Phosphorescence quenching has been used to image hypoxia in tumors using dendritic probes Oxyphor R2 [124] and G2 [125]. Dorsal window tumor model in small rodents is especially well suited for optical imaging [126], since the tumor grows between two thin glass slides in a chamber clipped to the back of a live animal, allowing for effective optical sectioning.

In recent experiments in collaboration with the Dewhirst laboratory (Duke), the dorsal window model was used to image hypoxia in mice using a gen 2 Pd tetrabenzoporphyrin–AG dendrimer (Figure 14.23), modified with PEG residues. The probe was injected iv, and the imaging was performed in wide-field epiluminescent

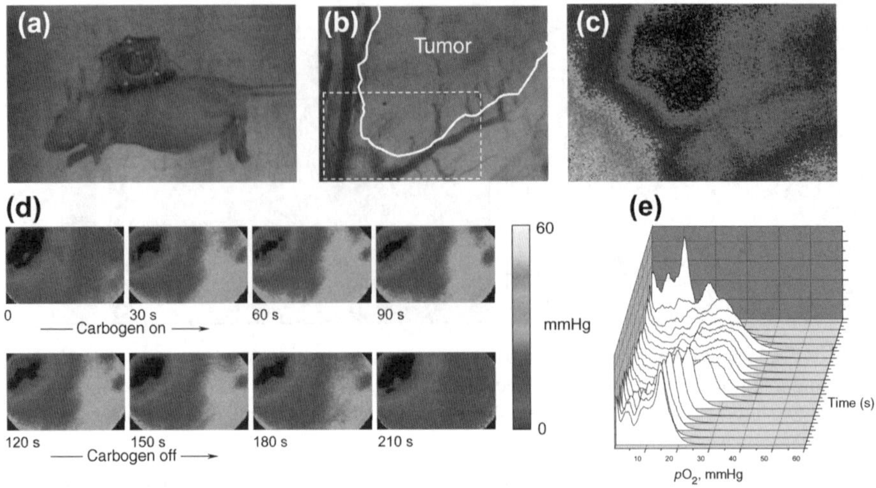

FIGURE 14.23 Imaging tumor hypoxia in a mouse dorsal window tumor model using Oxyphor G3 (gen 2 Pd tetrabenzoporphyrin–AG dendrimer). Anesthetized animal with dorsal window clipped to its back (a). The tumor is growing between the two glass plates, and it can be seen in the upper right area of the transmitted light image (b). pO_2 image of the tumor area (c). pO_2 images during carbogen (O_2:CO_2 = 95:5) breathing cycle (d). Histograms of oxygen images (e). (See the color version of this figure in Color Plates section.)

setup using LED-based (LumiLed, 635 nm, 1 W) for excitation and gated ICCD for detection of phosphorescence.

The tumor in the image (Figure 14.23b) is located in the upper right region, and it is severely hypoxic (c), with pO_2 values in the center close to zero. In order to determine how the tumor (in another animal) and the normal surrounding tissue would react to the changes in pO_2 in the inhaled gas, the mouse was given carbogen (O_2:CO_2 = 95:5) to breath, and pO_2 images were taken at ~15 s intervals (d). While the normal tissue reoxygenates quite effectively, the tumor remains severely hypoxic in spite of the high pO_2 in the inhaled gas mixture. The pO_2 histogram (e) reiterates this point, showing that regions of hypoxia are present in the imaged tissue regardless of the amount of oxygen in the inhaled gas mixture. This result suggests, among other things, that special attention must be devoted to tumor oxygenation in those cancer therapies that rely on oxygen as a cotreatment (e.g., PDT). The ability to quantify oxygen in tumors prior to and during the treatment using dendritic phosphorescent probes might be extremely useful asset in development of therapy protocols.

14.10.4 Imaging of Phosphorescence in the Brain

Brain activity is critically dependent on having an adequate supply of oxygen, but many aspects of the regulation of the oxygen supply are still not well understood. Phosphorescence quenching is well suited for measurements of oxygen in the

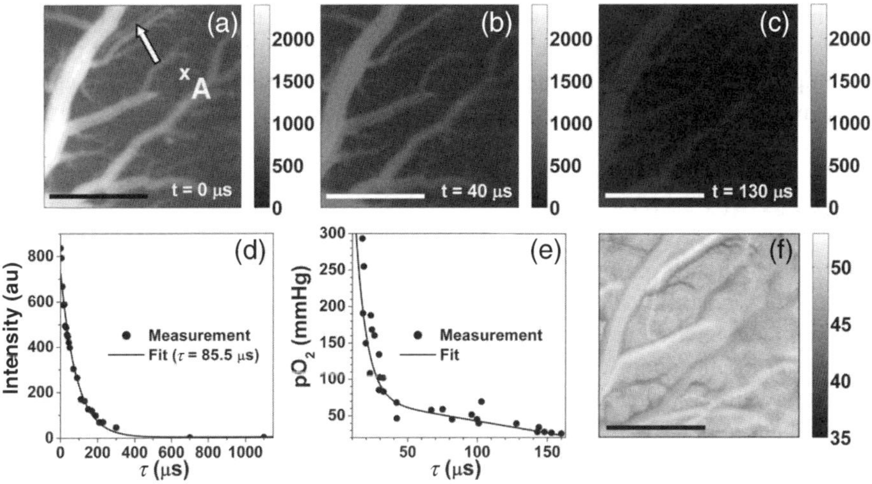

FIGURE 14.24 Formation of the pO_2 image (a), (b), and (c) phosphorescence intensity images at different time delays with respect to the laser pulse. (d) Measurement and single exponential decay lifetime (τ) fit of the phosphorescence intensity at point A marked with the cross in (a). (e) *In vitro* calibration of Oxyphor R2 at high concentration (4×10^{-5} M). (f) pO_2 image was obtained based on the probe calibration and lifetime values in each pixel. Arterial pO_2 measured during the experiment was $\approx$106 mmHg. Scale bar is 1 mm [128].

brain [106,107,127]. As an illustration, image of pO_2 in the rat brain, obtained using Oxyphor R2 and wide-field time-domain microscopy setup, is shown in Figure 14.24 [128].

More recent developments include high-resolution functional pO_2 measurements in the brain by way of confocal microscopy (Figure 14.25) [18].

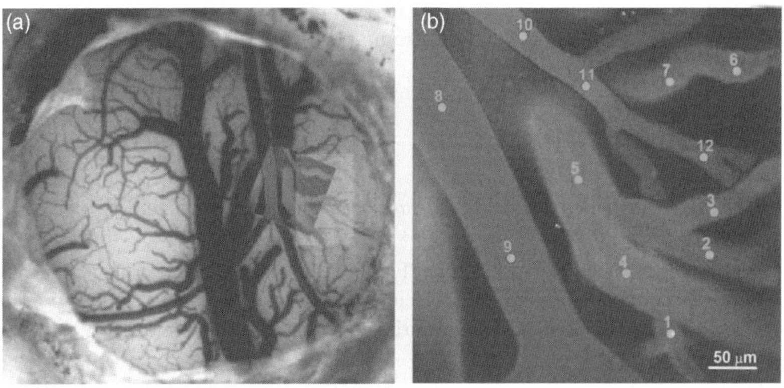

FIGURE 14.25 (a) CCD image of cranial window in a rat with confocal angiogram overlaid and region of functional activation identified in green. (b) Color-coded angiogram of microvessels in rat somatosensory cortex, with identified locations for pO_2 measurement [18]. (See the color version of this figure in Color Plates section.)

The above examples demonstrate applicability of the method to the functional studies involving hemodynamic response to neuronal stimulation, holding promise for improving our understanding of the physiological basis of functional magnetic resonance imaging.

14.11 CONCLUSION

The construction of porphyrin-based dendritic oxygen probes once again demonstrates the versatility of dendritic encapsulation as method for controlling the chromophore environment. Dendritic encapsulation has been used to optimize the properties of phosphorescent probes for oxygen measurements in tissue; including the sensitivity to quenching by oxygen, the permeability through biological membranes, and the solubility in biological media. Using a multilayered modular design, dendritic oxygen probes have been tailored to many different modes of oxygen measurement application, from point measurements to high-resolution microscopic imaging and 3-D tomography using infrared light. The described design provide a versatile framework for controlling the properties of phosphorescent oxygen sensors, tailoring them to meet the requirements of almost any applications requiring accurate measurement of oxygen.

ACKNOWLEDGMENTS

Support for this research was provided by grants HL081273, NS031465, EB3663, and EB007279 from the National Institutes of Health of the USA and a grant 02-5403-21-2 from the US Air Force. The authors are extremely grateful to all their collaborators, current group members and many postdoctoral researchers who over the years participated in the studies described in the chapter. SAV is especially thankful to Dr. Louise E. Sinks for proof reading the manuscript.

REFERENCES

[1] J. M. J. Fréchet, D. A. Tomalia, *Dendrimers and Other Dendritic Polymers*, Wiley, New York, **2001**.

[2] W. Krause, N. Hackmann-Schlichter, F. Maier, R. Muller, in Dendrimers II. Architecture, Nanostructure and Supramolecular Chemistry, Vol. *210*, *Topics in Current Chemistry ed.* (Ed.: F. Vögtle), Springer-Verlag, Berlin, Heidelberg, New York, **2000**, p. 261.

[3] S. Hecht, J. M. J. Fréchet, *Angew. Chem. Int. Ed.* **2001**, *40*, 74.

[4] C. B. Gorman, J. C. Smith, *J. Am. Chem. Soc.* **2000**, *122*, 9342.

[5] W. S. Li, T. Aida, *Chem. Rev.* **2009**, *109*, 6047.

[6] W. Maes, W. Dehaen, *Eur. J. Org. Chem.* **2009**, 4719.

[7] J. G. Rajendran, K. A. Krohn, *Radiol. Clin. N. Am.* **2005**, *43*, 169.

[8] D. F. Wilson, *Am. J. Physiol.-Heart Circ. Physiol.* **2008**, *294*, H11.

[9] C. J. Koch, Measurement of absolute oxygen levels in cells and tissues using oxygen sensors and 2-nitroimidazole EF5," Methods in Enzymology, v. 352 (Redox Cell Biology and Genetics, Part A), pp. 3–31 (2002).

[10] H. M. Swartz, R. B. Clarkson, *Phys. Med. Biol.* **1998**, *43*, 1957.

[11] M. C. Krishna, S. English, K. Yamada, J. Yoo, R. Murugesan, N. Devasahayam, J. A. Cook, K. Golman, J. H. Ardenkjaer-Larsen, S. Subramanian, J. B. Mitchell, *Proceedings of the National Academy of Sciences of the United States of America* **2002**, *99*, 2216.

[12] F. F. Jobsis, *Science* **1977**, *198*, 1264.

[13] L. S. Ziemer, S. M. Evans, A. Kachur, A. L. Shuman, C. A. Cardi, W. T. Jenkins, J. S. Karp, A. Alavi, W. R. Dolbier, C. J. Koch, *Eur. J. Nucl. Med. Mol. Imag.* **2003**, *30*, 259.

[14] J. M. Vanderkooi, G. Maniara, T. J. Green, D. F. Wilson, *J. Biol. Chem.* **1987**, *262*, 5476.

[15] M. Pawlowski, D. F. Wilson, *Adv. Expt. Med. Biol.* **1992**, *316*, 179.

[16] W. L. Rumsey, J. M. Vanderkooi, D. F. Wilson, *Science* **1988**, *241*, 1649.

[17] O. S. Finikova, A. Y. Lebedev, A. Aprelev, T. Troxler, F. Gao, C. Garnacho, S. Muro, R. M. Hochstrasser, S. A. Vinogradov, *ChemPhysChem* **2008**, *9*, 1673.

[18] M. A. Yaseen, V. J. Srinivasan, S. Sakadžic, W. Wu, S. Ruvinskaya, S. A. Vinogradov, D. A. Boas, *Opt. Express* **2009**, *17*, 22341.

[19] V. Soloviev, D. Wilson, S. Vinogradov, *Appl. Opt.* **2003**, *42*, 113.

[20] V. Y. Soloviev, D. F. Wilson, S. A. Vinogradov, *Appl. Opt.* **2004**, *43*, 564.

[21] S. V. Apreleva, D. F. Wilson, S. A. Vinogradov, *Appl. Opt.* **2006**, *45*, 8547.

[22] N. J. Turro, *Modern Molecular Photochemistry*, University Science Books, Sausalito, CA, **1991**.

[23] P.G.T. Fogg, W. Gerrard, *Solubility of Gases in Liquids*, Wiley, New York, **2001**.

[24] D. Eastwood, M. Gouterman, *J. Mol. Spectrosc.* **1970**, *35*, 359.

[25] D. H. Kim, D. Holten, M. Gouterman, J. W. Buchler, *J. Am. Chem. Soc.* **1984**, *106*, 4015.

[26] K. J. Morris, M. S. Roach, W. Y. Xu, J. N. Demas, B. A. DeGraff, *Anal. Chem.* **2007**, *79*, 9310.

[27] P. Caliceti, F. M. Veronese, *Adv. Drug Deliv. Rev.* **2003**, *55*, 1261–1277.

[28] S. A. Vinogradov, D. F. Wilson, *J. Chem. Soc., Perkin Trans. II* **1995**, 103.

[29] S. A. Vinogradov, L.-W. Lo, W. T. Jenkins, S. M. Evans, C. Koch, D. F. Wilson, *Biophys. J.* **1996**, *70*, 1609.

[30] M. Sibrian-Vazquez, T. J. Jensen, M. G. H. Vicente, *J. Photochem. Photobiol. B-Biol.* **2007**, *86*, 9.

[31] F. N. Castellano, J. R. Lakowicz, *Photochem. Photobiol.* **1998**, *67*, 179.

[32] V. Balzani, P. Ceroni, M. Maestri, C. Saudan, V. Vicinelli, *Dendrimers V: Functional and Hyperbranched Building Blocks, Photophysical Properties, Applications in Materials and Life Sciences*, Vol. *228*, Springer, Berlin, **2003**, p. 159.

[33] P. Ceroni, G. Bergamini, F. Marchioni, V. Balzani, *Prog. Polym. Sci.* **2005**, *30*, 453.

[34] J. S. Lindsey, I. C. Schreiman, H. C. Hsu, P. C. Kearney, A. M. Marguerettaz, *J. Org. Chem.* **1987**, *52*, 827.

[35] M. O. Senge, *Acc. Chem. Res.* **2005**, *38*, 733.

[36] J. W. Buchler, Static coordination chemistry of metalloporphyrins, in *Porphyrins and Metalloporphyrins* (Ed.: K. M. Smith), Elsevier, New York, **1975**, Chapter 5.

[37] For examples see: T. D. Lash, Synthesis of novel porphyrinoid chromophores, in *Porphyrin Handbook*, Vol. *2* (Eds.: K. M. Kadish, K. M. Smith, R. Guilard), Academic Press, New York, **2000**, Chapter 10, p. 125.

[38] D. B. Papkovsky, G. V. Ponomarev, W. Trettnak, P. Oleary, *Analytical Chemistry* **1995**, *67*, 4112.

[39] M. P. Tsvirko, V. V. Sapunov, K. N. Soloviyev, *Optika i Spektroskopiya (Russ)* **1973**, *34*, 1094.

[40] V. V. Rozhkov, M. Khajehpour, S. A. Vinogradov, *Inorg. Chem.* **2003**, *42*, 4253.

[41] P. Taroni, A. Pifferi, A. Torricelli, D. Comelli, R. Cubeddu, *Photochem. Photobiol. Sci.* **2003**, *2*, 124.

[42] U. Tripathy, D. Kowalska, X. Liu, S. Velate, R. P. Steer, *J. Phys. Chem. A* **2008**, *112*, 5824.

[43] V. N. Kopranenkov, S. N. Dashkevich, E. A. Luk'yanets, *J. Gen. Chem. (Russ.)* **1981**, *51*, 2165.

[44] L. Edwards, M. Gouterman, C. B. Rose, *J. Am. Chem. Soc.* **1976**, *98*, 7638.

[45] K. Ichimura, M. Sakuragi, H. Morii, M. Yasuike, M. Fukui, O. Ohno, *Inorg. Chim. Acta* **1990**, *176*, 31.

[46] V. N. Kopranenkov, E. A. Makarova, S. N. Dashkevich, E. A. Luk'janets, *Khimiya Geterotsiklicheskikh Soedinenii* **1988**, 773.

[47] D. Barton, J. Kervagoret, S. Zard, *Tetrahedron* **1990**, *46*, 7587.

[48] S. Ito, T. Murashima, H. Uno, N. Ono, *Chem. Commun.* **1998**, 1661.

[49] S. Ito, N. Ochi, H. Uno, T. Murashima, N. Ono, *Chem. Commun.* **2000**, 893.

[50] O. Finikova, A. Cheprakov, I. Beletskaya, S. Vinogradov, *Chem. Commun.* **2001**, 261.

[51] O. S. Finikova, A. V. Cheprakov, I. P. Beletskaya, P. J. Carroll, S. A. Vinogradov, *J. Org. Chem.* **2004**, *69*, 522.

[52] O. S. Finikova, A. V. Cheprakov, P. J. Carroll, S. A. Vinogradov, *J. Org. Chem.* **2003**, *68*, 7517.

[53] O. S. Finikova, S. E. Aleshchenkov, R. P. Brinas, A. V. Cheprakov, P. J. Carroll, S. A. Vinogradov, *J. Org. Chem.* **2005**, *70*, 4617.

[54] C. Borek, K. Hanson, P. I. Djurovich, M. E. Thompson, K. Aznavour, R. Bau, Y. R. Sun, S. R. Forrest, J. Brooks, L. Michalski, J. Brown, *Angew. Chem. Int. Ed.* **2007**, *46*, 1109.

[55] J. R. Sommer, R. T. Farley, K. R. Graham, Y. X. Yang, J. R. Reynolds, J. G. Xue, K. S. Schanze, *ACS Appl. Mater. Interfaces* **2009**, *1*, 274.

[56] R. Kumar, T. Y. Ohulchanskyy, I. Roy, S. K. Gupta, C. Borek, M. E. Thompson, P. N. Prasad, *ACS Appl. Mater. Interfaces* **2009**, *1*, 1474.

[57] S. Baluschev, V. Yakutkin, T. Miteva, Y. Avlasevich, S. Chernov, S. Aleshchenkov, G. Nelles, A. Cheprakov, A. Yasuda, K. Mullen, G. Wegner, *Angew. Chem. Int. Ed.* **2007**, *46*, 7693.

[58] O. S. Finikova, A. V. Cheprakov, S. A. Vinogradov, *J. Org. Chem.* **2005**, *70*, 9562.

[59] M. A. Filatov, A. V. Cheprakov, I. P. Beletskaya, *Eur. J. Org. Chem.* **2007**, 3468.

[60] M. A. Filatov, A. Y. Lebedev, S. A. Vinogradov, A. V. Cheprakov, *J. Org. Chem.* **2008**, *73*, 4175.

[61] A. Y. Lebedev, M. A. Filatov, A. V. Cheprakov, S. A. Vinogradov, *J. Phys. Chem. A* **2008**, *112*, 7723.

[62] P. J. Dandliker, F. Diederich, M. Gross, C. B. Knobler, A. Louati, E. M. Sanford, *Angew. Chem. Int. Ed. Engl.* **1994**, *33*, 1739.

[63] P. J. Dandliker, F. Diederich, J. P. Gisselbrecht, A. Louati, M. Gross, *Angew. Chem. Int. Ed. Engl.* **1996**, *34*, 2725.

[64] S. A. Vinogradov, D. F. Wilson, *Adv. Exp. Med. Biol.* **1997**, *428*, 657.

[65] S. A. Vinogradov, L. W. Lo, D. F. Wilson, *Chem.-Eur. J.* **1999**, *5*, 1338.

[66] I. B. Rietveld, E. Kim, S. A. Vinogradov, *Tetrahedron* **2003**, *59*, 3821.

[67] I. Dunphy, S. A. Vinogradov, D. F. Wilson, *Anal. Biochem.* **2002**, *310*, 191.

[68] R. H. Jin, T. Aida, S. Inoue, *J. Chem. Soc.-Chem. Commun.* **1993**, 1260.

[69] J. Issberner, F. Vögtle, L. De Cola, V. Balzani, *Chem.-Eur. J.* **1997**, *3*, 706.

[70] L. Twyman, A. E. Beezer, J. C. Mitchell, *Tetrahedron Lett.* **1994**, *35*, 4423.

[71] O. S. Finikova, A. S. Galkin, V. V. Rozhkov, M. C. Cordero, C. Hägerhäll, S. A. Vinogradov, *J. Am. Chem. Soc.* **2003**, *125*, 4882.

[72] Z. J. Kaminski, *Synthesis* **1987**, 917.

[73] A. D. Estrada, A. Ponticorvo, T. N. Ford, A. K. Dunn, *Opt. Lett.* **2008**, *33*, 1038.

[74] S. A. Vinogradov, D. F. Wilson, *Appl. Spectrosc.* **2000**, *54*, 849.

[75] V. Rozhkov, D. Wilson, S. Vinogradov, *Macromolecules* **2002**, *35*, 1991.

[76] C. J. Hawker, K. L. Wooley, J. M. J. Frechet, *J. Chem. Soc., Perkin Trans. I* **1993**, 1287.

[77] G. R. Newkome, X. F. Lin, *Macromolecules* **1991**, *24*, 1443.

[78] G. R. Newkome, X. F. Lin, J. K. Young, *Synlett* **1992**, 53.

[79] W. R. Dichtel, K. Y. Baek, J. M. J. Frechet, I. B. Rietveld, S. A. Vinogradov, *J. Polym. Sci. Part A-Polym. Chem.* **2006**, *44*, 4939.

[80] V. Gajbhiye, P. V. Kumar, R. K. Tekade, N. K. Jain, *Curr. Pharmaceut. Design* **2007**, *13*, 415.

[81] S. J. Guillaudeu, M. E. Fox, Y. M. Haidar, E. E. Dy, F. C. Szoka, J. M. J. Frechet, *Bioconjug. Chem.* **2008**, *19*, 461.

[82] M. Jikei, M. A. Kakimoto, *J. Polym. Sci. Part A-Polym. Chem.* **2004**, *42*, 1293.

[83] M. Okazaki, I. Washio, Y. Shibasaki, M. Ueda, *J. Am. Chem. Soc.* **2003**, *125*, 8120.

[84] I. Washio, Y. Shibasaki, M. Ueda, *Org. Lett.* **2003**, *5*, 4159.

[85] T. M. Miller, T. X. Neenan, *Chem. Mater.* **1990**, *2*, 346.

[86] S. C. E. Backson, P. M. Bayliff, W. J. Feast, A. M. Kenwright, D. Parker, R. W. Richards, *Macromol. Symp.* **1994**, *77*, 1.

[87] S. A. Vinogradov, *Org. Lett.* **2005**, *7*, 1761.

[88] J. Greenstein, M. Winitz, *Chem. Amino Acids*, John Wiley & Sons, New York, **1961**.

[89] A. Y. Lebedev, A. V. Cheprakov, S. Sakadzic, D. A. Boas, D. F. Wilson, S. A. Vinogradov, *ACS Appl. Mater. Interfaces* **2009**, *1*, 1292.

[90] Z. J. Kaminski, *Synthesis* **1987**, 917.

[91] S. J. Strickler, R. A. Berg, *J. Chem. Phys.* **1962**, *37*, 814.

[92] W. Denk, J. H. Strickler, W. W. Webb, *Science* **1990**, *248*, 73.

[93] W. Denk, D. W. Piston, W. W. Webb, in *Handbook of Biological Confocal Microscopy*, 3rd edition (Ed.: J. B. Pawley), Springer Science + Business Media, LLC, New York, **2006**, p. 535.

[94] E. G. Mik, T. G. van Leeuwen, N. J. Raat, C. Ince, *J. Appl. Physiol.* **2004**, *97*, 1962.

[95] For examples see: (a) M. Drobizhev, A. Karotki, M. Kruk, N. Z. Mamardashvili, A. Rebane, *Chem. Phys. Lett.* **2002**, *361*, 504; (b) A. Karotki, M. Drobizhev, M. Kruk, C. Spangler, E. Nickel, N. Mamardashvili, A. Rebane, *J. Opt. Soc. Am. B-Opt. Phys.* **2003**, *20*, 321; (c) M. Drobizhev, Y. Stepanenko, Y. Dzenis, A. Karotki, A. Rebane, P. N. Taylor, H. L. Anderson, *J. Phys. Chem. B* **2005**, *109*, 7223; (d) K. Ogawa, A. Ohashi, Y. Kobuke, K. Kamada, K. Ohta, *J. Phys. Chem. B* **2005**, *109*, 22003; (e) J. Dy, K. Ogawa, K. Kamada, K. Ohta, Y. Kobuke, *Chem. Commun.* **2008**, 3411; (f) D. Y. Kim, T. K. Alm, J. H. Kwon, D. Kim, T. Ikeue, N. Aratani, A. Osuka, M. Shigeiwa, S. Maeda, *J. Phys. Chem. A* **2005**, *109*, 2996; (g) T. K. Ahn, K. S. Kim, D. Y. Kim, S. B. Noh, N. Aratani, C. Ikeda, A. Osuka, D. Kim, *J. Am. Chem. Soc.* **2006**, *128*, 1700.

[96] (a) A. Karotki, M. Drobizhev, Y. Dzenis, P. N. Taylor, H. L. Anderson, A. Rebane, *Phys. Chem. Chem. Phys.* **2004**, *6*, 7–10; (b) M. Drobizhev, Y. Stepanenko, Y. Dzenis, A. Karotki, A. Rebane, P. N. Taylor, H. L. Anderson, *J. Am. Chem. Soc.* **2004**, *126*, 15352–15353.

[97] R. P. Briñas, T. Troxler, R. M. Hochstrasser, S. A. Vinogradov, *Journal of the American Chemical Society* **2005**, *127*, 11851.

[98] W. R. Dichtel, J. M. Serin, C. Edder, J. M. J. Fréchet, M. Matuszewski, L. S. Tan, T. Y. Ohulchanskyy, P. N. Prasad, *J. Am. Chem. Soc.* **2004**, *126*, 5380.

[99] M. A. Oar, J. M. Serin, W. R. Dichtel, J. M. J. Fréchet, *Chem. Mater.* **2005**, *17*, 2267.

[100] O. S. Finikova, T. Troxler, A. Senes, W. F. DeGrado, R. M. Hochstrasser, S. A. Vinogradov, *J. Phys. Chem. A* **2007**, *111*, 6977.

[101] C. Xu, W. W. Webb, *J. Opt. Soc. Am. B-Opt. Phys.* **1996**, *13*, 481.

[102] N. S. Makarov, M. Drobizhev, A. Rebane, *Opt. Express* **2008**, *16*, 4029.

[103] O. S. Finikova, P. Chen, Z. P. Ou, K. M. Kadish, S. A. Vinogradov, *J. Photochem. Photobiol. A-Chem.* **2008**, *198*, 75.

[104] A. Y. Lebedev, T. Troxler, S. A. Vinogradov, *J. Porphyrins Phthalocyanines* **2008**, *12*, 1261.

[105] For examples see: (a) B. J. Behnke, M. D. Delp, P. McDonough, S. A. Spier, D. C. Poole, T. I. Musch, *Cardiovasc. Res.* **2004**, *61*, 325; (b) D. C. Poole, B. J. Behnke, P. McDonough, R. M. McAllister, D. F. Wilson, *Microcirculation* **2004**, *11*, 317; (c) D. C. Poole, B. J. Behnke, D. J. Padilla, *Med. Sci. Sports Exerc.* **2005**, *37*, 1559; (d) L. F. Ferreira, P. McDonough, B. J. Behnke, T. I. Musch, D. C. Poole, *Respir. Physiol. Neurobiol.* **2006**, *153*, 237.

[106] For examples see: (a) W. M. DeCampli, G. Schears, R. Myung, S. Schultz, J. Creed, A. Pastuszko, D. F. Wilson, *J. Thorac. Cardiovasc. Surg.* **2003**, *125*, 472; (b) G. Schears, S. E. Schultz, J. Creed, W. J. Greeley, D. F. Wilson, A. Pastuszko, *Ann. Thorac. Surg.* **2003**, *75*, 560.

[107] I. Vanzetta and A. Grinvald, Increased cortical oxidative metabolism due to sensory stimulation: Implications for functional brain imaging, *Science*, **1999**, *286*, 1555–1558.

[108] For examples see: (a) A. S. Golub, M. C. Barker, R. N. Pittman, *Am. J. Physiol.-Heart Circ. Physiol.* **2007**, *293*, H1097; (b) A. S. Golub, R. N. Pittman, *Am. J. Physiol.-Heart Circ. Physiol.* **2008**, *294*, H2905; (c) A. S. Golub, M. C. Barker, R. N. Pittman, *Am. J. Physiol.-Heart Circ. Physiol.* **2008**, *294*, H21; (d) L. N. Torres, R. N. Pittman, I. P. Torres, *Microvasc. Res.* **2008**, *75*, 217.

[109] R. Pirow, C. Baumer, R. J. Paul, *J. Exp. Biol.* **2004**, *207*, 4393.

[110] N. M. Doliba, W. Qin, M. Z. Vatamaniuk, C. W. Buettger, H. W. Collins, M. A. Magnuson, K. H. Kaestner, D. F. Wilson, R. D. Carr, F. M. Matschinsky, *Am. J. Physiol.-Endocrinol. Metab.* **2006**, *291*, E525.

[111] For example see: (a) G. Schears, D. Antoni, S. Schultz, T. Zaitseva, W. Greeley, D. F. Wilson, A. Pastuszko, *Neurochem. Res.* **2005**, *30*, 1453; (b) G. Schears, T. Zaitseva, S. Schultz, W. Greeley, D. Antoni, D. F. Wilson, A. Pastuszko, *Eur. J. Cardio-Thorac. Surg.* **2006**, *29*, 168; (c) T. Zaitseva, S. Schultz, G. Schears, P. Pastuszko, S. Markowitz, W. Greeley, D. F. Wilson, A. Pastuszko, *Ann. Thorac. Surg.* **2006**, *82*, 2247.

[112] W. R. Rumsey, B. Abbott, L.-W. Lo, S. A. Vinogradov, D. F. Wilson, *Adv. Expt. Med. Biol.* **1997**, *411*, 591.

[113] T. Johannes, E. G. Mik, C. Ince, *J. Appl. Physiol.* **2006**, *100*, 1301.

[114] For examples see: (a) T. Johannes, E. G. Mik, B. Nohe, K. E. Unertl, C. Ince, *Am. J. Physiol.-Renal Physiol.* **2007**, *292*, F796; (b) E. G. Mik, T. Johannes, C. Ince, *Am. J. Physiol.-Renal Physiol.* **2008**, *294*, F676; (c) T. Johannes, E. G. Mik, C. Ince, *Shock* **2009**, *31*, 97; (d) T. Johannes, E. G. Mik, K. Klingel, H. J. Dieterich, K. E. Unertl, C. Ince, *Shock* **2009**, *31*, 521.

[115] L. S. Ziemer, W. M. F. Lee, S. A. Vinogradov, C. Sehgal, D. F. Wilson, *J. Appl. Physiol.* **2005**, *98*, 1503.

[116] D. F. Wilson, W. M. F. Lee, S. Makonnen, O. Finikova, S. Apreleva, S. A. Vinogradov, *J. Appl. Physiol.* **2006**, *101*, 1648.

[117] M. Ranji, D. L. Laggard, S. V. Apreleva, S. A. Vinogradov, B. Chance, *Opt. Lett.* **2006**, *31*, 2995.

[118] R. D. Shonat, D. F. Wilson, C. E. Riva, M. Pawlowski, *Appl. Opt.* **1992**, *33*, 3711.

[119] (a) S. R. Chamot, S. D. Cranstoun, B. L. Petrig, C. J. Pournaras, C. E. Riva, *J. Biomed. Opt.* **2003**, *8*, 63; (b) T. K. Stepinac, S. K. Chamot, E. Rungger-Brandle, P. Ferrez, J. L. Munoz, H. van den Bergh, C. E. Riva, C. J. Pournaras, G. A. Wagnieres, *Investig. Ophthalmol. Vis. Sci.* **2005**, *46*, 956.

[120] R. D. Shonat, A. C. Kight, *Ann. Biomed. Eng.* **2003**, *31*, 1084.

[121] D. F. Wilson, S. A. Vinogradov, P. Grosul, M. N. Vaccarezza, A. Kuroki, J. Bennett, *Appl. Opt.* **2005**, *44*, 5239.

[122] D. M. Brizel, G. L. Rosner, L. R. Prosnitz, M. W. Dewhirst, *Int. J. Radiat. Oncol. Biol. Phys.* **1995**, *32*, 1121.

[123] P. Vaupel, F. Kallinowski, P. Okunieff, *Cancer Res.* **1989**, *49*, 6449.

[124] K. Erickson, R. A. Braun, D. H. Yu, J. Lanzen, D. Wilson, D. M. Brizel, T. W. Secomb, J. E. Biaglow, M. W. Dewhirst, *Cancer Res.* **2003**, *63*, 4705.

[125] L. I. Cardenas-Navia, D. Mace, R. A. Richardson, D. F. Wilson, S. Shan, M. W. Dewhirst, *Cancer Res.* **2008**, *68*, 5812.

[126] D. F. Wilson, S. M. Evans, W. T. Jenkins, S. A. Vinogradov, E. Ong, M. W. Dewhirst, *Adv. Exp. Med. Biol.* **2008**, *454*, 603.

[127] B. M. Ances, D. F. Wilson, J. H. Greenberg, J. A. Detre, *J. Cerebral Blood Flow Metab.* **2001**, *21*, 511.

[128] S. Sakadzic, S. Yuan, E. Dilekoz, S. Ruvinskaya, S. A. Vinogradov, C. Ayata, D. A. Boas, *Appl. Opt.* **2009**, *48*, D169.

15

PEPTIDE DENDRIMERS AS ARTIFICIAL PROTEINS

TAMIS DARBRE AND JEAN-LOUIS REYMOND

Department of Chemistry and Biochemistry, University of Berne, Freiestrasse 3, CH-3012, Berne, Switzerland

15.1 INTRODUCTION

Dendrimers have been suggested early on as synthetic macromolecular analogs of proteins [1–4], in particular enzymes [5,6]. Indeed the tree-like dendrimer topology produces macromolecules that are expected to adopt a protein-like globular structure without the need for folding, thus circumventing an important obstacle in artificial protein design [7]. Most synthetic work with dendrimers has however remained quite distant from proteins for two reasons: (1) the building blocks used for dendrimers are quite different from the proteinogenic α-amino acids; (2) the use of fixed branching structures with changes possible only at the dendrimer core or at the periphery strongly limits structural variations compared to natural proteins where changes occur by amino acid substitution throughout the structure. For instance in early work by Denkewalter and coworkers with poly-lysine dendrimers [8] and in the multiple antigenic peptides by Tam and coworkers [9], a poly-lysine tree is used for the only purpose of providing multivalency without reference to artificial protein design.

In this chapter, we review our approach to peptide dendrimers as artificial proteins [10]. The work is based on the combinatorial solid-phase synthesis of peptide sequences featuring diamino acids branching points at every 2nd, 3rd, or 4th position to obtain peptide dendrimers that can be easily varied at the core, within

Designing Dendrimers, Edited by Sebastiano Campagna, Paola Ceroni, Fausto Puntoriero.
© 2012 John Wiley & Sons, Inc. Published 2012 by John Wiley & Sons, Inc.

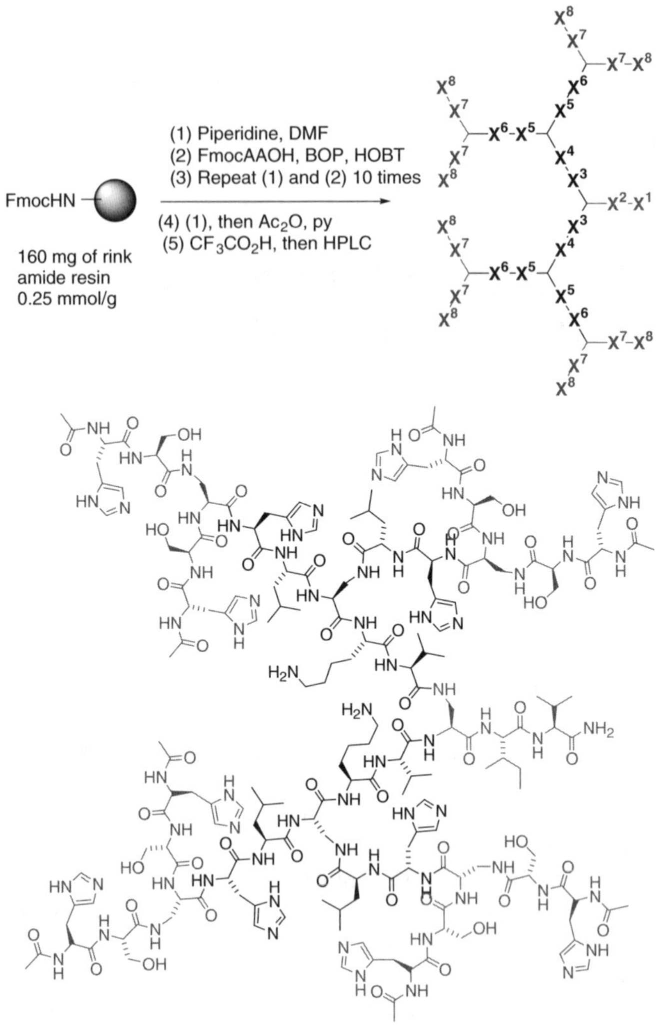

C11 (19.7 mg, 8% isolated yield)

FIGURE 15.1 Synthesis and structure of the third-generation peptide dendrimer **C11** (AcHisSer)$_8$(*Dap*HisLeu)$_4$(*Dap*LysVal)$_2$*Dap*IleValNH$_2$ (*Dap* = branching L-2,3-diaminopropanoic acid) identified by screening as an esterase dendrimer [14].

the branches, and at the branches' ends (Figure 15.1). These peptide dendrimers are identical to proteins in terms of composition and only differ in their topology. Furthermore, they can be optimized iteratively by amino acid substitutions in a manner similar to the natural evolution of proteins. We present our progress toward protein-like functions in the area of artificial enzymes, metal binding, and glycopeptide dendrimers, as well as structural studies by spectroscopic and molecular modeling methods.

15.2 SOLID PHASE AND COMBINATORIAL SYNTHESIS

15.2.1 SPPS of Peptide Dendrimers and Their Combinatorial Libraries

Solution phase synthesis and convergent syntheses are usually preferred for dendrimer assembly to minimize sterical hindrance and purification problems. Contrary to these principles, we found that peptide dendrimers up to the fourth generation can be obtained in good yields and purities by a divergent synthesis protocol on solid support under the standard conditions of solid phase peptide synthesis with Fmoc-protected amino acid building blocks (Fmoc-SPPS) [11,12]. The dendrimer core is constituted by the C-terminal amino acids which are coupled first to the solid support. Branching proceeds as diamino acids are added along the sequence. A typical dendrimer of third generation with branches, 2 amino acids in length is assembled in 11 coupling steps for a total of 37 amino acids and an average molecular weight of 4400 Da, as shown for the esterase dendrimer **C11** $(AcHisSer)_8$$(\textit{Dap}HisLeu)_4$ $(\textit{Dap}LysVal)_2$$\textit{Dap}IleValNH_2$ (Figure 15.1).

The isolated yields of peptide dendrimer obtained from Fmoc-SPPS after side-chain deprotection, cleavage from the solid support, and purification by reverse-phase HPLC are comparable to the yields for linear peptides of similar size. As for linear peptides, the isolated yields are influenced by the amino acid sequence and by the physico-chemical properties of the product. In particular strongly hydrophobic sequences may be difficult or impossible to purify due to their lack of solubility in any solvent. Overall, many "well-behaved" peptide dendrimers comprising a good balance of charged, polar, and hydrophobic amino acids, are readily accessible by SPPS. Peptide dendrimer synthesis is also facilitated by the commercial availability of amino acid building blocks for Fmoc-SPPS at reasonable prices, which renders the synthesis quite affordable and potentially scalable using the existing industrial infrastructure for SPPS.

The split-and-mix protocol for assembling peptide libraries by SPPS [13], which was a founding experiment in combinatorial chemistry, can be applied to prepare the corresponding peptide dendrimer libraries on solid support [14]. In these so-called "one-bead-one-compound" (OBOC) libraries, each bead of the solid support carries only one possible sequence, yet a large number of different sequences are assembled simultaneously in the course of the synthesis. The number of different sequences accessible in a split-and-mix synthesis is limited by the number of beads used in the SPPS synthesis batch. A typical synthesis batch of 500 mg of 90 μm diameter beads contains approximately one million beads, which imposes a limit of 100,000 sequences for the library size to ensure 10-fold coverage of sequence space.

15.2.2 Activity Screening

OBOC libraries are typically tested for activity using a bead-staining assay [15]. Such assays include binding assays with a labeled or colored ligand, for example, vitamin B_{12}, and catalysis assays looking for conversion of a chromogenic or fluorogenic substrate adsorbed on the beads, such as the hydrolysis of acyloxypyrene trisulfonate

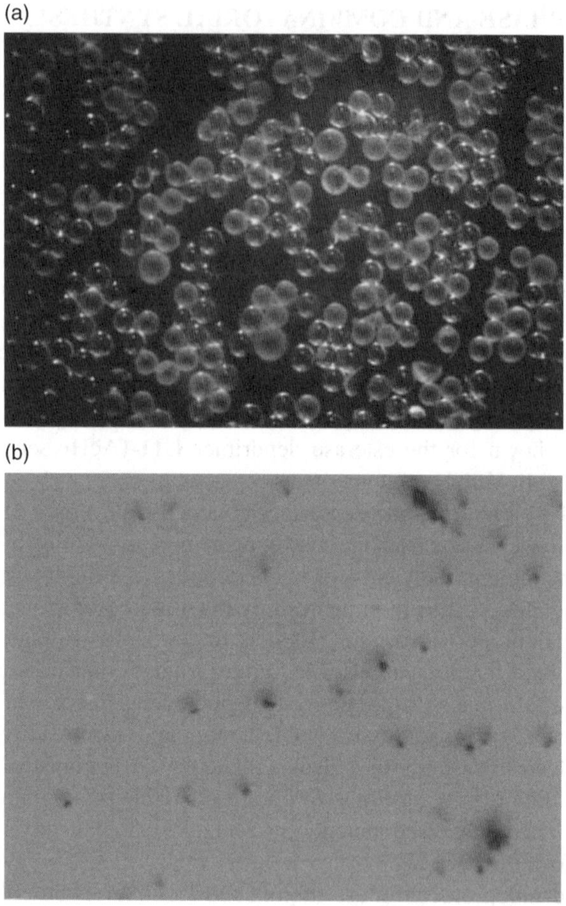

FIGURE 15.2 (a) On-bead assay with 8-butyryloxypyrene-1,3,6-trisulfonate **1** leading to the identification of dendrimer **C11** [14]. Microscopic picture under illumination with UV 365 nm. Beads were soaked with 80 μM **1** (structure in Figure 15.4) in aq. 20 mM bis-tris pH 6.0 and spread out in a petri-dish. The bead surrounded by a bright halo in the upper right center of the image contains a catalytic sequence. (b) Off-bead assay used for an SAR study of the single-site esterase dendrimer **RG3** $(AcTyrThr)_8$ $(\textit{Dap}TrpGly)_4(\textit{Dap}ArgSer)_2\textit{Dap}HisSerNH_2$. Photograph of a silica gel plate impregnated with 80 μM of substrate **1** in bis-tris buffer pH 6.0 in contact with photolyzed beads of a combinatorial library [17]. (See the color version of this figure in Color Plates section.)

1 producing the green fluorescent hydroxypyrene trisulfonate **2** (Figure 15.4) used for the discovery of dendrimer **C11** above (Figure 15.2a). The conditions of such assays must be carefully optimized to obtain a differentiated response among the library beads. Under optimized conditions, "positive" beads are those showing the strongest staining reaction even using dilute staining reagent for short assay times, while "negative" beads are those that remain unstained even under forcing conditions such as excess staining reagent for extended periods of time.

On-bead activity assays may be subject to artifacts presumably caused by the very high concentration of compound on the solid support [16]. Activities visible on the bead thus often disappear once the compound is cleaved from the support and tested in solution. To allow direct screening of our OBOC libraries in solution, we have established a so-called "off-bead" screening protocol in which the peptide dendrimers are cleaved from the solid support by photolysis of a photolabile linker. The cleaved dendrimer then diffuses in the underlying silica gel plate that is impregnated with the assay reagent, for example, a fluorogenic substrate [17]. In this manner, beads carrying active peptide dendrimers are identified by the halo formed on the matrix surrounding the beads, as shown for the identification of esterase peptide dendrimers (Figure 15.2b).

The results of OBOC library screening are analyzed by determining the structure of the compounds present on either the "positive" hits, the "negative" hits, or a random selection of beads. From such analysis, a structure–activity relationship may emerge leading to the identification of active compounds. Importantly, results from library screening must be confirmed by resynthesis, purification, and testing of pure substances. For example, the esterase dendrimer **RG3** (AcTyrThr)$_8$(*Dap*TrpGly)$_4$(*Dap*ArgSer)$_2$*Dap*HisSerNH$_2$ and its analog **HG3** (AcIlePro)$_8$(*Dap*IleThr)$_4$(*Dap*HisAla)$_2$*Dap*HisLeuNH$_2$ were initially identified by testing a focused "core-active site" combinatorial library featuring catalytic residues at the core positions only using an on-bead activity assay with the fluorogenic 8-butyryloxypyrene-1,3,6-trisulfonate **1** [18]. A subsequent investigation of the same combinatorial library using an off-bead activity assay confirmed the esterase activity of **RG3** and **HG3** with acyloxypyrene trisulfonates.[17] The off-bead assay also allowed to analyze the negative hits, which revealed that activity of the catalytic histidine residues at the dendrimer core is controlled by the nature of the residues in the dendrimer outer branches. Thus, histidine-containing peptide dendrimers were also found in the nonhits, but these contained anionic residues in the outer dendrimer branches. Resynthesis confirmed the absence of activity in these histidine-containing anionic dendrimers, for example, **D16** (AcGluThr)$_8$(*Dap*GluPhe)$_4$(*Dap*HisLeu)$_2$*Dap*HisAspNH$_2$, which simply lack any binding to the trianionic substrate, while partly anionic dendrimers such as **D14** (AcIleThr)$_8$(*Dap*IlePro)$_4$(*Dap*HisAsp)$_2$*Dap*ArgLeuNH$_2$ are active but have weak substrate binding (high K_M).

15.2.3 TAGSFREE Decoding of Combinatorial Libraries

One of the bottlenecks in testing OBOC libraries is the identification of the structure of the compounds on the beads. This can be realized by either reading an encoding tag that has been separately introduced on the bead or analyzing the bead directly by mass spectrometry or, in the case of linear peptide with a free N-terminus, by Edman microsequencing [19]. These methods are complex, expensive, and mostly not applicable to peptide dendrimers. To circumvent these limitations, we have developed a library design algorithm for OBOC libraries of peptides and peptide dendrimers called TAGSFREE, which allows direct sequence determination from the total amino acid analysis (AAA) of the bead [20]. AAA is a one-step quantification of amino acids by acidic hydrolysis and separation of their phenyl thiourea derivatives by HPLC.

For peptide dendrimers, the TAGSFREE algorithm simply imposes to use each amino acid only twice in the course of the library synthesis in two successive branches of the dendrimer.[14] The amount of an amino acid in the AAA of a synthesis bead then indicates if and in which branch the amino acid is present in the dendrimer. AAA also provides a composition analysis that is useful to identify beads that are empty or carry incomplete sequences, thus helping to avoid false positives. We have used

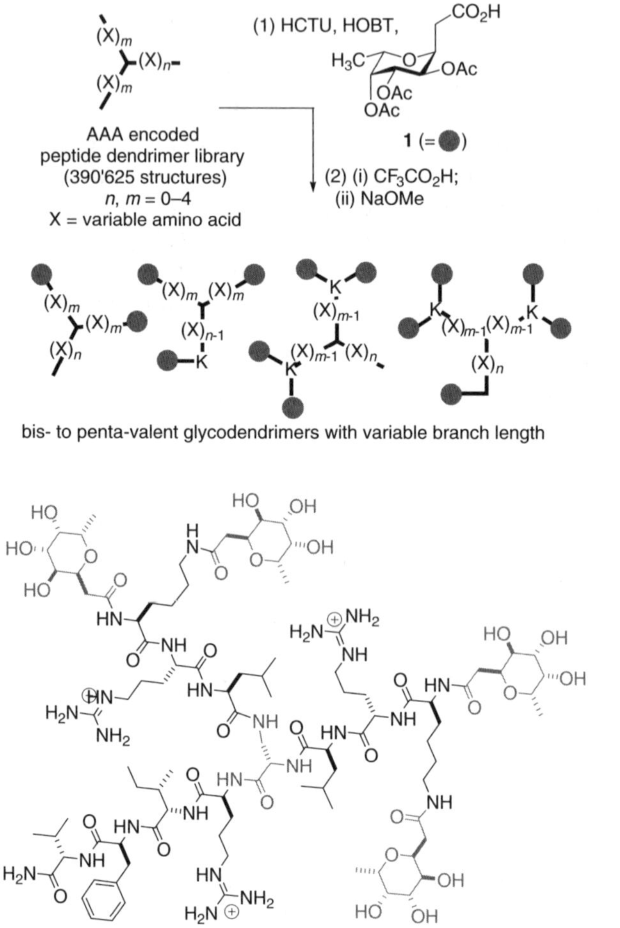

KT1

FIGURE 15.3 TAGSFREE encoded combinatorial library of glycopeptide dendrimers with variable branch length and multivalency and the lectin binding dendrimer **KT1** cFuc$_4$(*Lys*ArgLeu)$_2$*Dap*ArgIlePheLeuNH$_2$ identified by screening for binding to the fucose-specific plant lectin UEA-I from *Ulex europaeus*. Bis-Fmoc lysine was used as one of the four different amino acids at the first and last variable positions in the sequence to enable valency variations.

TAGSFREE for many different peptide dendrimer combinatorial libraries and thus tested probably the largest number of dendrimers ever prepared by any method so far. One of the most spectacular applications of the TAGSFREE design is a 390,625-member combinatorial library of glycopeptide dendrimers with variable multivalency and variable branch length based on the silent encoding of amino acid deletions made possible by the AAA, which lead to the identification of dendrimer **KT1** $cFuc_4$(*Lys*ArgLeu)$_2$*Dap*ArgIlePheLeuNH_2 as a potent lectin ligand (Figure 15.3) [21].

15.3 ENZYME MODELS

From the point of view of synthetic chemists, enzyme catalysis is one of the most fascinating properties of natural proteins, which calls for the study of synthetic models of enzymes [22]. Enzyme catalysis results from cooperativity between different amino acids and cofactors within the protein. Considering that arranging amino acids in a dendritic framework primarily enforces spatial proximity, peptide dendrimers seem particularly well suited to investigate such cooperativity effects. Within the SPPS combinatorial approach chosen for the assembly of peptide dendrimers, investigating catalysis should be realized by library screening, which is facilitated by numerous and efficient high-throughput assays available to measure enzyme-catalyzed reactions [23]. Dendrimer libraries for catalysis are designed to display amino acid residues with known catalytic properties for the reaction under study.

Our work so far has concerned histidine-containing ester hydrolysis catalysts as above, and amine-functionalized dendrimers that catalyze aldol reactions via enamine or enolate mechanisms.[10] We have also investigated peptide dendrimer models of metalloproteins, namely peptide dendrimers that bind vitamin B_{12} by coordination at cobalt of a nucleophilic residue at the dendrimer core [24], and the binding of Fe(II) to bipyridine-containing peptide dendrimers [25]. All of these systems operate in aqueous environment under conditions that are typical for enzyme-catalyzed processes. Catalytic and binding properties are governed by dendrimer size and the nature and position of the amino acid residues within the branches. Results indicate that multivalency, hydrophobic and electrostatic effects are key parameters influencing binding and catalysis in these peptide dendrimers. Optimization of catalysis and binding is possible by amino acid substitutions.

15.3.1 Dendritic Effects

Dendritic effects in catalysis are understood as a variation of catalytic efficiency as the dendrimer size increases. In the first approach toward catalytic dendrimers, catalytic groups are displayed in multiple copies at the dendrimer surface. Favorable cooperative effects between these groups might enhance catalysis compared to a monovalent or lower valency system. However, multivalency often also induces steric crowding that can block substrate access and decrease catalytic efficiency [10a].

Multivalent display of histidine residues favors catalysis in peptide dendrimers hydrolyzing fluorogenic ester substrates. The effect was directly apparent during our first study realized with a small library of peptide dendrimers featuring the catalytic triad amino acids aspartate, histidine, and serine in various combinations, which lead to peptide dendrimers catalyzing the hydrolysis of quinolium esters [26]. The dendritic effects were characterized in detail in a dendrimer series assembled by iteration of the $(HisSer)_2$*Dap* dendron (*Dap* = branching L-2,3-diaminopropanoic acid). These dendrimers catalyzed the fluorogenic hydrolysis of acyloxypyrene trisulfonates (Figure 15.4) [27]. The catalytic proficiency k_{cat}/K_M increased sharply

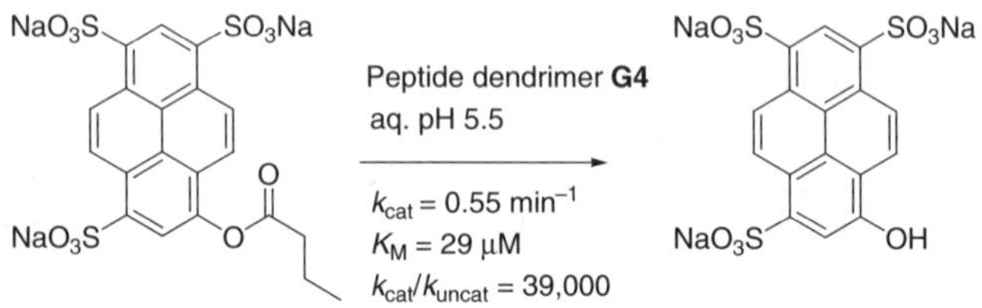

1 2

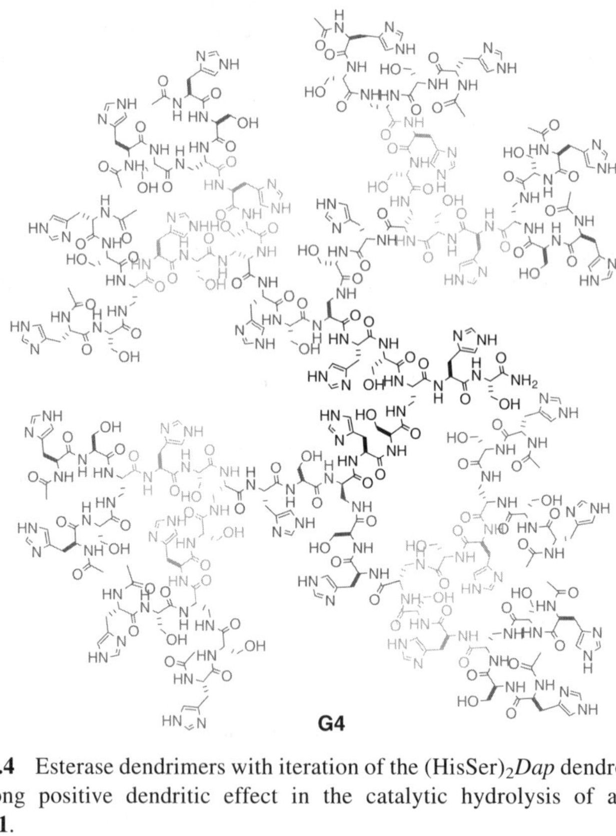

G4

FIGURE 15.4 Esterase dendrimers with iteration of the $(HisSer)_2$*Dap* dendron such as **G4** show a strong positive dendritic effect in the catalytic hydrolysis of acyloxypyrene trisulfonate **1**.

in higher generation dendrimers. The positive dendritic effect on catalysis lead to a remarkable rate acceleration of $k_{cat}/k_{uncat} = 39{,}000$ with the fourth-generation dendrimer **G4** (AcHisSer)$_{16}$(*Dap*HisSer)$_8$(*Dap*HisSer)$_4$(*Dap*HisSer)$_2$*Dap*HisSerNH$_2$ and substrate **1**. This corresponds to a 1200-fold increase in reactivity per histidine residue when compared to 4-methylimidazole as reference monovalent catalyst.

Isothermal titration calorimetry showed that the esterase dendrimer **G4** and its lower generation analogs bind one substrate molecule for three histidine residues. The ester substrate **1** binds the dendrimer approximately 10-fold stronger than the hydroxypyrene trisulfonate reaction product **2** by hydrophobic interactions between the acyl group and the dendrimer interior, which avoids product inhibition and allows multiple turnover to take place. Acid–base titration showed a significant spread in the pK_a values of histidine side chains toward lower values within the dendrimer [28]. This suggests that the positive dendritic effect is caused in part by allowing the coexistence of protonated and free base forms of the histidine side chains in the same dendrimer. The protonated histidine favors electrostatic binding of the anionic substrate **1**, while the free base form is necessary to promote nucleophilic or general base catalyzed cleavage of the ester.

A similar positive dendritic effect by multivalent display of catalytic groups was observed in aldolase peptide dendrimers with N-terminal proline residues, for example, **L2K7** (ProLys)$_8$(*Dap*SerArg)$_4$(*Dap*βAlaVal)$_2$*Dap*TyrLeuNH$_2$ (Figure 15.5), which were investigated for catalysis of the coupling of acetone and cyclohexanone to 4-nitrobenzaldehyde and dihydroxyacetone to 4-bromobenzaldehyde to form the corresponding aldols [29]. The catalytic sequences were identified by on-bead screening of a combinatorial library using either staining with a dye-labeled 1,3-diketone to mark enamine-reactive amino groups, or the fluorogenic substrate dihydroxyacetone coumarin ether to detect enolization catalysis [30]. Iteration of the (ProLys)$_2$*Dap* and (ProThr)$_2$*Dap* dendrons identified in active sequences to form dendrimers of first, second, and third generation showed that the activity per N-terminal proline for acetone aldolization was highest for the tetravalent second-generation dendrimers **R1G2** (ProThr)$_4$(*Dap*ProThr)$_2$*Dap*ProThrNH$_2$ and **R2G2** (ProLys)$_4$(*Dap*-ProLys)$_2$*Dap*ProLysNH$_2$. Trapping of the enamine intermediate with acetone by reduction gave no reaction of lysine side chains and alkylation of 50% of the N-terminal prolines, suggesting that the multivalency effect was caused by cooperativity between pairs of proline residues, with one residue forming the enamine and the other acting as general acid–base catalyst in the aldol reaction.

15.3.2 Hydrophobic and Electrostatic Effects

Hydrophobic and electrostatic effects controlled dendrimer binding to the acyloxypyrene trisulfonates **1–3** in the multivalent esterase dendrimers such as **G4** discussed above (Figure 15.4). These factors also determined the properties of three series of peptide dendrimers designed to mimic enzymes with a single active site. These dendrimers are equipped with catalytic residues at the dendrimer core, while amino acids in the outer dendrimer branches modulate global properties such as charges and hydrophobicity.

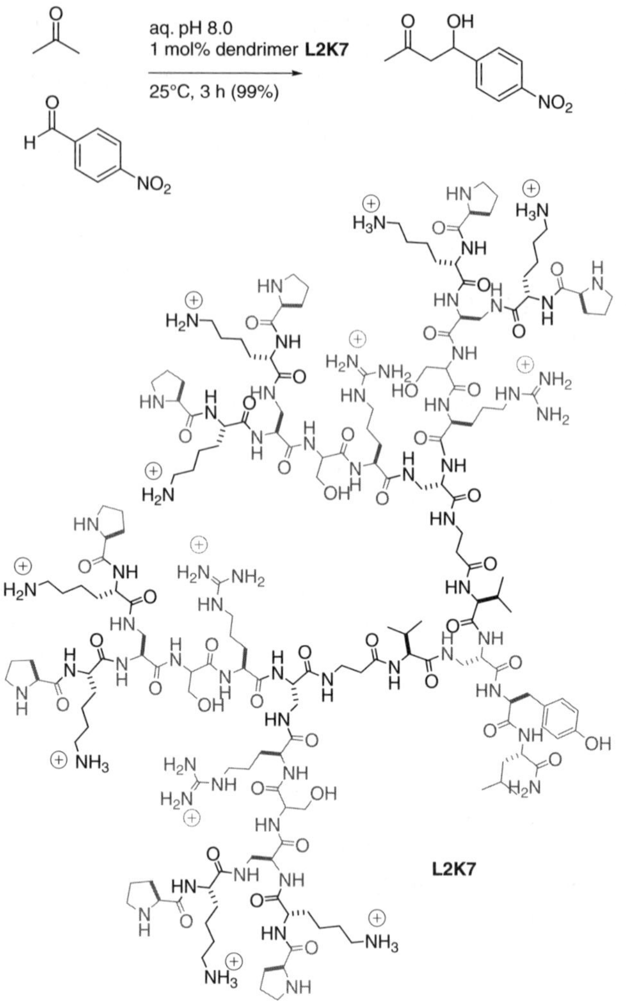

FIGURE 15.5 Aldolase dendrimers with N-terminal prolines catalyze aldol reaction with a multivalency effect.

The first case concerns peptide dendrimer **RG3** (AcTyrThr)$_8$(*Dap*TrpGly)$_4$ (*Dap*ArgSer)$_2$*Dap*HisSerNH$_2$ catalyzing the hydrolysis of acyloxypyrene trisulfonates. Its catalytic site consists of two cationic arginines at position X^4 for substrate binding and one nucleophilic histidine at the core position X^2 for ester cleavage (Figure 15.2) [18]. Structure–activity relationship studies showed that the catalytic proficiency k_{cat}/K_M for the hydrolysis of ester **1** increased in the series **RG0** AcArgSerAlaHisSerNH$_2$ → **RG1** (AcArgSer)$_2$*Dap*HisSerNH$_2$ → **RG2** (AcTrpGly)$_4$(*Dap*ArgSer)$_2$*Dap*HisSerNH$_2$ → **RG3**. Addition of the outer dendrimer branches containing aromatic residues lowers the K_M values (stronger substrate binding) without affecting the catalytic rate constant k_{cat}. Direct hydrophobic contacts between the aromatic residues in the outer dendrimer branches and the substrate are supported by substrate docking studies (Figure 15.11) [31]. On the other hand and as discussed above, catalysis by these dendrimers was entirely inhibited by the presence of anionic residues in the outer dendrimer branches due to reduced substrate binding, as in dendrimer **D16** $(AcGluThr)_8(\mathit{Dap}GluPhe)_4(\mathit{Dap}HisLeu)_2\mathit{Dap}HisAspNH_2$ [17].

In the second case, the effect of electrostatic charges on dendrimer properties was seen in peptide dendrimers selected as ligands for vitamin B_{12}. The third-generation peptide dendrimer **B1** $(AcGluSer)_8(\mathit{Dap}GluAla)_4(\mathit{Lys}AmbTyr)_2\mathit{Dap}CysAspNH_2$ (Amb = 4-aminomethyl benzoic acid) and its core-histidine analog **B1H** $(AcGluSer)_8(\mathit{Dap}GluAla)_4(\mathit{Lys}AmbTyr)_2\mathit{Dap}HisAspNH_2$ were identified as potent ligand for vitamin B_{12} through an on-bead binding assay of the "core-active site" library mentioned above, followed by secondary combinatorial optimization (Figure 15.6) [24a]. The primary binding interaction in these dendrimers is coordination of the core nucleophilic thiol or imidazole group to cobalt. Strikingly, binding is also controlled by the electrostatic charge in the outer dendrimer branches. Multiple anionic glutamate residues in the outer branches were clearly selected by the

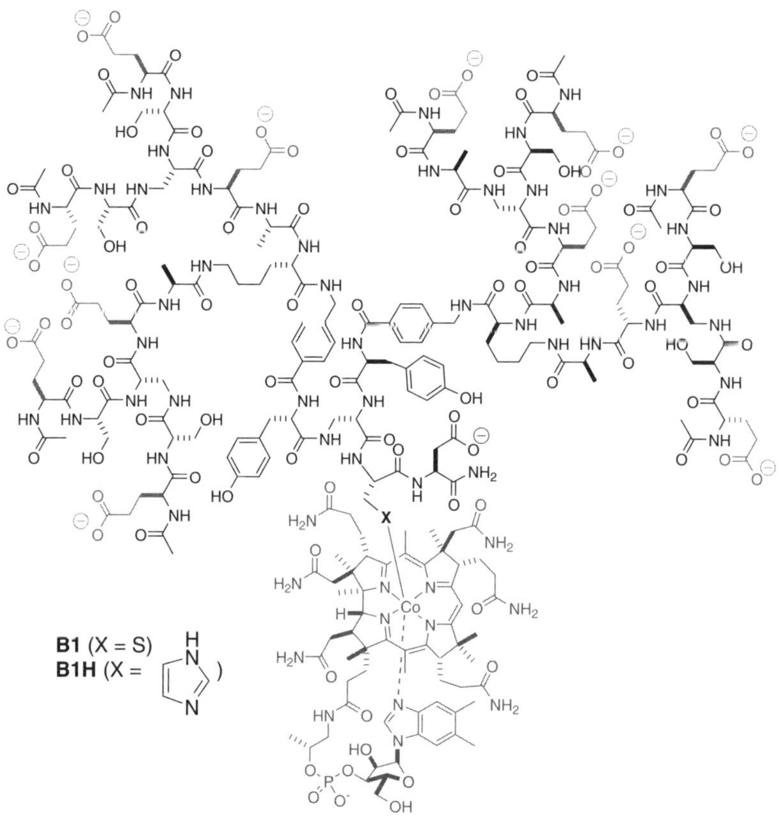

FIGURE 15.6 Peptide dendrimer ligands of vitamin B_{12}.

on-bead binding assay with vitamin B_{12}. The effects appeared to be kinetic in nature. Indeed, the cationic analog **B1K** $(AcLysSer)_8$(*Dap*LysAla)$_4$(*Lys*AmbTyr)$_2$-*Dap*CysLeuNH_2 where the 12 glutamate residues have been replaced by cationic lysines did not show measurable binding to vitamin B_{12} due to an extremely slow binding equilibrium, and the neutral glutamine analog **N3** $(AcGlnSer)_8$(*Dap*Gln-Ala)$_4$(*Dap*AmbTyr)$_2$*Dap*CysAspNH_2 showed stronger and slower binding compared to **B1** [24c].

In the third case, a similar control of metal coordination by charged residues was encountered in peptide dendrimers bearing a bipyridine ligand at their core [25]. In this system, multiple negative charges in the outer dendrimer branches of dendrimer **E1** $(AcGluThr)_4$(*Dap*GluVal)$_2$*Dap*BipβAlaAspNH_2 (Bip = 5′-amino-2,2′-bipyridine-5-carboxylic acid) (7 negative charges) and **G1** $(AcLysAla)_4$(*Dap*Glu-Pro)$_2$*Dap*GlyBipGlyLeuNH_2 (2 negative and 4 positive charges) lead to the formation of stable ternary complexes with Fe(II) Fe(**E1**)$_3$ ($K_a = 3.1 \times 10^{14}\,M^{-3}$) and Fe(**G1**)$_3$ ($K_a = 1.1 \times 10^{15}\,M^{-3}$). However, no complexation took place when multiple positive charges and no negative charges were present in the outer branches as in dendrimer **E2** $(AcArgLys)_4$(*Dap*HisVal)$_2$*Dap*BipAmbTyrNH_2 (8 positive charges), an effect which is probably also caused by kinetic inhibition of metal coordination.

15.3.3 Optimization by Amino Acid Substitution

A broad variety of Fmoc-protected amino acids, including proteinogenic and non-natural amino acids in both enantiomeric forms, are commercially available and can be used to vary the structure of any peptide dendrimer. This allows the straightforward design and synthesis of dendrimer combinatorial libraries and the preparation of analogs. In the absence of significant technical hurdles in the process, in particular by the use of the TAGSFREE design algorithm to facilitate analysis (see above) [20], iterative cycles of library design, synthesis, and functional selection offer a viable approach to discover and optimize peptide dendrimers for any given function. Dendrimer optimization by this process is related to the Darwinian evolution by mutagenesis and selection, which leads to functional proteins in nature.

Activity optimization by analog synthesis can be exemplified by the multivalent esterase dendrimers derived from the $(HisSer)_2$*Dap* dendron (Figure 15.4) [27]. A series of 32 analogs of the third-generation peptide dendrimer **G3** $(AcHisSer)_8$(*Dap*HisSer)$_4$(*Dap*HisSer)$_2$*Dap*HisSerNH_2 incorporating designed structural changes such as selected shuffling of amino acids between positions and addition or deletion of key residues uncovered structure–activity relationships that were not apparent from the initial study of dendritic effects in catalysis [28]. In particular, while the number of histidine residues primarily controlled substrate binding (K_M), the catalytic rate constant k_{cat} depend on the particular arrangement of histidine in the peptide dendrimer, and the nature of the other amino acids. The SAR study identified dendrimer **A3C** $(AcHisThr)_8$(*Dap*HisThr)$_4$(*Dap*HisThr)$_2$*Dap*HisThrNH_2 as the most active dendrimer. In this dendrimer, the serine residues have been exchanged throughout by threonine. This simple substitution lead to a fivefold increase in the catalytic rate constant up to $k_{cat}/k_{uncat} = 90{,}000$ ($k_{cat} = 1.3\ \mathrm{min}^{-1}$) for substrate **2**.

15.4 GLYCOPEPTIDE DENDRIMERS

Due to their size, peptide dendrimer analogs of proteins should be optimally suited to perturb protein–protein and carbohydrate–protein interactions. An initial study toward this goal was carried out focusing on lectin–glycoprotein interactions [32]. Lectins are ubiquitous proteins that bind to the terminal carbohydrates of glycoproteins or glycolipids by weak but specific binding sites. Lectins usually possess multiple binding sites and form oligomers, allowing avidity effects to enhance binding to carbohydrate epitopes displayed in polyvalent format on cell surfaces. Multivalent carbohydrate ligands generally bind strongly to lectins by virtue of such multivalency effects [33,34].

In the case of peptide dendrimers, one can expect that attaching carbohydrates at the end of the peptide dendrimer sequence should produce multivalent lectin-specific ligands in which binding by the carbohydrate moiety can be fine-tuned by the nature of the amino acid residues in the peptide dendrimer branches. These residues should at least determine the physicochemical properties of the dendrimers such as aqueous solubility, hydrophobicity, and charge, and at best engage in secondary binding interactions with the lectin. These hypotheses were borne out in practice in the study of fucosylated peptide dendrimer ligands to the bacterial lectin LecB as discussed below.

15.4.1 Synthesis of Glycopeptide Dendrimers

While the preparation of natural glycopeptides requires sophisticated chemical or enzymatic glycosylation chemistry, the decoration of synthetic macromolecules with carbohydrates is often more conveniently realized by reacting preassembled glycosidic building blocks by simpler coupling chemistries. In an initial synthetic study aimed at the dendrimer-mediated delivery of colchicine to cancer cells, three different strategies were investigated to obtain glycopeptide dendrimers [35]. The first approach used oxime bond formation between a multivalent aldehyde-terminated peptide dendrimer and an oxyamine-functionalized glycoside, as reported previously for the glycosylation of peptides [36] and peptide dendrimers [37]. The second approach used reductive alkylation of the free N-terminus with lactose [38], a method previously known for modifying whole proteins [39] and reported for the synthesis of glycodendrimers [40]. The third approach involved amide bond formation between an acetyl-protected glycosidic carboxylic acid building block and the N-termini of the peptide dendrimer directly on the solid support. Although this third approach required multigram quantities of carbohydrate building blocks, the synthesis was the most practical because it required only a single chromatographic purification and yielded products of excellent purity. For example, the *N*-acetyl-glucosamine terminated peptide dendrimer colchicine conjugate **30**, which shows selective cytotoxicity to cancer cells, was obtained by this approach (Figure 15.7).

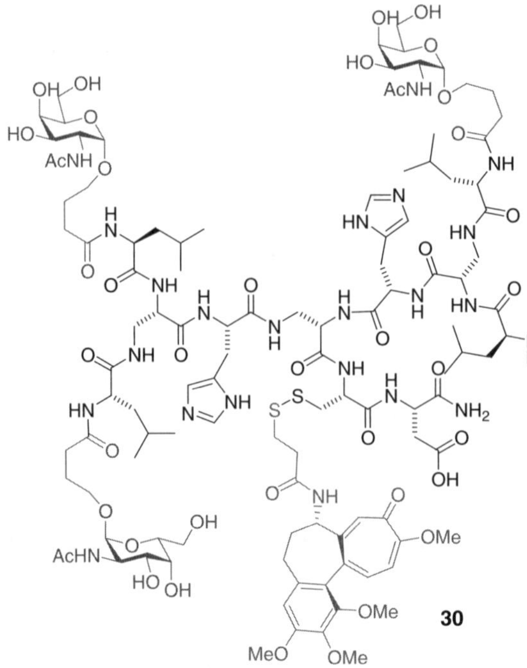

30

FIGURE 15.7 A cytotoxic colchicine glycopeptide dendrimer conjugate.

15.4.2 Lectin Binding

The selection of glycopeptide dendrimer ligands for lectins was applied to the case of the fucose-specific lectin B from the human pathogenic bacterium *Pseudomonas aeruginosa* [41]. In the first step, an on-bead lectin assay was used to screen combinatorial libraries of glycopeptide dendrimers bearing an N-terminal *c*-fucoside residue. In this assay, the solid support beads are incubated with the labeled lectin in the presence of fucose as a competitive ligand. The beads are then washed to remove unbound lectin, and stained by revealing the labeled lectin by a secondary reagent. Under optimized assay conditions, less than 5% of the beads showed strong staining. The control, nonfucosylated library showed no staining under the same conditions, and the staining intensity was reduced in the presence of high fucose concentrations, showing that the assay revealed specific binding between the dendrimers and the lectin. The assay was used to select lectin-binding fucosyl-peptide dendrimers such as **FD2** (cFucLysProLeu)$_4$(*Lys*PheLysIle)$_2$*Lys*HisIleNH$_2$ (Figure 15.8).

While on-bead assays are useful for the initial identification ligands, the quantification of the lectin–dendrimer interactions are carried out with purified ligands in solution. In our study of lecB-ligands, we used an enzyme-linked lectin assay (ELLA) in which the dendrimer competes in solution with a polymeric biotinylated fucose ligand for binding to the lectin adsorbed on the surface of microtiter plate wells [42]. The competitive binding is revealed by detection of immobilized biotin by secondary

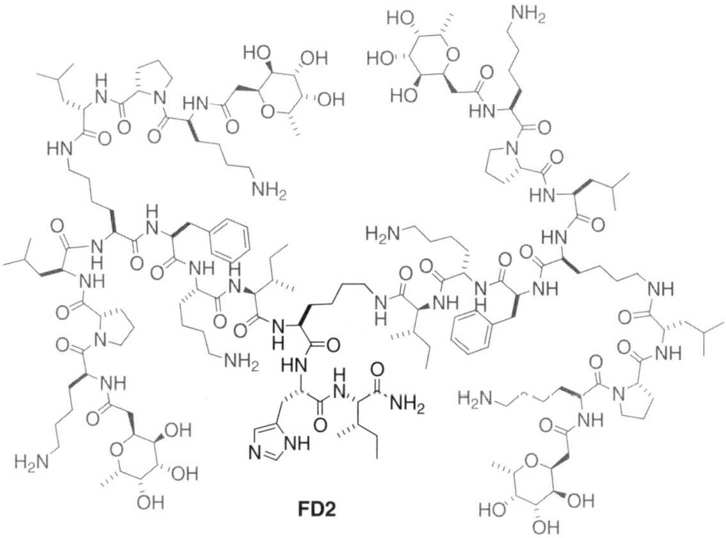

FD2

Lectin	IC_{50} (ELLA)
UEA-I	11 μM
LecB	0.14 μM

FIGURE 15.8 Fucosyl peptide dendrimer **FD2** (cFucLysProLeu)$_4$(*Lys*PheLysIle)$_2$*Lys*HisIleNH$_2$ is a potent inhibitor of fucose-specific lectins UEA-I and LecB.

staining. The ELLA was used to confirm the activity of positive hits from the on-bead screening, and to carry out more detailed SAR studies, in particular regarding a possible dendritic effect on binding. Thus, fragments of the most potent dendrimer **FD2** (cFucLysProLeu)$_4$(*Lys*PheLysIle)$_2$*Lys*HisIleNH$_2$ consisting of only the outer branches, such as **2G1** (cFuc-LysProLeu)$_2$*Lys*PheLysIleNH$_2$, showed much weaker binding than **FD2** itself. Increasing multivalency from tetravalent to octavalent by redesigning the sequence of **FD2** gave dendrimer **2G3** (cFucLysPro)$_8$(*Lys*LeuPhe)$_4$(*Lys*LysIle)$_2$*Lys*HisIleNH$_2$, which showed an even stronger binding to lecB ($IC_{50} = 25$ nM) [43].

15.4.3 Inhibition of *P. aeruginosa* Biofilms

P. aeruginosa causes lethal lung infections in immunocompromised patients and those suffering from cystic fibrosis or cancer. Tissue attachment and the formation of biofilms are essential steps in the process that also increase the antibiotic resistance of the pathogen [44]. The glycopeptide dendrimer **FD2** and several analogs inhibit biofilm formation and efficiently disperse established biofilms in cultures of several *P. aeruginosa* strains including antibiotic-resistant isolates (Figure 15.9) [45]. The effect is not mediated by cytotoxicity since bacterial growth is unaffected by the ligands. Control experiments with *lecB*-deletion mutants and dendrimers that are not ligands of lectin B show that biofilm inhibition requires both a potent ligand and lectin B expression, suggesting that the effect is indeed caused by binding of the ligand to the

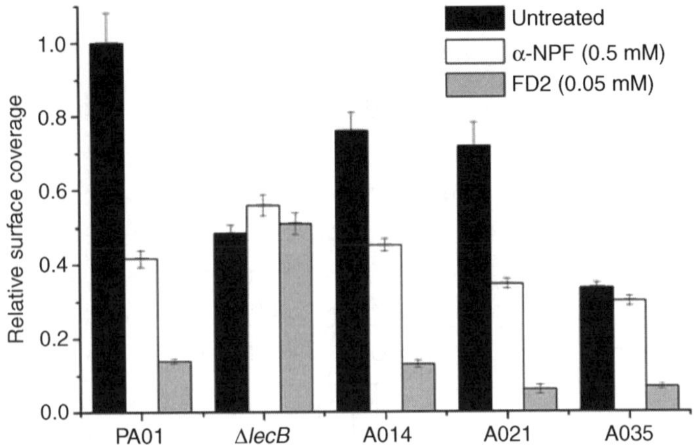

FIGURE 15.9 Biofilm inhibition with α-nitrophenyl fucoside (α-NPF, 0.5 mM) and **FD2** (0.05 mM) of *Pseudomonas aeruginosa* strains wild type (PAO1), ΔlecB mutant (PAO1 ΔlecB), and three clinical strains (A014, A021, A035).

lectin. This antibiofilm effect mediated by a lectin B ligand is unprecedented and suggests a new therapeutic approach to control *P. aeruginosa* infections.

15.5 STRUCTURE OF PEPTIDE DENDRIMERS

Almost all dendrimers described in the literature are assembled from building blocks containing freely rotatable bonds in their backbone, leading to conformationally flexible macromolecules whose structural determination by crystallography or NMR is difficult or impossible. Molecular dynamics simulations have been performed on various dendrimer types to understand their structural properties [46], in particular with poly(amidoamine) (PAMAM) dendrimers [47]. These investigations confirm the conformational flexibility of dendrimers and suggest substantial back folding of external branches inducing a high density located between the core and the periphery leaving some of the inner parts of the dendrimer accessible. Inspired by these reports, we have used molecular dynamics and spectroscopic methods to study the conformational behavior of peptide dendrimers and their relationship to protein structures in terms of compactness, folding, and stability. It should be noted that in contrast to most proteins, peptide dendrimers do not undergo irreversible denaturation, generally do not form precipitates or aggregates, and remain active even after prolonged heating in boiling water, as evidenced for esterolytic peptide dendrimers such as **C11** $(AcHisSer)_8(\mathit{Dap}HisLeu)_4(\mathit{Dap}LysVal)_2\mathit{Dap}IleValNH_2$ (Figure 15.1).

15.5.1 Peptide Dendrimers as Molten Globules

^{1}H-NMR spectra of peptide dendrimers generally show sharp signals without cross-peaks in ROESY experiments, suggesting the existence of multiple and rapidly equilibrating conformers. The conformational flexibility of peptide dendrimer was also apparent in studies of second- and third-generation dendrimers such as **3** $(AcAspSer)_8$$(DapSerHis)_4$$(DapPheLeu)_2$*Dap*CysPhe$NH_2$ as single dendrimers or as cystine-bridged dimers adsorbed on Au(111) surface and imaged by STM at air, under UHV, or in solution [48]. The images showed the dendrimers as globular features with dimensions suggesting an extended flattened conformation ($6 \times 9\ nm^2$ for **3**), forming both single globules and ordered arrays on the surface. Furthermore CD and FTIR data show that most peptide dendrimers lack typical protein-like secondary structures such as α-helices or β-sheets and consist mostly of random coils [31].

Systematic determination of hydrodynamic radii by diffusion NMR in single site esterase dendrimer **RG3** $(AcTyrThr)_8$(*Dap*TrpGly)$_4$(*Dap*ArgSer)$_2$*Dap*HisSerNH_2 and analogs showed that these dendrimers exist in a monomeric, nonaggregated state in aqueous solution at millimolar concentration, and that they have a compactness comparable to molten globules indicative of partial packing only [18]. A similar trend was observed in vitamin B_{12} binding dendrimers such as **B1** $(AcGluSer)_8$ (*Dap*GluAla)$_4$(*Lys*AmbTyr)$_2$*Dap*CysAspNH_2 (Figure 15.6) [24]. In this case, there was no increase in the hydrodynamic radius upon binding to the vitamin, suggesting that part of the binding affinity is caused by hydrophobic collapse of the dendrimer around the vitamin. The hydrophobic collapse is consistent with the formation of a more stable and compact vitamin B_{12} dendrimer complex by the glutamine dendrimer **N3** $(AcGlnSer)_8$(*Dap*GlnAla)$_4$(*Dap*AmbTyr)$_2$*Dap*CysAspNH_2 compared to its parent glutamate analog **N1** $(AcGluSer)_8$(*Dap*GluAla)$_4$(*Dap*AmbTyr)$_2$ *Dap*CysAspNH_2 [24c].

Proteolytic cleavage sites in proteins are found at exposed, disordered loops, while the protein interior is protected from degradation. Trypsin and chymotrypsin cleavage sites were found to be essentially stable toward proteolytic cleavage when placed in the first and second-generation branches of third-generation peptide dendrimer with only one amino acid per branch and a compact 2,3-diaminopropanoic acid branching unit, such as in peptide dendrimer **D18B** $(AcGlu)_8$(*Dap*Phe)$_4$(*Dap*Arg)$_2$*Dap*LeuNH_2, which is resistant to proteolysis while the corresponding linear peptide **L18** AcGluAlaPheAlaArgAlaLeuNH_2 is rapidly cleaved (Figure 15.10) [49]. By contrast, more open dendrimers containing di- and tripeptide branches, such as **D1** $(AcGlyPhePro)_4$(*Dap*HypArg$^{\downarrow}$Met)$_2$*Dap*-SerGlyβAlaNH_2, were cleaved by proteases, although much slower than the corresponding linear peptides, showing that these dendrimers are conformationally very flexible and that amino acids within the topological interior can be exposed at the surface and interact tightly with proteins.

15.5.2 Molecular Dynamics and Docking Studies

We have carried out a molecular dynamics (MD) and docking studies with the single-site esterase peptide dendrimer **RG3** $(AcTyrThr)_8$(*Dap*TrpGly)$_4$(*Dap*ArgSer)$_2$ *Dap*HisSerNH_2 and its analogs to better understand the origin of the increase in catalytic proficiency observed upon addition of the outer dendritic branches to the catalytic core [31]. The GROMACS [50] program was adapted to accept dendritic

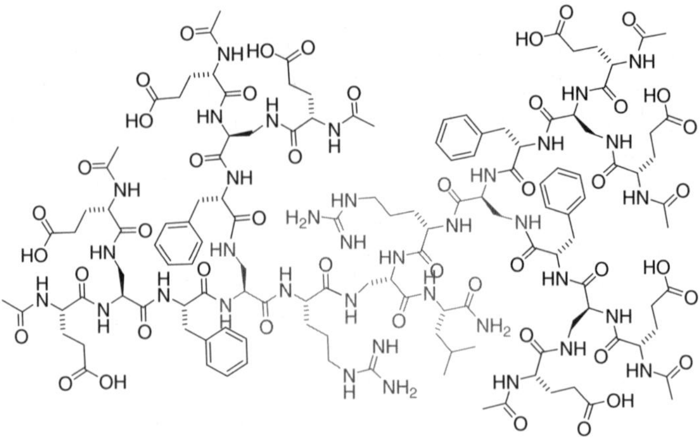

D18B

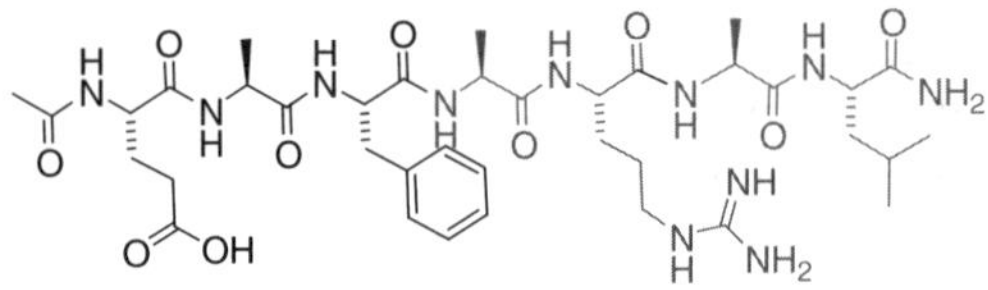

L18

FIGURE 15.10 The third-generation peptide dendrimer **D18B** is resistant to trypsin digestion, while its linear analog **L18** is rapidly cleaved.

peptide sequences using a customized topology manipulation kit. Low energy conformers were generated by gradual cooling from 800 to 200 K from 20 different starting high-energy conformations. The structural models obtained reproduced the hydrodynamic radius determined by diffusion NMR across dendrimers of various sizes and indicative of a molten globule state. Analysis of structural features in the low energy conformers generated by MD showed only 0.3 backbone hydrogen bonds per residue, which is much lower than the number of one backbone hydrogen bond per residue found in secondary structure elements of folded proteins, and might explain the conformational flexibility of peptide dendrimers.

The structural models of the single-site esterase peptide dendrimers **RG3** $(AcTyrThr)_8(\textit{Dap}TrpGly)_4(\textit{Dap}ArgSer)_2\textit{Dap}HisSerNH_2$ and **HG3** $(AcIlePro)_8(\textit{Dap}IleThr)_4(\textit{Dap}HisAla)_2\textit{Dap}HisLeuNH_2$ produced by MD showed that the pair of cationic residues at position X^3 (His or Arg) and the catalytic histidine residue at position X^1 appear on the dendrimer surface. Docking of butyryloxypyrene trisulfonate **1** showed the formation of at least one salt bridge between one of the substrate sulfonate groups and one of the cationic residues at position X^3 on the dendrimer. In addition, many low energy poses obtained by docking suitably positioned the reactive ester group of substrate **1** in proximity to the catalytic histidine residue at X^1 in an orientation suitable for esterolysis.

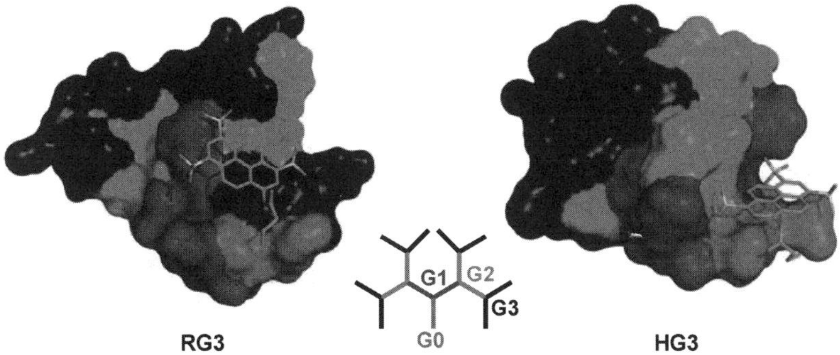

FIGURE 15.11 Docking poses of substrate **1** to peptide dendrimer **RG3** and **HG3**. The dendrimer is shown in surface representation color-coded by generation number (red = G0, blue = G1, green = G2, black = G3). Substrate carbons are represented in green, nitrogens in blue, oxygens in red, and sulfurs in yellow. (See the color version of this figure in Color Plates section.)

Interestingly, docking of substrate **1** to the esterase dendrimer **RG3**, which shows a positive dendritic effect in catalysis, shows frequent contacts between the pyrene trisulfonate and the aromatic side chains of the residues in the outer dendrimer layers. By contrast, substrate–dendrimer interactions mostly occur with the core residues in the case of dendrimer **HG3**, which showed essentially no dendritic effect in catalysis. Docking thus suggests that the increase in catalytic proficiency observed in dendrimer **RG3** upon addition of the aromatic outer dendrimer branches is caused by a direct binding interaction with the substrate. The difference in substrate–dendrimer interactions between **RG3** and **HG3** is illustrated by docking poses in Figure 15.11. The positive effect of the aromatic residues on catalysis in **RG3** was conserved when the pair of arginine residues at X^3 was exchanged for two histidine residues, resulting in a catalytically more efficient peptide dendrimer **RG3H** $(AcTyrThr)_8(\mathit{Dap}TrpGly)_4(\mathit{Dap}HisSer)_2\mathit{Dap}HisSerNH_2$, which a maximum rate acceleration of $k_{cat}/k_{uncat} = 1500$ for the hydrolysis of butyryloxypyrene trisulfonate **1** at its optimal pH 5.5 [51].

MD were also carried out for the antibiofilm glycopeptide dendrimer **FD2** [45]. The structural model obtained also suggests a molten globule state, as confirmed by diffusion NMR. The MD model was combined with the experimentally determined crystal structure of the complex between the terminal glycosylated tripeptide arm **2G0** $cFucLysProLeuNH_2$ of the dendrimer and lectin B. The resulting model of the protein–dendrimer complex showed that the dendrimers could only bind to one fucose binding site per dendrimer (Figure 15.12). Isothermal titration calorimetry data, however, showed binding by two of the four fucose residues per dendrimer. Divalent binding is probably realized by bridging two different lectins to form aggregates. This hypothesis is supported by the observation of such aggregates *in vitro* during attempted crystallization of the dendrimer–lectin complex. The formation of such aggregates could also explain the antibiofilm action of the dendrimer since they effectively neutralize the lectin.

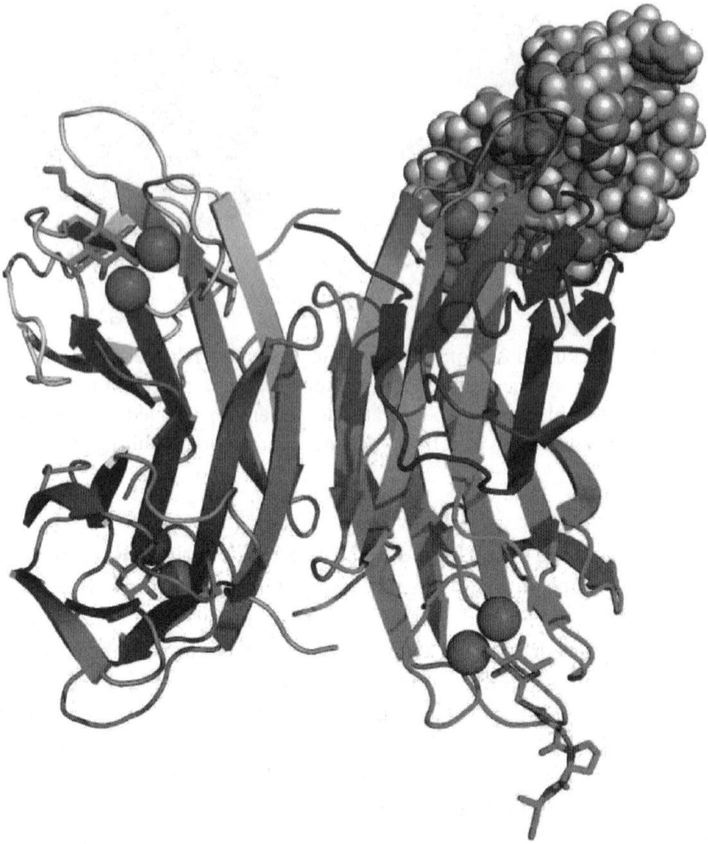

FIGURE 15.12 Structural model of the dendrimer **FD2**–LecB complex obtained by molecular dynamics and energy minimization. The Ca^{2+} ions are shown in magenta and indicate the location of the fucose-binding pocket. Three binding sites are shown with the resolved fucosyl residues of the tripeptide **2G0** cFucLysProLeuNH_2 and the fourth site is modeled with dendrimer **FD2**. (See the color version of this figure in Color Plates section.)

15.5.3 α-Helical Peptide Dendrimers

The structural models of peptide dendrimers obtained by MD showed a general low frequency of backbone H-bonds per residue, which might explain the conformational flexibility and the molten globule state of the dendrimers. These models suggest that the direct design of secondary structure elements might be necessary to obtain more compactly folded peptide dendrimers. This concept was tested by introducing branching lysine residues into the sequence of α-helical linear peptides, which lead to the identification of the second-generation peptide dendrimer **D3** $(AcAlaMetGluAla)_4(\mathit{Lys}LysLeuMetGlu)_2\mathit{Lys}MetLysLeuAlaNH_2$ existing in an α-helical conformation [52]. This peptide dendrimer shows higher α-helical content that its linear analog **L3** AcAlaMetgluAlaAlaLysLeuMetGluAlaMetLysLeuAlaNH_2, and a remarkable stability against pH-induced denaturation and thermal aggregation.

MD suggests that the dendrimer consists of a core α-helix that is surrounded and stabilized by the dendritic branches. This approach might be generally useful to design fully folded peptide dendrimer analogs of proteins.

15.6 CONCLUSION AND OUTLOOK

The investigation of peptide dendrimers as protein models offers a new insight into artificial protein-like systems. The approach is particularly straightforward since it capitalizes on solid-phase peptide synthesis and its combinatorial version for synthesis and functional optimization. The dendritic design produces macromolecules that adopt a protein-like globular conformation even without folding. Peptide dendrimers are quite stable against thermal denaturation and proteolytic degradation, properties that are quite difficult to obtain in proteins. Our work so far concerned enzyme model studies, including esterases, aldolases, and metal cofactor binding, and biofilm inhibition with glycopeptide dendrimers. Future challenges include the design of stable dendritic secondary structures, the synthesis of larger peptide dendrimers that are closer to proteins in terms of their size, and the continuing exploration of protein-like functions in peptide dendrimers.

ACKNOWLEDGEMENT

This work was supported financially by the University of Bern, the Swiss National Science Foundation, the Marie-Curie training program IBAAC, and the COST program Action D25, D34 and CM0801.

REFERENCES

[1] G. R. Newkome, C. N. Moorefield, F. Vögtle, *Dendrimers and Dendrons. Concepts, Synthesis and Perspectives*, Wiley-VCH, New York, **2001**.

[2] D. A. Tomalia, P. R. Dvornic, *Nature* **1994**, *372*, 617–618.

[3] J. K. M. Sanders, *Chem. Eur. J.* **1998**, *4*, 1378–1383.

[4] D. K. Smith, F. Diederich, *Chem. Eur. J.* **1998**, *4*, 1353–1361.

[5] L. J. Twyman, A. S. H. King, I. K. Martin, *Chem. Soc. Rev.* **2002**, *31*, 69–82.

[6] H. Brunner, *J. Organomet. Chem.* **1995**, *500*, 39–46.

[7] (a) M. H. Hecht, A. Das, A. Go, L. H. Bradley, Y. Wei, *Protein Sci.* **2004**, *13*, 1711–1723; (b) J. W. Bryson, S. F. Betz, H. S. Lu, D. J. Suich, H. X. Zhou, K. T. O'Neil, W. F. DeGrado, *Science* **1995**, *270*, 935–941.

[8] R. G. Denkewalter, J. F. Kolc, W. J. Lukasavage, Macromolecular highly branched α,ϖ-diamino carboxylic acids, U. S. Pat. 4,410,688, 10-18-1983.

[9] (a) J. P. Tam, *Proc. Natl. Acad. Sci. U.S.A.* **1988**, *85*, 5409–5413; (b) C. Rao, J. P. Tam, *J. Am. Chem. Soc.* **1994**, *116*, 6975–6976; (c) L. Zhang, J. P. Tam, *J. Am. Chem. Soc.* **1997**, *119*, 2363–2370; (d) J. P Tam, Y.A. Lu, J. L. Yang, *Eur. J. Biochem.* **2002**, *269*, 923–932.

[10] (a) J. Kofoed, J.-L. Reymond, *Curr. Opin. Chem. Biol.* **2005**, *9*, 656–664; (b) T. Darbre, J.-L. Reymond, *Acc. Chem. Res.* **2006**, *39*, 925–934; (c) T. Darbre, J.-L. Reymond, *Curr. Top. Med. Chem.* **2008**, *8*, 1286–1293.

[11] (a) R. B. Merrifield, *J. Am. Chem. Soc.* **1963**, *85*, 2149–2154; (b) S. S. Wang, R. B. Merrifield, *J. Am. Chem. Soc.* **1969**, *91*, 6488–6491; (c) W. C. Chan, P. D. White (Eds.), *Fmoc Solid Phase Peptide Synthesis: A Practical Approach*, Oxford University Press, **2000**; (d) M. Amblard, J. A. Fehrentz, J. Martinez, G. Subra, *Mol. Biotechnol.* **2006**, *33*, 239–254.

[12] D. Lagnoux, E. Delort, C. Douat-Casassus, A. Esposito, J.-L. Reymond, *Chem. Eur. J.* **2004**, *10*, 1215–1226.

[13] (a) K. S. Lam, S. E. Salmon, E. M. Hersh, V. J. Hruby, W. M. Kazmierski, R. J. Knapp, *Nature* **1991**, *354*, 82–84; (b) R. A. Houghten, C. Pinilla, S. E. Blondelle, J. R. Appel, C. T. Dooley, J. H. Cuervo, *Nature* **1991**, *354*, 84–86; (c) A. Furka, F. Sebestyén, M. Asgedom, G. Dibó, *Int. J. Pept. Protein Res.* **1991**, *37*, 487–493; (d) J. P. Kennedy, L. Williams, T. M. Bridges, R. N. Daniels, D. Weaver, C. W. Lindsley, *J. Comb. Chem.* **2008**, *10*, 345–354; (e) J. C. Kappel, Y. C. Fan, K. S. Lam, *J. Comb. Chem.* **2008**, *10*, 333–342.

[14] (a) A. Clouet, T. Darbre, J.-L. Reymond, *Angew. Chem. Int. Ed.* **2004**, *43*, 4612–4615; (b) A. Clouet, T. Darbre, J.-L. Reymond, *Biopolymers* **2006**, *84*, 114–123; (c) N. Maillard, A. Clouet, T. Darbre, J.-L. Reymond, *Nat. Protoc.* **2009**, *4*, 132–142.

[15] (a) R. Liu, J. Marik, K. S. Lam, *Methods Enzymol.* **2003**, *369*, 271–287; (b) M. Meldal, *Biopolymers* **2002**, *66*, 93–100.

[16] M. M. Marani, M. C. Martınez Ceron, S. L. Giudicessi, E. de Oliveira, S. Coté, R. Erra-Balsells, F. Albericio, O. Cascone, S. A. Camperi, *J. Comb. Chem.* **2009**, *11*, 146–150.

[17] N. Maillard, T. Darbre, J.-L. Reymond, *J. Comb. Chem.* **2009**, *11*, 667–675.

[18] S. Javor, E. Delort, T. Darbre, J.-L. Reymond, *J. Am. Chem. Soc.* **2007**, *129*, 13238–13246.

[19] (a) A. W. Czarnik, *Curr. Opin. Chem. Biol.* **1997**, *1*, 60–66; (b) A. W. Czarnik, *Proc. Natl. Acad. Sci. U.S.A.* **1997**, *94*, 12738–12739; (c) K. S. Lam, M. Lebl, V. Krchnak, *Chem. Rev.* **1997**, *97*, 411–448; (d) C. Barnes, S. Balasubramanian, *Curr. Opin. Chem. Biol.* **2000**, *4*, 346–350; (e) F. Guillier, D. Orain, M. Bradley, *Chem. Rev.* **2000**, *100*, 2091–2158; (g) R. L. Affleck, *Curr. Opin. Chem. Biol.* **2001**, *5*, 257–263; (g) N. J. Ede, Z. Wu, *Curr. Opin. Chem. Biol.* **2003**, *7*, 374–379.

[20] (a) J. Kofoed, J.-L. Reymond, *J. Comb. Chem.* **2007**, *9*, 1046–1052; (b) J. Kofoed, J.-L. Reymond, *Chem. Commun.* **2007**, 4453–4455; (c) V. S. Fluxa, J.-L. Reymond, *Bioorg. Med. Chem.* **2009**, *17*, 1018–1025.

[21] E. M. V. Johansson, E. Kolomiets, F. Rosenau, K.-E. Jaeger, T. Darbre, J.-L. Reymond, *New J. Chem.* **2007**, *31*, 1291–1299.

[22] (a) A. J. Kirby, *Angew. Chem. Int. Ed.* **1996**, *35*, 706–724; (b) W.-D. Woggon, *Acc. Chem. Res.* **2005**, *38*, 127–136; (c) W. B. Motherwell, M. J. Bingham, Y. Six, *Tetrahedron* **2001**, *57*, 4663–4686; (d) R. Breslow, Y. Huang, X. Zhang, J. Yang, *Proc. Natl. Acad. Sci. U.S.A.* **1997**, *94*, 11156–11158.

[23] J.-L. Reymond, V. S. Fluxà, N. Maillard, *Chem. Commun.* **2009**, 34–46.

[24] (a) P. Sommer, N. Uhlich, J.-L. Reymond, T. Darbre, *Chembiochem* **2008**, *9*, 689–693; (b) V. Duléry, N. A. Uhlich, N. Maillard, V. S. Fluxá, J. Garcia, P. Dumy, O. Renaudet, J.-L. Reymond, T. Darbre, *Org. Biomol. Chem.* **2008**, *6*, 4134–4141; (c) N. A. Uhlich, A.

Natalello, R. U. Kadam, S. M. Doglia, J.-L. Reymond, T. Darbre, *Chembiochem* **2010**, *11*, 358–365.

[25] N. A. Uhlich, P. Sommer, C. Bühr, S. Schürch, J.-L. Reymond, T. Darbre, *Chem. Commun.* **2009**, 6237–6239.

[26] (a) A. Esposito, E. Delort, D. Lagnoux, F. Djojo, J.-L. Reymond, *Angew. Chem. Int. Ed.* **2003**, *42*, 1381–1383; (b) C. Douat-Casassus, T. Darbre, J.-L. Reymond, *J. Am. Chem. Soc.* **2004**, *126*, 7817–7826; (c) A. Clouet, T. Darbre, J.-L. Reymond, *Adv. Synth. Catal.* **2004**, *346*, 1195–1204.

[27] E. Delort, T. Darbre, J.-L. Reymond, *J. Am. Chem. Soc.* **2004**, *126*, 15642–15643.

[28] E. Delort, N.-Q. Nguyen-Trung, T. Darbre, J.-L. Reymond, *J. Org. Chem.* **2006**, *71*, 4468–4480.

[29] J. Kofoed, T. Darbre, J.-L. Reymond, *Org. Biomol. Chem.* **2006**, 3268–3281.

[30] J. Kofoed, T. Darbre, J.-L. Reymond, *Chem. Commun.* **2006**, 1482–1484.

[31] S. Javor, J.-L. Reymond, *J. Org. Chem.* **2009**, *74*, 3665–3674.

[32] (a) N. Sharon, H. Lis, *Lectins*, 2nd edition, Kluwer Academic Publishers, Dordrecht, The Netherlands, **2003**; (b) Y. C. Lee, R. T. Lee, K. Rice, Y. Ichikawa, T.-C. Wong, *Pure Appl. Chem.* **1991**, *63*, 499–506.

[33] (a) Y. C. Lee, R. T. Lee, *Acc. Chem. Res.* **1995**, *28*, 321–327; (b) J. J. Lundquist, E. J. Toone, *Chem. Rev.* **2002**, *102*, 555–578; (c) J. D. Badjic, A. Nelson, S. J. Cantrill, W. B. Turnbull, J. F. Stoddart, *Acc. Chem. Res.* **2005**, *38*, 723–732.

[34] (a) M. Mammen, S.-K. Choi, G. M. Whitesides, *Angew. Chem. Int. Ed.* **1998**, *37*, 2754–2794; (b) C. R. Bertozzi, L. L. Kiessling, *Science* **2001**, *291*, 2357–2364; (c) P. I. Kitov, J. M. Sadowska, G. Mulvey, G. D. Armstrong, H. Ling, N. S. Pannu, R. J. Read, D. R. Bundle, *Nature* **2000**, *403*, 669–672; (d) H.-J. Gabius, H.-C. Siebert, S. André, J. Jiménez-Barbero, H. Rüdiger, *ChemBioChem* **2004**, *5*, 740–764; (e) M. Ambrosi, N. R. Cameron, B. G. Davis, *Org. Biomol. Chem.* **2005**, *3*, 1593–1608; (g) L. L. Kiessling, J. E. Gestwicki, L. E. Strong, *Angew. Chem. Int. Ed.* **2006**, *45*, 2348–2368; (g) D. Arosio, M. Fontanella, L. Baldini, L. Mauri, A. Bernardi, A. Casnati, F. Sansone, R. Ungaro, *J. Am. Chem. Soc.* **2005**, *127*, 3660–3661.

[35] D. Lagnoux, T. Darbre, M. L. Schmitz, J.-L. Reymond, *Chem. Eur. J.* **2005**, *11*, 3941–3950.

[36] (a) E. C. Rodriguez, K. A. Winans, D. S. King, C. R. Bertozzi, *J. Am. Chem. Soc.* **1997**, *119*, 9905–9906; (b) O. Renaudet, P. Dumy, *Org. Lett.* **2003**, *5*, 243–246.

[37] J. Shao, J. P. Tam, *J. Am. Chem. Soc.* **1995**, *117*, 3893–3899.

[38] (a) G. Gray, *Arch. Biochem. Biophys.* **1974**, *163*, 426–428; (b) B. A. Schwartz, G. R. Gray, *Arch. Biochem. Biophys.* **1977**, *181*, 542–549.

[39] (a) C. P. Stowell, Y. C. Lee, *Adv. Carbohydr. Chem. Biochem.* **1980**, *37*, 225–281; (b) H. Kataoka, M. Tavassoli, *Blood* **1985**, *65*, 1163–1171; (c) A. C. Roche, M. Barzilay, P. Midoux, S. Junqua, N. Sharon, M. Monsigny, *J. Cell Biol.* **1983**, *22*, 131–140.

[40] W. B. Turnbull, S. A. Kalovidouris, J. F. Stoddart, *Chem. Eur. J.* **2002**, *8*, 2988–2975.

[41] E. Kolomiets, E. M. V. Johansson, O. Renaudet, T. Darbre, J.-L. Reymond, *Org. Lett.* **2007**, *9*, 1465–1468.

[42] S. Perret, C. Sabin, C. Dumon, M. Pokorná, C. Gautier, O. Galanina, S. Ilia, N. Bovin, M. Nicaise, M. Desmadril, N. Gilboa-Garber, M. Wimmerová, E. P. Mitchell, A. Imberty, *Biochem. J.* **2005**, *389*, 325–332.

[43] E. Kolomiets, M. A. Swiderska, R. U. Kadam, E. M. V. Johansson, K.-E. Jaeger, Tamis Darbre, J.-L. Reymond, *ChemMedChem* **2009**, *4*, 562–569.

[44] V. E. Wagner, B. H. Iglewski, *Clin. Rev. Allergy Immunol.* **2008**, *35*, 124–134.

[45] E. M. V. Johansson, S. A. Crusz, E. Kolomiets, L. Buts, R. U. Kadam, M. Cacciarini, K.-M. Bartels, S. P. Diggle, M. Cámara, P. Williams, R. Loris, C. Nativi, F. Rosenau, K.-E. Jaeger, T. Darbre, J.-L. Reymond, *Chem. Biol.* **2008**, *15*, 1249–1257.

[46] (a) M. Ballauff, C. N. Likos, *Angew. Chem. Int. Ed.*, **2004**, *43*, 2998–3020; (b) N. W. Suek, M. H. Lamm, *Macromolecules* **2006**, *39*, 4247–4255; (c) R. L. Lescanec, M. Muthukumar, *Macromolecules* **1990**, *23*, 2280–2288.

[47] (a) S. T. Lin, P. K. Maiti, W. A. Goddard III, *J. Phys. Chem. B*, **2005**, *109*, 8663–8672; (b) P. K. Maiti, T. Çağin, S. T. Lin, W. A. Goddard III, *Macromolecules*, **2005**, *38*, 979–991; (c) M. L. Mansfield, L. I. Klushin, *Macromolecules* **1993**, *26*, 4262–4268.

[48] E. Delort, E. Szöcs, R. Widmer, H. Siegenthaler, J.-L. Reymond, *Macromol. Biosci.* **2007**, *7*, 1024–1031.

[49] P. Sommer, V. S. Fluxa, T. Darbre, J.-L. Reymond, *Chembiochem* **2009**, *10*, 1527–1536.

[50] (a) D. van der Spoel, E. Lindahl, B. Hess, G. Groenhof, A. E. Mark, H. J. C. Berendsen, *J. Comp. Chem.*, **2005**, *26*, 1701–1718; (b) E. Lindahl, B. Hess, D. van der Spoel, *J. Mol. Mod.* **2001**, *7*, 306–317.

[51] S. Javor, J.-L. Reymond, *Isr. J. Chem.* **2009**, *49*, 129–136.

[52] S. Javor, A. Natalello, S. M. Doglia, J.-L. Reymond, *J. Am. Chem. Soc.* **2008**, *130*, 17248–17249.

16

PHOSPHORUS-CONTAINING DENDRITIC ARCHITECTURES: SYNTHESIS AND APPLICATIONS

ANNE-MARIE CAMINADE[1,2] AND JEAN-PIERRE MAJORAL[1,2]

[1] *Laboratoire de Chimie de Coordination (LCC) CNRS, 205 route de Narbonne, 31077 Toulouse Cedex 4, France*

[2] *Université de Toulouse, UPS, INPT, LCC, F-31077 Toulouse, France*

16.1 INTRODUCTION

Most publications about dendrimers and dendritic structures concern organic dendrimers, in particular those having a nitrogen atom at each branching point such as poly (amidoamine) (PAMAM) [1] and poly(propyleneimine) (PPI) dendrimers [2]. However, heteroatom-containing dendrimers [3], possessing one heteroatom (in particular P and Si) at each branching point, attract an increasing interest since the pioneering works reported by Rebrov et al. in 1989 for poly(organosilane) dendrimers [4], and by Engel and coworkers in 1990 for poly(phosphonium) dendrimers [5]. We described in 1994 the first neutral phosphorus dendrimers (poly(thiophosphoryl)) [6]. Since that time we have published about 260 papers in this field, we have developed several methods of synthesis and we have found many applications in various fields. After a first part remembering important aspects concerning the synthesis of phosphorus-containing dendrimers, this chapter will emphasize their applications in three main fields: catalysis, (nano)materials, and biology.

Designing Dendrimers, Edited by Sebastiano Campagna, Paola Ceroni, Fausto Puntoriero.
© 2012 John Wiley & Sons, Inc. Published 2012 by John Wiley & Sons, Inc.

16.2 SYNTHESES OF PHOSPHORUS-CONTAINING DENDRIMERS

Besides our own work, a few other groups have reported the synthesis of phosphorus-containing dendrimers. We have mentioned in the introduction the poly(phosphonium) dendrimers proposed by Engel and coworkers [5]. This first two-step method consists of the cleavage of ether linkages by Me_3SiI to afford an alkyliodide, which is used in the second step for alkylating a phosphine possessing three ether groups. This process was applied up to the third generation (dendrimer **1-G$_3$**), and it allows multiplying by 3 the number of end groups at each generation (instead of 2 in most cases) (Figure 16.1). Small poly(phosphine) dendrimers were reported by DuBois and coworkers [7]; they were obtained by a radical addition of primary phosphines to the vinyl group of

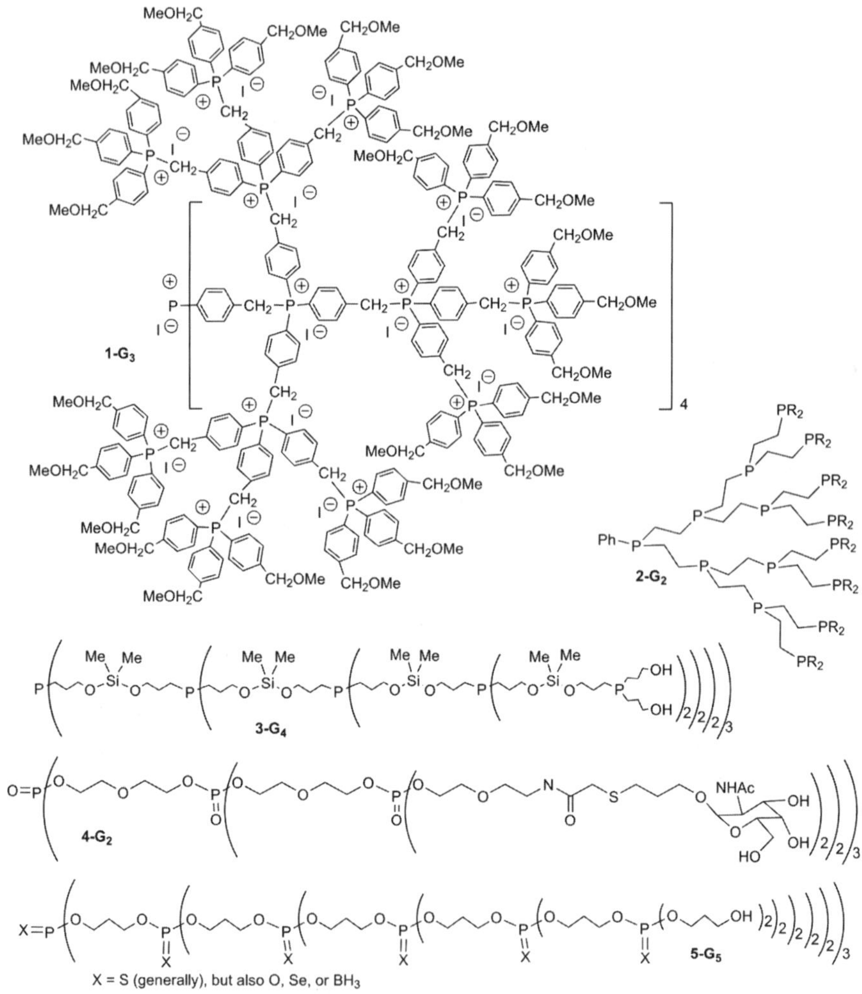

FIGURE 16.1 Chemical structures of various types of phosphorus dendrimers.

phosphonates. Reduction of the phosphonates by lithium aluminum hydride afforded again primary phosphines. This synthesis was carried out up to the second generation **2-G$_2$**. Larger poly(phosphine) dendrimers were proposed by Kakkar and coworkers [8]; the first step of the synthesis consisted of the aminosilanization of alcohol end groups; the second step is a controlled acid-based hydrolysis of the aminosilanes by a single OH of $P[(CH_2)_3OH]_3$. The synthesis was carried out up to the third generation **3-G$_3$**. Characterization of these compounds indicates dendrimer rather than hyperbranched growth, despite the possibility of multiple reactions of $P[(CH_2)_3OH]_3$. The use of phosphoramidites as building blocks for the synthesis of phosphorus dendrimers was developed by Roy [9] and by Salamonczyk et al. [10]; both methods afford poly (phosphate) dendrimers. The first method consists of coupling a phosphoramidite bearing siloxane groups with alcohol end groups, followed by oxidation. The second step of this method is the deprotection of the siloxane. Repeating twice both steps affords the second generation **4-G$_2$**, which can be functionalized by *N*-acetylgalactosamine [9]. The second method of the synthesis of phosphate dendrimers is related to the previous one; it consists first of the reaction of a triol with a phosphoramidite possessing acetate groups, followed by oxidation with elemental sulfur [10]. The second step is the deprotection of the acetates to afford the polyols of the next generation. The repetition of both steps was carried out up to the fifth generation **5-G$_5$**. The same type of method is also usable for the synthesis of dendrimers possessing selenium, oxygen [11] or boron hydride [12] instead of sulfur linked to phosphorus. It is even possible to build original dendrimers possessing a different type of phosphate at each generation, chosen between P=S, P=Se, and P=O, affording layered dendrimers [13] (Figure 16.1).

However, the most generally used method of synthesis of phosphorus-dendrimers is the one we described in 1994 [6]. It consists in applying a two-step reiterative process using successively 4-hydroxybenzaldehyde in basic conditions and $H_2NNMeP(S)Cl_2$ (Scheme 16.1). Both steps generate only NaCl and H_2O as by-products and are quantitative. This process gave rise to the highest generation well characterized obtained up to now for any type of dendrimers (generation 12, **6-G$_{12}$**) [14]. It is also the highest generation obtainable from a practical point of view (the twelfth generation is the last one soluble) and from a theoretical point of view, due to the fact that the number of end groups increases more rapidly than the free space available for them for high generations, preventing full substitutions.

These reactions can be applied starting from various cores, in particular $N_3P_3Cl_6$ instead of $P(S)Cl_3$ (compounds **7-G$_n$**, Scheme 16.1) [15], but also from a phthalocyanine core [16]. In fact any type of compound possessing aldehyde or PCl functions can be used as core, even an oligomer [17]. 4-Hydroxybenzaldehyde can be also replaced by any type of compound possessing both a phenol and an aldehyde, for instance the azobenzene derivative $HOC_6H_4{-}N{=}N{-}C_6H_4CHO$ [18].

Aldehydes and $P(S)Cl_2$ are the terminal groups of these series of dendrimers, both being among the most reactive groups in organic chemistry and phosphorus chemistry, respectively. Thus, we have carried out a lot of reactions starting from these end groups [19]; most of them were selected for obtaining special properties or applications, which will be emphasized later. These end groups allowed us to propose also

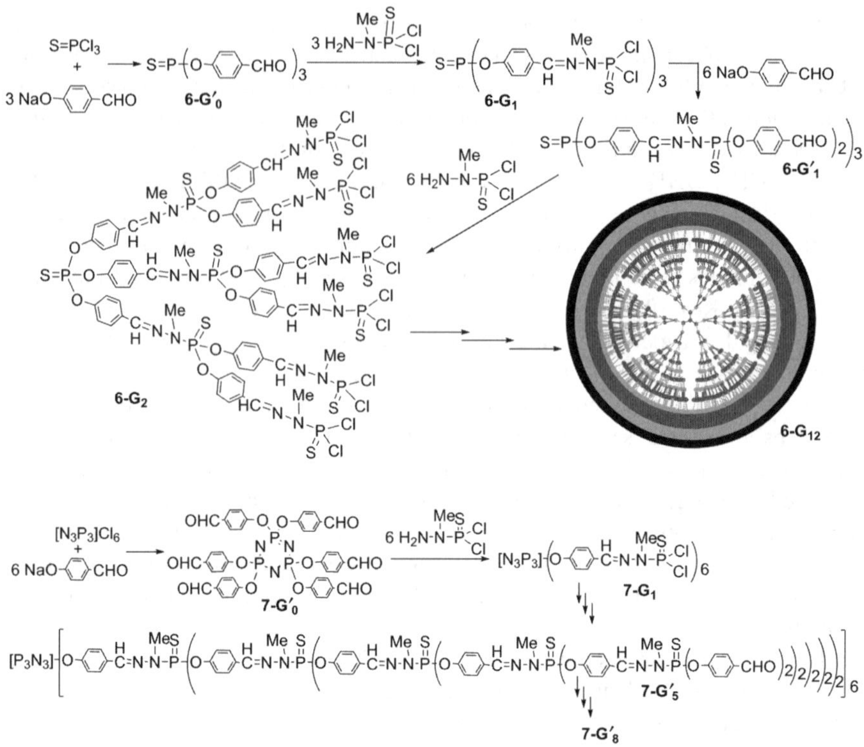

SCHEME 16.1

a new concept in dendrimer chemistry, that we called "multiplurifunctionalization." Indeed, a series of compounds having two, three, or even four different terminal groups linked to each branching point of the surface was obtained, thanks to the very special reactivity of the $P(X)Cl_2$ end groups ($X = O, S$). The grafting of several types of functional groups at each reactive site of the surface can be accomplished using the specific monosubstitution of functionalized amines to $P(X)Cl_2$ end groups. Dendrimers of type **8-G*n*** possessing three types of end groups (Cl, NH, propargyl) were obtained up to the seventh generation. The remaining Cl is able to react with phenols such as hydroxybenzaldehyde only when $X = O$, to afford another type of trifunctional dendrimers **9-G*n***, possessing NH, propargyl, and aldehyde end groups. The fourth terminal group is obtained from **9-G*n*** by a Wittig reaction between Ph_3PCHCN and the aldehydes, to afford dendrimers **10-G*n***. The four types of functional end groups are secondary amine, propargyl, alkene, and nitrile groups (Scheme 16.2) [20].

After our first method of synthesis shown in Scheme 16.1, we have also proposed several other types of synthesis of phosphorus dendrimers, including in particular the creation of P=N–P=S linkages at one, several, or all generations. These linkages are easily obtained by Staudinger reactions between phosphines and phosphorus azides. These reactions are quantitative and generate N_2 as the sole by-products, thus they are

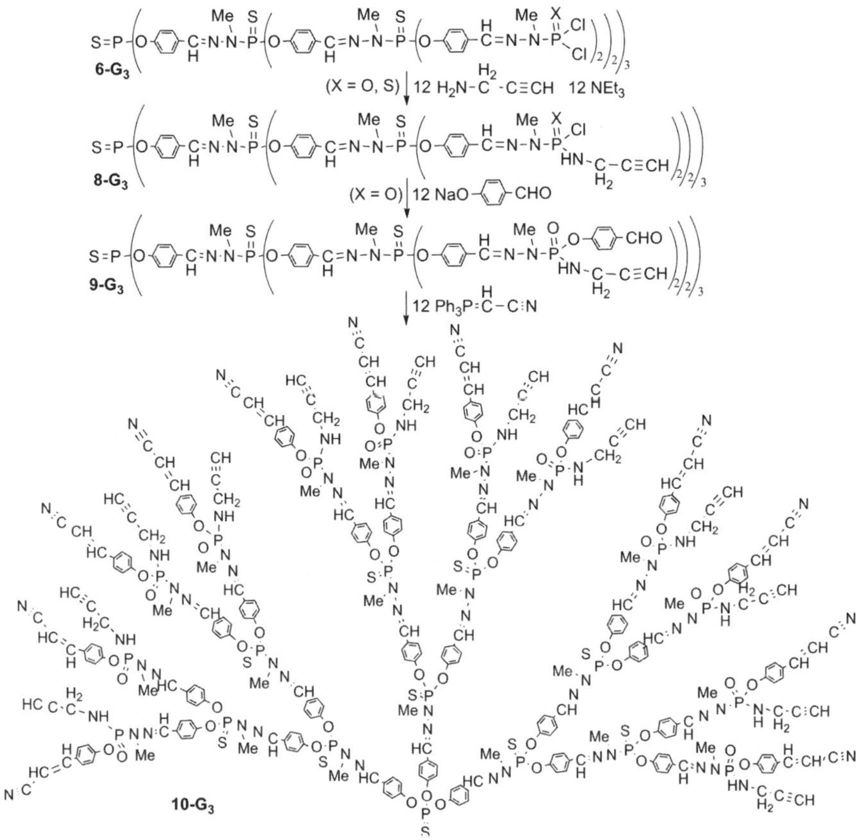

SCHEME 16.2

particularly attractive for the synthesis of dendrimers. In our first methods, phosphorus dendrimers including these linkages necessitate three steps to build one generation, starting from aldehyde groups: (i) condensation of methylhydrazine, (ii) reaction with either Ph_2PCl in basic conditions [21] or with Ph_2PCH_2OH [22], and (iii) Staudinger reaction between the phosphine end groups and an azido dialdehyde, which affords again aldehydes as end groups. The repetition of these three steps was carried out up to the third generation **11-G$_3$** (Scheme 16.3). It must be emphasized that this three-step method is fully compatible with the two-step method shown in Scheme 16.1, because both afford aldehyde terminal groups. It means that the P=N−P=S linkages can be introduced at one or several layers, where desired when applying the two-step process.

The presence of P=N−P=S linkages inside dendrimers led to the most original type of dendritic structures described up to now. Indeed, these linkages have a mesomeric form ($^+$P−N=P−S$^-$) in which a negative charge is located on sulfur. Due to this particular structure, these linkages are able to undergo a specific alkylation on sulfur with alkyl triflates [22], eventually functionalized [23]. This reaction is

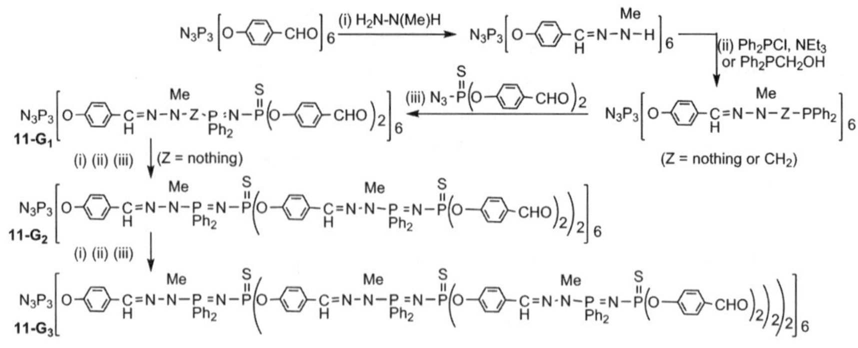

SCHEME 16.3

compatible with the presence of the other P=S groups in the dendrimer, which are not alkylated. The alkylation of sulfur of dendrimer **12-G$_3$**, having one layer of P=N−P=S groups, induces a weakening of the phosphorus–sulfur bond. The P−S bond can be easily cleaved by a nucleophilic phosphine such as $P(NMe_2)_3$, affording tricoordinated phosphorus atoms (P^{III}) inside the structure (dendrimer **13-G$_3$**). These P^{III} are able to react in Staudinger reactions with various functionalized azides, allowing the internal functionalization for instance with primary amines, or isothiocyanates [24], or pyrene derivatives [25]. However, the most interesting function to be grafted in this way is aldehyde, using $N_3P(S)(OC_6H_4CHO)_2$ (dendrimer **14c-G$_3$**); the condensation of the aldehydes with primary amines allowed the functionalization by macrocycles or azides (dendrimers **15-e-f-G$_3$**) [24] (Scheme 16.4).

A lot of reactions can be performed starting from these aldehyde internal groups, and in particular the methods of synthesis of phosphorus dendrimers shown in Schemes 16.1 and 16.3 can be used now inside these dendrimers. The two-step method was applied up to the growing of four generations inside the dendrimer **14c-G$_3$** (Scheme 16.5). The largest compound synthesized (**16-G$_3$G$_4$**) possesses 48 phenoxy end groups coming from the initial dendrimer, and 96 aldehyde end groups, coming from the reactivity inside the dendrimer, and the growing of new branches [26]. The three-step process can be applied to the dendrimer **14c-G$_3$** to afford dendrimer **17-G$_3$G$_4$** in which two different types of branches are forced to coexist, but keep their own properties (Scheme 16.6) [26]. Up to now, no other type of dendrimer (including organic dendrimers) has been able to afford so original structures. The reactivity of both types of branches is different, as illustrated by the reactivity of gold; only the newly synthesized branches, which possess P=N−P=S linkages are able to complex gold on P(S), whereas the other branches are not able, affording compound **18-G$_3$G$_4$** [27].

Another way for obtaining highly sophisticated dendritic structures consists in using dendrons (also called dendritic wedges), possessing as core a vinyl group activated by a P=N−P=S group (such as **19a,b-G$_3$**) with another dendron possessing a primary amine as core (such as **20c,d-G$_3$**). Michael addition can be performed with these dendrons. This reaction is compatible with several types of end groups, as shown by the obtaining of the “Janus” species **21ac-G$_3$G$_3$** and **21bd-G$_3$G$_3$**.

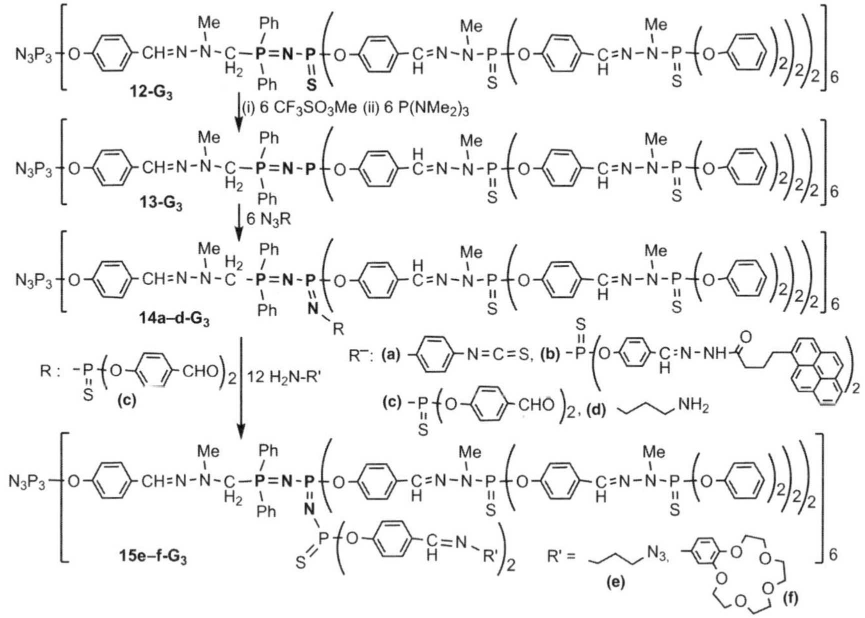

SCHEME 16.4

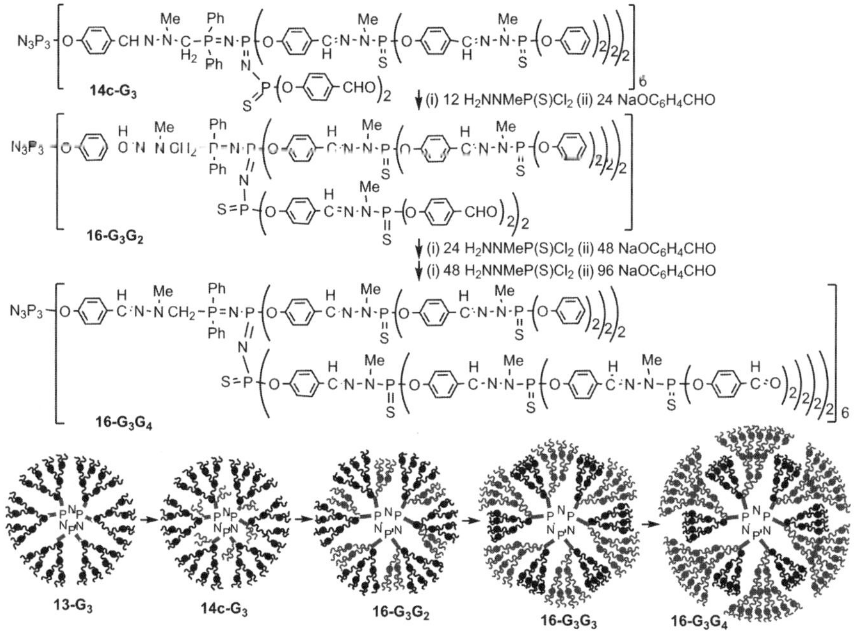

SCHEME 16.5

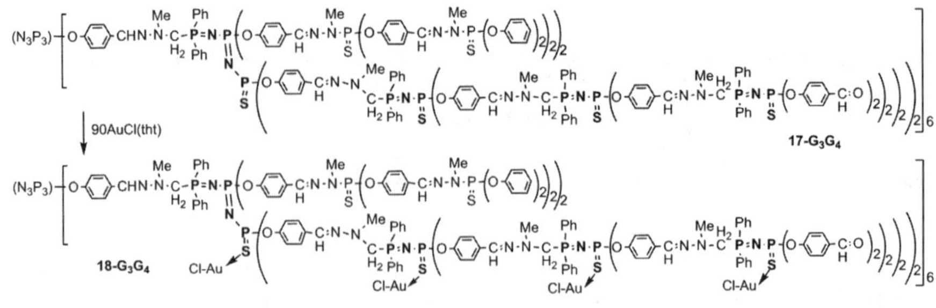

SCHEME 16.6

The former one is particularly interesting, because it possesses 16 diphenylphosphino groups on one side. The reaction of 16 equiv. of dendron **22-G$_3$** via Staudinger reactions afforded the dendritic species **23**, which possesses two types of end groups, that are 16 phenoxy groups on one side and 512 dimethylamino groups on the other side (Scheme 16.7) [28].

Figure 16.2 displays the various types of dendritic structures (dendrimers, dendrons, and special dendritic architectures) built with one phosphorus atom at each branching point synthesized up to now, thanks to the extraordinary versatility of the chemistry of phosphorus.

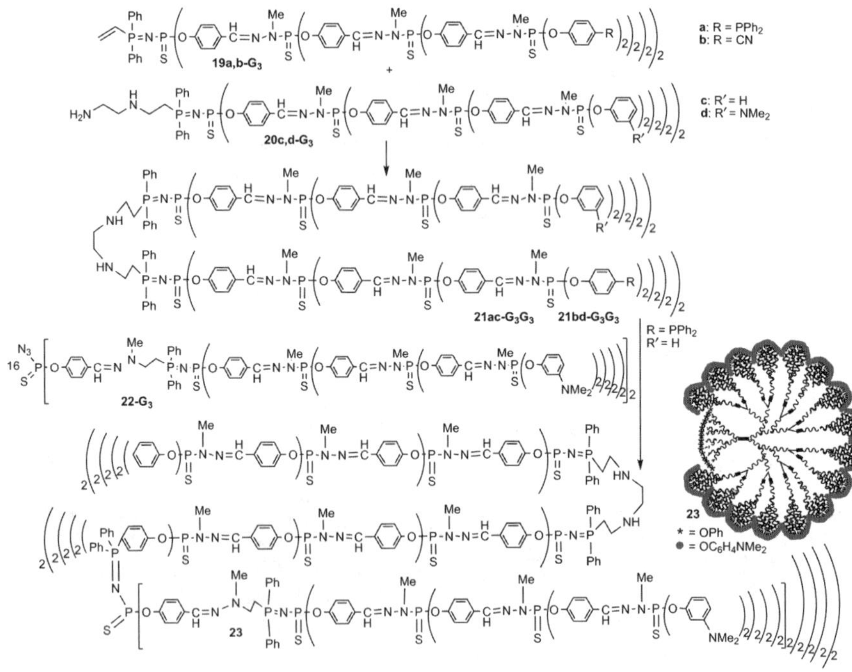

SCHEME 16.7

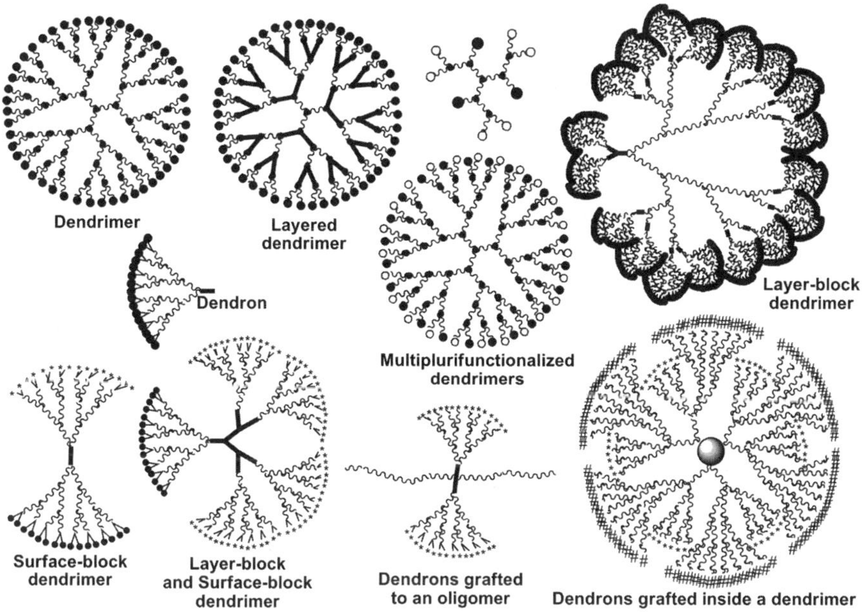

FIGURE 16.2 Various dendritic structures having phosphorus at each branching point.

Obviously, most of these dendritic structures are obtained by lengthy processes, thus we have tried to find more rapid methods of synthesis of dendrimers, in particular by multiplying the number of terminal groups at each synthetic step, and not every two or three steps as shown previously. We applied orthogonal coupling strategies, based on the use of two types of branched units bearing complementary functions, namely $\mathbf{AB_2}$ and $\mathbf{CD_2}$ monomers, where **A** reacts only with **D**, and **B** reacts only with **C**. In order to save time, quantitative reactions without protecting groups and activating agents are desirable. Two reactions are particularly useful for synthesizing phosphorus dendrimers, as shown previously: condensation reactions and Staudinger reactions. Thus, the goal is to synthesize compounds in which **A** is NH_2, **B** is PPh_2, **C** is N_3, and **D** is CHO. This goal was first achieved by designing compounds $H_2NNMeP(S)(OC_6H_4PPh_2)_2$ ($\mathbf{AB_2}$) and $N_3P(S)(OC_6H_4CHO)_2$ ($\mathbf{CD_2}$). Starting from a trialdehyde core ($\mathbf{D_3}$ type), the reaction of the $\mathbf{AB_2}$ monomer affords the first generation, then the reaction with the $\mathbf{CD_2}$ monomer affords directly the second generation. Using again the $\mathbf{AB_2}$, then the $\mathbf{CD_2}$ monomers affords the fourth generation of the layered dendrimers $\mathbf{24\text{-}G_4}$. In this series, the dendrimers have either aldehydes or phosphines as terminal groups. This method is very powerful and quantitative, and generates only N_2 and H_2O as by-products, thus it was even possible to synthesize the fourth generation in a one-pot (but multistep) process, adding periodically strictly stoichiometric amounts of reagents (Scheme 16.8) [29].

The same concept can be extended to monomers having a larger number of functional groups. In particular, $N_3P_3Cl_6$ offers the opportunity to differentiate the reactivity of one Cl among six, in order to graft specifically either one or five

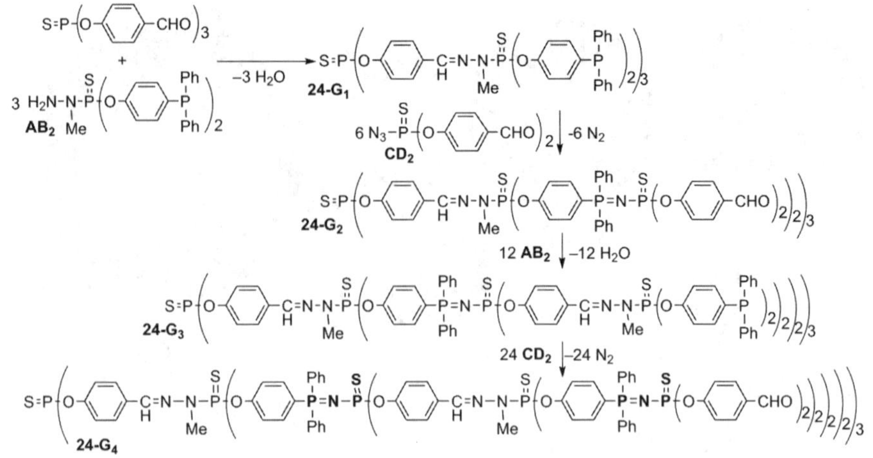

SCHEME 16.8

functions. $\mathbf{AB_5}$ and $\mathbf{CD_5}$ monomers based on the cyclotriphosphazene scaffold were synthesized in this way. Each step of the synthesis of dendrimers with $\mathbf{AB_5}$ and $\mathbf{CD_5}$ monomers needed several days to go to completion, and the synthesis was stopped at the third generation $\mathbf{25\text{-}G_3}$; no attempt was made to synthesize the next generation, due to the steric hindrance of the third generation outer shell (Scheme 16.9) [30].

In order to shorten more the synthetic process, hyperbranched polymers appear as a tempting alternative to dendrimers, due to their straightforward one-pot synthesis. Thus, we designed a protected branched monomer having two types of functions in a 1:2 ratio, in which the deprotected functions react readily intermolecularly. The Staudinger reaction is chosen, due to the absence of by-products and to its high yield. The monomer bears one azide and two protected phosphines that are deprotected by bases. Various hyperbranched polymers were obtained, with various molar masses and distribution, depending on the temperature, the reaction time, the solvent, the concentration, and the type of amine used for the deprotection. However, the degree of branching was found very high and practically identical (0.83–0.85), irrespective of

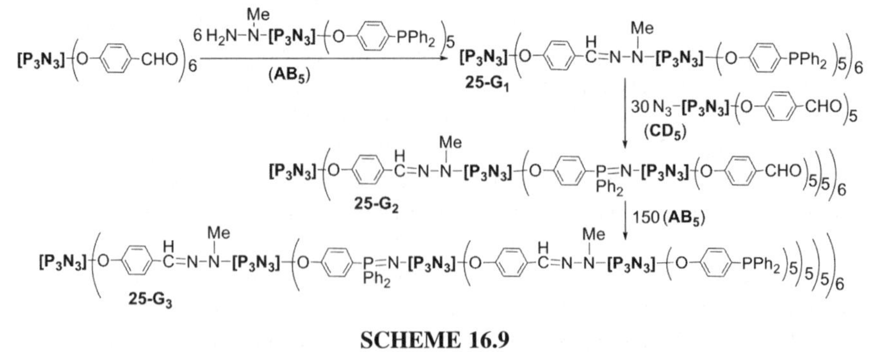

SCHEME 16.9

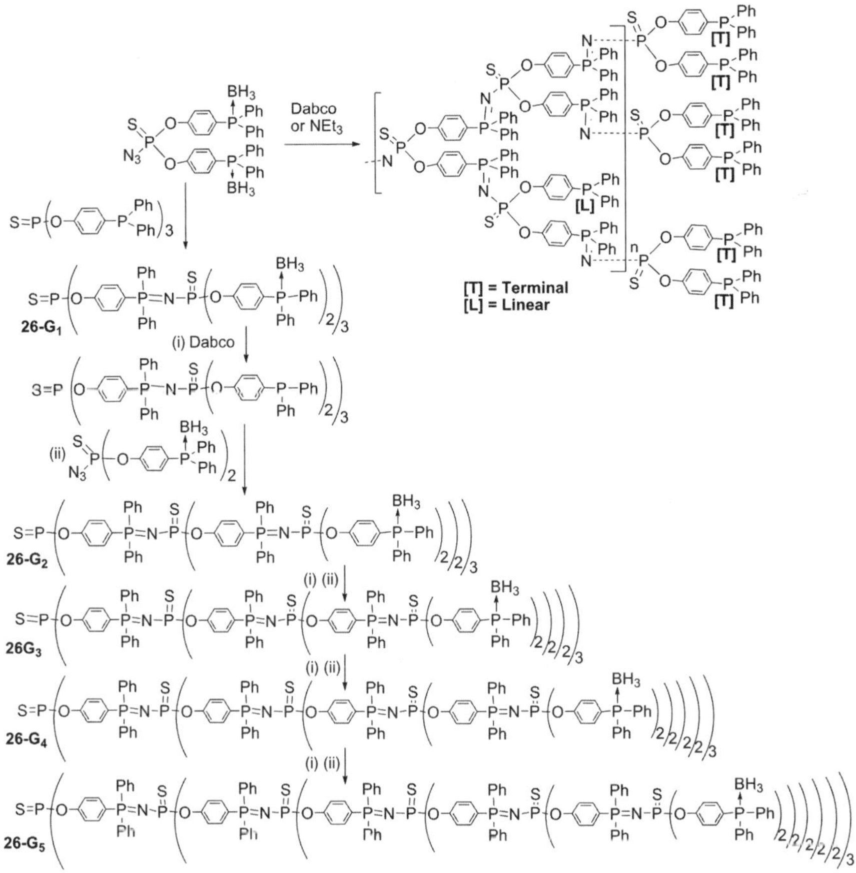

SCHEME 16.10

the conditions used (Scheme 16.10). The polydispersity is very high for these hyperbranched polymers (from 1.5 to 8, depending on the sample), and the properties, for instance concerning the intrinsic viscosity that slowly increases with the molecular weight, are different from those expected for dendrimers. In order to see if these differences were due to the topological structure or to the chemical structure, dendrimers issued from the same monomer were also synthesized by a stepwise process (Scheme 16.10). The series of compounds **26-G$_n$** have the normal behavior expected for dendrimers, in particular concerning the very low polydispersity (the highest value is 1.029 for **26-G$_5$**) and the variation of intrinsic viscosity with molecular weight (a bell-shaped curve). It can be inferred from these results that the replacement of dendrimers by hyperbranched polymers having the same chemical composition might not be a good solution in many cases, the physicochemical properties being too different [31].

After these synthetic aspects, we focused our interest on the potential uses of these phosphorus-containing dendrimers. In practically all cases, the dendrimers shown in Scheme 16.1 were used, mainly those built from the cyclotriphosphazene core (series **7-G**$_n$) suitably functionalized on the surface, depending on the properties desired [32]. Several properties that will be displayed in the following paragraphs are also known for other types of dendrimers. However in some cases, and especially concerning some biological properties, these phosphorus dendrimers display very unique properties.

16.3 ORGANOMETALLIC DERIVATIVES OF PHOSPHORUS DENDRIMERS AND THEIR USE AS CATALYSTS

The use of dendrimers as catalysts was recognized very early, with the aim of combining both the advantages of homogeneous catalysts (solubility) and heterogeneous catalysts (easy recovery), without their disadvantages. The first example dates back to 1994, using organic dendrimers [33]. Several reviews have covered this topic [34], but only very few concern phosphorus dendritic catalyst [35]. In many cases phosphorus derivatives (phosphines) are used to complex the metal, but the dendritic backbone does not contain phosphorus in their structure. In this chapter, we will focus only on the case of dendritic catalysts built from phosphorus-containing dendrimers. The first example in this field also dates back to 1994, when the palladium complex of the phosphorus dendrimers **2-G**$_n$ was used for the electrocatalyzed reduction of CO_2 to CO [7]. Later on, the rhodium complexes of the dendrimers **3-G**$_n$ were found efficient catalysts for olefin hydrogenation in a 1:200 metal-to-substrate ratio. The efficiency of the dendrimer was found similar to that of monomeric complexes, but the advantages of these dendritic catalysts are their easy separation and reuse [8].

We have also contributed to this field, but in first attempts we were mainly interested in organometallic chemistry and not in catalysis; several phosphorus dendrimers having metallic derivatives as end groups [36] or in their internal structure [37] were synthesized for this purpose. When using them later on as catalysts, we generally choose to graft the catalytic complexes at the periphery, as terminal groups of dendrimers. It must be emphasized that in all cases, the molar percentage of catalyst that is indicated is the molar percentage of metal, with one metal per ligating site on the dendrimer. In this way it is possible to compare directly the efficiency of a monomer and of a dendrimer. In our first example of catalysis, we wanted to compare the catalytic efficiency depending on the location of the catalytic sites. For this purpose, we grafted amino diphosphine ligands either at the surface of dendrimers or at the core of dendrons. The $RuH_2(PPh_3)_2$ complex **27-G**$_3$ was used for Knoevenagel condensations and diastereoselective Michael additions. In the latter case, the dendron **28-G**$_3$ having one $RuH_2(PPh_3)_2$ complex at the core was also tested. No difference in the diastereoselectivity was observed when compared to the dendrimer, and to the monomer (Figure 16.3). The $PdCl_2$ complex of the third-generation dendrimer shown in Figure 16.3 was used as catalyst in Stille couplings. In all these cases, it was possible to recover and reuse twice the dendritic catalyst, without any significant loss of activity in marked contrast with the behavior of the corresponding monomer [38].

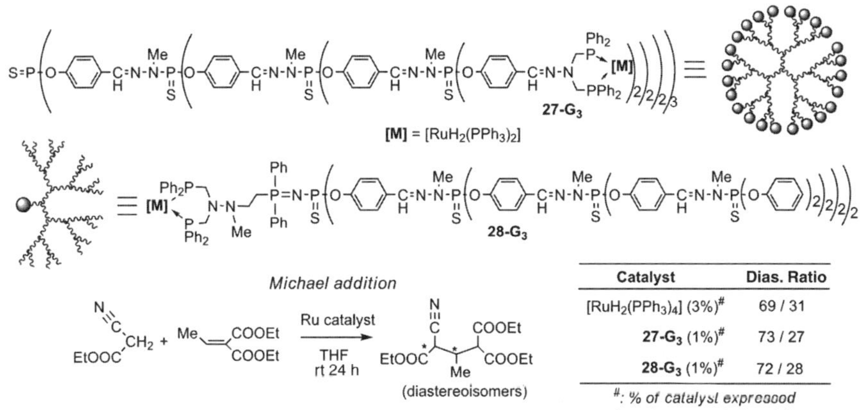

Catalyst	Dias. Ratio
$[RuH_2(PPh_3)_4]$ (3%)[#]	69 / 31
27-G$_3$ (1%)[#]	73 / 27
28-G$_3$ (1%)[#]	72 / 28

#: % of catalyst expressed in molar percent of metal.

FIGURE 16.3 Influence of the location of catalytic site(s) on Michael additions.

In another example of amino diphosphine ligands as terminal groups of phosphorus dendrimers, we wanted to compare the influence of a local steric hindrance on the efficiency of the catalysis, depending on the generation. For this purpose, the diphosphino dendrimers **29-G$_n$** and **30-G$_n$** were synthesized, and their palladium complexes used in Sonogashira reactions. In the case of the nonhindered dendrimer **29-G$_n$**, a slight increase of the efficiency of the catalysis was observed on going from the monomer **29** to the third generation **29-G$_3$** (slightly positive dendritic effect); in contrast, a detrimental influence of the generation was observed in the case of the more hindered compounds **30-G$_n$** from the first to the third generation (Figure 16.4). These experiments were carried out in mild conditions, and were found very tolerant to large amounts of water (more than 30% of water as solvent with acetonitrile) [39].

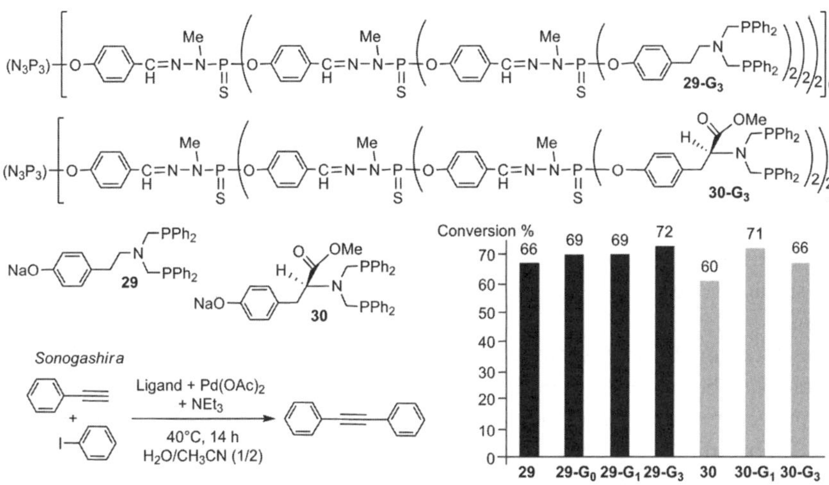

FIGURE 16.4 Influence of the generation on catalyzed Sonogashira reactions.

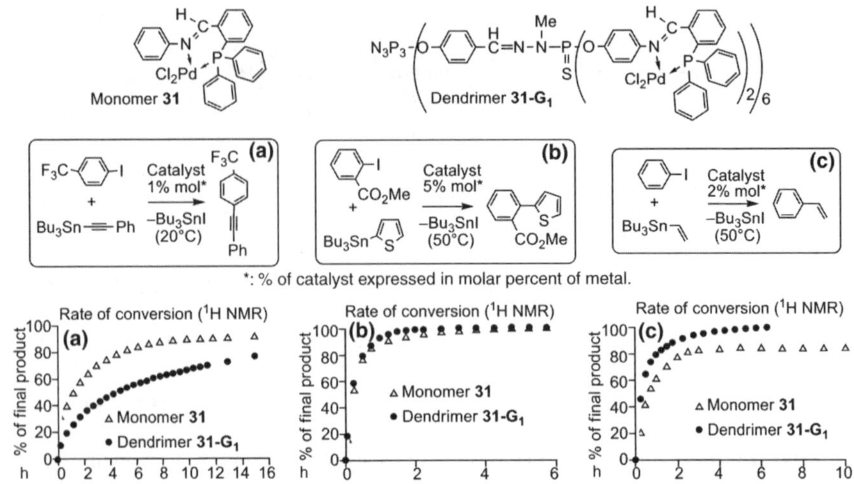

FIGURE 16.5 Monomer versus generation 1 dendrimer for catalyzing Stille couplings.

As shown above, the activity measured for the dendrimers was comparable or slightly higher than that of the monomers in many cases, but no general rules can be deduced concerning the efficiency of a dendritic catalyst. We have synthesized and complexed iminophosphine derivatives, in particular the monomeric palladium complex **31** and the first-generation dendrimer **31-G$_1$**. Comparison of their efficiency for various Stille couplings was really puzzling (see Figure 16.5). Measurements of the rate of conversion by ^{1}H NMR show that in case (a) the monomer is more efficient than the dendrimer, in case (b) the monomer and the dendrimer have the same efficiency, and in case (c) the dendrimer is more efficient. However, the real difference between **31** and **31-G$_1$** is that the latter can be recovered and reused, but not the former [40].

Another example of iminophosphine (compound **32-G$_3$**) derived from (2*S*)-2-amino-1-(diphenylphosphinyl)-3-methylbutane was complexed *in situ* by $[Pd(\eta^3\text{-}C_3H_5)Cl]_2$. This catalyst was used in asymmetric allylic alkylations of *rac*-(*E*)-diphenyl-2-propenyl acetate and pivalate. This reaction necessitates also a base (BSA) and an acetate salt; better results were obtained with LiOAc than with KOAc. The percentage of conversion, the yield of isolated 2-(1,3-diphenylallyl)-malonic acid dimethyl ester, and its enantiomeric excess were measured in each case, and were found good to very good (ee from 90 to 95%) (Figure 16.6). Furthermore, the dendritic catalyst can be recovered and reused at least two times, with almost the same efficiency [41].

The last example of phosphine dendrimers that we have synthesized was obtained by grafting the water-soluble PTA ligand (1,3,5-triaza-7-phosphaadamantane) as terminal groups of generations 1–3. Their ruthenium complexes (compounds **33-G$_n$**) were used as catalysts in aqueous media for two types of experiments. A slightly positive dendritic effect on the regioselectivity was observed for hydration of alkynes, and a large positive dendritic effect was observed in the biphasic

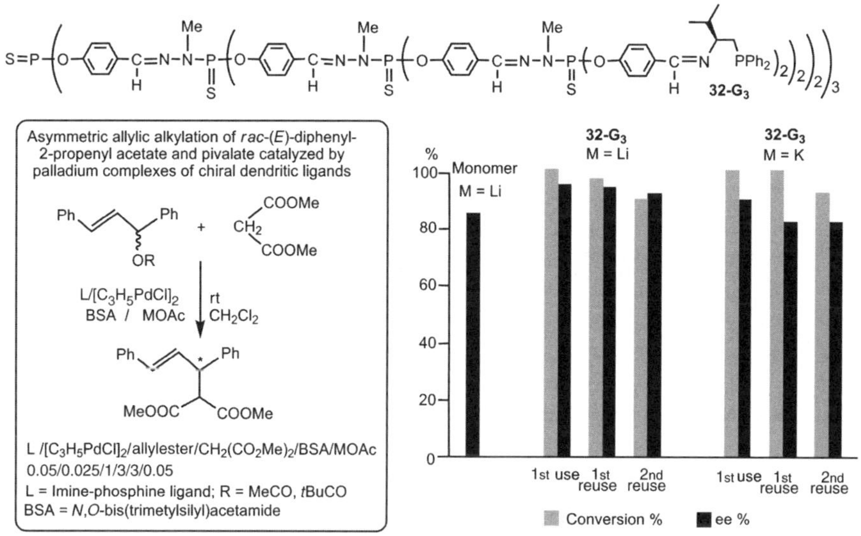

FIGURE 16.6 Influence of cocatalyst and generation for catalyzed asymmetric allylic alkylation.

(water/heptane)-catalyzed isomerization of allylic alcohols to ketones (from 38% of conversion with the monomer **33**, 63% with the first-generation dendrimer **33-G$_1$**, to 98% with the third generation **33-G$_3$**). The products were easily isolated and the catalyst could be reused several times, even the first generation (Figure 16.7) [42].

The best example of positive dendritic effect was obtained for the coupling of pyrazole, catalyzed by the copper complexes of pyridineimine-ended phosphorus dendrimers. Copper is a cheap and practically nontoxic metal, which is a very attractive alternative to transition metals. However, the monomer **34** is totally inefficient to catalyze this reaction, starting from either iodo- or bromobenzene. In sharp contrast, the dendrimers **34-G$_n$** enabled quantitative conversion of substrate into

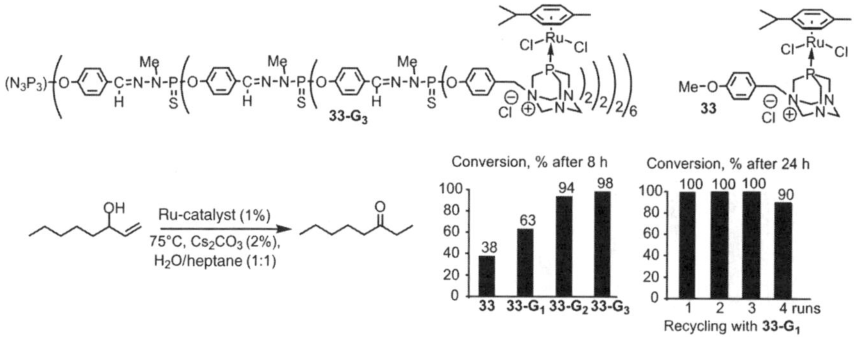

FIGURE 16.7 Role of the generation and recycling in aqueous media for alcohols isomerization.

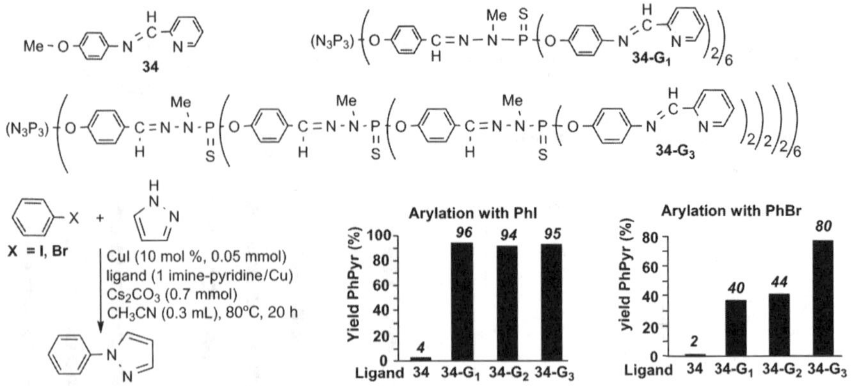

FIGURE 16.8 Copper(I) dendritic complexes for the catalyzed coupling of pyrazole.

product within 20 h at 80°C, when starting from iodobenzene. When starting from less reactive bromobenzene, the best catalytic activity was obtained in the presence of the third-generation dendritic ligand **34-G$_3$**, phenylpyrazole being obtained in 80% yield after 20 h at 80°C (Figure 16.8). We have also demonstrated specific advantages for copper(I) catalysis of the very important O- and N-arylation and vinylation of phenol and pyrazole using these copper dendritic complexes, for which the highest yields under the mildest conditions were used [43].

Another example of copper-catalyzed reactions was provided by dendrimers ended by bisoxazoline ligands attached via "click" reactions, from generation 1 to generation 3. These macromolecules were evaluated in copper(II)-catalyzed asymmetric benzoylations, starting from two different diols. Dendrimers **35-G$_1$** and **35-G$_2$** afford good yields and enantioselectivities in both cases, whereas the third generation has a detrimental influence on the enantioselectivity. The copper(II)-catalysts could be readily recovered and reused in several cycles (Figure 16.9) [44].

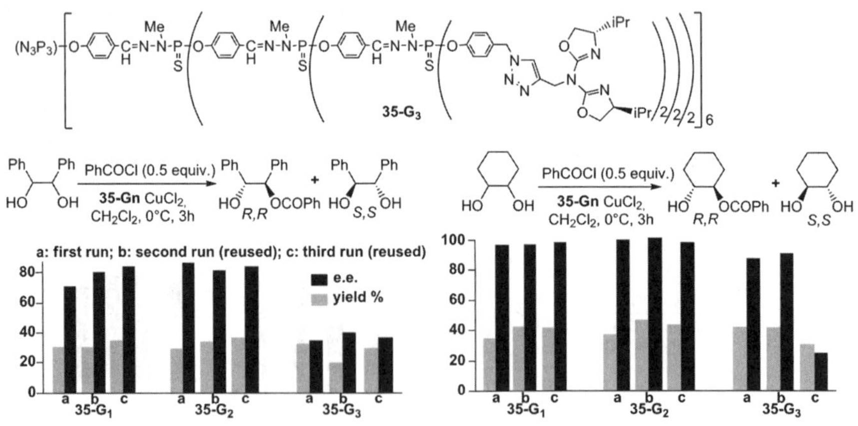

FIGURE 16.9 Cu(II) dendritic complexes and their reuse for catalyzed asymmetric benzoylations.

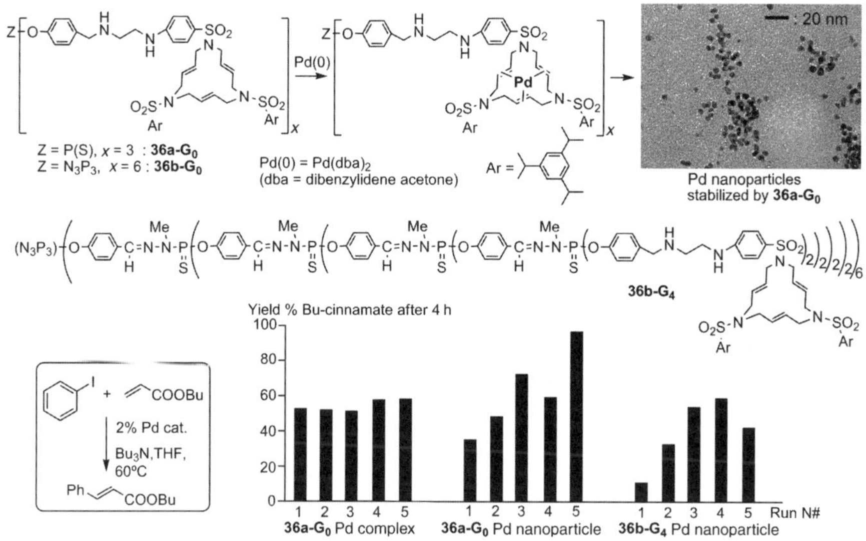

SCHEME 16.11

The last example in the field of catalysis is at the border with materials. Phosphorus dendrimers containing on their surfaces 15-membered azamacrocycles were synthesized up to the fourth generation (dendrimers **36a,b-G$_0$**, **36b-G$_1$**, **36b-G$_4$**), and their coordinating ability to palladium(0) atoms was tested. Depending mainly on the amount of Pd(0) used, either discrete complexes or new nanoparticulated materials were obtained, where nanoparticles are stabilized by the dendrimers of the zero, first, or fourth generation. No reduction process of palladium(II) salts was needed to prepare nanoparticles of 2.5–7.9 nm diameter. These complexes and nanoparticles were used for the Mizoroki–Heck reaction; in all cases, the catalyst could be recovered and reused several times. The Pd-complex of **36a-G$_0$** displays constant catalytic activity with recycling, whereas the Pd-nanoparticles stabilized with **36a-G$_0$** displays an increasing catalytic efficiency with recycling; this could be related to a decrease in the nanoparticles size, observed by electron microscopy (Scheme 16.11) [45].

16.4 PHOSPHORUS DENDRIMERS FOR THE ELABORATION OR MODIFICATION OF (NANO)MATERIALS

The development of nanomaterials is presently one of the major challenges for materials chemistry. In this perspective, nanosized molecules and particularly dendrimers are attracting a considerable attention. Indeed, dendrimers can play a key role in the chemical construction of periodic mesostructures synthesized by the cooperative assembly and organization of organic and inorganic components. We have already published two main reviews in this field [46]; the most important findings using phosphorus dendrimers will be highlighted in the next paragraphs.

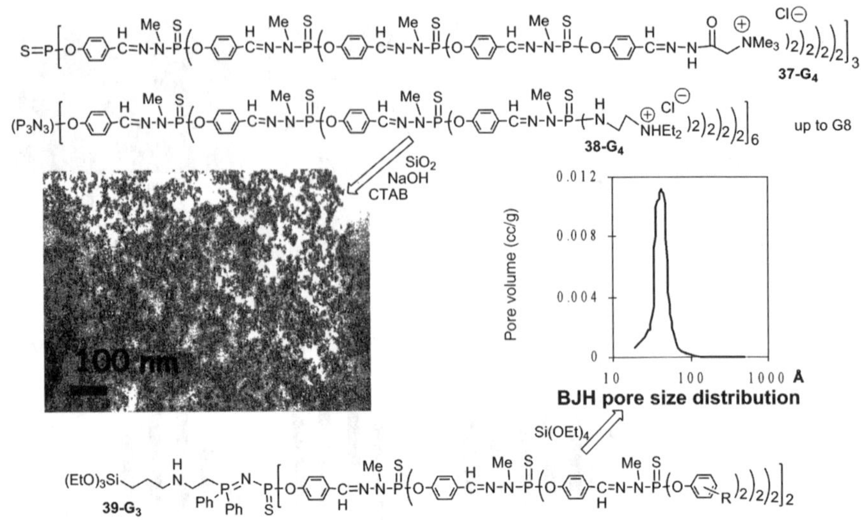

FIGURE 16.10 Positively charged dendrimers, and neutral dendrons, used for the elaboration of functionalized silica.

Phosphorus dendrimers are usable to create new silica materials controlled at the nanometric scale. The concomitant use of the positively charged dendrimers **37-G*n*** or **38-G*n*** (up to generation 8), cationic surfactants (cetyltrimethylammonium bromide (CTAB)) and sodium silicate in water allowed the synthesis of mesostructured nanoporous silica including dendrimers. We have shown that relatively important amounts of dendrimers (up to 26% in weight) can be incorporated into hexagonal silica phases during the structuring process. In these conditions, the inclusion of dendrimers does not modify the honeycomb structure characteristic of the MCM-41 phase, as well as its narrow pore size distribution of about 25 Å and its specific surface (Figure 16.10, left). These hybrid nanocomposites possess original properties, in particular the unprecedented possibility to selectively remove the surfactant while keeping the dendrimer inside the material, while noncovalently linked. Furthermore, all the dendrimers, and particularly their end groups are fully accessible [47].

In another approach to functionalized silica, we have covalently grafted dendrons **39-G$_3$** functionalized by $Si(OEt)_3$ at their core. The cohydrolysis and polycondensation of these macromolecules with a defined and varying number of equivalents of tetraethoxysilane (TEOS) was carried out via sol–gel protocol, giving rise to dendron-silica xerogels. The texture (porosity) of materials was determined by BET (Brunauer, Emmett, and Teller) measurements. A narrow pore size distribution was obtained in several cases (Figure 16.10, right). As the type of terminal groups of the dendrons can be easily varied, this process allowed the synthesis of numerous types of functionalized silica; we have in particular used as terminal groups of these dendrons phosphines, dyes, nitriles, and amines [48].

Besides silica, less classical periodic hybrid organic–inorganic materials with hierarchical structures and complex forms were obtained using as nanobuilding

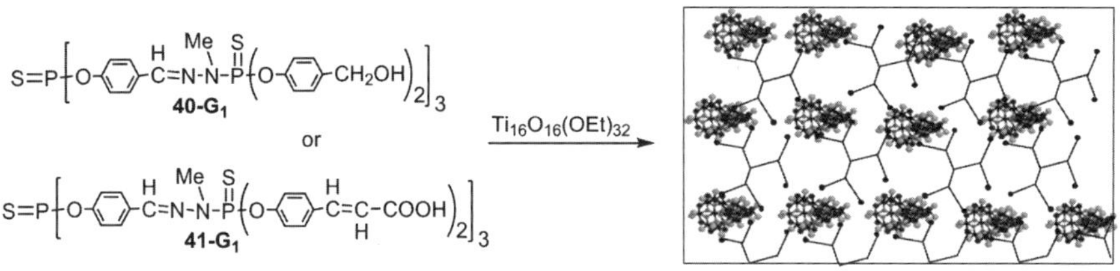

SCHEME 16.12

blocks the first-generation dendrimers **40-G$_1$** (alcohol terminal groups) or **41-G$_1$** (carboxylic acid terminal groups) and the structurally well-defined Ti-organo-oxo cluster $Ti_{16}O_{16}(OEt)_{32}$. The hybrid interface is created by *trans* alcoholysis in the case of the alcohol tips, and by bridging carboxylates in the case of acidic tips. In both cases a bicontinuous hybrid gel in which the titanium clusters are regularly spaced by the dendrimers was obtained (Scheme 16.12) [49].

Another type of organization was observed when some polycationic dendrimers such as **37-G$_n$** or **42-G$_n$** (both possessing amide terminal groups) were left for days in water. Indeed, solid hydrogels can be obtained in both cases, even when using amounts of dendrimers as small as 0.25% in weight in water; the gels thus obtained are solid at room temperature though they contain 99.75% of water (Figure 16.11, left). During the gelation process, large amounts of various types of hydrosoluble substances (for instance up to 30% in weight of nickel acetate) can be incorporated. Around 70,000 water molecules are estimated to be gelled by a molecule of dendrimer of generation 4 (**37-G$_4$**, **42-G$_4$**), that is gelation of about 1200 ± 1400 molecules of water per terminal ammonium or pyridinium unit, regardless of the generation considered [50]. The same gelation phenomenon was observed with nanolatexes coated with dendrons possessing the same types of terminal groups as **42-G$_n$** [51]. Dendrimers **42-G$_n$** were also used for obtaining macroscopic fibers. As shown in Figure 16.11 (right), a concentrated solution (10% in weight) of dendrimer in water was put in a syringe equipped with a kinked needle and linked to a step motor. The dendrimer solution contained in the syringe was continuously injected into a

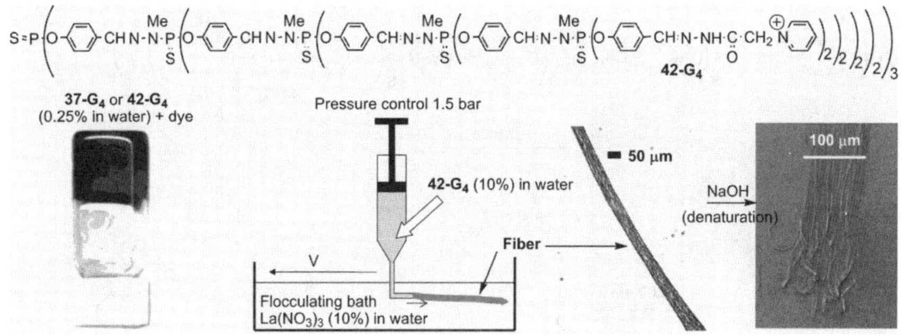

FIGURE 16.11 Hydrogels and fibers obtained from pyridinium-ended dendrimers.

40 cm-long cell containing a flocculating solution (10 wt% $La(NO_3)_3$ in water). The so-produced fiber was left 1 h in the solution then taken and analyzed. Mechanical measurements showed that dendrimer fibers exhibit an elastic behavior (reversible deformation of 3%, same order of magnitude than nylon and cotton fibers), followed by the rupture of the fiber above 9 GPa, whereas typical polymer fibers exhibit an irreversible deformation before rupture [52].

The stabilization of metal clusters by suitable ligands has allowed to make them isolable compounds, but the presence of this ligand shell modify the electronic properties, thus obtaining and organizing bare clusters appears highly desirable, but also highly challenging. The Au_{55} cluster is in particular very attractive for nanoelectronic devices, but numerous previous attempts to generate it failed and gave only crystals of Au_{13}. In sharp contrast, dendrimers **43-G*n*** possessing thiol terminal groups were able to peel off the PPh_3 ligands and the chlorine of $Au_{55}(PPh_3)_{12}Cl_6$, while keeping the Au_{55} structure, then to organize the bare Au_{55}, which coalesce in nanocrystals of $(Au_{55})\infty$, the dendrimer acting as template for crystallization. The best results were obtained with the fourth generation **43-G4** (Figure 16.12) [53]. The same phenomenon was observed later when the same dendrimer was deposited as thin layer on a silicon surface. Interaction of these dendrimer films with solutions of $Au_{55}(PPh_3)_{12}Cl_6$ also produced nanocrystals of $(Au_{55})\infty$ [54]. Very recently, we have used another type of dendrimer (**44-G*n***, ended by macrocycles closely related to **36b-G*n***) for performing also two different organizational roles. When interacting with $Pt(dba)_3$, these dendrimers first induce the formation of Pt(0) nanoparticles, then organize them in hyperbranched networks of coalesced Pt-nanoparticles, for which

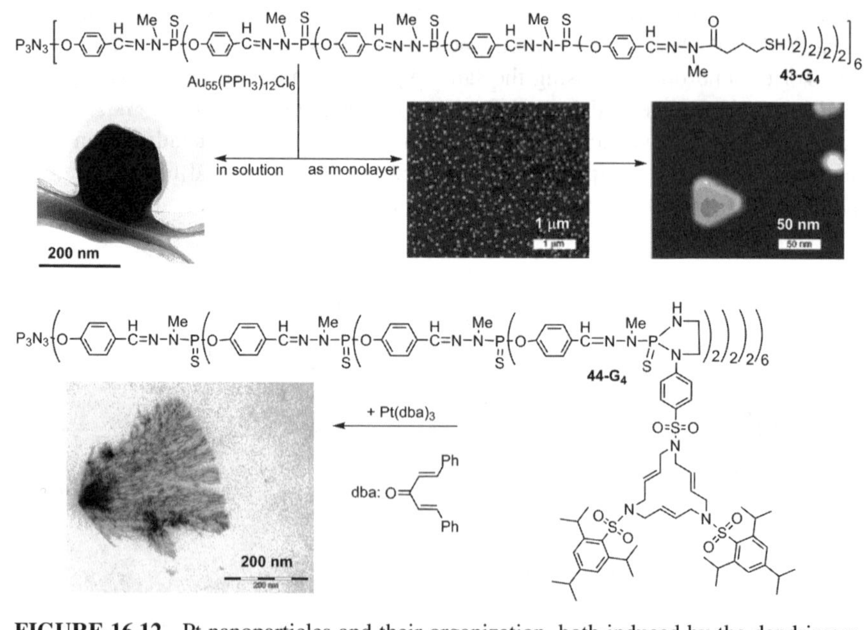

FIGURE 16.12 Pt nanoparticles and their organization, both induced by the dendrimers.

both the size and the degree of branching vary with the generation of the dendrimer. The largest networks were obtained with **44-G$_4$** (Figure 16.12) [55].

Another important field of research in materials chemistry consists in modifying the surface of an existing material by the deposition of organic thin films. Various techniques were already applied using phosphorus dendrimers; the first examples that we will see consists of obtaining modified electrodes, using dendrimers possessing electroactive groups in their structure. Dendrimers **45-G$_n$** ($n = 0$–4) possess bithiophene terminal groups that are electropolymerizable, leading to electrodes irreversibly modified by a dark blue film on the anode. This film is an electroactive conjugated polydendritic polymer, which remains electroactive in aqueous media, particularly when elaborated from the highest generations [56]. On the other hand, the electrodeposition of dendrimers **46-G$_3$**, possessing 24 TTF-crown ethers as terminal groups, is fully reversible. This dendrimer modified electrode is usable for the electrochemical sensing of a metal cation (i.e., Ba^{2+}), as shown in Figure 16.13 (upper part), which displays the changes of the electrochemical response upon increasing concentrations of Ba^{2+} [57]. However, one of the most popular electroactive derivatives is certainly ferrocene, thanks to its robustness and its full electrochemical reversibility in many cases. It has been linked very often to dendrimers, generally as terminal groups [58]. In the particular case of phosphorus dendrimers, we have grafted ferrocene derivatives as terminal groups, but also as core and in the interior; in all cases a blue film was reversibly deposited onto the electrode [59]. The most original examples are shown in Figure 16.13: ferrocenes functionalized by aldehydes were grafted as terminal groups

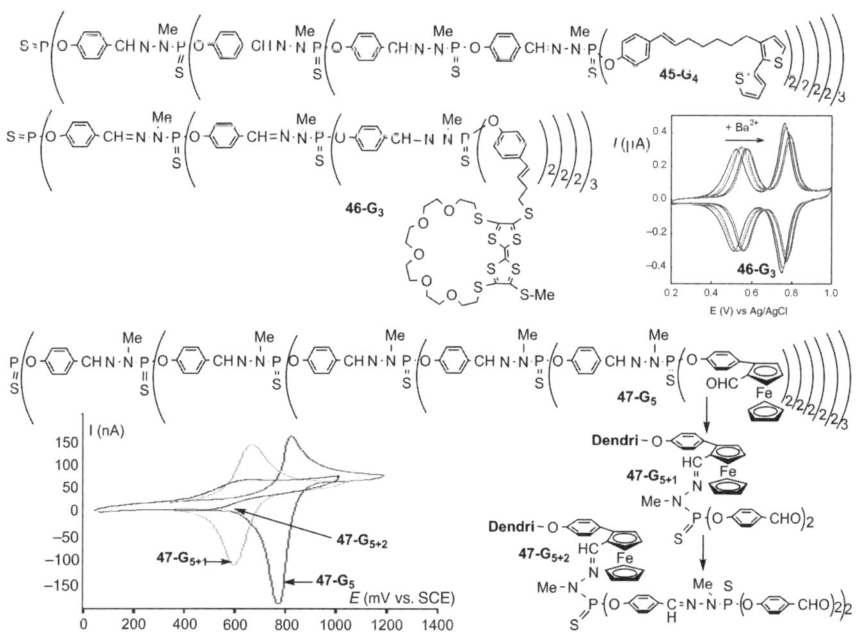

FIGURE 16.13 Various types of electrochemically active phosphorus dendrimers.

of the third, the fifth and the ninth-generation dendrimer **47-G$_n$**, then the aldehydes were reacted as shown previously in Scheme 16.1, to afford the generations $n+1$ and $n+2$. This concept is illustrated in Figure 16.13 (lower part) for the fifth generation **47-G$_5$**, leading then to sixth (**47-G$_{5+1}$**), then the seventh (**47-G$_{5+2}$**)-generation dendrimers. Progressive burying of the ferrocene inside the dendritic structure induces two differences in the electrochemical response: first, the rate of the electronic transfer decreases when the redox centers are more confined within the interior of the dendrimer; second, the reversibility of the system decreases with the increase of the generation of the dendrimer within the same series [60].

In connection with these electroactive dendrimers, we have synthesized the series of fluorescent dendrimers **48-G$_n$** ($n = 1$–4) bearing pyrene derivatives as terminal groups. The fluorescence intensity increased with the generation, and this series of compounds was used for the elaboration of organic light emitting diodes (OLEDs). The dendrimers are incorporated in poly(vinylcarbazole) and constitute the emitting (electroluminescent) layer, as shown in Figure 16.14 [61].

Another very important way to modify the surface of materials at the nanometric scale consists in using the "layer-by-layer" (LBL) deposition of positively and negatively charged entities [62]. Dendrimers having either positive charges (ammoniums, **38-G$_4$**) or negative charges (carboxylates, **49-G$_4$**) on their end groups were used for the coating of various surfaces such as microspheres to generate microcapsules [63], and glass in particular for the culture of neurons [64].

This LBL technique using ionic interactions was in particular applied for the elaboration of nanotubes of dendrimers. Ordered porous alumina templates (Figure 16.15a) were coated by immersion successively in water solutions of negatively charged dendrimers **49-G$_4$**, then positively charged dendrimers **38-G$_4$**. Bilayers of dendrimers were deposited by alternately immersing the templates in the corresponding solutions. Removal of the inorganic template was performed without destroying the dendrimers and afforded nanotubes that are the replica of the pores (400 nm width, 80 μm length, Figure 16.15b) [65]. The same methodology was used later with negatively charged quantum dots (QDs, fluorescent nanocrystals) emitting at 561, 594, and 614 nm, instead of the negatively charged dendrimers. Nanotubes

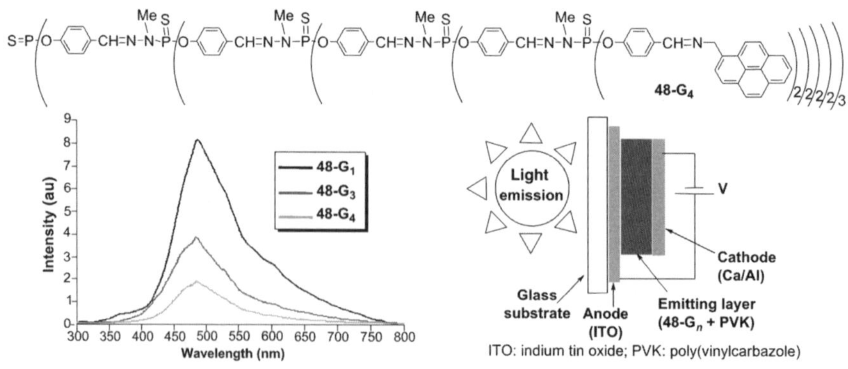

FIGURE 16.14 OLEDs elaborated from fluorescent dendrimers.

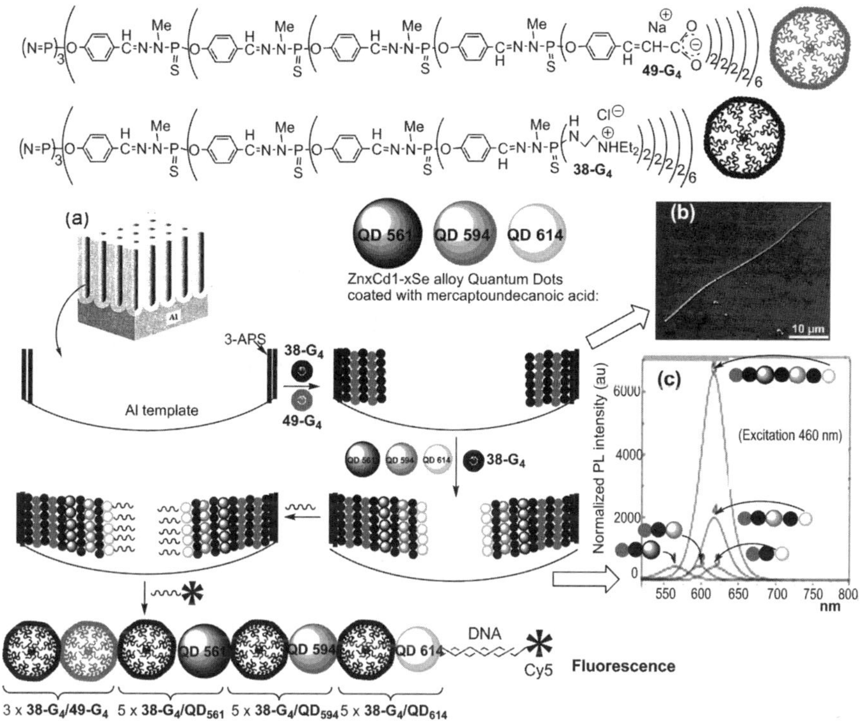

FIGURE 16.15 Nanotubes made by LBL deposition of charged dendrimers and quantum dots; their use for the detection of trace amounts of DNA.

containing a single type of QD were first prepared; as expected, their photoluminescence (PL) emission spectra appear at the expected wavelengths depending on the QDs used. Then, nanotubes containing multiple types of QDs were prepared. Their PL emission spectra show exclusively an emission peak centered at 614 nm, originating from QD614, indicating that an efficient excitation energy transfer takes place from the larger bandgap QDs to the ones with lower band energy (Figure 16.15c). This is a key feature for the enhanced detection sensitivity of DNA hybridization inside the nanotubes. Hybridization was carried out with amino group-functionalized probe DNA with Cy5-labeled complementary 15-mer DNA oligonucleotides. The measured normalized PL emission spectra display an intense shoulder at about 670 nm, originating from Cy5 that can be attributed to excitation energy transfers by FRET from the QDs to the Cy5. The results suggest that NTs containing a cascaded-energy-transfer architecture have potential utility for the detection of trace amounts of DNA [66].

Surfaces of materials can be modified also by covalently linking dendrimers. The main advantage of using dendrimers instead of simple monomers is the number of anchoring points afforded by the dendrimer, which enhances the stability of the interfaces; this is particularly interesting for detector applications. We have synthesized the fluorescent small dendrimer **50-G$_1$** incorporating reactive phosphonate

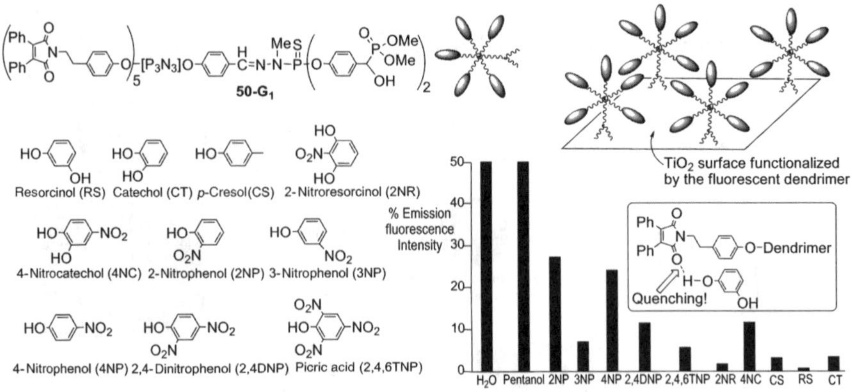

FIGURE 16.16 A chemical sensor for phenols, elaborated with a fluorescent dendrimer.

groups able to be anchored to a nanocrystalline titania mesostructured film. The resulting hybrid functionalized films are brightly fluorescent, which makes these devices quite attractive for use as sensors. Indeed, the formation of hydrogen bonds between alcohols and the carbonyl groups of the fluo-tag induces a quenching effect. Figure 16.16 shows that the dendrimers grafted to the films display high quenching efficiency toward phenolic OH moieties, especially those from resorcinol and 2-nitroresorcinol, with less than 1% of fluorescence remaining. This effect is attributed to the increased spatial proximity of the fluorescent molecules that makes the formation of hydrogen bonds between the hydroxyl moieties of the quenchers and the carbonyl groups of the dendrimer easy [67].

We have also elaborated very sensitive biological sensors by the covalent grafting of aldehyde terminated dendrimers onto aminosilanized glass slides. The contribution of phosphorus dendrimers to the field of bioarrays began in 1999, with the immobilization of a protein (human serum albumin (HSA)) [68]. AFM (atomic force microscopy) images of the slides showed a total coverage of the surface by dendrimers **6-G$_5$′**, and the presence of the HSA clusters (Figure 16.17).

Having in hand such methodology, it was later on developed for the elaboration of DNA microarrays that we called "dendrislides." Several generations of dendrimers **7-G$_n$′** were tested, and the best results concerning the signal-to-noise ratio were obtained with generations 4–7, thus generation 4 was chosen, because it is more easily synthesized than higher generations. To quantify the target/probe hybridization sensitivity, "dendrislides" and 12 commercially available glass slides were spotted with a 35 mer oligonucleotide, then hybridized with increasing concentrations of a Cy5-labeled 15 mer oligonucleotide complementary to the probe, from 0.001 to 100 nM. At target concentrations 0.001 nM of DNA, a fluorescence signal was still quantifiable only using the dendrislides elaborated with **7-G$_4$′** (Figure 16.17) [69]. Reusability of the dendrislides was also tested and was found excellent even after 10 hybridization/stripping cycles [70]. The same concept was developed later for liposomes arrays [71] and for the elaboration of sensitive piezoelectric mass-sensing devices [72].

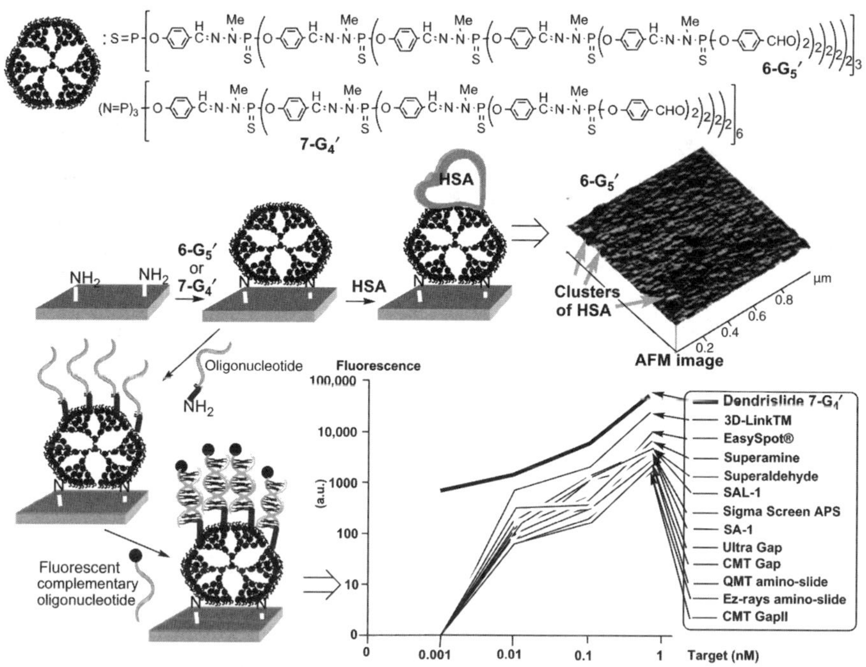

FIGURE 16.17 Very sensitive biochips elaborated with phosphorus dendrimers.

16.5 BIOLOGICAL PROPERTIES OF PHOSPHORUS DENDRIMERS

One of the most active areas of research about dendrimers concerns their biological properties, and numerous reviews in this field have been published recently [73]. Obviously, the first main point concerns the solubility in water [74], which is often achieved by grafting positively or negatively charged substituents as terminal groups, or PEG (polyethyleneglycol) chains. Two main different types of uses of dendrimers in biology can be distinguished: either the dendrimer is used for diagnosis (often by fluorescence) or the dendrimer is used for treatment (the dendrimer is the carrier of a drug or has its own activity). We have explored both types of properties using phosphorus dendrimers.

In the field of diagnosis, we have synthesized phosphorus dendrimers possessing as core a fluorophore specially engineered to be excited by two photons. Two-photon excited fluorescence is based on a nonlinear optical phenomenon, which is rapidly gaining popularity, due to a highly spatially confined excitation, an intrinsic three-dimensional resolution [75], and an increased penetration depth in tissues with reduced photodamages thanks to excitation in the near-infrared region. Even if most of the work in this field is done with quantum dots (inorganic crystals, for instance of type shown in Figure 16.15), phosphorus dendrimers linked to organic fluorophores possessing two-photon absorption (TPA) properties can be considered as an organic

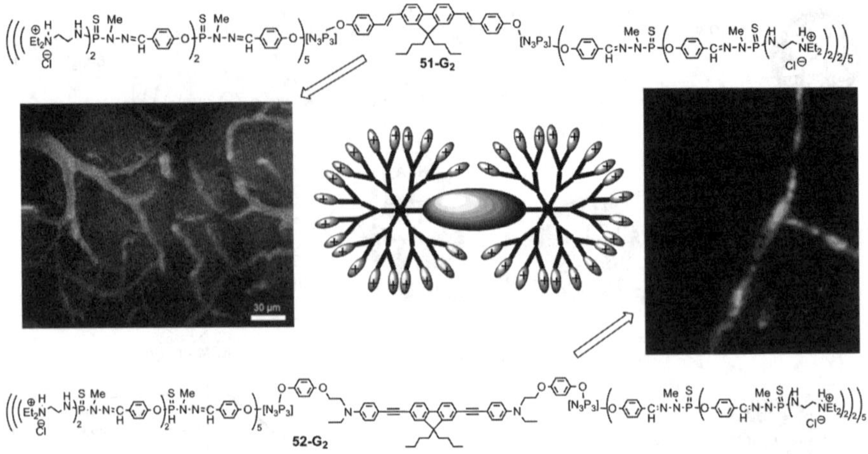

FIGURE 16.18 TPA fluorophores in the structure of dendrimers used for biological imaging.

alternative to inorganic quantum dots [76]. The second-generation dendrimer **51-G$_2$** injected intravenously to a rat allowed the imaging of the vascular network in the dorsal part of the rat olfactory bulb, at a depth of about 200 μm (left image in Figure 16.18) [77]. Analogously, the second-generation dendrimer **52-G$_2$** possessing a green emitter TPA fluorophore as core was used for intracardiac injection in a Xenopus tadpole, allowing imaging of the blood vessels of the tail (right image in Figure 16.18) [78].

Concerning the use of dendrimers for treating diseases, polycationic dendrimers have been recognized very early as useful synthetic vectors for transfection [79] and a huge quantity of publications was generated by this topic [80]. We have shown that the transfection efficiency of the phosphorus dendrimers having multiple ammonium terminal groups depends on the reversibility of protonation: tertiary ammonium groups are less toxic and more efficient than quaternary ammonium groups, presumably due to the adaptability of tertiary ammonium groups to pH (amine/ammonium). In addition, we have shown that the generation of the dendrimers has an important influence on the transfection efficiency, with the fourth generation being the best compromise between the transfection efficiency and the time needed to synthesize the dendrimer. The first experiments were done with dendrimers **38-G$_n$** as transfecting agents of the luciferase gene within 3T3 cells (Figure 16.19, left part) [81], then within human cells [82], and we have recently tested the efficiency of series of dendrimers with diverse types of ammonium terminal groups (pyrrolidine, morpholine, methyl piperazine, or phenyl piperazine) [83].

Dendrimers **38-G$_n$** were also found efficient antiprion agents. Prion disorders are fatal neurodegenerative disorders involving conformational changes from the normal cellular form of prion protein to an infectious scrapie isoform (PrP^{Sc}), leading to the formation of fibrils (amyloid-like structures). ScN2a cells infected with the scrapie strain 22L were treated with various amounts of dendrimers **38-G$_n$**, and analyzed after 3 days. The relative amount of PrP^{Sc} measured for treated cells compared to untreated

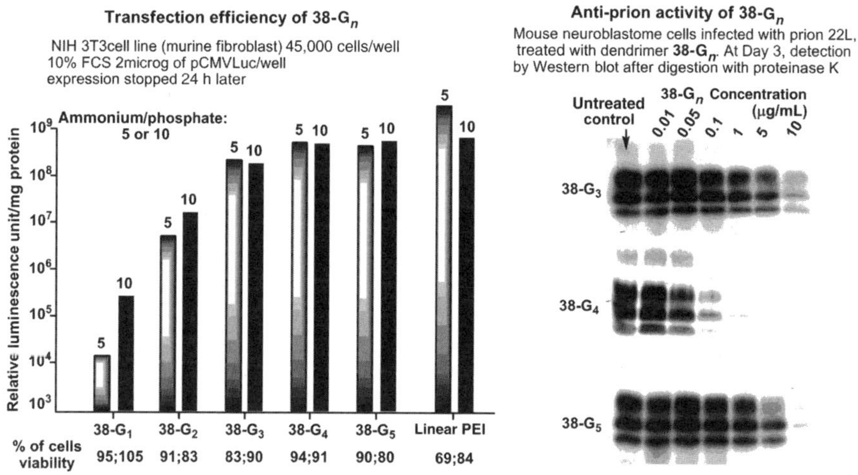

FIGURE 16.19 Transfection efficiency and antiprion activity of cationic phosphorus dendrimers.

cells, depending on the concentration and the generation of the dendrimer **38-G$_n$** is shown in Figure 16.19 (right part), by the disappearance of the black bars. The efficiency of the anti-PrP^{Sc} agents is **38-G$_4$** > **38-G$_5$** > **38-G$_3$**. The most efficient dendrimer (**38-G$_4$**) was then used against other strains of prions, including Bovine Spongiform Encephalopathy ("mad cow" disease) and was also found efficient in all cases. Furthermore, the cured cells show a persistent reduction in the PrP^{Sc} levels, and no transmission of infectivity was observed when the treated cells were used to "infect" healthy cells. In view of these important preliminary results, dendrimer **38-G$_4$** was tested with mice infected. Two days after contamination, a series of mice received an injection of 100 μg of **38-G$_4$**, which was repeated every 2 days for 1 month. The amount of PrP^{Sc} detected in the spleen of the treated mice compared to untreated mice shows a diminishing of 80% of the amount of PrP^{Sc} [84]. Recently, it was shown that dendrimer **38-G$_4$** interact with heparin, and that these interactions are responsible for the indirect mechanism of inhibiting the fibrils formation by dendrimers [85].

Besides polycationic dendrimers, polyanionic phosphorus dendrimers were also shown to possess interesting biological properties. The polyanionic dendrimers **53-G$_n$** were obtained by reaction of the carboxylic acid end groups of dendrimer **41-G$_n$** with an aminosugar analog of galactosylceramide. This species forms spontaneously bilayer vesicles in water by self-assembly [86], and possesses an anti-HIV activity. Indeed, these compounds were designed as multisite analogs of galactosylceramide, a glycolipid having a specific affinity for the V3 loop region of the gp120 viral envelope protein of HIV-1, with the aim of blocking HIV infection prior the entry of the virus into human cells. Antiviral assays on these supra-molecular entities (generations 0, 1, and 2), built from either a trifunctional (**53-G$_1$**, **53-G$_2$**) or a hexafunctional core (**53-G$_0$**, **53-G$_1$**), have shown the crucial role played

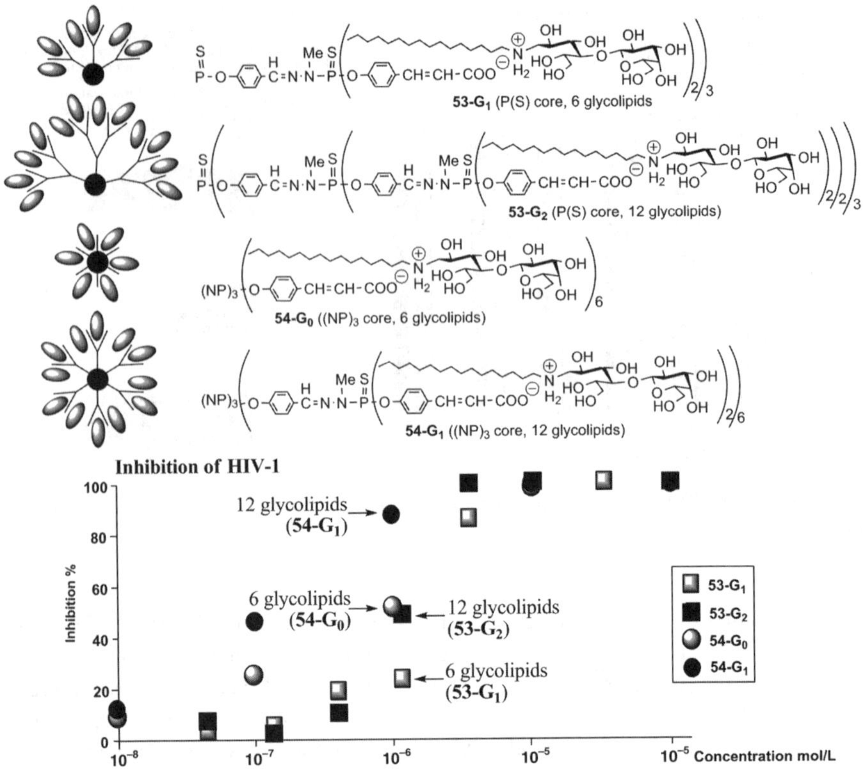

FIGURE 16.20 Anti-HIV properties of dendrimers depending on the generation and the shape.

by the multivalency, the lipophilicity, but also the shape on the biological activity (Figure 16.20) [87].

The last example of biological uses of phosphorus dendrimers concerns their influence toward immune blood cells. Peripheral blood immune cells are easily accessible and widespread in the whole body, thus they are a target of choice for trying to increase their fighting efficiency against bacterial, viral, or parasitic infections, and even against cancers. We have demonstrated that some phosphorus-containing dendrimers are able to promote human monocytes activation [88], possess anti-inflammatory properties for the human myeloid cell lineage [89], and are able to increase the efficiency of the IL2-dependent proliferation of human Natural Killer (NK) cells [90]. Indeed, phosphonates grafted to the surface of phosphorus-containing dendrimers possess the unexpected property to dramatically and selectively promote the multiplication of human NK cells. Depending on the size of the dendrimers, the type, the number, and even the geometry of the end groups they bear, large differences in their bioactivity were observed. The multiplication up to 500-fold in certain cases was obtained, particularly using dendrimer **55-G$_1$**, and is unprecedented (Figure 16.21). Furthermore, the bioactivity of the NK cells generated in the

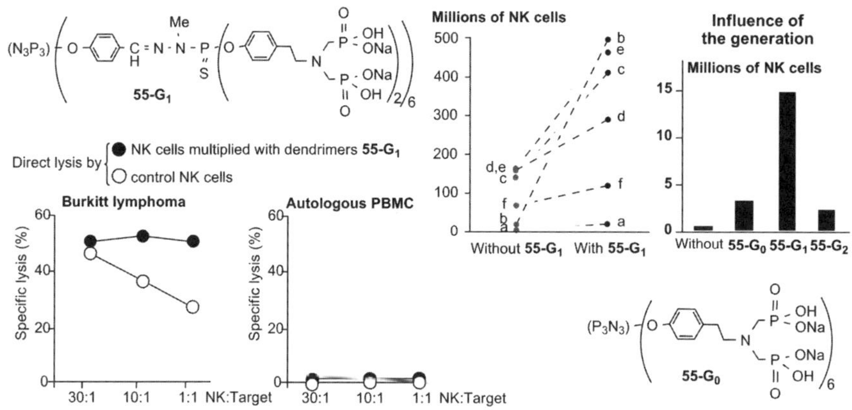

FIGURE 16.21 Multiplication of Natural Killer cells using phosphorus dendrimers and the biological efficiency of the NK cells thus generated.

presence of dendrimers is not modified. Cultures with these dendrimers preserve autologous lymphocytes, and did not induce activation or inhibition of the NK cells lytic response, nor compromise direct toxicity for their target cells, as shown in Figure 16.21 for the Burkitt lymphoma, but also for dozens of lymphoma and carcinoma cells strains. In view of the importance of these findings, we began a detailed QSAR (quantitative structure–activity relationship) study, by varying in particular the type of phosphonate terminal groups [91], or their density [92].

16.6 CONCLUSION

This chapter demonstrates that phosphorus-containing dendrimers constitute a specific class of dendrimers easily tunable due to the extraordinary versatility of phosphorus chemistry, and having a high potential of properties and uses. The versatility is demonstrated in particular with the synthesis of series of dendritic molecules whose complexity is unattainable up to now using other types of synthetic methodologies, without phosphorus; it is in particular illustrated by the reactivity inside the branches of phosphorus dendrimers after their synthesis. The reactivity of the end groups of these phosphorus dendrimers has also allowed the grafting of a tremendous number of type of functional groups. The variety of these end groups allowed studying potential applications of these dendrimers in three main fields: (i) catalysis, (ii) (nano)materials, and (iii) biology.

(i) The main advantage of using dendrimers in catalysis in general concerns their easy recovery, which can be an important criterion when pure products are needed, or when the catalyst is expensive; efforts are now focused on finding positive dendritic effects, on using cheap and nontoxic metals (or even no metals), and on using ecofriendly solvents (ethanol, water, super critical fluids).

(ii) A lot of work concerning the organization of matter at the nanometer scale has been carried out using phosphorus dendrimers, either to create new and organized nanomaterials or to modify the surface of pre-existing materials at the nanometric scale. In particular, functionalized silica, hydrogels, modified electrodes, chemical sensors, OLEDs, original organic nanotubes and microcapsules, exclusively made of dendrimers were obtained. In between materials and biology, very sensitive, stable, and reusable DNA-chips elaborated from phosphorus dendrimers were obtained and their industrial production is currently under development.

(iii) Tests in the field of biology concerning the cytotoxicity were reassuring, and allowed the use of phosphorus dendrimers both for diagnosis and for treatment of various diseases. Especially engineered fluorophore having two-photon absorption properties used as core of dendrimers allowed imaging vascular networks, whereas suitably functionalized phosphorus dendrimers are usable as transfection agents, as antiprion and anti-HIV agents, and for promoting the multiplication of human immune blood cells. Thus, we have demonstrated that phosphorus dendrimers have a very important potential of development for biological uses that will be one of the major fields of applications of phosphorus dendrimers in the next years.

REFERENCES

[1] D. A. Tomalia, H. Baker, J. Dewald, M. Hall, G. Kallos, S. Martin, J. Roeck, J. Ryder, P. Smith, *Polym. J.* **1985**, *17*, 117.

[2] (a) E. Buhleier, W. Wehner, F. Vögtle, *Synthesis* **1978**, 155; (b) C. Wörner, R. Mülhaupt, *Angew. Chem. Int. Ed. Engl.* **1993**, *32*, 1306–1308; (c) E. M. M. de Brabander-van den Berg, E. W. Meijer, *Angew. Chem. Int. Ed. Engl.* **1993**, *32*, 1308–1311.

[3] J.-P. Majoral, A.-M. Caminade, *Chem. Rev.* **1999**, *99*, 845.

[4] E. A. Rebrov, A. M. Muzafarov, V. S. Papkov, A. A. Zhdanov, *Doklady Akademii Nauk SSSR* **1989**, *309*, 376–380.

[5] K. Rengan, R. Engel, *J. Chem. Soc., Chem. Commun.* **1990**, 1084–1085.

[6] N. Launay, A.-M. Caminade, R. Lahana, J.-P. Majoral, *Angew. Chem. Int. Ed. Engl.* **1994**, *33*, 1589–1592.

[7] A. Miedaner, C. J. Curtis, R. M. Barkley, D. L. DuBois, *Inorg. Chem.* **1994**, *33*, 5482–5490.

[8] M. Petrucci-Samija, V. Guillemette, M. Dasgupta, A. K. Kakkar, *J. Am. Chem. Soc.* **1999**, *121*, 1968.

[9] R. Roy, *Topics Curr. Chem.* **1997**, *187*, 241.

[10] G. Salamonczyk, M. Kuznikowski, A. Skowronska, *Tetrahedron Lett.* **2000**, *41*, 1643.

[11] G. Salamonczyk, M. Kuznikowski, E. Poniatowska, *Chem. Commun.* **2001**, 2202.

[12] E. Poniatowska, G. Salamonczyk, *Tetrahedron Lett.* **2003**, *44*, 4315.

[13] G. Salamonczyk, M. Kuznikowski, E. Poniatowska, *Tetrahedron Lett.* **2002**, *43*, 1747.

[14] (a) N. Launay, A.-M. Caminade, J.-P. Majoral, *J. Am. Chem. Soc.* **1995**, *117*, 3282; (b) M.-L. Lartigue, B. Donnadieu, C. Galliot, A.-M. Caminade, J.-P. Majoral, J.-P. Fayet, *Macromolecules* **1997**, *30*, 7335–7337.

[15] N. Launay, A.-M. Caminade, J.-P. Majoral, *J. Organomet. Chem.* **1997**, *529*, 51.

[16] (a) J. Leclaire, Y. Coppel, A.-M. Caminade, J.-P. Majoral, *J. Am. Chem. Soc.* **2004**, *126*, 2304–2305; (b) J. Leclaire, R. Dagiral, S. Fery-Forgues, Y. Coppel, B. Donnadieu, A.-M. Caminade, J.-P. Majoral, *J. Am. Chem. Soc.* **2005**, *127*, 15762–15770.

[17] G. Magro, B. Donnadieu, A.-M. Caminade, J.-P. Majoral, *Chem. Eur. J.* **2003**, *9*, 2151–2159.

[18] R.-M. Sebastian, J.-C. Blais, A.-M. Caminade, J.-P. Majoral, *Chem. Eur. J.* **2002**, *8*, 2172–2183.

[19] (a) J.-P. Majoral, A.-M. Caminade, *Top. Curr. Chem.* **1998**, *197*, 79; (b) J.-P. Majoral, A.-M. Caminade, V. Maraval, *Chem. Commun.* **2002**, 2929–2942.

[20] M.-L. Lartigue, M. Slany, A.-M. Caminade, J.-P. Majoral, *Chem. Eur. J.* **1996**, *2*, 1417.

[21] C. Galliot, D. Prévoté, A.-M. Caminade, J.-P. Majoral, *J. Am. Chem. Soc.* **1995**, *117*, 5470.

[22] C. Larré, A.-M. Caminade, J.-P. Majoral, *Angew. Chem. Int. Ed. Engl.* **1997**, *36*, 596.

[23] C. Larré, B. Donnadieu, A.-M. Caminade, J.-P. Majoral, *J. Am. Chem. Soc.* **1998**, *120*, 4029.

[24] C. Larré, D. Bressolles, C. Turrin, B. Donnadieu, A.-M. Caminade, J.-P. Majoral, *J. Am. Chem. Soc.* **1998**, *120*, 13070.

[25] L. Brauge, A.-M. Caminade, J.-P. Majoral, S. Slomkowski, M. Wolszczak, *Macromolecules*, **2001**, *34*, 5599.

[26] C. Galliot, C. Larré, A.-M. Caminade, J.-P. Majoral, *Science* **1997**, *277*, 1981.

[27] C. Larré, B. Donnadieu, A.-M. Caminade, J.-P. Majoral, *Chem. Eur. J.* **1998**, *4*, 2031.

[28] V. Maraval, R. Laurent, B. Donnadieu, M. Mauzac, A.-M. Caminade, J.-P. Majoral, *J. Am Chem. Soc.* **2000**, *122*, 2499.

[29] L. Brauge, G. Magro, A.-M. Caminade, J.-P. Majoral, *J. Am. Chem. Soc.* **2001**, *123*, 6698–6699; correction: *J. Am. Chem. Soc.* **2001**, *123*, 8446.

[30] V. Maraval, A.-M. Caminade, J.-P. Majoral, J.-C. Blais, *Angew. Chem. Int. Ed.* **2003**, *42*, 1822–1826.

[31] S. Merino, L. Brauge, A.-M. Caminade, J.-P. Majoral, D. Taton, Y. Gnanou, *Chem. Eur. J.* **2001**, *7*, 3095–3105.

[32] J.-P. Majoral, A.-M. Caminade, *Top. Curr. Chem.* **2003**, *223*, 111–159.

[33] J. W. J. Knapen, A. W. van der Made, J. C. de Wilde, P. W. N. M. van Leeuwen, P. Wijkens, D. M. Grove, G. van Koten, *Nature* **1994**, *372*, 659.

[34] See for instance: (a) B. Helms, J. M. J. Fréchet, *Adv. Synth. Catal.* **2006**, *348*, 1125–1148; (b) J. K. Kassube, L. H. Gade, *Top. Organomet. Chem.* **2006**, *20*, 61–96; (c) A. Dahan, M. Portnoy, *J. Polym. Sci. A*, **2005**, *43*, 235–262; (d) R. van Heerbeek, P. C. J. Kamer, P. W. N. M. van Leeuwen, J. N. H. Reek, *Chem. Rev.* **2002**, *102*, 3717; (e) B. Romagnoli, W. Hayes, *J. Mater. Chem.* **2002**, *12*, 767; (f) D. Astruc, F. Chardac, *Chem. Rev.* **2001**, *101*, 2991; (g) R. Kreiter, A. W. Kleij, R. J. M. Gebbink, G. van Koten, *Top. Curr. Chem.* **2001**, *217*, 163.

[35] (a) A.-M. Caminade, V. Maraval, R. Laurent, J.-P. Majoral, *Curr. Org. Chem.* **2002**, *6*, 739; (b) A.-M. Caminade, J.-P. Majoral, *Coord. Chem. Rev.* **2005**, *249*, 1917–1926;

(c) J.-P. Majoral, A.-M. Caminade, R. Laurent, *ACS Sympos. Ser.*, **2006**, *928*, 230–243; (d) A.-M. Caminade, P. Servin, R. Laurent, J.-P. Majoral, *Chem. Soc. Rev.* **2008**, *37*, 56–67.

[36] (a) M. Slany, M. Bardají, M.-J. Casanove, A.-M. Caminade, J.-P. Majoral, B. Chaudret, *J. Am. Chem. Soc.* **1995**, *117*, 9764; (b) M. Bardají, M. Kustos, A.-M. Caminade, J.-P. Majoral, B. Chaudret, *Organometallics* **1997**, *16*, 403; (c) A.-M. Caminade, R. Laurent, B. Chaudret, J.-P. Majoral, *Coord. Chem. Rev.* **1998**, *178–180*, 793.

[37] A.-M. Caminade, J.-P. Majoral, *Coord. Chem. Rev.* **2005**, *249*, 1917–1926.

[38] V. Maraval, R. Laurent, A.-M. Caminade, J.-P. Majoral, *Organometallics* **2000**, *19*, 4025–4029.

[39] P. Servin, R. Laurent, A. Romerosa, M. Peruzzini, J.-P. Majoral, A.-M. Caminade, *Organometallics* **2008**, *27*, 2066–2073.

[40] M. Koprowski, R.-M. Sebastian, V. Maraval, M. Zablocka, V. Cadierno-Menendez, B. Donnadieu, A. Igau, A.-M. Caminade, J.-P. Majoral, *Organometallics* **2002**, *21*, 4680–4687.

[41] R. Laurent, A.-M. Caminade, J.-P. Majoral, *Tetrahedron Lett.* **2005**, *46*, 6503–6506.

[42] P. Servin, R. Laurent, L. Gonsalvi, M. Tristany, M. Peruzzini, J.-P. Majoral, A.-M. Caminade, *Dalton Trans.* **2009**, 4432–4434.

[43] A. Ouali, R. Laurent, A.-M. Caminade, J.-P. Majoral, M. Taillefer, *J. Am. Chem. Soc.* **2006**, *128*, 15990–15991.

[44] A. Gissibl, C. Padié, M. Hager, F. Jaroschick, R. Rasappan, E. Cuevas-Yanez, C.-O. Turrin, A.-M. Caminade, J.-P. Majoral, O. Reiser, *Org. Lett.* **2007**, *9*, 2895–2898.

[45] E. Badetti, A.-M. Caminade, J.-P. Majoral, M. Moreno-Manas, R.-M. Sebastian, *Langmuir* **2008**, *24*, 2090–2101.

[46] (a) A.-M. Caminade, J.-P. Majoral, *Accounts Chem. Res.* **2004**, *37*, 341–348; (b) A.-M. Caminade, J.-P. Majoral, *J. Mater. Chem.* **2005**, *15*, 3643–3649.

[47] P. Reinert, J.-Y. Chane-Ching, L. Bull, R. Dagiral, P. Batail, R. Laurent, A.-M. Caminade, J.-P. Majoral, *New J. Chem.* **2007**, *31*, 1259–1263.

[48] C.-O. Turrin, V. Maraval, A.-M. Caminade, J.-P. Majoral, A. Mehdi, C. Reyé, *Chem. Mat.* **2000**, *12*, 3848–3856.

[49] M. K. Boggiano, G. J. A. A. Soler-Illia, L. Rozes, C. Sanchez, C.-O. Turrin, A.-M. Caminade, J.-P. Majoral, *Angew. Chem. Int. Ed. Engl.* **2000**, *39*, 4249–4254.

[50] C. Marmillon, F. Gauffre, T. Gulik-Krzywicki, C. Loup, A.-M. Caminade, J.-P. Majoral, J.-P. Vors, E. Rump, *Angew. Chem. Int. Ed.* **2001**, *40*, 2626–2629.

[51] C. Larpent, C. Geniès, A. P. De Sousa Delgado, A.-M. Caminade, J.-P. Majoral, J.-F. Sassi, F. Leising, *Chem. Commun.* **2004**, 1816–1817.

[52] A. El Ghzaoui, F. Gauffre, A.-M. Caminade, J.-P. Majoral, H. Lannibois-Drean, *Langmuir* **2004**, *20*, 9348–9353.

[53] G. Schmid, W. Meyer-Zaika, R. Pugin, T. Sawitowski, A.-M. Caminade, J.-P. Majoral, C.-O. Turrin, *Chem. Eur. J.* **2000**, *6*, 1693–1697.

[54] G. Schmid, E. Emmrich, J.-P. Majoral, A.-M. Caminade, *Small* **2005**, *1*, 73–75.

[55] G. Franc, E. Badetti, V. Collière, J.-P. Majoral, R.-M. Sebastián, A.-M. Caminade, *Nanoscale* **2009**, *1*, 233–237.

[56] R.-M. Sebastián, A.-M. Caminade, J.-P. Majoral, E. Levillain, L. Huchet, J. Roncali, *Chem. Commun.* **2000**, 507–508.

[57] F. Le Derf, E. Levillain, A. Gorgues, M. Sallé, R.-M. Sebastian, A.-M. Caminade, J.-P. Majoral, *Angew. Chem. Int. Ed.* **2001**, *40*, 224–227.

[58] B. Alonso, E. Alonso, D. Astruc, J.-C. Blais, L. Djakovitch, J.-L. Fillaut, S. Nlate, F. Moulines, S. Rigaut, J. Ruiz, C. Valério, *Adv. Dendritic Macromol.* **2002**, *5*, 89–127.

[59] C.-O. Turrin, J. Chiffre, D. de Montauzon, J.-C. Daran, A.-M. Caminade, E. Manoury, G. Balavoine, J.-P. Majoral, *Macromolecules* **2000**, *33*, 7328–7336.

[60] C.-O. Turrin, J. Chiffre, D. de Montauzon, G. Balavoine, E. Manoury, A.-M. Caminade, J.-P. Majoral, *Organometallics* **2002**, *21*, 1891–1897.

[61] L. Brauge, G. Vériot, G. Franc, R. Deloncle, A.-M. Caminade, J.-P. Majoral, *Tetrahedron* **2006**, *62*, 11891–11899.

[62] G. Decher, *Science* **1997**, *277*, 1232–1237.

[63] (a) B. S. Kim, O. V. Lebedeva, D. H. Kim, A.-M. Caminade, J.-P. Majoral, W. Knoll, O. I. Vinogradova, *Langmuir* **2005**, *21*, 7200–7206; (b) B. S. Kim, O. V. Lebedeva, K. Koynov, H. Gong, A.-M. Caminade, J.-P. Majoral, O. I. Vinogradova, *Macromolecules* **2006**, *39*, 5479–5483.

[64] J.-L. Hernandez-Lopez, H. L. Khor, A.-M. Caminade, J.-P. Majoral, S. Mittler, W. Knoll, D. H. Kim, *Thin Solid Films* **2008**, *516*, 1256–1264.

[65] (a) D. H. Kim, P. Karan, P. Göring, J. Leclaire, A.-M. Caminade, J.-P. Majoral, U. Gösele, M. Steinhart, W. Knoll, *Small* **2005**, *1*, 99–102; (b) F. Yu, S. Ahl, A.-M. Caminade, J.-P. Majoral, W. Knoll, J. Erlebacher, *Anal. Chem.* **2006**, *78*, 7346–7350.

[66] (a) C. L. Feng, X. H. Zhong, M. Steinhart, A.-M. Caminade, J.-P. Majoral, W. Knoll, *Adv. Mater.* **2007**, *19*, 1933–1936; (b) C. L. Feng, X. H. Zhong, M. Steinhart, A.-M. Caminade, J.-P. Majoral, W. Knoll, *Small* **2008**, *4*, 566–571.

[67] E. Martinez-Ferrero, G. Franc, S. Mazères, C.-O. Turrin, A.-M. Caminade, J.-P. Majoral, C. Sanchez, *Chem. Eur. J.* **2008**, *14*, 7658–7669.

[68] S. Slomkowski, B. Miksa, M.M. Chehimi, M. Delamar, E. Cabet-Deliry, J.-P. Majoral, A.-M. Caminade, *React. Funct. Polym.* **1999**, *41*, 45–57.

[69] V. Le Berre, E. Trévisiol, A. Dagkessamanskaia, S. Sokol, A.-M. Caminade, J.-P. Majoral. B. Meunier, J. François, *Nucleic Acids Res.* **2003**, *31*, e88.

[70] E. Trévisiol, V. Leberre-Anton, J. Leclaire, G. Pratviel, A.-M. Caminade, J.-P. Majoral, J. M. François, B. Meunier, *New. J. Chem.* **2003**, *27*, 1713–1719.

[71] B. Chaize, M. Nguyen, T. Ruysschaert, V. Le Berre, A.-M. Caminade, J.-P. Majoral, G. Pratviel, B. Meunier, M. Winterhalter, D. Fournier, *Bioconjug. Chem.*, **2006**, *17*, 245–247.

[72] L. Nicu, M. Guirardel, F. Chambosse, P. Rougerie, S. Hinh, E. Trévisiol, J. M. François, J.-P. Majoral, A.-M. Caminade, E. Cattan, C. Bergaud, *Sens. Actuators B* **2005**, *110*, 125–136.

[73] See in particular: (a) S. E. Stiriba, H. Frey, R. Haag, *Angew. Chem. Int. Ed. Engl.* **2002**, *41*, 1329–1334; (b) M. J. Cloninger, *Curr. Opin. Chem. Biol.* **2002**, *6*, 742–748; (c) U. Boas, P. M.H. Heegaard, *Chem. Soc. Rev.* **2004**, *33*, 43–63; (d) S. Svenson, D. A. Tomalia, *Adv. Drug. Deliv. Rev.* **2005**, *57*, 2106–2129; (e) R. Dunca, L. Izzo, *Adv. Drug. Deliv Rev.* **2005**, *57*, 2215–2237; (f) C. C. Lee, J. A. MacKay, J. M. J. Fréchet, F. C. Szoka, *Nature Biotech.* **2005**, *23*, 1517–1526; (g) A.-M. Caminade, C.-O. Turrin, J.-P. Majoral, *Chem. Eur. J.* **2008**, *14*, 7422–7432; (h) O. Rolland, C.-O. Turrin, A.-M. Caminade, J.-P. Majoral, *New. J. Chem.* **2009**, *33*, 1809–1824.

[74] A.-M. Caminade, J.-P. Majoral, *Prog. Polym. Sci.* **2005**, *30*, 491–505.

[75] W. Denk, J. H. Strickler, W. W. Webb, *Science* **1990**, *248*, 73–76.

[76] (a) O. Mongin, T. Rama Krishna, M. H. V. Werts, A.-M. Caminade, J.-P. Majoral, M. Blanchard-Desce, *Chem. Commun.* **2006**, 915–917; (b) O. Mongin, A. Pla-Quintana, F. Terenziani, D. Drouin, C. Le Droumaguet, A.-M. Caminade, J.-P. Majoral, M. Blanchard-Desce, *New J. Chem.* **2007**, *31*, 1354–1367.

[77] T. R. Krishna, M. Parent, M. H. V. Werts, L. Moreaux, S. Gmouh, S. Charpak, A.-M. Caminade, J.-P. Majoral, M. Blanchard-Desce, *Angew. Chem. Int. Ed.* **2006**, *45*, 4645–4648.

[78] (a) O. Mongin, C. Rouxel, A.-C. Robin, A. Pla-Quintana, T. Rama Krishna, G. Recher, F. Tiaho, A.-M. Caminade, J.-P. Majoral, M. Blanchard-Desce, *Proc. SPIE-Int. Soc. Opt. Eng.*, **2008**, *7040*, 704006; (b) O. Mongin, C. Rouxel, J.-M. Vabre, Y. Mir, A. Pla-Quintana, Y. Wei, A.-M. Caminade, J.-P. Majoral, M. Blanchard-Desce, *Proc. SPIE-Int. Soc. Opt. Eng.*, **2009**, *7403*, 740303.

[79] J. Haensler, F.C. Szoka Jr., *Bioconjug. Chem.* **1993**, *4*, 372–379.

[80] C. Dufès, I. F. Uchegbu, A. G. Schätzlein, *Adv. Drug. Deliv. Rev.* **2005**, *57*, 2177–2202.

[81] (a) C. Loup, M.-A. Zanta, A.-M. Caminade, J.-P. Majoral, B. Meunier, *Chem. Eur. J.* **1999**, *5*, 3644–3650; (b) A. V. Maksimenko, V. Mandrouguine, M.B. Gottikh, J. R. Bertrand, J.-P. Majoral, C. Malvy, *J. Genet. Med.* **2003**, *5*, 61–71.

[82] M. Maszewska, J. Leclaire, M. Cieslak, B. Nawrot, A. Okruszek, A.-M. Caminade, J.-P. Majoral, *Oligonucleotides* **2003**, *13*, 193–207.

[83] C. Padié, M. Maszewska, K. Majchrzak, B. Nawrot, A.-M. Caminade, J.-P. Majoral, *New J. Chem.* **2009**, *33*, 318–326.

[84] J. Solassol, C. Crozet, V. Perrier, J. Leclaire, F. Béranger, A.-M. Caminade, B. Meunier, D. Dormon, J.-P. Majoral, S. Lehmann, *J. Gen. Virol.* **2004**, *85*, 1791–1799.

[85] B. Klajnert, M. Cangiotti, S. Calici, M. Ionov, J.-P. Majoral, A.-M. Caminade, J. Cladera, M. Bryszewska, M.F. Ottaviani, *New J. Chem.* **2009**, *33*, 1087–1093.

[86] M. Blanzat, C.-O. Turrin, E. Perez, I. Rico-Lattes, A.-M. Caminade, J.-P. Majoral, *Chem. Commun.* **2002**, 1864–1865.

[87] M. Blanzat, C.-O. Turrin, A.-M. Aubertin, C. Vidal, A.-M. Caminade, J.-P. Majoral, I. Rico-Lattes, A. Lattes, *ChemBioChem* **2005**, *6*, 2207–2213.

[88] M. Poupot, L. Griffe, P. Marchand, A. Maraval, O. Rolland, L. Martinet, F. E. L'Faqihi-Olive, C.-O. Turrin, A.-M. Caminade, J.-J. Fournié, J.-P. Majoral, R. Poupot, *FASEB J.* **2006**, *20*, 2339–2351.

[89] S. Fruchon, M. Poupot, L. Martinet, C.-O. Turrin, J.-P. Majoral, J.-J. Fournié, A.-M. Caminade, R. Poupot, *J. Leukocyte Biol.* **2009**, *85*, 553–562.

[90] L. Griffe, M. Poupot, P. Marchand, A. Maraval, C.-O. Turrin, O. Rolland, P. Métivier, G. Bacquet, J.-J. Fournié, A.-M. Caminade, R. Poupot, J.-P. Majoral, *Angew. Chem. Int. Ed.* **2007**, *46*, 2523–2526.

[91] P. Marchand, L. Griffe, M. Poupot, C.-O. Turrin, G. Bacquet, J.-J. Fournié, J.-P. Majoral, R. Poupot, A.-M. Caminade, *Bioorg. Med. Chem. Lett.* **2009**, *19*, 3963–3966.

[92] O. Rolland, L. Griffe, M. Poupot, A. Maraval, A. Ouali, Y. Coppel, J.-J. Fournié, G. Bacquet, C.-O. Turrin, A.-M. Caminade, J.-P. Majoral, R. Poupot, *Chem. Eur. J.* **2008**, *14*, 4836–4850.

INDEX

Designing Dendrimers, Edited by Sebastiano Campagna, Paola Ceroni, Fausto Puntoriero.

© 2012 John Wiley & Sons, Inc. Published 2012 by John Wiley & Sons, Inc.